GUIDE TO FLOWERING PLANT FAMILIES

GUIDE TO FLOWERING

PLANT FAMILIES

WENDY B. ZOMLEFER

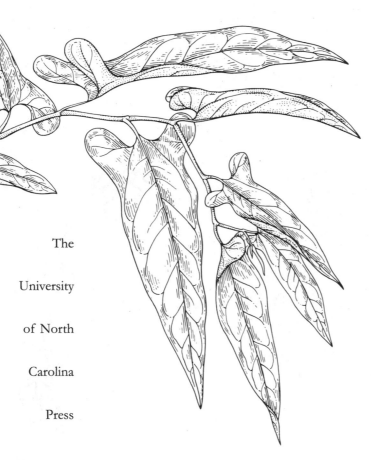

The

University

of North

Carolina

Press

Chapel Hill

& London

Manufactured in the United States of America

Portions copyright © 1980, 1983, 1986, 1989.

The paper in this book meets the guidelines for permanence and
durability of the Committee on Production Guidelines for Book
Longevity of the Council on Library Resources.

Library of Congress Cataloging-in-Publication Data

Zomlefer, Wendy B.

   Guide to flowering plant families / Wendy B. Zomlefer.

     p.   cm.

   Includes bibliographical references (p.   ) and index.

   ISBN 0-8078-2160-8 (cloth : alk. paper).

—ISBN 0-8078-4470-5 (pbk. : alk. paper)

   1. Angiosperms—United States—Classification.

2. Angiosperms—Canada—Classification.   3. Angiosperms—
Classification.   4. Botanical illustration—United States.

5. Botanical illustration—Canada.   6. Botanical illustration.   I. Title.

QK115.Z65   1994

582.13—dc20   94-5712

                                         CIP

Wendy B. Zomlefer, a plant taxonomist and professional scientific
illustrator, is the author of *Flowering Plants of Florida: A Guide to Common
Families*.

99 98 97 96   6 5 4 3

IN LOVING MEMORY OF MY MOTHER,

DOROTHY D. ZOMLEFER

But flowers distilled though they with winter meet,

Leese but their show, their substance still lives sweet.

—William Shakespeare, *Sonnet 5*

CONTENTS

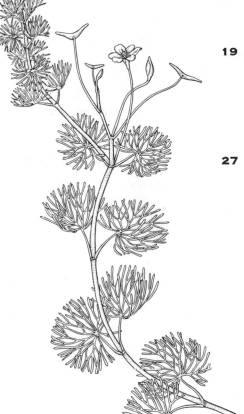

PREFACE

Guide to Flowering Plant Families is a combination of detailed illustration and family descriptions, a product of my dual professional backgrounds as scientist and illustrator. My interest in textbooks on flowering plants originated during my own student years, when I discovered that existing texts were out of date and/or contained poor-quality illustrations. Over the past fifteen years I have perfected my skills in a series of manuals adopted for use in college courses throughout the United States. The popularity of these publications indicated a great need for a comprehensive new reference with modern commentaries and accurate drawings. The response of colleagues and their students to my work has inspired this book, which includes 130 families, 23 comparative charts, and 158 plates of botanical illustrations. My hope is that it will serve botanists, horticulturists, students, and other plant enthusiasts as a guide to promote the appreciation of flowering plants everywhere.

I assume all editorial responsibility and welcome any corrections, comments, and constructive criticism, as well as appropriate reprints.

Wendy B. Zomlefer
Gainesville, Florida
August 20, 1993

ACKNOWLEDGMENTS

I am greatly indebted to the many individuals who assisted me in this long-term venture, especially during the years when I endeavored to balance book-associated activities with concurrent employment and doctoral studies (on a completely unrelated topic). I am grateful first to Walter S. Judd for his enthusiastic involvement in this project from day one. Our hours of animated discussion and debate, ranging in topic from the principles of phylogenetics and family circumscription to intriguing general gossip, have been instrumental in my development as a plant taxonomist. My colleague and good friend Kent D. Perkins deserves a very special thank-you for his help in all aspects of this study, including his information on word processing, references, and plant locations, as well as for proofreading numerous versions of the manuscript and cheerfully accompanying me on countless plant-collecting trips. Our friendship, based on our common love and appreciation of plants, has been deepened by our experiences and fond memories of working together on this project over the past fifteen years. The high quality of my illustrations in this work would not have been possible without the diligent tutelage of Marion R. Sheehan, my artistic mentor, who carefully cultivated my talents in the initial stages of the project. I am very grateful to Scott Zona, who critically reviewed the entire manuscript, updated the Arecaceae treatment, and patiently instructed me in the proper placement of adverbs. I thank also Linda Chandler, a careful and conscientious editor/scientist, who greatly improved the flow and consistency of the text.

Alone, I could not possibly have assembled all of the live plant material required for the illustrations, the most important part of the book. Robert L. Dressler and Brenda J. Herring top the list of the many friends and colleagues who provided me with plants and/or helped me collect them. Bob, an internationally renowned orchidologist, is also a local flora expert, and he gleefully supplied numerous species in prime condition, including custom-potted weeds and a 2 m tall *Yucca* plant complete with taproot-like rhizome. Brenda provided front-door service with refrigerated samples from her study site and also allowed me to accompany her on many collecting forays. Jack Fisher, Nina M. Woessner, and Roger W. Sanders graciously extended a helping hand during my several collecting efforts at Fairchild Tropical Garden; Nina, in particular, went out of her way to check on flowers for me between my trips. I thank also the following individuals for their help in locating certain plants: Howard

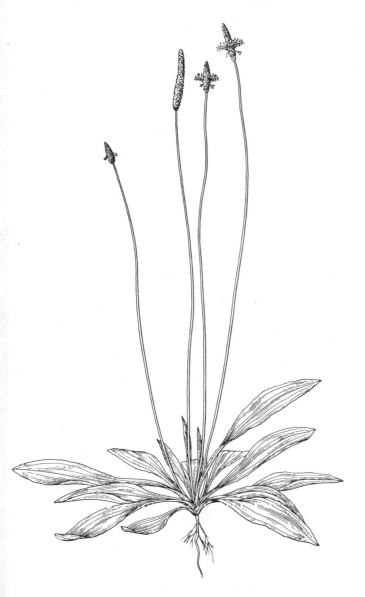

Adams, James Cobb, Ronald Determan, William J. Dunn, Mark Elliott, Angus K. Gholson, Jr., Jack L. Gibson, Dana G. Griffin III, Nancy Griffin, David W. Hall, Jeff Hillard, Walter S. Judd, Shirley M. Kooyman, Terry A. Lott, Darcy MacMahon, Richard Moyroud, John Nowak, David Pais, Scott Raybuck, Donna Ruhl, J. Dan Skean, Jr., Stuart Skeate, E. Eugene Spears, Jr., Stephen Sundlof, Bian Tan, Daniel B. Ward, Debbie L. White, W. Mark Whitten, Susan W. Williams, Tom Wood, and Scott Zona.

I benefited greatly from Phillip Martin's thorough and professional critique of my book proposal. I also very much appreciate all the members of the administration and staff at the University of North Carolina Press who were involved with this book. In particular, David Perry has had unwavering faith in this project since first perusing the proposal, and he patiently guided me through the long contractual phase and acceptance process. Pamela Upton expertly directed the book through production. Julia A. McVaugh painstakingly copyedited the text, and April Leidig-Higgins designed the attractive cover and text layout. I am also very impressed by the Marketing Department team, who carefully planned promotion for the book-to-be. I am grateful to Rudolph Schmid and an anonymous reviewer for their favorable readings of both the proposal and the manuscript for the Press.

I thank William L. Stern for his help in some anatomical aspects (including slide preparation); Robert F. Thorne for our lively correspondence concerning updates on his classification system; John Kartesz for suggestions on the format of plant common names; Gerald L. Smith for annotation of my *Hymenocallis* specimen; William J. Hess for identification of the *Sansevieria* plant; G. F. ("Stinger") Guala for help on the Poaceae treatment and specimens; Fred H. Utech for the pickled *Trillium* fruit with mature seeds; Susan Verhoek for information on *Manfreda*; Florence Sergile for translating several articles from French; Paloma Ibarra, Greg H. Cunningham, and John R. Hermsdorfer for the excellent photographic reductions of the illustrations; and Don Herring and Dianna C. Carver for computer assistance. The collections and library of the University of Florida herbarium (FLAS) have been crucial resources for this project.

I also extend sincere thanks and appreciation to my sister, Kayla S. Zomlefer, who has cheered me on through moments of doubt. Finally, I thank my parents, Jack and Dorothy Zomlefer, who always wholeheartedly supported my academic pursuits, even in my formative years when females were neither expected nor encouraged to excel in the sciences.

GUIDE TO FLOWERING PLANT FAMILIES

INTRODUCTION

The flowering plants (class Angiospermae) have dominated the land in most of the world for more than 100 million years and form the major component of the vegetation in most ecosystems. The estimated 233,885 species in 12,650 genera (Thorne 1992a) are extremely diverse in both habit and habitat. They include plants on which humans depend for sustenance, such as grains and legumes, as well as for numerous important products, including drugs, paper, and building materials. This large group usually is divided into 300 or more families—for example, into 437 by Thorne (1992a) and into 387 by Cronquist (1988). A knowledge of plant families is applicable anywhere in the world, whereas a familiarity with genera and species tends to be more regional in scope. Therefore, an understanding of flowering plants or even of the flora of any region begins with the recognition of families, rather than the much greater diversity of genera and species. Because of their predictive power, families are stressed as diagnostic tools in teaching, research, and floristics.

In North American taxonomy texts on angiosperm families, the examples and illustrations tend to emphasize horticultural and/or temperate (northeastern) plants. To counteract this trend, my earlier manuals (Zomlefer 1980, 1983, 1986, 1989) stressed the major temperate to tropical plant families characteristic of the southeastern United States. In this book, each family treatment includes a diagnosis and summary of characters, distribution data, important economic members, pollination ecology, and other relevant information, including mention of taxonomic problems within the family. The content of these discussions is based on the assumption that the reader has completed at least one introductory biology or botany course. The original plates of detailed illustrations depict dissections of familiar plants, which were chosen to represent both the diagnostic features and the diversity of each family. The various sections below further detail the intent and format of this book.

CHOICE OF CLASSIFICATION

The history of angiosperm classification shows a progression from purely artificial approaches (Linnaeus 1735, 1753) to attempts toward more "natural" phylogenetic considerations (see Stevens 1984 and 1986 for an excellent summary). Two general classification systems prevail in taxonomy texts today, those of (or based upon) Engler and Prantl (1887–1915) and Bessey (1915). The well-known Englerian system, once much more widely accepted, is the standard classification scheme used in many herbaria and floras. Today, most systematists disagree with Engler's concepts of "primitive" characters because he failed to recognize the significance of the reduction of parts in phylogenetic relationships; the major shortcoming of his system, therefore, was the tendency to equate "simple" with "primitive" (Cronquist 1965).

In contrast to Engler, Bessey (1915) regarded "primitive" families (comprising a "Ranalian complex") as being characterized by flowers with many, free, equal, and spirally arranged parts. Conversely, he considered "advancement" in floral evolution to be represented by the aggregation, fusion, reduction, and loss of parts. Although his taxonomy adopted concepts from other natural schemes (including those of de Jussieu and de Candolle), the principles of his work (outlined in detail) were mostly original. With some modifications, most modern classifications—for example, those of Cronquist (1981, 1983, 1988), Takhtajan (1969, 1980, 1983, 1991), Stebbins (1974), R. Dahlgren (1975, 1980, 1983; R. Dahlgren et al. 1981; R. Dahlgren and Rasmussen 1983; R. Dahlgren and Bremer 1985; G. Dahlgren 1989), and Thorne (1976, 1981, 1983, 1992a,b)—follow the Bessey tradition.

In Becker (1973) and Goldberg (1986, 1989), comparisons of the Englerian families with several versions of the systems mentioned above demonstrate that many of Engler's families are still basically accepted. Unfortunately, there previously has been no set criterion for family recognition: families historically have been circumscribed intuitively as "conceptually useful" units primarily based on phenetic considerations (similarities vs. differences in certain characters), coupled with "tradition" (Cronquist 1981). Flowering plant families usually are characterized by assemblages of reproductive features (flowers and fruits), plus some subsidiary data, such as vegetative and chemical characters (Stuessy 1990). Ideally, the most rigorous and objective circumscriptions of families and their relationships would result from a classification scheme built upon a series of phylogenetic analyses based strictly upon shared derived characters (monophyletic groupings); unfortunately, however, very few such analyses have been published on families of the class Angiospermae. The next chapter, " 'Dicotyledons' and Monocotyledons: An Example of Paraphyly," serves as an example of the application of cladistic methodology and concepts.

Until phylogenetic evidence regarding monophyletic circumscription for each of the angiosperm families is presented in a comprehensive form, this manual must be considered experimental, in that I use the family delineations of Thorne (1958, 1963, 1968, 1973a,b, 1974, 1975, 1976, 1977, 1981, 1983, 1992a,b, pers. comm.), whose system, thus far, has been presented mostly in synoptic listings. Taking a generalist approach, Thorne attempts to emphasize similarities and relationships between groups rather than differences; he avoids splitting "natural" groupings (especially families and orders) by utilizing an expanded hierarchy of superorders, suborders, and subfamilies. Thus, his family delimitations stress phenetic concerns, as well as phylogenetic relationships. He has found that some closely related groups, standardly treated as distinct families, have a phylogenetic gap too small to justify separate familial status. Moreover, he avoids multiplying taxa unnecessarily where common ancestry seems evident. As a result, Thorne's families (and other taxa) often tend to be more inclusive than those of other workers, while his use of supercategories and subcategories allows some emphasis on differences. Although he himself is not overly concerned about avoiding paraphyletic groups (Thorne pers. comm.), his taxa frequently are monophyletic since he constantly adjusts his system to incorporate the most recent available studies, which often are based on cladistic analyses of morphological and/or molecular data.

The following list of families included in this book is adapted from Thorne (1992a, pers. comm.), with the exception of the expanded delimitations of the Urticaceae (including the Moraceae) and Apiaceae (including the Araliaceae) maintained from Thorne's earlier (1983) scheme. His original broader circumscriptions of these two families probably more accurately reflect the true phylogeny (see the commentary under these families, and Judd et al. 1994). Other notable examples of Thorne's expanded, nontraditional family concepts are: Apocynaceae (including the Asclepiadaceae), Celastraceae (including the Hippocrateaceae), Cornaceae (including the Nyssaceae), Ericaceae (including the Pyrolaceae and Monotropaceae), Lythraceae (including the Punicaceae and Sonneratiaceae), Papaveraceae (including the Fumariaceae), Sapindaceae (including the Aceraceae and Hippocastanaceae), Scrophulariaceae (including the Orobanchaceae), and Typhaceae (including the Sparganiaceae). However, several families, such as the Capparaceae-Brassicaceae and Malvaceae-Bombacaceae, even though phenetically close, have been maintained separately by Thorne. Recent cladistic analyses (Judd et al. 1994) on certain family pairs support the combination of some of these families (see my commentary under Malvaceae and

Brassicaceae). Other unconventional circumscriptions by Thorne (1992a) incorporating recent cladistic studies include a restricted Verbenaceae (basically only the subfamily Verbenoideae) plus an expanded Lamiaceae (including much of the remainder of the traditional Verbenaceae *s.l.*). *Sambucus* and *Viburnum* have been transferred to the Adoxaceae, leaving a more narrowly defined Caprifoliaceae. The Loganiaceae include only the original subfamily Loganioideae (Thorne pers. comm.), and the Liliaceae comprise only 22 genera. For comparison, see Appendix A, a list of the approximately equivalent families covered here as arranged and delineated by Cronquist (1988).

Although disconcerting to some taxonomists, Thorne's unconventional circumscriptions emphasize taxonomy as a growing discipline, with classifications changing as new data become available. In particular, new and exciting molecular studies (specifically those based on the sequencing of the chloroplast gene, *rbc*L) have the potential to support or greatly alter traditional perspectives in plant systematics (Zurawski and Clegg 1993). Thorne's scheme is elastic, periodically updated in detailed lists; this book, however, is static and cannot include new data that are becoming available even as this manuscript goes to press. Because Thorne has yet to publish a detailed explanation for most of his circumscriptions, his scheme has not been widely adopted. This book attempts to highlight briefly some of these areas of controversy, piecing together evidence from the voluminous literature cited by Thorne in his numerous publications.

Thorne (1992a) briefly mentions the near-impossibility of constructing an "angiospermous family tree," although recent cladistic work certainly has contributed to our understanding of the phylogenetic relationships of angiosperm families. Therefore, this book does not stress higher ranks, such as subclass, superorder, order, and suborder, for they imply a level of knowledge that currently does not exist.

CHOICE OF FAMILIES AND FAMILY LIST

One hundred and thirty families have been chosen on the basis of floristic dominance, phylogenetic interest, and economic importance. They vary greatly in size, are distributed throughout Thorne's (1992a) superorders, and are usually covered in introductory or graduate taxonomy courses. The list attempts to strike a balance between predominantly herbaceous and predominantly woody families, as well as between those characteristic of either temperate or tropical areas. Although families with native and naturalized species predominate, a few with only horticultural representa-

tives in our area are included, such as the Strelitziaceae. Monocotyledonous families are well represented, especially the segregate families formerly included in the Liliaceae *s.l.* One hundred and fifteen families are covered by detailed treatments (plus illustrations); 15 families, indicated by an asterisk (*) in the following list, are described in the various comparative charts (also accompanied by separate plates of illustrations).

3

Class Angiospermae (Magnoliopsida)

Subclass Dicotyledoneae (Magnoliidae)

   Superorder Magnolianae

      Order Magnoliales
      (1) Illiciaceae
      (2) Magnoliaceae
      (3) Annonaceae
      (4) Aristolochiaceae
      (5) Lauraceae
      (6) Piperaceae

      Order Nelumbonales
      (7) Nelumbonaceae*

      Order Berberidales
      (8) Menispermaceae
      (9) Ranunculaceae
      (10) Papaveraceae (incl. Fumariaceae)

   Superorder Nymphaeanae

      Order Nymphaeales
      (11) Nymphaeaceae
      (12) Cabombaceae*

   Superorder Caryophyllanae (Centrospermae)

      Order Caryophyllales (Chenopodiales)
      (13) Caryophyllaceae
      (14) Portulacaceae
      (15) Cactaceae
      (16) Phytolaccaceae
      (17) Nyctaginaceae
      (18) Chenopodiaceae
      (19) Amaranthaceae

   Superorder Theanae

      Order Theales
      (20) Theaceae
      (21) Aquifoliaceae
      (22) Sarraceniaceae
      (23) Clusiaceae (Guttiferae; incl. Hypericaceae)

      Order Ericales
      (24) Ericaceae (incl. Pyrolaceae and Monotropaceae)

Order Styracales (Ebenales)
(25) Sapotaceae

Order Primulales
(26) Myrsinaceae

Order Polygonales
(27) Polygonaceae

4 Superorder Celastranae

Order Celastrales
(28) Celastraceae (incl. Hippocrateaceae)

Superorder Malvanae

Order Malvales
(29) Malvaceae
(30) Bombacaceae*
(31) Sterculiaceae
(32) Tiliaceae

Order Urticales
(33) Ulmaceae
(34) Urticaceae (incl. Moraceae)

Order Rhamnales
(35) Rhamnaceae

Order Euphorbiales
(36) Euphorbiaceae

Superorder Violanae

Order Violales
(37) Cistaceae
(38) Violaceae
(39) Salicaceae
(40) Passifloraceae
(41) Turneraceae
(42) Cucurbitaceae
(43) Begoniaceae

Order Brassicales (Capparales)
(44) Brassicaceae (Cruciferae)
(45) Capparaceae*

Superorder Santalanae

Order Santalales
(46) Viscaceae

Superorder Geranianae

Order Linales
(47) Zygophyllaceae

Order Rhizophorales
(48) Rhizophoraceae

Order Geraniales
(49) Oxalidaceae
(50) Geraniaceae

Order Polygalales (Malpighiales)
(51) Malpighiaceae
(52) Polygalaceae

Superorder Rutanae

Order Rutales
(53) Rutaceae
(54) Meliaceae
(55) Anacardiaceae
(56) Sapindaceae (incl. Aceraceae and Hippocastanaceae)
(57) Fabaceae (Leguminosae)

Superorder Rosanae

Order Hamamelidales
(58) Platanaceae
(59) Hamamelidaceae

Order Juglandales
(60) Juglandaceae
(61) Myricaceae

Order Betulales (Fagales)
(62) Betulaceae
(63) Fagaceae

Order Rosales
(64) Rosaceae

Superorder Cornanae

Order Hydrangeales
(65) Hydrangeaceae
(66) Escalloniaceae*

Order Cornales
(67) Vitaceae
(68) Cornaceae (incl. Nyssaceae)

Order Araliales
(69) Apiaceae (Umbelliferae; incl. Araliaceae)

Order Dipsacales
(70) Caprifoliaceae
(71) Adoxaceae (incl. *Sambucus* and *Viburnum*)*

Superorder Asteranae

Order Asterales
(72) Asteraceae (Compositae)

Order Campanulales
(73) Campanulaceae

Superorder Solananae

Order Solanales
(74) Solanaceae
(75) Convolvulaceae
(76) Boraginaceae
(77) Polemoniaceae

Superorder Myrtanae

Order Myrtales
(78) Lythraceae
(79) Melastomataceae
(80) Combretaceae
(81) Onagraceae
(82) Myrtaceae

Superorder Gentiananae

Order Gentianales
(83) Loganiaceae (only Loganioideae)
(84) Rubiaceae
(85) Apocynaceae (incl. Asclepiadaceae)
(86) Gentianaceae

Order Scrophulariales (Lamiales, Bignoniales)
(87) Oleaceae
(88) Bignoniaceae
(89) Scrophulariaceae (incl. Orobanchaceae)
(90) Plantaginaceae
(91) Lentibulariaceae
(92) Acanthaceae
(93) Verbenaceae (incl. only Verbenoideae)
(94) Lamiaceae (Labiatae; incl. Caryopteridoideae, Viticoideae, Chloanthoideae)

Subclass Monocotyledoneae (Liliidae)

Superorder Lilianae

Order Liliales
(95) Melanthiaceae*
(96) Alstroemeriaceae*
(97) Liliaceae (incl. Uvulariaceae and Calochortaceae)
(98) Trilliaceae*
(99) Iridaceae

Order Asparagales
(100) Asparagaceae (incl. Convallariaceae, Herreriaceae, and Ruscaceae)
(101) Hemerocallidaceae*
(102) Agavaceae
(103) Dracaenaceae (incl. Asteliaceae and Nolinaceae)*
(104) Amaryllidaceae
(105) Alliaceae*

Order Dioscoreales
(106) Smilacaceae
(107) Dioscoreaceae

Order Orchidales
(108) Orchidaceae

Superorder Alismatanae

Order Alismatales
(109) Alismataceae
(110) Hydrocharitaceae

Superorder Aranae

Order Arales
(111) Araceae
(112) Lemnaccae

Superorder Arecanae

Order Arecales
(113) Arecaceae (Palmae)

Superorder Commelinanae

Order Bromeliales
(114) Bromeliaceae
(115) Pontederiaceae
(116) Haemodoraceae

Order Typhales
(117) Typhaceae (incl. Sparganiaceae)

Order Zingiberales
(118) Zingiberaceae
(119) Costaceae
(120) Musaceae*
(121) Strelitziaceae*
(122) Heliconiaceae*
(123) Cannaceae
(124) Marantaceae

Order Commelinales
(125) Xyridaceae
(126) Commelinaceae
(127) Eriocaulaceae

Order Juncales (Cyperales)
(128) Juncaceae
(129) Cyperaceae

Order Poales
(130) Poaceae (Gramineae)

FAMILY TREATMENTS

The family treatments are divided into convenient sections (each explained in detail below) and accompanied by intricate plates of illustrations and comparative charts. The same basic terminology is used throughout the text. These technical terms (defined in the glossary) generally follow *Taxonomy of Vascular Plants* (Lawrence 1951), excepting some inflorescence and fruit definitions for which I favor a less restrictive application (see the introduction to the glossary). The family descriptions and discussions were compiled from standard texts (listed below), monographs, other specialized studies, and, most importantly, from many years of detailed personal observations of field and herbarium specimens. I have stressed in this book the extensive documentation of references. As the project has progressed over the past fifteen years, I have attempted to update the text by adding pertinent new information whenever possible.

Family diagnoses: Each family treatment begins with a very detailed description (diagnosis) of the major plant parts: habit, leaves, inflorescence, flowers, perianth, androecium, gynoecium, and fruit. These diagnoses are as consistent and parallel as possible. Care has been taken to include only those characters appropriate to Thorne's (1992a) particular circumscription of the family. The diagnoses do not include the many exceptions to the general family characteristics, and they stress the features of the species in North America north of Mexico.

The framework of the family diagnoses was assembled from the following primary sources: Rendle (1925), Lawrence (1951), Wood (1958 and subsequent papers of this series), Melchior (1964), Cronquist (1968, 1981), Hutchinson (1969, 1973), Willis (1973), Heywood (1978), Benson (1979), Dahlgren and Clifford (1982), Dahlgren et al. (1985), and family monographs (listed at the end of each family treatment). Additional data were added, when necessary, from Porter (1967), Smith (1977), Baumgardt (1982), Jones and Luchsinger (1986), Radford (1986), Mabberley (1987), and Hickey and King (1988); most of these references, however, are synopses of Cronquist (1968, 1981).

Family characterization: For easy reference, each diagnosis is followed by this brief summary of distinguishing family features that emphasizes plants in the United States. Occasionally, as for the Haemodoraceae, the morphology of a family on a worldwide basis differs significantly from typical plants in North America. In these cases, a discussion of this variation is included under the "Commentary" section.

The chemical and anatomical features of the family are listed following the summary of distinguishing morphological characters. Unless referenced by particular specialized publications, general information on plant chemistry comes from Willaman (1961), Raffauf (1970), Willaman and Li (1970), Gibbs (1974), Jensen et al. (1975), and Robinson (1983), and anatomical details come from Metcalfe and Chalk (1950, 1979, 1985) and Esau (1965, 1977).

Genera/species: Worldwide genera and species totals are approximate, because the taxonomy of many groups is not well understood; the numbers probably are most accurate for small families and recently monographed taxa. The estimates are usually the numbers given by Thorne (1992a), who based them on Willis (1973), Mabberley (1987), and appropriate monographs. In several families (e.g., Orchidaceae), however, these estimates are taken directly from other recent publications (cited parenthetically following the numbers) or from Thorne (pers. comm.), who reevaluated some of the totals in his 1992a publication. When a family comprises six or fewer genera (e.g., Typhaceae), all genera (with the number of estimated species) are listed in this section.

Distribution: The worldwide distribution of each family is summarized from Willis (1973), Mabberley (1987), and Thorne (1992a).

Major genera: Unless otherwise indicated, the largest genera (and accompanying ranges in number of species) are based on Willis (1973) and Mabberley (1987), used in conjunction with Brummitt (1992), who conveniently lists all genera in each family. In a few instances (e.g., Orchidaceae), this genus list with species numbers is from a cited reference. This section is deleted in those small families (six or fewer genera) where all genera (with approximate species numbers) are included under "Genera/species."

U.S./Canadian representatives: The checklist compiled by Kartesz and Kartesz (1980) is the origin for estimates of genera and species in "the United States and Canada" (all the continental United States and Canada plus Greenland, Hawaii, Puerto Rico, and the Virgin Islands). These numbers have been adjusted to reflect Thorne's (1992a) circumscription of the particular family. The largest genera are given in order of size. Every genus is listed (with species number estimates) for families represented by six or fewer genera.

Economic plants and products: Examples of the most important economic plants in each family, including edible, medicinal, poisonous, and weedy species, are noted here. This section concludes with examples of plants commonly grown as ornamentals. Uphof (1968), Willis (1973), Mabberley (1987), and Facciola (1990) are the primary sources used to verify economic plants; Muenscher (1951), Kingsbury (1964), and Morton (1977) are major references for poisonous species. Approximate numbers of horticulturally important genera, examples of commonly cultivated species, and their common names are derived from Bailey and Bailey (1976) and Huxley (1992). The format for common names here (and in the Commentary section) is based on the conventions suggested by Kartesz and Thieret (1991). Some exceptions have been made for some names in common use: in particular, the designations "plant," "shrub," "tree," and "vine" in a common name are treated as separate words rather than as hyphenated or unhyphenated suffixes.

Commentary: This book differs from other similar manuals in having this type of commentary section, modeled after the expository style of Rendle (1925), who emphasized interesting features and taxonomic history of the families. The discussions are based on my personal observations in conjunction with the cited publications. I have attempted to emphasize the relevant morphological and anatomical features included on the accompanying detailed plate(s) of illustrations for each family.

Although the content of the commentary differs among families, pollination biology is mentioned consistently since the primary function of flowers is sexual reproduction—flowers do not exist solely as a source of characters for taxonomists! Knuth (1906, 1908, 1909), Proctor and Yeo (1972), Faegri and van der Pijl (1980), and Bertin and Newman (1993) were consulted as general references on pollination, in addition to the cited specialized studies. Unfortunately, for several families (especially those of tropical areas) little or no information is available (Maas and Westra 1993).

References cited: The reference list after each family summary is a compilation of the supplemental, often current, sources of information cited directly in the various sections of the family treatments. Many major and/or standard works on particular families are included. These listings are not, however, intended to be complete literature reviews for each family. The abbreviations for the journals are standardized according to *Botanico-Periodicum-Huntianum/supplementum* (Bridson and Smith 1991), the revised and updated version of the original "*B-P-H*" (Lawrence et al. 1968).

TABLES

Twenty-two detailed charts, found throughout the text, compare pertinent characteristics of certain taxa, usually subfamilies or related families. In some cases (e.g., Table 13, comparing the Caprifoliaceae *s.s.* and Adoxaceae), the tables aid in highlighting the features of certain nontraditional family delineations. The generalized table constituting Appendix B summarizes the important features of all 130 families covered in this text. Keys to all of the families of angiosperms are beyond the scope of this text and probably would not work successfully at local levels due to the numerous exceptions. Recommended are keys to families (or gen-

era) in appropriate floras for particular local geographical areas; there are also several excellent wide-ranging, higher-level keys, including Hutchinson (1967), Simpson and Janos (1974), Geesink et al. (1981), Batson (1984a,b), and Davis and Cullen (1989). In addition, due to new technology, several interactive computer keys such as MEKA/MEKAEDIT (Duncan and Meacham 1986) and INTKEY (Watson and Dallwitz 1991a,b) are now available.

ILLUSTRATIONS AND PLANT MATERIAL

The drawings of plants in this book are my own original artwork prepared exclusively for this series. The plants illustrated for each family have the important morphological characters mentioned in the discussion and usually are widespread species characteristic of the native and naturalized flora of the eastern United States. Often these particular species (or similar species) are common throughout the remainder of the United States, as well. A very few horticultural species (such as *Heliconia rostrata* and *Strelitzia reginae*) are also included, however. My location in north-central Florida is ideal for collecting a representative range of temperate to tropical species. The plates often depict several plants, because a single species usually cannot portray adequately the scope of variation within a family.

In compiling the collecting list for the initial stages of the project (Zomlefer 1980, 1983), I first consulted *Manual of the Southeastern Flora* (Small 1933) and *Manual of the Vascular Flora of the Carolinas* (Radford et al. 1968) to determine possible species that best illustrated the familial characters; for subsequent families (Zomlefer 1986, 1989), I used *The Biology of Trees Native to Tropical Florida* (Tomlinson 1980), *Guide to The Vascular Plants of Central Florida* (Wunderlin 1982), and *Guide to the Vascular Plants of the Florida Panhandle* (Clewell 1985) to determine appropriate plants. The flora of Long and Lakela (1976) was consulted occasionally as a reference for the more tropical species available in this area. *Aquatic and Wetland Plants of Southeastern United States* (Godfrey and Wooten 1979, 1981) and *Trees, Shrubs, and Woody Vines of Northern Florida and Adjacent Georgia and Alabama* (Godfrey 1988) were valuable for supplemental information. The collections of the University of Florida herbarium (FLAS) were indispensable in determining plant locations.

The 312 species of primarily native and naturalized plants depicted in these plates were drawn mostly from live material collected by me (frequently with the assistance of the botanists and horticulturists cited in the acknowledgments). Herbarium specimens were used, however, for a very few of the fruit illustrations (and to supplement the fresh material). Whenever feasible, freehand habit sketches were completed in the field, and the fresh flowers were dissected as soon as possible. Even with the best of care, cut (or even potted) wild flowers often wilt quickly, or at best last only a few days. Certain plants (such as the weedy *Sambucus canadensis*) wilt immediately; other flowers (such as those of *Thalia geniculata*) are naturally ephemeral. Fragile flowers like these were best transported in inflated zip-closure plastic bags placed on ice in an insulated cooler. Not all flowers are open all day: the ephemeral flowers of *Portulaca*, for example, are expanded for only a few hours daily in full sunlight, and those of *Canna flaccida* and *Mirabilis jalapa* open at dusk. The inflorescences of other species, such as those of *Costus speciosus* and *Xyris* species, produce only one new flower per day.

Thus, many species required numerous trips to the same and/or additional location(s) for replacement material—and later, for fresh fruiting specimens. These return forays could be frustrating, for I often found that species had vanished due to habitat destruction (e.g., roadside mowing, herbicide application, drainage, and construction). Due to the rise in violent crime over the years, personal safety became a serious concern that thwarted my collecting on numerous occasions. The problems of obtaining a legal permit also delayed or prevented my access to certain sites. In addition, the time and effort spent in locating both staminate and carpellate plants of the 20 dioecious and polygamodioecious species were equivalent to searching for two different species.

For the purpose of future reference, many plants were photographed (35 mm color slides; SX-70 Polaroid prints) and pressed, and flowers and fleshy fruits were preserved in 60% ethanol. Vouchers are deposited at FLAS. I have attempted to consider the natural variability of each taxon and to illustrate the most typical individuals based on fieldwork.

The chapter "Observing, Dissecting, and Drawing Flowering Plants" summarizes some fundamental aspects of both examining and sketching live material, based on my experiences in preparing the illustrations for this text. All flower dissections were done with the aid of a Wild M5A Stereomicroscope, and sketching was facilitated by a drawing tube attachment. Anatomical and palynological illustrations were drawn using a Wild M20 compound microscope equipped with a camera lucida. In addition to the numerous standard sources cited previously (under Family Treatments), the drawings in the following publications also were valuable resources: Engler and Prantl (1887–1915),

Sargent (1890), Swingle (1928), Johnson (1931), Pool (1941), Erdtman (1966, 1969), Wood (1974), and Corner (1976).

For inking the plates, I developed a style based on the methods of Marion Ruff Sheehan, who illustrated such classic works as *Taxonomy of Vascular Plants* (Lawrence 1951) and *Hortus Third* (Bailey and Bailey 1976). Excessive shading, which can obliterate detail, has been avoided. Cut material is indicated by a dashed line, and different tissue types are separated by dotted lines. The shading of xylem (radiating lines), phloem (stippling), and sclerenchyma (solid black) in the anatomical drawings follows the conventions of Metcalfe and Chalk (1950). In fruit and seed dissections, light stippling indicates endosperm or perisperm, and a darker screen represents fleshy tissue.

The final composition of each family plate required much preliminary planning. The parts are arranged as logically as possible while still taking aesthetics into consideration. The darkest and heaviest drawings (usually the habit and fruit) are placed near the bottom whenever possible. Lighter flowers and their components, appearing in the upper portion, are arranged in a systematic sequence of dissection. When more than one species is included on a plate, parts of each species are grouped together. The final illustrations were inked with Gillott 659 crowquill nibs and Pelikan drawing ink on Bienfang Satin Design (No. 150H) drafting vellum. Each figure was reduced to 50% of its original size for this publication.

REFERENCES CITED

Bailey, L. H., and E. Z. Bailey. 1976. *Hortus third: A concise dictionary of plants cultivated in the United States and Canada.* Revised by the Bailey Hortorium. Macmillan, New York.

Batson, W. T. 1984a. *A guide to the genera of the plants of eastern North America.* University of South Carolina Press, Columbia.

———. 1984b. *A guide to the genera of the plants of western North America.* University of South Carolina Press, Columbia.

Baumgardt, J. P. 1982. *How to identify flowering plant families: A practical guide for horticulturists.* Timber Press, Beaverton, Ore.

Becker, K. M. 1973. A comparison of angiosperm classification systems. *Taxon* 22:19–50.

Benson, L. 1979. *Plant classification.* 2d ed. D. C. Heath, Lexington, Mass.

Bertin, R. I., and C. M. Newman. 1993. Dichogamy in angiosperms. *Bot. Rev.* (Lancaster) 59:112–152.

Bessey, C. E. 1915. The phylogenetic taxonomy of flowering plants. *Ann. Missouri Bot. Gard.* 2:109–164.

Bridson, G. D. R., and E. R. Smith, eds. 1991. *Botanico-Periodicum-Huntianum/supplementum.* Hunt Institute for Botanical Documentation, Carnegie Mellon University, Pittsburgh.

Brummitt, R. K. 1992. *Vascular plant families and genera.* Royal Botanic Gardens, Kew.

Clewell, A. F. 1985. *Guide to the vascular plants of the Florida panhandle.* University Presses of Florida, Gainesville.

Corner, E. J. H. 1976. *The seeds of dicotyledons.* 2 vols. Cambridge University Press, London.

Cronquist, A. 1965. The status of the general system of classification of flowering plants. *Ann. Missouri Bot. Gard.* 52:281–303.

——. 1968. *The evolution and classification of flowering plants.* Houghton Mifflin, Boston.

——. 1981. *An integrated system of classification of flowering plants.* Columbia University Press, New York.

——. 1983. Some realignments in the dicotyledons. *Nordic J. Bot.* 3:75–83.

——. 1988. *The evolution and classification of flowering plants.* 2d ed. New York Botanical Garden, Bronx.

Dahlgren, G. 1989. The last Dahlgrenogram: System of classification of the dicotyledons. In *Plant taxonomy, phytogeography and related subjects,* ed. K. Tan, R. R. Mill, and T. S. Elias, pp. 249–260. Edinburgh University Press, Edinburgh.

Dahlgren, R. M. T. 1975. A system of classification of angiosperms to be used to demonstrate the distribution of characters. *Bot. Not.* 128:119–147.

——. 1980. A revised system of the angiosperms. *Bot. J. Linn. Soc.* 80:91–124.

——. 1983. General aspects of angiosperm evolution and macrosystematics. *Nordic J. Bot.* 3:119–149.

—— and K. Bremer. 1985. Major clades of the angiosperms. *Cladistics* 1:349–368.

—— and H. T. Clifford. 1982. *The monocotyledons: A comparative study.* Academic Press, London.

——, ——, and P. F. Yeo. 1985. *The families of the monocotyledons.* Springer-Verlag, Berlin.

—— and F. N. Rasmussen. 1983. Monocot evolution: Characters and phylogenetic estimation. In *Evolutionary biology,* vol. 16, ed. M. K. Hecht, B. Wallace, and G. T. Prance, pp. 255–395. Plenum Press, New York.

——, S. Rosendal-Jensen, and B. J. Nielsen. 1981. A revised classification of the angiosperms with comments on correlation between chemical and other characters. In *Phytochemistry and angiosperm taxonomy,* ed. D. A. Young and D. S. Seigler, pp. 149–204. Praeger Publishers, New York.

Davis, P. H., and J. Cullen. 1989. *The identification of flowering plant families.* 3d ed. Cambridge University Press, Cambridge.

Duncan, T., and C. A. Meacham. 1986. MEKA/MEKA-EDIT. University of California, Berkeley.

Engler, A., and K. Prantl. 1887–1915. *Die natürlichen Pflanzenfamilien,* parts II–IV. G. Kreysing, Leipzig.

Erdtman, G. 1966. *Pollen morphology and plant taxonomy.* Hafner, New York.

——. 1969. *Handbook of palynology.* Hafner, New York.

Esau, K. 1965. *Plant anatomy.* John Wiley and Sons, New York.

——. 1977. *Anatomy of seed plants.* 2d ed. John Wiley and Sons, New York.

Facciola, S. 1990. *Cornucopia: A source book of edible plants.* Kampong Publications, Vista, Calif.

Faegri, K., and L. van der Pijl. 1980. *The principles of pollination ecology.* 3d ed. Pergamon Press, Oxford.

Geesink, L., A. J. M. Leeuwenberg, C. E. Ridsdale, and V. F. Veldkamp. 1981. *Thonner's analytical key to the families of flowering plants.* Leiden University Press, The Hague.

Gibbs, R. D. 1974. *Chemotaxonomy of flowering plants.* 4 vols. McGill-Queen's University Press, Montreal.

Godfrey, R. K. 1988. *Trees, shrubs, and woody vines of northern Florida and adjacent Georgia and Alabama.* University of Georgia Press, Athens.

—— and J. W. Wooten. 1979. *Aquatic and wetland plants of southeastern United States: Monocotyledons.* University of Georgia Press, Athens.

—— and ——. 1981. *Aquatic and wetland plants of southeastern United States: Dicotyledons.* University of Georgia Press, Athens.

Goldberg, A. 1986. Classification, evolution, and phylogeny of the families of dicotyledons. *Smithsonian Contr. Bot.* 58:1–314.

——. 1989. Classification, evolution, and phylogeny of the dicotyledons. *Smithsonian Contr. Bot.* 71:1–73.

Heywood, V. H., ed. 1978. *Flowering plants of the world.* Mayflower Books, New York.

Hickey, M., and C. J. King. 1988. *100 families of flowering plants.* 2d ed. Cambridge University Press, Cambridge.

Hutchinson, J. 1967. *Key to the families of flowering plants of the world.* Oxford University Press, Oxford.

——. 1969. *Evolution and phylogeny of flowering plants. Dicotyledons: Facts and theory.* Academic Press, London.

——. 1973. *The families of flowering plants.* 3d ed. Oxford University Press, Oxford.

Huxley, A., ed. 1992. *The new Royal Horticultural Society dictionary of gardening.* 4 vols. Macmillan, London.

Jensen, S. R., B. J. Nielsen, and R. Dahlgren. 1975. Iridoid compounds, their occurrence and systematic importance in the angiosperms. *Bot. Not.* 128:148–180.

Johnson, A. M. 1931. *Taxonomy of flowering plants.* Century, New York.

Jones, S. B., and L. E. Luchsinger. 1986. *Plant systematics.* McGraw-Hill, New York.

Judd, W. S., R. W. Sanders, and M. J. Donoghue. 1994. Angiosperm family pairs: Preliminary cladistic analyses. *Harvard Pap. Bot.* 5:1–51.

Kartesz, J. T., and R. Kartesz. 1980. *A synonymized checklist of the vascular flora of the United States, Canada, and Greenland.* Vol. 2, *The biota of North America.* University of North Carolina Press, Chapel Hill.

—— and J. W. Thieret. 1991. Common names for vascular plants: Guidelines for use and application. *Sida* 14:421–434.

Kingsbury, J. M. 1964. *Poisonous plants of the United States and Canada.* Prentice-Hall, Englewood, N.J.

Knuth, P. 1906. *Handbook of flower pollination.* Trans. J. R. A. Davis. Vol. 1, *Introduction and literature.* Oxford University Press, Oxford.

——. 1908. Ibid., vol. 2, *Ranunculaceae to Stylidieae.* Oxford University Press, Oxford.

——. 1909. Ibid., vol. 3, *Goodenovieae to Cyadeae.* Oxford University Press, Oxford.

Lawrence, G. H. M. 1951. *Taxonomy of vascular plants.* Macmillan, New York.

——, A. F. G. Buchheim, G. S. Daniels, and H. Dolezal, eds. 1968. *Botanico-Periodicum-Huntianum.* [*B-P-H*]. Hunt Botanical Library, Pittsburgh.

Linnaeus, C. 1735. *Systema naturae*. Lugduni Batavorum, Leiden.

———. 1753. *Species plantarum*. 2 vols. Holmiae Impensis Laurentii Salvii, Stockholm.

Long, R. W., and O. Lakela. 1976. *A flora of tropical Florida*. Banyan Books, Miami.

Maas, P. J. M., and L. Y. T. Westra. 1993. *Neotropical plant families*. Koeltz Scientific Books, Königstein, Germany.

Mabberley, D. J. 1987. *The plant-book*. Cambridge University Press, Cambridge.

Melchior, H., ed. 1964. *A. Engler's Syllabus der Pflanzenfamilien*. 12th ed. Vol. 2. Gebrüder Borntraeger, Berlin.

Metcalfe, C. R., and L. Chalk. 1950. *Anatomy of the dicotyledons*. 2 vols. Oxford University Press, Oxford.

——— and ———, eds. 1979. *Anatomy of the dicotyledons*. 2d ed. Vol. 1, *Systematic anatomy of leaf and stem, with a brief history of the subject*. Oxford University Press, Oxford.

——— and ———, eds. 1985. Ibid., vol. 2, *Wood structure and conclusion of the general introduction*. Oxford University Press, Oxford.

Morton, J. F. 1977. Poisonous and injurious higher plants and fungi. In *Forensic medicine*, vol. 3, *Environmental hazards*, ed. C. G. Tedeschi, W. G. Eckert, and L. G. Tedeschi, pp. 1456–1567. W. B. Saunders, Philadelphia.

Muenscher, W. C. 1951. *Poisonous plants of the United States*. Macmillan, New York.

Pool, R. J. 1941. *Flowers and flowering plants*. McGraw-Hill, New York.

Porter, C. L. 1967. *Taxonomy of flowering plants*. 2d ed. W. H. Freeman, San Francisco.

Proctor, M., and P. Yeo. 1972. *The pollination of flowers*. Taplinger, New York.

Radford, A. E. 1986. *Fundamentals of plant systematics*. Harper and Row, New York.

———, H. E. Ahles, and C. R. Bell. 1968. *Manual of the vascular flora of the Carolinas*. University of North Carolina Press, Chapel Hill.

Raffauf, R. F. 1970. Table of alkaloids by family; Index of genera. In *A handbook of alkaloids and alkaloid-containing plants*. Wiley-Interscience, New York.

Rendle, A. B. 1925. *The classification of flowering plants*. 2 vols. Cambridge University Press, Cambridge.

Robinson, T. 1983. *The organic constituents of higher plants*. Cordus Press, North Amherst, Mass.

Sargent, C. S. 1890. *Silva of North America*. Vols. 1–14. Murray Printing, Cambridge, Mass.

Simpson, D. R., and D. Janos. 1974. *Punch card key to the families of dicotyledons of the Western Hemisphere south of the United States*. Field Museum of Natural History, Chicago.

Small, J. K. 1933. *Manual of the southeastern flora*. "Publ. by the author," New York.

Smith, J. P. 1977. *Vascular plant families*. Mad River Press, Eureka, Calif.

Stebbins, G. L. 1974. *Flowering plants: Evolution above the species level*. Belknap Press of Harvard University Press, Cambridge, Mass.

Stevens, P. F. 1984. Metaphors and typology in the development of botanical systematics 1690–1960, or the art of putting new wine in old bottles. *Taxon* 33:169–211.

———. 1986. Evolutionary classification in botany, 1960–1985. *J. Arnold Arbor.* 67:313–339.

Stuessy, T. F. 1990. *Plant taxonomy*. Columbia University Press, New York.

Swingle, D. B. 1928. *A textbook of systematic botany*. McGraw-Hill, New York.

Takhtajan, A. 1969. *Flowering plants: Origin and dispersal*. Smithsonian Institution Press, Washington, D.C.

———. 1980. Outline of the classification of flowering plants (Magnoliophyta). *Bot. Rev.* (Lancaster) 46:225–359.

———. 1983. The systematic arrangement of dicotyledonous families. In *Anatomy of the dicotyledons*, 2d ed., vol. 2, ed. C. R. Metcalfe and L. Chalk, pp. 180–201. Oxford University Press, Oxford.

———. 1991. *Evolutionary trends in flowering plants*. Columbia University Press, New York.

Thorne, R. F. 1958. Some guiding principles of angiosperm phylogeny. *Brittonia* 10:72–77.

———. 1963. Some problems and guiding principles of angiosperm phylogeny. *Amer. Naturalist* 97:287–305.

———. 1968. Synopsis of a putatively phylogenetic classification of the flowering plants. *Aliso* 6:57–66.

———. 1973a. The "Amentiferae" or Hamamelidae as an artificial group: A summary statement. *Brittonia* 25:395–405.

———. 1973b. Inclusion of the Apiaceae (Umbelliferae) in the Araliaceae. *Notes Roy. Bot. Gard. Edinburgh* 32:161–165.

———. 1974. A phylogenetic classification of the Annoniflorae. *Aliso* 8:147–209.

———. 1975. Angiosperm phylogeny and geography. *Ann. Missouri Bot. Gard.* 62:362–367.

———. 1976. A phylogenetic classification of the Angiospermae. *Evol. Biol.* 9:35–106.

———. 1977. Some realignments in the Angiospermae. *Pl. Syst. Evol.*, Suppl. 1:299–319.

———. 1981. Phytochemistry and angiosperm phylogeny: A summary statement. In *Phytochemistry and angiosperm phylogeny*, ed. D. A. Young and D. S. Seigler, pp. 233–295. Praeger, New York.

———. 1983. Proposed new realignments in the angiosperms. *Nordic J. Bot.* 3:85–117.

———. 1992a. Classification and geography of the flowering plants. *Bot. Rev.* (Lancaster) 58:225–348.

———. 1992b. An updated phylogenetic classification of the flowering plants. *Aliso* 13:365–389.

Tomlinson, P. B. 1980. *The biology of trees native to tropical Florida*. "Publ. by the author," Petersham, Mass.

Uphof, J. C. T. 1968. *Dictionary of economic plants*. 2d ed. J. Cramer, Lehre, Germany.

Watson, L., and M. J. Dallwitz. 1991a. The families of angiosperms: Automated descriptions, with interactive identification and information retrieval. *Austral. Syst. Bot.* 4:681–695.

——— and ———. 1991b. INTKEY. Taxonomy Laboratory, Research School of Biological Sciences, Australian National University, Canberra.

Willaman, J. J. 1961. Alkaloid-bearing plants and their contained alkaloids. *Techn. Bull. U.S.D.A.* 1234:1–239.

——— and H.-L. Li. 1970. Alkaloid-bearing plants and their contained alkaloids, 1957–1968. *Lloydia* 33, Suppl. 3A:1–286.

Willis, J. C. 1973. *A dictionary of flowering plants and ferns*. 8th ed., rev. H. K. A. Shaw. University Printing House, Cambridge.

Wood, C. E. 1958. The genera of woody Ranales in the southeastern United States. *J. Arnold Arbor.* 39:296–346.

———. 1974. *A student's atlas of flowering plants: Some dicotyledons of eastern North America.* Harper and Row, New York.

Wunderlin, R. P. 1982. *Guide to the vascular plants of central Florida.* University Presses of Florida, Gainesville.

Zomlefer, W. B. 1980. Common Florida angiosperm families. M.S. thesis, University of Florida, Gainesville.

———. 1983. *Common Florida angiosperm families*, part 1. Biological Illustrations, Gainesville.

———. 1986. *Common Florida angiosperm families*, part 2. Biological Illustrations, Gainesville.

———. 1989. *Flowering plants of Florida: A guide to common families.* Biological Illustrations, Gainesville.

Zurawski, G., and M. T. Clegg. 1993. *rbc*L sequence data and phylogenetic reconstruction in seed plants: Foreword. *Ann. Missouri Bot. Gard.* 80:523–525.

"Dicotyledons" and Monocotyledons

An Example of Paraphyly

As emphasized in the introduction, the most objective delineation of angiosperm families and their relationships would be based on shared derived characters, resulting in groupings described as "monophyletic." The discussion below, using the "dicot" vs. monocot dilemma as an example, explains some of the particular concepts and terminology associated with cladistics, a philosophical and methodological approach that has been accepted in much of systematic zoology for more than three decades but has only relatively recently been adopted by certain botanists, who face much criticism and opposition from traditionalists (see Cronquist 1987). The main difference in the traditionalists' approach is their acceptance of paraphyletic taxa (explained below), coupled with the often inexplicit assignment of character states at inappropriate levels ("levels of universality," also discussed below).

John Ray (1686–88) was the first botanist to establish a classification system recognizing two main subdivisions of flowering plants: the dicotyledons or dicots ("two cotyledons"), and the monocotyledons or monocots ("one cotyledon"). These subgroupings are widely accepted by recent systematists (as in, e.g., the subclasses Dicotyledoneae and Monocotyledoneae of Thorne 1992, and the classes Magnoliopsida and Liliopsida of Cronquist 1988), who generally acknowledge that the monocots actually have been derived from some group of dicots. Besides cotyledon number, the two groups traditionally have been distinguished by combinations of the characters listed in Table 1, a standard comparison chart of these two groups. However, there are exceptions for all of these features. In particular, certain families of dicots are characterized by suites of character states common among the monocots but rare or uncommon in most dicots (Burger 1977; Huber 1977). According to Dahlgren and Clifford (1982), the dicot families linked to the monocots compose two groups: the "Nymphaeiflorae" (Cabombaceae, Nymphaeaceae, Ceratophyllaceae, Saururaceae, and Piperaceae) and the "Magnoliiflorae" (Annonaceae, Aristolochiaceae, Magnoliaceae, Illiciaceae, Winteraceae, and other families).

The difficulty in circumscribing the dicots may be easily explained in the context of cladistics. The details of Hennigian cladistics, succinctly summarized by Dahlgren et al. (1985), are briefly outlined here. The basic tenet of cladistic analysis is the principle that only shared derived ("advanced") character states ("synapomorphies") can provide evidence of phylogenetic relationship and be used to define taxa; shared ancestral ("primitive") character states ("symplesiomorphies") are not considered a valid basis for grouping taxa. Figure 1 depicts a series of cladograms (phy-

TABLE 1. Traditional chart comparing "dicotyledons" with monocotyledons. The features listed for the "dicotyledons" actually most accurately characterize the tricolpate-pollen group in Fig. 2 (see text for explanation). Chart modified from Porter (1967), Cronquist (1988), and Schmid (1992).

CHARACTER	"DICOTYLEDONS"	MONOCOTYLEDONS
GENERA/SPECIES	9,882/175,690	2,768/58,195
COTYLEDON(S)	usually 2 (rarely 1, 3, 4, or embryo undifferentiated) usually develop above ground	usually 1 (sometimes embryo undifferentiated) usually develops underground
MATURE ROOT SYSTEM	primary and/or adventitious	entirely adventitious
LEAF VENATION	usually net	usually parallel
STEM VASCULAR BUNDLE ARRANGEMENT	usually in a ring pith and cortex usually distinct	scattered, or in 2 or more rings pith and cortex usually not distinct
INTRAFASCICULAR CAMBIUM (true secondary growth)	usually present	lacking (typically, no cambium of any kind)
HABIT	woody or herbaceous	usually herbaceous
SIEVE TUBE PLASTIDS	S- or (seldom) P-type protein bodies variously shaped	P-type protein bodies exclusively wedge-shaped
NUMBER OF FLORAL PARTS	typically 5-, less often 4- or (seldom) 3-merous (carpels often fewer); or indefinite number	typically 3- or (seldom) 4-merous (carpels 3 or fewer); or indefinite number
POLLEN TYPE	typically tricolpate (3 furrows or apertures), but often monocolpate in *Magnolia*-like group	basically monocolpate (1 furrow or pore)

logenetic trees) with taxa shown in various groupings. The construction of such a phylogenetic tree is based on changes in character states (from ancestral to derived). Each branching point (bifurcation) represents a hypothesis of sister group relationship—thus, the cladogram itself represents a hypothesis of genealogical relationships of the included taxa. Relationships based on overall similarity (= phenetics) are ignored. A group composed of descendants from two or more ancestral sources is a polyphyletic group (Fig. 1A). All taxonomists reject a polyphyletic taxon, because it does not represent a single line of descent. Monophyletic groups (Fig. 1B) ideally comprise *all* derivatives of a single common ancestor and are defined solely by shared derived features. Paraphyletic groups (Fig. 1C) include the most recent ancestor and *some, but not all*, of its descendants. Cladists accept only strict monophyly, while traditionalists (such as Cronquist) use both monophyletic and paraphyletic taxa if they are phenetically cohesive (Ashlock 1979)—that is, if all members of a group are generally more similar to one another than to members of another group. However,

a paraphyletic group represents an incomplete system, usually maintained on the grounds that the constituents have diverged from the remainder of the descendants in some significant way as ascertained subjectively by a taxonomist (Donoghue and Cantino 1988). Thus, in these "natural" or "evolutionary" classifications, the degree of similarity, based on both ancestral and derived character states, is the primary criterion for establishing relationships, and many groups, thus circumscribed, are paraphyletic. Paraphyletic taxa are therefore typically difficult to characterize, have no reality in nature, and are less predictive than monophyletic groups (Bremer and Wanntorp 1978).

Because no complete phylogenetic analysis (based on shared derived characters) has been completed for all of the Angiospermae, paraphyletic taxa are quite common in current classification systems, including several families as delineated by Thorne (1992) and presented in this book (for example, see the commentary under Caprifoliaceae). An obvious, higher-level example, common to all current classification systems, is the monocot-"dicot" situation. Figure 2 represents a

A. Polyphyletic Groups

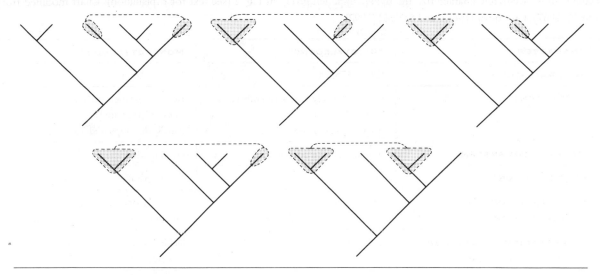

B. Monophyletic Groups

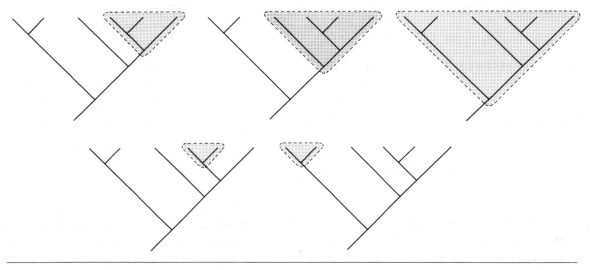

C. Paraphyletic Groups

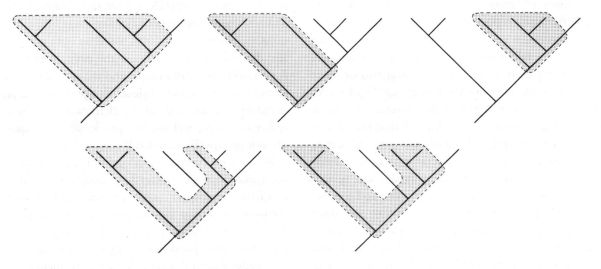

Figure 1. Different approaches to grouping taxa. **A,** Examples of polyphyletic groups; **B,** Examples of monophyletic groups; **C,** Examples of paraphyletic groups.

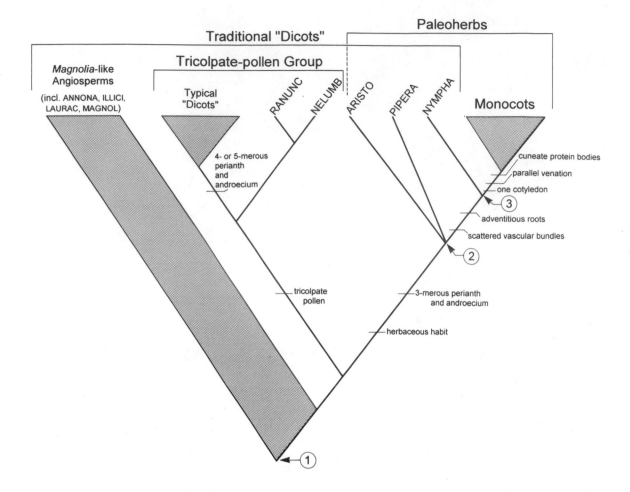

Figure 2. Relationships of "dicotyledons" and monocotyledons. Hypothesis of general phylogenetic relationships of all angiosperms from Donoghue and Doyle (1989a,b), modified by W. S. Judd (pers. comm.) and Zomlefer. Included are some characters from the standard "dicot"-monocot chart (Table 1). Arrows indicate possible positions for rooting the cladogram (see summary in Qiu et al. 1994). Although shown unrooted, the configuration is equivalent to the root at **1** (Donoghue and Doyle 1989a); hypothesized roots at **2** and **3** approximate those supported by Taylor and Hickey (1992) and Hamby and Zimmer (1992), respectively. The traditional "dicots" are paraphyletic; the monocots, monophyletic; and the so-called primitive angiosperms (*sensu* Cronquist 1988 or Thorne 1992), paraphyletic. See text for discussion. Abbreviations: ANNONA (Annonaceae), ARISTO (Aristolochiaceae), ILLICI (Illiciaceae), LAURAC (Lauraceae), MAGNOL (Magnoliaceae), NELUMB (Nelumbonaceae), NYMPHA (Nymphaeaceae), PIPERA (Piperaceae), RANUNC (Ranunculaceae).

hypothesis of relationships of all angiosperms, simplified from the preliminary, morphological and anatomical cladistic analyses by Donoghue and Doyle (1989a,b) with modifications by Walter S. Judd (pers. comm.) and myself. Additional morphological studies incorporated into this diagram include those by Loconte and Stevenson (1991) and Taylor and Hickey (1992), as well as the molecular analyses by Zimmer et al. (1989) and Hamby and Zimmer (1992). Included are the features from the standard monocot-dicot chart (Table 1) for comparison; most of these characters (herbaceous habit, three-merous perianth, scattered bundles, and parallel venation), however, have evolved more than once.

Figure 2 clearly illustrates the "dicots" as a paraphyletic taxon, and the monophyletic monocots are defined by several derived features: single cotyledon, cuneate protein bodies in sieve element plastids (found also in *Asarum* of the Aristolochiaceae), and parallel venation. Three main assemblages are evident: the *Magnolia*-like angiosperms (monophyletic or paraphyletic), the tricolpate-pollen dicots (probably monophyletic), and a monophyletic or paraphyletic "paleoherb" group (monocots plus Aristolochiaceae, Nymphaeaceae, and Piperaceae). It should be noted that most characters usually cited for the "dicots" apply mainly to the tricolpate-pollen group. Also, this hypothesis suggests that the so-called primitive angio-

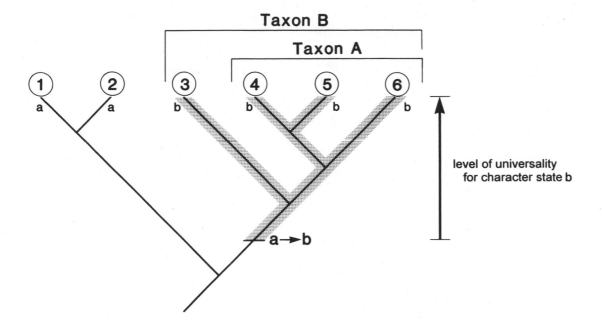

Figure 3. Level of universality. Cladogram illustrating that the level of universality for derived character state "b" is Taxon B, and not any subsets of Taxon B (such as Taxon A). See text for discussion.

sperms of many taxonomists (Thorne's superorder Magnolianae, Cronquist's subclass Magnoliidae), comprising the *Magnolia*-like group, the Ranunculaceae-Nelumbonaceae clade, and some (Thorne) or all (Cronquist) of the paleoherb dicots, are likely paraphyletic (see also the summary in Qiu et al. 1993). Recent molecular data (Olmstead et al. 1992; Chase et al. 1993) support the tricolpate clade and the paraphyly of the dicots. Thus, a provisional cladogram may more accurately reflect probable evolutionary relationships than does the conventional "dicot" vs. monocot dichotomy.

Cladists disregard paraphyletic groups because such artificial taxa are created to emphasize certain characters or phenetic gaps (differences between taxa). The acceptance of paraphyly is related to an uncritical view of characters at undefined levels of universality (Humphries and Chappill 1988). At a certain point in the evolutionary history of a group, a character may evolve from an ancestral to a derived state, defining a monophyletic group inclusive only at that point (upwards on a cladogram). This shared derived feature (synapomorphy) cannot be applied again as a defining character of taxa embedded within this monophyletic group (i.e., later or lower-level taxa), because at those less inclusive points the character is a shared ancestral character state (symplesiomorphy). Figure 3 graphically depicts this concept, with a character evolving from ancestral state *a* to derived state *b*. Character state *b* is a synapomorphy defining the inclusive Taxon B

(taxa 3+4+5+6) but a symplesiomorphy for any less inclusive, lower-level taxon higher on the cladogram (e.g., Taxon A or 4+5+6). Thus, the cladistic approach does not discard ancestral ("primitive") character states, but emphasizes that the character has already been employed as a synapomorphy for a higher, more inclusive taxon (Wiley 1981). Confusion arises when traditional taxonomists often describe characters (actually character states) as "primitive" or "advanced" (e.g., throughout Cronquist 1981)—terms that are meaningless when there is little or no reference to the levels at which the characters are employed.

Table 1 lists many such examples of synapomorphic and symplesiomorphic characters used to define groups (as do all comparison charts in this book). The "dicots" are characterized by several ancestral characters; the primary root, for example, actually arose as a derived character at least at the level of universality of seed plants and progymnosperms. In comparison, the characters traditionally used to define the monophyletic monocots are at their level of universality (e.g., cuneate protein bodies in the sieve element plastids), at a slightly higher level (e.g., scattered vascular bundles—including a few dicot groups), or at a much higher level (e.g., three-merous perianth; see Fig. 2).

Therefore, the main disadvantage of the traditional approach is that conventional taxonomists differ in the extent to which phenetic (overall similarity, gaps, divergence) vs. phylogenetic criteria are incorporated into their schemes, resulting in the great differences among current classification systems mentioned in the introductory chapter. In comparison, the primary benefits of cladistic analysis are (1) the care with which characters are presented: the characters, their separation into states, their polarity (assignment of ancestral vs. advanced state), and their level of universality must be stated explicitly; and (2) the ability to construct an explicit and reproducible hypothesis of genealogical relationships using homologous characters at their appropriate level (Zomlefer 1993). The problems with cladistics (e.g., dealing with character reversal and loss) that are emphasized by critics exist in "traditional" taxonomic work as well (Donoghue and Cantino 1988; Humphries and Chappill 1988). Phylogenetic hypotheses based on phenetic (or "overall similarity") concepts do not accurately reflect genealogical relationships and are misleading and irreproducible due to the subjectivity involved (Humphries and Chappill 1988). The choice is whether to retain phenetically defined taxa (even when in conflict with the best assessment of phylogeny), or to update classifications in order to portray accurately our current knowledge of phylogenetic relationships—even if this means altering or discarding traditional circumscriptions of taxa (Donoghue and Cantino 1988).

REFERENCES CITED

Ashlock, P. D. 1979. An evolutionary systematist's view of classification. *Syst. Zool.* 28:441–450.

Bremer, K., and H.-E. Wanntorp. 1978. Phylogenetic systematics in botany. *Taxon* 27:317–329.

Burger, W. C. 1977. The Piperales and the monocots—Alternate hypotheses for the origin of monocotyledonous flowers. *Bot. Rev.* (Lancaster) 43:345–393.

Chase, M. W., D. E. Soltis, R. G. Olmstead, D. Morgan, D. H. Les, et al. 1993. Phylogenetics of seed plants: An analysis of nucleotide sequences from the plastid gene *rbc*L. *Ann. Missouri Bot. Gard.* 80:528–580.

Cronquist, A. 1981. *An integrated system of flowering plants.* Columbia University Press, New York.

——. 1987. A botanical critique of cladism. *Bot. Rev.* (Lancaster) 53:1–52.

——. 1988. *The evolution and classification of flowering plants*, 2d ed., p. 262. New York Botanical Garden, Bronx.

Dahlgren, R. M. T., and H. T. Clifford. 1982. *The monocotyledons: A comparative study*, pp. 323, 330–345. Academic Press, London.

——, ——, and P. F. Yeo. 1985. *The families of the monocotyledons*, pp. 23–25, 44–47. Springer-Verlag, Berlin.

Donoghue, M. J., and P. D. Cantino. 1988. Paraphyly, ancestors, and the goals of taxonomy: A botanical defense of cladism. *Bot. Rev.* (Lancaster) 54:107–128.

——, and J. A. Doyle. 1989a. Phylogenetic analysis of angiosperms and the relationships of Hamamelidae. In *Evolution, systematics, and fossil history of the Hamamelidae*, vol. 1, *Introduction and "lower" Hamamelidae*, ed. P. R. Crane and S. Blackmore, pp. 17–45. Syst. Assoc. Special Vol. 40A. Clarendon Press, Oxford.

—— and ——. 1989b. Phylogenetic studies of seed plants and angiosperms based on morphological characters. In *The hierarchy of life*, ed. B. Fernholm, K. Bremer, and H. Jörnvall, pp. 181–193. Elsevier Science Publishers, Amsterdam.

Hamby, R. K., and E. A. Zimmer. 1992. Ribosomal RNA as a phylogenetic tool in plant systematics. In *Molecular systematics of plants*, ed. P. S. Soltis, D. E. Soltis, and J. J. Doyle, pp. 50–91. Chapman and Hall, New York.

Huber, H. 1977. The treatment of the monocotyledons in an evolutionary system of classification. *Pl. Syst. Evol.*, Suppl. 1:285–298.

Humphries, C. J., and J. A. Chappill. 1988. Systematics as science: A response to Cronquist. *Bot. Rev.* (Lancaster) 54:129–144.

Loconte, H., and D. W. Stevenson. 1991. Cladistics of the Magnoliidae. *Cladistics* 7:267–296.

Olmstead, R. G., H. J. Michaels, K. M. Scott, and J. D. Palmer. 1992. Monophyly of the Asteridae and identification of their major lineages inferred from DNA sequences of *rbc*L. *Ann. Missouri Bot. Gard.* 79:249–265.

Porter, C. L. 1967. *Taxonomy of flowering plants*, p. 137. W. H. Freeman, San Francisco.

Qiu, Y.-L., M. W. Chase, D. H. Les, and C. R. Parks. 1993. Molecular phylogenetics of the Magnoliidae: Cladistic analyses of the nucleotide sequences of the plastid gene *rbc*L. *Ann. Missouri Bot. Gard.* 80:587–606.

Ray, J. 1686–88. *Historia plantarum.* 2 vols. Mariae Clark, London.

Schmid, R. 1992. *Diversity of plants and fungi*, pp. 153–154. Burgess International Group, Edina, Minn.

Taylor, D. W., and L. J. Hickey. 1992. Phylogenetic evidence for the herbaceous origin of angiosperms. *Pl. Syst. Evol.* 180:137–156.

Thorne, R. F. 1992. Classification and geography of the flowering plants. *Bot. Rev.* (Lancaster) 58:225–348.

Wiley, E. O. 1981. *Phylogenetics*, pp. 126–130. John Wiley and Sons, New York.

Zimmer, E. A., R. K. Hamby, M. L. Arnold, D. A. LeBlanc, and E. C. Theriot. 1989. Ribosomal RNA phylogenies and flowering plant evolution. In *The hierarchy of life*, ed. B. Fernholm, K. Bremer, and H. Jörnvall, pp. 205–214. Elsevier Science Publishers, Amsterdam.

Zomlefer, W. B. 1993. A revision of *Rigodium* (Musci: Rigodiaceae). *Bryologist* 96:1–72.

OBSERVING, DISSECTING, AND DRAWING FLOWERING PLANTS

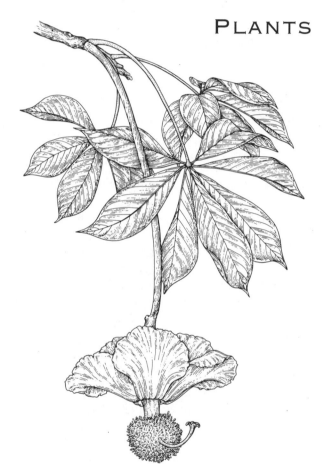

EXAMINATION IN THE FIELD AND LABORATORY

The most important tool for learning plant morphology is careful observation, whether in the field or in the laboratory. Since some variability occurs in nearly every plant group, several members of the same population always should be examined. The most effective scrutiny utilizes all the senses. For example, the scent of a crushed leaf can demonstrate the presence of aromatic oils, and the prevalence of certain acrid, bitter, or pungent compounds can be verified by a brief taste of the sap of a broken stem or branch. (A novice, however, must be extremely wary of tasting an unfamiliar plant!) Touching the leaves and stems can confirm the degree of pubescence or scabrousness. In addition to the above-ground habit, the type of root system should be noted, along with the features of the sap of a broken root.

In the laboratory, freehand sections with a scalpel or razor blade are adequate for observing nodal anatomy. The stem should be cut immediately below the point of leaf attachment. Phloroglucinal with a few drops of HCl stains the xylem (and any other tissues containing lignin) a bright red, and toluidine blue may aid in distinguishing laticifers in tissues.

The careful observation of floral morphology requires a dissecting microscope (or, in the field, a 10× handlens). Several flowers must be checked in order to determine the most typical individuals of the population. All dissections should be made under the microscope, and several flowers are needed for the various sections. Other essential tools are a scalpel or single-edged razor blade, dissecting needles (with a sharp or spear-shaped point), and forceps. Certain double-edged razor blades (broken in half lengthwise) also have a fine cutting edge. Water applied with an eyedropper will help keep the material from dehydrating under the microscope lamp, and a combination of clay and insect pins can be used to hold the flower and other parts in position.

This type of study requires patience and care. With practice, small flowers can be easily manipulated under the microscope with a needle (or forceps) in each hand. Certain features, such as symmetry, must be analyzed carefully before dissection. Starting at the outermost whorl and working inward, the number of parts and their degree of fusion should be noted. The degree of fusion is very important: parts that superficially appear to be free may actually be fused at the base. The number of styles and/or stigmas (which may indicate carpel number) should also be noted at this stage. The attachment of parts and the ovary position are best shown by a longitudinal section of the flower. A flower cross section will reveal the arrangement of parts in relation to one another, the carpel and locule number, and the placentation type.

Floral formulas and floral diagrams are two convenient shorthand notations for recording all of these features. A floral formula is a numerical notation of floral whorls that also summarizes the relationships of floral parts. The basic numerals (and accompanying symbols) identify the number of parts composing the calyx, corolla, androecium, and gynoecium, respectively (for examples see, Fig. 4: 1c, 2c, 3d, 4c, 5c). Complete connation is usually depicted by a circle around the appropriate floral whorl number, and basal connation, by a semicircle subtending the base of the number. Fusion between floral whorls is typically shown by lines connecting the pertinent numbers. A line above the carpel number indicates an inferior ovary (i.e., floral parts "above" the ovary), while a line below represents a superior ovary. The format varies among authors, however, because the abbreviations and symbols are not standardized. For example, unlike those in Fig. 4, the whorls are sometimes represented by letters with the number of constituent parts in super- or subscript (see Baumgardt 1982 and Jones and Luchsinger 1986 for examples). The floral formula method works best for studies of lower taxonomic levels (genera and species), because generalized formulas for families usually include ranges of numbers and degrees of fusion reflecting the variation of floral structure.

A floral diagram (Fig. 4: 4b, 5b) is a diagrammatic floor plan of a flower representing the floral whorls as viewed from the top. A cross section of the bud, just prior to anthesis (Fig. 4: 3c), is often the best way to see these relationships between the floral whorls. The floral diagram frequently is idealized as though all floral parts were at the same level, thus showing the number in each whorl and the degree of fusion within and between them, as well as the floral symmetry and relative position of associated structures (such as nectaries). However, a separate longitudinal section of the pistil may be necessary to show the exact ovary position. Each taxonomist devises conventions for stylizing the parts, such as shading to differentiate between the calyx and corolla, or various symbols for the anthers (see Porter 1967; Baumgardt 1982; Jones and Luchsinger 1986). Floral diagrams were used more commonly in classic taxonomy works (e.g., Engler and Prantl 1887–1915; Rendle 1925; Swingle 1928); they still occasionally appear in modern manuals (e.g., Porter 1967; Baumgardt 1982) and floras (Correll and Correll 1982).

IMPORTANT ASPECTS OF BOTANICAL ILLUSTRATION

Laboratory sketches provide an excellent record for studying plant morphology. Although the following discussion emphasizes techniques for creating accurate illustrations suitable for publication, the outlined principles for drawing plants and flowers can aid in quick, rough sketching as well. Accurate botanical drawings require technical knowledge and patience in addition to artistic skills. Illustrations should be precise and drawn to scale, with correct perspective and properly executed dissections; overshading and unnecessary detail should be avoided. The following discussion (originally in Zomlefer 1977, 1980) outlines some helpful tips to avoid several common inaccuracies. Additional aspects of botanical illustration are thoroughly covered by Sheehan (= Ruff 1950; Sheehan 1989), West (1983), and Holmgren and Angell (1986).

Figure 4. Floral diagrams and floral formulas. 1, *Polygonum punctatum* (*Persicaria punctata*, Polygonaceae; see also Fig. 37: 1d–h): **a,** flower, ×10; **b,** cross section through middle of bud, ×15; **c,** floral formula (note: styles free). **2,** *Tilia americana* var. *caroliniana* (*T. caroliniana*, Tiliaceae; see also Fig. 42: e–m): **a,** flower, ×3½; **b,** cross section of flower, ×10; **c,** floral formula. **3,** *Crotalaria spectabilis* (Fabaceae; see also Fig. 71: 2c–i): **a,** two views of flower, ×⅘; **b,** expanded corolla, ×1; **c,** cross section of bud, ×5; **d,** floral formula. **4,** *Lantana camara* (Verbenaceae; see also Fig. 119: 1f–l): **a,** lateral view of flower, ×3; **b,** floral diagram (schematic; note one-locular ovary); **c,** floral formula (see **b**: two ovary locules reduced to one). **5,** *Canna flaccida* (Cannaceae; see also Fig. 156: d–q): **a,** flower, ×⅓; **b,** floral diagram (schematic; basal connation of androecium, corolla-androecium adnation, and stamen-style attachment not shown): 1 = one "wing" (an outer staminode), 2 = expanded outer staminode (behind stamen), 3 = one "wing" (an inner staminode), 4 = lip (inner staminode), 5 = petaloid stamen with half-anther; **c,** floral formula. Floral diagrams and bud/flower cross sections: dark screen = calyx; unshaded perianth = tepals or corolla; spaced stipples = staminodes. Floral formula format: floral symmetry: -tepal number- or sepal number, petal number; androecial element (stamen ± staminode) number; carpel number. Symbols (see also Appendix B): a = actinomorphic, z = zygomorphic, i = irregular; circle = complete connation, semicircle = partial (apical or basal) connation; curved lines connecting whorls = adnation; * = staminodes; line above carpel number = inferior ovary, line below number = superior ovary.

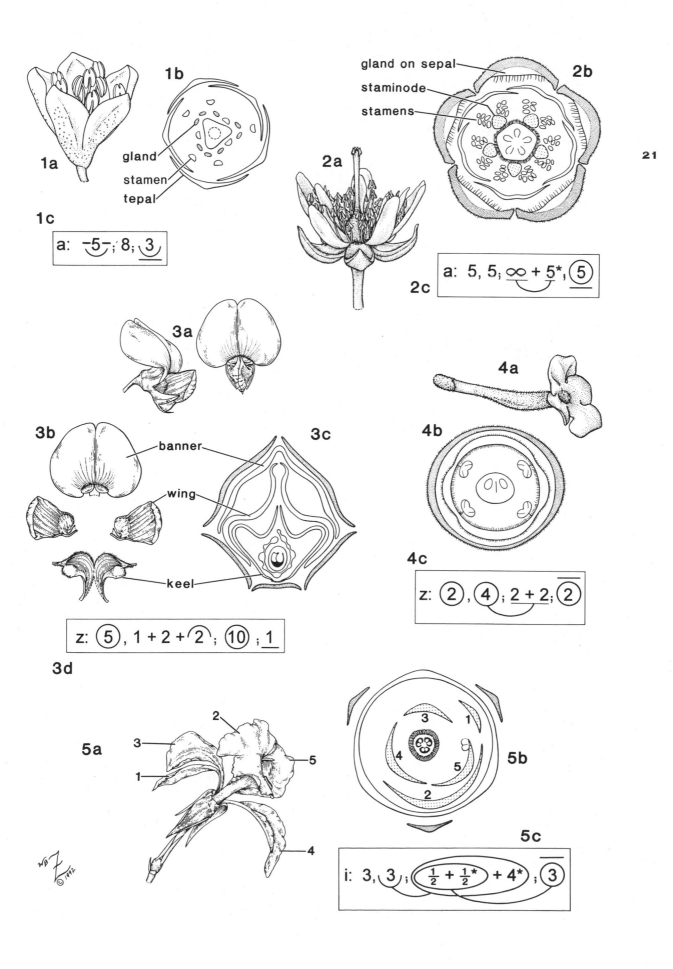

1a

1b

gland
stamen
tepal

1c

a: $\underline{-5-}$; 8; $\underline{\textcircled{3}}$

2b

gland on sepal
staminode
stamens

21

2a

2c

a: 5, 5; $\infty + \underline{5}^*$, $\textcircled{5}$

3a

3b

banner

3c

wing

keel

3d

z: $\textcircled{5}$, $1 + 2 + 2$; $\textcircled{10}$; $\underline{1}$

4a

4b

4c

z: $\textcircled{2}$, $\textcircled{4}$; $2 + 2$; $\overline{\textcircled{2}}$

5a

3
2
1
5
4

5b

3 1
4
5
2

5c

i: 3, 3; $(\frac{1}{2} + \frac{1}{2}^*) + 4^*$; $\overline{\textcircled{3}}$

WBZ ©1992

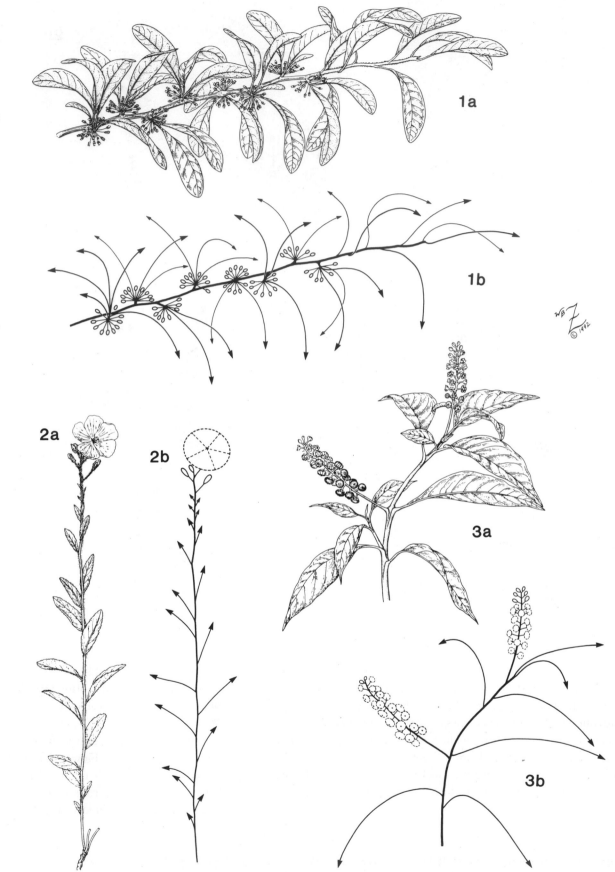

22

1a

1b

2a

2b

3a

3b

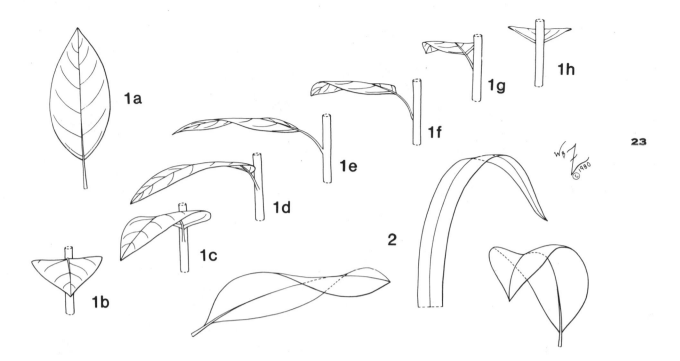

Figure 6. Perspective in leaves. 1, Foreshortening in leaves: **a,** flat leaf (in plane of stem); **b–h,** changes in leaf length as it is rotated around the stem. **2,** Various leaves in perspective.

Simplification is important when beginning to sketch. First, the best angle to view the plant must be ascertained, so that the essential features are clearly visible. The complexity of plants can be confusing (and overwhelming), but even an entire plant can be subdivided into manageable components (see Fig. 5). At first glance, a stem (or branch) may appear to be covered in an incomprehensible array of foliage, when actually the underlying arrangement is quite logical, with the stem composing a basic framework upon which the leaves and flowers are attached at specific points (as in Fig. 5: 1b). To capture the true appearance or "habit" of the plant, the angle at which the leaves (or inflorescences, flowers, or fruits) are held in relation to the axis also must be carefully noted. Once the fundamental construction is established, the leaves (and other major parts) can be roughly blocked in. Attention to detail then becomes more important. For example, one obvious feature, the pinnate venation of many species, frequently is inaccurately portrayed by novices, who fail to notice the exact number of secondary veins and the angle at which they diverge from the midvein.

Inattention to perspective, especially of leaves, is a common error in botanical drawings (Fitch 1869). Because of their curvature, many leaves seen in perspective must be drawn foreshortened to depict an illusion of projection into space (see Fig. 6: 1a–h). "Foreshortening" is a proportional contracting of the depth of an object. For example, a flat leaf in top view clearly shows its entire width and length (Fig. 6: 1a), but in a lateral view (Fig. 6: 1e) only the entire length can be seen because the width is foreshortened. The length of a leaf becomes more and more foreshortened as the stem is rotated toward or away from the viewer. Ultimately, in an end view, the length essentially becomes totally obscured (Fig. 6: 1b,h). In addition, the midvein of a well-drawn curved or flat leaf should be a continuous smooth line, with the edge of the blade represented by two continuous lines, one on each side of the midvein (Fig. 6: 2).

Errors in perspective are also common in flower (and fruit) drawings (see Fig. 7: 1a–e). Flower shapes basically are geometrical forms—usually circles, ellipses, cones, or cylinders (Fig. 7: 1a–c); fruits are often fundamentally spheres, cones, or cylinders. Relating

Figure 5. Plant habit structure. 1, *Sideroxylon lanuginosum* (*Bumelia lanuginosa*; Sapotaceae): **a,** flowering branch, ×¼; **b,** preliminary sketch. **2,** *Piriqueta caroliniana* (Turneraceae): **a,** habit, ×⅜; **b,** preliminary sketch. **3,** *Phytolacca americana* var. *rigida* (*P. rigida*; Phytolaccaceae): **a,** portion of plant with flowers and fruit, ×⅓; **b,** preliminary sketch.

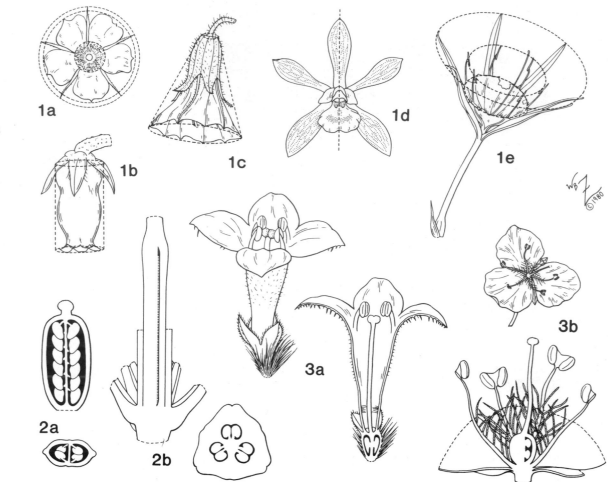

Figure 7. Perspective in flowers and flower dissections. 1, Flower shape: **a,** circle, *Rosa palustris* (Rosaceae), ×½; **b,** cylinder, *Lyonia lucida* (Ericaceae), ×3; **c,** cone, *Physalis heterophylla* (Solanaceae), ×2 ¼; **d,** bilateral symmetry, *Encyclia tampensis* (Orchidaceae), ×1; **e,** parallel ellipses, *Hymenocallis rotata* (Amaryllidaceae), ×⅕. **2,** Sections of pistils: **a,** even-carpellate, *Descurainia pinnata* (Brassicaceae), longitudinal and cross sections, ×25; **b,** odd-carpellate, *Lilium catesbaei* (Liliaceae), longitudinal section, ×2¼, and cross section, ×6. **3,** Longitudinal sections of flowers: **a,** even-merous, *Diodia teres* (Rubiaceae), flower and longitudinal section of flower, ×6; **b,** odd-merous, *Callisia graminea* (*Cuthbertia graminea*; Commelinaceae), flower, ×1½, and longitudinal section of flower, ×6.

the flower or fruit to a fundamental shape is extremely helpful when translating the three-dimensional object into a two-dimensional sketch. For example, as a circular corolla is tipped toward or away from the viewer, it becomes an ellipse with a decreasing width. Flower parts, such as the stamens and calyx, often align along ellipses parallel to the ellipse formed by the tipped corolla (Fig. 7: 1e). A corolla limb should line up with the tube, with the tube apex diameter equal to the diameter of the opening in the limb. The equal halves of zygomorphic flowers are essentially mirror images of each other (Fig. 7: 1d).

Certain dissection conventions are helpful. Anther shape is usually best depicted before dehiscence. Placentation and locule number generally are most easily

seen in an ovary cross section taken at the widest part of the pistil. A longitudinal section of an entire flower is very complex and difficult; such a section should cut the flower exactly in half (e.g., Fig. 7: 3a,b). The alignment of the perianth in relationship to the ovary locules is very important. Often even-merous flowers are cut along two opposite petals (Fig. 7: 3a). A bisected odd-merous flower must have a bisected petal on one side of the flower and a bisected sepal on the other (Fig. 7: 3b). The total stamen number should be halved, with a half-stamen next to its corresponding opposite perianth segment (usually a sepal). Ovary locules also should be depicted correctly in a longitudinal section. With odd-carpellate pistils, usually only one locule should be revealed by the cut (as in Fig 7:

2b), and this locule should be opposite a bisected petal, tepal, or sometimes sepal (Fig. 7: 3b). In a bisected even-carpellate pistil, two locules usually should be exposed (Fig. 7: 2a). Because of these complexities, a longitudinal section of a flower should be somewhat stylized and simplified whenever possible to eliminate unnecessary and confusing detail.

REFERENCES CITED

Baumgardt, J. P. 1982. *How to identify flowering plant families: A practical guide for horticulturists*, pp. 39–51. Timber Press, Beaverton, Ore.

Correll, D. S., and H. B. Correll. 1982. *Flora of the Bahama archipelago*. J. Cramer, Vaduz, Germany.

Engler, A., and K. Prantl. 1887–1915. *Die natürlichen Pflanzenfamilien*, parts II–IV. G. Kreysing, Leipzig.

Fitch, W. H. 1869. Botanical drawings—Nos. I–VII. *Gard. Chron.* 1:7; 3:51; 5:110; 7:165; 9:221; 12:305; 15:389; 19:499.

Holmgren, N. H., and B. Angell. 1986. *Botanical illustration: Preparation for publication*. New York Botanical Garden, Bronx.

Jones, S. B., and L. E. Luchsinger. 1986. *Plant systematics*, pp. 249–252. McGraw-Hill, New York.

Porter, C. L. 1967. *Taxonomy of flowering plants*, pp. 103–105. W. H. Freeman, San Francisco.

Rendle, A. B. 1925. *The classification of flowering plants*. 2 vols. Cambridge University Press, Cambridge.

Ruff, M. E. (= M. R. Sheehan). 1950. Methods and techniques in botanical illustration. M.S. thesis, Cornell University, Ithaca, N.Y.

Sheehan, M. R. 1989. Illustrating plants. In *The guild handbook of illustration*, ed. E. R. S. Hodges, pp. 189–220. Van Nostrand Reinhold, New York.

Swingle, D. B. 1928. *A textbook of systematic botany*. McGraw-Hill, New York.

West, K. 1983. *How to draw plants*. Watson-Guptill, New York.

Zomlefer, W. B. 1977. The challenges of botanical illustrating. *Bull. Marie Selby Bot. Gard.* 4:12–13.

———. 1980. Common Florida angiosperm families. M.S. thesis, University of Florida, Gainesville.

ILLICIACEAE
Star Anise Family

Shrubs to small trees, aromatic (due to ethereal oils). **Leaves** simple, entire, alternate (often clustered and forming pseudowhorls), persistent, exstipulate. **Flowers** solitary or 2 to 3 per cluster, axillary, actinomorphic, perfect, hypogynous, with short convex to conical receptacle, small to showy. **Perianth** of numerous tepals in several whorls, not clearly differentiated into calyx and corolla, with outermost whorl(s) generally sepaloid and the inner whorls petaloid, green, cream, yellow to red, distinct, imbricate. **Androecium** of numerous stamens spirally arranged on the receptacle; filaments distinct, short, thick, poorly differentiated from the anthers; anthers with enlarged connective, basifixed, dehiscing longitudinally, introrse. **Gynoecium** apocarpous and of 5 to numerous compressed carpels in one whorl; ovaries superior, 1-locular; ovules 1 in each carpel, anatropous, placentation subbasal; style 1, incompletely fused, not well differentiated; stigma decurrent along inner margins of style. **Fruit** a radially spreading aggregate of 1-seeded follicles (each splitting along the adaxial suture); seeds flattened, glossy; endosperm copious, oily; embryo minute.

Family characterization: Shrubs to trees with aromatic parts; simple, coriaceous leaves with entire margins; flowers with numerous distinct parts spirally arranged on a short receptacle; thick stamens not clearly differentiated into filament and anther; uniovulate carpels with conduplicate style and decurrent stigma; and star-like aggregate of follicles. Anatomical features: ethereal oil cells in the parenchymatous tissues; sclereids in the mesophyll; and 1-trace unilacunar nodes (Fig. 8: 1b; Metcalfe 1987).

Genus/species: *Illicium* (37 spp.); a monotypic family.

Distribution: Disjunct: tropical montane and temperate areas of North America (southeastern United States, West Indies, Mexico) and southeastern Asia.

U.S./Canadian representatives: *Illicium* (2 spp.)

Economic plants and products: Anise oil (anethole) used in perfumes, flavoring, and medicines from bark, fruit, and seeds of a few *Illicium* species (anise tree, star anise). Several ornamental shrubs, including *Illicium verum* (Chinese anise) and *I. floridanum* (purple anise).

Commentary: The placement of this small monotypic family has been debated extensively (see Metcalfe 1987 for a summary). The Illiciaceae have been closely allied with the Schisandraceae (two genera of southeastern United States and Asia), and both families are characterized by a unique type of tricolpate pollen (Doyle and Hotton 1991). They often are included with the Magnoliaceae complex of families (Magnoliales of Thorne 1992) and have even been submerged formerly within the Magnoliaceae (Smith 1947). However, recent molecular data (e.g., Qiu et al. 1993) indicate that the Illiciaceae (and Schisandraceae) may be closely allied with the Nymphaeales rather than with such families as the Magnoliaceae and Annonaceae. The two species of *Illicium* that occur in the United States, *I. floridanum* (Florida anise; Fig. 8: 2a,b) and *I. parviflorum* (yellow anise; Fig. 8: 1a–l), are restricted to the southeast (Wood 1958).

The floral structure comprises numerous free and distinct, spirally arranged parts. As in *Nymphaea*, the calyx and corolla are not clearly differentiated, and in some species the innermost whorl of petaloid tepals also intergrades with the fleshy stamens that are not clearly differentiated into filament and anther. The apocarpous gynoecium develops into a characteristic star-like fruit (Fig. 8: 1j), a more or less coherent, radiating ring of follicles that often differ in size or shape. Each segment dehisces elastically along the upper side, squeezing out a single, smooth, glossy seed (Roberts and Haynes 1983).

The protogynous flowers of *Illicium* generally are pollinated by beetles, but those of our species attract a variety of small insects (mostly flies; Thien et al. 1983; White and Thien 1985). *Illicium floridanum* has a fishy odor, and *I. parviflorum*, a sweet fragrance. The flies pollinate the flowers as they walk across the stamens and carpels while searching for the small amounts of nectar produced at the base of the stamens. Both species of *Illicium* are self-incompatible.

REFERENCES CITED

Doyle, J. A., and C. L. Hotton. 1991. Diversification of early angiosperm pollen in a cladistic context. In *Pollen and*

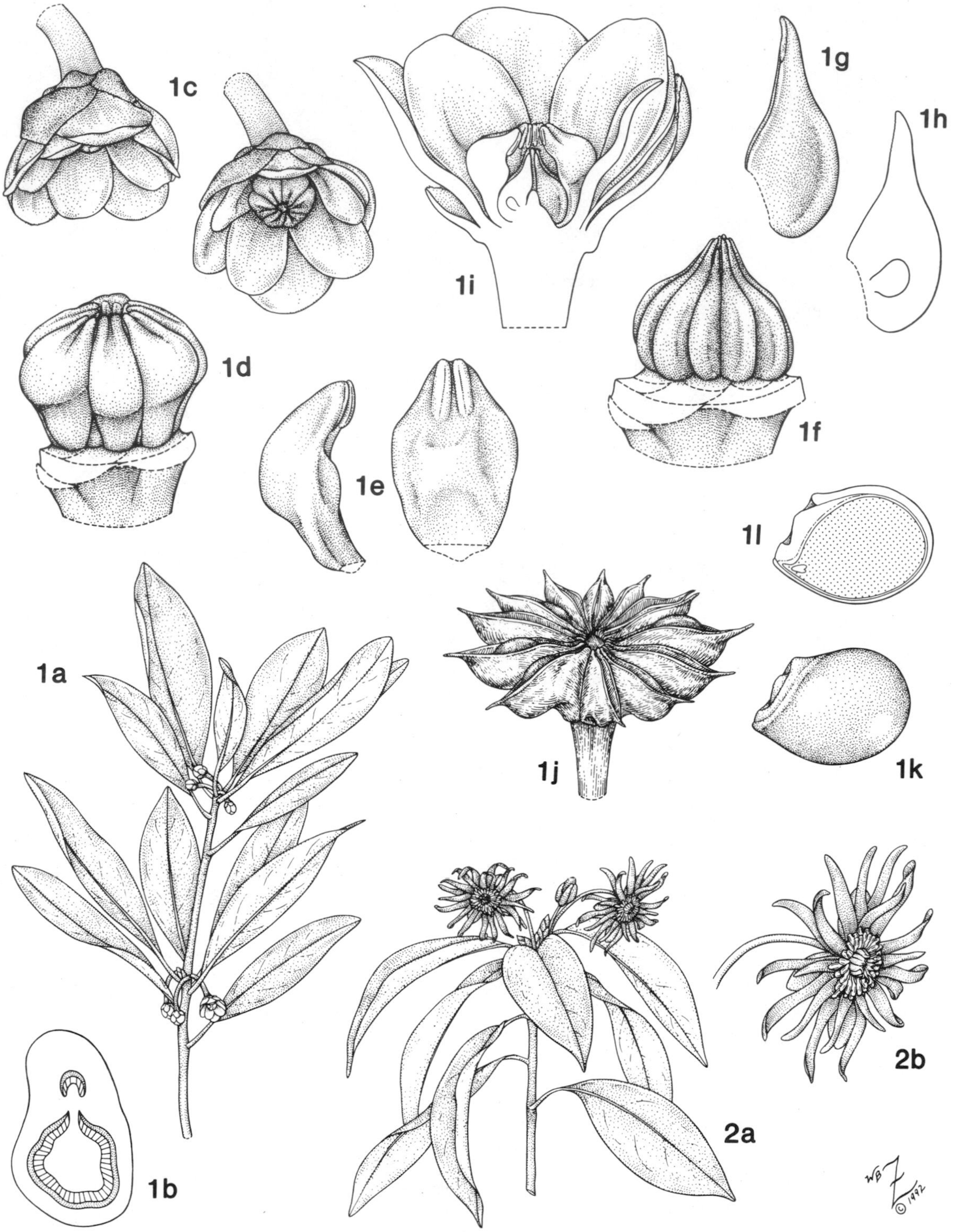

28

Figure 8. Illiciaceae. 1, *Illicium parviflorum*: **a,** flowering branch, ×⅖; **b,** cross section of node (one-trace unilacunar), ×7; **c,** two views of flower, ×3; **d,** androecium and gynoecium, ×7; **e,** two views of stamen, ×10; **f,** gynoecium, ×9; **g,** carpel, ×15; **h,** longitudinal section of carpel, ×15; **i,** longitudinal section of flower, ×6; **j,** fruit (aggregate of follicles), ×2¼; **k,** seed, ×5; **l,** longitudinal section of seed, ×5. **2,** *Illicium floridanum*: **a,** flowering branch, ×⅜; **b,** flower, ×1.

spores, ed. S. Blackmore and S. H. Barnes, pp. 169–195. Syst. Assoc. Special Vol. 44. Clarendon Press, Oxford.

Metcalfe, C. R. 1987. *Anatomy of the dicotyledons.* 2d ed. Vol. 3, *Magnoliales, Illiciales, and Laurales*, pp. 75–89. Oxford University Press, Oxford.

Qiu, Y.-L., M. W. Chase, D. H. Les, and C. R. Parks. 1993. Molecular phylogenetics of the Magnoliidae: Cladistic analyses of nucleotide sequences of the plastid gene *rbc*L. *Ann. Missouri Bot. Gard.* 80:587–606.

Roberts, M. L., and R. R. Haynes. 1983. Ballistic seed dispersal in *Illicium* (Illiciaceae). *Pl. Syst. Evol.* 143:227–232.

Smith, A. C. 1947. The families Illiciaceae and Schisandraceae. *Sargentia* 7:1–224.

Thien, L. B., D. A. White, and L. Y. Yatsu. 1983. The reproductive biology of a relict—*Illicium floridanum* Ellis. *Amer. J. Bot.* 70:719–727.

Thorne, R. F. 1992. Classification and geography of the flowering plants. *Bot. Rev.* (Lancaster) 58:225–348.

White, D. A., and L. B. Thien. 1985. The pollination of *Illicium parviflorum* (Illiciaceae). *J. Elisha Mitchell Sci. Soc.* 101:15–18.

Wood, C. E. 1958. The genera of the woody Ranales in the southeastern United States. *J. Arnold Arbor.* 39:296–346.

MAGNOLIACEAE
Magnolia Family

Trees or shrubs, aromatic (due to ethereal oils). **Leaves** simple, usually entire, alternate, deciduous or persistent, stipulate; stipules large, enclosing the young flower and leaf buds, caducous and leaving a scar around the node. **Flowers** solitary, usually terminal, actinomorphic, perfect, hypogynous, with elongated receptacles, showy. **Perianth** of 6 to many tepals (but usually 3 whorls of 3), undifferentiated and petaloid or outer whorl sepaloid, distinct, green, yellow, or white, imbricate. **Androecium** of numerous laminar stamens spirally arranged on the receptacle, filaments poorly differentiated from the anthers, distinct; anthers dehiscing longitudinally, introrse or latrorse. **Gynoecium** apocarpous and of numerous conduplicate carpels spirally arranged on the upper portion of the receptacle, coherent and closely packed; ovaries superior, 1-locular; ovules 1 to several in each carpel, anatropous, placentation parietal on the adaxial suture; style 1, recurved, usually not well differentiated; stigma 1, decurrent along inner margin of style. **Fruit** a cone-like aggregate of follicles (each splitting along the abaxial suture), samaras, or berries; seeds often with reddish sarcotesta, suspended by thread-like funicles (when carpel dehiscent); endosperm copious, oily; embryo minute.

Family characterization: Trees or shrubs with aromatic parts; simple, stipulate leaves with entire margins; annular stipular scars; perfect, actinomorphic, showy, solitary flowers; laminar stamens undifferenti-

ated into filament and anther; numerous distinct carpels and stamens spirally arranged on an elongated receptacle; conduplicate carpels with decurrent stigmas; and cone-like aggregate fruit. Pollen grains consistently monocolpate (Fig. 9: m; Canright 1953). Alkaloids and cyanogenic glycosides commonly present. Tissues with calcium oxalate crystals. Anatomical features: ethereal oil cells in the tissues; and multilacunar nodes (Fig. 9: c; Metcalfe 1987).

Genera/species: 7/220 (Thorne pers. comm.)

Distribution: Disjunct: mostly temperate to tropical southeastern Asia and eastern North America; a few in South America.

Major genera: *Magnolia* (120 spp.) and *Michelia* (45–50 spp.)

U.S./Canadian representatives: *Magnolia* (10 spp.) and *Liriodendron* (1 sp.)

Economic plants and products: Timber from *Magnolia* and *Liriodendron*. Ornamental trees (species of 5 genera), including *Liriodendron* (tulip tree), *Magnolia*, and *Michelia* (banana shrub).

Commentary: The circumscription of most genera of the Magnoliaceae has been problematical due to the homogeneity in morphology (Nooteboom 1985, 1987). A notable exception is the very distinctive genus *Liriodendron*, which was segregated into its own family by Barkley (1975); its inclusion in the Magnoliaceae, however, is supported by molecular data (Qiu et al. 1993).

The Magnoliaceae have often been considered to be the archetypal "primitive" taxon of extant angiosperms ("living fossils") due to the fossil record, the extant discontinuous distribution, and especially their "unspecialized" morphology, including the woody habit, perfect actinomorphic flowers, numerous and free spirally arranged stamens and carpels, laminar stamens, and monocolpate pollen (Maneval 1914; Takhtajan 1969). These features generally characterize the "woody Ranales" or "Magnoliales-Laurales" assemblage of dicots (see "*Magnolia*-like angiosperms" in Fig. 2), which includes families such as the Illiciaceae, Annonaceae, and Lauraceae.

The characteristic flower of the Magnoliaceae has very showy tepals and is protected in bud by prominent stipules (Fig. 9: d). Sometimes, as in *Magnolia virginiana*, the outermost whorl of tepals is calyx-like. Another prominent feature is the elongate receptacle bearing many spirally arranged stamens and carpels (Fig. 9: f,g). The stamens in our Magnoliaceae are three-veined with little differentiation into filament, connective, and anther (Canright 1952). The locules

Figure 9. Magnoliaceae. *Magnolia virginiana*: **a,** flowering branch, ×²⁄₅; **b,** stem showing annular stipular scars, ×1; **c,** cross section of node (multilacunar), ×6; **d,** flower bud subtended by stipules, ×½; **e,** flower, ×½; **f,** androecium and gynoecium, ×2⅓; **g,** receptacle with gynoecium (stamens removed), ×2⅓; **h,** detail of carpels, ×5; **i,** longitudinal section of flower, ×2⅓; **j,** two views of stamen, ×3; **k,** cross section of "anther" showing vascular bundles, ×12; **l,** cross section of "filament" showing vascular bundles, ×12; **m,** pollen grain (monocolpate), ×260; **n,** fruit (aggregate of follicles), ×½; **o,** seed, ×2⅓; **p,** cross section of seed, ×2⅓.

occur either laterally (two on each side) or on the adaxial surface (Fig. 9: j). The carpels are incompletely fused along the adaxial suture, with the stigmatic surface decurrent along the inner margins of the recurved style (Fig. 9: h; Canright 1960). The closely packed, overlapping carpels, which occur on the upper part of the receptacle, are actually coherent (in our members) but appear to be connate (Wood 1958).

Magnoliaceous flowers do not produce nectar, but pollinators (beetles) are attracted to the strong fragrance. Generally, pollination occurs as the beetle moves among the floral parts and its underside touches the anthers and stigmas. Self-pollination generally is prevented by protogyny.

The distinctive magnoliaceous fruits are cone-like structures with aggregates of follicles (*Magnolia*; Fig. 9: n) or samaras (*Liriodendron*) on the elongated receptacle. In *Magnolia* the follicles split along the abaxial (outer) suture to release the seeds, which hang by delicate silky threads (unrolled spiral vessels of the funicles). The seeds, each with a bright red sarcotesta, are dispersed by birds.

REFERENCES CITED

Barkley, F. A. 1975. A note concerning two flowering plants. *Phytologia* 32:304.

Canright, J. E. 1952. The comparative morphology and relationships of the Magnoliaceae. I. Trends in specialization in the stamens. *Amer. J. Bot.* 39:484–497.

——. 1953. Ibid., II. Significance of the pollen. *Phytomorphology* 3:355–365.

——. 1960. Ibid., III. Carpels. *Amer. J. Bot.* 47:145–155.

Maneval, W. E. 1914. The development of *Magnolia* and *Liriodendron*, including a discussion of the primitiveness of the Magnoliaceae. *Bot. Gaz.* 57:1–31.

Metcalfe, C. R. 1987. *Anatomy of the dicotyledons.* 2d ed. Vol. 3, *Magnoliales, Illiciales, and Laurales*, pp. 24–33. Oxford University Press, Oxford.

Nooteboom, H. P. 1985. Notes on Magnoliaceae with a revision of *Pachylarnax* and *Elmerrillia* and the Malesian species of *Mangelietia* and *Michelia*. *Blumea* 31:65–121.

——. 1987. Notes on Magnoliaceae II. Revision of *Magnolia* sections *Maingola* (Malesian species), *Aromadendron* and *Blumiana*. *Blumea* 32:343–382.

Qiu, Y.-L., M. W. Chase, D. H. Les, and C. R. Parks. 1993. Molecular phylogenetics of the Magnoliidae: Cladistic analysis of nucleotide sequences of the plastid gene *rbc*L. *Ann. Missouri Bot. Gard.* 80:587–606.

Takhtajan, A. 1969. *Flowering plants: Origin and dispersal*, pp. 69–94. Smithsonian Institution Press, Washington, D.C.

Wood, C. E. 1958. The genera of the woody Ranales in the southeastern United States. *J. Arnold Arbor.* 39:296–346.

ANNONACEAE
Annona Family

Trees or shrubs, aromatic (due to ethereal oils). **Leaves** simple, entire, alternate, distichous, deciduous or persistent, often minutely punctate, exstipulate. **Flowers** solitary or in small cymose clusters, axillary, actinomorphic, perfect, hypogynous, with short convex receptacles, nodding, showy. **Calyx** of 3 sepals, distinct or basally connate, valvate. **Corolla** of 6 petals, biseriate (inner whorl usually smaller than outer whorl), distinct, maroon, white, or yellowish, imbricate or valvate. **Androecium** of numerous laminar stamens spirally arranged on the receptacle; filaments short and poorly differentiated from the anthers, distinct; anthers 4-locular, with pronounced peltate connective, dehiscing longitudinally, extrorse or latrorse. **Gynoecium** apocarpous and of usually numerous carpels; ovaries superior, 1-locular; ovules usually numerous in each carpel, anatropous, placentation usually parietal on the adaxial suture; style 1 and very short and thick or absent; stigma 1, terminal. **Fruit** an aggregate of berries or a syncarp formed by connation of carpels and adnation to the receptacle; seeds large, sometimes arillate; endosperm copious, oily, ruminate; embryo small.

Family characterization: Trees or shrubs with aromatic parts; actinomorphic, nodding flowers; perianth of 3 sepals and 6 biseriate petals with the inner whorl of petals smaller than the outer; numerous, peltate, laminar stamens spirally compacted on a short, convex receptacle; numerous distinct carpels; baccate fruit; and seeds with ruminate endosperm. Pollen grains monocolpate. Various alkaloids and tannins usually present. Tissues commonly with calcium oxalate crystals. Anatomical features: ethereal oil cells in the tissues; septate pith; and trilacunar nodes (Fig. 10: lb; Metcalfe 1987).

Genera/species: 132/2,300

Distribution: Almost exclusively in lowland evergreen forests of the tropics; a notable exception is the temperate American genus *Asimina* (see Walker 1971 and Thorne 1974).

Major genera: *Guatteria* (250 spp.), *Uvaria* (100–150 spp.), *Xylopia* (100–150 spp.), *Annona* (100–120 spp.), *Polyalthia* (120 spp.), *Goniothalamus* (115 spp.), and *Artabotrys* (100+ spp.)

U.S./Canadian representatives: *Asimina* (8 spp.), *Annona* (5 spp.), *Deeringothamnus* (2 spp.), *Guatteria* (2 spp.), *Oxandra* (2 spp.), and *Rollinia* (1 sp.)

Economic plants and products: Edible fruit from *Annona* (custard-apple, cherimoya, soursop, sweetsop) and *Asimina* (pawpaw). Perfumes (ethereal oils) from *Cananga odorata* (ilang-ilang or ylang-ylang). Ornamental trees and shrubs (species of 9 genera), including *Annona*, *Cananga*, *Polyalthia*, and *Rollinia*.

Commentary: The Annonaceae, a family of the "woody Ranales," are considered closely related to the Magnoliaceae, sharing such features as numerous and free stamens and carpels, laminar stamens, and monocolpate pollen (see Sastri 1969 and Thorne 1974). This so-called primitive floral morphology contrasts with some derived features, such as vessel elements with simple perforations (Vander Wyk and Canright 1956).

The various morphological and anatomical features are unusually consistent within this well-defined and

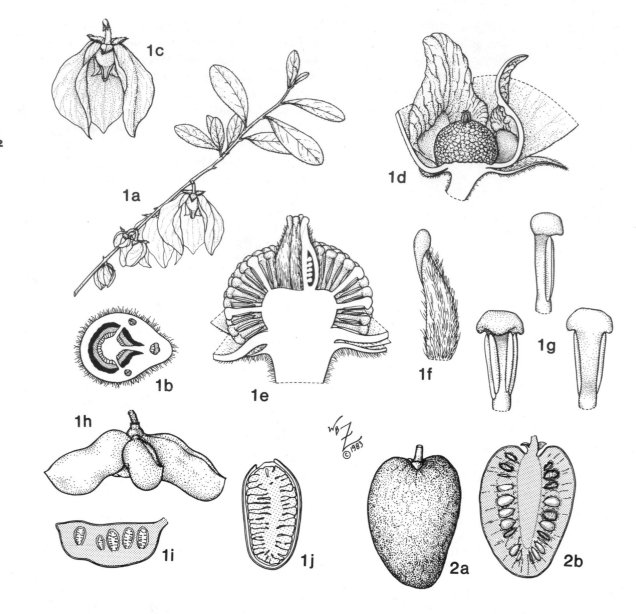

Figure 10. Annonaceae. 1, *Asimina incana*: **a,** flowering branch, ×⅓; **b,** cross section of node (three-trace trilacunar), ×10; **c,** flower, ×½; **d,** androecium and gynoecium (with half of perianth), ×2⅓; **e,** longitudinal section of flower, ×5; **f,** carpel, ×10; **g,** three views of stamen, ×12; **h,** fruit (aggregate of berries), ×½; **i,** longitudinal section of berry, ×½; **j,** longitudinal section of seed, ×2⅓. **2,** *Annona glabra*: **a,** fruit (syncarp formed by fusion of carpels and receptacle), ×⅓; **b,** longitudinal section of fruit, ×⅓.

presumably monophyletic group. Due to the pronounced homogeneity within the family, subfamilial taxa (especially genera) tend to be difficult to delimit (Leboeuf et al. 1982). Tribes and genera are classically distinguished by certain floral features, such as petal aestivation and anther connective shape, as well as fruit type (Walker 1971).

A conspicuous feature of the showy, nodding flower is the short and rounded receptacle packed with many laminar stamens, and at the apex, several to many distinct carpels (Fig. 10: 1e). The peculiar stamens have short, thick filaments and very conspicuous peltate connectives (Fig. 10: 1g). The carpels have poorly differentiated thick styles (Fig. 10: 1f) and develop typically into an aggregate of berries. When the berries are many-seeded, as in *Asimina*, the pericarp often forms fleshy ingrowths between the seeds. In some genera, such as *Annona*, the maturing carpels become united with each other and also with the fleshy receptacle to form an aggregate fruit (Fig. 10: 2a,b).

Annonaceous seeds are characterized by a complex morphology due to the unusual pronounced basipetal growth of the ovule and developing seed at the end opposite the micropyle (see Corner 1949). The seeds of some genera are arillate; other seeds develop a "middle integument," which sometimes outgrows the integuments and forms the covering of the seed. The ruminations of the endosperm are transverse infoldings caused by overgrowth of the integuments (Fig. 10: 1j).

The flowers of the family, generally pollinated by beetles, are protogynous. For example, in *Annona* the stigmas and styles become stuck together with exudate (Wood 1958), and the sticky stigmas are receptive before the surrounding extrorse anthers dehisce (Vithanage 1982). *Asimina* flowers have a fragrant or unpleasant scent and secrete nectar at the base of the three small inner petals, which envelop the receptacle (Norman and Clayton 1986). The nectar cannot be reached easily without touching the mature stigmas. Later, as the anthers mature, the petals expand and insects become covered with pollen on the way to the nectar, insuring cross-pollination.

REFERENCES CITED

Corner, E. J. H. 1949. The annonaceous seed and its four integuments. *New Phytol.* 48:332–364.

Leboeuf, M., A. Cavé, P. K. Bhaumik, B. Mukherjee, and R. Mukherjee. 1982. The phytochemistry of the Annonaceae. *Phytochemistry* 21:2783–2813.

Metcalfe, C. R. 1987. *Anatomy of the dicotyledons.* 2d ed. Vol. 3, *Magnoliales, Illiciales, and Laurales*, pp. 34–48. Oxford University Press, Oxford.

Norman, E. M., and D. Clayton. 1986. Reproductive biology of two Florida pawpaws: *Asimina obovata* and *A. pygmaea* (Annonaceae). *Bull. Torrey Bot. Club* 113:16–22.

Sastri, R. L. N. 1969. Comparative morphology and phylogeny of the Ranales. *Biol. Rev. Cambridge Philos. Soc.* 44: 291–319.

Thorne, R. F. 1974. A phylogenetic classification of the Annoniflorae. *Aliso* 8:147–209.

Vander Wyk, R., and J. E. Canright. 1956. The anatomy and relationships of the Annonaceae. *Trop. Woods* 104:1–24.

Vithanage, H. I. M. V. 1982. Pollen-stigma interactions: Development and cytochemistry of stigma papillae and their secretions in *Annona squamosa* L. *Ann. Bot.* (Oxford) 54: 153–167.

Walker, J. W. 1971. Pollen morphology, phytogeography, and phylogeny of the Annonaceae. *Contr. Gray Herb.* 202:1–131.

Wood, C. E. 1958. The genera of the woody Ranales in the southeastern United States. *J. Arnold Arbor.* 39:296–346.

ARISTOLOCHIACEAE
Birthwort or Dutchman's-pipe Family

Woody vines, shrubs, or (less often) perennial herbs, with aromatic stems and leaves (due to ethereal oils), with bitter or peppery-tasting yellow sap. **Leaves** simple, entire or sometimes lobed, alternate, palmately veined, often with cordate bases, exstipulate. **Flowers** solitary or in cymose or racemose clusters, usually axillary, zygomorphic or sometimes actinomorphic, perfect, usually epigynous, showy to inconspicuous, with patches of secretory hairs (nectaries) in perianth tube, often fetid. **Calyx** synsepalous with 1 to 3 lobes, with tubular, often curved base and spreading limb, often large and petaloid, usually lurid (dull red and mottled). **Corolla** absent or seldom vestigial. **Androecium** of 6 to numerous stamens, free or often adnate to the style forming a gynostemium; filaments distinct, free or adnate to style, short, thick; anthers free or adnate to style, dehiscing longitudinally, extrorse. **Gynoecium** of 1 pistil, typically 4- to 6-carpellate; ovary usually inferior, with as many locules as carpels (occasionally septa incomplete), often twisted during development; ovules numerous in each locule, anatropous, placentation typically axile; style 1, stout, free or usually adnate to androecium; stigmas as many as carpels, lobed, spreading. **Fruit** usually a septicidal capsule, often dehiscing basally; seeds numerous, three-sided or flattened, frequently immersed in pulpy endocarp; endosperm copious, oily; embryo minute, weakly dicotyledonous or undifferentiated.

Family characterization: Herbaceous to often woody vines with bitter or peppery-tasting, yellow sap; palmately veined leaves with cordate bases; fetid, 3-merous flowers with bizarre coloration and campanulate or trumpet-like shape; style and androecium often fused into a gynostemium; 3- or 6-locular inferior ovary (often twisted during development); a parachute-like, septicidal capsule as the fruit type; and poorly differentiated embryo. Aristolochic acid (bitter-tasting, yellow, nitrogenous compounds) or various alkaloids present. Anatomical features: spherical ethereal oil cells in parenchymatous tissues of stems and leaves; and extremely broad medullary rays that split the xylem in the stem (sometimes erroneously interpreted as anomalous secondary growth).

Genera/species: 8/400

Distribution: Primarily pantropical (excluding Australia); a few representatives in temperate areas.

Major genera: *Aristolochia* (300–350 spp.) and *Asarum* (70 spp.)

U.S./Canadian representatives: *Aristolochia* (21 spp.), *Hexastylis* (9 spp.), and *Asarum* (4 spp.)

Economic plants and products: Herbal medicines from dried rhizomes, roots, and stems of *Aristolochia* spp. (birthwort, snakeroot). Ornamental plants: sev-

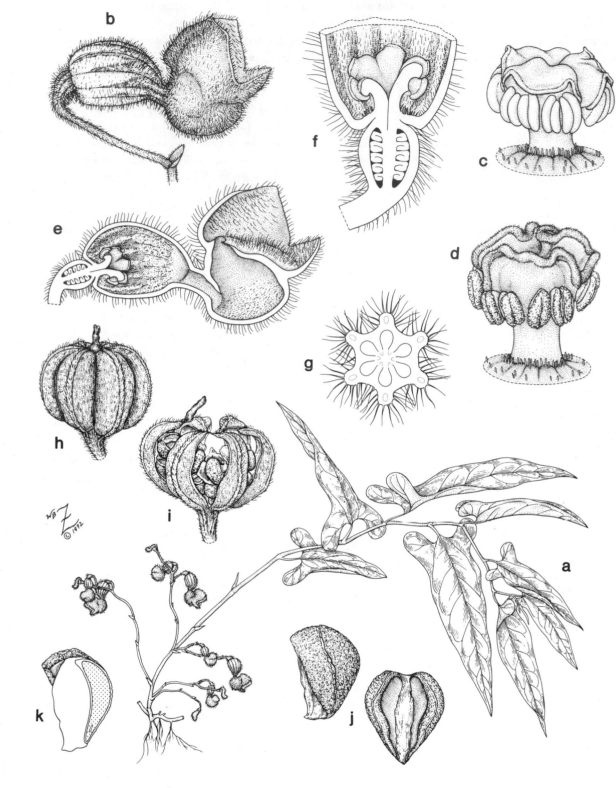

34

Figure 11. Aristolochiaceae. *Aristolochia serpentaria*: **a,** habit, ×½; **b,** flower, ×3; **c,** gynostemium from flower at anthesis, ×12; **d,** gynostemium (anthers dehisced, stigmas folded upward), ×12; **e,** longitudinal section of flower, ×3½; **f,** longitudinal section of gynoecium and androecium, ×8; **g,** cross section of ovary, ×12; **h,** young capsule, ×3; **i,** dehiscing capsule, ×3; **j,** two views of seed, ×7; **k,** longitudinal section of seed, ×7.

eral species of *Aristolochia* (Dutchman's-pipe, pelican-flower) and *Asarum* (asarabacca, wild-ginger).

Commentary: The affinities of the Aristolochiaceae have been much debated, and the family is commonly allied with those of the Magnoliales (e.g., Thorne 1992). Taxonomists also have considered the Aristolochiaceae as a "link" between the monocots and dicots, due to characters such as the three-merous flower parts; the similarity to the Dioscoreaceae has been especially stressed (Huber 1977; Dahlgren and Clifford 1982; Al-Shehbaz and Schubert 1989). A more accurate representation of this relationship is depicted in Fig. 2, which is based on the hypothesis of Donoghue and Doyle (1989), as discussed in the introductory chapter on monocots and dicots. Here the family, as a member of the "paleoherbs," is close to the monocot clade and not associated with the "woody Magnoliales" group.

At least three-fourths of the species in the Aristolochiaceae belong to the genus *Aristolochia*. The family is divided into three tribes (or in Thorne 1992, two subfamilies) based primarily upon the development of the perianth and the morphology of the androecium (Gregory 1956). The petaloid perianth, which likely represents the calyx, sometimes has been interpreted as involucral in origin (see Cronquist 1981). Of our representatives, *Aristolochia* is characterized by the zygomorphic perianth (and gynostemium formed by the fusion of androecium and style; Fig. 11: c,e). The perianth of *Asarum* and *Hexastylis* is actinomorphic (and the free stamens are connivent with the styles).

The highly specialized perianth of *Aristolochia* is trumpet- or pipe-shaped (Fig. 11: b): an elongated, curved tube (swollen at the base, constricted at the neck) and often an abruptly expanded limb. Small flies and gnats are attracted to the bizarre flowers whose coloration (dull red and mottled) often mimics carrion (Faegri and van der Pijl 1980). Nectar is secreted by patches of hairs in the perianth tube; some species also emit heat. The perianth acts as a trap, with a polished "slip-zone" and/or retrorse hairs in the tube (Fig. 11: e). The stigmas crown the gynostemium, curving over the anthers before pollination (Fig. 11: c). Later the anthers are exposed as the lobes shrivel and become erect (Fig. 11: d). Trapped insects escape when the stamens have matured, and the retrorse hairs wither (or in some species, fenestrations at the tube base expand). The campanulate flowers of *Asarum* are darkly colored and reportedly resinously scented (Knuth 1909). Pollination is more straightforward in these inconspicuous flowers, which also are protogynous and probably attract flies.

REFERENCES CITED

Al-Shehbaz, I., and B. G. Schubert. 1989. The Dioscoreaceae in the southeastern United States. *J. Arnold Arbor.* 70:57–95.

Cronquist, A. 1981. *An integrated system of classification of flowering plants*, pp. 90–93. Columbia University Press, New York.

Dahlgren, R. M. T., and H. T. Clifford. 1982. *The monocotyledons: A comparative study*, pp. 323, 330–345. Academic Press, London.

Donoghue, M. J., and J. A. Doyle. 1989. Phylogenetic analysis of the angiosperms and the relationships of the Hamamelidae. In *Evolution, systematics, and fossil history of the Hamamelidae*, ed. P. R. Crane and S. Blackmore, vol. 1, *Introduction and "lower" Hamamelidae*, pp. 17–45. Syst. Assoc. Special Vol. 40A. Clarendon Press, Oxford.

Faegri, K., and L. van der Pijl. 1980. *The principles of pollination ecology*, pp. 103–105. Pergamon Press, Oxford.

Gregory, M. P. 1956. A phyletic rearrangement in the Aristolochiaceae. *Amer. J. Bot.* 43:110–122.

Huber, H. 1977. The treatment of the monocotyledons in an evolutionary system of classification. *Pl. Syst. Evol.*, Suppl. 1:285–298.

Knuth, P. 1909. *Handbook of flower pollination.* Trans. H. R. A. Davis. Vol. 3, *Goodenovieae to Cyadeae*, pp. 355–356. Oxford University Press, Oxford.

Thorne, R. F. 1992. Classification and geography of the flowering plants. *Bot. Rev.* (Lancaster) 58:225–348.

LAURACEAE
Laurel Family

Trees or shrubs, aromatic (due to ethereal oils). **Leaves** simple, entire or occasionally lobed, alternate or sometimes opposite, usually persistent, minutely punctate, exstipulate. **Inflorescence** determinate, basically cymose and appearing racemose, paniculate, or umbellate, axillary, sometimes subtended by involucre. **Flowers** actinomorphic, perfect or sometimes imperfect (then plants usually dioecious), hypogynous to perigynous, small. **Perianth** of 6 tepals, biseriate, undifferentiated or outer whorl smaller, distinct or basally connate, sepaloid, greenish, yellowish, or white, imbricate, often persistent and accrescent. **Androecium** of basically 12 stamens in 4 whorls of 3, 1 or more whorls reduced to staminodes or absent; filaments distinct, those of at least 1 whorl (usually innermost) bearing a pair of lateral glandular outgrowths; anthers 2- or 4-locular at anthesis, basifixed, dehiscing by upturning flap-like valves, introrse (outer 2 whorls) or extrorse (innermost). **Gynoecium** of 1 pistil, 1-carpellate; ovary superior, 1-locular; ovule solitary, anatropous, pendulous, placentation parietal; style 1, simple; stigma 1 or sometimes 2- or 3-lobed. **Fruit** a berry or drupe, often subtended or completely surrounded by a fleshy

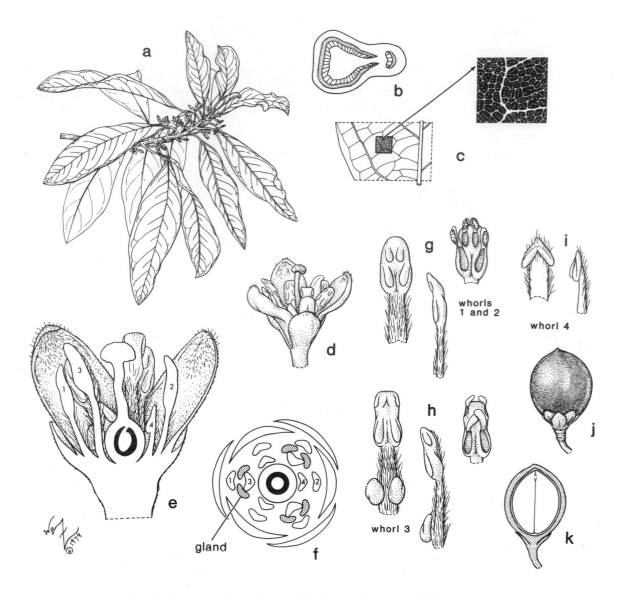

36

Figure 12. Lauraceae. *Persea borbonia*: **a,** flowering branch, ×²⁄₅; **b,** cross section of node (one-trace unilacunar), ×6; **c,** adaxial side of leaf, ×3, with detail showing finely punctate surface, ×12; **d,** flower, ×6; **e,** longitudinal section of flower, ×12; **f,** cross section of flower, ×12; **g,** three views of stamen (first and second whorls), ×12; **h,** three views of stamen with glands (third whorl), ×12; **i,** two views of staminode (fourth whorl), ×12; **j,** drupe, ×1½; **k,** longitudinal section of drupe, ×1½.

cupule; endosperm absent; embryo large with thick and fleshy cotyledons.

Family characterization: Trees or shrubs with aromatic leaves and bark; small, actinomorphic flowers with 6 tepals; whorled stamens; anthers dehiscing by valves; filaments (of 1 whorl) often with glandular outgrowths; 1-carpellate pistil with a single pendulous ovule; drupe or berry subtended by a cupule as the fruit type; and nonendospermous seeds. Tannins and often alkaloids present. Anatomical features: ethereal oil cells in the tissues (Janssonius 1926); and 1-trace unilacunar nodes (Fig. 12: b; Metcalfe 1987).

Genera/species: 31/2,490

Distribution: Tropical and subtropical areas; centers of diversity in Brazil and southeast Asia.

Major genera: *Litsea* (400 spp.), *Ocotea* (200–400 spp.), *Cryptocarya* (200–250 spp.), *Cinnamomum* (250 spp.), and *Persea* (150 spp.)

U.S./Canadian representatives: 13 genera/35 spp.; largest genera: *Ocotea*, *Nectandra*, and *Persea*

Economic plants and products: Spices and herbs from *Cinnamomum* (cinnamon, camphor), *Laurus* (bay leaf), *Lindera* (benzoil), and *Sassafras*. Fruit from *Persea americana* (avocado). Fragrant wood for cabinet work from *Nectandra* and *Sassafras*. Timber from *Ocotea*. Ornamental trees (species of 12 genera), including *Cinnamomum* (camphor tree) and *Laurus* (laurel).

Commentary: The Lauraceae are usually divided into two subfamilies, the large Lauroideae and the monotypic Cassythoideae (*Cassytha*, about 20 species of parasitic climbers); the latter group is sometimes treated as a segregate family (Cassythaceae). Genera (and tribes) of the family are determined primarily by androecial morphology, inflorescence type, and the nature of the perianth in fruit (Wood 1958). Artificial generic and specific boundaries are common due to convergent evolution of certain characters (van der Werff 1991).

The unusual androecial characters in the family are particularly stressed in the literature. Basically, the lauraceous flower has twelve stamens in four whorls of three (see Fig. 12: f), but often the innermost whorl is reduced to staminodes or is absent. The staminodes are commonly glandular. The anthers are two- or four-sporangiate and dehisce by upwardly recurving flaps that pull out all of the pollen from each locule (Fig. 12: g,h). Those of the outer two whorls are commonly introrse, and those of the inner whorl, extrorse (see Fig. 12: e).

The perfect flowers of the family are protogynous, with the stamens elongating and dehiscing after the stigma is receptive. The pollen presentation mechanism of the Lauraceae is complex (see Tomlinson 1980). Basically, the entire contents of each anther locule is presented to a pollinator as a mass stuck to the upturned flaps. Insects (often flies) are attracted to the nectar produced by glandular outgrowths that usually occur on the filaments of the third whorl (Fig. 12: h). Anatomical studies indicate that these glands represent modified stamens (Sastri 1965).

Usually the persistent perianth or perianth tube forms a cupule that partially subtends the baccate fruit. The cupule may be quite showy, as in *Nectandra coriacea* (cupule red-orange), but those of *Persea* species (Fig. 12: j) are inconspicuous. The one-seeded berries or drupes are usually dispersed by birds.

REFERENCES CITED

Janssonius, H. 1926. Mucilage cells and oil cells in the woods of the Lauraceae. *Trop. Woods* 6:3, 4.

Metcalfe, C. R. 1987. *Anatomy of the dicotyledons.* 2d ed. Vol. 3, *Magnoliales, Illiciales, and Laurales*, pp. 152–173. Oxford University Press, Oxford.

Sastri, R. L. N. 1965. Studies in Lauraceae. V. Comparative morphology of the flower. *Ann. Bot.* (Oxford) 29:39–44.

Tomlinson, P. B. 1980. *The biology of trees native to tropical Florida*, pp. 183–192. "Publ. by the author," Petersham, Mass.

van der Werff, H. 1991. A key to the genera of Lauraceae in the New World. *Ann. Missouri Bot. Gard.* 78:377–387.

Wood, C. E. 1958. The genera of woody Ranales in the southeastern United States. *J. Arnold Arbor.* 39:296–346.

PIPERACEAE
Pepper Family

Perennial herbs or sometimes shrubs to small trees, erect or prostrate, terrestrial or epiphytic, aromatic, commonly succulent, with jointed and/or swollen nodes. **Leaves** simple, entire, usually alternate, palmately or pinnately veined, stipulate (stipules adnate to petiole) or exstipulate. **Inflorescence** indeterminate, spicate and composed of numerous flowers packed onto a cylindrical fleshy axis (spadix) with these spicate units solitary or sometimes in umbellate arrangements, terminal, axillary, or arising on the stem opposite the leaves. **Flowers** actinomorphic or zygomorphic, perfect or sometimes imperfect (then plants monoecious or dioecious), hypogynous, very minute, each subtended by a peltate bract. **Perianth** absent. **Androecium** often of 6 stamens (biseriate) or reduced to 2, 3, or 4; filaments usually distinct; anthers basically 2-locular with locules distinct or confluent (then appearing 1-locular), dehiscing longitudinally, extrorse. **Gynoecium** of 1 pistil, 1-, 3-, or 4-carpellate; ovary superior, 1-locular; ovule solitary, orthotropous, erect, placentation basal; style 1 and short or absent, stigmas 1 to 4, capitate, lobed or brush-like. **Fruit** a small drupe with thin mesocarp; seed 1; perisperm copious, starchy, mealy; endosperm scanty, fleshy; embryo tiny, scarcely differentiated, embedded in endosperm.

Family characterization: Aromatic, fleshy herbs to shrubs with jointed nodes; reduced, naked flowers, each subtended by a peltate bract and crowded onto a fleshy axis (spadix); 1-locular and -ovulate ovary; a drupelet as the fruit type; and seeds with perisperm and poorly differentiated embryo. Various kinds of alkaloids present. Tissues with calcium oxalate crystals. Anatomical features: spherical ethereal oil cells embedded in parenchymatous tissues of leaves and stems; vascular bundles of stem scattered or organized into 2 or more rings (the outer ring often becoming continuous due to intrafascicular cambium; Fig. 13: 1c,d; 2b); and trilacunar or multilacunar nodes.

Genera/species: 8/2,000+

Distribution: Pantropical, most diverse in northern South America, Central America, and Malaysia. Com-

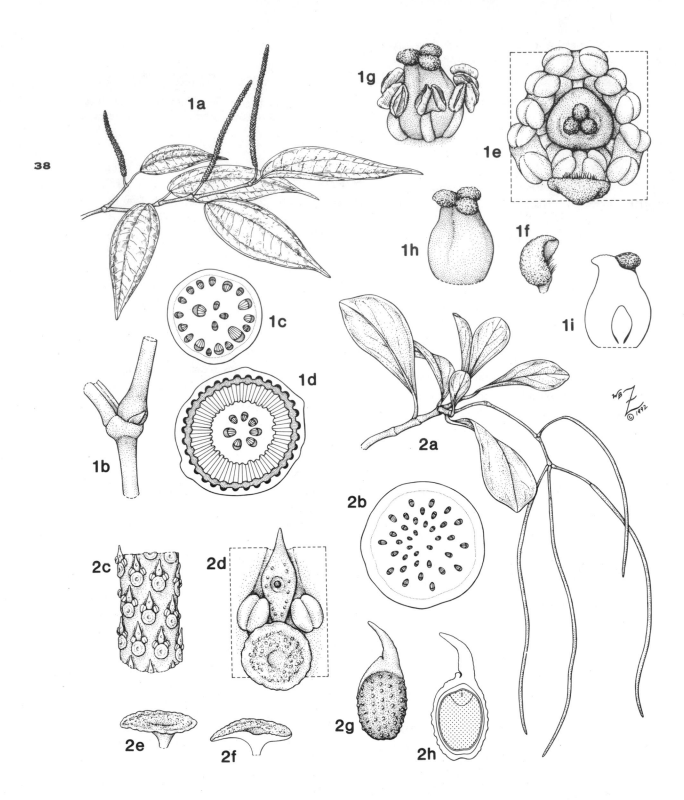

Figure 13. Piperaceae. 1, *Piper aduncum*: **a,** habit, ×½; **b,** node, ×8; **c,** cross section of young stem, ×25; **d,** cross section of older stem with secondary growth, ×9; **e,** top view of flower and subtending bract, ×30; **f,** bract, ×30; **g,** flower (anthers dehisced), ×30; **h,** pistil, ×30; **i,** longitudinal section of pistil, ×30. **2,** *Peperomia magnoliifolia*: **a,** habit, ×¼; **b,** cross section of stem, ×5; **c,** portion of spadix, ×8; **d,** flower and subtending bract, ×35; **e,** bract, ×35; **f,** longitudinal section of bract, ×35; **g,** drupe, ×20; **h,** longitudinal section of drupe, ×25.

monly in mesic, shady forests or in disturbed, sunny habitats.

Major genera: *Peperomia* (1,000+ spp.) and *Piper* (1,000+ spp.)

U.S./Canadian representatives: *Peperomia* (73 spp.), *Piper* (13 spp.), and *Pothomorphe* (2 spp.)

Economic plants and products: White and black pepper (*Piper nigrum* fruits); numerous *Piper* species sources of similar condiments (e.g., cubebs), as well as medicinal preparations (e.g., kava, a narcotic beverage). Species of *Macropiper*, *Peperomia* (radiator plant), and *Piper* cultivated as ornamentals, often as houseplants.

Commentary: The species of *Piper* (1,000+) and *Peperomia* (1,000+) make up most of the Piperaceae, which contain an uncertain number of genera (8 to 15 or more) and species (2,000+). The circumscription of *Piper*, in particular, has been problematic, with several subgenera or sections sometimes treated as segregate genera. In addition, the 600 to 1,000 (or more) described "species" of *Peperomia* often have been based on variable characters (e.g., leaf size and shape).

The Piperaceae usually are divided into two tribes or subfamilies, Piperoideae and Peperomioideae (Thorne 1992), but *Peperomia* and allied genera sometimes are treated as a segregate family, due to differences in several anatomical (see the discussion below), morphological, and embryological features (see Bornstein 1991). The Peperomioideae are characterized by herbaceous, often epiphytic habit; exstipulate leaves with multilayered epidermis; two stamens; one-locular anthers; unicarpellate ovary; and unitegmic ovules. The Piperoideae are defined by woody, usually terrestrial habit; stipulate leaves with one-layered epidermis; one to ten stamens; two-locular anthers; two- to six-carpellate ovary; and bitegmic ovules.

The Piperaceae share a suite of characters that are common among monocots but rare in the other dicots, such as the trimerous perianth, scattered vascular bundles (actually not homologous with those of the monocots), and monosulcate pollen (Burger 1977). The family thus has been considered a "link" between monocots and dicots (Dahlgren and Clifford 1982), although probably a more accurate representation of this relationship is shown in Fig. 2, where the Piperaceae are hypothesized as a member of the paleoherb clade of Donoghue and Doyle (1989).

The most outstanding anatomical feature of the family is the arrangement of the vascular tissue in the stem (Metcalfe and Chalk 1950). The widely spaced and often scattered vascular bundles frequently have xylem vessels arranged in U-shaped groups, as in the monocots. Sometimes the scattered bundles are organized into a vague circular pattern (e.g., species of *Peperomia*; Fig. 13: 2b) or into two or more distinct rings, with the outermost bundles often becoming a continuous cylinder due to intrafascicular cambial growth (e.g., *Piper*; Fig. 13: 1c,d).

Numerous, tiny, naked flowers, each subtended by a peltate bract (Fig. 13: 2e,f), are sunken into a fleshy axis (spadix). The gynoecium matures before the androecium. Little is known about pollination of the family, although authors sometimes mention the similarity of the reduced flowers and dense, spicate inflorescences to those of the wind-pollinated "Amentiferae" (e.g., Juglandaceae, Betulaceae). However, anemophily is rare in tropical plants. Pollination in this family is at least sometimes entomophilous, as reported by Semple (1974) for *Piper*. Pollen in the family is commonly glutinous, and several species (e.g., *Peperomia*) have scented flowers.

The small drupes are probably animal dispersed. Those of *Piper*, with a thin fleshy layer, are likely consumed by animals. The viscid drupelets of *Peperomia* (Fig. 13: 2g) often are somewhat tuberculate and cling to passers-by.

REFERENCES CITED

Bornstein, A. J. 1991. The Piperaceae in the southeastern United States. *J. Arnold Arbor*, Suppl. Ser. 1:349–366.

Burger, W. C. 1977. The Piperales and the monocots: Alternative hypothesis for the origin of the monocotyledon flower. *Bot. Rev.* (Lancaster) 43:345–393.

Dahlgren, R. M. T., and H. T. Clifford. 1982. *The monocotyledons: A comparative study*, pp. 323, 330–345. Academic Press, London.

Donoghue, M. J., and J. A. Doyle. 1989. Phylogenetic analysis of the angiosperms and the relationships of the Hamamelidae. In *Evolution, systematics, and fossil history of the Hamamelidae*, ed. P. R. Crane and S. Blackmore, vol. 1, *Introduction and "lower" Hamamelidae*, pp. 17–45. Syst. Assoc. Special Vol. 40A. Clarendon Press, Oxford.

Metcalfe, C. R., and L. Chalk. 1950. *Anatomy of the dicotyledons*, vol. 2, pp. 1120–1127. Oxford University Press, Oxford.

Semple, K. S. 1974. Pollination in Piperaceae. *Ann. Missouri Bot. Gard.* 61:868–871.

Thorne, R. F. 1992. Classification and geography of the flowering plants. *Bot. Rev.* (Lancaster) 58:225–348.

MENISPERMACEAE
Moonseed Family

Usually perennial herbaceous vines, woody (at least at base). **Leaves** simple, entire or occasionally palmately lobed, sometimes peltate, alternate, usually palmately veined, deciduous or persistent, with petiole sometimes swollen distally and/or basally, exstipulate. **Inflorescence** usually determinate, cymose, paniculate, racemose, or umbellate, commonly supra-axillary. **Flowers** actinomorphic, imperfect (plants dioecious), hypogynous, minute. **Calyx** of usually 6 sepals, biseriate, distinct, imbricate or valvate. **Corolla** of usually 6 petals, biseriate, distinct, usually smaller than the sepals and encircling the filaments, whitish, greenish, or yellowish, imbricate. **Androecium** of 6 stamens, biseriate, opposite the petals; filaments distinct to sometimes variously connate; anthers basifixed, 4-locular but becoming 1- or 2-locular at anthesis, dehiscing longitudinally, introrse or sometimes extrorse; staminodes often present in carpellate flowers. **Gynoecium** apocarpous and of usually 3 or 6 carpels in 1 whorl, often on a gynophore; ovaries superior, 1-locular, gibbose; ovules 2 in each carpel (1 aborting), anatropous, placentation parietal; style 1, short or absent; stigma 1, entire to laciniate; rudimentary gynoecium occasionally present in staminate flowers. **Fruit** an aggregate of drupes or sometimes nutlets, with persistent style appearing basal due to asymmetric growth of maturing carpel, with curved and sculptured endocarp; seeds horseshoe-shaped; endosperm oily, sometimes ruminate, or sometimes absent; embryo curved or sometimes coiled.

Family characterization: Dioecious perennial vines; leaves with palmate venation; inconspicuous, imperfect, 3-merous flowers with biseriate calyx, corolla, and androecium; apocarpous carpels in 1 whorl; aggregate of drupes with elaborately sculptured endocarp; and horseshoe-shaped seed with curved/coiled embryo. Sesquiterpenoids (bitter, poisonous compounds) and alkaloids (e.g., berberine) present. Anatomical features: broad medullary rays in young stems; and anomalous secondary growth (fascicular or extrafascicular cambia forming concentric rings of bundles, resulting in alternating layers of xylem and phloem; see Amaranthaceae).

Genera/species: 65/350

Distribution: Mainly tropical rain forests (especially in the Old World), with some representatives in subtropical to temperate areas.

Major genera: *Stephania* (40 spp.), *Tinospora* (32–40 spp.), and *Abuta* (31–35 spp.)

U.S./Canadian representatives: *Cocculus* (6 spp.), *Calycocarpum* (1 sp.), *Cissampelos* (1 sp.), *Hyperbaena* (1 sp.), and *Menispermum* (1 sp.)

Economic plants and products: Many medicinal/ poisonous plants (due to alkaloids and terpenoids), including *Anamirta* (fish-berry—with picrotoxin, used as a fish poison, a parasiticide, and an ingredient in ointments); *Chondrodendron* (with tubocurarine chloride, a muscle relaxant used medicinally and as an arrow poison [curare]); *Cissampelos* (false pareira root—used as a snake-bite antidote and a contraceptive); and *Jateorhiza* (calumba—used as a medicinal tonic). Ornamental plants: species of *Cocculus* (coralbeads, red-berried moonseed) and *Menispermum* (moonseed).

Commentary: The Menispermaceae usually are included in the Ranunculales/Berberidales complex of families. Recent cladistic analyses on families of this group support the monophyly of Menispermaceae (e.g., dioecism, large embryo; Loconte and Estes 1989). The large and complex family has been divided traditionally into four to eight tribes based upon characters of the fruit (endocarp) and seeds (endosperm and embryo; Ernst 1964).

The typical fruit type of the family (Fig. 14: s) is an aggregate of drupes (usually blue or black) that develop from an apocarpous gynoecium (Fig. 14: k). Each maturing carpel enlarges asymmetrically, so that the style appears to be nearly basal on the more or less spherical ripe drupe (Fig. 14: t). The bony endocarp (the outer part of the stone or "pit" surrounding the seed) is curved, laterally compressed, and attractively sculptured with ribs and/or tubercles (Fig. 14: v). The name "moonseed" for the family is in reference to the shape of the seed (like a crescent). The curved to coiled embryo is molded around an adaxial intrusion of the endocarp (Fig. 14: u), called a condyle (Thanikaimoni 1986). This ingrowth appears on the surface of the endocarp as a variously shaped depression; internally the condyle may be solid, hollow, or perforated.

Little has been reported on the pollination of these small, imperfect flowers. Barneby and Krukoff (1971) postulate that insect pollinators are likely since the dioecious vines are widely separated in tropical forests.

REFERENCES CITED

Barneby, R. C., and B. A. Krukoff. 1971. Supplementary notes on American Menispermaceae. VIII. A generic survey of the American Triclisieae and Anomospermeae. *Mem. New York Bot. Gard.* 22:1–89.

Ernst, W. R. 1964. The genera of Berberidaceae, Lardizabalaceae and Menispermaceae in the southeastern United States. *J. Arnold Arbor.* 45:1–35.

Loconte, H., and J. R. Estes. 1989. Phylogenetic systematics of Berberidaceae and Ranunculales (Magnoliidae). *Syst. Bot.* 14:565–579.

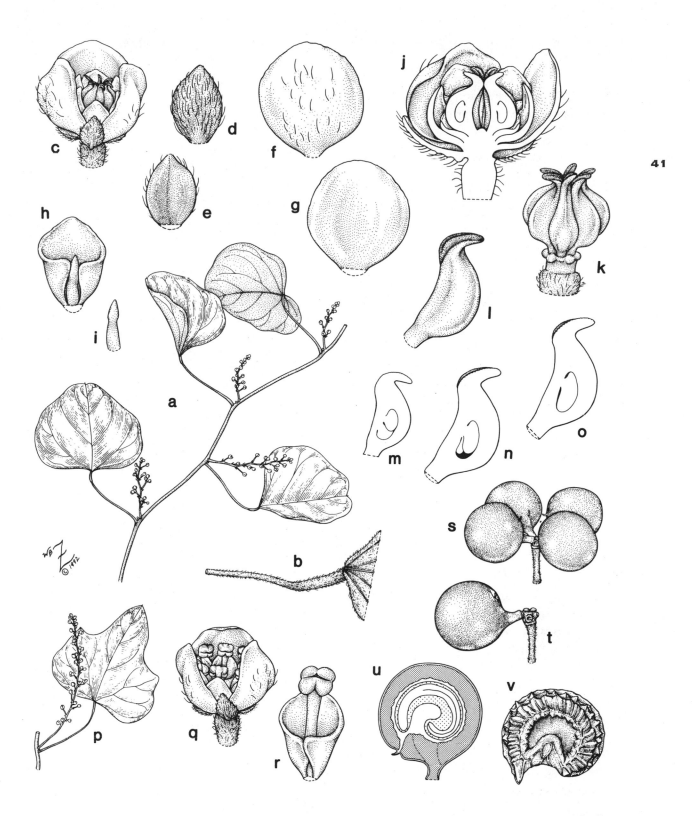

Figure 14. Menispermaceae. *Cocculus carolinus:* **a,** carpellate flowering plant, ×⅖; **b,** detail of petiole apex showing pulvinus, ×2¼; **c,** carpellate flower, ×9; **d,** outer sepal (abaxial side), ×15; **e,** outer sepal (adaxial side), ×15; **f,** inner sepal (abaxial side), ×12; **g,** inner sepal (adaxial side), ×12; **h,** staminode with subtending petal, ×15; **i,** staminode, ×15; **j,** longitudinal section of carpellate flower, ×15; **k,** gynoecium, ×15; **l,** carpel, ×25; **m,** longitudinal section of young carpel (two-ovulate), ×25; **n,** longitudinal section of older carpel (one ovule aborting), ×25; **o,** mature carpel (one-ovulate), ×25; **p,** staminate flowering plant, ×⅖; **q,** staminate flower, ×9; **r,** stamen with subtending petal, ×15; **s,** fruit (aggregate of drupelets), ×2¼; **t,** drupelet, ×2½; **u,** longitudinal section of drupelet, ×4; **v,** pyrene, ×5.

41

Thanikaimoni, G. 1986. Evolution of Menispermaceae. *Canad. J. Bot.* 64:3130–3133.

RANUNCULACEAE
Buttercup or Crowfoot Family

Annual or perennial herbs, sometimes shrubs or vines, terrestrial or sometimes aquatic, often with rhizomes or tuberous roots. **Leaves** simple or variously compound or dissected, serrate to lobed, usually alternate, exstipulate, often with sheathing petioles. **Inflorescence** determinate, cymose and often appearing racemose or paniculate, or flower sometimes solitary, terminal. **Flowers** actinomorphic or sometimes zygomorphic, usually perfect, hypogynous, often showy, with short and globose to elongate receptacle. **Calyx** of typically 5 sepals, distinct, caducous, often showy and petaloid, variously colored, imbricate. **Corolla** of typically 5 petals or often absent, distinct, with modified nectariferous bases or reduced to small nectariferous sacs or scales, imbricate. **Androecium** of usually numerous stamens spirally arranged on the receptacle, with the outer stamens sometimes reduced to staminodes; filaments distinct; anthers basifixed, dehiscing longitudinally, extrorse. **Gynoecium** apocarpous and of usually several to numerous carpels spirally arranged on the receptacle; ovaries superior, 1-locular; ovules 1 to numerous in each carpel, anatropous, placentation parietal along the ventral suture or nearly basal; style 1; stigma 1, often bilobed. **Fruit** typically an aggregate of follicles, achenes, or berries; endosperm copious, oily; embryo minute, straight.

Family characterization: Herbaceous plants commonly with rhizomes or tubers; compound or dissected leaves with sheathing petiole bases; perianth often of showy petaloid sepals and reduced or modified petals; numerous distinct stamens and carpels spirally arranged on the receptacle; aggregate fruit of follicles, achenes, or berries; and seeds with minute embryo and copious endosperm. Various alkaloids, glycosides, and/or saponins characteristically present in sap. Anatomical feature: vascular bundles of stem often scattered or in concentric rings.

Genera/species: 46/1,900

Distribution: Primarily in temperate to boreal regions of the Northern Hemisphere; especially diverse in eastern Asia and eastern North America.

Major genera: *Ranunculus* (250–400 spp.), *Aconitum* (100–300 spp.), *Clematis* (230–250 spp.), *Delphinium* (250 spp.), *Anemone* (120–150 spp.), and *Thalictrum* (85–150 spp.)

U.S./Canadian representatives: 24 genera/323 spp.; largest genera: *Ranunculus, Delphinium, Clematis,* and *Thalictrum*

Economic plants and products: Several poisonous and medicinal plants (due to various alkaloids, glycosides, and/or saponins), including *Aconitum* (wolfbane, monk's-hood—with aconite), *Actaea* (baneberry—with berberine), *Delphinium* (larkspur—with delphinine), *Hydrastis* (goldenseal—with berberine), and *Ranunculus* (buttercup—with ranunculin). Ornamental plants (species of 29 genera), including *Anemone* (windflower), *Aquilegia* (columbine), *Clematis* (virgin's-bower), *Delphinium* (larkspur), *Helleborus* (hellebore), *Ranunculus* (buttercup, crowfoot), *Thalictrum* (meadowrue), and *Trollius* (globeflower).

Commentary: The Ranunculaceae traditionally have been regarded as the most "primitive" of herbaceous angiosperms, due to such features as numerous, distinct carpels and stamens spirally arranged on an enlarged receptacle, incomplete carpel fusion, and follicular fruits (see the commentary for Magnoliaceae, and Leppik 1964). More accurately stated, however, the family is probably somewhat basal in the "tricolpate pollen group" of dicots (see Fig. 2), according to the hypothesis of Donoghue and Doyle (1989). The heterogeneous family has historically been divided into two (or three) subfamilies (and several tribes) based primarily on characters of the ovules (number per ovary) and of the fruit (aggregates of follicles, berries, or achenes) and, more recently, on chromosome morphology (see Hoot 1991 for a brief summary). Generic limits (often based on perianth features) tend to be indistinct. The cladistic studies of Loconte and Estes (1989) support the monophyly of the family and the close relationship to families such as the Menispermaceae and Berberidaceae.

The differentiation of the perianth varies considerably within the family. For example, the flowers of *Ranunculus* (Fig. 16: 2b) have a greenish calyx and showy corolla, as is typical of most flowers (Keener 1976). Many genera, however, such as *Anemone* and *Clematis* (Fig. 15: 1; 2b,d), are characterized by apetalous flowers with petaloid sepals (Keener 1975a,b). The perianth forms spurs in the more specialized groups, such as *Aquilegia* (spurred petals; Fig. 16: 1c) and *Delphinium* (spurred sepals and petals).

The conspicuous flowers of this family usually are pollinated by insects. Nectar often is secreted by the petals, as in *Ranunculus* (nectaries at petal bases; Fig. 16: 2c) or *Aquilegia* (nectaries within spurs). The flowers of several taxa (e.g., *Clematis* and *Anemone* species) that do

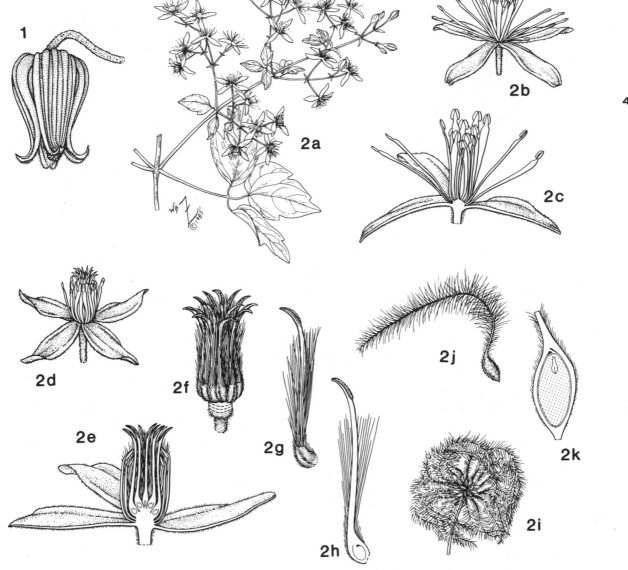

Figure 15. Ranunculaceae. 1, *Clematis reticulata*: flower, ×1½. **2,** *Clematis catesbyana*: **a,** staminate flowering plant, ×½; **b,** staminate flower, ×2; **c,** longitudinal section of staminate flower, ×3; **d,** carpellate flower, ×2; **e,** longitudinal section of carpellate flower, ×4; **f,** gynoecium, ×5; **g,** carpel, ×8; **h,** longitudinal section of carpel, ×9; **i,** fruit (aggregate of achenes), ×1; **j,** achene, ×2½; **k,** longitudinal section of achene, ×8.

not produce nectar are visited by insects that gather the pollen. Details of the pollination mechanisms vary greatly depending upon the particular floral morphology. The more open flowers (e.g., those of *Ranunculus*) are visited by various insects, while more elaborate flowers with spurs (e.g., those of *Aquilegia*) are pollinated by insects with long proboscises (or by hummingbirds). Flowers generally are protandrous, although self-pollination is possible when the anthers of the inner whorl dehisce immediately before or during the maturation of the stigmas. The flowers of *Thalictrum* are anemophilous.

REFERENCES CITED

Donoghue, M. J., and J. A. Doyle. 1989. Phylogenetic analysis of the angiosperms and the relationships of the Hamamelidae. In *Evolution, systematics, and fossil history of the Hamamelidae*, ed. P. R. Crane and S. Blackmore, vol. 1, *Introduction and "lower" Hamamelidae*, pp. 17–45. Syst. Assoc. Special Vol. 40A. Clarendon Press, Oxford.

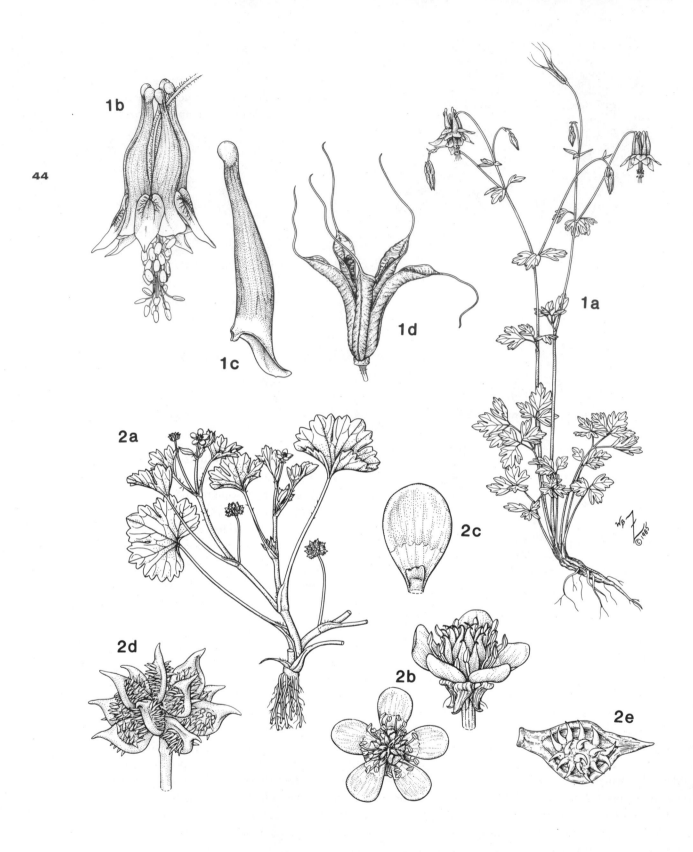

Figure 16. Ranunculaceae (continued). 1, *Aquilegia canadensis*: **a,** habit, ×⅖; **b,** flower, ×1¼; **c,** petal (note spur), ×2¼; **d,** fruit (aggregate of follicles), ×1¾. **2,** *Ranunculus muricatus*: **a,** habit, ×⅓; **b,** two views of flower, ×3½; **c,** adaxial view of petal showing nectary at base, ×6; **d,** young fruit (aggregate of achenes), ×2¾; **e,** mature achene, ×6.

Hoot, S. B. 1991. Phylogeny of the Ranunculaceae based on epidermal microcharacters and macromorphology. *Syst. Bot.* 16:741–755.

Keener, C. S. 1975a. Studies in the Ranunculaceae of the southeastern United States: I: *Anemone* L. *Castanea* 40:36–44.

——. 1975b. Ibid., III: *Clematis* L. *Sida* 6:33–47.

——. 1976. Ibid., V: *Ranunculus* L. *Sida* 6:266–283.

Leppik, E. E. 1964. Floral evolution in the Ranunculaceae. *Iowa State Coll. J. Sci.* 39:1–101.

Loconte, H., and J. R. Estes. 1989. Phylogenetic systematics of Berberidaceae and Ranunculales (Magnoliidae). *Syst. Bot.* 14:565–579.

PAPAVERACEAE

Poppy Family

Including the Fumariaceae: Fumitory Family

Annual or perennial herbs, sometimes vines, or occasionally shrubs or trees; sap acrid, watery or colored (reddish-orange, yellow, to nearly white). **Leaves** simple to compound, typically pinnately or palmately lobed or dissected, usually alternate, in basal rosettes and/or cauline, exstipulate. **Inflorescence** determinate or indeterminate, often cymose and appearing racemose or umbellate, or flower solitary, terminal and/or axillary. **Flowers** actinomorphic to zygomorphic, perfect, hypogynous, small to large and showy. **Calyx** of 2 or sometimes 3 sepals, distinct, large (enclosing the bud) or small (then bract-like and peltate), caducous, imbricate. **Corolla** of 4 or sometimes 6 (or more) petals, biseriate, distinct, spreading, and equal, or more often coherent and closed (then 1 or both outer petals basally saccate/spurred and inner 2 petals narrower and apically coherent/connate, forming a cap over the stigma[s]), typically white or brightly colored, imbricate and sometimes crumpled in bud. **Androecium** of 6 (diadelphous) or numerous (whorled) stamens; filaments winged, distinct (when numerous) or connate/coherent in 2 groups of 3 (then often basally spurred and/or nectariferous); anthers basifixed, all 2-locular (when stamens numerous) or dimorphic (when stamens diadelphous) with the central one of each group 2-locular and the 2 laterals 1-locular, dehiscing longitudinally, introrse or extrorse. **Gynoecium** of 1 pistil, 2- to many-carpellate; ovary superior, 1-locular; ovules few to many, anatropous, amphitropous, or campylotropous, placentation parietal (sometimes on intruding placentae); style 1 or absent; stigma(s) 1 to numerous, capitate, lobed. **Fruit** a septicidal, poricidal, or transversely septate capsule or occasionally a nut; seeds obovoid to reniform, generally arillate; endosperm copious, oily or fleshy; embryo minute, straight or slightly curved.

Family characterization: Herbaceous plants with watery or colored, acrid sap; pinnately lobed/dissected leaves; 2- or 3-merous, hypogynous flowers; 2 or 3 caducous sepals; biseriate corolla commonly of 4 or 6 petals; 6 diadelphous stamens or numerous whorled stamens; 1-locular compound ovary with parietal placentation; and variously dehiscent capsular fruit with arillate seeds. Various alkaloids (e.g., protopine) within laticifers or elongate sacs (containing colored or milky latex) or elongate secretory cells (containing watery sap).

Genera/species: 42/660

Distribution: Primarily in north temperate to subtropical regions, with a few representatives in southern montane Africa and eastern Australia; particularly diverse in western North America and temperate Eurasia.

Major genera: *Corydalis* (300–320 spp.), *Papaver* (50–100 spp.), and *Fumaria* (55 spp.)

U.S./Canadian representatives: 23 genera/93 spp.; largest genera: *Argemone*, *Papaver*, *Eschscholzia*, and *Corydalis*

Economic plants and products: Drugs (due to alkaloids) from sap of unripe capsules of *Papaver somniferum* (opium poppy—opium and derivatives morphine, heroin, and codeine) and from rhizomes of *Sanguinaria canadensis* (bloodroot, puccoon—various alkaloids). Edible seeds and oil also from *Papaver somniferum*. Ornamental plants (species of 28 genera), including *Argemone* (argemony, prickly-poppy), *Corydalis*, *Dendromecon* (bush-poppy, tree-poppy), *Dicentra* (bleedingheart, Dutchman's-breeches), *Eschscholzia* (California-poppy), *Hunnemannia* (Mexican tulip-poppy), *Macleaya* (plume-poppy), and *Meconopsis* (Asiatic-poppy).

Commentary: The Papaveraceae are here defined broadly (Thorne 1992) to include the large subfamily Fumarioideae, which comprises such familiar genera as *Corydalis*, *Fumaria*, and *Dicentra*. With the family thus circumscribed, about half of the species belong to the genus *Corydalis*. Although most European botanists treat the Fumarioideae as a subfamily of the Papaveraceae *s.l.*, American botanists tend to accept the group as a distinct family (Fumariaceae) for purposes of convenience. The characters often used to separate these two groups are summarized in Table 2; however, the chart is misleading since both taxa share numerous vegetative, floral, anatomical, chemical, and cytological characters. For example, the laticifers (of Papaveraceae *s.s.*) and the secretory cells (of Fumarioideae) ac-

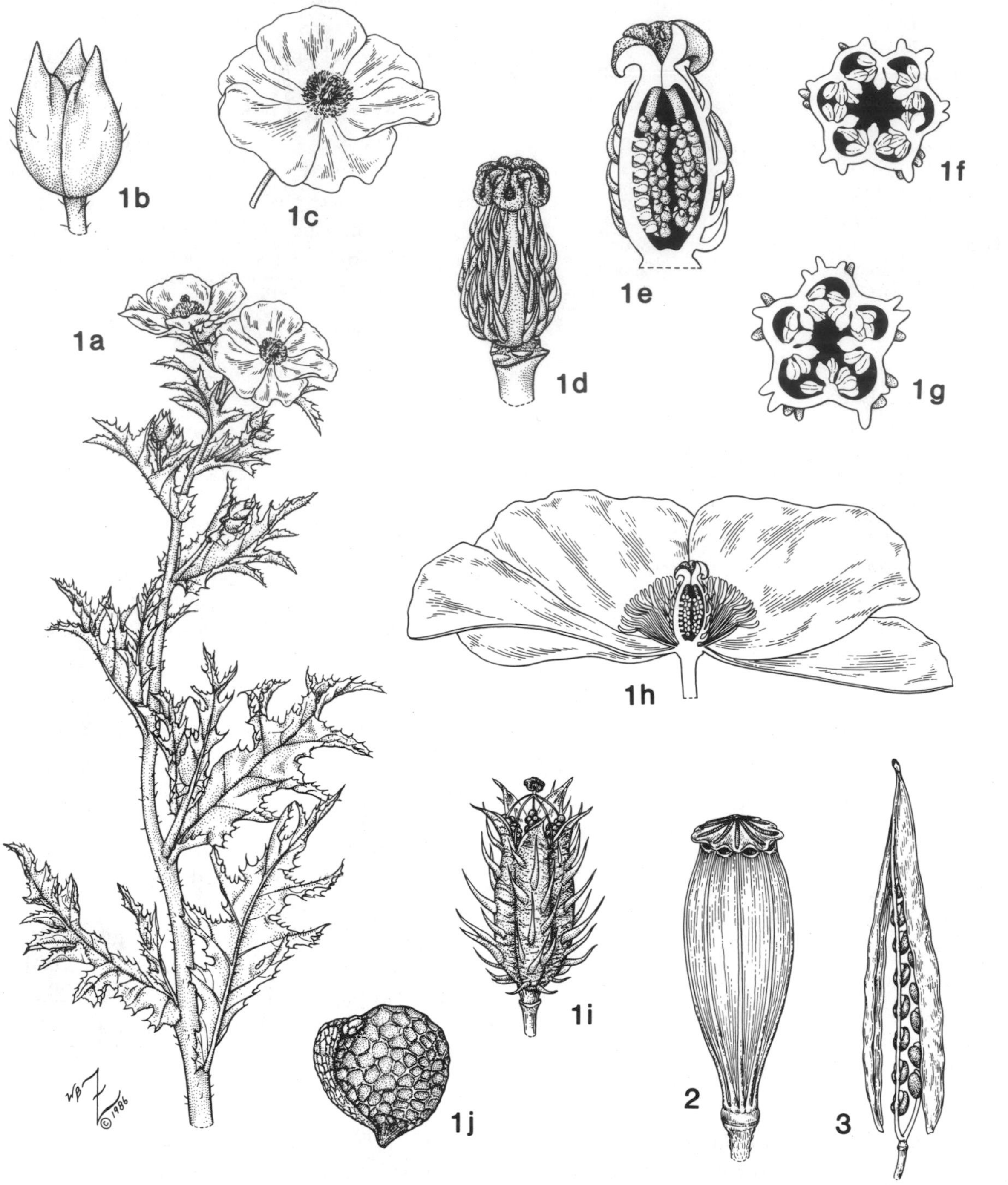

Figure 17. Papaveraceae. 1, *Argemone albiflora*: **a,** upper portion of plant, ×¼; **b,** bud, ×1½; **c,** flower, ×⅖; **d,** pistil, ×3; **e,** longitudinal section of pistil, ×4; **f,** cross section of ovary with six carpels, ×5; **g,** cross section of ovary with five carpels, ×5; **h,** longitudinal section of flower, ×1¼; **i,** capsule, ×1¼; **j,** seed (with aril), ×10. **2,** *Papaver dubium*: capsule, ×2½. **3,** *Chelidonium majus*: capsule, ×2.

tually are homologous structures. The latter view of Fumarioideae as a segregate family also ignores the transitional genera *Hypecoum* and *Pteridophyllum*,.characterized by weakly zygomorphic flowers with four distinct stamens (Arber 1932). These two genera have

been included as one (Hypecoideae) or two (Hypecoideae and Pteridophylloideae) subfamilies within the Papaveraceae *s.s.* or Fumariaceae, or each genus has been segregated in its own monotypic family (Hypecoaceae and Pteridophyllaceae). Thorne (1974) sug-

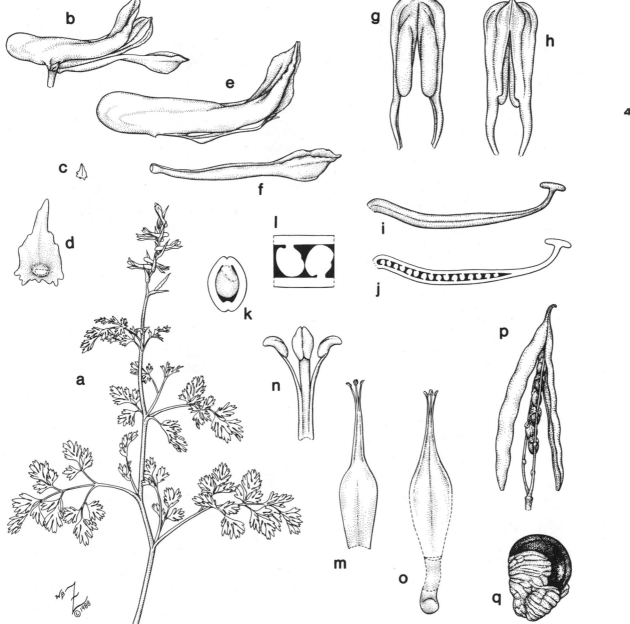

Figure 18. Papaveraceae (continued). *Corydalis micrantha*: **a,** habit, ×²/₅; **b,** flower, ×3; **c,** sepal (to same scale as **e** and **f**), ×4 ½; **d,** adaxial side of sepal, showing peltate attachment, ×25; **e,** upper petal (and androecium), ×4½; **f,** lower petal, ×4½; **g,** lateral petals (connate at apices, adaxial side), ×4½; **h,** lateral petals (abaxial side), ×4½; **i,** pistil, ×6; **j,** longitudinal section of pistil, ×6; **k,** cross section of ovary, ×22; **l,** portion of longitudinal section of ovary, ×22; **m,** upper staminal fascicle, ×6; **n,** apex of upper staminal bundle (from bud) showing dimorphic anthers, ×12; **o,** lower staminal fascicle (abaxial side) with spur, ×6; **p,** capsule, ×3; **q,** seed (with aril), ×10.

gests that American botanists prefer the segregate "Fumariaceae" because they are unfamiliar with these intermediate genera that are restricted to the Old World. A recent cladistic summary discussing characters of the major taxa of this Papaveraceae complex (Judd et al. 1994) also supports the treatment of all the above-mentioned taxa as a single family, Papaveraceae *s.l.,* and not as two (or more) families.

The extremes in the floral structure of the family, particularly of the perianth and androecium, are summarized in Table 2 (although transitional forms of these structures, mentioned briefly above, occur). The

TABLE 2. Major morphological differences traditionally used to separate the Papaveraceae *s.s.* and the Fumarioideae ("Fumariaceae"). Many of these character states actually are found in both taxa (see text for discussion).

CHARACTER	PAPAVERACEAE *S.S.* (SEVERAL SUBFAMILIES)	FUMARIOIDEAE ("FUMARIACEAE")
GENERA/SPECIES	24/201	18/461
HABIT	herbs to (sometimes) soft-wooded arborescent shrubs	delicate herbs with brittle stems, often with swollen rootstocks
SAP	usually milky or colored latex in articulated laticifers or elongate latex cells	colorless and watery, often in elongate secretory cells
INFLORESCENCE	flowers usually solitary, or less often in cymose or umbellate arrangements	racemose, or sometimes cymose
FLORAL SYMMETRY	actinomorphic	zygomorphic to occasionally actinomorphic
NECTARIES	none	present (filament base)
SEPALS	2 or 3 large and fully enclosing bud before anthesis	2 small (bract-like) and peltate; not enclosing bud
PETALS	2 + 2, 3 + 3, or more spreading equal distinct wrinkled, crumpled in bud	2 + 2 closed unequal coherent/connate not wrinkled
STAMENS	numerous distinct, in several whorls	3 + 3 diadelphous
ANTHERS	all 2-loculed	central 2-loculed, laterals 1-loculed
OVARY	2- to many-carpellate	2-carpellate
STIGMA	large, discoid, and lobed dry	capitate or lobed wet
STYLE	short to absent	elongate
FRUIT TYPE	usually a poricidal or septicidal capsule	usually a septicidal capsule, or nut-like

48

bud may be enclosed by cap-like sepals that fall off when the flower expands, as in *Argemone*, where each hooded sepal is terminated by a sharp prickle (Fig. 17: 1b). Often, as in *Corydalis*, the sepals are inconspicuous, scale-like, and attached by a more or less peltate stub (Fig. 18: c,d). The petals are biseriate, usually twice the sepals in number, and vary from equal ("open," actinomorphic flower; e.g., *Argemone*) to modified and unequal ("closed," zygomorphic flower; *Corydalis*). In the Fumarioideae, the large outer pair of petals more or less completely encloses the laterally compressed, somewhat tubular flower (Fig. 18: e,f). A prominent basal pouch or spur characterizes one (*Corydalis*) or both (*Dicentra*) of these petals. The inner, narrower petals are fused apically (or coherent) into a crested hood over the anthers and stigma (Fig. 18: g,h);

the combined structure blocks the entrance to the flower.

The androecium may consist of four to numerous stamens, all with typical two-locular anthers. However, the six stamens of the androecium of the Fumarioideae are fused into two bundles (diadelphous), each with a common, broad, strap-like base terminating in three branches (Fig. 18: m–o). The central branch of each bundle bears a two-locular anther, but the anthers on the lateral filaments are one-locular (Fig. 18: n). Each bundle is opposite the outer petals and encloses the style and stigma, to which the anthers are appressed. One (*Corydalis*; Fig. 18: o) or both (*Dicentra*) bundle(s) often is/are basally spurred and/or nectariferous.

The showy flowers with white or brightly colored

petals (and often a fragrance) usually attract insect pollinators. The open actinomorphic flowers of the family with conspicuous numerous stamens (e.g., *Argemone*) do not produce nectar, and bees, flies, and beetles visit to collect the copious pollen (Vogel 1978). The outermost stamens mature before the innermost, and the flowers vary from protogynous to protandrous. The broad, flat-topped, and radiating stigma (Fig. 17: 1d) serves as a convenient landing platform (resulting in pollination as a pollen-dusted insect alights on the lobed disc).

In the specialized flowers of the Fumarioideae, nectar is secreted into the saccate portion of the outer petal(s) by the basally spurred androecial bundle(s). The pollination mechanism is similar to that of the papilionaceous Fabaceae whose keel, which encloses and protects the anthers and stigma, serves the same purpose as the apically connate, inner petals of the Fumarioideae. A nectar-seeking bee, probing down into the spur, must depress the hooded inner petals, thereby becoming dusted by the exposed, pollen-covered anthers (Ryberg 1960). Pollen is then transferred to the stigmas of other visited flowers. In many species the hood springs back when the pressure is removed; other flowers are explosive, as in some of the Fabaceae. Despite the proximity of the anthers to the stigma in both the open, pollen flowers and the closed, nectar flowers, self-pollination is not necessarily automatic, since many species of the Papaveraceae evidently are self-sterile.

The fruit of the family often is a capsule with variable (and often incomplete) apical or basal dehiscence (Ernst 1962). In *Argemone*, for example, the several flap-like septicidal valves are apical (Fig. 17: 1i), while the two valves of *Corydalis* are basipetally dehiscent (Fig. 18: p). The arillate seeds of capsular fruits frequently are dispersed by ants. In many of the true poppies (*Papaver* species) the capsules are poricidal, with a series of small holes below the flat, persistent stigma (Fig. 17: 2). The Papaveraceae often have been placed near the Brassicaceae-Capparaceae partly due to the convergent development of a "silique" with a frame-like "replum," as in *Chelidonium* and *Corydalis*, where the seeds are attached to the rim-like, persistent parietal placentae (Figs. 17: 3; 18: p). The fruit of *Fumaria* is a nut.

REFERENCES CITED

Arber, A. 1932. Studies in floral morphology. IV. On the Hypecoideae, with special reference to the androecium. *New Phytol.* 30:145–173.

Ernst, W. R. 1962. The genera of Papaveraceae and Fumariaceae in the southeastern United States. *J. Arnold Arbor.* 43:315–343.

Judd, W. S., R. W. Sanders, and M. J. Donoghue. 1994. Angiosperm family pairs: Preliminary cladistic analyses. *Harvard Pap. Bot.* 5:1–51.

Ryberg, M. 1960. A morphological study of the Fumariaceae and the taxonomic significance of the characters examined. *Acta Horti Berg.* 19:121–248.

Thorne, R. F. 1974. A phylogenetic classification of the Annoniflorae. *Aliso* 8:147–209.

———. 1992. Classification and geography of the flowering plants. *Bot. Rev.* (Lancaster) 58:225–348.

Vogel, S. 1978. Evolutionary shifts from reward to deception in pollen flowers. In *The pollination of flowers by insects*, ed. A. J. Richards, pp. 89–96. Linn. Soc. Symp. Ser. No. 6. Academic Press, London.

NYMPHAEACEAE
Water-lily Family

Perennial aquatic herbs, scapose, with milky sap; rhizome large, with massive shoot apices. **Leaves** simple, usually entire, alternate, with palmate to pinnate venation, long-petiolate with usually floating blade, large, exstipulate. **Flowers** solitary, axillary, actinomorphic, perfect, hypogynous, perigynous to sometimes epigynous, large and showy, floating or raised above the water surface, with long peduncles. **Calyx** of generally 4 or 6 sepals, distinct, sometimes petaloid, imbricate. **Corolla** of 8 to many petals in several whorls, staminodial in origin, distinct, often not clearly differentiated from the stamens, with the inner whorl(s) transitional or with all petals stamen-like, sometimes with abaxial nectaries, yellow, white, red, or blue, imbricate. **Androecium** of numerous laminar and spirally arranged stamens; filaments poorly differentiated from the anthers, distinct; anthers dehiscing longitudinally, introrse; sometimes inner whorl staminodial. **Gynoecium** of generally 1 pistil (with degree of carpel connation various), 5- to many-carpellate; ovary superior to sometimes inferior, with as many locules as carpels; ovules numerous in each locule, anatropous, pendulous, placentation parietal or lamellate; styles absent; stigma(s) large and discoid, with receptive area(s) over the entire surface or restricted to radiating lines. **Fruit** berry-like, spongy, irregularly dehiscent (due to swelling of mucilage surrounding seeds); seeds small, usually operculate, often arillate; endosperm scanty; perisperm copious; embryo straight.

Family characterization: Perennial aquatic herbs with large rhizomes and milky sap; large, long-petiolate leaves with floating, cordate to orbicular (peltate) blades; large, long-peduncled, solitary flowers with numerous petals, stamens, and carpels; laminar stamens undifferentiated into filament and anther; ex-

50

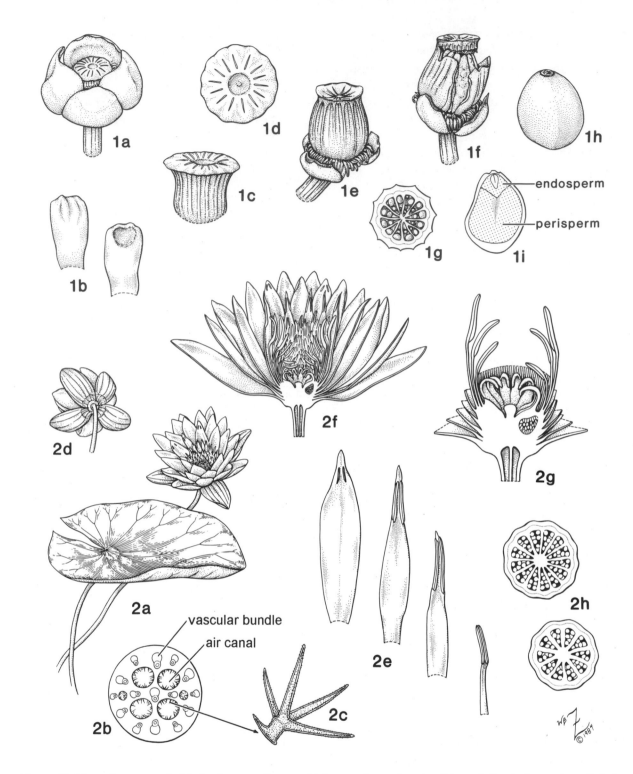

Figure 19. Nymphaeaceae. 1, *Nuphar luteum*: **a,** flower, ×½; **b,** two views of petal, ×2¼; **c,** pistil, ×1; **d,** top view of pistil showing radiating stigmatic lines, ×1; **e,** fruit, ×½; **f,** dehiscing fruit, ×⅖; **g,** cross section of fruit, ×½; **h,** seed, ×4; **i,** cross section of seed, ×4. **2,** *Nymphaea odorata*: **a,** flower and floating leaf, ×¼; **b,** cross section of petiole, ×6; **c,** branched sclereid from petiole, ×80; **d,** view of flower showing the four sepals, ×¼; **e,** sequence showing transition from petaloid staminode of the outer whorl (top) to stamen of the inner whorl (bottom), ×1½; **f,** longitudinal section of flower, ×1; **g,** longitudinal section of flower (perianth removed), ×2; **h,** cross section of ovaries from two flowers showing varying carpel number among flowers, ×2.

TABLE 3. Comparison of the Nymphaeaceae, Nelumbonaceae, and Cabombaceae. The Nymphaeaceae are treated in a somewhat restricted sense here, *sensu* Thorne (1974, 1992), who does not consider the Nelumbonaceae to be closely related to the Nymphaeaceae and Cabombaceae (see Qiu et al. 1993).

CHARACTER	NYMPHAEACEAE	NELUMBONACEAE	CABOMBACEAE
GENERA (SPECIES)	Barclaya (2), *Euryale* (1), *Nuphar* (20), *Nymphaea* (35), *Ondinea* (1), and *Victoria* (2)	*Nelumbo* (2)	*Brasenia* (1) and *Cabomba* (7)
DISTRIBUTION	widespread: tropical to northern cold-temperate regions	eastern Asia and eastern North America	tropical to warm-temperate areas
HABIT	scapose	scapose	caulescent
VESSELS	absent	present in roots	absent
LEAVES	alternate, long-petiolate, and cordate to peltate	alternate, long-petiolate, and peltate	alternate, long-petiolate, and cordate to peltate *or* opposite or whorled, short petiolate, and blades dissected
PERIANTH	4 to 6 sepals (sometimes petaloid) + 8 to numerous petals (not clearly differentiated from stamens)	2 "sepals" + numerous petals	3 + 3 tepals
STAMENS	numerous laminar	numerous filament + anther with prolonged connective	3 to numerous filament + anther
POLLEN TYPE	monocolpate	tricolpate	usually monocolpate
CARPELS	5 to numerous basically fused	numerous distinct, individually embedded in enlarged obconical receptacle	2 to 18 distinct
OVULES PER CARPEL OR LOCULE	numerous	1	usually 2 or 3
STIGMA	enlarged, discoid, and radiating	capitate	terminal or decurrent
FRUIT TYPE	berry-like, spongy, and irregularly dehiscent	aggregate of nuts, each free in cavities of enlarged receptacle	achene-like (nutlets)
SEEDS	operculate often arillate	—— no arils	operculate no arils
ENDOSPERM	scanty	essentially none	scanty
PERISPERM	copious	none	copious
EMBRYO SIZE	small	large (cotyledons fleshy)	small

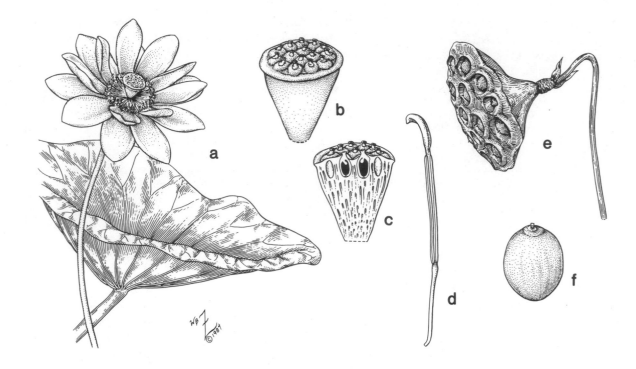

Figure 20. Nelumbonaceae. *Nelumbo lutea*: **a,** flower and emergent leaf, ×¼; **b,** receptacle of flower (note sunken carpels), ×1; **c,** longitudinal section of receptacle, ×1; **d,** stamen, ×2; **e,** fruiting receptacle, ×⅖; **f,** nutlet, ×1½. See Table 3 for family characteristics.

panded discoid and radiating stigma; spongy, dehiscent, berry-like fruit; and operculate seeds with perisperm. Pollen grains monocolpate. Alkaloids often present. Anatomical features: parenchymatous tissues with schizogenous intercellular spaces and articulated laticifers; branched sclereids (especially in the leaves; Fig. 19: 2c); scattered vascular bundles (in the axes, petioles, and peduncles; Fig. 19: 2b) with no vessels or cambium; and root hairs that arise from specialized cells (as in many monocots).

Genera/species: *Nymphaea* (35–50 spp.), *Nuphar* (20–25 spp.), *Barclaya* (2–4 spp.), *Victoria* (2–3 spp.), *Euryale* (1 sp.), and *Ondinea* (1 sp.)

Distribution: Widespread from tropical to northern cold temperate regions in quiet freshwaters (e.g., ponds, streams, and lakes).

U.S./Canadian representatives: *Nymphaea* (11 spp.) and *Nuphar* (2 spp.)

Economic plants and products: Edible seeds from species of *Victoria* and *Nymphaea* (rhizomes also edible). Ornamental plants for pools and aquaria: *Euryale* (prickly water-lily), *Nuphar* (yellow water-lily, spatterdock), *Nymphaea* (water-lily), and *Victoria* (royal water-lily, Amazon water-lily).

Commentary: The Nymphaeaceae are treated in a rather restricted sense here (*sensu* Thorne 1974, 1992); other authors also include species here referred to the Cabombaceae and Nelumbonaceae (see Table 3; Wood 1959; Simon 1971; Ito 1987). These families share such morphological features as the aquatic perennial habit; long-petiolate, cordate to peltate leaves; solitary, long-peduncled flowers; and parietal placentation—as well as several anatomical features. However, numerous characters, especially gynoecial features, have been used to separate these three groups (see Table 3 and Figs. 19, 20, and 21). Molecular data (Qiu et al. 1993) indicate that the Nelumbonaceae, in particular, are not closely related to the Nymphaeaceae.

All three families were considered traditionally to be allies of the "primitive" families of dicots (Li 1955). In addition, the Nymphaeaceae have several distinctive anatomical characteristics that "link" them to the monocots (Huber 1977; Dahlgren and Clifford 1982). For example, the plants lack a cambium and true vessels and have long tracheids with spiral or annular thickenings. As in the monocots, the vascular bundles are scattered within the ground tissue, and the root hairs originate from specialized cells. The family is a member of the "paleoherb" group of Donoghue and Doyle (1989), and Fig. 2 shows its hypothesized position in relation to the monocots.

The flowers of *Nuphar* (Fig. 19: 1a) and *Nymphaea*

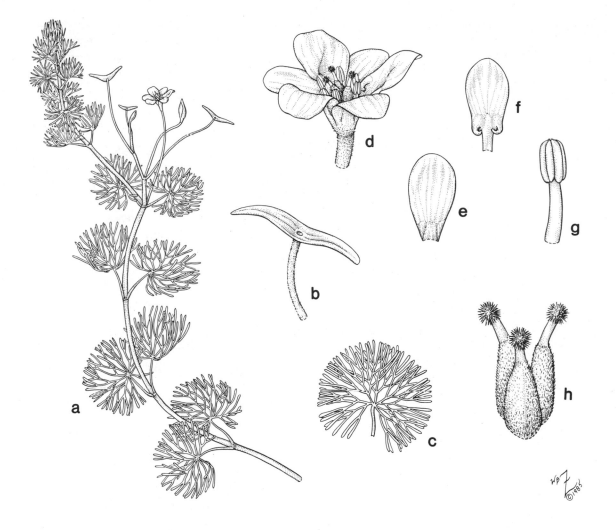

Figure 21. Cabombaceae. *Cabomba caroliniana*: **a,** habit, ×½; **b,** floating leaf, ×2½; **c,** submersed leaf, ×⅔; **d,** flower, ×4; **e,** outer tepal, ×3; **f,** inner tepal, ×3; **g,** stamen, ×9; **h,** gynoecium, ×8. See Table 3 for family characteristics.

(Fig. 19: 2a) are distinctive with numerous petals and stamens and a conspicuous stigma (Moseley 1961, 1965). The petals blend into the laminar stamens with transitional forms. In *Nuphar*, the petals are scale-like, and each has a nectary on the abaxial side (Fig. 19: 1b). The receptive stigmatic surface is represented by radiating lines on the discoid stigma (Fig. 19: 1c,d). The carpels of *Nymphaea* form a broad concave stigma, with each carpel having an incurved elongation (see Fig. 19: 2g). In this case, the entire upper surface of each carpel is receptive.

The flowers are protogynous and are visited mainly by pollen-collecting insects (flies, beetles). Self-pollination also may occur in some species as the flowers open. Although the flowers of *Nymphaea* do not produce nectar, the receptive stigmatic cup fills up with a sweet liquid secreted by the stigma, and some species are also fragrant.

Water-lily seeds and fruits, which generally mature underwater, are adapted for water dispersal. The ripe fruit eventually bursts due to the swelling of the mucilage surrounding the seeds. After the outer rind of a *Nuphar* fruit ruptures, each protruding carpel separates (like a segment of an orange) and floats away (Fig. 19: 1f). The released seeds of *Nymphaea* rise to the surface of the water due to the air enclosed in their arils.

REFERENCES CITED

Dahlgren, R. M. T., and H. T. Clifford. 1982. *The monocotyledons: A comparative study*, pp. 323, 330–345. Academic Press, London.

Donoghue, M. J., and J. A. Doyle. 1989. Phylogenetic analysis of the angiosperms and the relationships of the Hamamelidae. In *Evolution, systematics, and fossil history of the Hamamelidae*, ed. P. R. Crane and S. Blackmore, vol. 1, *Introduction and "lower" Hamamelidae*, pp. 17–45. Syst. Assoc. Special Vol. 40A. Clarendon Press, Oxford.

Huber, H. 1977. The treatment of the monocotyledons in an

evolutionary system of classification. *Pl. Syst. Evol.*, Suppl. 1:285–298.

Ito, M. 1987. Phylogenetic systematics of the Nymphaeales. *Bot. Mag.* (Tokyo) 100:17–35.

Li, H.-L. 1955. Classification and phylogeny of Nymphaeaceae and allied families. *Amer. Midl. Naturalist* 54: 33–41.

Moseley, M. F. 1961. Morphological studies of the Nymphaeaceae. II. The flower of *Nymphaea*. *Bot. Gaz.* 122: 233–259.

——. 1965. Ibid., III. The floral anatomy of *Nuphar*. *Phytomorphology* 15:54–84.

Qiu, Y.-L., M. W. Chase, D. H. Les, and C. R. Parks. 1993. Molecular phylogenetics of the Magnoliidae: Cladistic analyses of nucleotide sequences of the plastid gene *rbc*L. *Ann. Missouri Bot. Gard.* 80:587–606.

Simon, J.-P. 1971. Comparative serology of the order Nymphaeales. II. Relationships of Nymphaeaceae and Nelumbonaceae. *Aliso* 7:325–350.

Thorne, R. F. 1974. A phylogenetic classification of the Annoniflorae. *Aliso* 8:147–209.

——. 1992. Classification and geography of the flowering plants. *Bot. Rev.* (Lancaster) 58:225–348.

Wood, C. E. 1959. The genera of Nymphaeaceae and Ceratophyllaceae in the southeastern United States. *J. Arnold Arbor.* 40:94–112.

CARYOPHYLLACEAE

Pink Family

Usually annual or perennial herbs; stems typically with swollen nodes. **Leaves** simple, entire, opposite, decussate, usually narrow, appearing parallel-veined, usually narrow, often basally connate or connected by a transverse line on the stem, usually exstipulate. **Inflorescence** determinate, cymose, flower sometimes solitary, usually terminal. **Flowers** actinomorphic, usually perfect, frequently showy, typically hypogynous. **Calyx** of 5 or occasionally 4 sepals, distinct to connate, imbricate, often with membranous margins, persistent. **Corolla** of 5 or occasionally 4 petals, sometimes reduced or absent, distinct, often differentiated into claw and limb with appendages on the inner surface of the claw-limb junction, often apically notched, commonly white or pink, imbricate. **Androecium** of 5 (uniseriate) to 10 (biseriate with outer whorl apparently opposite the petals) or occasionally 1 to 4 stamens; filaments distinct, sometimes basally adnate to the petals or calyx, often with nectaries at base; anthers basifixed, dehiscing longitudinally. **Gynoecium** of 1 pistil, 2- to 5-carpellate; ovary superior, usually 1-locular (at least above), sometimes 3- to 5-locular at the base, often stalked; ovules usually numerous, sometimes solitary to few, usually campylotropous, placentation free-central (in 1-locular ovary), free-central above and axile below (in ovary partitioned at base), or occasionally basal (when ovule solitary); styles 2 to 5;

stigmas 2 to 5, minute. **Fruit** usually a capsule dehiscing apically by valves or teeth, or sometimes a utricle; seeds with ornamented seed coat; perisperm hard; embryo curved, surrounding the perisperm (peripheral).

Family characterization: Herbaceous plants with swollen nodes; simple, opposite, and narrow leaves with entire margins and connate or connected bases; pink or white petals (when present) often with an apical notch; 2- to 5-carpellate 1-locular ovary with free-central placentation (at least above); a capsule (dehiscing by apical valves or teeth) or utricle as the fruit type; and seeds with perisperm and a peripheral, curved embryo. Saponins and lychnose (unusual storage carbohydrate) present. Anatomical features: sieve tubes with a special kind of P-type plastid (Behnke 1976); and unilacunar nodes.

Genera/species: 70/1,750

Distribution: Primarily in north temperate regions, with a few representatives in south temperate zones, montane tropics, and the Arctic; especially diverse in the Mediterranean region.

Major genera: *Silene* (500 spp.), *Dianthus* (300 spp.), *Arenaria* (150–250 spp.), and *Gypsophila* (125 spp.)

U.S./Canadian representatives: 35 genera/326 spp.; largest genera: *Silene*, *Stellaria*, and *Minuartia*

Economic plants and products: Many widespread weedy plants, such as *Cerastium* and *Stellaria* (chickweeds). Garden ornamentals (species of 22 genera), including *Arenaria* (sandwort), *Cerastium* (mouse-ear chickweed), *Dianthus* (carnation, pink, sweet William), *Gypsophila* (baby's-breath), *Lychnis* (Maltese-cross), *Saponaria* (soapwort), and *Silene* (catchfly).

Commentary: The Caryophyllaceae usually are divided into three distinct subfamilies primarily based upon the presence of stipules, fusion of the sepals, and the morphology of the petals. The phylogenetic relationships of the family have been scrutinized closely because the Caryophyllaceae differ from most allied families (in the "Centrospermae" or Caryophyllidae/Caryophyllanae) in having anthocyanins (as in most flowering plants) rather than the betalains found in their allies (see the summary in Giannasi et al. 1992, and references and commentary under the Chenopodiaceae). Although this is a uniform and easily recognizable family, the generic limits are somewhat difficult and controversial.

The flowers of the family demonstrate a range of morphological complexity in such characters as fusion of the calyx, presence and morphology of the corolla, and number of stamens and carpels. Variation may occur even in the same species; for example, *Stellaria*

54

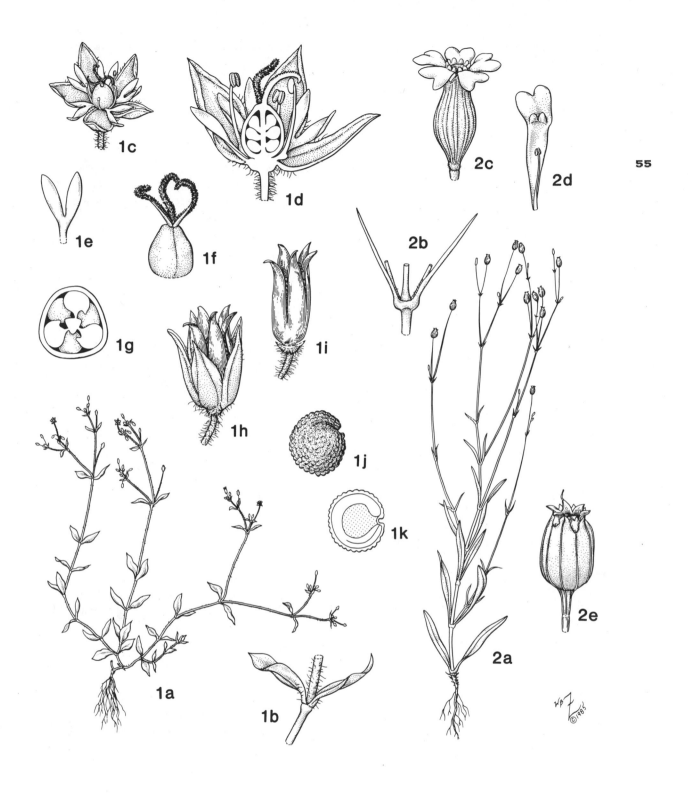

55

Figure 22. Caryophyllaceae. 1, *Stellaria media*: **a,** habit, ×⅓; **b,** node, ×3; **c,** flower, ×6; **d,** longitudinal section of flower, ×12; **e,** petal, ×12; **f,** pistil, ×12; **g,** cross section of ovary, ×22; **h,** capsule with persistent calyx, ×6; **i,** capsule (calyx removed), ×6; **j,** seed, ×15; **k,** longitudinal section of seed, ×15. **2,** *Silene antirrhina*: **a,** habit, ×½; **b,** node, ×2; **c,** flower, ×4½; **d,** petal and adnate stamen, ×6; **e,** capsule, ×4½.

media may have five, ten, or even three stamens. The petals of the Caryophyllaceae are likely derived from the stamens (Thomson 1942; Cronquist 1981) and often are bifid at the apex and/or clawed with appendages ("corona") at the claw-limb junction (Fig. 22: 1e, 2d).

Pollinators visit for the nectar secreted at the base of the stamens. Flies and bees generally pollinate the more open-type flowers, such as *Stellaria* (Fig. 22: 1c). Others (*Silene*; Fig. 22: 2c) conceal the nectar in the tube formed by the synsepalous calyx and the long-clawed and appendaged petals, and these flowers are probed by larger bees and Lepidoptera. The flowers commonly are protandrous, although self-pollination occurs in several species.

An unusual capsule (Fig. 22: 2e), which opens by the recurving of its apical teeth, is characteristic of most species. Wind or animals are required to shake the capsules in order to disperse the seeds. In a few others (such as *Paronychia*), the fruit is a utricle.

REFERENCES CITED

Behnke, H.-D. 1976. Ultrastructure of sieve-element plastids in Caryophyllales (Centrospermae), evidence for delimitation and classification of the order. *Pl. Syst. Evol.* 126:31–54.

Cronquist, A. 1981. *An integrated system of classification of flowering plants*, pp. 272–276. Columbia University Press, New York.

Giannasi, D. E., G. Zurawski, G. Learn, and M. T. Clegg. 1992. Evolutionary relationships of the Caryophyllidae based on comparative *rbc*L sequences. *Syst. Bot.* 17:1–15.

Thomson, B. F. 1942. The floral morphology of the Caryophyllaceae. *Amer. J. Bot.* 29:333–349.

PORTULACACEAE

Purslane Family

Annual or perennial herbs or suffrutescent shrubs, succulent, often prostrate, with mucilaginous tissues, sometimes with bitter sap, with taproot or fleshy roots. **Leaves** simple, entire, opposite or alternate, sometimes in basal rosettes, succulent, cylindrical to flat, usually stipulate; stipules scarious or of tufted hairs. **Inflorescence** determinate, cymose and often appearing racemose, paniculate, or capitate, terminal or lateral, or sometimes flowers solitary in leaf axils. **Flowers** actinomorphic, perfect, hypogynous to half-epigynous, small and showy, subtended by 2 unequal bracts ("calyx"), often with extrastaminal nectariferous disc. **Perianth** of 4 to 6 tepals, uniseriate (appearing biseriate due to involucre), petaloid, distinct or basally connate, ephemeral, variously colored, with shiny surface, imbricate. **Androecium** of 4 to 6 stamens or sometimes numerous and fascicled, opposite the tepals, sometimes epitepalous; filaments distinct, sometimes with basal glands; anthers dehiscing longitudinally, introrse. **Gynoecium** of 1 pistil, 2- to 9-carpellate; ovary superior to half-inferior, 1-locular (but septate early in development); ovules few to usually many, usually campylotropous or amphitropous, placentation free-central or basal; styles as many as carpels, variously basally connate; stigmas linear to capitate. **Fruit** a circumscissile or loculicidal capsule; seeds lenticular, compressed, sometimes arillate; perisperm copious, mealy; embryo curved, surrounding the perisperm (peripheral).

Family characterization: Succulent herbs to shrubs with mucilaginous tissues; thick leaves with scarious or tufted-hair stipules; involucre of 2 unequal bracts ("calyx"); uniseriate perianth of petaloid, ephemeral tepals ("corolla"); 1-locular ovary with basal, central placenta; branched style; circumscissile or loculicidal capsular fruit; and perisperm surrounded by curved embryo. Betalains (nitrogen-containing pigments), oxalic acid, and nitrates present. Anatomical features: mucilaginous parenchyma in leaves and stems; sieve tubes with special P-type plastid (Behnke 1976b); and unilacunar nodes. Leaves of many species also with Kranz anatomy (see the discussion below).

Genera/species: 19/500

Distribution: Widespread in warm and temperate climates; most diverse in western North America, southern South America, and southern Africa.

Major genera: *Portulaca* (40–200 spp.), *Calandrinia* (150 spp.), *Anacampseros* (70 spp.), and *Talinum* (50 spp.)

U.S./Canadian representatives: 8 genera/108 spp.; largest genera: *Claytonia*, *Portulaca*, *Talinum*, and *Lewisia*

Economic plants and products: Several poisonous plants (due to oxalic acid) and weeds, including species of *Calandrinia* and *Portulaca*. Edible greens from a few species, such as *Portulaca oleracea* (common purslane). Ornamental plants (species of 9 genera), including *Calandrinia* (rock-purslane), *Lewisia* (bitter-root), *Portulaca* (purslane, rose-moss), and *Talinum* (fameflower).

Commentary: Taxonomists generally link the Portulacaceae to other families in the "Centrospermae" complex (or Caryophyllales) due to characters such as betalains, basal placentation, and curved peripheral embryo (Behnke 1976a; Eckardt 1976). Anomalous secondary growth, however, has not been reported for the purslane family. Many genera formerly included within the Portulacaceae have been transferred to

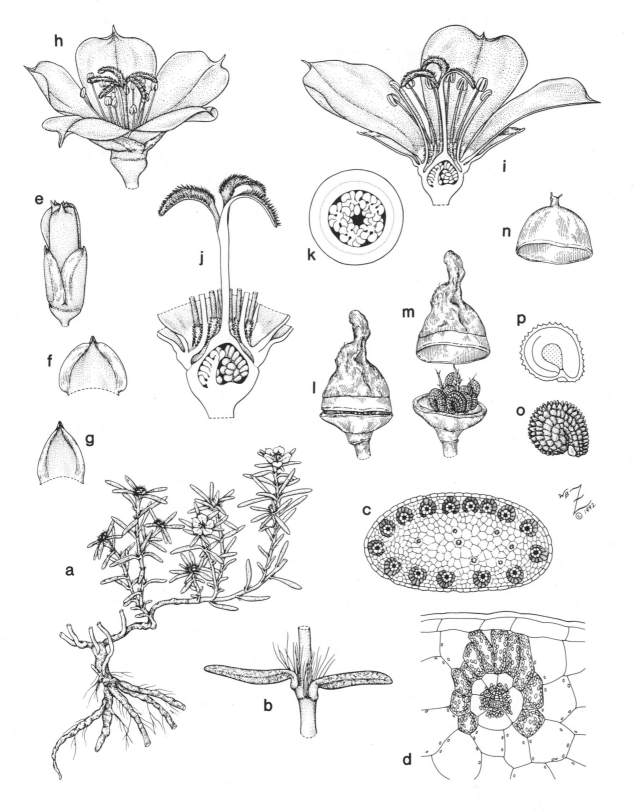

Figure 23. Portulacaceae. *Portulaca pilosa*: **a,** habit, ×³⁄₄; **b,** node (note hair-tuft stipules), ×5; **c,** cross section of leaf, ×25; **d,** detail of **c** showing vascular bundle and surrounding mesophyll cells (Kranz anatomy), ×100; **e,** bud, ×6; **f,** abaxial bract ("sepal"), ×6; **g,** adaxial bract ("sepal"), ×6; **h,** flower, ×6; **i,** longitudinal section of flower, ×8; **j,** longitudinal section of basal portion of flower, ×12; **k,** cross section of ovary (five-carpellate; varies among flowers from four- to six-carpellate), ×20; **l,** pyxis (beginning to dehisce) with persistent perianth, ×10; **m,** dehiscing pyxis, ×10; **n,** operculum of pyxis with persistent perianth removed, ×10; **o,** seed, ×30; **p,** longitudinal section of seed, ×30.

other families such as the Basellaceae, Aizoaceae, and Molluginaceae (Rodman 1990). The remaining genera in the family have been variously grouped into subfamilies and/or tribes (Bogle 1969; Carolin 1987; Nyananyo 1990).

These succulent plants are adapted for dry areas of high light intensity. The mesophyll in the leaves of many species has a particular organization ("Kranz anatomy") associated with the C_4 photosynthetic pathway, an adaptation for these stressful habitats. Some palisade cells are arranged in a ring around a layer of large bundle sheath cells, resulting in two concentric cell wreaths around the vascular bundle (Fig. 23: d). The leaf mesophyll also has large water-storage cells (see Fig. 23: c).

The characteristic showy flower of the family (Fig. 23: h) appears to have a biseriate perianth consisting of two unequal "sepals" and four to six variously colored "petals" with a satiny surface. Morphologically, the perianth is actually uniseriate, with a modified involucre (two bracts; Fig. 23: e–g) subtending the petaloid calyx (Eckardt 1976). Presumably to avoid confusion, most texts (and monographs) adopt the floral interpretation of "calyx" and "corolla"; however, tradition and convenience are not valid reasons to perpetuate misconceptions concerning floral structure.

The ephemeral flowers, which are protandrous or homogamous, usually open for a short period in full sunlight. Small pollinators (e.g., flies and ants) are attracted to the brightly colored perianth and the nectar (produced by the glands or disc at the base of the stamens). Self-pollination may occur as the anthers are pressed against the stigmas when the flowers close (or fail to expand in inclement weather; Knuth 1908).

REFERENCES CITED

Behnke, H.-D. 1976a. A tabulated survey of some characters of systematic importance in Centrospermous families. *Pl. Syst. Evol.* 126:95–98.

———. 1976b. Ultrastructure of sieve-element plastids in Caryophyllales (Centrospermae), evidence for delimitation and classification of the order. *Pl. Syst. Evol.* 126:31–54.

Bogle, A. L. 1969. The genera of Portulacaceae and Basellaceae in the southeastern United States. *J. Arnold Arbor.* 50:566–598.

Carolin, R. 1987. A review of the family Portulacaceae. *Austral. J. Bot.* 35:383–412.

Eckardt, T. 1976. Classical morphological features of centrospermous families. *Pl. Syst. Evol.* 126:5–25.

Knuth, P. 1908. *Handbook of flower pollination.* Trans. J. R. A. Davis. Vol. 2, *Ranunculaceae to Stylidieae*, pp. 201–202. Oxford University Press, Oxford.

Nyananyo, B. L. 1990. Tribal and generic relationships in the Portulacaceae (Centrospermae). *Feddes Repert.* 101:237–241.

Rodman, J. E. 1990. Centrospermae revisited, part I. *Taxon* 39:383–393.

CACTACEAE
Cactus Family

Perennial herbs, vines, shrubs, or sometimes small trees, often growing in xeric habitats, sometimes epiphytic, succulent, with leaves generally reduced to spines or quickly deciduous, with watery or mucilaginous sap; stems usually greatly enlarged and cylindrical, conical, or flattened, often tuberculate or ribbed, simple or branched, often jointed; roots usually superficial, slender, fleshy. **Leaves** simple, alternate, fleshy, sometimes large and persistent to usually rudimentary (scale-like) and caducous, with axillary buds (or branches) specialized into cushion-like areas (areoles) bearing spine clusters (modified axillary shoot leaves) and sometimes bristles (glochids). **Flowers** usually solitary, borne upon or near the areoles, actinomorphic to sometimes slightly zygomorphic, perfect, epigynous, often large and showy, sessile, with nectariferous ring along inner surface of hypanthium. **Perianth** of numerous intergrading sepaloid to petaloid tepals, spirally arranged, basally connate (forming a hypanthium), generally red, purple, orange, yellow, or white. **Androecium** of numerous stamens, spirally arranged or clustered, arising from hypanthium; filaments distinct; anthers basifixed or dorsifixed, dehiscing longitudinally, introrse or latrorse. **Gynoecium** of 1 pistil, 3- to many-carpellate; ovary inferior (see Boke 1964), 1-locular, usually embedded into the stem, commonly covered with hairs, bristles, or spines; ovules numerous, campylotropous to anatropous, with funicles often connate at base and forming a bundle, placentation parietal; style 1; stigmas as many as carpels, usually radiating, thick, soft papillose. **Fruit** a berry, often spiny or bristly; seeds numerous, immersed in pulp; perisperm absent or sometimes abundant, starchy and mealy; embryo curved or sometimes straight.

Family characterization: Succulent herbs to small trees usually growing in warm and dry habitats; enlarged stems that often are tuberculate, ribbed and/or jointed; superficial roots; reduced to absent leaves with specialized axillary bud areas (areoles) bearing spines and sometimes bristles (glochids); solitary and showy flowers with intergrading sepals and petals; numerous spirally arranged tepals and stamens fused basally to form a hypanthium; 1-locular, inferior ovary sunken into the stem; parietal placentation; and baccate, often spiny fruit. Betalains (nitrogen-containing pigments) present (see references under Caryophyllaceae and

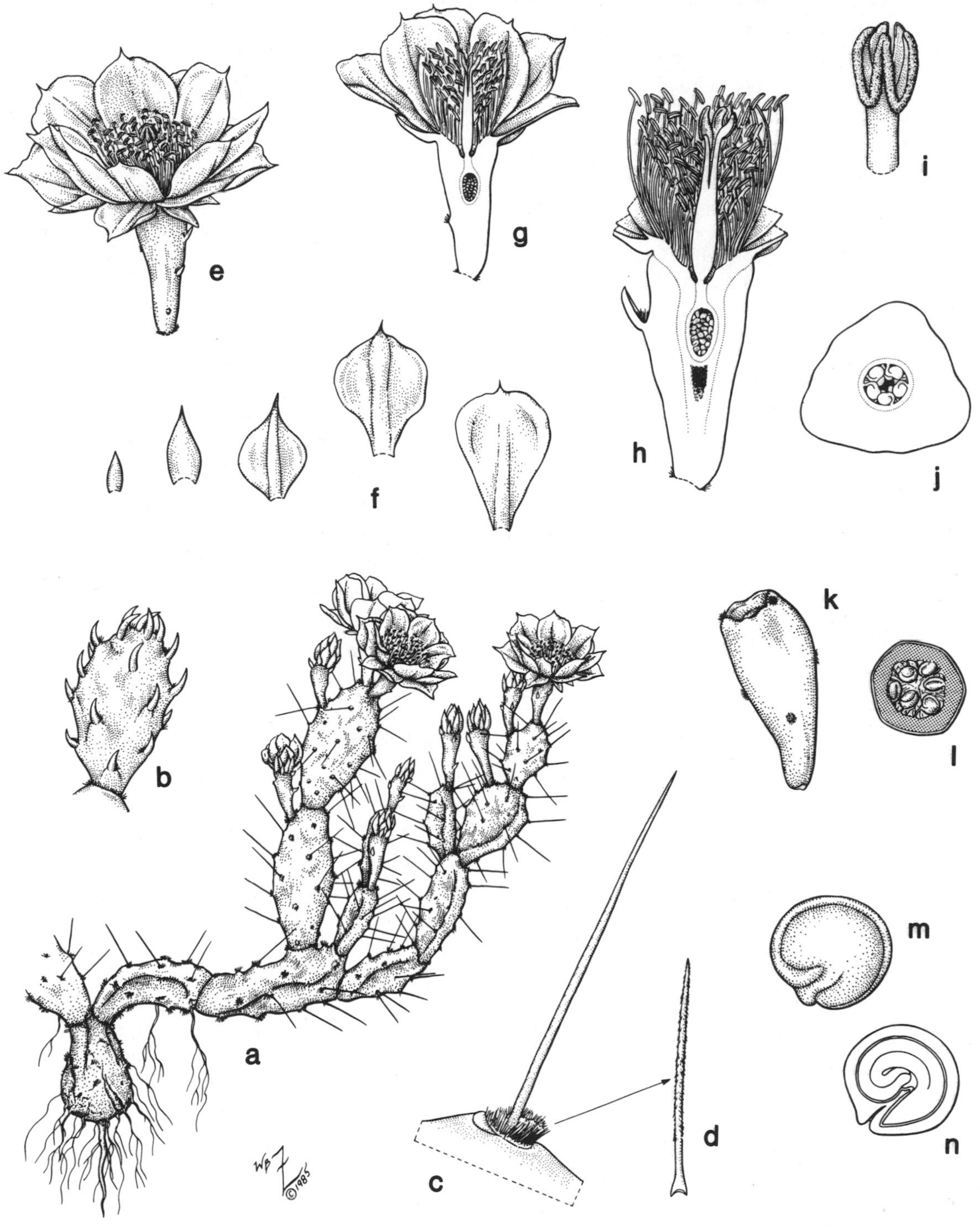

Figure 24. Cactaceae. *Opuntia humifusa*: **a,** habit, ×¼; **b,** young pad with scale-like leaves, ×⅔; **c,** detail of areole, ×2; **d,** glochid, ×14; **e,** flower, ×⅔; **f,** sequence showing transition from sepaloid tepal of the outer whorl (left) to petaloid tepal of the inner whorl (right), ×⅔; **g,** longitudinal section of flower, ×⅔; **h,** longitudinal section of flower (perianth removed), ×1⅓; **i,** stigma, ×3; **j,** cross section of ovary, ×2; **k,** berry, ×⅔; **l,** cross section of berry, ×¾; **m,** seed, ×4½; **n,** longitudinal section of seed, ×4½.

Chenopodiaceae). Alkaloids or saponins commonly present. Anatomical features: parenchymatous tissues commonly with scattered mucilage cells and calcium oxalate crystals; sieve tubes with a special kind of P-type plastid; and several adaptations of the stem related to water storage and retention (see the discussion below).

60

Genera/species: 93/1,488 (Hunt and Taylor 1990)

Distribution: Primarily localized in desert and semi-desert areas of North, Central, and South America; a few genera (probably all introduced) in other areas of the world (e.g., Africa, Ceylon, India, and Australia).

Major genera: *Opuntia* (220 spp.), *Mammillaria* (150 spp.), *Cleistocactus* (64 spp.), and *Echinopsis* (60 spp.) (Hunt and Taylor 1990)

U.S./Canadian representatives: 19 genera/174 spp.; largest genera: *Opuntia* and *Cereus*

Economic plants and products: Timber from several, such as species of *Cephalocereus* and *Pereskia*. Medicinal and ritual drugs from a few, such as *Lophophora* (peyote or mescal-button—with mescaline, a narcotic alkaloid). Edible fruits from species of *Opuntia* (prickly-pear). Ornamental plants (species of at least 100 genera), including *Cereus* (hedge cactus), *Echinopsis* (sea-urchin cactus), *Epiphyllum* (orchid cactus), *Hylocereus* (night-blooming cereus), *Lobivia* (cob cactus), *Mammillaria* (pincushion cactus), *Notocactus* (ball cactus), *Opuntia* (prickly-pear, cholla), *Rebutia* (crown cactus), *Rhipsalis* (mistletoe cactus), and *Schlumbergera* (Christmas cactus, crab cactus, Thanksgiving cactus).

Commentary: The Cactaceae are divided generally into three subfamilies (or tribes) based on seed characters and the presence or absence of leaves, glochids, and flower stalks (see the classifications in Britton and Rose 1923, Benson 1982, and Gibson et al. 1986). Generic and specific delimitations have been difficult, due in part to the difficulty in preserving the succulent plants for comparative herbarium studies. Genera and species tend to be overdescribed, resulting in a range of 30 to 300 genera and 1,000 to 2,000+ species cited in the literature. Collectors and commercial growers probably have also contributed to the proliferation of new "genera" and "species." The latest consensus listing by the International Organization for Succulent Plant Study (IOS) totals 93 accepted genera with an estimated 1,488 species (Hunt and Taylor 1990).

Most cacti are highly specialized stem-succulent herbs and rarely resemble the usual leaf-bearing type of plant. The enlarged stems, often with tubercles or prominent ribs, vary in shape from globose to cylindrical or flattened. Leaves, when present, usually are ephemeral. Except in a few taxa with well-developed leaves (*Pereskia*), photosynthesis (with "crassulacean acid metabolism" or CAM, a variant of the C_4 pathway) takes place primarily in the stems (see Benson 1982). Specializations of the stem structure for water storage and retention include the thick epidermis with sunken stomata and strong cuticle. Beneath this outer layer occurs a collenchymatous hypodermis surrounding a well-developed ground tissue of water-storage cells. The vascular system forms a cylindrical network embedded in the ground tissue.

The spines on the stem may help promote the accumulation of dew or water droplets around the plant. Spines are restricted to hairy, cushion-like areas called areoles (Fig. 24: c), which usually occur at the tips of the tubercles or along the ridges or edges of the stem. An areole is regarded as a modified axillary bud (or short branch), and a spine represents a specialized leaf (or bud scale) of that axillary branch. Besides affording protection from herbivores, the spines assist in vegetative propagation by animal dispersal. In some cacti (*Opuntia*), tufts of barbed bristles (glochids; Fig. 24: d) also occur in the areoles.

The showy flowers of many species often are rapidly produced during or at the end of a rainy season, and they usually last only one or a few days. Several (*Cereus* and relatives) are night-blooming. Bees, beetles, birds, bats, and sphinx moths visit many species for the nectar secreted along the inner surface of the hypanthium and/or for the copious pollen. Some cacti (e.g., certain *Opuntia* species) have sensitive stamens that incurve when stimulated by crawling insects. With the anthers closer to the stigmas, the passageway for the pollinator becomes narrower. This does not necessarily promote self-pollination because most cacti are protandrous (and self-sterile). The stigmas are appressed when the anthers dehisce and later become erect or spread out over the stamens.

The baccate fruits (Fig. 24: k) are often sweet and tasty and dispersed by animals.

REFERENCES CITED

Benson, N. L. 1982. *The cacti of the United States and Canada.* Stanford University Press, Stanford.

Boke, N. H. 1964. The cactus gynoecium: A new interpretation. *Amer. J. Bot.* 51:598–610.

Britton, N. L., and J. N. Rose. 1923. *The Cactaceae.* 4 vols. Publ. Carnegie Inst. Wash. No. 248.

Gibson, A. C., K. C. Spencer, R. Bajaj, and J. L. McLaughlin. 1986. The ever-changing landscape of cactus systematics. *Ann. Missouri Bot. Gard.* 73:532–555.

Hunt, D., and N. Taylor, eds. 1990. The genera of Cactaceae: Progress towards consensus. Report of the IOS working party, 1987–90. *Bradleya* 8:85–107.

PHYTOLACCACEAE
Pokeweed Family

Perennial herbs, shrubs, to sometimes trees, somewhat succulent. **Leaves** simple, entire, alternate, typically exstipulate. **Inflorescence** usually indeterminate, often appearing racemose, paniculate, or spicate, axillary or arising on the stem opposite the leaves. **Flowers** actinomorphic, usually perfect, hypogynous, small and inconspicuous, often with nectariferous disc. **Perianth** of usually 4 or 5 tepals, uniseriate, distinct or basally connate, succulent or leathery, greenish, white, or somewhat colored, imbricate, persistent. **Androecium** various, usually of either 4 or 5 (uniseriate) or 8 or 10 (biseriate) stamens, inserted on the disc; filaments distinct or basally connate; anthers dorsifixed, dehiscing longitudinally, introrse. **Gynoecium** more or less syncarpous (ovaries connate), 1- to many-carpellate; ovary superior, with as many locules as carpels; ovules 1 in each locule, campylotropous, placentation axile (syncarpous ovary) or basal (1-carpellate pistil); styles as many as carpels, distinct or basally connate, short; stigmas distinct, linear or peltate. **Fruit** a berry, drupe, samara, or achene; seeds reniform, often arillate; perisperm copious, hard or starchy and mealy; embryo curved, surrounding the perisperm (peripheral).

Family characterization: Somewhat succulent herbs or shrubs; small, hypogynous flowers in racemose/spicate inflorescences; minute, 4- or 5-merous, uniseriate perianth; stamens on a nectariferous disc; 1 ovule in each locule; usually a berry-like fruit with colored, juicy sap; and arillate seed with perisperm surrounded by curved embryo. Accumulating free oxalates and nitrates. Saponins and betalains (nitrogen-containing pigments) present. Tissues with calcium oxalate crystals (usually raphides). Anatomical features: anomalous secondary growth (similar to that in the Amaranthaceae; Mikesell 1979); sieve tubes with a special kind of P-type plastid (Behnke 1976); and unilacunar nodes (Fig. 25: 1b).

Genera/species: 14/97

Distribution: Widespread in tropical and subtropical regions; especially diverse in the American tropics and subtropics.

Major genus: *Phytolacca* (25–35 spp.)

U.S./Canadian representatives: 9 genera/15 spp.; largest genus: *Phytolacca*

Economic plants and products: Edible greens (young shoots) and red dyes (from berries) from species of *Phytolacca* (inkberry) and *Rivina* (bloodberry, rouge plant). Several poisonous plants (e.g., *Phytolacca* spp.). Medicinal preparations from roots of *Agdestis*, *Petiveria*, and *Phytolacca*. Ornamental plants (species of at least 6 genera), including *Agdestis*, *Phytolacca* (pokeberry, pokeweed), *Rivina*, and *Trichostigma*.

Commentary: The Phytolaccaceae are here defined rather broadly (Thorne unpublished update [1992]) with five subfamilies; some or all of these have sometimes been treated as small, often monotypic, segregate families as delimited previously in Thorne (1992) and summarized, with a key, in Brown and Varadarajan (1985). Depending upon the various combinations of these groups, the resultant "Phytolaccaceae" may be polyphyletic or paraphyletic (Brown and Varadarajan 1985; Rettig et al. 1992). Basically these taxa are distinguished by differences in the gynoecium, androecium, and fruit type. For example, the Petiveriaceae (which includes *Rivina*) are characterized by features such as a uniovulate ovary and more or less lenticular seed. The Phytolaccaceae, as variously circumscribed, are commonly considered allied with the Nyctaginaceae within the "Centrospermae" complex of families (Rettig et al. 1992).

The carpel and stamen numbers vary considerably in the family and may even differ in flowers of the same species (Rogers 1985). The flowers of *Rivina*, for example, each have a single carpel and uniseriate androecium (Fig. 25: 2a,b); the compound pistil of *Phytolacca* is multi-carpellate, and the stamens, biseriate (Fig. 25: 1c,d,f). The great variation in fruit type includes multi-seeded (*Phytolacca*; Fig. 25: 1g,h) or one-seeded (*Rivina*; Fig. 25: 2d,e) berries with a strongly pigmented juice. The single carpel of *Petiveria alliacea* develops into an achene with a terminal hook; that of *Trichostigma*, into a fleshy drupe-like fruit. Frugivorous birds frequently are important dispersal agents (McDonnell et al. 1984).

The small flowers often are aggregated into conspicuous racemose or spicate inflorescences. The nectariferous disc (Fig. 25: 1d,f), either on the receptacle between the stamens and pistils or in a ring around the base of the stamens, produces easily accessible nectar, which presumably attracts various insect pollinators.

REFERENCES CITED

Behnke, H.-D. 1976. Ultrastructure of sieve-element plastids in Caryophyllales (Centrospermae), evidence for delimitation and classification of the order. *Pl. Syst. Evol.* 126:31–54.

Brown, G. K., and G. S. Varadarajan. 1985. Studies in Caryophyllales I: Re-evaluation of classification of Phytolaccaceae s.l. *Syst. Bot.* 10:49–63.

McDonnell, M. J., E. W. Stiles, G. P. Cheplick, and J. J.

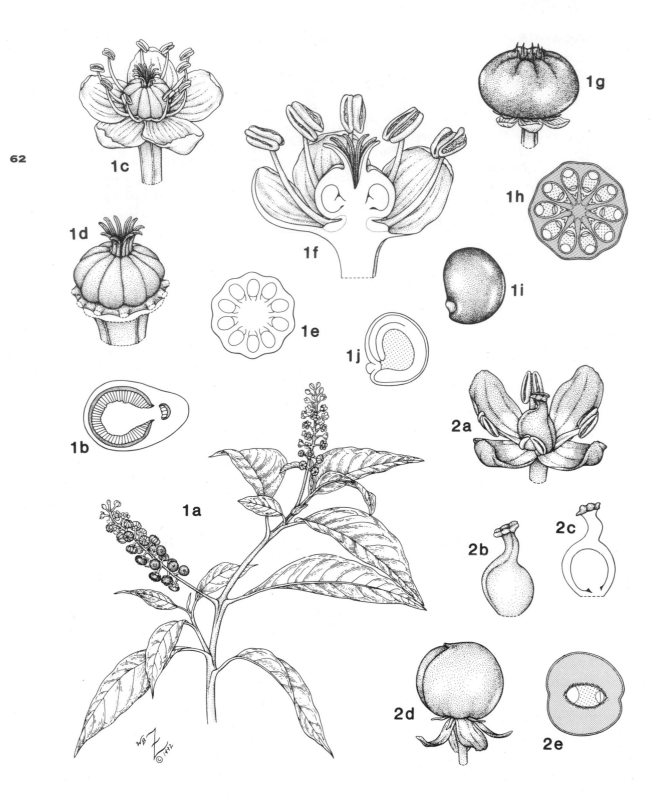

Figure 25. Phytolaccaceae. 1, *Phytolacca americana* var. *rigida* (*P. rigida*): **a,** portion of plant with flowers and fruit, ×²⁄₅; **b,** cross section of node (one-trace unilacunar), ×3; **c,** flower, ×7; **d,** pistil and disc, ×12; **e,** cross section of ovary (carpel number varies among flowers), ×12; **f,** longitudinal section of flower, ×12; **g,** berry, ×3½; **h,** cross section of berry, ×3½; **i,** seed, ×8; **j,** longitudinal section of seed, ×8. **2,** *Rivina humilis*: **a,** flower, ×11; **b,** pistil, ×15; **c,** longitudinal section of pistil, ×15; **d,** berry, ×5; **e,** cross section of berry, ×5.

Armesto. 1984. Bird-dispersal of *Phytolacca americana* L. and the influence of fruit removal on subsequent fruit development. *Amer. J. Bot.* 71:895–901.

Mikesell, J. E. 1979. Anomalous secondary thickening in *Phytolacca americana* L. (Phytolaccaceae). *Amer. J. Bot.* 66: 997–1005.

Rettig, J. H., H. D. Wilson, and J. R. Manhart. 1992. Phylogeny of the Caryophyllales—Gene sequence data. *Taxon* 41:201–209.

Rogers, G. K. 1985. The genera of Phytolaccaceae in the southeastern United States. *J. Arnold Arbor.* 66:1–37.

Thorne, R. F. 1992. Classification and geography of the flowering plants. *Bot. Rev.* (Lancaster) 58:225–348.

NYCTAGINACEAE
Four-o'clock Family

Annual or perennial herbs, shrubs, or trees. **Leaves** simple, entire, usually opposite, exstipulate. **Inflorescence** determinate, cymose and sometimes capitate, often subtended by conspicuous foliaceous or petaloid involucre, terminal or axillary. **Flowers** actinomorphic, perfect or seldom imperfect (then plants monoecious or dioecious), hypogynous, showy or inconspicuous, with intrastaminal nectariferous disc. **Perianth** syntepalous with typically 5 lobes, uniseriate, salverform to funnelform, often petaloid and colorful, with persistent base, induplicate-valvate and plicate or contorted in bud. **Androecium** of usually 5 stamens (but varying from 1 to numerous); filaments distinct or basally connate, often unequal; anthers basifixed, dehiscing longitudinally, latrorse. **Gynoecium** of 1 carpel; ovary superior, 1-locular; ovule solitary, campylotropous or sometimes hemitropous, placentation basal; style 1, long and slender; stigma capitate. **Fruit** an achene, often enveloped by accrescent calyx base; endosperm scanty or absent; perisperm copious, mealy or hard; embryo straight or curved, peripheral.

Family characterization: Herbs or woody plants with opposite leaves; hypogynous flowers subtended by conspicuous involucre of bracts; uniseriate petaloid perianth with induplicate-valvate aestivation; unicarpellate pistil with a single, basal ovule; fruit an achene enclosed within accrescent perianth base; and seeds with perisperm. Commonly accumulating free oxalates. Betalains (nitrogen-containing pigments) present (see Chenopodiaceae). Tissues with calcium oxalate crystals (raphides). Anatomical features: anomalous secondary growth (similar to that of the Amaranthaceae; Fig. 26: 1b,c; Esau and Cheadle 1969); sieve tubes with a special kind of P-type plastid (Behnke 1976); and unilacunar nodes.

Genera/species: 30/290

Distribution: Predominantly in tropical and subtropical areas, with a few representatives in temperate regions; particularly well represented in the American tropics.

Major genera: *Neea* (80 spp.), *Mirabilis* (45–60 spp.), *Pisonia* (35–50 spp.), *Boerhavia* (40 spp.), and *Abronia* (35 spp.)

U.S./Canadian representatives: 16 genera/119 spp.; largest genera: *Mirabilis, Abronia,* and *Boerhavia*

Economic plants and products: Various edible plants such as species of *Boerhavia* (leaves and roots) and *Pisonia* (leaves), as well as medicinal plants (e.g., *Mirabilis* spp. roots, the source of a jalap substitute, a purgative). Ornamental plants (species of at least 5 genera), including *Abronia* (sand-verbena), *Bougainvillea, Mirabilis* (four-o'clock), *Nyctaginia* (scarlet muskflower), and *Pisonia.*

Commentary: The Nyctaginaceae, generally included in the "Centrospermae" complex of families (see the commentary and references under the Chenopodiaceae), are considered to be closely allied with the Phytolaccaceae (Rettig et al. 1992). The family has been divided into a varying number of subfamilies and/or tribes, often based on the habit type, flower gender (perfect or imperfect), fruit morphology, and embryo characters (Bogle 1974). At least half of the genera in the family are monotypic, while the circumscription of several of the large genera (e.g., *Boerhavia, Mirabilis,* and *Pisonia*) is problematic.

The characteristic flower of the family consists of a uniseriate perianth, usually five stamens, and a unicarpellate pistil with a long, slender style. The cymose inflorescences are subtended by an involucre of bracts. Interpretation of the involucre and perianth in relation to floral structure may be confusing. The tubular perianth, usually petaloid and corolla-like, actually represents a modified calyx; the flowers lack a corolla. The involucre of bracts subtending the inflorescence may mimic a calyx in some species: in *Mirabilis jalapa* (Fig. 26: 3a,b), for example, the involucre of connate bracts ("calyx") subtends a reduced inflorescence of a single flower with a showy, salverform calyx ("corolla"). The involucre may be modified variously, as in *Boerhavia* (Fig. 26: 1d,e): the minute flowers have a colored perianth, and the involucre is reduced to scales. In *Bougainvillea* (Fig. 26: 2a,b), an inflorescence consists of three large, colorful, and petaloid bracts, each subtending a relatively inconspicuous flower.

The salverform to funnelform perianth is conspicuously constricted just above the ovary. The upper, expanded portion withers after anthesis, but the base

64

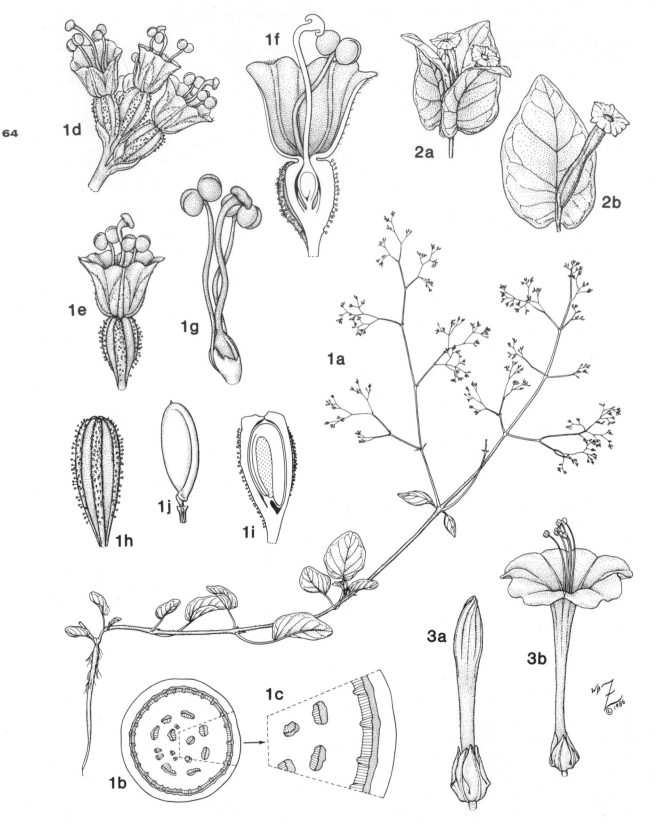

Figure 26. Nyctaginaceae. 1, *Boerhavia diffusa*: **a,** habit, ×¼; **b,** cross section of stem showing anomalous secondary growth pattern, ×10; **c,** detail of **b,** ×20; **d,** portion of inflorescence, ×10; **e,** flower, ×12; **f,** longitudinal section of flower, ×20; **g,** androecium, gynoecium, and disc, ×20; **h,** fruit ("anthocarp": achene enclosed by persistent perianth), ×9; **i,** longitudinal section of fruit, ×9; **j,** achene (persistent perianth removed), ×10. **2,** *Bougainvillea spectabilis*: **a,** inflorescence, ×1¼; **b,** flower and subtending bract, ×1⅔. **3,** *Mirabilis jalapa*: **a,** bud with subtending involucre ("calyx"), ×1 ¾; **b,** flower with subtending involucre ("calyx"), ×1.

persists and encloses the mature fruit (achene or utricle)—thus, simulating a pericarp (Fig. 26: 1i). The accessory perianth base plus fruit often is called an "anthocarp" in the literature on the Nyctaginaceae. The accrescent perianth is ribbed and/or grooved and may be hard, leathery, or fleshy. Various modifications to facilitate dispersal include glandular hairs that secrete mucilage (e.g., *Boerhavia*, Fig. 26: 1h; species of *Pisonia*). However, in *Bougainvillea*, the persistent bracts form parachute-like wings for wind dispersal, and the developing fruits of *Okenia* are pushed underground by their elongating pedicels.

The attractive flowers often are fragrant, and the nectariferous disc or ring (Fig. 26: 1f,g) secretes nectar at the base of the long perianth tube. Pollinators include various insects (bees, Lepidoptera) and birds (e.g., hummingbirds in *Bougainvillea*). Night-blooming species (such as those of *Mirabilis*) are visited by nocturnal insects, such as hawkmoths (Cruden 1973). The flowers of the family generally are protogynous, but self-pollination may occur as the style and anthers curl up together when the flowers close.

REFERENCES CITED

Behnke, H.-D. 1976. Ultrastructure of sieve-element plastids in Caryophyllales (Centrospermae), evidence for delimitation and classification of the order. *Pl. Syst. Evol.* 126:31–54.

Bogle, A. L. 1974. The genera of Nyctaginaceae in the southeastern United States. *J. Arnold Arbor.* 55:1–37.

Cruden, R. W. 1973. Reproductive biology of weedy and cultivated *Mirabilis* (Nyctaginaceae). *Amer. J. Bot.* 60:802–809.

Esau, K., and V. I. Cheadle. 1969. Secondary growth in *Bougainvillea. Ann. Bot.* (Oxford) 33:807–819.

Rettig, J. H., H. D. Wilson, and J. R. Manhart. 1992. Phylogeny of the Caryophyllales—Gene sequence data. *Taxon* 41:201–209.

CHENOPODIACEAE
Goosefoot Family

Annual or perennial herbs or sometimes shrubs, generally growing in disturbed, saline or xeric habitats, often succulent. **Leaves** simple, entire or sometimes serrate or lobed, usually alternate, generally succulent, sometimes terete or reduced to scales, often covered with hairs (causing a "mealy" appearance), exstipulate. **Inflorescence** determinate, cymose (with flowers congested in leaf axils) and often appearing spicate, racemose, or paniculate, axillary. **Flowers** usually actinomorphic, perfect or (less often) imperfect (then plants dioecious to monoecious), hypogynous, minute, bracteate. **Perianth** of usually 5 tepals, uniseriate,

distinct or basally connate, herbaceous to membranous, green or greenish, more or less imbricate, persistent, generally accrescent. **Androecium** of usually 5 stamens (opposite the tepals), arising from receptacle, inserted on disc, or adnate to perianth; filaments usually distinct, incurved in bud; anthers basifixed, dehiscing longitudinally, introrse or latrorse. **Gynoecium** of 1 pistil, 2- or 3-carpellate; ovary superior, 1-locular; ovule solitary, campylotropous or sometimes amphitropous, placentation basal; style usually 1; stigmas 2 or 3, filiform. **Fruit** an achene, utricle, or occasionally a circumscissile capsule (pyxis), often subtended by persistent tepals and bracts, sometimes forming a multiple fruit by connation of tepals of several flowers; seeds lenticular; perisperm copious, starchy, usually hard; embryo curved or spirally twisted, surrounding the perisperm (peripheral).

Family characterization: More or less fleshy herbs to shrubs growing in weedy, xeric, or saline habitats; minute flowers in dense cymose inflorescences; greenish, herbaceous to membranous, uniseriate perianth of 5 tepals; 5 distinct stamens; 1-locular ovary with a solitary basal ovule; and perisperm surrounded by curved or coiled embryo. Betalains (nitrogen-containing pigments) present (see the references below). Organic acids and often also free nitrates and/or oxalates, various alkaloids, and saponins present. Tissues with calcium oxalate crystals. Anatomical features: anomalous secondary growth (similar to that of the Amaranthaceae; Fig. 27: 1b); sieve tubes with a special kind of P-type plastid (Behnke 1976); and unilacunar nodes. Leaves of many species often with Kranz anatomy (associated with C_4 photosynthesis; see the discussion under Portulacaceae).

Genera/species: 104/1,510

Distribution: Cosmopolitan; especially abundant in weedy, xeric or saline areas.

Major genera: *Atriplex* (100–200 spp.), *Chenopodium* (100–150 spp.), *Salsola* (150 spp.), *Suaeda* (110 spp.), and *Obione* (100+ spp.)

U.S./Canadian representatives: 25 genera/187 spp.; largest genera: *Atriplex* and *Chenopodium*

Economic plants and products: Food plants: *Beta vulgaris* (beet, Swiss chard), *Spinacia oleracea* (spinach), and several species of *Chenopodium* (edible greens and quinoa, pseudo-grains from mealy perisperm of seeds). Several weedy plants, such as species of *Chenopodium* (goosefoot, lamb's-quarters) and *Salsola* (Russian-thistle). Ornamental plants (species of 7 genera), including *Atriplex* (saltbush), *Chenopodium*, *Kochia* (summer-cypress), and *Salicornia* (glasswort).

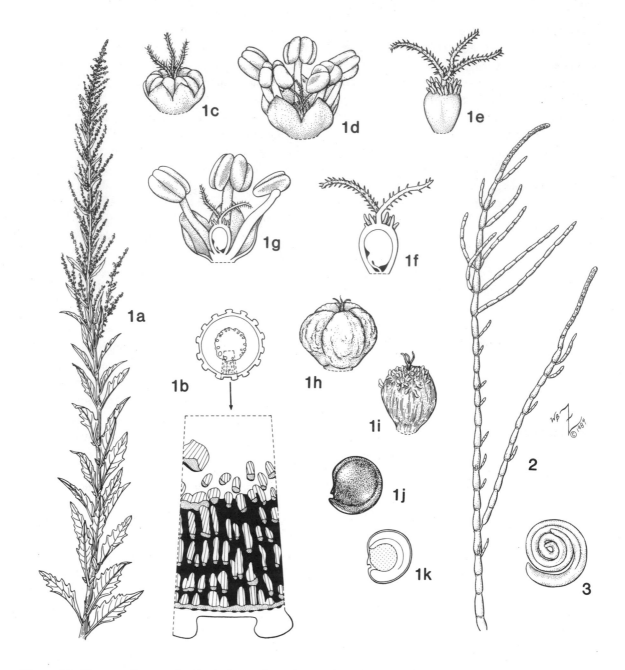

Figure 27. Chenopodiaceae. 1, *Chenopodium ambrosioides*: **a,** habit, ×⅓; **b,** cross section of stem, ×2, with detail showing anomalous secondary growth pattern, ×15; **c,** flower before extension of stamens, ×12; **d,** flower, ×12; **e,** pistil, ×25; **f,** longitudinal section of pistil, ×25; **g,** longitudinal section of flower, ×15; **h,** fruit (utricle subtended by accrescent perianth), ×18; **i,** utricle, ×25; **j,** seed, ×25; **k,** longitudinal section of seed, ×25. **2,** *Salicornia virginica*: habit, ×½. **3,** *Suaeda maritima*: embryo, ×12.

Commentary: The Chenopodiaceae usually are divided into two or three subfamilies (see Williams and Ford-Lloyd 1974; Blackwell 1977) primarily based on characters of the embryo (curved or spirally coiled). The family generally is considered to be closely allied with the Amaranthaceae (see Hershkovitz 1989; Rettig et al. 1992); the two taxa may be sister groups, or the Amaranthaceae may be derived within the Chenopodiaceae (Rodman 1990). These two families form the core of the so-called Centrospermae ("free-central") complex of families (order Caryophyllales in Thorne 1992). Numerous studies and commentaries highlight the controversies concerning the taxonomy of this complex (e.g., Mabry et al. 1963; Mabry et al. 1972; Mabry 1974, 1977; Eckardt 1976; Rodman et al. 1984; Hershkovitz 1989; Rodman 1990; Giannasi et al. 1992; and Rettig et al. 1992). Most of these families produce the red pigment betacyanin, in contrast to the anthocyanin pigments found in most other flowering plants. Other characters of the complex include the basal placentation and the curved, peripheral embryo.

The varied vegetative habits within the Chenopodiaceae include adaptations for habitats with soils containing a large percentage of inorganic salts. The species of seashores (and similar habitats) generally are succulent and brittle plants with reduced (*Salsola*) to nearly absent (*Salicornia*; Fig. 27: 2) leaves. Others in the family, such as *Chenopodium* (Fig. 27: 1a), are leafy weeds inhabiting salt-rich soils around dwellings and disturbed areas. Species of such varied genera as *Chenopodium* and *Salsola* often have a vestiture of hairs with distended, thin-walled apices containing water and oxalates. The desiccated hairs appear as white flakes on mature plant parts, causing a "mealy" appearance. Water also is retained in water-storage tissue often present in the mesophyll.

The inconspicuous flowers reportedly are rarely visited by insects and are assumed to be generally anemophilous (Proctor and Yeo 1972; Frankel and Galun 1977). Imperfect flowers lack nectaries, but nectariferous glands or a disc may occur at the filament bases of perfect flowers. Protandry or protogyny promotes cross-pollination, although self-pollination does occur in several members, such as *Atriplex* species.

REFERENCES CITED

Behnke, H.-D. 1976. Ultrastructure of sieve-element plastids in Caryophyllales (Centrospermae), evidence for delimitation and classification in the order. *Pl. Syst. Evol.* 126:31–54.

Blackwell, W. H. 1977. The subfamilies of the Chenopodiaceae. *Taxon* 26:395–397.

Eckardt, T. 1976. Classical morphological features of centrospermous families. *Pl. Syst. Evol.* 126:5–25.

Frankel, R., and E. Galun. 1977. *Pollination mechanisms, reproduction and plant breeding*, pp. 17, 22. Monogr. Theor. & Appl. Genet. 2. Springer-Verlag, Berlin.

Giannasi, D. E., G. Zurawski, G. Learn, and M. T. Clegg. 1992. Evolutionary relationships of the Caryphyllidae based on comparative *rbc*L sequences. *Syst. Bot.* 17:1–15.

Hershkovitz, M. A. 1989. Phylogenetic studies in Centrospermae: A brief appraisal. *Taxon* 38:602–610.

Mabry, T. J. 1974. Is the order Centrospermae monophyletic? In *Chemistry in botanical classification*, ed. G. Bendz and J. Santesson, pp. 275–285. Nobel Symp. 25. Academic Press, London.

———. 1977. The order Centrospermae. *Ann. Missouri Bot. Gard.* 64:210–220.

———, L. Kimler, and C. Chang. 1972. The betalains: Structure, function, and biogenesis, and the plant order Centrospermae. In *Recent advances in phytochemistry*, vol. 5, *Structural and functional aspects of phytochemistry*, ed. V. C. Runeckles and T. C. Tso, pp. 105–134. Academic Press, London.

———, A. Taylor, and B. L. Turner. 1963. The betacyanins and their distribution. *Phytochemistry* 2:61–64.

Proctor, M., and P. Yeo. 1972. *The pollination of flowers*, p. 273. Taplinger, New York.

Rettig, J. H., H. D. Wilson, and J. R. Manhart. 1992. Phylogeny of the Caryophyllales—Gene sequence data. *Taxon* 41:201–209.

Rodman, J. E. 1990. Centrospermae revisited, part I. *Taxon* 39:383–393.

———, M. K. Oliver, R. R. Nakamura, J. U. McClammer, and A. H. Bledsoe. 1984. A taxonomic analysis and revised classification of Centrospermae. *Syst. Bot.* 9:297–323.

Thorne, R. F. 1992. Classification and geography of the flowering plants. *Bot. Rev.* (Lancaster) 58:225–348.

Williams, J. T., and B. V. Ford-Lloyd. 1974. The systematics of the Chenopodiaceae. *Taxon* 23:353–354.

AMARANTHACEAE
Amaranth or Pigweed Family

Annual or perennial herbs or sometimes shrubs; stems reddish. **Leaves** simple, entire, alternate or opposite, sometimes succulent, exstipulate. **Inflorescence** determinate, cymose and usually appearing racemose, spicate, or capitate, or flowers solitary, axillary. **Flowers** actinomorphic, perfect or sometimes imperfect (then plants dioecious or polygamodioecious), hypogynous, minute, subtended by a persistent scarious bract and 2 similar bracteoles. **Perianth** of 3, 4, or most commonly 5 tepals, uniseriate, distinct or basally connate, scarious, minute, white, pink, or red, imbricate. **Androecium** of 5 stamens (opposite the tepals), often alternating with simple or fringed enations; filaments partially to totally connate (monadelphous) and forming a membranous tube; anthers 2- or 4-locular at anthesis, dorsifixed, dehiscing longitudinally. **Gynoecium** of 1 pistil, 2- or 3-carpellate; ovary superior, 1-locular, compressed; ovule usually solitary, campylotropous, pendulous or erect, placentation basal, styles 1 to 3; stigmas 1 to 3, capitate, bifid or trifid. **Fruit** a utricle, achene, or circumscissile capsule (pyxis); seeds globose or lenticular; perisperm copious, mealy; embryo curved, surrounding the perisperm (peripheral).

Family characterization: Herbaceous to shrubby plants with reddish stems; dense cymose inflorescences; minute flowers subtended by scarious bracts; uniseriate perianth of scarious tepals; monadelphous stamens often alternating with enations; 1-locular ovary with a solitary basal ovule; and perisperm surrounded by curved embryo. Betalains (nitrogen-containing pigments) present (see the references under Chenopodiaceae), and also often free oxalates (nitrates) and saponins. Tissues commonly with calcium oxalate crystals. Anatomical features: anomalous sec-

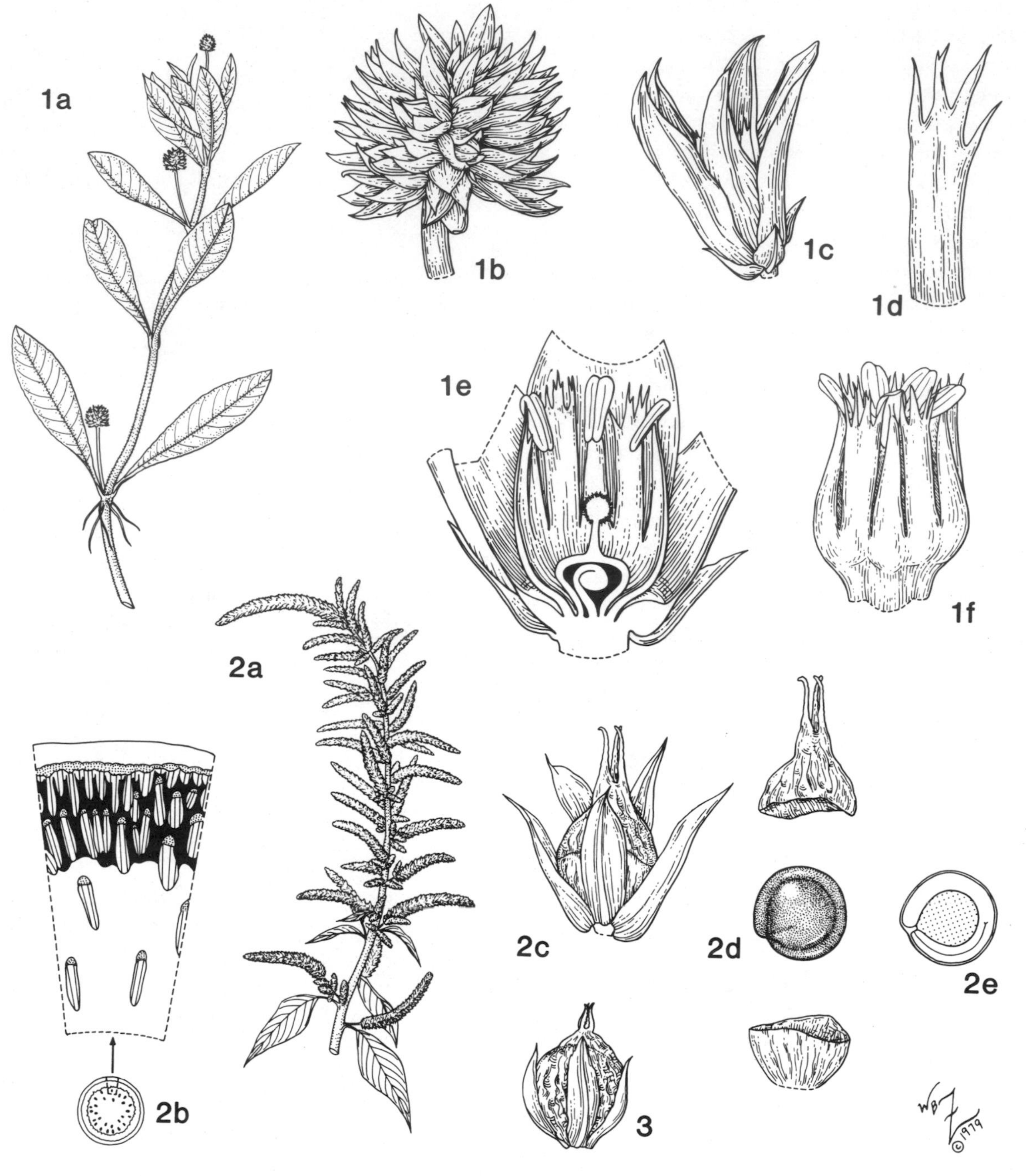

Figure 28. Amaranthaceae. 1, *Alternanthera philoxeroides*: **a,** habit, ×2/5; **b,** inflorescence, ×3; **c,** flower, ×6; **d,** staminal enation, ×20; **e,** longitudinal section of flower, ×12; **f,** androecium (monadelphous), ×12. **2,** *Amaranthus hybridus*: **a,** flowering branch, ×2/5; **b,** cross section of stem, ×1/2, with detail showing anomalous secondary growth pattern, ×12; **c,** pyxis, ×12; **d,** dehiscing pyxis and seed, ×12; **e,** longitudinal section of seed, ×12. **3,** *Amaranthus viridis*: utricle, ×12.

ondary growth (see the discussion below); sieve tubes with a special kind of P-type plastid (Behnke 1976); and unilacunar nodes. Leaves of many species with Kranz anatomy (associated with C_4 photosynthesis; see the discussion under Portulacaceae).

Genera/species: 65/850

Distribution: Widespread in tropical to temperate zones; especially diverse in tropical America and Africa.

Major genera: *Alternanthera* (80–200 spp.), *Ptilotus* (100 spp.), and *Gomphrena* (100 spp.)

U.S./Canadian representatives: 18 genera/108 spp.; largest genera: *Amaranthus* and *Iresine*

Economic plants and products: Edible greens and

"cereal" (pseudo-grains) from *Amaranthus* spp. Many weedy species, including *Alternanthera* (alligator-weed), *Amaranthus* (pigweed), and *Iresine*. Ornamental plants: *Alternanthera* (copperleaf), *Amaranthus* (prince's-feather, foxtail), *Celosia* (cock's-comb), *Froelichia* (cottonweed), *Gomphrena* (globe-amaranth), and *Iresine* (bloodleaf).

Commentary: The Amaranthaceae are divided into two subfamilies primarily based on the number of anther locules at anthesis (two or four) and the number of ovules per ovary (one, or one to many; see Robertson 1981). The family is allied with the Chenopodiaceae (Hershkovitz 1989; Rettig et al. 1992). The floral structure of the two families is very similar (e.g., one whorl of tepals and superior ovary with one ovule), but the flower parts of the Amaranthaceae are scarious and the stamens are monadelphous (Fig. 28: 1f) and frequently appendaged (Fig. 28: 1d).

Often the stems are tinged with red due to the betacyanin pigments (see the references concerning these pigments and the "Centrospermae" in the discussions under Caryophyllaceae and Chenopodiaceae). The stems of many species undergo anomalous secondary growth with the vascular bundles in several concentric rings or sometimes in irregular patterns (Balfour 1965). Secondary growth begins normally, but later an additional vascular cambium differentiates within or outside the phloem. New xylem is produced toward the inside, and phloem, toward the outside. Series of vascular cambia then arise successively farther out from the center. The production of new secondary tissue by each cambium results in many alternating layers of xylem and phloem (Fig. 28: 2b).

The coarse weeds are recognized easily in the field due to their scarious reduced flowers in congested inflorescences. The small flowers, with often a white, red, or pink perianth, apparently are entomophilous and probably attract various unspecialized insect visitors. The flowers of most genera are nectariferous with a more or less conspicuous annular nectary at the adaxial base of the staminal tube (e.g., species of *Alternanthera* and *Gomphrena*) or layer of glandular tissue lining the tube's adaxial surface (*Froelichia*; Zandonella 1967). The notable exception is *Amaranthus*, whose flowers presumably are anemophilous: the flowers lack nectaries and produce relatively small, unornamented pollen grains (Zandonella and Lecocq 1977).

REFERENCES CITED

Balfour, E. 1965. Anomalous secondary thickening in Chenopodiaceae, Nyctaginaceae and Amaranthaceae. *Phytomorphology* 15:111–112.
Behnke, H.-D. 1976. Ultrastructure of sieve-element plastids in Caryophyllales (Centrospermae), evidence for delimitation and classification of the order. *Pl. Syst. Evol.* 126:31–54.
Hershkovitz, M. A. 1989. Phylogenetic studies in Centrospermae: A brief appraisal. *Taxon* 38:602–610.
Rettig, J. H., H. D. Wilson, and J. R. Manhart. 1992. Phylogeny of the Caryophyllales—Gene sequence data. *Taxon* 41:201–209.
Robertson, K. R. 1981. The genera of Amaranthaceae in the southeastern United States. *J. Arnold Arbor.* 62:267–313.
Zandonella, P. 1967. Les nectaires des Amaranthaceae. *Compt. Rend. Hebd. Séances Acad. Sci.*, Sér. D 264:2559–2562.
—— and M. Lecocq. 1977. Morphologie pollinique et mode de pollinisation chez des Amaranthaceae. *Pollen & Spores* 19:119–141.

THEACEAE
Camellia or Tea Family

Trees or shrubs. **Leaves** simple, serrate or less commonly entire, alternate, often coriaceous, usually persistent, exstipulate. **Flowers** generally solitary and axillary, occasionally fasciculate, actinomorphic, perfect, usually hypogynous to somewhat perigynous, showy, subtended by 1 or more bract(s) (often similar to the sepals). **Calyx** of usually 5 sepals, distinct to more often basally connate, usually persistent, imbricate. **Corolla** of usually 5 petals, distinct to more often slightly basally connate, often with wrinkled edges, typically white, imbricate or seldom contorted/convolute. **Androecium** of numerous stamens; filaments distinct to usually basally connate into a ring (monadelphous) or 5 fascicles (opposite the petals), basally adnate to the corolla; anthers basifixed or dorsifixed and versatile, dehiscing longitudinally, introrse or latrorse. **Gynoecium** of 1 pistil, 3- to 5-carpellate; ovary superior, with as many locules as carpels; ovules 2 to numerous in each locule, anatropous or somewhat campylotropous, placentation axile; styles as many as carpels, distinct to basally connate; stigma(s) 1 and lobed or sometimes 3 to 5, distinct and capitate. **Fruit** usually a loculicidal capsule with a persistent central column, woody; seeds often ovoid and compressed or globose; endosperm scanty or absent; embryo large, straight or curved.

Family characterization: Trees or shrubs with simple, leathery, serrate leaves; solitary, actinomorphic, showy flowers subtended by conspicuous bracts; 5-merous perianth; often white, slightly wrinkled petals; numerous stamens basally connate into fascicles or a ring; and woody loculicidal capsule with persistent central column as the fruit type. Saponins and alkaloids present. Tissues with tannins and calcium oxalate crystals. Anatomical features: abundant sclereids

70

Figure 29. Theaceae. *Gordonia lasianthus*: **a,** flowering branch, ×½; **b,** cross section of node (one-trace unilacunar), ×5; **c,** branched sclereids from leaf, ×135; **d,** flower, ×½; **e,** petal and adnate stamen fascicle, ×¾; **f,** longitudinal section of flower, ×⅘; **g,** longitudinal section of flower (perianth removed), ×2½; **h,** cross section of ovary, ×6; **i,** capsule with persistent calyx, ×1½; **j,** seed, ×4½; **k,** longitudinal section of seed, ×4½.

(often much-branched) in stems, leaves, and floral parts (Fig. 29: c); mucilage cells often in the leaf epidermis; and unilacunar nodes (Fig. 29: b).

Genera/species: 28/500

Distribution: Primarily in tropical to subtropical regions, with a few representatives in warm temperate areas; especially diverse in tropical America and Asia.

Major genera: *Eurya* (70–130 spp.), *Ternstroemia* (85–100 spp.), *Camellia* (82 spp.), and *Gordonia* (40–70 spp.)

U.S./Canadian representatives: 7 genera/12 spp.; largest genus: *Ternstroemia*

Economic plants and products: Commercial tea leaves and tea-seed oil from *Camellia* (*Thea*) *sinensis* (tea). Ornamental plants (species of at least 9 genera), including *Camellia*, *Cleyera*, *Eurya*, *Franklinia* (Franklin tree), *Gordonia* (loblolly bay), *Stewartia*, and *Ternstroemia*. The notable cultivated species native to the southeastern United States, *Franklinia alatamaha*, has been extirpated from the wild (Harper and Leeds 1938).

Commentary: Two large subfamilies (Theoideae and Ternstromioideae) or tribes form the core of the Theaceae *s.s.* (Keng 1962). Some authorities have defined the family more broadly by including several other subfamilies (or tribes) that now are generally recognized as small (e.g., Actinidiaceae) or monotypic (e.g., Pelliciereaceae) segregate families. The circumscription of the large, probably over-described genera of the Theaceae also is problematic (Wood 1959).

A five-merous perianth and numerous stamens compose the characteristic flowers of the family (Keng 1962). The subtending floral bract(s) often blend into the five sepals (e.g., *Camellia*). The petals (often white) have wrinkled edges, and the outermost usually is somewhat cupped and smaller than the other four. Stamens are adnate to the corolla base and aggregated into five fascicles or united into a ring (monadelphous). The degree of filament fusion varies considerably. When fascicled, the filaments may be distinct (*Franklinia*) or basally connate (*Gordonia*) with each stamen cluster opposite a petal (Fig. 29: e). In *Stewartia*, the androecium is united into a shallow ring, and in *Camellia*, the monadelphous filaments form a well-developed fleshy cup.

The handsome, showy flowers usually are fragrant and often secrete nectar from the base of the filaments or ovary. The fragrant flowers of *Camellia*, however, do not produce nectar and probably are visited by pollen-collecting insects.

REFERENCES CITED

Harper, F., and A. N. Leeds. 1938. A supplementary chapter on *Franklinia alatamaha*. *Bartonia* 19:1–13.

Keng, H. 1962. Comparative morphological studies in Theaceae. *Univ. Calif. Publ. Bot.* 33:269–383.

Wood, C. E. 1959. The genera of Theaceae of the southeastern United States. *J. Arnold Arbor.* 40:413–419.

AQUIFOLIACEAE
Holly Family

Trees and shrubs, usually growing on acid soil. **Leaves** simple, entire to serrate, commonly spiny, alternate, coriaceous, sometimes persistent, minutely stipulate. **Inflorescence** determinate, cymose, fasciculate, or flowers solitary, axillary. **Flowers** actinomorphic, perfect or frequently imperfect (then plants dioecious or polygamodioecious), hypogynous, small. **Calyx** of usually 4 or 5 sepals, basally connate, imbricate, persistent. **Corolla** of usually 4 or 5 petals, usually basally connate, white or greenish, imbricate. **Androecium** of usually 4 or 5 stamens, often basally adnate to corolla; filaments distinct; anthers dorsifixed, dehiscing longitudinally, introrse; staminodes present in carpellate flowers. **Gynoecium** of 1 pistil, 4- to 7-carpellate; ovary superior, with as many locules as carpels; ovules 1 in each locule, anatropous, pendulous, often with funicular protuberance, placentation axile; style 1, very short or absent; stigma capitate, discoid, or lobed; rudimentary pistil usually present in staminate flowers. **Fruit** a drupe with 4 to 7 pyrenes (stones), frequently brightly colored (red, yellow, orange-red, pink, or purple-black); endosperm copious, fleshy; embryo minute, straight.

Family characterization: Acidophilic woody plants often with spiny or toothed leaves; often 4-merous, imperfect, axillary flowers; superior ovary with 1 pendulous ovule per locule; and brightly colored drupe with often 4 pyrenes. Anatomical features: unitegmic and tenuinucellate ovules; and 1-trace unilacunar nodes (Fig. 30: 1e).

Genera/species: *Ilex* (at least 400 spp.); a monotypic family.

Distribution: Generally widespread throughout tropical and temperate areas; relatively uncommon in Africa and Australia.

U.S./Canadian representatives: *Ilex* (31 spp.)

Economic plants and products: Teas with high caffeine content from *I. paraguariensis* (yerbe maté) and *I. vomitoria* (yaupon, the reputed emetic of the Coastal Plains Indians). Fine white hardwood from several *Ilex* species. Several ornamental shrubs, including *Ilex aquifolium* (English holly, European holly) and *I. opaca* (American holly).

Commentary: As defined here, the Aquifoliaceae are a monotypic family, comprising only the genus *Ilex* (Thorne 1992). Other authors (e.g., Brizicky 1964) recognize the segregate genus *Nemopanthus* and sometimes also include the closely related *Phelline* (10 species) and *Sphenostemon* (7 species) for an expanded circumscription of the family. Thorne (1992) and other taxonomists (Baas 1975) maintain the latter two genera as segregate families (the Phellinaceae and Sphenostemonaceae, respectively). Taxonomists also disagree on the affinities of the Aquifoliaceae to other families. Generally, the hollies have been allied with the Celastraceae (and other families in the Celastrales), but Thorne considers the family (plus the two related segregates) with the Theaceae and allies (Theales).

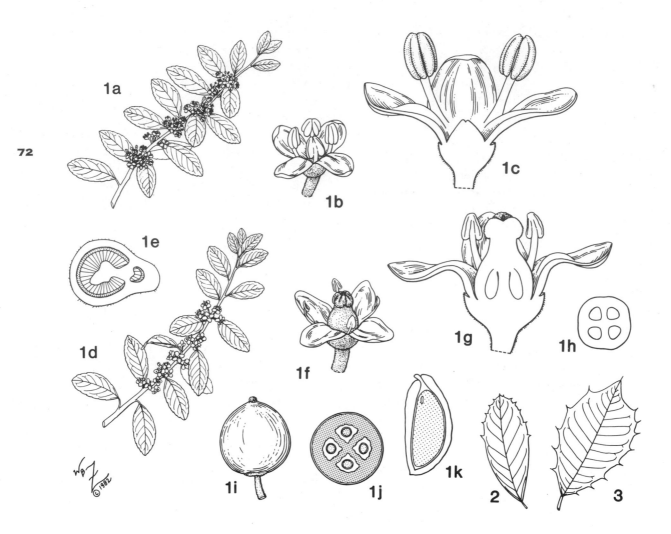

Figure 30. Aquifoliaceae. 1, *Ilex vomitoria*: **a,** staminate flowering branch, ×½; **b,** staminate flower, ×6; **c,** longitudinal section of staminate flower, ×12; **d,** carpellate flowering branch, ×½; **e,** cross section of node (one-trace unilacunar), ×10; **f,** carpellate flower, ×6; **g,** longitudinal section of carpellate flower, ×12; **h,** cross section of ovary, ×12; **i,** drupe, ×3; **j,** cross section of drupe, ×3; **k,** longitudinal section of pyrene, ×7. **2,** *Ilex cassine*: leaf, ×½. **3,** *Ilex opaca*: leaf, ×½.

The specific and intraspecific delimitations of *Ilex*, a nomenclaturally complex group, also is problematical, and species generally are separated on leaf and fruit characters. The popularity of the cultivated hollies, *Ilex aquifolium* (English holly) and *I. opaca* (American holly), probably has led to the misconception that all hollies are characterized by spiny-margined leaves (Fig. 30: 2, 3). However, many species of *Ilex* have leaves with entire to serrate margins lacking spines (Fig. 30: 1a,d).

The small and inconspicuous flowers of most hollies are entomophilous, although wind may aid in the pollination of some species (Brizicky 1964). Cross-pollination is reinforced by dioecism. Generally, the carpellate flowers are relatively few to solitary in the leaf axils (Fig. 30: 1d), while the staminate flowers tend to be much more numerous and obviously fascicled (Fig. 30: 1a). Glands on the adaxial surface of the petals secrete some nectar, and the inconspicuous flowers commonly attract bees as principal pollinators.

The brightly colored drupes are disseminated by birds. At the time of dispersal the embryo is commonly undifferentiated, and it requires a long duration to mature before germination of the seed (Tomlinson 1980).

REFERENCES CITED

Baas, P. 1975. Vegetative anatomy and the affinities of Aquifoliaceae, *Sphenostemon*, *Phelline*, and *Onotheca*. *Blumea* 22: 311–407.

Brizicky, G. K. 1964. The genera of Celastrales in the southeastern United States. *J. Arnold Arbor.* 45:206–234.

Thorne, R. F. 1992. Classification and geography of the flowering plants. *Bot. Rev.* (Lancaster) 58:225–348.

Tomlinson, P. B. 1980. *The biology of trees native to tropical*

Florida, pp. 99–103. "Publ. by the author," Petersham, Mass.

SARRACENIACEAE
Pitcher Plant Family

Perennial herbs growing in marshes or bogs, insectivorous, with rhizomes. **Leaves** simple, alternate, in basal rosettes, highly modified, tubular or trumpet-shaped with ridge or laminar wing on adaxial side and a terminally expanded hood, often with retrorse hairs on the inner surface, liquid-filled at base, green, yellow, maroon, or variegated, exstipulate. **Flowers** solitary and terminal, actinomorphic, perfect, hypogynous, large, nodding. **Calyx** of usually 5 sepals, distinct, often colored and petaloid, imbricate, persistent. **Corolla** of usually 5 petals, distinct, caducous, usually yellow, maroon, or red, imbricate. **Androecium** of numerous stamens; filaments distinct; anthers basifixed or versatile, dehiscing longitudinally, extrorse. **Gynoecium** of 1 pistil, usually 5-carpellate; ovary superior, usually 5-locular and -lobed; ovules numerous, anatropous, placentation basically axile (with partitions above often not meeting or joined); style 1, usually with greatly expanded peltate apex with 5 lobes or occasionally shortly 5-branched, persistent; stigmas 5, small and restricted to area beneath each style lobe tip or occasionally terminal. **Fruit** a loculicidal capsule; seeds numerous, small, often winged; endosperm copious, oily, firm-fleshy; embryo minute, linear.

Family characterization: Perennial insectivorous herbs growing in marshes or bogs; highly modified and specialized tubular leaves in basal rosettes; solitary, nodding, perfect flowers with 5-merous perianth and gynoecium; numerous stamens; usually dilated peltate style with a restricted stigmatic area under each of 5 lobes; and loculicidal capsule as the fruit type. Tannins commonly present. Several anatomical features related to specialized leaves (see Metcalfe and Chalk 1950, and the discussion below).

Genera/species: *Sarracenia* (8–10 spp.), *Heliamphora* (5–6 spp.), and *Darlingtonia* (1 sp.)

Distribution: In marshy and boggy habitats of eastern North America, coastal Oregon and northern California, and northern South America.

U.S./Canadian representatives: *Sarracenia* (8 spp.) and *Darlingtonia* (1 sp.)

Economic plants and products: Novelty house plants: several species of *Sarracenia* (pitcher plant) and *Darlingtonia californica* (cobra plant, California pitcher plant).

Commentary: The Sarraceniaceae, a distinctive insectivorous group, have been allied with various families (see Wood 1960; DeBuhr 1975). The family primarily is American, with several species of *Sarracenia* in eastern North America and the monotypic *Darlingtonia* in the Pacific northwest. The morphology of the leaves ("pitchers") provides the most important taxonomic characters used in distinguishing species of *Sarracenia*.

The tubular leaves of *Sarracenia* consist of gracefully curved pitchers, each covered by a hood or lid that represents a prolongation of the dorsal side (Fig. 31: b). A ridge or laminar wing is present along the ventral surface. Insects are attracted to the bright coloring and often strong odors of the leaves. Several species also have translucent, window-like spots (fenestrations) around the necks of the pitchers. In addition, the outer surface of the leaf and the inner surface of the lid and mouth generally have secreting glandular hairs (Fig. 31: c,e). After entering the pitcher, an insect may slip downward into the leaf along a slick zone composed of epidermal cells with downwardly directed projections (covering the surface like fish scales; Fig. 31: f); it next encounters long, recurved hairs (which cover most of the tube base; Fig. 31: g) and becomes trapped within the leaf, later to drown in the liquid accumulated at the base. In most species of *Sarracenia*, the prey is digested by acids and enzymes and then absorbed by the plant. Despite the insectivorous nature of the leaves, several insect larvae (of certain flies and mosquitoes) may inhabit the pitchers (Lloyd 1942; Juniper et al. 1989). The tubular leaves of *Darlingtonia californica* are similar to those of *Sarracenia*, but the apex of the curved hood terminates in a conspicuous, forked, glandular appendage that attracts prey. The prey is digested inside the tube by bacterial action, not by secreted enzymes as in *Sarracenia*.

Little has been reported about pollination mechanisms in the family. Many *Sarracenia* flowers have a musty or sometimes sweet odor and are visited by bees (Juniper et al. 1989). Hybridization occurs within *Sarracenia*, although species that occur in the same area tend to be isolated temporally by flowering periods (McDaniel 1971).

REFERENCES CITED

DeBuhr, L. E. 1975. Phylogenetic relationships of the Sarraceniaceae. *Taxon* 24:297–306.

Juniper, B. E., B. J. Robbins, and D. M. Joel. 1989. *The carnivorous plants*, pp. 37–41, 54–64, 108–112, 252–254, 270–271. Academic Press, London.

Lloyd, F. E. 1942. *The carnivorous plants*, pp. 17–50. Chronica Botanica, Waltham, Mass.

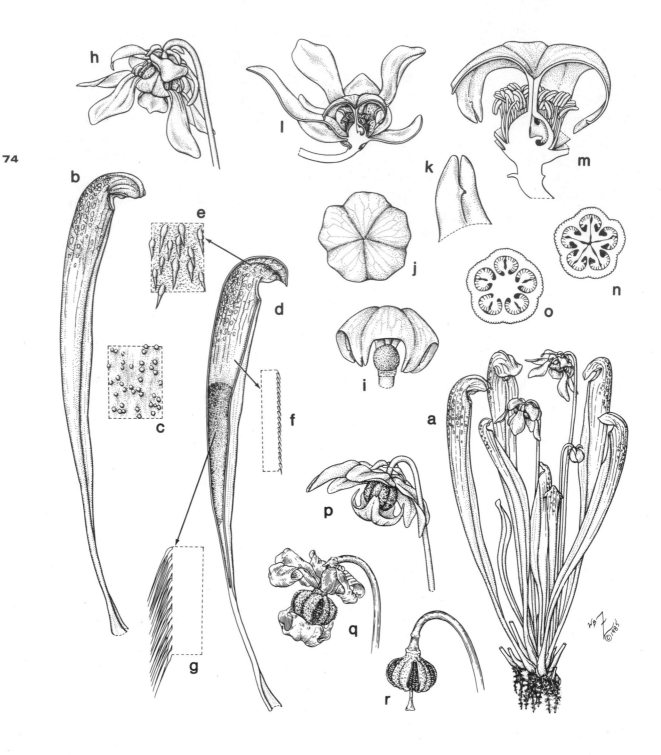

Figure 31. Sarraceniaceae. *Sarracenia minor*: **a,** habit, ×⅕; **b,** leaf, ×⅓; **c,** glands on outer surface of leaf, ×25; **d,** longitudinal section of leaf, ×⅓; **e,** glandular hairs along inner surface of hood, ×25; **f,** detail of upper section of leaf ("smooth zone"), ×25; **g,** detail of lower section of leaf, ×25; **h,** flower, ×½; **i,** pistil, ×1; **j,** top view of expanded style, ×1; **k,** tip of style branch with stigma, ×3; **l,** longitudinal section of flower, ×⅔; **m,** longitudinal section of flower (perianth removed), ×1¾; **n,** cross section of ovary near middle, ×3½; **o,** cross section of ovary at base, ×3½; **p,** young capsule, ×½; **q,** mature capsule with persistent calyx, ×⅔; **r,** capsule (calyx removed), ×⅔.

McDaniel, S. T. 1971. The genus *Sarracenia* (Sarraceniaceae). *Bull. Tall Timbers Res. Sta.* 9:1–36.

Metcalfe, C. R., and L. Chalk. 1950. *Anatomy of the dicotyledons*, vol. 2, pp. 71–74. Oxford University Press, Oxford.

Wood, C. E. 1960. The genera of Sarraceniaceae and Droseraceae in the southeastern United States. *J. Arnold Arbor.* 41:152–163.

CLUSIACEAE OR GUTTIFERAE

Garcinia or St. John's-wort Family
Including the Hypericaceae

Trees, shrubs, or annual or perennial herbs, with clear or colored resinous sap. **Leaves** simple, entire, opposite or sometimes whorled, often coriaceous and persistent, pellucid or black punctate, exstipulate. **Inflorescence** determinate, cymose and often appearing umbellate, terminal. **Flowers** actinomorphic, perfect or imperfect (then plants usually dioecious), typically showy, hypogynous. **Calyx** of 2 to 10 sepals, distinct, often unequal in size, imbricate, persistent. **Corolla** of 4 to 12 petals, distinct, often asymmetrical, yellow, pink, or white, imbricate or contorted. **Androecium** of 4 to numerous stamens; filaments distinct to usually connate into 3 to 5 fascicles; anthers dehiscing longitudinally; staminodes sometimes present. **Gynoecium** of 1 pistil, generally 3- to 5-carpellate; ovary superior, 3- to 5- or sometimes 1-locular; ovules numerous, anatropous, placentation parietal (in 1-locular ovaries), apparently axile due to intrusions of placentae, or axile; styles absent or 3 to 5, distinct to partially connate; stigmas broad and radiating (when sessile) or capitate (when styles present). **Fruit** typically a septicidal capsule or sometimes a berry or drupe; seeds numerous, sometimes arillate; endosperm absent; embryo straight.

Family characterization: Herbaceous to shrubby plants with resinous sap; opposite, translucent or black punctate leaves; 4- or 5-merous perianth; often yellow petals; numerous fascicled stamens; 3- to 5-carpellate superior ovary; septicidal capsule; and nonendospermous seeds. Tannins and calcium oxalate crystals commonly present. Anatomical features: schizogenous secretory cavities and/or canals containing oils and/or resins (often appearing on the leaves and perianth as transparent or black dots); tenuinucellate ovules; and 1-trace unilacunar nodes.

Genera/species: 45/1,010

Distribution: Widespread in tropical regions; shrubby and herbaceous *Hypericum* species predominate in temperate areas.

Major genera: *Hypericum* (370–400 spp.), *Garcinia* (200–400 spp.), and *Clusia* (145 spp.)

U.S./Canadian representatives: *Hypericum* (55 spp.), *Clusia* (4 spp.), *Triadenum* (4 spp.), *Calophyllum* (2 spp.), *Rheedia* (2 spp.), and *Mammea* (1 sp.)

Economic plants and products: Edible fruits from *Garcinia* (mangosteen) and *Mammea* (mammey-apple). Gums, resins, oils, and dyes from *Calophyllum*, *Clusia*, and *Garcinia*. Ornamental plants (species of 6 genera), including *Clusia* (pitch-apple) and *Hypericum* (St. Andrew's-cross, St. John's-wort).

Commentary: The Clusiaceae generally are divided into five closely interrelated subfamilies based on variations in the degree of stamen connation, carpel number, and fruit type. Some authors such as Lawrence (1951) and Hutchinson (1969) have separated the Hypericoideae (herbaceous to shrubby habit, fascicled stamens, distinct styles) into a distinct family, the Hypericaceae. Anatomical studies of Vestal (1937) and others have concluded that the arborescent Hypericoideae do not differ significantly from the rest of the Clusiaceae. In addition, the elevation of the Hypericoideae to the rank of family probably necessitates the equivalent segregation of the other subfamilies as well (Adams 1962; Wood and Adams 1976). In the literature, "Guttiferae" also has been included erroneously as a synonym for "Hypericaceae." According to the *International Code of Botanical Nomenclature* (Greuter et al. 1988), Guttiferae is the correct alternate name for Clusiaceae only; Hypericaceae, a segregate family, is not a proper alternative for either Guttiferae or Clusiaceae.

The name Guttiferae, meaning "gum-bearing," refers to the characteristic resinous sap that is white or yellowish in many arborescent members (e.g., *Clusia*) or clear (*Hypericum tetrapetalum*) or with black pigment (*H. perforatum*) in other species. The secretory cavities and/or canals occur in most plant parts and appear as clear or blackish dots (or short lines) on the leaves (Fig. 32: 2b) and perianth (Fig. 32: 2c).

The showy flowers with numerous stamens usually lack nectariferous structures and probably are visited by insects (various bees) for the copious pollen. The numerous staminodes of the carpellate flowers of *Clusia rosea* (Fig. 32: 3a) are fused into a ring that produces a sticky, brownish fluid that attracts bees. However, staminate plants of this species evidently no longer exist, and viable seeds (Fig. 32: 3d) are produced apomictically (Maguire 1976).

REFERENCES CITED

Adams, P. 1962. Studies in the Guttiferae. II. Taxonomic and distributional observations on North American taxa. *Rhodora* 64:231–242.

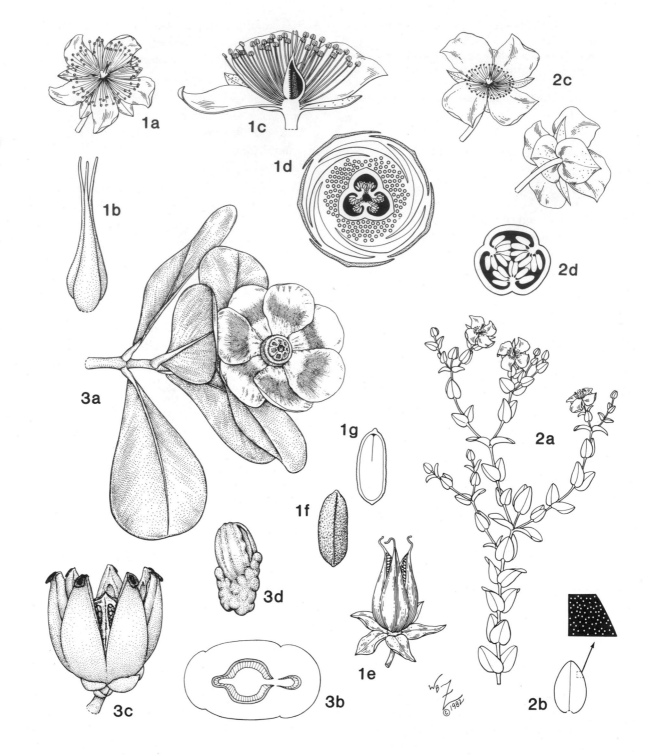

Figure 32. Clusiaceae. 1, *Hypericum myrtifolium*: **a,** flower, ×1½; **b,** pistil, ×5; **c,** longitudinal section of flower, ×3; **d,** cross section of flower, ×12; **e,** septicidal capsule (with persistent calyx), ×3; **f,** seed, ×25; **g,** longitudinal section of seed, ×25. **2,** *Hypericum tetrapetalum*: **a,** habit, ×⅖; **b,** leaf, ×1, with detail showing punctate surface, ×6; **c,** two views of flower, ×1⅓; **d,** cross section of ovary, ×12. **3,** *Clusia rosea*: **a,** carpellate flowering branch, ×⅔; **b,** cross section of node (one-trace unilacunar), ×3½; **c,** capsule, ×½; **d,** seed (note aril), ×3½.

Greuter, W., H. M. Burdet, W. G. Chaloner, V. Demoulin, R. Grolle, et al., eds. 1988. *International code of botanical nomenclature*, p. 23. Koeltz Scientific Books, Königstein.

Hutchinson, J. 1969. *Evolution and phylogeny of flowering plants*, pp. 314–321. Academic Press, London.

Lawrence, G. H. M. 1951. *Taxonomy of vascular plants*, pp. 603–605. Macmillan, New York.

Maguire, B. 1976. Apomixis in the genus *Clusia* (Clusiaceae)—A preliminary report. *Taxon* 25:241–244.

Vestal, P. A. 1937. The significance of comparative anatomy in establishing the relationship of the Hypericaceae to the Guttiferae and their allies. *Philipp. J. Sci.* 64:199–256.

Wood, C. E., and P. Adams. 1976. The genera of Guttifcrae (Clusiaceae) in the southeastern United States. *J. Arnold Arbor.* 57:74–90.

ERICACEAE
Heath Family
Including the Pyrolaceae and Monotropaceae

Shrubs to small trees or sometimes perennial herbs, usually growing on acid soil, sometimes parasitic; roots mycorrhizal. **Leaves** simple, entire to serrate, with flat to recurved margins, usually alternate, usually coriaceous and persistent, exstipulate. **Inflorescence** usually indeterminate, racemose, paniculate, corymbose, or fasciculate, or flowers solitary, axillary or terminal. **Flowers** actinomorphic or slightly zygomorphic, perfect, hypogynous to epigynous, generally showy, with hypogynous or epigynous nectariferous disc. **Calyx** usually synsepalous with typically 4 or 5 lobes, imbricate to valvate, persistent. **Corolla** usually sympetalous with usually 4 or 5 lobes, funnelform, campanulate, to urceolate, usually white or brightly colored and conspicuous, convolute or imbricate. **Androecium** of usually 8 or 10 stamens, biseriate with the outer whorl opposite the petals, inserted on edge of disc; filaments usually distinct, flattened, dilated, or S-shaped, sometimes spurred at or near anther junction; anthers appearing dorsifixed, often with terminal awns or tubules, dehiscing by pores or sometimes longitudinally, appearing introrse or apical due to inversion during development; pollen grains in tetrads. **Gynoecium** of 1 pistil, usually 4- or 5-carpellate; ovary superior to inferior, 5-locular; ovules numerous, anatropous to campylotropous, placentation axile on protruding placentae; style 1, conical to filiform, inserted in depression at ovary apex; stigma 1, capitate and often lobed. **Fruit** a loculicidal or septicidal capsule, berry, or drupe; seeds small, sometimes winged; endosperm copious, fleshy; embryo small, straight.

Family characterization: Acidophilic shrubby plants with mycorrhizal roots; coriaceous, evergreen, alternate, and simple leaves; sympetalous, 4- or 5-merous, campanulate to urceolate corolla; 8 or 10 biseriate, often appendaged stamens with outer whorl opposite the petals; inverted anthers often dehiscing by "apical" pores; pollen in tetrads; ovary with numerous ovules and axile placentation on ovary wall intrusions; and single style inserted into depressed ovary apex. Iridoid compounds and various di- and triterpenoid compounds present. Tissues with calcium oxalate crystals. Anatomical features: unitegmic and tenuinucellate ovules; 1-trace unilacunar nodes (Fig. 33: 1b); and a radiating stylar canal (thc rays continuous with each locule; Fig. 33: 1e).

Genera/species: 99/2,245

Distribution: Widespread on acidic soils throughout temperate zones; centers of diversity in southern Africa and western China. Also conspicuous in the Arctic and in the montane tropics; uncommon in Australia.

Major genera: *Rhododendron* (500–850 spp.), *Erica* (500–665 spp.), *Vaccinium* (300–450 spp.), *Gaultheria* (150–200 spp.), and *Cavendishia* (100 spp.)

U.S./Canadian representatives: 44 genera/219 spp.; largest genera: *Arctostaphylos*, *Vaccinium*, and *Rhododendron*

Economic plants and products: Fruit from *Vaccinium* spp. (blueberries and cranberries). Toxic resins (diterpenes or andromedo-toxin) in many, such as *Kalmia* (lambkill, mountain-laurel). Medicinal plants: *Ledum* (Labrador-tea) and *Gaultheria* (lemonleaf—with oil of wintergreen: high methylsalicylate content). Ornamental trees and shrubs (species of 40+ genera), including *Andromeda* (bog-rosemary), *Arctostaphylos* (bearberry, manzanita), *Calluna* (heather), *Erica* (heath), *Kalmia* (mountain-laurel), *Leucothoë* (fetterbush), *Lyonia* (staggerbush), *Pieris*, and *Rhododendron* (azalea, rhododendron).

Commentary: The intrafamilial classification of the Ericaceae has been a subject of dispute. The genera often are either small and distinct or large and complex (see Wood 1961). The delineation of the family also is not consistent in the literature, with the subfamilies Vaccinioideae, Monotropoideae, and Pyroloideae often raised to familial status; however, Thorne (1992) maintains these taxa as subfamilies. Table 4 summarizes the characters used to separate the subfamilies as delineated by Stevens (1971), who includes the three, sometimes segregated, taxa. The traditional Vaccinioideae (or Vacciniaceae) have been distinguished from the rest of the Ericaceae by the partially to completely inferior ovary (Fig. 33: 2c) and the baccate fruit (Fig. 33: 2d). Stevens's expanded concept of this group in-

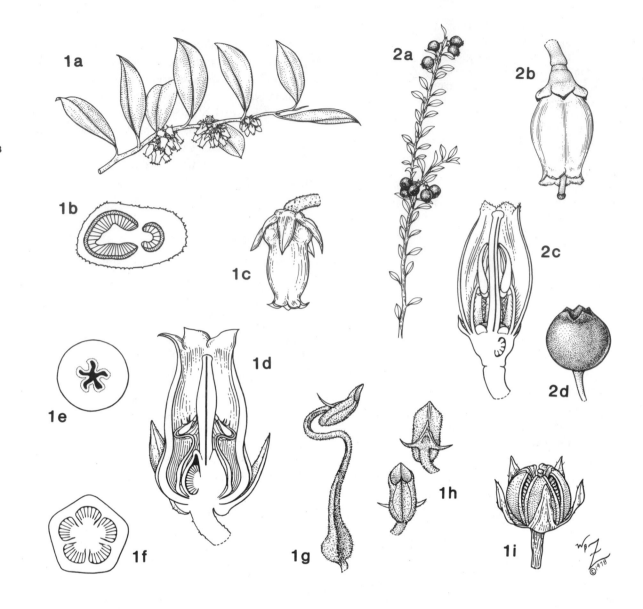

Figure 33. Ericaceae. 1, *Lyonia lucida*: **a,** flowering branch, ×½; **b,** cross section of node (one-trace unilacunar), ×9; **c,** flower, ×3; **d,** longitudinal section of flower, ×6; **e,** cross section of style showing stylar canal, ×18; **f,** cross section of ovary, ×12; **g,** stamen, ×12; **h,** two views of anther, ×12; **i,** capsule, ×4. **2,** *Vaccinium myrsinites*: **a,** fruiting branch, ×½; **b,** flower, ×4; **c,** longitudinal section of flower, ×6; **d,** berry, ×3.

cludes some members with superior ovaries (the Arbutoideae), which have numerous features in common with Vaccinioideae *s.s.* The Monotropoideae and Pyroloideae have been separated from the Ericaceae *s.s.* by characters including herbaceous habit, apopetalous corolla, loculicidal capsule, and reduced embryo. The Monotropoideae (see Fig. 34: 3a–d) are further specialized, with an achlorophyllous herbaceous habit, scale-like leaves, and extremely minute embryos. According to Wallace (1976), numerous transitional features indicate that the distinctive characters of these two groups are due to the parasitic

dependence on mycorrhizal fungi, and that the Monotropoideae represent the culmination of fungal symbiont relationships found throughout the family.

Recent cladistic studies of morphological, phytochemical, and molecular characters (Anderberg 1992, 1993; Judd and Kron 1993; Kron and Chase 1993) have examined the subfamilies and the phylogenetic position of the family. The currently accepted five subfamilies may not all be monophyletic, and the hypothesis of all these authors indicates that the Ericaceae, as defined here, may be paraphyletic. They suggest the inclusion of the closely related Epacridaceae

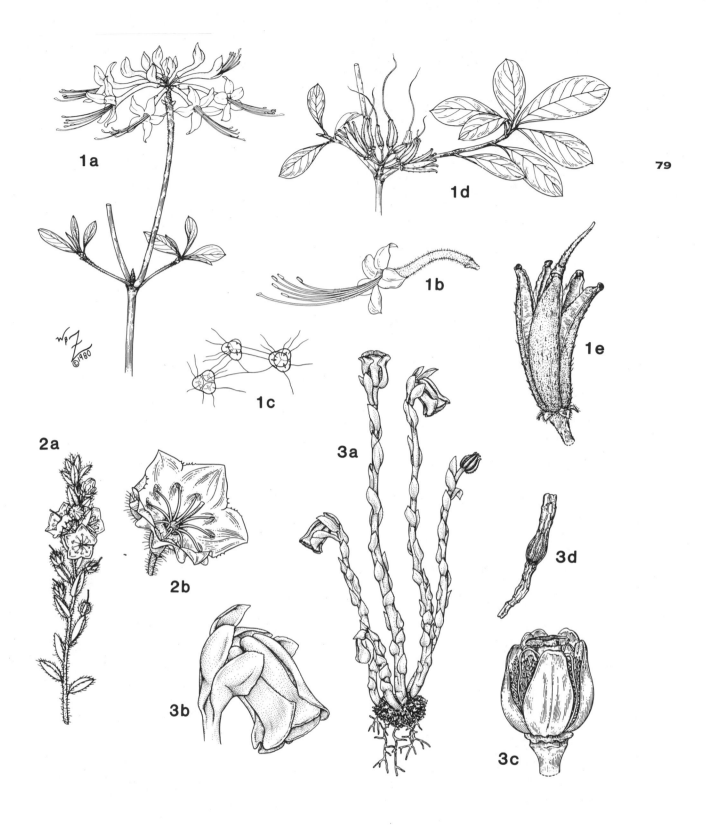

Figure 34. Ericaceae (continued). 1, *Rhododendron canescens*: **a,** flowering branch (with immature leaves), ×½; **b,** flower, ×1; **c,** pollen grains (in tetrads) and viscin strands, ×150; **d,** fruiting branch with mature leaves, ×½; **e,** capsule, ×2½. **2,** *Kalmia hirsuta*: **a,** flowering branch, ×¾; **b,** flower, ×2⅓. **3,** *Monotropa uniflora*: **a,** habit, ×½; **b,** flower, ×1¾; **c,** capsule (persistent perianth and androecium removed), ×3; **d,** seed, ×30.

TABLE 4. Morphological characters used to distinguish the subfamilies of the Ericaceae. Features from Stevens (1971).

CHARACTER	ERICOIDEAE	VACCINIOIDEAE (INCL. ARBUTOIDEAE) ("VACCINIACEAE")	RHODODENDROIDEAE	PYROLOIDEAE ("PYROLACEAE")	MONOTROPOIDEAE ("MONOTROPACEAE")
GENERA/SPECIES	17/865	54/660	15/700	3/10	10/12
REPRESENTATIVE GENERA	*Calluna, Erica*	*Arbutus, Gaultheria, Lyonia, Vaccinium*	*Andromeda, Kalmia, Rhododendron*	*Chimaphila, Pyrola*	*Monotropa*
HABIT	shrubs to small trees	shrubs, woody vines, or small trees	shrubs to small trees	herbs (sometimes achlorophyllous)	fleshy herbs (achlorophyllous)
LEAVES	scale-like or needle-like ("ericoid")	flat and well-developed	flat and well-developed, or occasionally ericoid	well-developed, or reduced to scales	scale-like
FLOWERS	± pendulous	pendulous	pendulous or erect	± pendulous	pendulous
COROLLA	urceolate to tubular, sometimes campanulate / persistent (in fruit)	usually urceolate / deciduous	tubular, campanulate, to funnelform, occasionally urceolate / deciduous	± rotate (petals ± distinct) / ± deciduous	urceolate (petals distinct or fused) / deciduous or persistent
ANTHER APPENDAGES	present (flattened spurs) or absent	present (awns, spurs)	absent	absent	absent, or sometimes present (spurs)
OVARY POSITION	superior	superior or inferior	superior	superior	superior
FRUIT TYPE	usually a loculicidal capsule	loculicidal capsule or berry	septicidal capsule	loculicidal capsule	loculicidal capsule
OTHER	all native to Old World		pollen connected by viscin strands	embryo reduced	embryo reduced

(30 genera/400 spp.) and Empetraceae (3 genera/6 spp.) for an expanded, monophyletic Ericaceae *s.l.*

Stamen characters have been stressed, particularly as taxonomic indicators within the family (Watson 1965). The filaments usually are flattened, and in some genera, such as *Lyonia*, distinctly S-shaped (Fig. 33: 1g). Prominent spurs may occur at or near the junction of the filament and anther. Ericaceous anthers usually become inverted on the filaments during development and appear introrse. Anther dehiscence is by "apical" or morphologically basal pores, or sometimes longitudinally. In *Vaccinium* and related genera, the anther apex is attenuated into long tubules terminating in pores (Fig. 33: 2c).

Various insects are attracted to the conspicuous white or brightly colored corollas, as well as to the scent and concealed nectar. Species with urceolate corollas (*Vaccinium*, *Lyonia*) often have appendaged anthers (Fig. 33: 1g); when the visiting insect reaches for the nectar at the corolla base, it brushes against the appendages and receives a shower of pollen. Some flowers (*Rhododendron*; Fig. 34: 1b) have flaring corollas, exserted stamens, unappendaged anthers, and sticky projecting stigmas. The pollen tetrads (Fig. 34: 1c) often are coalesced into masses by cobwebby viscin strands (see the commentary under Onagraceae); the entire anther contents may then be pulled out by the pollinator and deposited on the protruding stigma of the next flower. The flowers of *Kalmia* (Fig. 34: 2b) have an interesting pollination specialization with bow-like filaments and anthers tucked into deep corolla depressions: as an insect probes for nectar, it triggers the filaments. The anthers then strike against the pollinator and powder it with pollen.

REFERENCES CITED

Anderberg, A. A. 1992. The circumscription of the Ericales, and their cladistic relationships to other families of "higher" dicotyledons. *Syst. Bot.* 17:660–675.

——. 1993. Cladistic relationships and major clades of the Ericales. *Pl. Syst. Evol.* 184:207–231.

Judd, W. S., and K. A. Kron. 1993. Circumscription of Ericaceae (Ericales) as determined by preliminary cladistic analyses based on morphological, anatomical, and embryological features. *Brittonia* 45:99–114.

Kron, K. A., and M. W. Chase. 1993. Systematics of the Ericaceae, Empetraceae, Epacridaceae and related taxa based on *rbc*L sequence data. *Ann. Missouri Bot. Gard.* 80: 735–741.

Stevens, P. F. 1971. A classification of the Ericaceae: Subfamilies and tribes. *Bot. J. Linn. Soc.* 64:1–53.

Thorne, R. F. 1992. Classification and geography of the flowering plants. *Bot. Rev.* (Lancaster) 58:225–348.

Wallace, G. D. 1976. Interrelationships of the subfamilies of the Ericaceae and the derivation of the Monotropoideae. *Bot. Not.* 128:286–298.

Watson, L. 1965. The taxonomic significance of certain anatomical variations among Ericaceae. *J. Linn. Soc., Bot.* 59: 111–126.

Wood, C. E. 1961. The genera of Ericaceae in the southeastern United States. *J. Arnold Arbor.* 42:10–80.

SAPOTACEAE

Sapodilla or Sapote Family

Trees or shrubs, with milky sap. **Leaves** simple, entire, alternate, sometimes pseudoverticillate, coriaceous, usually exstipulate. **Inflorescence** determinate, cymose with flowers often in fascicles and appearing umbellate, or sometimes flower solitary, axillary, sometimes cauliflorous and occurring at nodes on old wood. **Flowers** actinomorphic, usually perfect, hypogynous, small. **Calyx** of often 5 (uniseriate) or 4, 6, or 8 (biseriate) sepals, distinct or basally connate, imbricate. **Corolla** sympetalous, usually with as many lobes as sepals, sometimes with paired petaloid appendages, often white or cream-colored, imbricate. **Androecium** of basically 8 to 15 stamens in 2 or 3 whorls of 4 or 5, usually with the outer whorl(s) reduced to staminodes (often petaloid) or absent, epipetalous; filaments distinct; anthers basifixed, dehiscing longitudinally, extrorse. **Gynoecium** of 1 pistil, usually 4- or 5-carpellate; ovary superior, usually 4- or 5-locular, usually hirsute; ovules 1 in each locule, anatropous (to hemitropous), placentation axile or axile-basal; style 1; stigma 1, capitate or slightly lobed, inconspicuous. **Fruit** a berry, often with a thin, leathery to bony outer layer; seeds 1 to few, large, often with thick and hard seed coat and conspicuous hilum (scar); endosperm scanty to copious, oily, fleshy, or hard, or absent; embryo large.

Family characterization: Trees or shrubs with laticiferous parts; thick, sympetalous corolla, each lobe often with paired appendages; 2 or 3 whorls of epipetalous stamens and petaloid staminodes; a berry as the fruit type; and seeds with hard and thick testa and large hilum. Steroids, terpenoids, and tannins commonly present (Waterman and Mahmoud 1991). Tissues with calcium oxalate crystals. Anatomical features: well-developed latex-sacs (in leaves, bark, and pith); vesture of unicellular 2-armed hairs ("malpighian hairs"; Fig. 35: c; sometimes one branch suppressed); 3-trace trilacunar nodes (Fig. 35: b); and unitegmic and tenuinucellate ovules.

Genera/species: 53/1,100 (Pennington 1991)

Distribution: Pantropical; especially in lowland and montane rain forests.

Major genera: *Pouteria* (325 spp.), *Palaquium* (110 spp.), *Planchonella* (100 spp.), *Madhuca* (100 spp.), *Side-*

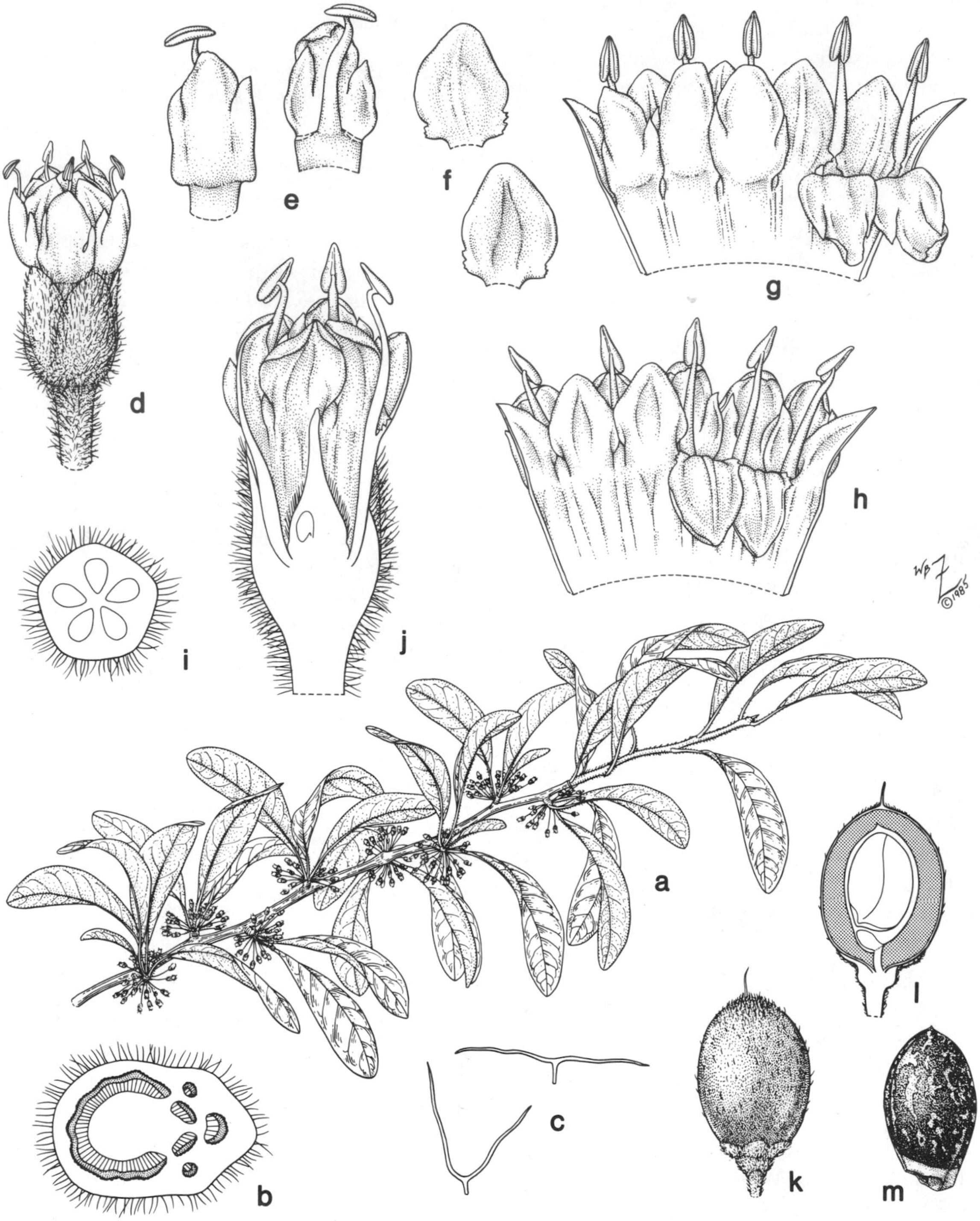

Figure 35. Sapotaceae. *Sideroxylon lanuginosum* (*Bumelia lanuginosa*): **a,** flowering branch, ×⅓; **b,** cross section of node (three-trace trilacunar), ×12; **c,** malpighian hairs from abaxial surface of leaf, ×25; **d,** flower, ×7; **e,** two views of corolla lobe and adnate stamen, ×12; **f,** two views of staminode, ×12; **g,** outer (abaxial) surface of expanded corolla and androecium with two corolla lobes folded down, ×10; **h,** inner (adaxial) surface of expanded corolla and androecium with two staminodia folded down, ×10; **i,** cross section of ovary, ×25; **j,** longitudinal section of flower, ×12; **k,** berry, ×3; **l,** longitudinal section of berry, ×3; **m,** seed, ×3½.

roxylon (74 spp.), and *Chrysophyllum* (71 spp.) (Pennington 1991)

U.S./Canadian representatives: 9 genera/39 spp.; largest genera: *Pouteria*, *Sideroxylon* (incl. *Bumelia*), and *Manilkara*

Economic plants and products: Chicle (essential chewing gum ingredient) from *Manilkara* (= *Achras*). Gutta-percha (rubbery compounds) from several, such as species of *Mimusops*, *Palaquium*, and *Payena*. Edible fruits from *Chrysophyllum* (star-apple), *Manilkara* (sapodilla), and *Pouteria* (eggfruit or canistel; incl. *Calocarpum*, mamey sapote). Edible oils (from seeds) of *Madhuca* and *Vitellaria* (incl. *Butyrospermum*, sheabutter). Very durable timber from *Manilkara* (bulletwood), *Mimusops* (cherry-mahogany), and *Sideroxylon* (ironwood). Ornamental plants (species of 10 genera), including *Chrysophyllum* (satinleaf), *Manilkara*, *Pouteria*, and *Sideroxylon* (incl. *Bumelia*, buckthorn; incl. *Dipholis*).

Commentary: The delimitation of the genera (35–75) and species (600–1,100) of the Sapotaceae has varied considerably, with various authors splitting and/or combining groups, resulting in confusing nomenclature (see Wood and Channell 1960 and Pennington 1991 for historical summaries). Seed characters (presence of endosperm, morphology of hilum) and the presence of staminodia and/or of appendages on the corolla are important features for separating genera. According to Pennington (1991), however, very few genera are defined by unique characters.

The perianth is extremely variable within the family (Pennington 1991). For example, the sepals vary in number and arrangement: five and uniseriate (*Sideroxylon*, incl. *Bumelia*), or biseriate with two (*Pouteria*), three (*Manilkara*), or four (*Mimusops*) sepals in each whorl. The number of corolla lobes usually is equal to that of the sepals. In addition, two lateral petaloid appendages may occur on each corolla lobe (Fig. 35: e), as well as petaloid staminodes that alternate with the lobes (Fig. 35: f,h).

The pollination biology of the family has not been thoroughly examined. In several species, such as *Sideroxylon celastrinum* (= *Bumelia celastrina*) and *Sideroxylon salicifolium* (= *Dipholis salicifolia*), the receptive stigma protrudes before the flower opens and before the anthers dehisce (Tomlinson 1980). In others (*S. reclinatum*, = *Bumelia reclinata*), the style remains enclosed by the corolla appendages, which diverge later to expose the stigma after the pollen is shed. Flowers in some groups (such as *Chrysophyllum* species) are sweet-scented. Many flowers are nocturnal and bat-pollinated.

The distinctive seed of the family (Fig. 35: m) has a hard, smooth, and glossy seed coat (usually brown) that contrasts with the pale, rough-textured hilum, the scar representing the point of attachment of the seed to the ovary wall. Hilum shape and position are used for distinguishing genera.

REFERENCES CITED

Pennington, T. D. 1991. *The genera of Sapotaceae*, pp. 75–261. Royal Botanic Gardens, Kew, and New York Botanical Gardens, Bronx.

Tomlinson, P. B. 1980. *The biology of trees native to tropical Florida*, pp. 382–396. "Publ. by the author," Petersham, Mass.

Waterman, P. G., and E. N. Mahmoud. 1991. Chemical taxonomy of the Sapotaceae: Patterns in distribution of some simple phenolic compounds. In T. D. Pennington, *The genera of Sapotaceae*, pp. 51–74. Royal Botanic Gardens, Kew, and New York Botanical Gardens, Bronx.

Wood, C. E., and R. B. Channell. 1960. The genera of Ebenales in the southeastern United States. *J. Arnold Arbor.* 41:1–35.

MYRSINACEAE
Myrsine Family

Trees, shrubs, or sometimes woody vines. **Leaves** simple, entire or less commonly serrate, alternate, coriaceous, punctate and with linear secretory lines, often with glandular trichomes, usually persistent, exstipulate. **Inflorescence** determinate, cymose with flowers often in fascicles or appearing paniculate or corymbose, axillary or terminal. **Flowers** actinomorphic, perfect or sometimes imperfect (then plants dioecious or polygamodioecious), usually hypogynous, often small. **Calyx** of typically 5 sepals, distinct to basally connate, usually glandular dotted and streaked, persistent, imbricate, convolute or valvate. **Corolla** sympetalous with typically 5 lobes, rotate to shortly salverform, greenish or white to pinkish-red, usually glandular dotted and streaked, imbricate or convolute. **Androecium** of 5 stamens opposite the corolla lobes, epipetalous; filaments distinct or occasionally basally connate; anthers basifixed, dehiscing longitudinally or by apical slits or pores, introrse. **Gynoecium** of 1 pistil, usually 4- or more often 5-carpellate; ovary usually superior, 1-locular; ovules 1 to several, anatropous to hemitropous or nearly campylotropous, embedded in an expanded free-central placenta that fills the locule; style often short; stigma capitate, simple or lobed. **Fruit** usually a drupe with a 1- to few-seeded pyrene, often glandular dotted and streaked; seeds small, dark brown to black; endosperm copious, oily, fleshy or hard; embryo cylindrical, straight or slightly curved, embedded in endosperm.

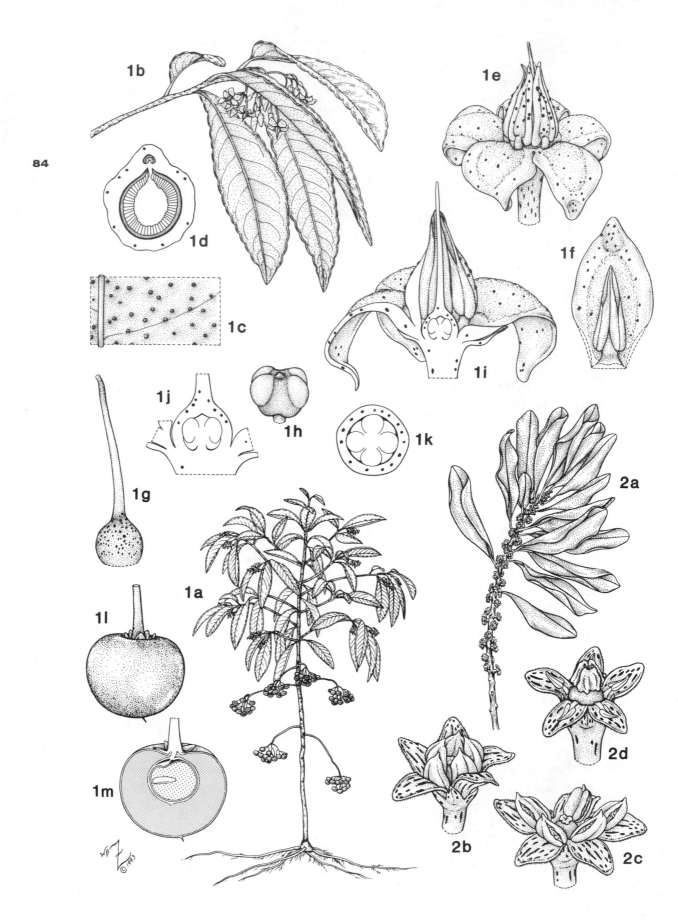

Family characterization: Trees or shrubs with resinous dotted and streaked vegetative and floral parts; persistent, coriaceous, punctate leaves with peltate or capitate glandular trichomes; 5 epipetalous stamens opposite the 5 fused petals; hypogynous 1-locular ovary with ovules embedded in an enlarged placenta that fills the locule; and a drupe as the fruit type. Saponins and tannins present. Tissues with calcium oxalate crystals. Anatomical features: schizogenous secretory ducts, cavities, or canals containing resinous material (appearing as translucent, reddish or black dots and streaks) in various plant parts (stems, leaves, flower parts); tenuinucellate ovules; and unilacunar nodes (Fig. 36: 1d).

Genera/species: 33/1,000

Distribution: Widely distributed in tropical to warm-temperate areas.

Major genera: *Ardisia* (250–400 spp.), *Myrsine* (*Rapanea, Suttonia*; 155–207 spp.), *Maesa* (100–200 spp.), *Embelia* (130 spp.), *Cybianthus* (40–124 spp.), and *Oncostemum* (100 spp.)

U.S./Canadian representatives: 8 genera/35 spp.; largest genera: *Myrsine* and *Ardisia*

Economic plants and products: Ornamental plants, including species of: *Ardisia* (coral-berry, marlberry), *Maesa,* and *Myrsine* (*Rapanea, Suttonia*).

Commentary: The Myrsinaceae often are divided into two subfamilies, the large Myrsinoideae (superior ovary, one-seeded fruit) and the monotypic Maesoideae (inferior to half-inferior ovary, many-seeded fruit). The former subfamily is divided further into two tribes based upon ovule number and arrangement in the placenta (Channell and Wood 1959). The genus *Aegiceras*, comprising two species of mangroves with septate anther chambers, sometimes has been separated as a segregate family, and in some classification schemes the closely related Theophrastaceae (*Theophrastus* and allies) are included within the Myrsinaceae. Both of these families (of the order Primulales) are characterized by usually a five-merous perianth, epipetalous stamens, and a superior, four- or five-carpellate ovary with free-central placentation (Frohne and John 1978).

Vegetatively, the Myrsinaceae are characterized by the glandular, reddish or blackish dots, streaks, or lines found in most plant parts (e.g., stems, leaves, perianth, ovary, fruit; see Fig. 36: 1b–g,i–k; 2b–d). These macroscopically visible areas represent secretory cells, cavities, or canals that contain yellow to reddish-brown resinous substances. When held up to the light, leaves may appear translucent dotted (punctate; Tomlinson 1980). Glandular trichomes, often peltate or capitate (heads multicellular), also are common on the leaf surface (Fig. 36: 1c).

The distinctive placentation type of the family consists of one to several ovules embedded in a fleshy, proliferated placenta (Fig. 36: 1h) that is attached to the base of the unilocular ovary (thus, "free-central" placentation; Fig. 36: 1j,k). The large placenta completely fills the locule. Most ovules abort as the ovary matures, resulting in a typically one-seeded pyrene in the drupe (Fig. 36: 1m).

Little has been reported on the pollination biology of the flowers of the family, which are entomophilous. In *Ardisia*, the large, yellow, upright anthers (Fig. 36: 1e) contrast in color with the petals (white, pink to purple), and the flowers are fragrant. Flowers are perfect (at least in *A. escallonioides*) and probably protogynous (Tomlinson 1980). Before anthesis, the presumably receptive stigma protrudes from the tube formed by the erect anthers and closed corolla; pollen is released from the anthers later when the flower opens. Other members of the family are characterized by imperfect flowers. For example, plants of *Myrsine* basically are dioecious (Fig. 36: 2b–d), although perfect flowers also may be present. The dehisced anthers of the staminate flowers become exposed as the petals spread (Fig. 36: 2b,c).

REFERENCES CITED

Channell, R. B., and C. E. Wood. 1959. The genera of the Primulales of the southeastern United States. *J. Arnold Arbor.* 40:268–288.

Frohne, D., and J. John. 1978. The Primulales: Serological contributions to the problem of their systematic position. *Biochem. Syst. & Ecol.* 6:315–322.

Tomlinson, P. B. 1980. *The biology of trees native to tropical Florida,* pp. 260–264. "Publ. by the author," Petersham, Mass.

Figure 36. Myrsinaceae. 1, *Ardisia crenata*: **a,** plant with flowers and fruit, ×1/7; **b,** flowering branch, ×2/3; **c,** adaxial side of leaf showing punctate surface, ×5; **d,** cross section of node (one-trace unilacunar), ×10; **e,** flower, ×6; **f,** petal and epipetalous stamen, ×7; **g,** pistil, ×10; **h,** placenta with ovules, ×20; **i,** longitudinal section of flower, ×9; **j,** longitudinal section of ovary, ×15; **k,** cross section of ovary, ×18; **l,** drupe, ×2½; **m,** longitudinal section of drupe, ×2¾. **2,** *Myrsine floridana* (*Rapanea punctata*): **a,** staminate flowering branch, ×2/5; **b,** staminate flower at anthesis, ×6; **c,** fully expanded staminate flower, ×6; **d,** carpellate flower, ×8.

POLYGONACEAE

Buckwheat, Knotweed, or Smartweed Family

Mainly annual or perennial herbs or shrubs, sometimes vines; stems with swollen nodes. **Leaves** simple, usually entire, usually alternate, with a sheathing membranous stipule (ocrea) at petiole base. **Inflorescence** determinate, cymose and often appearing racemose, paniculate, spicate, or capitate, terminal or axillary. **Flowers** actinomorphic, usually perfect, hypogynous, small, subtended by persistent bract(s). **Perianth** of usually 5 or 6 tepals (5 from the union of 1 tepal from each whorl), biseriate, distinct or basally connate, typically petaloid, white, greenish, pinkish, or reddish, often with wings, spines, or hooks, imbricate, persistent, often becoming enlarged and accrescent or membranous in fruit. **Androecium** of 6 to 9 stamens, biseriate, generally paired (those opposite the inner tepals longer); filaments distinct or basally connate, sometimes basally adnate to perianth, sometimes alternating with nectariferous glands; anthers versatile or basifixed, dehiscing longitudinally, introrse (outer whorl) or extrorse (inner whorl). **Gynoecium** of 1 pistil, usually 3- or sometimes 2-carpellate; ovary superior, 1-locular; ovule solitary, orthotropous, placentation basal, style(s) 1 to 3; stigmas capitate or feathery. **Fruit** an achene, lenticular (2 carpels) or trigonous (3 carpels), often enclosed by persistent, membranous tepals or accrescent hypanthium; endosperm copious, mealy or horny; embryo curved or straight.

Family characterization: Herbaceous to shrubby plants with swollen nodes and sheathing stipules (ocreae); small flowers with 5 or 6 petaloid tepals; dense cymose inflorescences; 1-locular ovary with a solitary basal ovule; and an achene subtended by persistent perianth as the fruit type. Oxalic acid (detectable by the sour taste of the sap) and tannins present. Tissues with calcium oxalate crystals.

Genera/species: 49/1,100

Distribution: Mainly in northern temperate regions; a few genera in tropical and subtropical areas.

Major genera: *Polygonum* (150–300 spp.), *Rumex* (200 spp.), *Eriogonum* (150–200 spp.), and *Coccoloba* (150 spp.)

U.S./Canadian representatives: 24 genera/446 spp.; largest genera: *Eriogonum*, *Polygonum*, and *Rumex*

Economic plants and products: Food plants: *Fagopyrum* (buckwheat), *Rheum* (rhubarb), and *Rumex* (dock, sorrel). Jelly from fruits of *Coccoloba* spp. Orna-mental plants (species of 15 genera), including *Antigonon* (coral vine), *Coccoloba* (sea-grape), and *Homalocladium* (ribbonbush).

Commentary: Although the interfamilial relationships and some generic delineations have been disputed (see Graham and Wood 1965; Nowicke and Skvarla 1977), the Polygonaceae comprise a very homogeneous and apparently monophyletic family with many distinctive vegetative and floral features. One striking character is the conspicuous membranous sheath at the petiole base called the ocrea (ochrea; Fig. 37: 1b), which is, at least partially, stipular in origin (Mitra 1945). Polygonaceous floral morphology also has received much attention (Laubengayer 1937). The basic floral plan (*Rumex*, *Eriogonum*) is three-merous, while the pentamerous perianth of some species of *Polygonum* and *Polygonella* is derived by the fusion of a tepal from the outer whorl with one from the inner. The origin of this tepal is evident from its intermediate position (one edge aligned with the inner whorl, and one edge, with the outer; Fig. 37: 1f) and often also from its two veins.

The flowers of the Polygonaceae generally are entomophilous (bees and flies). Many have attractive petaloid tepals and nectar (secreted by glands at the filament bases, as in some *Polygonum* species; Fig. 37: 1g); others also have a strong scent, as in *Coccoloba* flowers that have a fetid odor. Anemophily also occurs in the family, as in species of *Rumex*, which have pendulous greenish flowers with large feathery stigmas (Fig. 37: 2b).

The persistent tepals often aid in fruit dispersal. In *Rumex*, for example, the inner whorl enlarges and forms three membranous wings (Fig. 37: 2c) that often are armed with hooks. In *Coccoloba*, the achene is surrounded completely by an accrescent perianth (Fig. 37: 3a,b).

REFERENCES CITED

Graham, S. A., and C. E. Wood. 1965. The genera of Polygonaceae in the southeastern United States. *J. Arnold Arbor.* 46:91–121.

Laubengayer, R. A. 1937. Studies in anatomy and morphology of the polygonaceous flower. *Amer. J. Bot.* 24:329–343.

Mitra, G. C. 1945. The origin, development and morphology of the ochrea in *Polygonum orientale* L. *J. Indian Bot. Soc.* 24:191–200.

Nowicke, J. W., and J. J. Skvarla. 1977. Pollen morphology and the relationship of the Plumbaginaceae, Polygonaceae, and Primulaceae to the order Centrospermae. *Smithsonian Contr. Bot.* 37:1–64.

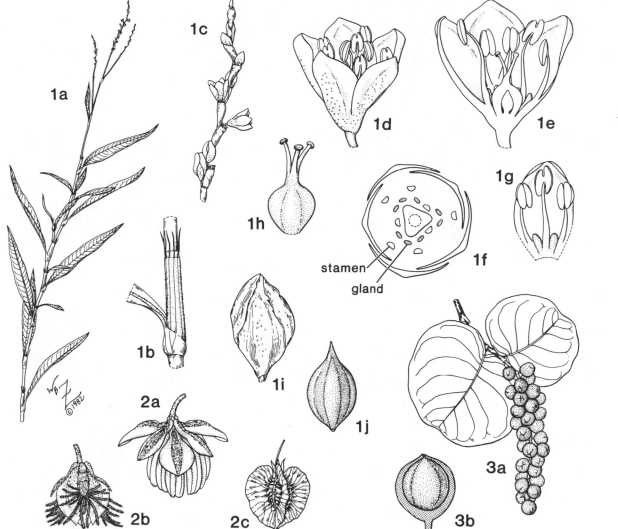

stamen

gland

Figure 37. Polygonaceae. 1, *Polygonum punctatum* (*Persicaria punctata*): **a,** habit, ×¼; **b,** node showing ocrea, ×1½; **c,** inflorescence, ×2⅓; **d,** flower, ×10; **e,** longitudinal section of flower, ×12; **f,** cross section through middle of bud, ×15; **g,** one inner tepal with portion of androecium and basal glands, ×12; **h,** pistil, ×17; **i,** achene with persistent perianth, ×9; **j,** achene, ×9. **2,** *Rumex hastatulus*: **a,** staminate flower, ×10; **b,** carpellate flower, ×14; **c,** achene with winged perianth, ×6. **3,** *Coccoloba uvifera*: **a,** fruiting branch, ×⅕; **b,** achene with half of persistent perianth removed, ×1. For floral formula of *Polygonum punctatum*, see Fig. 4: 1c.

CELASTRACEAE

Bittersweet or Staff Tree Family
Including the Hippocrateaceae

Trees, shrubs, or woody vines. **Leaves** simple, entire to serrate, opposite or alternate, often coriaceous, deciduous or persistent, stipulate (stipules minute and caducous) or exstipulate. **Inflorescence** determinate, cymose and sometimes appearing fasciculate, racemose, or paniculate, or sometimes flowers solitary, terminal or axillary. **Flowers** actinomorphic, perfect or sometimes imperfect (then plants monoecious, dioecious, or polygamodioecious), hypogynous, perigynous, to sometimes half-epigynous (sometimes appearing epigynous due to adnation of disc to ovary), small, with well-developed intrastaminal or extrastaminal disc (often adnate to ovary). **Calyx** of 4 or 5 sepals, distinct to basally connate, generally persistent, usually imbricate. **Corolla** of 4 or 5 petals, distinct, spreading to reflexed, greenish to white, usually imbricate. **Androecium** of 3 to 5 stamens, inserted on, within, or under disc; filaments distinct to sometimes

88

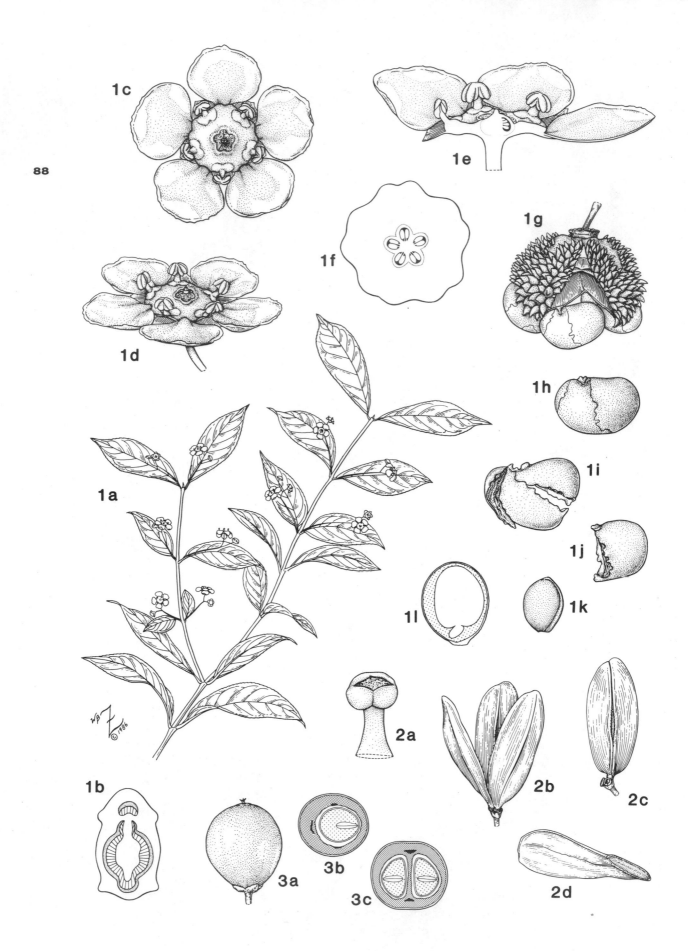

basally connate; anthers basifixed or dorsifixed, dehiscing longitudinally or transversely, introrse or sometimes extrorse; staminodes sometimes present in carpellate flowers. **Gynoecium** of 1 pistil, 2- to 5-carpellate; ovary superior to half-inferior (or sometimes appearing inferior due to adnation to disc), with as many locules as carpels; ovules 2 to numerous in each locule, anatropous, erect or sometimes pendulous, placentation axile; style short; stigma capitate, often 2- to 5-lobed; rudimentary pistil present in staminate flowers. **Fruit** usually a loculicidal capsule or a drupe; seeds often arillate or winged; endosperm copious to sometimes absent, oily; embryo large, straight, surrounded by endosperm.

Family characterization: Trees, shrubs, or lianas; small greenish flowers with 4- or 5-merous perianth and androecium; ovary surrounded by and/or adnate to conspicuous glandular disc; a capsule or a drupe as the fruit type; and seeds (of capsules) covered with brightly colored (red or orange) arils. Tissues commonly with tannins and calcium oxalate crystals. Anatomical features: laticiferous canals or sacs in phloem or parenchymatous tissues of stems and leaves; and unilacunar nodes (Fig. 38: 1b).

Genera/species: 55/855

Distribution: Mainly tropical to subtropical and extending into temperate regions.

Major genera: *Maytenus* (225 spp.), *Salacia* (150–200 spp.), *Euonymus* (177 spp.), *Hippocratea* (1–120 spp.), and *Cassine* (40–80 spp.)

U.S./Canadian representatives: 13 genera/32 spp.; largest genera: *Euonymus*, *Maytenus*, and *Mortonia*

Economic plants and products: Carving wood and rubber-like latex from several *Euonymus* species (spindle tree). Several medicinal plants, such as species of *Catha* (Arabian-tea and khat from leaves and twigs), *Celastrus* (medicinal oil from seeds), and *Maytenus* (medicinal substances from bark). Ornamental plants (species of at least 13 genera), including *Cassine*, *Catha*, *Celastrus* (bittersweet), *Euonymus* (spindle tree), *Maytenus*, and *Schaefferia*.

Commentary: The circumscription of the Celastraceae has varied somewhat depending on the authority,

with several genera sometimes segregated as small, often monotypic, families (e.g., Goupiaceae and Lophopyxidaceae; Thorne 1992). The family is here defined rather broadly (five subfamilies); noteworthy is the inclusion of the subfamily Hippocrateoideae (4 genera/100 spp.), distinguished by transverse anther dehiscence (Fig. 38: 2a) and nonarillate and nonendospermous seeds (Fig. 38: 2d). Authors who recognize the Hippocrateaceae readily acknowledge the very close relationship to the Celastraceae *s.s.* (Brizicky 1964). Generic circumscription within the family is problematical, with several of the large genera (e.g., *Maytenus*, *Hippocratea*, *Salacia*) variously divided into many other segregate genera (den Hartog and Baas 1978).

The most conspicuous feature of the small flowers is the fleshy nectariferous disc that is flat to cup-shaped, often with crenulate to deeply lobed margins (Fig. 38: 1c–f). The superior ovary, variously fused to the disc, may appear inferior, especially when almost completely immersed (as in *Euonymus*, Fig. 38: 1e; *Hippocratea*). Similar disc positions occur in the flowers of the Rhamnaceae (but stamens are opposite the petals in that family). The dilated bases of the stamen filaments often are inserted on or under the outer margin of the disc between the lobes (intrastaminal disc, as in *Euonymus*), or sometimes the stamens arise from the inner margins of the disc (extrastaminal disc, as in *Hippocratea*).

The inconspicuous, greenish-white (occasionally purplish) flowers are pollinated by various small insects (bees, flies, ants, beetles) attracted to the easily accessible nectar secreted by the disc surrounded by spreading to recurved petals. Protandry is prevalent in the family.

The fruit type of the Celastraceae varies but often is a leathery and colorful (orange, yellow, red, purple) capsule (Fig. 38: 1g) opening to expose the red arils (Fig. 38: 1h–k) that surround the seeds (e.g., *Euonymus*, *Maytenus*, *Celastrus*). The red to purple capsules of several *Euonymus* species often are also tuberculate. Colorful drupes are characteristic of many genera, such as *Gyminda* (blue-black drupe) and *Schaefferia* and *Crossopetalum* (red drupes; Fig. 38: 3a–c). The attractive drupes and arillate seeds of capsular fruits are dispersed by

Figure 38. Celastraceae. 1, *Euonymus americanus*: **a,** flowering branch, ×⅜; **b,** node (one-trace unilacunar), ×10; **c,** top view of flower, ×4½; **d,** lateral view of flower, ×4½; **e,** longitudinal section of flower, ×6; **f,** cross section of ovary and disc, ×8; **g,** capsule, ×2; **h,** cluster of arillate seeds from one locule, ×2½; **i,** cluster of seeds from one locule (separating), ×2½; **j,** seed with aril, ×2½; **k,** seed (aril removed), ×2½; **l,** longitudinal section of seed, ×4½. **2,** *Hippocratea volubilis*: **a,** stamen, ×25; **b,** capsule, ×½; **c,** dehiscing capsule valve, ×⅓; **d,** seed, ×¾. **3,** *Crossopetalum rhacoma*. **a,** drupe, ×5; **b,** cross section of drupe with one pyrene, ×5; **c,** cross section of drupe with two pyrenes, ×5.

birds. Disproportionate growth in the maturing carpels of *Hippocratea* results in an unusual modified capsule with three widely divergent, vertically compressed segments or lobes (Fig. 38: 2b,c). Each section dehisces along a median suture to release basally winged seeds that are wind dispersed (Fig. 38: 2d).

90 **REFERENCES CITED**

Brizicky, G. K. 1964. The genera of Celastrales in the southeastern United States. *J. Arnold Arbor.* 45:206–234.

den Hartog, R. M., and P. Baas. 1978. Epidermal characters of the Celastraceae sensu lato. *Acta Bot. Neerl.* 27:355–388.

Thorne, R. F. 1992. Classification and geography of the flowering plants. *Bot. Rev.* (Lancaster) 58:225–348.

MALVACEAE
Mallow Family

Annual, biennial, or perennial herbs or sometimes shrubs to small trees, with mucilaginous sap and fibrous bark; pubescence lepidote or stellate. **Leaves** simple, entire or serrate to variously lobed, alternate, usually palmately veined, stipulate. **Inflorescence** determinate, cymose, or often flowers solitary and axillary. **Flowers** actinomorphic, perfect, hypogynous, showy, frequently with an epicalyx of distinct or connate bracts. **Calyx** of 5 sepals, distinct or basally connate, valvate. **Corolla** of 5 petals, distinct, obovate, asymmetrical, adnate to the base of the staminal column, large and showy, variously colored, usually convolute. **Androecium** of numerous stamens; filaments connate into a tube (monadelphous); anthers 1-locular, reniform, dehiscing by longitudinal slits, extrorse; pollen grains large, spherical, echinulate. **Gynoecium** of 1 pistil, 2- to many-carpellate; ovary superior, 2- to many-locular with locules often forming a ring; ovules 1 to many in each locule, anatropous, usually ascending, placentation axile on the inner angle of each locule; style 1 and apically branched or as many as carpels; stigma capitate to discoid or decurrent along inner side of style branch. **Fruit** a loculicidal capsule or schizocarp; seeds often pubescent or comose; endosperm scanty, oily; embryo straight or curved.

Family characterization: Herbaceous to shrubby plants with mucilaginous sap and stellate or lepidote pubescence; stipulate leaves with palmate venation; flowers often with epicalyx (whorl of bracts); convolute corolla of 5 distinct petals; numerous monadelphous stamens with 1-locular anthers; large echinulate pollen grains; few to numerous carpels arranged in a ring; and a loculicidal capsule or a schizocarp as the fruit type. Sterculic and malvalic acids commonly present. Tissues with calcium oxalate crystals. Anatomical feature: mucilage receptacles (cells, cavities, and/or canals) in the parenchymatous tissues.

Genera/species: 75/1,000+

Distribution: Generally cosmopolitan; particularly diverse in the American tropics.

Major genera: *Hibiscus* (300 spp.), *Sida* (200 spp.), and *Pavonia* (200 spp.)

U.S./Canadian representatives: 41 genera/266 spp.; largest genera: *Hibiscus, Abutilon, Sida,* and *Sidalcea*

Economic plants and products: Cotton from seedhairs of *Gossypium.* Many fiber plants, such as *Abutilon* (China jute). Okra from *Hibiscus esculentus* (immature fruit). Ornamental plants (species of 40 genera), including *Althaea* (hollyhock), *Hibiscus* (rose-of-Sharon), *Malva* (mallow, cheeseweed), *Malvaviscus* (turk's-cap), and *Thespesia* (Portia tree).

Commentary: The circumscription of the Malvaceae and other closely allied families of the order Malvales (e.g., Bombacaceae, Tiliaceae, and Sterculiaceae) is arbitrary, and genera frequently have been shifted among them (Edlin 1935). Basically all of these families are characterized by mucilage cells/canals; stellate pubescence; palmately-veined, stipulate leaves; hypogynous, showy flowers; five-merous perianth; and numerous, fascicled to monadelphous stamens. Table 5 lists the character states traditionally used to distinguish the Malvaceae, Bombacaceae, Sterculiaceae, and Tiliaceae. In particular, the tropical Bombacaceae (Fig. 40) often have been separated from the Malvaceae on the basis of the large tree habit and "smooth" pollen. Tiliaceae and Sterculiaceae are especially difficult to circumscribe (see the commentary under these two families). Judd et al. (1994) suggest that the Bombacaceae likely are paraphyletic and probably should be merged with the Malvaceae; they additionally emphasize the need for much more critical review of the Tiliaceae and Sterculiaceae.

Many taxonomic problems arise in the intrafamilial classification of the Malvaceae, with botanists offering conflicting opinions on the delineation of tribes, genera, and species. Characters of the epicalyx are among those useful for distinguishing genera (Fryxell 1988). This calyx-like whorl (Fig. 39: 1c,k), which protects the delicate flower bud, may represent an aggregation of bracts or of stipules (Kearney 1951). The segments of the epicalyx often are subulate or lanceolate (as in Fig. 39: 1c,k), but vary from broad and foliar (*Gossypium gossypioides*) to inconspicuous (certain *Hibiscus* spp.).

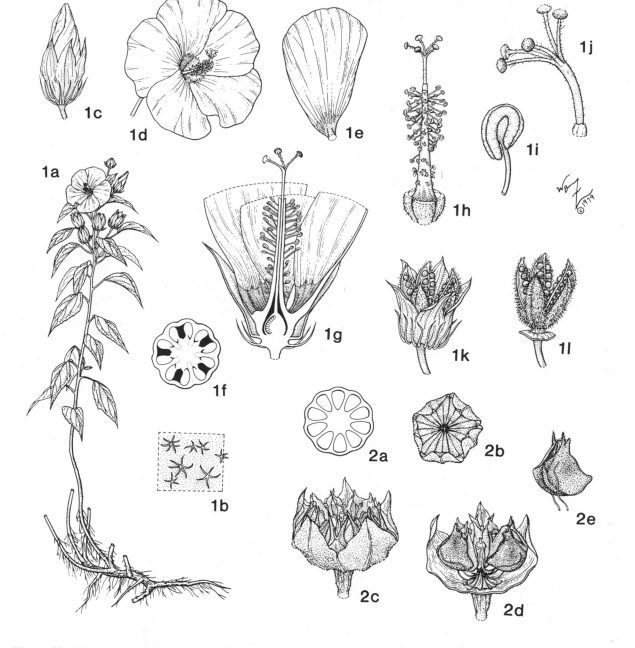

Figure 39. Malvaceae. **1,** *Hibiscus incanus* (*H. moscheutos* subsp. *incanus*): **a,** habit, ×¹⁄₁₀; **b,** stellate hairs on leaf, ×25; **c,** flower bud, ×½; **d,** flower, ×²⁄₅; **e,** petal, ×²⁄₅; **f,** cross section of ovary, ×3; **g,** longitudinal section of flower, ×³⁄₄; **h,** androecium and gynoecium, ×³⁄₄; **i,** anther, ×6; **j,** style branches and stigmas, ×1²⁄₃; **k,** capsule with persistent calyx, ×³⁄₄; **l,** capsule (calyx removed), ×³⁄₄. **2,** *Sida rhombifolia*: **a,** cross section of ovary, ×12; **b,** top view of immature schizocarp, ×3; **c,** lateral view of schizocarp with persistent calyx, ×4; **d,** schizocarp with half of calyx removed, ×4; **e,** segment of schizocarp, ×6.

The lobes of the epicalyx also vary in number (two to twenty) and degree of fusion (distinct to connate into a cup). The epicalyx is lacking in some genera.

The mode of fruit dehiscence also is used as a generic (as well as tribal) character. Genera such as *Sida* and *Malva*, for example, have loosely coherent carpels arranged in a ring (Fig. 39: 2a,b), which separate from a persistent central column at maturity (Fig. 39: 2c,d).

The common name "cheeseweed" for *Malva* refers to the resemblance of the schizocarp to a round of cheese cut into wedges. Other members of the family, such as *Hibiscus*, have loculicidal capsules (Fig. 39: 1k,l). Seed dispersal can be facilitated by apical awns on the carpels (*Sida*; Fig. 39: 2e) or hair on the seeds (some *Hibiscus* spp.).

The typical malvaceous flower is recognized easily

TABLE 5. Morphological characters used to distinguish the four poorly defined families composing the core of the order Malvales (see text for discussion).

CHARACTER	MALVACEAE	BOMBACACEAE	STERCULIACEAE	TILIACEAE
GENERA/SPECIES	75/1,000	20/180	60/700	49/450
DISTRIBUTION	cosmopolitan	tropical	almost exclusively pantropical, a few in warm-temperate areas	tropical to subtropical areas, a few in temperate regions
HABIT	herbs, or sometimes shrubs to small trees	large trees	trees, shrubs, or sometimes herbs	trees, shrubs, or occasionally herbs
LEAVES	simple (but may be palmately lobed)	simple, or palmately compound	simple, or palmately compound	simple, or sometimes palmately compound
FLOWERS	actinomorphic perfect	usually actinomorphic perfect	actinomorphic, or sometimes zygomorphic perfect or imperfect	actinomorphic usually perfect
EPICALYX	often present	often present	sometimes present	sometimes present
PETALS	present	present, or occasionally absent	often reduced to absent	present, or sometimes absent
STAMEN NUMBER	numerous	5 to numerous (often some staminodial)	5 + 5 (5 often staminodial), usually on androgynophore	usually numerous (sometimes 5 or more staminodial), usually on short androgynophore
FILAMENT FUSION	monadelphous	connate into 5 to 15 fascicles or ± monadelphous (divided tube)	monadelphous	distinct, or connate into 5 or 10 fascicles
ANTHER LOCULE NUMBER	1	1	2 (parallel or divergent)	2 (contiguous, or separated on bifurcated filament)
POLLEN	spiny	"smooth" to minutely spiny	"smooth" to spiny	"smooth"
FRUIT TYPE	loculicidal capsule, schizocarp, or occasionally indehiscent pod or berry	loculicidal capsule, or sometimes indehiscent pod	capsule, schizocarp, an aggregate of follicles, or sometimes a berry	nut, loculicidal capsule, schizocarp, or drupe
OTHER		seeds often arillate and embedded in pithy/hairy tissue	seeds sometimes arillate	

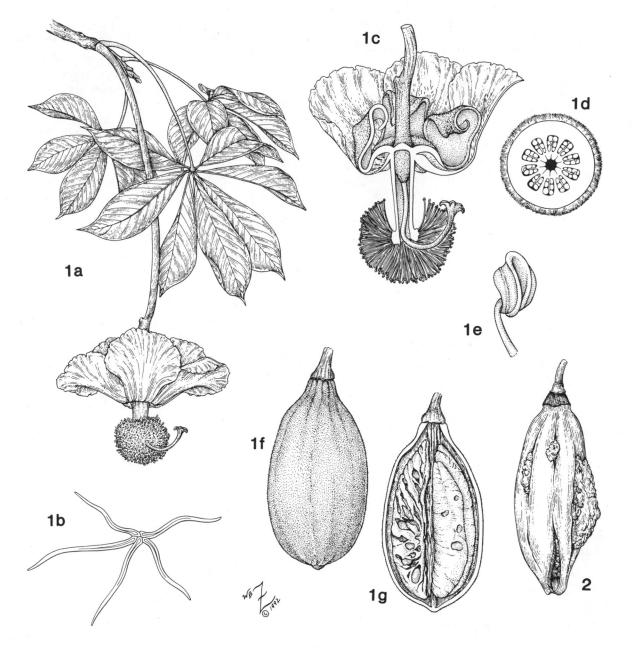

Figure 40. Bombacaceae. 1, *Adansonia digitata*: **a,** flowering branch, ×⅓; **b,** stellate hair from leaf, ×100; **c,** flower with half of perianth and androecium removed, ×⅖; **d,** cross section of ovary, ×2¼; **e,** dehisced stamen (one-locular), ×6; **f,** fruit (indehiscent pod), ×¼; **g,** longitudinal section of fruit showing seeds embedded in pithy pulp, ×¼. **2,** *Ceiba pentandra*: dehiscing capsule, ×½. See Table 5 for family characteristics.

in the field. At first glance, the corolla appears to be sympetalous. Actually, the distinct petals are adnate to the base of the staminal column, and in this way are indirectly connected to one another. When the flower first opens, the immature stigmas remain within the staminal tube, and the anthers spread out in the center of the flower. The single locule of each anther (actually a half-anther resulting from split stamens; Fig. 39: 1i) dehisces by a slit across the top that divides the anther in half (van Heel 1966). After the filaments of the dehisced anthers recurve, the receptive stigmas expand (Fig. 39: 1j). Insects (and for some genera, hummingbirds) are attracted to the conspicuous flowers and to the nectar, which is secreted by hairs on the adaxial basal surface of the sepals, or sometimes, the petals (van Heel 1966).

REFERENCES CITED

Edlin, H. L. 1935. A critical revision of certain taxonomic groups of the Malvales. *New Phytol.* 34:1–20, 122–143.

Fryxell, P. A. 1988. Malvaceae of Mexico. *Syst. Bot. Monogr.* 25:1–522.

Heel, W. A. van. 1966. Morphology of the androecium in Malvales. *Blumea* 13:177–394.

Judd, W. S., R. W. Sanders, and M. J. Donoghue. 1994. Angiosperm family pairs: Preliminary cladistic analyses. *Harvard Pap. Bot.* 5:1–51.

Kearney, T. H. 1951. The American genera of Malvaceae. *Amer. Midl. Naturalist* 46:93–131.

STERCULIACEAE
Cocoa Family

Trees, shrubs, or sometimes herbs, with mucilaginous sap; pubescence stellate or lepidote. **Leaves** simple or sometimes palmately compound, entire, serrate, or sometimes palmately lobed, alternate, usually palmately veined, stipulate. **Inflorescence** determinate, basically cymose and in complex racemose and paniculate arrangements, usually axillary, sometimes cauliflorous. **Flowers** actinomorphic or sometimes zygomorphic, perfect or imperfect (then plants monoeious or polygamous), hypogynous, generally showy, sometimes with epicalyx. **Perianth** biseriate or frequently uniseriate. **Calyx** of typically 5 sepals, usually basally connate, sometimes petaloid (yellow, orange, maroon), with tufts of nectariferous hairs at base, valvate. **Corolla** usually of 5 petals when present, often reduced or absent, distinct, sometimes basally adnate to androecium, usually clawed, variously colored, small, convolute. **Androecium** of typically 10 stamens, basically biseriate with the outer whorl opposite the sepals and often reduced to staminodes, usually arising from a raised stalk (androgynophore); filaments connate into a tube (monadelphous); anthers dorsifixed, with parallel or divergent locules, dehiscing longitudinally, extrorse; staminodes present in carpellate flowers. **Gynoecium** of 1 pistil (syncarpous) or with only styles connate, usually 5-carpellate, stipitate (often raised on an androgynophore); ovary (ovaries) superior, of distinct or connate carpels, 5-locular (syncarpous ovary) or 1-locular (ovaries distinct); ovules 2 to many in each locule (carpel), anatropous or hemitropous, axile (syncarpous pistil) or parietal (ovaries distinct); styles usually 5, distinct or variously basally connate; stigmas usually 5; rudimentary pistil present in staminate flowers. **Fruit** often a capsule, a schizocarp, an aggregate of follicles, or indehiscent, sometimes a berry; seeds sometimes arillate; endosperm usually copious, oily; embryo straight or curved.

Family characterization: Trees or shrubs with mucilaginous sap and stellate or lepidote pubescence; stipulate leaves with palmate venation; usually 5-merous flowers, often with reduced or absent corolla; sepals with fringe of glandular hairs on adaxial surface; monadelphous, basically biseriate stamens with outer whorl often staminodial; androecium and gynoecium raised on a stalk (androgynophore); and various fruit types (usually capsules, schizocarps, or follicles). Tannins, sterculic and malvalic acids, and various alkaloids (e.g., caffeine, theobromine) present. Anatomical features: mucilage receptacles (cells, cavities and/or canals) in the parenchymatous tissues; and trilacunar nodes.

Genera/species: 60/700

Distribution: Almost exclusively pantropical, with a few representatives in warm-temperate areas.

Major genera: *Sterculia* (200–300 spp.), *Dombeya* (200–300 spp.), and *Cola* (125 spp.)

U.S./Canadian representatives: 11 genera/26 spp.; largest genera: *Ayenia* and *Melochia*

Economic plants and products: Chocolate, cocoa, and cocoa butter from seeds of *Theobroma cacao* (cacao). Cola (kola) nuts, cola (a beverage), and kolanine (glucoside used as a stimulant) from *Cola* spp. Lumber from species of several genera, such as *Sterculia* and *Cola*. Ornamental plants (species of at least 10 to 20 genera), including *Dombeya*, *Firmiana* (Chinese-parasol tree), *Fremontodendron* ("Fremontia," flannelbush), and *Sterculia* (flame tree).

Commentary: Depending on the authority, the Sterculiaceae comprise 60 to 73 genera (700 to 1,500 species). The family, in the most inclusive concept, is characterized by considerable floral and vegetative

Figure 41. Sterculiaceae. 1, *Firmiana simplex*: **a,** carpellate flowering branch, ×⅙; **b,** stellate hair from leaf, ×90; **c,** carpellate flower, ×3; **d,** androecium (staminodial) and gynoecium on androgynophore (carpellate flower), ×6; **e,** longitudinal section of androecium and gynoecium, ×6; **f,** cross section of ovaries, ×10; **g,** longitudinal section of carpellate flower, ×4; **h,** staminate flower, ×3; **i,** androecium and gynoecium (carpels rudimentary) on androgynophore (staminate flower), ×4½; **j,** anthers (from bud), ×5; **k,** longitudinal section of androecium and gynoecium (carpels rudimentary; staminate flower), ×5; **l,** longitudinal section of staminate flower, ×4; **m,** fruit (aggregate of follicles), ×⅖; **n,** leaf-like follicle with seeds attached on margins, ×⅖; **o,** seed, ×2½; **p,** longitudinal section of seed, ×3. 2, *Melochia corchorifolia*: **a,** schizocarp, ×6; **b,** segment of schizocarp with seeds, ×6.

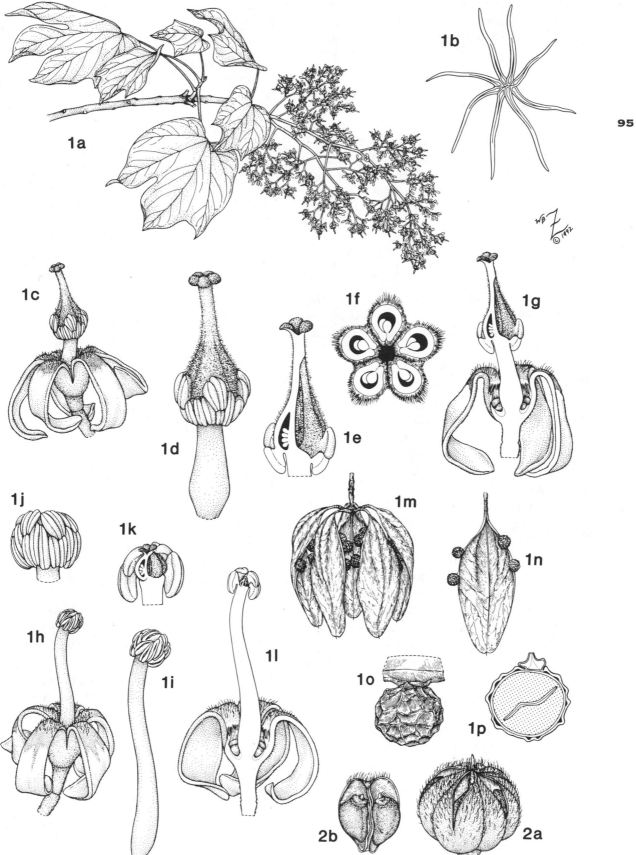

95

variation and has been divided into two subfamilies (and eight to ten tribes): the well-defined Sterculioideae (with only one tribe, Sterculieae), and the heterogeneous Byttnerioideae (with the remaining tribes). Several monographers, such as Edlin (1935), have advocated a restricted definition of the family, including only the relatively homogeneous subfamily Sterculioideae, with the remainder of the genera/tribes placed in the Byttneriaceae. The Sterculiaceae are closely related to the Tiliaceae, Malvaceae, and Bombacaceae; the circumscription of these families in relation to one another is problematical (see Table 5 and the commentary under Malvaceae). Several genera formerly placed in the Sterculiaceae have been transferred to the Bombacaceae and/or the Malvaceae.

The flowers of the Sterculiaceae are usually five-merous with a reduced to absent corolla. The monadelphous androecium is basically biseriate (van Heel 1966). The elements of the inner whorl, which develop first, alternate with the sepals (i.e., opposite the petals, if present) and consist of five fertile stamens (or five fascicles of two or three stamens). The outer whorl of the androecium, opposite the sepals, often is staminodial or even absent. The typical five-carpellate gynoecium is either syncarpous or with carpels connate only by the styles (e.g., *Firmiana*, Fig. 41: 1e, and *Sterculia*; Saunders 1931). Both androecium and gynoecium often arise from the apex of a well-developed stalk, the androgynophore (Fig. 41: 1d,i).

The fruit (e.g., Fig. 41: 1m, 2a) also is extremely variable (fleshy to leathery, and dehiscent or indehiscent), but most often consists of five follicles derived directly from distinct, individual carpels or from the segmentation of an originally syncarpous, compound pistil. The segments often split open early, exposing the developing seeds along the margins (Fig. 41: 1n).

The flowers attract various pollinators (small birds, wasps, bees, moths, ants, midges). Nectar is produced by glandular hairs forming a fringe on the adaxial side of the sepal bases and/or an extrastaminal disc lining the base of the calyx and receptacle around the androgynophore (Fig. 41: 1g,l; van Heel 1966). The protandry of many species apparently favors cross-pollination (Brizicky 1966).

REFERENCES CITED

Brizicky, G. K. 1966. The genera of Sterculiaceae in the southeastern United States. *J. Arnold Arbor.* 47:60–74.

Edlin, H. L. 1935. A critical revision of certain taxonomic groups of the Malvales. *New Phytol.* 34:1–20, 122–143.

Heel, W. A. van. 1966. Morphology of the androecium in Malvales. *Blumea* 13:177–394.

Saunders, E. R. 1931. On carpel polymorphism. IV. *Ann. Bot.* (Oxford) 45:91–110.

TILIACEAE
Linden Family

Trees, shrubs, or occasionally annual herbs; pubescence stellate or lepidote. **Leaves** simple or sometimes palmately lobed, serrate, often with obliquely based blades, alternate, distichous, usually palmately veined, stipulate; stipules caducous or persistent. **Inflorescence** determinate, cymose or sometimes flowers paired or solitary, axillary or terminal. **Flowers** actinomorphic, usually perfect, hypogynous, with an extrastaminal nectariferous disc and/or glands (inside base of sepals or petals), small, sometimes with an epicalyx. **Calyx** of usually 5 sepals, distinct or occasionally basally connate, sometimes with basal nectaries on adaxial surface, valvate. **Corolla** of typically 5 petals or sometimes absent, distinct, usually white to yellowish, sometimes with basal nectaries on adaxial surface, imbricate, convolute, or valvate. **Androecium** of usually numerous stamens, 5 or more sometimes modified into petaloid staminodes, commonly arising on short androgynophore; filaments distinct or basally connate into 5 or 10 fascicles, sometimes bifid at apex; anthers dorsifixed, with contiguous or separated locules (on bifurcated filament), dehiscing longitudinally or sometimes by apical pores, introrse or extrorse. **Gynoecium** of 1 pistil, typically 2- to 5-carpellate, commonly on short androgynophore; ovary superior, with as many locules as carpels; ovules 2 to several in each locule, anatropous to hemitropous, placentation axile; style 1; stigmas as many as carpels, capitate or lobed. **Fruit** a nut, loculicidal capsule, schizocarp, or drupe; endosperm scanty to copious, oily; embryo straight or sometimes with folded cotyledons.

Family characterization: Trees or shrubs with stellate or lepidote pubescence; stipulate, 2-ranked leaves with oblique bases and palmate venation; numerous distinct stamens more or less in 5 fascicles opposite the petals; stipitate ovary on short androgynophore; and a nut or loculicidal capsule as the fruit type. Tannins present. Anatomical features: mucilage cells or cavities in parenchymatous tissues of stems, leaves, and flowers; and trilacunar nodes.

Genera/species: 49/450

Distribution: Widespread in tropical and subtropical regions, with relatively few representatives in temperate areas.

Major genera: *Grewia* (150 spp.), *Triumfetta* (100–150 spp.), *Corchorus* (40–100 spp.), and *Tilia* (45–50 spp.)

U.S./Canadian representatives: *Tilia* (6 spp.), *Triumfetta* (6 spp.), and *Corchorus* (4 spp.)

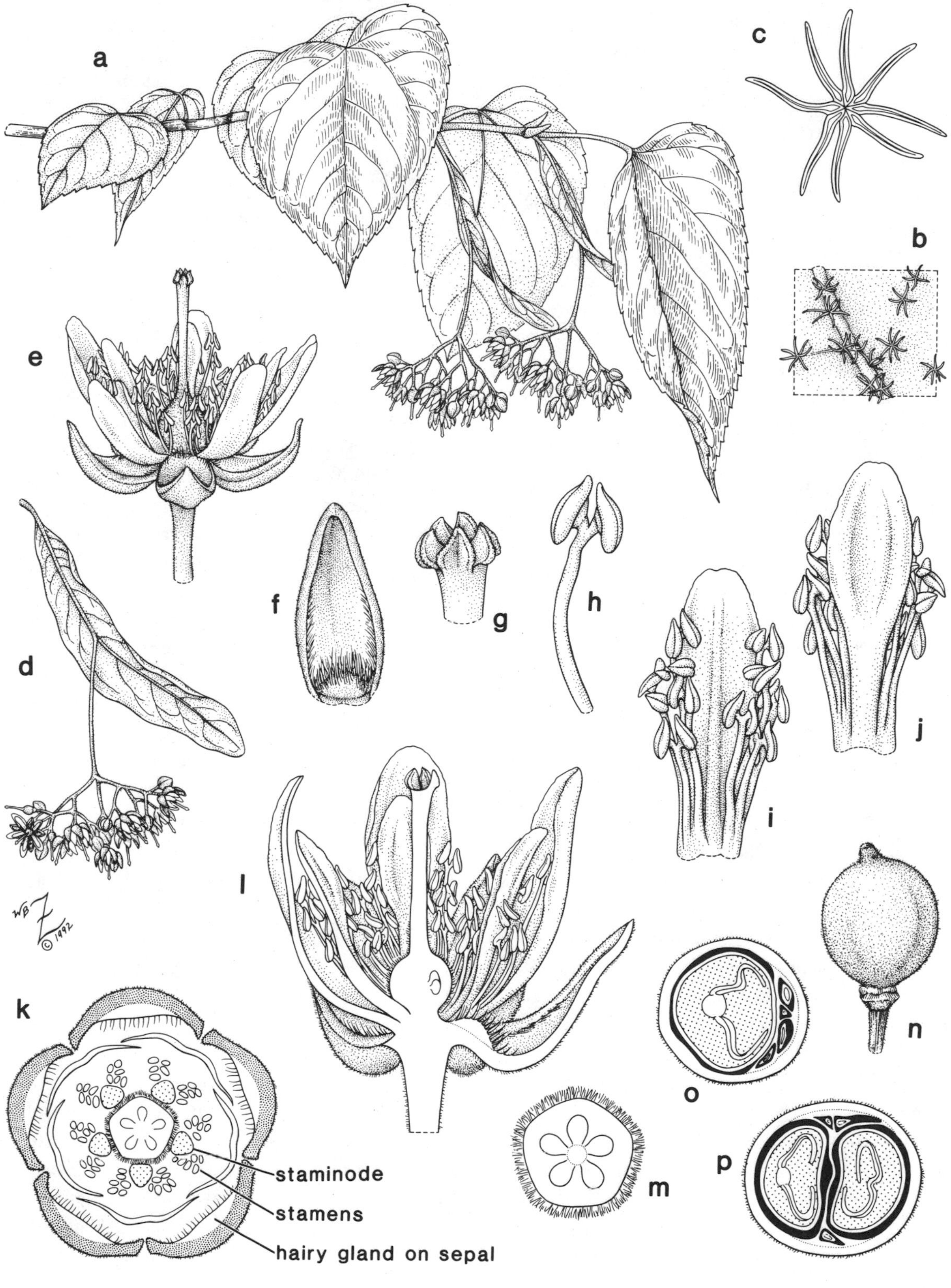

97

staminode

stamens

hairy gland on sepal

Figure 42. Tiliaceae. *Tilia americana* var. *caroliniana* (*T. caroliniana*): **a,** flowering branch, ×½; **b,** abaxial surface of leaf showing stellate hairs, ×12; **c,** stellate hair from leaf, ×85; **d,** inflorescence and subtending bract, ×½; **e,** flower, ×4; **f,** adaxial side of sepal showing hairy gland at base, ×6; **g,** stigma, ×12; **h,** stamen, ×12; **i,** abaxial view of petaloid staminode and attached stamens, ×10; **j,** adaxial view of petaloid staminode and attached stamens, ×10; **k,** cross section of flower, ×10; **l,** longitudinal section of flower, ×7; **m,** cross section of ovary, ×12; **n,** nutlet, ×3½; **o,** cross section of nutlet with one seed, ×4½; **p,** cross section of nutlet with two seeds, ×4½. For floral formula, see Fig. 4: 2c.

Economic plants and products: Lumber from species of several genera, notably *Tilia* (basswood, linden). Jute from phloem fibers of *Corchorus* spp. Ornamental trees and shrubs (species of about 10 genera), including *Corchorus* (jute), *Grewia*, *Sparmannia* (indoorlinden), and *Tilia* (linden, lime tree).

Commentary: The Tiliaceae generally are divided into four subfamilies and sixteen tribes, and the genera usually are well defined (Brizicky 1965). However, the family is very closely allied to the Elaeocarpaceae, Sterculiaceae, Malvaceae, and Bombacaceae—all families with weak and uncertain circumscription (see Table 5, the commentary under Malvaceae, and Edlin 1935). Several genera have been transferred repeatedly from one family to another (e.g., *Muntingia*: Tiliaceae–Elaeocarpaceae; *Corchoropsis*: Tiliaceae–Sterculiaceae).

The variable androecium of the Tiliaceae fundamentally is biseriate (but one whorl suppressed) and typically is composed of numerous stamens (sometimes including five or more staminodes) that are derived from the splitting of an original five. Although more or less distinct, the filaments may be coalesced into five (or ten) fascicles (Fig. 42: k) that are opposite the petals (van Heel 1966). In flowers of *Tilia*, the main representative of the family in the north temperate zone, the outermost stamen in each fascicle is modified into a petaloid staminode (Fig. 42: i,j). In addition, the filaments of the stamens are split at the apex, resulting in the separation of the anther locules (Fig. 42: h).

The genera of the Tiliaceae often are distinguished on the basis of fruit morphology. The fruit type varies from indehiscent (*Grewia*—drupe, *Tilia*—nutlet; Fig. 42: n–p) to dehiscent (*Corchorus*—loculicidal capsule). In *Tilia*, a conspicuous wing-like bract, adnate to the primary axis of the inflorescence (Fig. 42: d), aids in fruit dispersal when the mature infructescence becomes detached as a unit.

The small flowers, with various types of nectaries and often also a strong scent, are adapted for insect pollinators (frequently various bees and flies). For example, the flowers of *Triumfetta* have an extrastaminal disc; those of *Tilia* secrete nectar from a hairy gland (Fig. 42: f) at the base of the adaxial surface of each sepal (Anderson 1976).

REFERENCES CITED

Anderson, G. 1976. The pollination biology of *Tilia*. *Amer. J. Bot.* 63:1203–1212.

Brizicky, G. K. 1965. The genera of Tiliaceae and Elaeocarpaceae in the southeastern United States. *J. Arnold Arbor.* 46:286–307.

Edlin, H. L. 1935. A critical revision of certain taxonomic groups of the Malvales. *New Phytol.* 34:1–20, 122–143.

Heel, W. A. van. 1966. Morphology of the androecium in Malvales. *Blumea* 13:177–394.

ULMACEAE
Elm Family

Trees and shrubs, with watery to mucilaginous sap. **Leaves** simple, entire to variously serrate, usually with obliquely based blades, alternate, commonly distichous, stipulate (stipules caducous). **Inflorescence** determinate, cymose and of congested fascicles, or flower solitary, axillary. **Flowers** actinomorphic to slightly zygomorphic, perfect or imperfect (then plants usually monoecious), hypogynous, small and inconspicuous. **Perianth** of generally 4 to 8 tepals, distinct to connate, campanulate, foliaceous, imbricate, persistent. **Androecium** with as many stamens as tepals and opposite them, free or adnate to the perianth base; filaments distinct, erect in bud; anthers basifixed, dehiscing longitudinally, extrorse or introrse. **Gynoecium** of 1 pistil, 2-carpellate; ovary superior, usually 1-locular, sessile to stalked; ovule solitary, anatropous or amphitropous, pendulous, placentation apical; styles 2, linear, stigmas decurrent and along upper inner surface of styles; rudimentary pistil often present in staminate flowers. **Fruit** a nutlet, samara, or drupe; endosperm usually absent or scanty; embryo straight or curved.

Family characterization: Trees or shrubs with watery to slightly mucilaginous sap; distichous, stipulate leaves with oblique bases; reduced flowers with uniseriate perianths; stamens with erect filaments in bud; a samara or drupe as the fruit type; and seeds with little or no endosperm. Anatomical features: mucilage cells and/or canals in the tissues; often calcification or silicification of certain cell walls (epidermal hairs of stem and leaves); and trilacunar nodes (Fig. 43: 1b).

Figure 43. Ulmaceae. **1,** *Celtis laevigata*: **a,** flowering branch (with immature leaves), ×1¼; **b,** cross section of node (three-trace trilacunar), ×9; **c,** perfect flower, ×7; **d,** carpellate flower, ×7; **e,** longitudinal section of carpellate flower, ×12; **f,** longitudinal section of staminate flower bud, ×12; **g,** staminate flower, ×7; **h,** fruiting branch, ×⅗; **i,** mature leaf, ×½; **j,** drupe, ×3½; **k,** longitudinal section of drupe, ×3½. **2,** *Ulmus alata*: **a,** leaf, ×1; **b,** flower, ×8; **c,** samara, ×5½.

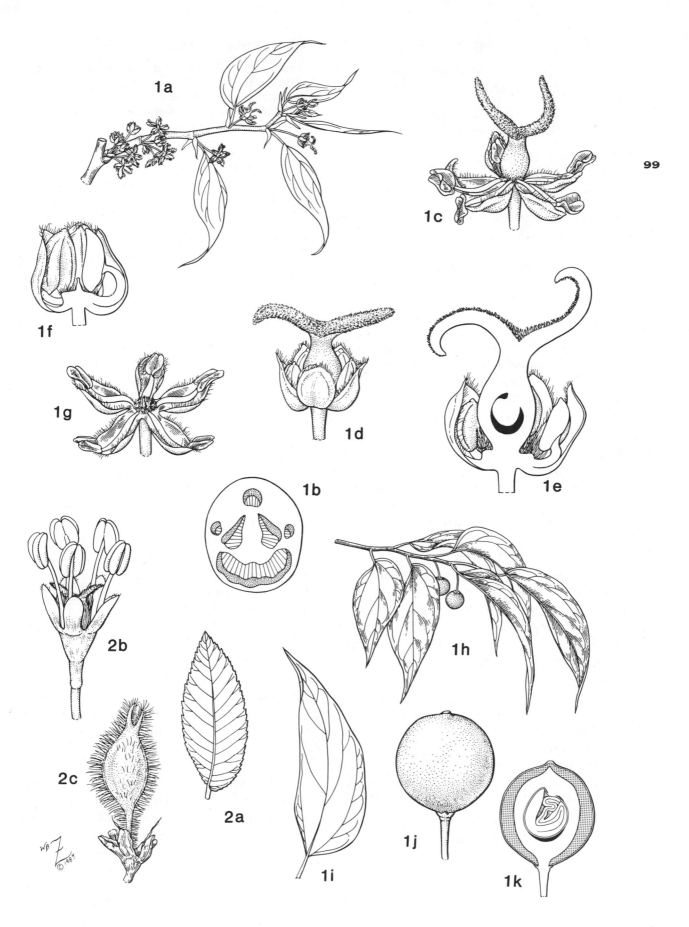

1a

1c

1f

1g

1d

1e

1b

2b

1h

2c

2a

1i

1j

1k

WB🗲
©1987

Genera/species: 15/200

Distribution: Primarily throughout temperate and tropical regions of the Northern Hemisphere.

Major genera: *Celtis* (60–80 spp.), *Ulmus* (18–45 spp.), and *Trema* (10–30 spp.)

U.S./Canadian representatives: *Ulmus* (11 spp.), *Celtis* (8 spp.), *Trema* (3 spp.), *Planera* (1 sp.), and *Zelkova* (1 sp.)

Economic plants and products: Timber from several, such as species of *Ulmus* (elm). Medicinal bark from *Ulmus rubra* (slippery elm—with high mucilage content). Edible fruit from *Celtis* (hackberries, sugarberries). Ornamental trees and shrubs (species of 8 genera), including *Celtis*, *Planera* (water-elm, planer tree), *Ulmus*, and *Zelkova*.

Commentary: The Ulmaceae are divided into two very distinctive subfamilies, the Ulmoideae (incl. *Ulmus*, Fig. 43: 2; *Planera*) and the Celtidoideae (incl. *Celtis*, Fig. 43: 1; *Trema*), on the basis of numerous morphological, chemical, and anatomical features (see the summaries in Elias 1970, Cronquist 1981, and Manchester 1989). For example, leaf, fruit, and seed characters differ in the two groups as follows: Ulmoideae—pinnately veined leaves with secondary veins running to the teeth, dry (often winged) fruit, flat seeds, straight embryo with flat cotyledons, and no endosperm; and Celtidoideae—leaf venation usually with three main veins diverging from the base and secondary veins forming a series of arches, baccate fruit (drupes), round seeds with folded or rolled cotyledons, and some endosperm usually present. In addition, the two groups are characterized by different (but overlapping) distributions, with the Ulmoideae mainly north temperate in range, and the Celtidoideae, typically tropical to subtropical (Berg 1989). Some taxonomists advocate Celtidoideae as a segregate family, a view supported by preliminary cladistic analyses of morphological characters of the Moraceae-Urticaceae complex by Judd et al. (1994).

The reduced and relatively inconspicuous flowers of the Ulmaceae are anemophilous and bloom early in the season (Berg 1977). Most species have imperfect flowers and are monoecious. A notable exception is *Ulmus*, in which the flowers usually are perfect (Fig. 43: 2b). In monoecious genera, often the carpellate flowers on a tree will develop before the staminate flowers (and also before the perfect flowers, if also present). The mechanisms for preventing possible self-pollination in *Ulmus* flowers are not well understood.

REFERENCES CITED

Berg, C. C. 1977. Urticales, their differentiation and systematic position. *Pl. Syst. Evol.*, Suppl. 1:349–374.

———. 1989. Systematics and phylogeny of the Urticales. In *Evolution, systematics, and fossil history of the Hamamelidae*, ed. P. R. Crane and S. Blackmore, vol. 2, "Higher" Hamamelidae, pp. 193–200. Syst. Assoc. Special Vol. 40B. Clarendon Press, Oxford.

Cronquist, A. 1981. *An integrated system of classification of flowering plants*, pp. 189–193. Columbia University Press, New York.

Elias, T. S. 1970. The genera of Ulmaceae in the southeastern United States. *J. Arnold Arbor.* 51:18–40.

Judd, W. S., R. W. Sanders, and M. J. Donoghue. 1994. Angiosperm family pairs: Preliminary cladistic analyses. *Harvard Pap. Bot.* 5:1–51.

Manchester, S. R. 1989. Systematics and fossil history of the Ulmaceae. In *Evolution, systematics, and fossil history of the Hamamelidae*, ed. P. R. Crane and S. Blackmore, vol. 2, "Higher" Hamamelidae, pp. 221–251. Syst. Assoc. Special Vol. 40B. Clarendon Press, Oxford.

URTICACEAE

Nettle Family

Including the Moraceae: Mulberry Family

Trees, shrubs, or herbs, with milky or watery sap, sometimes with glandular or stinging hairs. **Leaves** simple, entire, serrate, or lobed, alternate or sometimes opposite, pinnately to more frequently palmately veined, deciduous or persistent, stipulate (stipules caducous, small or cap-like and leaving a circular scar). **Inflorescence** determinate, cymose and often appearing racemose, spicate (erect or pendulous), umbellate, or capitate/globose, or flowers along inner surface of involuted or invaginated hollow receptacle, axillary. **Flowers** actinomorphic, imperfect (then plants monoecious, dioecious, or polygamous), hypogynous to epigynous, minute, sessile to subsessile. **Perianth** of 3 to 5 tepals or reduced or absent, uniseriate or biseriate, distinct to connate, often green or yellowish, imbricate or valvate, often persistent and accrescent. **Androecium** with as many stamens as tepals and opposite them; filaments distinct, erect or inflexed in bud; anthers versatile, dehiscing longitudinally; staminodes sometimes present in carpellate flowers. **Gynoecium** of 1 pistil, basically 2-carpellate but often appearing 1-carpellate (pseudomonomerous) with usually 1 carpel aborting; ovary superior to inferior, usually 1-locular; ovule solitary, anatropous or orthotropous, pendulous or erect, placentation basal, apical, or subapical; style(s) 1 or 2, filiform; stigma(s) 1 or 2; rudimentary pistil often present in staminate

100

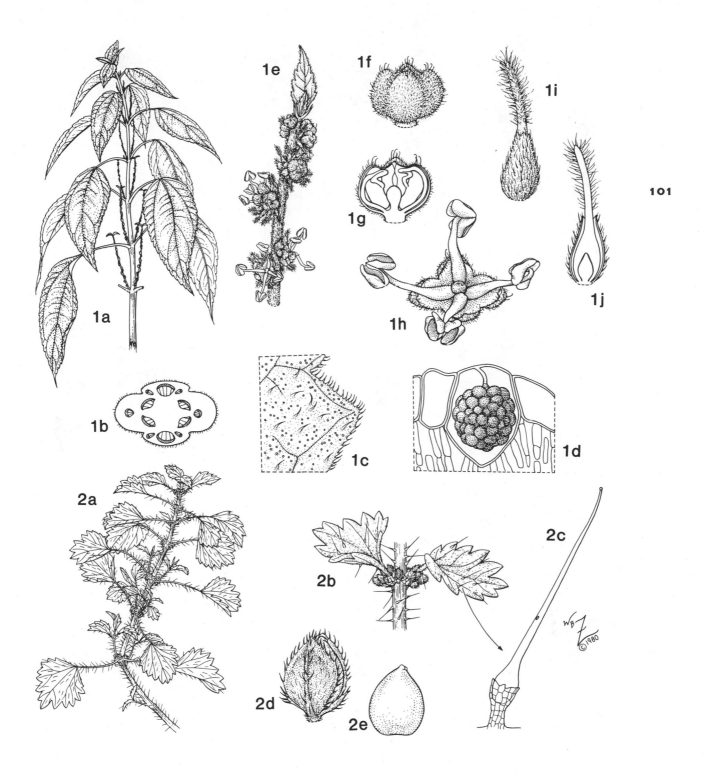

Figure 44. Urticaceae. 1, *Boehmeria cylindrica*: **a,** habit, ×⅖; **b,** cross section of node (trilacunar), ×6; **c,** surface of dried leaf showing cystoliths (circular raised areas), ×12; **d,** cross section of fresh leaf showing cystolith in epidermal cell, ×375; **e,** inflorescence, ×4½; **f,** staminate flower before anthesis, ×12; **g,** longitudinal section of staminate flower before anthesis (filaments coiled), ×12; **h,** staminate flower after anthesis (filaments expanded), ×12; **i,** carpellate flower, ×25; **j,** longitudinal section of carpellate flower, ×25. **2,** *Urtica chamaedryoides*: **a,** habit, ×½; **b,** detail of node showing young achenes, staminate flowers, and stinging hairs, ×3; **c,** stinging hair on leaf, ×36; **d,** achene with persistent calyx, ×12; **e,** achene (mucilage dried), ×12.

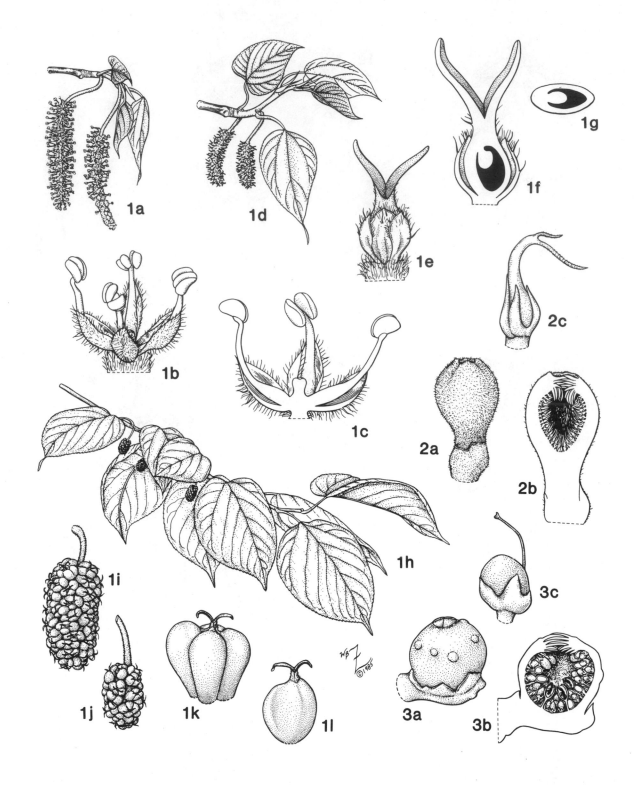

Figure 45. Urticaceae (continued). 1, *Morus rubra*: **a,** staminate flowering branch, ×¾; **b,** staminate flower, ×8; **c,** longitudinal section of staminate flower, ×9; **d,** carpellate flowering branch, ×¾; **e,** carpellate flower, ×12; **f,** longitudinal section of carpellate flower, ×20; **g,** cross section of ovary, ×25; **h,** fruiting branch, ×⅓; **i,** syncarp (from cultivated specimen), ×1½; **j,** syncarp (from wild specimen), ×1½; **k,** drupe with accrescent perianth, ×6; **l,** drupe (perianth removed), ×6. **2,** *Ficus carica*: **a,** inflorescence (syconium), ×3; **b,** longitudinal section of syconium, ×4; **c,** carpellate flower, ×25. **3,** *Ficus aurea*: **a,** syncarp ("fig"), ×3; **b,** longitudinal section of syncarp (with many wasps removed), ×4; **c,** drupe with persistent perianth, ×20.

TABLE 6. Major morphological differences often used to distinguish the four main groupings of the Moraceae-Urticaceae complex. See Berg (1989) for a more detailed summary of these characters. In this text, the Urticaceae *s.s.*, "Moraceae," and "Cecropiaceae" are combined into the Urticaceae *s.l.*, *sensu* Thorne (1981, 1983).

CHARACTER	URTICOIDEAE (URTICACEAE S.S.)	MOROIDEAE ("MORACEAE")	CECROPIOIDEAE ("CECROPIACEAE," CONOCEPHALOIDEAE)	CANNABACEAE
GENERA/SPECIES	39/800	53/1,400	6/275	3/3
DISTRIBUTION	mainly tropical to subtropical, with some representatives in temperate areas	mainly tropical to subtropical, less common in temperate areas	tropical	north temperate regions
HABIT	herbs, occasionally shrubs, or rarely small trees / usually monoecious	trees, shrubs, woody vines, or seldom herbs / monoecious	trees, shrubs, or woody vines / dioecious	herbs / dioecious
SAP	watery	usually milky	watery to slightly milky	watery
NODAL ANATOMY	three-trace trilacunar	multilacunar	three-trace trilacunar to multilacunar	three-trace trilacunar
FILAMENT CONFIGURATION IN BUD	inflexed	straight or inflexed	straight or inflexed	straight
CARPEL NUMBER	1 (pseudomonomerous)	2 (but 1 may be rudimentary)	1 (pseudomonomerous)	2
PLACENTATION	basal	± apical	± basal	± apical
FRUIT TYPE	achene	drupe (often multiple)	achene, drupe (often multiple)	achene
EMBRYO SHAPE	straight	curved	straight	curved
OTHER	plants often with stinging hairs	stipules often broadly sheathing	often rapidly growing weeds of disturbed areas	plants aromatic

flowers. **Fruit** an achene or a drupe (sometimes dehiscent), often aggregated into a multiple fruit arising from the union of fruits of different flowers, their perianth, and common receptacle; endosperm fleshy and oily or mealy, or sometimes absent; embryo curved or straight.

Family characterization: Herbs to trees with milky or watery sap; reduced imperfect flowers in modified cymose inflorescences; reduced or absent perianth; 1-locular gynoecium with solitary ovule; and an achene or a multiple fruit of drupes or achenes as the fruit type. Anatomical features: often calcification or silicification of certain cell walls (epidermal hairs and tissues of leaves); cystoliths in leaf epidermal cells (discussed below); and trilacunar to multilacunar nodes.

Genera/species: 98/2,475

Distribution: Widespread in tropical and subtropical regions and also well-represented in temperate areas.

Major genera: *Ficus* (800 spp.), *Pilea* (250–400 spp.), *Elatostema* (200 spp.), *Dorstenia* (170 spp.), and *Cecropia* (75–100 spp.)

U.S./Canadian representatives: 30 genera/54 spp.; largest genera: *Pilea*, *Pipturus*, and *Ficus*.

Economic plants and products: Fiber for cordage and textiles from *Boehmeria* (ramie) and *Urtica*. Edible greens from young shoots of some temperate members, such as *Urtica* spp. Edible fruits from *Artocarpus* (breadfruit, jackfruit), *Ficus* (figs), and *Morus* (mulberries). Timber from *Chlorophora* (fustic, iroko-wood) and *Maclura* (osage-orange). Rubber from latex of several, such as species of *Castilla* and *Ficus*. Ornamental plants (species of at least 20 genera), including *Broussonetia* (paper-mulberry), *Cecropia*, *Chlorophora*, *Cudrania*, *Dorstenia* (pick-a-back plant), *Ficus* (various figs, India rubber plant, banyan), *Maclura*, *Pellionia*, *Pilea*, and *Soleirolia* (baby's-tears).

104

Commentary: The delimitation of the Urticaceae and Moraceae varies considerably in the literature (see Tippo 1938; Berg 1977; Friis 1989). Several genera (e.g., *Poikilospermum*) have been shifted back and forth between the two families; in addition, the Cannabaceae and Cecropiaceae have been considered as segregate families or as subfamilies within either the Moraceae or Urticaceae (Berg 1978; Friis 1989; Humphries and Blackmore 1989). These four major groupings (see Table 6) are distinguished by such characters as herbaceous or woody habit, clear or milky sap, one or two style(s), and basal or apical placentation (Berg 1989).

The broad definition of the Urticaceae here follows that of Thorne (1981, 1983), who maintained the Cannabaceae as a segregate family and defined the Urticaceae *s.l.* with three subfamilies: Urticoideae, Moroideae, and Cecropioideae. Although he recently (Thorne 1992) split the group into three component families based on a narrower phylogenetic gap (Thorne pers. comm.), the original expanded circumscription, based on phenetic criteria, has been supported by subsequent phylogenetic data (Humphries and Blackmore 1989; Judd et al. 1994). These studies demonstrate the monophyletic Urticaceae *s.s.* as derived within the "Moraceae," and the variously circumscribed "Moraceae" and "Cecropiaceae" as probably paraphyletic. Therefore, the authors of both analyses suggest the inclusion of the Moraceae and Cecropiaceae within an expanded Urticaceae *s.l.* Much more critical analysis is needed, however, to further define the relationships between the various genera of this complex, including those of the Cannabaceae.

Of note is the tribe Urticeae (stinging nettles), from which the family receives its common name. This group is characterized by stinging hairs (embedded in epidermal cells; Fig. 44: 2b,c) that have a special "hypodermic needle" mechanism for releasing their irritating contents (histamines and acetocholines): when the tip of the tubular hair is broken, pressure upon the sac-like base forces the irritants out (Thurston and Lersten 1969; Woodland 1989).

Other interesting vegetative characters of taxonomic importance within the family include cystoliths, concretions of calcium carbonate attached to a cel-lulose stalk within an enlarged specialized cell (lithocyst; Fig. 44: 1d). Prominent cystoliths usually are present in epidermal cells on the abaxial or adaxial leaf surface (Fig. 44: 1c). Although they are difficult to see in fresh material, cystoliths usually are obvious in dried leaves as elevated areas with distinctive shapes that may vary in different species and higher taxa. Miller (1971) reports that cystolith shape may at least be correlated with tribes within the Urticoideae.

The inflorescence type is important in classification within the Urticaceae. For example, *Morus* is characterized by staminate and carpellate catkins (Fig. 45: 1a,d), while in *Boehmeria* the staminate and carpellate flowers are arranged together in erect spicate inflorescences (Fig. 44: 1e). The carpellate flowers of *Maclura* and *Broussonetia* are congested into globose heads. In *Ficus*, an involuted receptacle becomes a hollow and fleshy structure (syconium) that bears the flowers along the interior surface (Fig. 45: 2a,b).

The minute, reduced flowers of the family often are well adapted for anemophily. In many species, such as *Boehmeria cylindrica*, the inflexed filaments of the staminate flower buds extend elastically, thereby causing the anthers to release clouds of pollen in a sudden burst (see Fig. 44: 1g,h). On the other hand, the specialized and very complex entomophilous pollination of *Ficus* by gall-wasps has received much attention in the literature (Proctor and Yeo 1972; Faegri and Pijl 1980; Berg 1990). Basically, the flowers (enclosed in the hollow, fleshy axis) are pollinated when the wasp enters to lay eggs in the ovaries of the carpellate flowers; newly emergent wasps carry pollen from that inflorescence to new syconia. The development of the carpellate flowers before the staminate flowers within the same inflorescence encourages cross-pollination.

The fruit type of the family varies from an achene enclosed by the persistent perianths (Fig. 44: 2d,e) to a multiple fruit of drupes (from adjacent flowers), accrescent perianth, and the fleshy common receptacle. The individual drupes and subtending perianths may be distinguished easily in a mulberry (*Morus*; Fig. 45: 1i–k), which superficially resembles a blackberry (aggregate fruit of species of *Rubus*). In a breadfruit (*Artocarpus*), the drupes and axis are well-united into one massive structure. A fleshy receptacle with drupes inside forms a fig (fruit of *Ficus*; Fig. 45: 3a–c).

REFERENCES CITED

Berg, C. C. 1977. Urticales, their differentiation and systematic position. *Pl. Syst. Evol.*, Suppl. 1:349–374.

——. 1978. Cecropiaceae, a new family of the Urticales. *Taxon* 27:39–44.

——. 1989. Systematics and phylogeny of the Urticales. In *Evolution, systematics, and fossil history of the Hamamelidae*, ed. P. R. Crane and S. Blackmore, vol. 2, *"Higher" Hamamelidae*, pp. 193–220. Syst. Assoc. Special Vol. 40B. Clarendon Press, Oxford.

——. 1990. Reproduction and evolution in *Ficus* (Moraceae): Traits connected with the adequate rearing of pollinators. *Mem. New York Bot. Gard.* 55:169–185.

Faegri, K., and L. van der Pijl. 1980. *The principles of pollination ecology*, pp. 176–178. Pergamon Press, Oxford.

Friis, I. 1989. The Urticaceae: A systematic review. In *Evolution, systematics, and fossil history of the Hamamelidae*, ed. P. R. Crane and S. Blackmore, vol. 2, *"Higher" Hamamelidae*, pp. 285–308. Syst. Assoc. Special Vol. 40B. Clarendon Press, Oxford.

Humphries, C. J., and S. Blackmore. 1989. A review of the classification of the Moraceae. In *Evolution, systematics, and fossil history of the Hamamelidae*, ed. P. R. Crane and S. Blackmore, vol. 2, *"Higher" Hamamelidae*, pp. 267–277. Syst. Assoc. Special Vol. 40B. Clarendon Press, Oxford.

Judd, W. S., R. W. Sanders, and M. J. Donoghue. 1994. Angiosperm family pairs: Preliminary cladistic analyses. *Harvard Pap. Bot.* 5:1–51.

Miller, N. G. 1971. The genera of Urticaceae in the southeastern United States. *J. Arnold Arbor.* 52:40–68.

Proctor, M., and P. Yeo. 1972. *The pollination of flowers*, pp. 312–316. Taplinger, New York.

Thorne, R. F. 1981. Phytochemistry and angiosperm phylogeny: A summary statement. In *Phytochemistry and angiosperm phylogeny*, ed. D. A. Young and D. S. Seigler, pp. 233–295. Praeger, New York.

——. 1983. Proposed new realignments in the angiosperms. *Nordic J. Bot.* 3:85–117.

——. 1992. Classification and geography of the flowering plants. *Bot. Rev.* (Lancaster) 58:225–348.

Thurston, E. L., and N. R. Lersten. 1969. The morphology and toxicology of plant stinging hairs. *Bot. Rev.* (Lancaster) 35:393–412.

Tippo, O. 1938. Comparative anatomy of the Moraceae and their presumed allies. *Bot. Gaz.* 100:1–99.

Woodland, D. W. 1989. Biology of temperate Urticaceae (nettle) family. In *Evolution, systematics, and fossil history of the Hamamelidae*, ed. P. R. Crane and S. Blackmore, vol. 2, *"Higher" Hamamelidae*, pp. 309–318. Syst. Assoc. Special Vol. 40B. Clarendon Press, Oxford.

RHAMNACEAE
Buckthorn Family

Trees, shrubs, or sometimes woody vines (then with hooks, tendrils, or twining stems), sometimes armed with thorns or stipular spines. **Leaves** simple, entire or serrate, usually alternate, deciduous or persistent, stipulate; stipules minute, caducous, sometimes modified into spines. **Inflorescence** determinate, cymose and appearing corymbose, umbellate, paniculate, spicate, racemose, or fasciculate, or sometimes reduced to a solitary flower, axillary or occasionally terminal. **Flowers** actinomorphic, usually perfect or occasionally imperfect (then plants usually monoecious), perigynous to sometimes appearing epigynous due to adnation of disc to ovary, small and inconspicuous, with a hypanthium bearing a nectariferous intrastaminal disc. **Calyx** of 4 or usually 5 sepals, distinct, valvate. **Corolla** of 4 or usually 5 petals or sometimes absent, distinct, frequently clawed, usually hooded to concave and enclosing the stamens at emergence, greenish, white, or pink to blue. **Androecium** of 4 or usually 5 stamens, opposite the petals and encircled by them, epipetalous, inserted on or below disc margin; filaments distinct; anthers dorsifixed, dehiscing longitudinally, introrse. **Gynoecium** of 1 pistil, usually 2- or 3-carpellate; ovary superior or sometimes appearing inferior due to adnation to disc, with as many locules as carpels; ovules usually 1 in each locule, anatropous, erect, placentation basically axile but appearing basal; style 1, often 2- or 3-lobed or -branched; stigmas capitate to elongate. **Fruit** typically a dehiscent or indehiscent drupe with 1 to several pyrenes; seeds smooth, convex; endosperm scanty to copious; embryo straight, large.

Family characterization: Trees, shrubs, or woody vines with simple stipulate leaves; small, perigynous, cup-shaped flowers with well-developed hypanthium and conspicuous intrastaminal nectar disc; concave and clawed petals opposite to and enclosing the stamens; a solitary basal ovule in each ovary locule; and drupaceous fruits. Saponins (glycosides), certain alkaloids, and tannins present. Tissues with calcium oxalate crystals. Anatomical features: mucilage cells (and sometimes cavities) in leaves and stems; and trilacunar nodes.

Genera/species: 45/850

Distribution: Widespread in temperate to tropical regions, but most common in subtropical to tropical areas.

Major genera: *Rhamnus* (125–160 spp.), *Phylica* (150 spp.), *Ziziphus* (86–100 spp.), and *Ceanothus* (55 spp.)

U.S./Canadian representatives: 14 genera/123 spp.; largest genera: *Ceanothus*, *Gouania*, and *Rhamnus*

Economic plants and products: Several medicinal plants, such as species of *Rhamnus* (purgatives; e.g., cascara sagrada); green and yellow dyes also extracted from fruits and bark of certain *Rhamnus* species. Edible fruits (*Ziziphus* spp.: Chinese-date, jujube, lotus) and edible fruit pedicels (*Hovenia dulcis*: Japanese raisin tree). Ornamental plants (species of 17 genera), including *Berchemia* (rattan vine, supplejack), *Ceanothus*

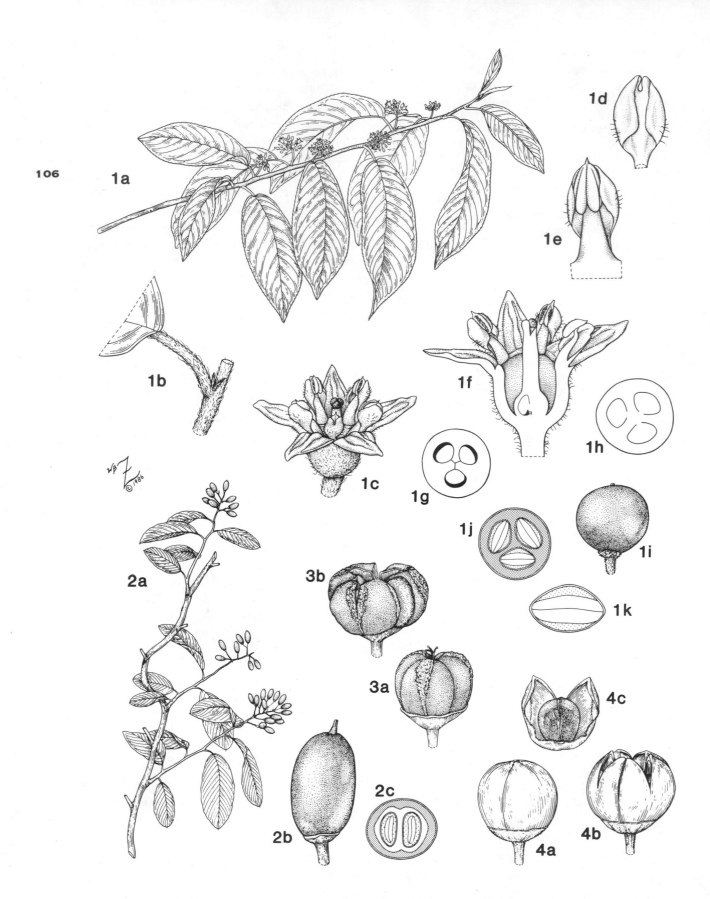

(California-lilac, redroot), *Colletia, Hovenia, Noltea, Paliurus* (Jerusalem-thorn), *Pomaderris, Rhamnus* (buck-thorn), and *Spyridium*.

Commentary: The Rhamnaceae often are associated with the Vitaceae in various classification schemes, as in Cronquist (1988). Preliminary molecular data (Chase et al. 1993), however, place the family as members of the sister group to the Urticaceae *s.l.* and Ulmaceae.

The flowers of the family resemble those of the Rosaceae and are characteristically perigynous, with the hypanthium bearing an intrastaminal nectariferous disc (Prichard 1955). The disc usually is a conspicuous, fleshy, and lobed structure (as in *Colubrina*), and may form a fleshy lining on the inner wall of the hypanthium (as in *Rhamnus*; Fig. 46: 1f), sometimes covering the inside of the sepals as well. The degree of fusion of the disc to the ovary varies, and when immersed in the disc, as in *Ziziphus*, the ovary appears inferior. The filaments of the stamens are inserted at or below the outer disc margin and are opposite to and enclosed by the concave to hooded portions of the clawed petals arising from the hypanthium (Fig. 46: 1e).

The small, inconspicuous flowers often are fragrant and produce copious nectar. Many species of *Rhamnus*, for example, are important "honey plants." Pollinators generally include bees, beetles, and flies, and protandry promotes cross-pollination. At anthesis, the filaments are erect, and the stigma is unreceptive. The filaments generally recurve after anther dehiscence as the stigma matures.

Authors variously describe the unusual dehiscing fruits of the family as "drupes," "capsules," or "schizocarps." The fruit basically is drupaceous: a more or less fleshy mesocarp surrounds the hard and bony endocarp that forms one to three separate stones (pyrenes) or one fused plurilocular stone (Brizicky 1964). The pyrenes may be indehiscent, as in a typical drupe (e.g., *Berchemia*, Fig. 46: 2b,c, and *Rhamnus caroliniana*, Fig. 46: 1i,j). Often, however, the mesocarp is thin, and the pyrenes dehisce along the adaxial suture when the fruit dehydrates, thereby resembling a schizocarp segment or a capsule valve. Frequently, the seeds are explosively ejected from the spitting endocarp (e.g., *Colu-*

brina, Fig 46: 4a–c, and *Ceanothus*, Fig. 46: 3a,b). The dehiscent fruits of some species (e.g., some *Sageretia* species) sometimes are erroneously described as indehiscent. The dry fruits of *Gouania* (and allied genera) are more correctly described as true schizocarps.

REFERENCES CITED

Brizicky, G. K. 1964. The genera of Rhamnaceae in the southeastern United States. *J. Arnold Arbor.* 45:439–463.

Chase, M. W., D. E. Soltis, R. G. Olmstead, D. Morgan, D. H. Les, et al. 1993. Phylogenetics of seed plants: An analysis of nucleotide sequences from the plastid gene *rbc*L. *Ann. Missouri Bot. Gard.* 80:528–580.

Cronquist, A. 1988. *The evolution and classification of flowering plants*, 2d ed., pp. 396–397. New York Botanical Garden, Bronx.

Prichard, E. C. 1955. Morphological studies in Rhamnaceae. *J. Elisha Mitchell Sci. Soc.* 71:82–106.

EUPHORBIACEAE
Spurge Family

Trees, shrubs, herbs, or vines, sometimes succulent and cactus-like, often with colored or milky sap, often with glands on vegetative parts. **Leaves** simple to variously compound, entire or serrate to lobed, alternate or sometimes opposite or whorled, usually stipulate. **Inflorescence** determinate, cymose and appearing spicate, capitate, or highly specialized and forming cyathia, or sometimes flower solitary, sometimes subtended by showy bracts. **Flowers** actinomorphic, imperfect (plants monoecious or dioecious), hypogynous, often very reduced due to suppression of parts, with extrastaminal nectariferous disc or glands present in one or both sexes. **Calyx** absent or of typically 5 sepals, distinct, valvate or imbricate. **Corolla** usually absent or of typically 5 petals, distinct or connivent, valvate or imbricate. **Androecium** of 1 to numerous stamens; filaments distinct to completely connate (monadelphous); anthers basifixed, dehiscing longitudinally; staminodes sometimes present in carpellate flowers. **Gynoecium** of 1 pistil, usually 3-carpellate; ovary superior, usually 3-locular and 3-lobed; ovules 1 or sometimes 2 in each locule, anatropous, pendulous, often with thickened funicle, placentation axile; styles

Figure 46. Rhamnaceae. 1, *Rhamnus caroliniana*: **a,** flowering branch, ×²⁄₅; **b,** detail of node showing stipules, ×2¼; **c,** flower, ×9; **d,** petal, ×25; **e,** stamen and subtending petal, ×25; **f,** longitudinal section of flower, ×12; **g,** cross section of ovary at base, ×25; **h,** cross section of ovary near middle, ×25; **i,** drupe, ×2 ½; **j,** cross section of drupe, ×2½; **k,** cross section of pyrene, ×5. **2,** *Berchemia scandens*: **a,** portion of plant with fruit, ×½; **b,** drupe, ×4; **c,** cross section of drupe, ×4. **3,** *Ceanothus americanus*: **a,** fruit (dehiscent drupe), ×4½; **b,** dehiscing fruit, ×4½. **4,** *Colubrina elliptica*: **a,** fruit (dehiscent drupe), ×3½; **b,** dehiscing fruit, ×3½; **c,** segment of fruit with seed, ×3½.

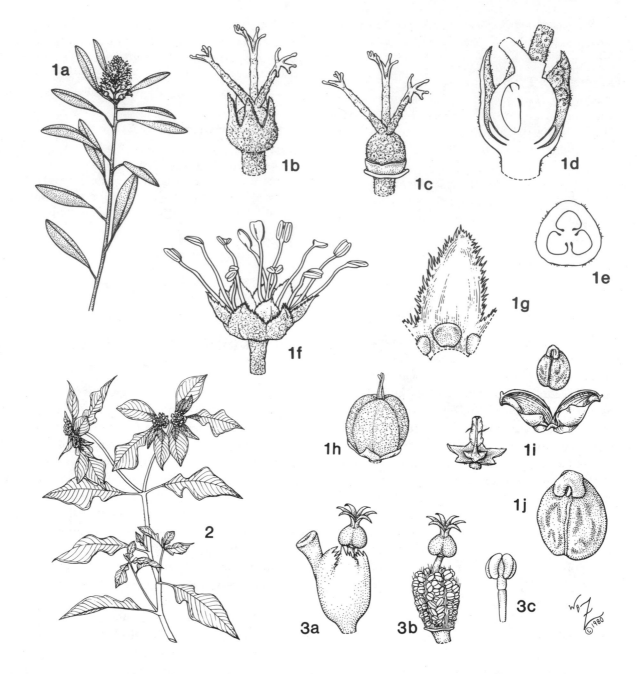

Figure 47. Euphorbiaceae. 1, *Croton argyranthemus*: **a,** habit, ×½; **b,** carpellate flower, ×6; **c,** carpellate flower with perianth removed, ×6; **d,** longitudinal section of carpellate flower, ×12; **e,** cross section of ovary, ×12; **f,** staminate flower, ×6; **g,** adaxial side of calyx lobe showing gland, ×12; **h,** schizocarp before dehiscence, ×3; **i,** persistent calyx and central column, dehiscing segment of the schizocarp, and seed, ×3; **j,** seed, ×6. **2,** *Euphorbia* (Subg. *Poinsettia*) *cyathophora*: habit, ×⅕. **3,** *Euphorbia* (Subg. *Poinsettia*) *heterophylla*: **a,** cyathium, ×6; **b,** cyathium with involucre removed, ×6; **c,** staminate flower, ×12.

3, distinct or basally connate, each often bifid; stigmas linear or broad, frequently lacerate; rudimentary pistil frequently present in staminate flowers. **Fruit** a schizocarp separating elastically into usually 3 segments that split ventrally; seeds 1 or sometimes 2 per segment, often carunculate, sometimes mottled; endosperm copious, fleshy; embryo straight or bent.

Family characterization: Herbs to shrubs with milky or resinous sap; alternate stipulate leaves; imperfect, hypogynous, reduced flowers that are often clustered into cyathia; extrastaminal nectariferous disc or glands; 3-carpellate and -lobed ovary with distinct bifid styles; schizocarp with 3 segments; and carunculate seeds. Various alkaloids, cyanogenic glycosides, and tannins commonly present. Tissues with calcium oxalate crystals. Anatomical feature: secretory tubes, cells, or sacs (containing latex, tannins, or other substances) in stems and leaves.

Genera/species: 307/7,030

Distribution: Mainly tropical; also many representatives in temperate regions.

Major genera: *Euphorbia* (1,600–2,000+ spp.), *Croton* (750 spp.), *Phyllanthus* (600 spp.), *Acalypha* (430–450 spp.), *Glochidion* (300 spp.), *Macaranga* (240–280 spp.), *Drypetes* (200 spp.), and *Jatropha* (170–175 spp.)

U.S./Canadian representatives: 47 genera/358 spp.; largest genera: *Chamaesyce, Euphorbia,* and *Croton*

Economic plants and products: Various commercial products from *Aleurites* (tung oil, waxes), *Croton* (oils, resins), *Euphorbia* (resins, waxes), *Hevea* (rubber), *Manihot* (arrowroot starch, cassava, rubber, tapioca), *Ricinus* (castor oil), and *Sapium* (oils, fats, waxes). Many poisonous plants (due to toxic alkaloids and cyanogenic glycosides), such as *Ricinus communis* (castorbean). Ornamental plants (species of 30 genera), including *Acalypha* (chenille plant, copperleaf), *Codiaeum* (croton), *Euphorbia* (crown-of-thorns, pencil tree, poinsettia), *Jatropha* (coral plant, physic-nut), *Phyllanthus* (emblic, gooseberry tree), and *Synadenium* (African milkbush).

Commentary: The phylogenetic relationships of the Euphorbiaceae to other families have been much disputed (see Webster 1967, 1987). Since this diverse family is difficult to characterize, other families (such as the Buxaceae) sometimes have been included. Intrafamilial classification of the Euphorbiaceae has also been a controversial topic due to the great variation in vegetative and floral structure. For example, euphorbs vary in habit from cosmopolitan weedy herbs, to desert succulents, to tropical trees, and the flowers can vary considerably in degree of specialization (see below). Since the range of morphological variation is so great, many authors hypothesize a polyphyletic origin for the family. The recent classification of Webster (1975) recognizes five subfamilies based upon such characters as number of ovules per locule, basic chromosome number, pollen grain type, and latex characters.

Floral morphology varies greatly throughout the family (Venkata and Ramalakshmi 1968). Generally, euphorbiaceous flowers are imperfect, and the carpellate flowers have a three-carpellate and -locular superior ovary that matures into a schizocarp. The androecium of the staminate flowers consists of five or ten (sometimes numerous or reduced to three) stamens that are distinct to variously connate. The least specialized flower types have a five-merous uniseriate

perianth (*Manihot*) or sometimes a biseriate perianth (some *Croton* species; Fig. 47: 1f). When uniseriate, the perianth lacks the innermost whorl.

The greatest floral reduction and specialization occur in *Euphorbia* and allied genera. The fascinating flowers and inflorescence of *Euphorbia* have been studied extensively (Haber 1925; Rao 1971). The carpellate flower consists of a single ovary elevated upon a pedicel. The staminate flower is reduced to a solitary stamen and has a joint between the stamen and its pedicel (Fig. 47: 3c). Typically, one carpellate and several staminate flowers are arranged inside a cup-like structure (involucre) composed of four or five bracts (Fig. 47: 3a,b). This entire highly reduced cymose inflorescence is called a "cyathium." Often the margins of the involucral bracts bear colored glands that may also have petaloid appendages, and sometimes the involucral bracts themselves are colorful and conspicuous. With all of these structures, the compact cyathium resembles a complete flower (or pseudanthium) with a "calyx" (involucre of bracts), a "corolla" (gland appendages, involucre), one "pistil" (carpellate flower), and several "stamens" (staminate flowers).

Despite the wide floral diversity, the fruit usually is a schizocarp (Fig. 47: 1h,i). The three segments separate elastically from the persistent central column (Fig. 47: 1i); simultaneously, each segment splits lengthwise into two valves, thereby releasing the seed(s). In this way, the seeds may be dispersed up to several meters. The seeds (Fig. 47: 1j) sometimes are mottled and often have a caruncle, which is a spongy, tumor-like outgrowth of the integuments. The caruncles of seeds (as in the Violaceae) often function as oil bodies and attract ants, which aid in dispersal. However, Webster (1967) reports that the field data to support ant dispersal in the Euphorbiaceae are insufficient.

Most euphorbs easily attract pollinators (mostly flies) with the nectar secreted by the extrastaminal disc or glands (Fig. 47: 1g) or, in the most specialized members, by involucral glands (Fig. 47: 3a). The flowers of a few genera, such as those of *Acalypha* and *Ricinus*, are anemophilous.

109

REFERENCES CITED

Haber, J. M. 1925. The anatomy and morphology of the flower of *Euphorbia*. Ann. Bot. (Oxford) 39:657–707.

Rao, C. V. 1971. Anatomy of the inflorescence of some Euphorbiaceae with a discussion on the phylogeny and evolution of the inflorescence including the cyathium. Bot. Not. 124:39–64.

Venkata, R. C., and T. Ramalakshmi. 1968. Floral anatomy of some Euphorbiaceae. I. Non-cyathium taxa. J. Indian Bot. Soc. 47:278–300.

Webster, G. L. 1967. The genera of Euphorbiaceae in the southeastern United States. *J. Arnold Arbor.* 48:303–430.

——. 1975. Conspectus of a new classification of the Euphorbiaceae. *Taxon* 24:593–601.

——. 1987. The saga of the spurges: A review of classification and relationships in the Euphorbiales. *Bot. J. Linn. Soc.* 94:3–46.

110

CISTACEAE
Rock-rose Family

Herbs to shrubs, often growing in exposed areas with alkaline or sandy soils, typically with pubescence of glandular hairs and/or nonglandular tufted hairs. **Leaves** simple, entire, often with inrolled margins, often scale-like, usually opposite, stipulate or exstipulate. **Inflorescence** determinate, cymose and often appearing racemose, or flower solitary, terminal or axillary. **Flowers** actinomorphic, perfect, hypogynous, showy, sometimes with nectariferous disc, sometimes cleistogamous. **Calyx** of basically 5 sepals, unequal with 2 outer sepals reduced (bract-like) to absent, distinct, convolute, generally persistent. **Corolla** of 5 or occasionally 3 petals, absent in cleistogamous flowers, distinct, fugacious or ephemeral, often yellow or reddish, convolute (twisted in opposite direction of calyx) or sometimes imbricate, often crumpled in bud. **Androecium** of numerous stamens; filaments distinct; anthers basifixed, dehiscing longitudinally, introrse. **Gynoecium** of 1 pistil, 3- or occasionally 5-carpellate; ovary superior, 1-locular, sometimes appearing 5- to 10-locular due to intruding placentae; ovules 2 to numerous on each placenta, orthotropous, with well-developed funicles, placentation parietal often on intruded placentae; style 1; stigma usually 1, capitate, discoid, or occasionally plumose, often lobed. **Fruit** a loculicidal capsule, leathery or woody; seeds small, angular, with rough surface, with outer seed coat often gelatinous when moistened; endosperm copious, mealy or often cartilaginous; embryo curved to coiled.

Family characterization: Herbs or shrubs growing in exposed areas with alkaline or sandy soils; pubescence of tufted and/or glandular hairs; small to scale-like leaves; calyx of 5 unequal sepals with the outer 2 smaller and bract-like; fugacious/marcescent petals; numerous stamens; 1-locular ovary with parietal placentation; capsular fruits with mucilaginous seeds (when moistened); and curved to circinate embryo. Tissues commonly with tannins and calcium oxalate crystals. Anatomical features: "combretaceous hairs" (described below); xylem and phloem forming a continuous ring; and unilacunar nodes.

Genera/species: 8/200

Distribution: Primarily north temperate regions (although with a few representatives in South America); particularly diverse in the Mediterranean region and the eastern United States.

Major genus: *Helianthemum* (100–110 spp.)

U.S./Canadian representatives: *Lechea* (17 spp.), *Helianthemum* (14 spp.), *Hudsonia* (3 spp.), and *Cistus* (1 sp.)

Economic plants and products: Ladanum (or labdanum), fragrant resin used in incense, perfumes, and medicines, from leaves and branches of *Cistus* spp. Ornamental plants (species of 7 genera), including *Cistus* (rock-rose), *Halimium*, *Helianthemum* (sun-rose, rock-rose), and *Hudsonia* (beach-heather).

Commentary: Due to superficial resemblance the Cistaceae have been allied with the Clusiaceae, but now they are generally placed in the Violales (e.g., Cronquist 1988; Thorne 1992). With 100+ species, the genus *Helianthemum* constitutes at least half of the family; however, the delimitation of this genus is problematic (probably paraphyletic), because some authorities recognize several segregates (e.g., *Crocanthemum* and *Heteromeris*; Brizicky 1964).

Vegetatively, the family is characterized by the pubescence of the stem, leaf, and calyx. The so-called stellate hairs of the family (Fig. 48: b,c) actually are several unicellular hairs clustered in tufts (Fig. 48: d). These nonglandular hairs are rigid at the base, where each appears to have a double structure, as though a second hair or cell were enclosed in the basal portion. These "combretaceous hairs" also characterize the Combretaceae, where, however, they occur singly. Various kinds of multicellular, peltate or gland-tipped hairs also are common. These glandular hairs may be uniseriate or sometimes capitate, and contain various substances (such as resins and ethereal oils). The various hair types are useful in distinguishing species in the family (e.g., Daoud and Wilbur 1965). The tufted (and/or peltate) hairs often give a silvery appearance to the plants.

The ephemeral flowers are homogamous to slightly protogynous and open only in full sunlight, usually only for a few hours. The brightly colored petals and numerous stamens attract bees, flies, and beetles that visit to collect the copious pollen. In some flowers (e.g., some species of *Helianthemum*), the stamens are "sensitive," moving outwards when touched by insects and dusting them with pollen. Self-pollination may also occur as the anthers are pressed against the stigma when the flower closes and probably is prevalent in the flowers of *Lechea*, which rarely expand (Hodgdon 1938).

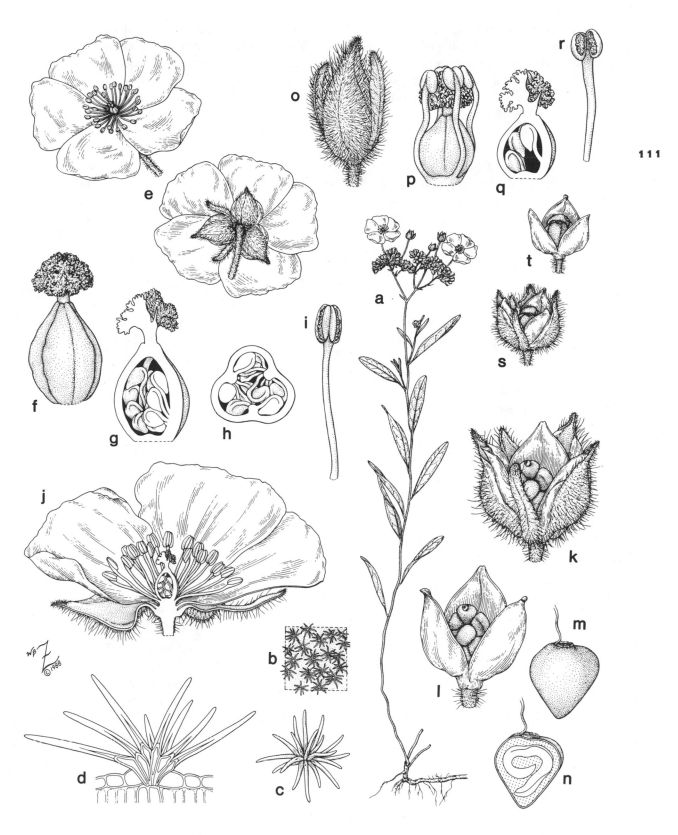

Figure 48. Cistaceae. *Helianthemum corymbosum*: **a,** habit, ×½; **b,** adaxial leaf surface showing tufted hairs, ×30; **c,** hair tuft from leaf, ×135; **d,** cross section of hair tuft ("combretaceous hairs"), ×215; **e–n,** chasmogamous flower: **e,** two views of flower, ×2½; **f,** pistil, ×20; **g,** longitudinal section of pistil, ×20; **h,** cross section of ovary, ×23; **i,** stamen, ×15; **j,** longitudinal section of flower, ×6; **k,** capsule with persistent calyx, ×6; **l,** capsule (persistent calyx removed), ×6; **m,** seed, ×20; **n,** longitudinal section of seed, ×20; **o–t,** cleistogamous flower: **o,** cleistogamous flower, ×15; **p,** androccium and gynoecium, ×25; **q,** longitudinal section of pistil, ×25; **r,** stamen, ×28; **s,** capsule with persistent calyx, ×6; **t,** capsule (persistent calyx removed), ×6.

The occurrence of cleistogamous ("closed") flowers is distributed widely in the family (e.g., many American species of *Helianthemum*). Cleistogamous flowers, which self-pollinate within a closed perianth, often occur on the same plant as the chasmogamous forms. For example, in *Helianthemum corymbosum*, the long pedicels of the few chasmogamous flowers protrude above the numerous cleistogamous flowers in the corymb (Fig. 48: a). In this particular species, the two flower types develop more or less at the same time; in others, the chasmogamous flowers are vernal, while the cleistogamous flowers develop much later in the autumn. The cleistogamous flowers (Fig. 48: o) are apetalous, and the calyx and pistil (Fig. 48: p) are much smaller than those in the chasmogamous type (Fig. 48: f); also, the stamens (Fig. 48: p) and ovules (Fig. 48: q) are much reduced in number (Barnhart 1900). In chasmogamous flowers, the numerous long filaments far exceed the pistil in height (Fig. 48: j), and the anthers dehisce longitudinally along their entire length (Fig. 48: i). The few anthers of the cleistogamous forms, in contrast, are fused to the stigma and dehisce only at the point of adherence (Fig. 48: p,r); the mature capsules (Fig. 48: s,t) are smaller than those of the chasmogamous flower (Fig. 48: k,l) and contain fewer and smaller seeds (Wilbur and Daoud 1964; Daoud and Wilbur 1965).

REFERENCES CITED

Barnhart, J. H. 1900. Heteromorphism in *Helianthemum*. *Bull. Torrey Bot. Club* 27:589–592.

Brizicky, G. K. 1964. The genera of Cistaceae in the southeastern United States. *J. Arnold Arbor.* 45:346–357.

Cronquist, A. 1988. *The evolution and classification of flowering plants*, 2d ed., pp. 339–345. New York Botanical Garden, Bronx.

Daoud, H. S., and R. L. Wilbur. 1965. A revision of the North American species of *Helianthemum* (Cistaceae). *Rhodora* 67:63–82, 201–216, 255–312.

Hodgdon, A. R. 1938. A taxonomic study of *Lechea. Rhodora* 40:29–69, 87–131.

Thorne, R. F. 1992. Classification and geography of the flowering plants. *Bot. Rev.* (Lancaster) 58:225–348.

Wilbur, R. L., and H. S. Daoud. 1964. The genus *Helianthemum* (Cistaceae) in the southeastern United States. *J. Elisha Mitchell Sci. Soc.* 80:38–43.

VIOLACEAE
Violet Family

Perennial herbs or shrubs to small trees. **Leaves** simple, sometimes lobed, alternate or forming basal rosettes, stipulate. **Inflorescence** indeterminate, racemose or flowers solitary to few in leaf axils. **Flowers** zygomorphic or sometimes actinomorphic, perfect, hypogynous to slightly perigynous (with short hypanthium), showy, usually nodding, sometimes cleistogamous. **Calyx** of 5 sepals, distinct, usually unequal and auricled at base, imbricate, persistent. **Corolla** of 5 petals, distinct, usually unequal with large spurred anterior petal, white, yellow, green, or purple, imbricate or contorted. **Androecium** of 5 stamens, connivent or syngenesious in a ring close around the gynoecium; filaments very short, dilated, free or often basally connate; anthers basifixed, dehiscing longitudinally, introrse, with a membranous appendage extending from the broad connective, abaxial pair (or all) with a spur-like nectary at the base. **Gynoecium** of 1 pistil, typically 3-carpellate; ovary superior, 1-locular; ovules 1 to many, anatropous, placentation parietal; style simple, columnar or clavate, flexuous; stigma head-like with restricted receptive area. **Fruit** usually a 3-valved elastic loculicidal capsule; seeds obovoid or globose, often carunculate; endosperm copious, fleshy; embryo straight.

Family characterization: Perennial herbs; zygomorphic flowers; spurred anterior petal and anthers; coherent stamens with short filaments; 3-carpellate pistil with parietal placentae and unusual style and stigma; and explosive capsule with carunculate seeds. Saponins and alkaloids often present. Tissues with calcium oxalate crystals.

Genera/species: 22/900

Distribution: Cosmopolitan; the temperate species mainly are perennial herbs, and the tropical species mainly are trees and shrubs.

Major genera: *Viola* (500 spp.), *Rinorea* (200–340 spp.), and *Hybanthus* (150 spp.)

U.S./Canadian representatives: *Viola* (87 spp.), *Isodendrion* (14 spp.), and *Hybanthus* (5 spp.)

Economic plants and products: Perfume from hybrids of *Viola odorata*. Medicinal compounds (saponins and alkaloids used as expectorants and emetics) from the roots of *Hybanthus* and *Viola*. Ornamental plants: species of *Hybanthus* (green-violet), *Hymenanthera*, *Melicytus*, and *Viola* (violet, pansy).

Commentary: At least one-half of the species in the Violaceae belong to the genus *Viola* (violets), a temperate group of usually perennial herbs. The family is represented in the continental United States and Canada by at least 80 species of *Viola* and 3 of *Hybanthus* (green-violets).

The flowers of the family generally are entomophilous, but pollination biology is well documented only in *Viola* (Brizicky 1961). The characteristic flowers of *Viola* are highly specialized for insect pollination. The large spurred petal (Fig. 49: 1c) stores the nectar pro-

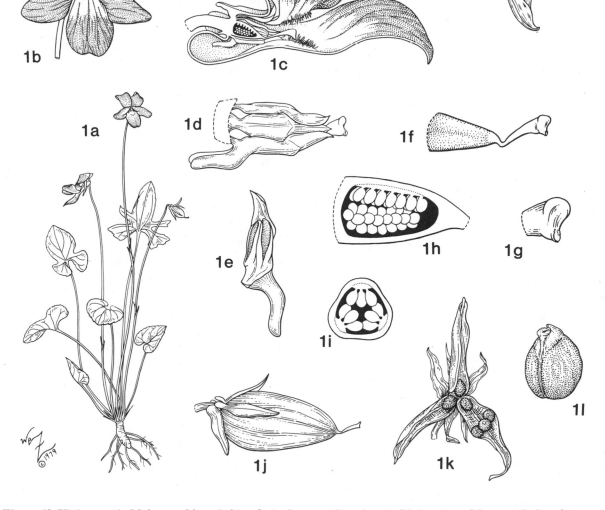

Figure 49. Violaceae. 1, *Viola septemloba*: **a,** habit, ×⅖; **b,** flower, ×1½; **c,** longitudinal section of flower, ×3; **d,** androecium and gynoecium, ×6; **e,** spurred stamen, ×6; **f,** pistil, ×6; **g,** stigma, ×12; **h,** longitudinal section of ovary, ×12; **i,** cross section of ovary, ×12; **j,** capsule before dehiscence, ×3; **k,** dehisced capsule, ×3; **l,** seed, ×12. **2,** *Viola sororia*: cleistogamous flower, ×6.

duced by the gland-like appendages (spurs) on the lowermost anthers (Fig. 49: 1e). Nectar guides (colored streaks on the petals) lead to the nectaries. The introrse connivent anthers form a tight ring around the ovary (Fig. 49: 1d). When an insect (usually a bee) probes for nectar, it brushes against the spurred anthers and receives a shower of pollen; as it enters another flower, the pollen is deposited on the receptive stigma, a restricted area on the stigmatic head (Fig. 49: 1g).

Delicate violet flowers are characteristic of the spring flora, often growing in shaded, moist soils. After the normal flowering season is past, many *Viola* species produce cleistogamous flowers (Fig. 49: 2). Research on *Viola* (Russell 1960) has indicated that cleistogamous flowers are produced in response to longer photoperiods. These inconspicuous, bud-like, apetalous flowers are concealed beneath the leaves on short pedicels. The anthers, often reduced to two, are closely appressed to the short style and stigma. Little pollen is produced. Self-pollination occurs when the pollen grains germinate in the anthers, and the tubes grow through the upper anther walls directly into the stigma (Madge 1929). Similar cleistogamous flowers have been reported for species of *Hybanthus* (Brizicky 1961).

The loculicidal capsule typically splits elastically into three boat-shaped valves (Fig. 49: 1k). In some species of *Viola*, the valves close when drying, which forces the seeds out one by one—sometimes dispersing them for a distance of several meters. Ants, which disperse the seeds of many species, are attracted to an appendage (caruncle; Fig. 49: 1l) on the raphe of the seed (Gates 1943).

REFERENCES CITED

Brizicky, G. K. 1961. The genera of Violaceae in the southeastern United States. *J. Arnold Arbor.* 42:321–333.

Gates, B. N. 1943. Carunculate seed dissemination by ants. *Rhodora* 45:438–445.

Madge, M. 1929. Spermatogenesis and fertilization in the cleistogamous flowers of *Viola odorata* var. *praecox* Gregory. *Ann. Bot.* (Oxford) 43:545–577.

Russell, N. H. 1960. Studies in the photoperiodic responses of violets (*Viola*). *SouthW. Naturalist* 5:177–186.

SALICACEAE

Willow Family

Trees, shrubs, or sometimes subshrubs. **Leaves** simple, often serrate, alternate, deciduous, stipulate; stipules often conspicuous, caducous. **Inflorescence** indeterminate, spicate, erect or pendulous (catkin), lateral. **Flowers** more or less actinomorphic, imperfect (plants dioecious), hypogynous, usually precocious, very reduced, each subtended by a fringed or hairy bract. **Perianth** absent or reduced to a cupular disc or 1 or 2 nectariferous glands. **Androecium** of 2 to numerous stamens; filaments distinct to basally connate; anthers basifixed, dehiscing longitudinally. **Gynoecium** of 1 pistil, 2- to 4-carpellate; ovary superior, 1-locular, sometimes stalked; ovules generally numerous, anatropous, placentation parietal or basal; style 1 or absent; stigmas 2 to 4; often bifid or irregularly lobed. **Fruit** a 2- to 4-valved loculicidal capsule; seeds small, numerous, usually with an apical tuft of hairs (coma); endosperm absent or scanty and oily; embryo small, straight.

Family characterization: Fast-growing and often clonal dioecious trees and shrubs; deciduous, stipulate leaves; reduced imperfect flowers subtended by fringed bracts and arranged in erect or pendent, spicate inflorescences (catkins); absent or vestigial perianth represented by cup-like disc or gland(s); and loculicidal capsule with comose seeds. Certain glucosides (e.g., salicin and populin), cyanogenic glycosides, and tannins present. Tissues with calcium oxalate crystals. Anatomical features: periderm with superficial origin (from epidermis or the next lower layer); distal end of petioles with closed ring(s) of xylem and phloem (Fig. 50: 1d); often unitegmic ovules (or second integument very thin); and trilacunar nodes (Fig. 50: 1b).

Genera/species: *Salix* (300–500 spp.), *Populus* (35 spp.), and *Chosenia* (1 sp.)

Distribution: Generally widespread, with centers of diversity in north temperate and subarctic regions; absent in Australia and the Malay Archipelago. Common in moist habitats.

U.S./Canadian representatives: *Salix* (106 spp.) and *Populus* (11 spp.)

Economic plants and products: Slender stems ("osiers") used in basketry from *Salix*. Lumber from species of *Populus* (wood pulp, boxes, matches) and *Salix* (boxes, matches). Medicinal bark from *Salix* (due to salicylic acid). Ornamental trees: species of *Salix* (willow) and *Populus* (aspen, cottonwood, poplar).

Commentary: The 300 to 500 species of *Salix* constitute most of the family. Species delimitations are difficult in this genus (and also in *Populus*) due to extensive hybridization. The affinities of the Salicaceae to other groups have been disputed, but the family generally is considered closely related to the very heterogeneous Flacourtiaceae (see Meeuse 1975; Thorne 1973, 1992; Cronquist 1981). Critical review of these and other allied families of the Violales is needed (see Judd et al. 1994). In the past, the Salicaceae have sometimes been included in the "Amentiferae" (see the discussions under Fagaceae and Betulaceae), a polyphyletic assemblage of woody plants with reduced flowers arranged in aments (Hjelmqvist 1948; Stern 1973; Dilcher and Zavada 1986). Actually, the apparent "simplicity" of the flowers is due to extreme reduction (Fisher 1928).

The minute and imperfect flowers, each in the axil of a bract, are congested into erect or pendulous spicate inflorescences (aments, catkins) that are unisexual (Fig. 50: 1e,i). The trees are dioecious. In *Populus*, the inflorescences are pendulous, and the flowers are entirely wind-pollinated. Although a disc- or cup-shaped gland (reduced perianth) occurs at the base of each flower (Fig. 50: 2), no nectar or scent is produced. The flowers of *Salix*, in contrast, attract various insects (especially bees and moths) that visit to collect pollen and/or nectar. Both the staminate and carpellate flowers secrete abundant and sweetly scented nectar from glands at the flower bases (Fig. 50: 1f,j). Because the flowers often appear before or at the same time as the leaves, the erect spikes of reduced flowers are somewhat conspicuous, especially the staminate inflorescences with their long filaments and bright yellow an-

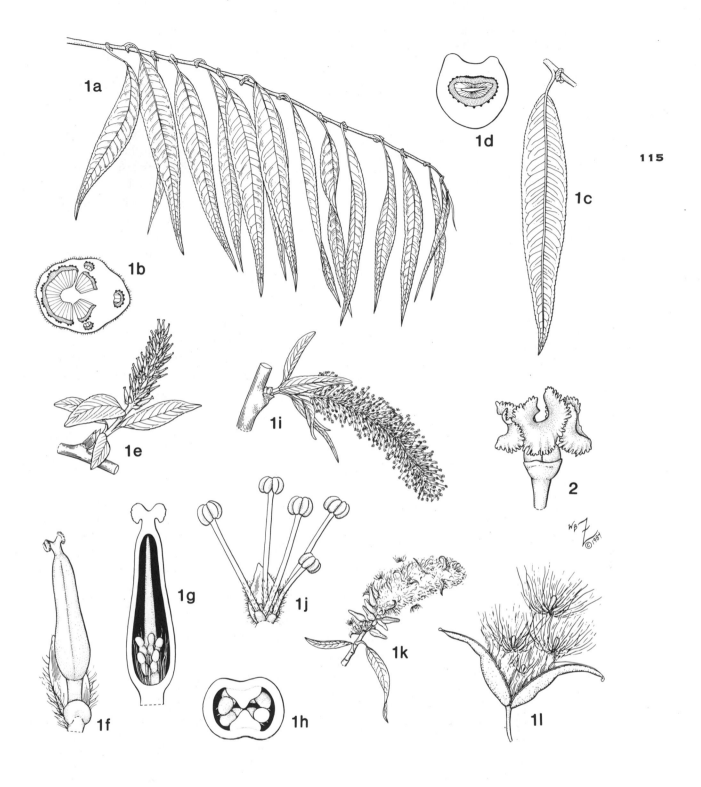

115

Figure 50. Salicaceae. 1, *Salix caroliniana*: **a,** sterile branch with mature leaves, ×⅓; **b,** cross section of node (three-trace trilacunar), ×6; **c,** leaf, ×½; **d,** cross section of petiole, ×18; **e,** carpellate inflorescence, ×1¼; **f,** carpellate flower and bract, ×10; **g,** longitudinal section of pistil, ×15; **h,** cross section of ovary, ×25; **i,** staminate inflorescence, ×1¼; **j,** staminate flower and bract, ×8; **k,** fruiting branch, ×¾; **l,** capsule, ×4½. **2,** *Populus deltoides*: carpellate flower, ×4½.

thers. Although entomophily is prevalent in *Salix*, much pollen also is spread by the wind, which probably is also important in pollination (Meeuse 1978; Argus 1986).

REFERENCES CITED

Argus, G. W. 1986. The genus *Salix* (Salicaceae) in the southeastern United States. *Syst. Bot. Monogr.* 9:1–170.

Cronquist, A. 1981. *An integrated system of classification of flowering plants*, pp. 432–435. Columbia University Press, New York.

Dilcher, D. L., and M. S. Zavada. 1986. Phylogeny of the Hamamelidae: An introduction. *Ann. Missouri Bot. Gard.* 73:225–226.

Fisher, M. J. 1928. The morphology and anatomy of flowers of Salicaceae. I. and II. *Amer. J. Bot.* 15:307–326, 372–394.

Hjelmqvist, H. 1948. Studies on the floral morphology and phylogeny of the Amentiferae. *Bot. Not.*, Suppl. 2:1–171.

Judd, W. S., R. W. Sanders, and M. J. Donoghue. 1994. Angiosperm family pairs: Preliminary cladistic analyses. *Harvard Pap. Bot.* 5:1–51.

Meeuse, A. D. J. 1975. Taxonomic relationships of Salicaceae and Flacourtiaceae. *Acta Bot. Neerl.* 24:437–457.

———. 1978. Entomophily in *Salix*: Theoretical considerations. In *The pollination of flowers by insects*, ed. A. J. Richards, pp. 47–50. Linn. Soc. Symp. Ser. No. 6. Academic Press, London.

Stern, W. L. 1973. Development of the amentiferous concept. *Brittonia* 25:316–333.

Thorne, R. F. 1973. The "Amentiferae" or Hamamelidae as an artificial group: A summary statement. *Brittonia* 25:395–405.

———. 1992. Classification and geography of the flowering plants. *Bot. Rev.* (Lancaster) 58:225–348.

PASSIFLORACEAE
Passion-flower Family

Herbaceous or woody vines with axillary tendrils, or sometimes shrubs. **Leaves** simple or sometimes compound, entire or serrate, often palmately lobed and veined, alternate, often with nectariferous glands on the petiole, stipulate; stipules small, deciduous. **Inflorescence** determinate, cymose and often reduced to 1 or 2 flower(s), bracteate, axillary. **Flowers** actinomorphic, usually perfect, perigynous, with cupulate to tubular hypanthium, showy, usually with nec-

tariferous disc. **Calyx** of typically 5 sepals, distinct or basally connate, often petaloid and/or fleshy, sometimes apically appendaged on abaxial side, persistent, imbricate. **Corolla** of typically 5 petals, distinct or basally connate, spreading, with fringed corona of 1 to several whorls of filaments and/or scales, greenish, white, or brightly colored, imbricate. **Androecium** of typically 5 stamens, often arising from a raised stalk (androgynophore); filaments distinct or basally connate; anthers dorsifixed, versatile, dehiscing longitudinally, introrse then extrorse (when mature). **Gynoecium** of 1 pistil, usually 3-carpellate; ovary superior, 1-locular, stipitate (often raised on androgynophore or sometimes a gynophore); ovules numerous, anatropous, with well-developed funicles, placentation parietal with intruding placentae; styles usually 3, distinct or basally connate; stigmas 3, often capitate, clavate, or discoid. **Fruit** a berry or sometimes a loculicidal capsule; seeds compressed, arillate, with pitted or reticulate seed coat; endosperm copious, oily, fleshy; embryo large, straight, spatulate, embedded in endosperm.

Family characterization: Vines with axillary tendrils; palmately lobed and veined leaves with glandular petioles; axillary, perigynous flowers with conspicuous fringed corona; androecium and gynoecium elevated on a stalk (androgynophore); ovary with 3 intruded parietal placentae; and a berry with arillate seeds as the fruit type. Alkaloids (e.g., passiflorine) and cyanogenic glucosides (hydrocyanic acid) present. Tissues commonly with calcium oxalate crystals. Anatomical features: stems frequently with anomalous secondary growth (Ayensu and Stern 1964); and trilacunar nodes.

Genera/species: 18/630

Distribution: Widespread in tropical to warm-temperate regions; most diverse in tropical America and Africa.

Major genera: *Passiflora* (350–500 spp.) and *Adenia* (93 spp.)

U.S./Canadian representatives: *Passiflora* (26 spp.)

Economic plants and products: Edible fruits from several species of *Passiflora* (granadilla, passion-fruit). Ornamental plants (species of 4 genera): *Adenia, Male-*

Figure 51. Passifloraceae. 1, *Passiflora incarnata*: **a,** habit, ×²/₅; **b,** petiole-leaf blade junction showing paired extrafloral nectaries, ×2½; **c,** flower, ×¾; **d,** pistil and one stamen showing relative positions of styles and anther at anthesis ("male stage"), ×2¼; **e,** androecium, gynoecium, androgynophore, and limen ("female stage"), ×2¼; **f,** longitudinal section of ovary, ×5; **g,** cross section of ovary, ×9; **h,** longitudinal section of flower, ×2½; **i,** berry, ×½; **j,** cross section of berry, ×½; **k,** seed (enclosed by aril), ×2½; **l,** seed (completely enclosed by sarcotesta; one-half of aril removed), ×2¼; **m,** seed (aril and sarcotesta removed), ×3. **2,** *Passiflora multiflora*: cross section of mature stem showing anomalous secondary growth pattern (stippled area represents parenchymatous cells of the xylem, phloem, and pith), ×2½.

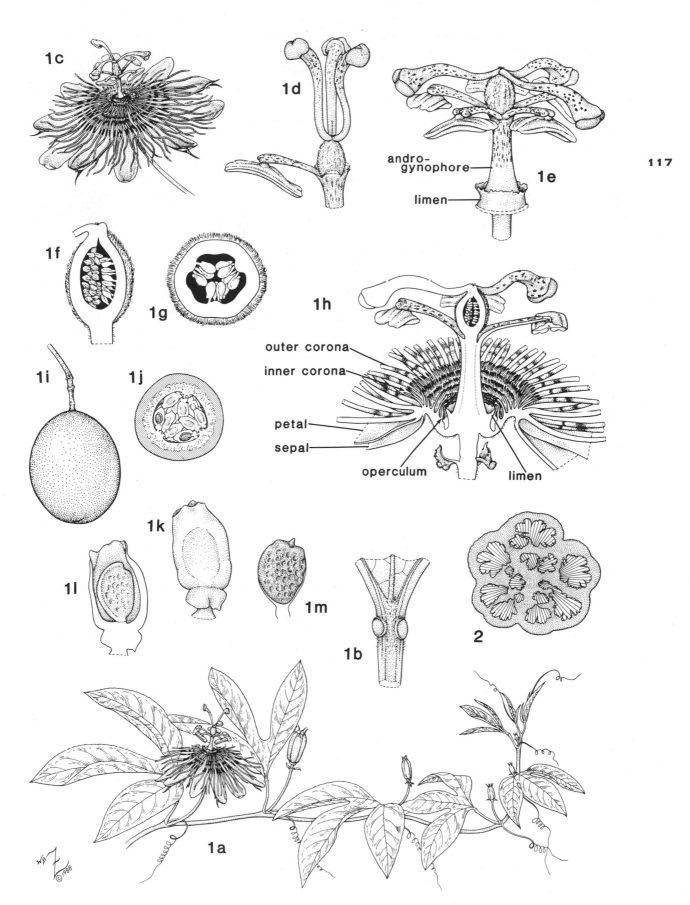

117

sherbia, *Passiflora* (passion-flower), and *Tetrapathaea* (New Zealand passion-flower, New Zealand passion-fruit).

Commentary: The genus *Passiflora* comprises at least one-half to two-thirds of the species of the family. *Paropsia* and allied genera, characterized by an arborescent habit, often are included in the Flacourtiaceae (de Wilde 1971).

Vegetatively, the Passifloraceae in our area are characterized as vines with axillary tendrils and often nectariferous petioles (Brizicky 1961). The tendrils represent modified inflorescences (or parts of inflorescences). Extrafloral nectaries, consisting of sessile or shortly stalked glands, usually occur at the petiole base (Fig. 51: 1b). Although not emphasized in the literature, anomalous stem development occurs in the family (e.g., species of *Passiflora*; Fig. 51: 2): included phloem (patches of phloem embedded in the xylem cylinder), alternating wedge-shaped sections of xylem and phloem with step-like margins, and irregularly shaped, scattered strands of xylem and phloem associated with vascular cambium fragments (Ayensu and Stern 1964).

Certain aspects of the complex floral structure are the most distinctive features of the family. The perigynous flowers have a cup-shaped to tubular hypanthium bearing the perianth and fringed corona at the rim (Fig. 51: 1h); in the center of the floral cup, the androecium and gynoecium are conspicuously elevated on a stalk (androgynophore; Fig. 51: 1e). The characteristic corona is brightly colored and occurs within the corolla. This structure varies from a single membrane or whorl of filaments to a complex system of several series of filaments, scales, and/or membranes (as in *Passiflora*; de Wilde 1974). In the broadest interpretation, the term "corona" refers to all accessory structures between the corolla and androecium. However, monographers have developed various terminology defining the several zones of the complex corona of *Passiflora* (Puri 1948; Killip 1938). The "outer corona" (or "radii") consists of the long filaments immediately within the petals, and "inner corona" (or "pali") refers to the inner whorls of shorter filaments or scales (see Fig. 51: 1h). The innermost whorl, the "operculum," is membranous and often dentate or fringed and serves as a cover over the nectar-secreting chamber. The outer and inner corona and operculum are derived from the perianth. A lobed rim or cup ("limen"; Fig. 51: 1e) occurring at or above the base of the androgynophore probably is staminodial in origin.

The conspicuous attractive flowers, often strongly scented, are pollinated by various bees, moths, butterflies, hummingbirds, and even bats (Sazima and Sazima 1978). An annular nectariferous region ("annulus") within the hypanthium secretes nectar at the base of the androgynophore, and the corona, with the various filament whorls appearing as colored concentric zones, serves as a nectar guide. The flowers often are protandrous and sometimes functionally staminate (May and Spears 1988). Before anthesis, the anthers are introrse and parallel to their filaments, and both stamens and styles are initially erect (Fig. 51: 1d). As the filaments radiate and incline when the flower opens, the versatile anthers rotate to become perpendicular to their filaments and are horizontal and extrorse in the final position (male stage in protandrous flowers; Fig. 51: 1e). The styles also recurve, eventually becoming situated between the anthers. Pollinators touch the inclined anthers and/or stigmas as they probe for the nectar in the floral cup (Janzen 1968).

In *Passiflora*, the fruit is a berry with a leathery to fleshy pericarp (Fig. 51: 1i,j). Each seed is embedded in a mucilaginous, pulpy aril that often is brightly colored (Fig. 51: 1k,l). The seed coat consists of an outer fleshy layer (sarcotesta) surrounding an inner hard layer. The seeds probably are dispersed by birds.

REFERENCES CITED

Ayensu, E. S., and W. L. Stern. 1964. Systematic anatomy and ontogeny of the stem in Passifloraceae. *Contr. U.S. Natl. Herb.* 34:45–73.

Brizicky, G. K. 1961. The genera of Turneraceae and Passifloraceae in the southeastern United States. *J. Arnold Arbor.* 42:204–218.

Janzen, D. H. 1968. Reproductive behavior in the Passifloraceae and some of its pollinators in Central America. *Behaviour* 32:33–48.

Killip, E. P. 1938. The American species of Passifloraceae. *Field Mus. Nat. Hist., Bot. Ser.* 19 (part 1): 1–331.

May, P. G., and E. E. Spears. 1988. Andromonoecy and variation in phenotypic gender of *Passiflora incarnata* (Passifloraceae). *Amer. J. Bot.* 75:1830–1841.

Puri, V. 1948. Studies in floral anatomy. V. On the structure and nature of the corona in certain species of the Passifloraceae. *J. Indian Bot. Soc.* 27:130–149.

Sazima, M., and I. Sazima. 1978. Bat pollination of the passion flower, *Passiflora mucronata*, in southeastern Brazil. *Biotropica* 10:100–109.

de Wilde, W. J. J. O. 1971. The systematic position of the tribe Paropsieae, in particular the genus *Ancistrothyrsus*, and a key to the genera of Passifloraceae. *Blumea* 19:99–104.

——. 1974. The genera of the tribe Passifloreae (Passifloraceae) with special reference to flower morphology. *Blumea* 22:37–50.

TURNERACEAE

Turnera Family

Usually perennial herbs or shrubs, with pubescence of diverse hairs (glandular, nonglandular, simple, multicellular). **Leaves** simple, entire or serrate (teeth sometimes glandular), occasionally lobed, alternate, usually with glands at blade base, exstipulate or sometimes stipulate (stipules minute). **Inflorescence** determinate, cymose and appearing fasciculate or racemose, axillary or terminal, or often flowers solitary and axillary, with pedicel sometimes fused to petiole of subtending leaf. **Flowers** actinomorphic, perfect, hypogynous, showy, usually heterostylous (dimorphic), often subtended by 2 bracteoles, with short to frequently tubular hypanthium, often with 5 glands between androecium and corolla. **Calyx** of 5 sepals, distinct, imbricate. **Corolla** of 5 petals, distinct, color various (often white, yellow, or purple), ephemeral, convolute. **Androecium** of 5 stamens, inserted inside the hypanthium; filaments distinct; anthers dorsifixed, dehiscing longitudinally, introrse. **Gynoecium** of 1 pistil, 3-carpellate; ovary superior, 1-locular; ovules usually numerous, anatropous, placentation parietal; styles 3, often flattened, sometimes bifid; stigmas fringed or brush-like. **Fruit** usually a loculicidal capsule; seeds arillate, with reticulate seed coat; endosperm copious, fleshy; embryo straight to slightly curved, embedded in endosperm.

Family characterization: Shrubs or perennial herbs with pubescence of diverse hair types (glandular, nonglandular, simple, multicellular, etc.) and often extrafloral nectaries; dimorphic, heterostylous flowers with tubular hypanthium; 5 stamens arising from inside the hypanthium; 1-locular ovary with 3 parietal placentae; 3 styles with fimbriate stigmas; and loculicidal capsule with arillate seeds. Cyanogenic glycosides often present. Tissues commonly with calcium oxalate crystals. Anatomical feature: tanniferous cells in leaves, pith, and cortex.

Genera/species: 8/120

Distribution: Primarily tropical and subtropical regions of America, Africa, Madagascar, and Mascarene Islands, with a few representatives in warm-temperate areas; most diverse in the American tropics.

Major genera: *Turnera* (60 spp.) and *Piriqueta* (19–20 spp.)

U.S./Canadian representatives: *Piriqueta* (4 spp.) and *Turnera* (3 spp.)

Economic plants and products: Herbal medicines (laxatives, stimulants) and flavorings from the leaves of a few *Turnera* spp. *Turnera ulmifolia* (sage-rose, yellow-alder, West Indian–holly) sometimes cultivated as an ornamental.

Commentary: On the basis of anatomy, palynology, and embryology, taxonomists generally agree that the Turneraceae are very closely related to the Passifloraceae (Brizicky 1961; Vijayaraghavan and Kaur 1966).

Vegetatively, the family is characterized by the pubescence of the stem, leaf, pedicel, calyx, and fruit. The various hair types include both nonglandular (simple and unicellular or variously stellate) and glandular (multicellular and stalked with various apices or tapering). Stalked glands (referred to as "extrafloral nectaries") with concave heads (which become convex at the onset of secretion) may occur at the base or margins of the leaf blade or on the petiole (e.g., leaves of *Piriqueta*, Fig. 52: b,c, and *Turnera* spp.; Metcalfe and Chalk 1950).

The majority of species in the family have dimorphic heterostyly, a type of flower polymorphism in which the flowers of the same species have two distinct style lengths that are associated with differences in other floral parts (see Fig. 52: e–k, and Hardin 1953). This outcrossing mechanism is discussed in detail in the commentary under Rubiaceae. Insects (usually bees) are attracted to the colorful, ephemeral flowers that often have extrastaminal nectaries (five glands, between the petals and stamens; e.g., *Piriqueta*, Fig. 52: i). Self-pollination may occur in the flowers of homostylous and incompletely heterostylous species when the petals shrivel and press the anthers and stigmas together.

The loculicidal capsules (Fig. 52: n) usually contain numerous seeds characterized by a reticulate seed coat plus a membranous unilateral aril (Fig. 52: o,p). The arils attract ants that disperse the seeds.

REFERENCES CITED

Brizicky, G. K. 1961. The genera of Turneraceae and Passifloraceae in the southeastern United States. *J. Arnold Arbor.* 42: 204–218.

Hardin, J. W. 1953. Heterostyly in *Piriqueta caroliniana. Castanea* 18:103–107.

Metcalfe, C. R., and L. Chalk. 1950. *Anatomy of the dicotyledons*, pp. 669–674. Oxford University Press, Oxford.

Vijayaraghavan, M. R., and D. Kaur. 1966. Morphology and embryology of *Turnera ulmifolia* L. and affinities of the family Turneraceae. *Phytomorphology* 19:539–553.

120

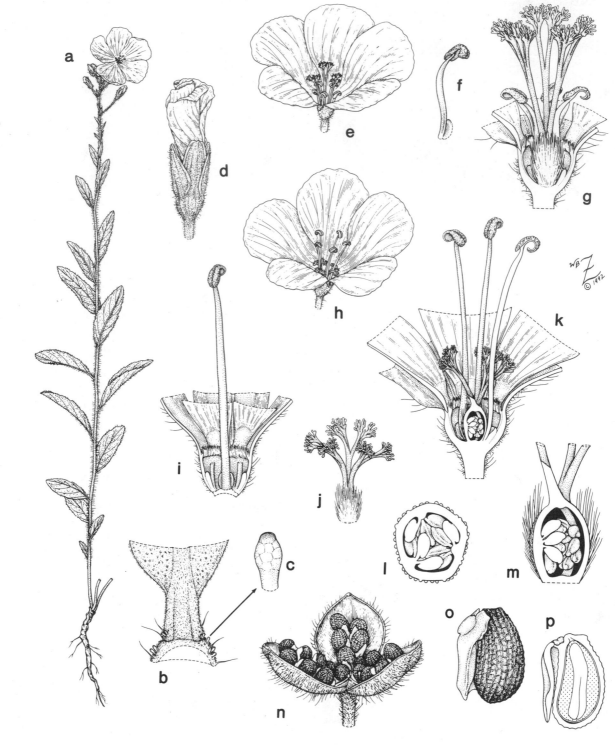

Figure 52. Turneraceae. *Piriqueta caroliniana*: **a,** habit, ×½; **b,** adaxial surface of leaf base showing extrafloral nectaries, ×8; **c,** extrafloral nectary (glandular hair), ×60; **d,** bud, ×2½; **e–g,** long-styled flower: **e,** long-styled flower, ×2; **f,** stamen (short) and basal gland, ×5; **g,** long-styled flower with half of perianth removed, ×5; **h–l,** short-styled flower: **h,** short-styled flower, ×2; **i,** stamen (long) and portion of hypanthium showing corona (hair fringe) at perianth bases and glands at base of filaments, ×5; **j,** pistil (short-styled), ×5; **k,** longitudinal section of short-styled flower, ×5; **l,** cross section of ovary, ×12; **m,** longitudinal section of ovary, ×10; **n,** capsule, ×5; **o,** seed (with aril), ×12; **p,** longitudinal section of seed, ×12.

CUCURBITACEAE
Pumpkin or Squash Family

Annual or sometimes perennial vines, sometimes softly woody, climbing or prostrate and trailing, often coarse and scabrous, sometimes somewhat succulent with abundant watery sap, with swollen tuberous rootstock; stems often 5-angled in cross section. **Leaves** simple and palmately 5-lobed or sometimes palmately compound, alternate, palmately veined, with spirally coiled tendrils at petiole base, exstipulate. **Inflorescence** determinate, cymose or flowers solitary, axillary. **Flowers** actinomorphic, imperfect (plants monoecious or dioecious), epigynous, showy, with cup- or tube-like expanded hypanthium, typically with variously developed nectary, ephemeral. **Calyx** synsepalous with 5 lobes, tubular, imbricate or open. **Corolla** sympetalous with 5 lobes, campanulate, rotate, to salverform, yellow or sometimes greenish, orange-yellow, or white, valvate or induplicate-valvate. **Androecium** of basically 5 stamens but often apparently 3 or sometimes 1 due to various modifications (cohesion, connation, and/or convolution); filaments distinct or coherent to connate in 2 pairs (with the fifth one distinct) to completely connate (monadelphous); anthers 1- or seemingly 2-locular, distinct or coherent to connate (in 2 pairs or completely connate), straight or often variously bent to convoluted, dehiscing longitudinally, extrorse; staminodes sometimes present in carpellate flowers. **Gynoecium** of 1 pistil, usually 3-carpellate; ovary inferior, typically 1-locular; ovules numerous, anatropous, placentation fundamentally parietal (with 3 enlarged and bifid placentae almost completely filling the locule); style usually 1, columnar; stigmas usually 3, thick, each bilobed or bifid; rudimentary pistil often present in staminate flowers. **Fruit** usually a berry or a modified berry (pepo) with leathery or hard pericarp or sometimes a variously dehiscing capsule; seeds many, large, usually compressed, with many-layered seed coat; endosperm absent; embryo straight, oily, with large and flat cotyledons.

Family characterization: Prostrate or climbing vines often with coarse and scabrous parts; 5-angled stems; palmately lobed or compound leaves with tendrils at the petiole base; yellow or white imperfect flowers; unusual stamens highly modified by displacement, reduction, fusion, and/or folding; inferior ovary with 3 greatly enlarged parietal placentae; and a berry (often a pepo) or fleshy capsule as the fruit type. Cucurbitacins (tetracyclic triterpenoids, bitter-tasting substances) and various alkaloids typically present. Anatomical features: trichomes often with calcified walls and with cystoliths (at the base and in nearby cells); bicollateral vascular bundles in the stem (often arranged in 2 concentric rings; Fig. 53: 1c) and petiole; and trilacunar nodes.

Genera/species: 118/825

Distribution: Pantropical and subtropical; a few representatives in temperate to cooler climates.

Major genera: *Cayaponia* (60 spp.), *Momordica* (45 spp.), *Gurania* (40 spp.), *Sicyos* (40 spp.), and *Trichosanthes* (40 spp.) (Jeffrey 1990a)

U.S./Canadian representatives: 26 genera/76 spp.; largest genera: *Sicyos* and *Cucurbita*

Economic plants and products: Food plants: *Citrullus* (watermelon), *Cucumis* (cantaloupe, cucumber, gherkin, honeydew, musk melon), *Cucurbita* (gourds, pumpkins, squashes, vegetable spaghetti), *Lagenaria* (calabash), *Momordica* (balsam-apple, bitter-melon), and *Sechium* (chayote). Loofah sponges (dried vascular system of fruit) from *Luffa*. Ornamental plants (species of 22 genera), including *Benincasa* (Chinese-watermelon), *Coccinea* (ivy-gourd), *Ecballium* (squirting-cucumber), *Lagenaria*, *Luffa* (loofah, vegetable-sponge), *Sicana* (casa-banana or cassabanana, musk-cucumber), *Sicyos* (burr-cucumber), and *Trichosanthes* (snake-gourd).

Commentary: The Cucurbitaceae compose a distinctive family in habit, floral structure, and biochemistry. Many of the major genera lack modern monographic revision (Jeffrey 1990a,b). There have been several opinions concerning the phylogenetic position of the family (see Cronquist 1981), which is now generally allied with the Begoniaceae (Jeffrey 1990b; Thorne 1992).

The unusual androecium of the cucurbits basically consists of five stamens that vary in the connation of the filaments and anthers (Chakravarty 1958). Key characters of genera often are based upon the morphology of the androecium (Jeffrey 1962, 1967, 1980). The most common situation (e.g., *Melothria*) is an androecium of apparently "three" stamens (Fig. 53: 1e), two with four pollen sacs (Fig. 53: 1g) and one with two pollen sacs (Fig. 53: 1h). Each two-locular stamen (with four pollen sacs; Fig. 53: 1g) actually is a compound stamen formed by the fusion of two adjacent stamens. Other variations include contorted "capitate" anthers (as in *Momordica*; Fig. 53: 2a) and/or monadelphous stamens, as in species of *Cucurbita* or *Sicyos*.

The flowers attract various insects (often bees) that visit to collect the nectar or sometimes also the copious pollen. Cross-pollination is favored by the mono-

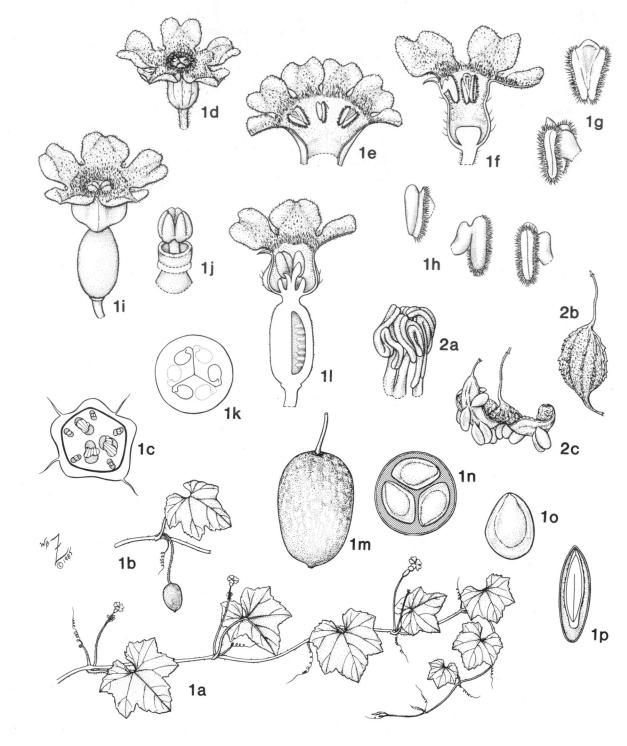

Figure 53. Cucurbitaceae. 1, *Melothria pendula*: **a,** flowering plant, ×½; **b,** portion of plant with mature fruit, ×½; **c,** cross section of stem showing bicollateral bundles in two rings, ×18; **d,** staminate flower, ×5; **e,** expanded corolla and epipetalous androecium of staminate flower, ×6; **f,** longitudinal section of staminate flower, ×7½; **g,** two views of a pair of connate stamens, ×15; **h,** three views of the single stamen, ×15; **i,** carpellate flower, ×5; **j,** detail of apical portion of pistil showing style, stigma, and nectary, ×7½; **k,** cross section of ovary, ×10; **l,** longitudinal section of carpellate flower, ×6; **m,** berry, ×2; **n,** cross section of berry, ×2; **o,** seed with mucilaginous outer seed coat, ×3; **p,** longitudinal section of seed, ×4½. **2,** *Momordica charantia*: **a,** androecium, ×6; **b,** fruit, ×⅔; **c,** dehisced fruit, ×⅔.

ecious or dioecious nature of the plants, and in monoecious species the staminate flowers often appear (on the younger parts of the stem) before the carpellate flowers on the same plant. As in the Begoniaceae, pollen-collecting insects mistake the thick, bright yellow-orange stigmas for an androecium (fused anthers) and thus visit carpellate flowers and pollinate them.

The Cucurbitaceae also are characterized by distinctive fruits and seeds. The fruit usually is a berry (Fig. 53: 1m,n), although other types occur (such as fleshy capsules; Fig. 53: 2b,c). When modified with a firm-walled epicarp (rind), as in *Citrullus* (watermelon), the berry is termed a "pepo." The seed has a several-layered seed coat (see Fig. 53: 1p), and the outer layers often swell in water (Corner 1976). The outermost covering is derived from the carpel wall (endocarp), and the inner layers, from the outer integument.

REFERENCES CITED

Chakravarty, H. L. 1958. Morphology of the staminate flowers in the Cucurbitaceae with special reference to the evolution of the stamens. *Lloydia* 21:49–87.

Corner, E. J. H. 1976. *The seeds of dicotyledons*, pp. 112–114. Cambridge University Press, Cambridge.

Cronquist, A. 1981. *An integrated system of classification of flowering plants*, pp. 422–425. Columbia University Press, New York.

Jeffrey, C. 1962. Notes on Cucurbitaceae, including a proposed new classification of the family. *Kew Bull.* 15:337–371.

——. 1967. On the classification of the Cucurbitaceae. *Kew Bull.* 20:417–426.

——. 1980. A review of the Cucurbitaceae. *Bot. J. Linn. Soc.* 81:233–247.

——. 1990a. Appendix: An outline classification of the Cucurbitaceae. In *Biology and utilization of the Cucurbitaceae*, ed. D. M. Bates, R. W. Robinson, and C. Jeffrey, pp. 449–463. Cornell University Press, Ithaca, N.Y.

——. 1990b. Systematics of the Cucurbitaceae: An overview. In *Biology and utilization of the Cucurbitaceae*, ed. D. M. Bates, R. W. Robinson, and C. Jeffrey, pp. 3–9. Cornell University Press, Ithaca, N.Y.

Thorne, R. F. 1992. Classification and geography of the flowering plants. *Bot. Rev.* (Lancaster) 58:225–348.

BEGONIACEAE
Begonia Family

Perennial herbs or low shrubs, erect or creeping, sometimes climbing or acaulescent, succulent, with thick rhizomes or tubers; stems jointed. **Leaves** simple, entire or serrate, often palmately lobed, commonly with obliquely based blades, alternate, usually palmately veined, usually distichous, fleshy, stipulate; stipules large, membranous, persistent. **Inflorescence**

determinate, cymose, axillary. **Flowers** usually zygomorphic, imperfect (plants monoecious), epigynous, showy. **Perianth** of petaloid tepals, differing in staminate and carpellate flowers, distinct, usually white, pink, or red; of staminate flowers: 4 tepals, biseriate with the 2 outer larger than the 2 inner, valvate; of carpellate flowers: 2 to usually 5 sepals, uniseriate, imbricate. **Androecium** of numerous stamens in many whorls; filaments distinct or basally connate; anthers basifixed, with pronounced connective separating and extending beyond the locules, dehiscing longitudinally, more or less latrorse. **Gynoecium** of 1 pistil, usually 3-carpellate; ovary inferior, 3-locular, 3-winged (wings equal or unequal); ovules numerous, anatropous, usually axile on intruded forked or lobed placentae; styles 3, distinct or basally connate, deeply bifid; stigmas 6, twisted, strongly papillose on all sides. **Fruit** a loculicidal capsule, winged; seeds numerous, minute, with reticulate seed coat and collar (operculum); endosperm absent; embryo straight, oily.

Family characterization: Succulent herbs or subshrubs with jointed stems; asymmetric, palmately veined leaves with large membranous stipules; imperfect, zygomorphic flowers; petaloid tepals with number varying in staminate (2 whorls of 2) and carpellate (usually 5) flowers; androecium of numerous stamens with pronounced connectives; inferior, winged ovary with deeply intruded forked/lobed placentae; bifid styles with twisted stigmas; winged loculicidal capsule as the fruit type; and nonendospermous, operculate seeds. Free organic acids (e.g., oxalic and malic acid) accumulating in the cells. Anatomical features: cystoliths in the leaves and sometimes the stem; large water-storage cells in the leaf hypodermis; and a transverse ring of specialized elongated cells of the testa (collar cells; Fig. 54: l,m) that rupture as an operculum (de Lange and Bouman 1992). Photosynthesis in many species via "succulent metabolism" (crassulacean acid metabolism, CAM).

Genera/species: *Begonia* (900–920 spp.) and *Hillebrandia* (1 sp.) (Thorne pers. comm.)

Distribution: Widespread in tropical to subtropical regions (excluding Polynesia and Australia); most diverse in northern South America. Commonly in damp, shady forests.

U.S./Canadian representatives: *Begonia* (2 spp.)

Economic plants and products: More than 100 species and 10,000 hybrids and cultivars of *Begonia* cultivated as ornamentals grown for attractive foliage and flowers.

Commentary: The 900+ species of *Begonia* compose

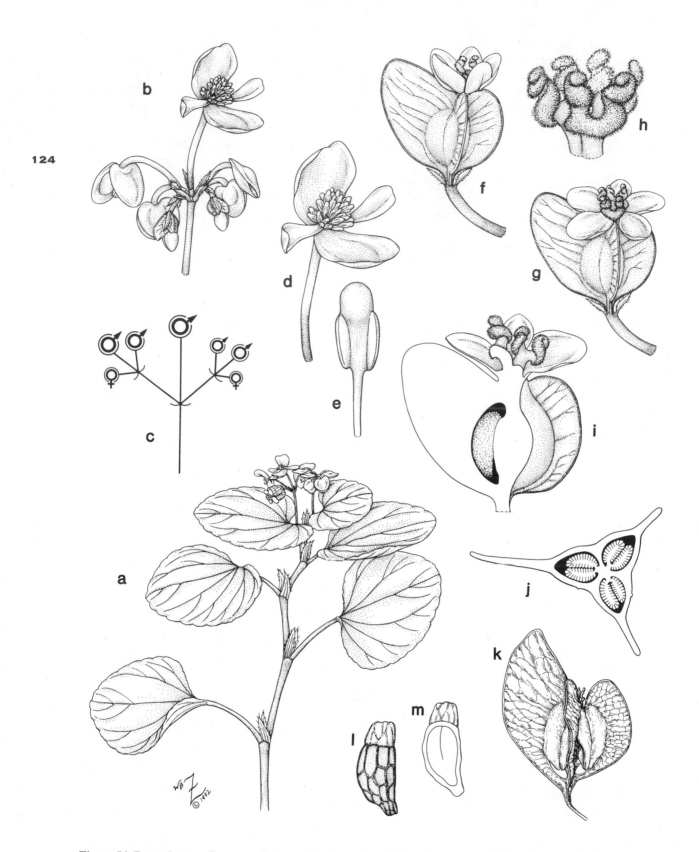

Figure 54. Begoniaceae. *Begonia cucullata* var. *hookeri*: **a,** habit, ×½; **b,** inflorescence, ×1¾; **c,** diagram of inflorescence (cyme) showing relative positions of staminate and carpellate flowers and sequence of development; **d,** staminate flower, ×2½; **e,** stamen, ×15; **f,** carpellate flower at anthesis, ×3; **g,** carpellate flower, ×3; **h,** stigmas, ×7; **i,** longitudinal section of flower, ×4½; **j,** cross section of ovary, ×6; **k,** capsule, ×2½; **l,** seed (note "operculum"), ×50; **m,** longitudinal section of seed, ×50.

the majority of this small family, resulting in a relatively homogeneous group (Smith and Wasshausen 1986). Some authors accept several small segregate genera, such as *Symbegonia* (Golding and Karegeannes 1986). The Begoniaceae usually have been associated with the Violales (e.g., Cronquist 1988; Thorne 1992).

The species are widely distributed in the tropics, mostly in damp, wooded areas. The more or less succulent plants reproduce vegetatively from groups of small tubers (often found in the leaf axils) or by adventitious buds that readily form on detached leaves in contact with moist soil. The buds arise on the upper leaf surface from a meristem that develops within the callus formed over the wound.

The plants are monoecious, with the staminate and carpellate flowers in the same cymose inflorescence (Fig. 54: b,c). Typically, the staminate flowers are more numerous and open first (on the primary axis), with the last axis (or axes) of the cyme bearing the few carpellate flowers (Fig. 54: c). The petaloid perianth is brightly colored and differs in morphology depending on the sex of the flower (Baranov 1981). The perianth of the staminate flower (Fig. 54: d) consists of a larger, outer pair of tepals that surround the smaller, inner pair; the whorl of numerous stamens form a conspicuous yellow globose mass in the center. The carpellate flower (Fig. 54: f,g) has a more or less uniseriate perianth (two to five tepals) that top a winged ovary and surround the twisted, bright-yellow stigmas (Fig. 54: h). The showy flowers in spreading inflorescences have no nectaries: insects visit to gather pollen. The bright-yellow, papillose stigmas mimic a cluster of anthers, which persuade the pollen-collecting insects to visit (and thereby pollinate) the carpellate flowers (Faegri and van der Pijl 1980). Cross-pollination is reinforced by the earlier development of the staminate flowers in the cyme.

REFERENCES CITED

Baranov, A. I. 1981. Studies in the Begoniaceae. *Phytologia Mem.* 4:1–88.

Cronquist, A. 1988. *The evolution and classification of flowering plants*, 2d ed., pp. 339–345. New York Botanical Garden, Bronx.

Faegri, K., and L. van der Pijl. 1980. *The principles of pollination ecology*, p. 58. Pergamon Press, Oxford.

Golding, J., and C. E. Karegeannes. 1986. Begoniaceae, part II: Annotated species list. *Smithsonian Contr. Bot.* 60:131–584.

de Lange, A., and F. Bouman. 1992. Seed micromorphology of the genus *Begonia* in Africa: Taxonomic and ecological implications. In *Studies in Begoniaceae III*, ed. J. J. F. E. de Wilde, pp. 1–82. Wageningen Agric. Univ. Pap. Wageningen, The Netherlands.

Smith, L. B., and D. C. Wasshausen. 1986. Begoniaceae, part I: Illustrated key. *Smithsonian Contr. Bot.* 60:1–129.

Thorne, R. F. 1992. Classification and geography of the flowering plants. *Bot. Rev.* (Lancaster) 58:225–348.

BRASSICACEAE OR CRUCIFERAE
Mustard Family

Annual, biennial, or perennial herbs with watery, acrid sap. **Leaves** simple or sometimes pinnately lobed, alternate and often forming basal rosettes, exstipulate. **Inflorescence** indeterminate, racemose or corymbose, terminal. **Flowers** actinomorphic, perfect, hypogynous, ebracteate, usually inconspicuous. **Calyx** of 4 sepals, distinct, biseriate, imbricate. **Corolla** of 4 petals, distinct, arranged diagonally (cruciform), long-clawed with horizontally spreading limbs, generally white, yellow, lavender, or pink, imbricate or contorted. **Androecium** of 6 stamens, tetradynamous; filaments free or connate in pairs (usually the inner), often with basal glands; anthers basifixed, dehiscing longitudinally, introrse. **Gynoecium** of 1 pistil, 2- (or 4: see Eames and Wilson 1928, 1930) carpellate; ovary superior, 2-locular by means of membranous partition (false septum); ovules usually many, anatropous or campylotropous, placentation parietal in ovary wall-partition angle; style absent or short; stigma 2-lobed and commissural (at right angles to partition) or capitate. **Fruit** a 2-valved silique (longer than broad) or silicle (wider than long), opening from the base toward the apex, sometimes transversely dehiscent or indehiscent, flattened parallel or perpendicular to the partition, often thin and membranous; seeds (in dehiscent fruit) attached to persistent exposed rim (replum); endosperm scanty or absent; embryo large, curved, bent, or folded.

Family characterization: Herbaceous plants with acrid juice; ebracteate flowers; 4 clawed petals arranged in a "cross"; tetradynamous stamens; 2-locular ovary with membranous false partition and parietal placentae; commissural stigma; distinctive fruit (silique or silicle) with seeds attached to persistent rim (replum); and strongly curved to folded embryo. Glucosinolates ("mustard oils," sulfur-containing compounds) present (Rodman 1981). Anatomical feature: secretory cells (myrosin cells; Jørgensen 1981) containing myrosinase (Björkman 1976) in all plant organs.

Genera/species: 376/3,200

Distribution: Primarily in temperate and cold regions of the Northern Hemisphere; centers of diversity in

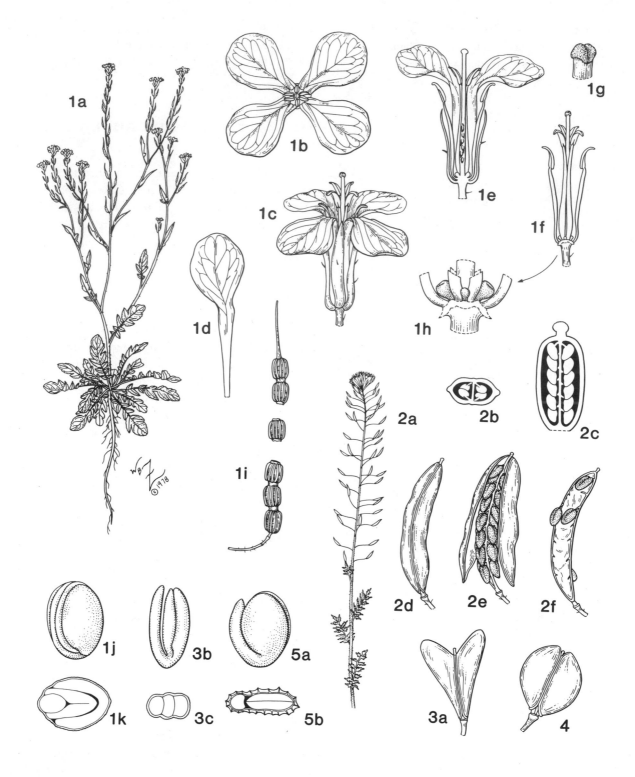

Figure 55. Brassicaceae. 1, *Raphanus raphanistrum*: **a,** habit, ×1/10; **b,** top view of flower, ×3; **c,** lateral view of flower, ×3; **d,** clawed petal, ×3; **e,** longitudinal section of flower, ×3; **f,** androecium (tetradynamous) and gynoecium, ×3; **g,** stigma, ×12; **h,** glands at base of filaments, ×12; **i,** dehiscing fruit, ×1; **j,** embryo (cotyledons conduplicate), ×9; **k,** cross section of seed, ×9. **2,** *Descurainia pinnata*: **a,** habit, ×1/2; **b,** cross section of ovary, ×25; **c,** longitudinal section of pistil, ×25; **d,** silique before dehiscence, ×6; **e,** dehiscing silique, ×6; **f,** replum with a few attached seeds, ×6. **3,** *Capsella bursa-pastoris* (*C. rubella*): **a,** silicle, ×3; **b,** embryo (cotyledons incumbent), ×25; **c,** cross section of seed, ×25. **4,** *Lepidium virginicum*: silicle, ×6. **5,** *Rorippa nasturtium-aquaticum* (*Nasturtium officinale*): **a,** embryo (cotyledons accumbent), ×25; **b,** cross section of seed, ×25.

TABLE 7. Major morphological differences traditionally used to separate the Brassicaceae and Capparaceae. Recent cladistic analyses based on molecular (Rodman et al. 1994) and morphological (Judd et al. 1994) data suggest that the two families should be combined (see text for discussion).

CHARACTER	BRASSICACEAE (CRUCIFERAE)	CAPPARACEAE
GENERA/SPECIES	376/3,200	46/930
REPRESENTATIVE GENERA	*Alyssum, Arabis, Cardamine, Draba, Lepidium*	*Capparis, Cleome, Polanisia*
DISTRIBUTION	mostly temperate and cold regions of the Northern Hemisphere	widespread in tropical to subtropical regions, with a few in the temperate zone
HABIT	herbs	shrubs, or sometimes herbs or trees
LEAVES	simple, sometimes pinnately lobed exstipulate	simple, or often palmately compound stipulate (stipules minute, modified) or exstipulate
INFLORESCENCE	racemose or corymbose	racemose, or occasionally flowers solitary
FLOWERS	usually actinomorphic usually ebracteate	actinomorphic, or more often ± zygomorphic usually bracteate
SEPALS	4	2 to 6
PETALS	4, forming a cross	2 to 6
STAMEN NUMBER	2 + 4 (tetradynamous)	usually 6 to many; occasionally 4
OVARY	usually sessile (no gynophore) 2-locular (false septum)	usually stipitate (with gynophore) usually 1-locular (no false septum)
FRUIT TYPE	usually a 2-valved silique/silicle with replum + thin partition, sometimes transversely dehiscent, or indehiscent	usually a berry, sometimes indehiscent, or a silique with frame-like replum (no partition), occasionally a capsule, nut, or drupe
EMBRYO SHAPE	strongly curved to folded	± curved

the Mediterranean region and southwestern and central Asia.

Major genera: *Draba* (300 spp.), *Alyssum* (150–168 spp.), *Cardamine* (130–160 spp.), *Lepidium* (150 spp.), and *Arabis* (120 spp.)

U.S./Canadian representatives: 94 genera/634 species; largest genera: *Draba, Arabis,* and *Lesquerella*

Economic plants and products: Food and condiments from *Armoracia* (horse-radish), *Brassica* (black mustard, broccoli, Brussels sprouts, cabbage, cauliflower, Chinese cabbage, kale, kohlrabi, rape, rutabaga, turnip), *Raphanus* (radish), *Rorippa* (incl. *Nasturtium,* watercress), and *Sinapis* (yellow mustard). Canola oil extracted from seeds of *Brassica* spp. cultivars. Many weedy species (see Rollins 1981), including *Capsella* (shepherd's-purse), *Descurainia* (tansy-mustard), and *Lepidium* (pepper-grass). Ornamental plants (species of about 57 genera), including *Arabis* (rockcress), *Ery-*

simum (wallflower), *Hesperis* (rocket), *Iberis* (candytuft), and *Lunaria* (honesty, money plant).

Commentary: The Brassicaceae, a predominantly north temperate family, are considered to be a well-defined taxon because of the distinctive flowers and fruits (Al-Shehbaz 1984), discussed in more detail below. Taxonomists have long agreed that the family is very closely related to the tropical to subtropical Capparaceae (caper family; Fig. 56), according to numerous shared characters (see Rodman 1991a,b), especially the (apparently) two-carpellate ovary with parietal placentae. Table 7 lists the major morphological differences between the two groups; many character states overlap, however. Recent cladistic analyses (Rodman et al. 1993; Judd et al. 1994) indicate that the Capparaceae, as circumscribed in most classification schemes (such as Thorne 1992), are paraphyletic, and that the sister group Brassicaceae are monophyletic. Combining the two families is the taxonomic solution

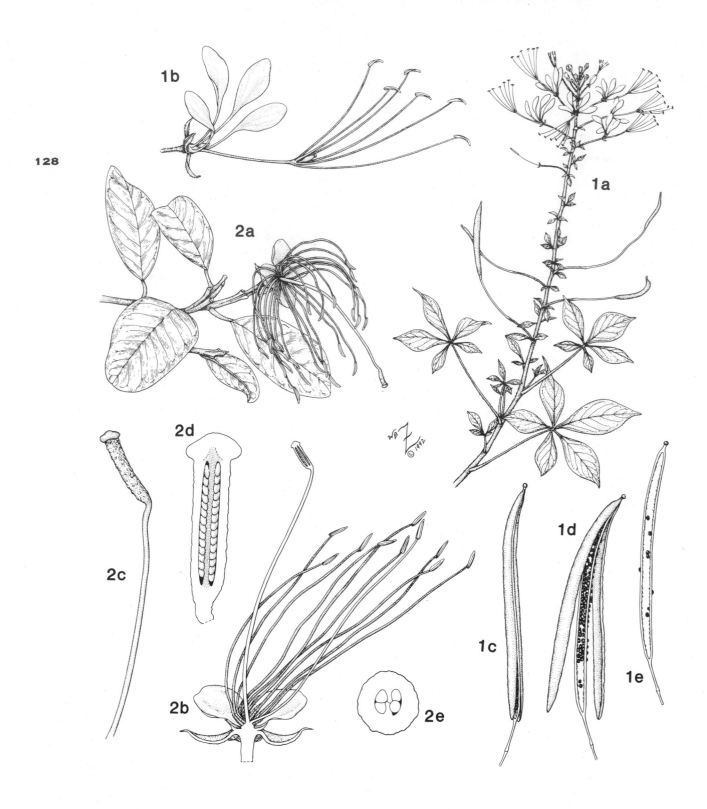

Figure 56. Capparaceae. 1, *Cleome gynandra*: **a,** habit, ×½; **b,** flower, ×2½; **c,** fruit, ×¾; **d,** dehiscing fruit showing frame-like replum, ×¾; **e,** frame-like replum with attached seeds, ×¾. **2,** *Capparis cynophallophora*: **a,** flowering branch, ×1; **b,** longitudinal section of flower, ×1½; **c,** gynophore and pistil, ×3½; **d,** longitudinal section of pistil, ×9; **e,** cross section of ovary, ×15. See Table 7 for family characteristics.

proposed by Judd et al. (1994), but this expanded circumscription has not yet been adopted by Thorne.

The morphology of the gynoecium of the Brassicaceae (and Capparaceae) has received much attention in the literature (see Al-Shehbaz 1984 for a summary). A basic carpel number of four (Eames and Wilson 1928, 1930) or two (Puri 1951) has been proposed. Although often considered unique to the Brassicaceae, the gynoecium with replum (plus false partition, discussed below under fruit) evidently is homologous to those in the Capparaceae with a replum (but lacking a membranous septum; Al-Shehbaz 1984).

The most obvious floral character of the Brassicaceae is the cruciform arrangement of the petals (Fig. 55: 1b); "Cruciferae," meaning "cross-bearing," is a conserved family synonym allowed by the *International Code of Botanical Nomenclature* (Greuter et al. 1988). The often inconspicuous flowers exhibit certain adaptations for entomophily; those of many genera are protogynous, and several are protandrous (see Rollins 1993). The glands at the base of the filaments (Fig. 55: 1f,h) secrete nectar, which accumulates in the gibbose bases of the inner sepals. The long-clawed petals (Fig. 55: 1d) with spreading limbs protect the nectar from dilution by rain. Furthermore, in many genera the flowers are arranged in corymbs, resulting in many flowers massed together on one level (Fig. 55: 1a). Some or all of the anthers become twisted so that a visiting insect can touch an anther with one side of its body and the stigma with the other. The stigma is in such close proximity to the anthers that self-pollination also is prevalent in the family.

Since fruit morphology is commonly used in artificial keys for practical identification, crucifers should be collected late in the season with both flowers and mature fruits (Hedge 1976). Traditionally, the two-valved fruit has arbitrarily been termed "silique" when "long" (two or three times longer than wide; Fig. 55: 2d,e) or "silicle" when "short" (two or three times wider than long; Fig. 55: 3a, 4). The silique or silicle is divided into two chambers by a membranous partition (Fig. 55: 2f) that is formed by the union of outgrowths of the placentae. This septum is bounded by the persistent rim-like placentae called the replum; some authors, however, use the term "replum" to refer specifically to the partition itself. The fruit may be either flattened parallel to a broadened partition (*Lunaria*) or perpendicular to a narrow partition (*Capsella*; Fig. 55: 3a). Generally, each chamber of the fruit has two rows of seeds. In elongated fruits, the seeds of the two rows in each chamber tend to alternate with each other, resulting in an apparent single row of seeds per chamber. After the fruit dehisces, the exposed seeds adhere to the replum and eventually become detached by the wind (Fig. 55: 2f). However, several species have indehiscent fruit (Zohary 1948), which may break into single-seeded portions, as in *Raphanus* (Fig. 55: 1i). Although the silicle/silique with replum often is cited as a unique fruit of the Brassicaceae, several of the Capparaceae (e.g., *Cleome*; Fig. 56: 1c–e) share this derived type. In these cases, however, the silique-like fruits have a thickened, frame-like, persistent replum that lacks the membranous partition (Fig. 56: 1e)—a fruit type also found in some of the Papaveraceae (Ernst 1963; also see the commentary under Papaveraceae concerning fruit types).

The embryo characteristically is strongly curved to folded. The position of the radicle relative to the cotyledons is an important generic feature (Vaughan et al. 1976), although it is almost impossible to use in the field. There are three main types of embryo forms: cotyledons incumbent—when the dorsal sides (faces) of the cotyledons are parallel to the radicle (Fig. 55: 3b,c); cotyledons accumbent—when the edges of both cotyledons are adjacent to the radicle (Fig. 55: 5a,b); and cotyledons conduplicate—when the dorsal sides of the cotyledons are folded lengthwise over the radicle and more or less enclose it (Fig. 55: 1j,k).

REFERENCES CITED

Al-Shehbaz, I. A. 1984. The tribes of Cruciferae (Brassicaceae) in the southeastern United States. *J. Arnold Arbor.* 65:343–373.

Björkman, R. 1976. Properties and function of plant myrosinases. In *The biology and chemistry of the Cruciferae*, ed. J. G. Vaughan, A. J. MacLeod, and B. M. G. Jones, pp. 191–205. Academic Press, London.

Eames, A. J., and C. L. Wilson. 1928. Carpel morphology in the Cruciferae. *Amer. J. Bot.* 15:251–270.

—— and ——. 1930. Crucifer carpels. *Amer. J. Bot.* 17:638–656.

Ernst, W. R. 1963. The genera of Capparaceae and Moringaceae in the southeastern United States. *J. Arnold Arbor.* 44:81–95.

Greuter, W., H. M. Burdet, W. G. Chaloner, V. Demoulin, R. Grolle, et al., eds. 1988. *International code of botanical nomenclature*, p. 23. Koeltz Scientific Books, Königstein, Germany.

Hedge, I. C. 1976. A systematic and geographical survey of Old World Cruciferae. In *The biology and chemistry of the Cruciferae*, ed. J. G. Vaughan, A. J. MacLeod, and B. M. G. Jones, pp. 1–45. Academic Press, London.

Jørgensen, L. B. 1981. Myrosin cells and dilated cisternae of the endoplasmic reticulum in the order Capparales. *Nordic J. Bot.* 1:433–445.

Judd, W. S., R. W. Sanders, and M. J. Donoghue. 1994. Angiosperm family pairs: Preliminary cladistic analyses. *Harvard Pap. Bot.* 5:1–51.

Puri, V. 1951. The role of floral anatomy in the solution of morphological problems. *Bot. Rev.* (Lancaster) 17:471–553.

Rodman, J. E. 1981. Divergence, convergence, and parallelism in phytochemical characters: The glucosinolate-myrosinase system. In *Phytochemistry and angiosperm phylogeny*, ed. D. A. Young and D. S. Seigler, pp. 43–79. Praeger, New York.

——. 1991a. A taxonomic analysis of glucosinolate-producing plants. Part 1: Phenetics. *Syst. Bot.* 16:598–618.

——. 1991b. A taxonomic analysis of glucosinolate-producing plants. Part 2: Cladistics. *Syst. Bot.* 16:619–629.

——, R. A. Price, K. Karol, E. Conti, K. J. Sytsma, et al. 1993. Nucleotide sequences of the *rbc*L gene indicate monophyly of the mustard oil plants. *Ann. Missouri Bot. Gard.* 80:686–699.

Rollins, R. C. 1981. Weeds of the Cruciferae (Brassicaceae) in North America. *J. Arnold Arbor.* 62:517–540.

——. 1993. *The Cruciferae of continental North America*, pp. 21–79. Stanford University Press, Stanford.

Thorne, R. F. 1992. Classification and geography of the flowering plants. *Bot. Rev.* (Lancaster) 58:225–348.

Vaughan, J. G., J. R. Phelan, and K. E. Denford. 1976. Seed studies in the Cruciferae. In *The biology and chemistry of the Cruciferae*, ed. J. G. Vaughan, A. J. MacLeod, and B. M. G. Jones, pp. 119–144. Academic Press, London.

Zohary, M. 1948. Follicular dehiscence in Cruciferae. *Lloydia* 11:226–228.

VISCACEAE

Mistletoe Family

Perennial herbs or small shrubs, hemiparasitic on tree branches, brittle; stems with sympodial growth, dichotomously branched, with jointed nodes; roots modified into haustoria (penetrating the host). **Leaves** simple, entire, opposite, decussate, parallel-veined, leathery to somewhat succulent, persistent, sometimes reduced to scales, exstipulate. **Inflorescence** determinate, cymose and often appearing spicate, axillary. **Flowers** actinomorphic, imperfect (plants usually dioecious), epigynous, with cup-shaped receptacle, minute, sessile. **Perianth** of 3 or 4 tepals in 1 whorl, sepaloid, distinct or basally connate, usually yellowish or greenish, valvate. **Androecium** of 3 or 4 stamens, opposite and adnate to the tepals; filaments very short to usually absent; anthers variously modified, often sessile on tepal, 2- or 1-locular (due to confluence), dehiscing by pores or transverse slits, introrse. **Gynoecium** of 1 pistil, 3- or 4-carpellate; ovary inferior, obscurely 1-locular with large placenta-like structure (mamelon) filling the space; ovules undifferentiated as 2 embryo sacs (female gametophytes) within the placenta; style short or absent; stigma capitate; rudimentary pistil present in staminate flowers. **Fruit** a berry, with viscous layer surrounding the seed, sometimes explosive; seed 1, with no seed coat; endosperm copious, starchy, chlorophyllous (green); embryo large, straight, embedded in endosperm.

Family characterization: Hemiparasitic, brittle shrubs attached to tree branches with haustoria; dichotomously branched, jointed stems; leathery, opposite leaves; minute, imperfect flowers (plants monoecious); anthers sessile on the uniseriate, sepaloid tepals; inferior ovary with 1 inconspicuous locule filled with a large placenta containing undifferentiated ovules (embryo sacs); a berry with viscous tissue as the fruit type; and a single seed with chlorophyllous (green) endosperm and no seed coat. Tissues with calcium oxalate crystals. Anatomical feature: scattered groups of silicified cells and often also sclereids commonly in the mesophyll.

Genera/species: 8/440

Distribution: Cosmopolitan; most abundant in tropical to subtropical regions.

Major genera: *Phoradendron* (190 spp.) and *Viscum* (60–100 spp.)

U.S./Canadian representatives: *Phoradendron* (17 spp.), *Arceuthobium* (16 spp.), *Korthalsella* (7 spp.), *Dendrophthora* (2 spp.), and *Viscum* (1 sp.)

Economic plants and products: Several plantation tree parasites, such as *Arceuthobium* spp. (dwarf-mistletoe). Poisonous berries from *Phoradendron* spp. Decorative Christmas mistletoe from *Viscum album* (European) and *Phoradendron leucarpum* (= *P. serotinum*, North American).

Commentary: Traditionally, the Viscaceae have been treated as a subfamily (Viscoideae) within the Loranthaceae because the two taxa share certain modified characters adaptive for parasitism. Detailed embryological studies, however, have established that the two groups should be maintained as separate families (see the summary in Barlow 1964). In addition to a different embryology, the Loranthaceae *s.s.* ("showy mistletoes") lack jointed stems and are characterized by conspicuous, usually perfect flowers with a biseriate perianth (Barlow 1964; Kuijt 1969).

The Viscaceae are hemiparasites on a variety of tree hosts that provide water and inorganic nutrients. With chlorophyllous leaves (and stems), the mistletoes apparently are able to assimilate carbon into organic compounds (via photosynthesis). The absorptive structures (haustoria or suckers) usually are interpreted as modified adventitious roots (Kuijt 1969). These root-like strands penetrate the host cortex and phloem and spread laterally, forming an anastomosing system that connects the parasite directly to the host xylem (Fisher 1983). Conspicuous swelling of the host

130

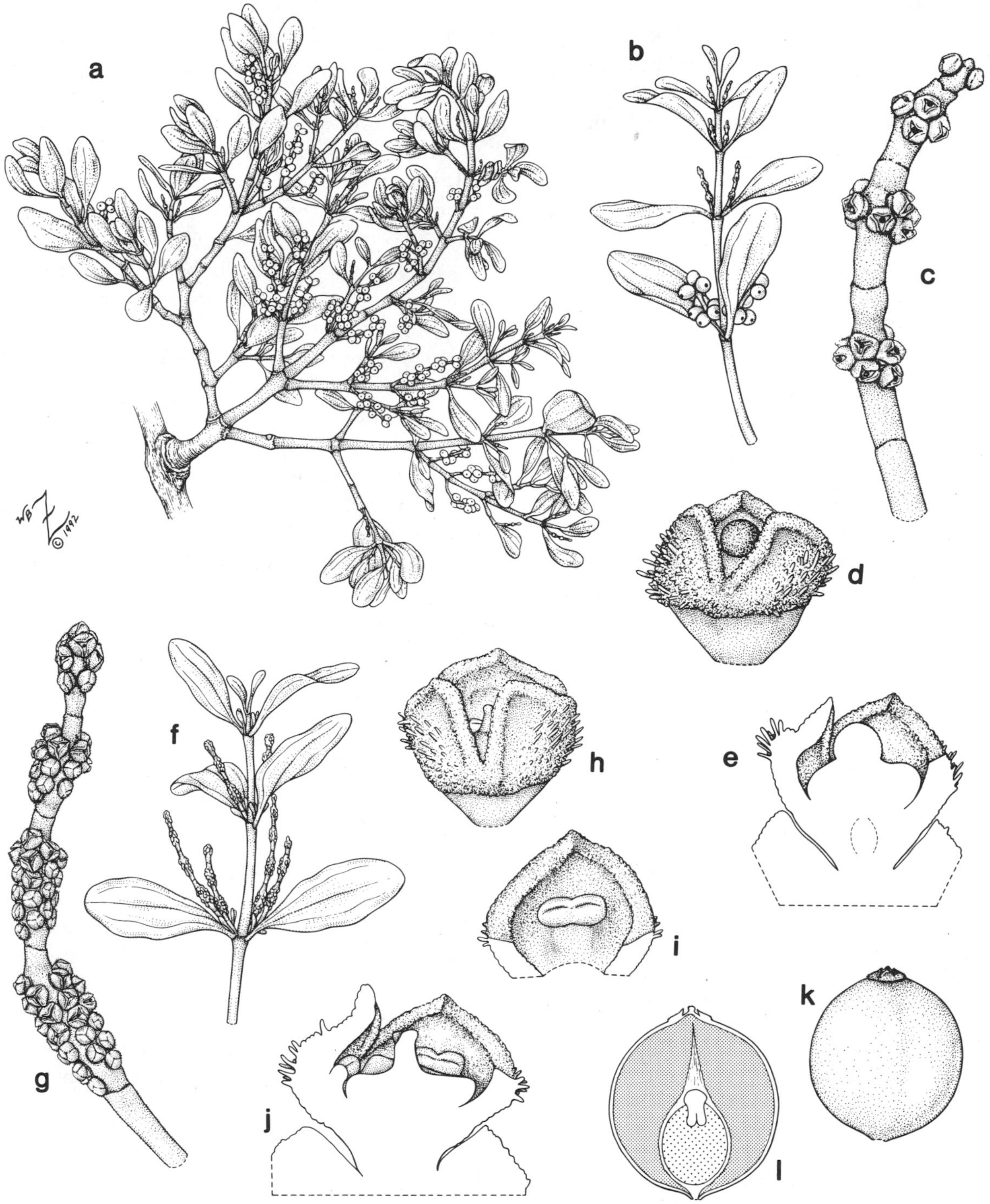

Figure 57. Viscaceae. *Phoradendron leucarpum* (*P. serotinum*): **a,** carpellate flowering plant (with mature fruit from previous year's flowers) on *Fraxinus caroliniana*, ×⅓; **b,** carpellate flowering branch (with mature fruit from previous year's flowers), ×⅔; **c,** carpellate inflorescence, ×5; **d,** carpellate flower, ×30; **e,** longitudinal section of carpellate flower, ×30; **f,** staminate flowering branch, ×⅔; **g,** staminate inflorescence, ×3; **h,** staminate flower, ×25; **i,** adaxial side of tepal (from staminate flower) with adnate anther, ×30; **j,** longitudinal section of staminate flower, ×30; **k,** berry, ×5; **l,** longitudinal section of berry, ×6.

tissue occurs at the point of attachment, and the base of the mistletoe itself becomes enlarged as well (Fig. 57: a). Infected, leafed-out trees may be detected easily after a storm, when the stems and leaves of the brittle mistletoes accumulate on the ground under the tree.

The distinctive embryology of the Viscaceae also is complex (Bhandari and Vohra 1983). A placenta-like structure (mamelon; Fig. 57: e), filling the locule of the ovary, contains two embryo sacs (female gametophytes). A complete ovule (= embryo sac, megasporangium, and integument[s]) is not developed; the single seed, therefore, has no seed coat (Fig. 57: l), which ordinarily is derived from the integument(s) of an ovule.

The flowers of *Phoradendron* bloom in late autumn and winter in the southeastern United States (Allard 1943). The fruit may take more than a year to mature, so carpellate flowers (of the current year) and mature fruit (from the previous year) may be found on the same plant (Fig. 57: a,b). The viscid layer surrounding the seed adheres to fruit-eating birds that subsequently disperse them, often by rubbing the sticky seeds off on a new tree branch. The berries of *Arceuthobium* are explosive.

The minute flowers are entomophilous or anemophilous. Kuijt (1982) reports Hymenoptera as likely pollinators.

REFERENCES CITED

Allard, H. A. 1943. The eastern false mistletoe (*Phoradendron flavescens*): When does it flower? *Castanea* 8:72–78.

Barlow, B. A. 1964. Classification of the Loranthaceae and Viscaceae. *Proc. Linn. Soc. New South Wales* 89:268–272.

Bhandari, N. N., and S. C. A. Vohra. 1983. Embryology and affinities of Viscaceae. In *The biology of mistletoes*, ed. M. Calder and P. Bernhardt, pp. 69–86. Academic Press, Sydney.

Fisher, J. T. 1983. Water relations of mistletoes and their hosts. In *The biology of mistletoes*, ed. M. Calder and P. Bernhardt, pp. 161–184. Academic Press, Sydney.

Kuijt, J. 1969. *The biology of parasitic flowering plants*, pp. 13–52, 193–194. University of California Press, Berkeley.

—— 1982. The Viscaceae in the southeastern United States. *J. Arnold Arbor.* 63:401–410.

ZYGOPHYLLACEAE
Caltrop, Creosote Bush, or *Lignum Vitae* Family

Shrubs, annual or perennial herbs, or seldom trees, generally growing in saline or xeric habitats; stems with sympodial growth, often dichotomously branched, with jointed nodes. **Leaves** usually even-pinnately compound (terminal leaflet absent), entire, typically opposite, often with one leaf at a node larger that the other, fleshy or leathery, often resinous, stipulate; stipules slender, firm, sometimes modified into spines, persistent. **Inflorescence** determinate, cymose or flowers often solitary, terminal or pseudo-axillary. **Flowers** actinomorphic or sometimes slightly zygomorphic, perfect, hypogynous, showy, with conspicuous intra- or extrastaminal disc. **Calyx** of usually 5 sepals, distinct or sometimes basally connate, imbricate or valvate, persistent. **Corolla** of usually 5 petals, distinct, often clawed, sometimes twisted, variously colored, imbricate or convolute. **Androecium** of usually 10 stamens, biseriate with the outer whorl opposite the petals; filaments distinct, often with basal glands or appendages; anthers basifixed to dorsifixed (versatile), dehiscing longitudinally, introrse or latrorse. **Gynoecium** of 1 pistil, usually 5-carpellate; ovary superior, usually 5-locular, ridged to winged, sessile or sometimes with gynophore; ovules 2 to many in each locule, usually anatropous, pendulous, placentation axile; style 1, angular to furrowed; stigma usually 1, capitate or lobed. **Fruit** a septicidal or loculicidal capsule or a schizocarp; seeds sometimes arillate; endosperm usually copious, hard, oily; embryo straight or slightly curved, enveloped by endosperm.

Family characterization: Shrubs to herbs growing in xeric habitats; stems with jointed nodes; fleshy or leathery, even-pinnately compound leaves with persistent stipules; 5-merous flowers with conspicuous nectariferous disc; often clawed petals; biseriate stamens with basal glands/appendages; ridged or winged ovary; and a capsule or schizocarp as the fruit type. Saponins and often alkaloids or mustard oils present. Anatomical features: woody members with characteristic storied xylem (see commentary); leaves of several species with Kranz anatomy (see the commentary under Portulacaceae); and trilacunar nodes.

Genera/species: 28/255

Distribution: Primarily arid, tropical to subtropical areas, with some representatives in temperate regions; especially abundant in dry areas of the Mediterranean region (major component of scrub vegetation).

Major genera: *Zygophyllum* (80–100 spp.), *Fagonia* (30–40 spp.), and *Tribulus* (20–25 spp.)

U.S./Canadian representatives: *Kallstroemia* (7 spp.), *Fagonia* (4 spp.), *Guaiacum* (3 spp.), *Peganum* (2 spp.), *Tribulus* (2 spp.), *Larrea* (1 sp.), and *Zygophyllum* (1 sp.)

Economic plants and products: Strong wood and medicinal resin (guaiacum) from *Guaiacum* spp. (*lignum*

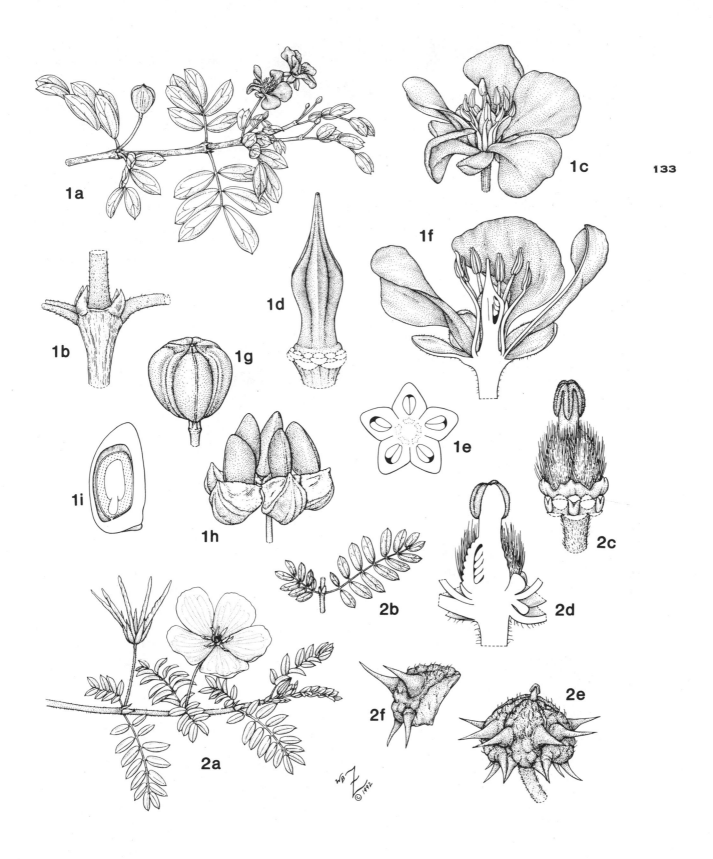

Figure 58. Zygophyllaceae. 1, *Guaiacum sanctum*: **a,** flowering branch, ×²⁄₃; **b,** node showing stipules, ×3; **c,** flower, ×2; **d,** pistil and disc, ×7; **e,** cross section of ovary, ×15; **f,** longitudinal section of flower, ×3½; **g,** capsule before dehiscence, ×1½; **h,** dehisced capsule with exposed arillate seeds, ×1¼; **i,** longitudinal section of seed, ×2. **2,** *Tribulus cistoides*: **a,** habit, ×²⁄₃; **b,** node showing unequal opposite leaves, ×¾; **c,** pistil, disc, and extrastaminal glands, ×6; **d,** longitudinal section of pistil and flower base, ×6; **e,** schizocarp, ×2½; **f,** segment of schizocarp, ×2½.

vitae). Spices from *Zygophyllum* spp. (caper-bean). Several weedy and poisonous plants, such as species of *Tribulus* (burrnut, puncture-weed). Ornamental plants (species of about 6 genera), including *Larrea* (creosote bush), *Tribulus* (caltrop, Jamaican-buttercup), and *Zygophyllum* (caper-bean).

Commentary: Discerning taxonomic affinities of the Zygophyllaceae to other families has been problematical, and the family has been placed in several orders (e.g., Sapindales by Cronquist 1988; Linales by Thorne 1992). The family itself has been divided into seven or eight subfamilies distinguished primarily by differences in fruit types, leaf characters, and carpel number. Several subfamilies that are small or monotypic (e.g., Nitrarioideae) often have been elevated as segregate families (Saleh and El-Hadidi 1977).

A unifying feature of the family is the characteristic wood anatomy, which has been studied extensively (Metcalfe and Chalk 1950). The elements of the wood (vessels, tracheids, fibers) are arranged in horizontally aligned tiers (seen on the tangential surfaces), forming a distinct stratified pattern ("storied wood"). This wood of the Zygophyllaceae is notable also for the high number of stories or layers (i.e., 8–17 stories per mm). The extremely dense wood of *Guaiacum officinale* (and also *G. sanctum*; *lignum vitae*) is reportedly the hardest and strongest of commercial timbers.

The fruit type of the family typically is a capsule (e.g., *Guaiacum*; Fig. 58: 1g,h) or schizocarp (e.g., *Tribulus*; Fig. 58: 2e,f). The spiny schizocarp of *Tribulus* (puncture-weed) clings to animals. In *Guaiacum*, the capsules dehisce to expose seeds (each covered by a thin, bright-red aril; Fig. 58: 1i) probably dispersed by birds (Tomlinson 1980).

Insect pollinators are attracted to the variously colored flowers producing nectar from the disc (Fig. 58: 1d, 2c) and/or glands (Fig. 58: 2c,d) at the base of the stamens. Self-pollination also has been noted in the family (*Kallstroemia*, Porter 1972).

REFERENCES CITED

Cronquist, A. 1988. *The evolution and classification of flowering plants*, 2d ed., pp. 403–407. New York Botanical Garden, Bronx.

Metcalfe, C. R. and L. Chalk, eds. 1950. *Anatomy of the dicotyledons*, vol. 1, pp. 286–287, 291. Oxford University Press, Oxford.

Porter, D. M. 1972. The genera of Zygophyllaceae in the southeastern United States. *J. Arnold Arbor.* 53:531–552.

Saleh, N. A. M., and M. N. El-Hadidi. 1977. An approach to the chemosystematics of the Zygophyllaceae. *Biochem. Syst. & Ecol.* 5:121–128.

Thorne, R. F. 1992. Classification and geography of the flowering plants. *Bot. Rev.* (Lancaster) 58:225–348.

Tomlinson, P. B. 1980. *The biology of trees native to tropical Florida*, pp. 426–428. "Publ. by the author," Petersham, Mass.

RHIZOPHORACEAE
Red Mangrove Family

Shrubs or trees, frequently having a "mangrove habit" (with conspicuous prop roots) and growing in flooded (tidal) swamps; branches with swollen nodes. **Leaves** simple, entire, usually opposite, coriaceous, persistent, stipulate; stipules usually conspicuous, interpetiolar, caducous, sometimes with colleters at inner surface of base. **Inflorescence** determinate, cymose or less often appearing racemose, axillary. **Flowers** actinomorphic, usually perfect, hypogynous to more commonly epigynous, with hypanthium (in epigynous flowers) sometimes prolonged beyond the ovary, small. **Calyx** of typically 4 or 5 to sometimes many sepals, usually basally connate, thick, usually fleshy or coriaceous, valvate, persistent. **Corolla** with as many petals as sepals, distinct, often clawed and lacerate or emarginate, usually fleshy or coriaceous, white or yellowish white, convolute or inflexed in bud. **Androecium** of usually 8 or 10 to sometimes many stamens, uniseriate, often in pairs opposite the petals; filaments distinct or connate at base, very short, or absent; anthers basifixed, 4-locular or cross-partitioned, dehiscing longitudinally or by a flap, introrse. **Gynoecium** of 1 pistil, typically 2- to 4-carpellate; ovary superior, half-inferior to inferior, usually 2- to 4-locular; ovules 2 in each locule, anatropous to hemitropous, pendulous, placentation axile or axile-basal; style 1; stigma 1, lobed. **Fruit** usually a leathery berry or drupe, or sometimes dry and indehiscent, terminated by the persistent calyx; seeds often viviparous (in mangrove species); endosperm copious or scanty, fleshy, oily, sometimes forming aril-like outgrowth; embryo straight, linear, often chlorophyllous, with enlarged hypocotyl.

Family characterization: Woody plants with mangrove habit (prop roots) growing in muddy tidal shores, brackish streams, or lagoons; opposite, abaxially punctate, coriaceous leaves with large interpetiolar stipules; leathery 4-merous flowers; 8 uniseriate stamens in pairs opposite the adaxially hairy petals; sessile and chambered anthers, each dehiscing by a flap; leathery berry containing 1 seed (due to abortion); and viviparous and chlorophyllous embryo with enlarged hypocotyl. Tissues (especially the bark) with tannins and calcium oxalate crystals. Anatomical feature: trilacunar nodes.

Genera/species: 12/84

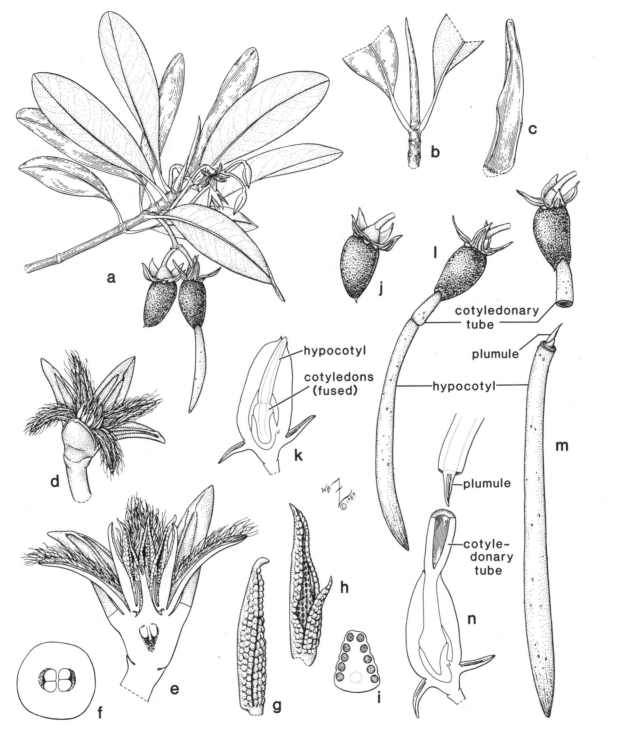

Figure 59. Rhizophoraceae. *Rhizophora mangle*: **a,** flowering and fruiting branch (note vivipary), ×½; **b,** apex of branch showing interpetiolar stipules, ×½; **c,** one stipule, ×1¼; **d,** flower, ×3; **e,** longitudinal section of flower, ×4½; **f,** cross section of ovary, ×6; **g,** stamen (anther sessile), ×9; **h,** dehiscing anther, ×9; **i,** cross section of anther, ×15; **j,** fruit (seed in early stages of germination), ×⅔; **k,** longitudinal section of fruit, ×1; **l,** fruit with developing cotyledonary tube and hypocotyl, ×⅔; **m,** fruit and detached hypocotyl, ×⅔; **n,** longitudinal section of fruit and detached hypocotyl, ×1.

Distribution: Tropical to subtropical rain forests, shorelines, and muddy tidal flats; several genera composing a major component of the world's mangrove vegetation.

Major genus: *Cassipourea* (62–80 spp.)

U.S./Canadian representatives: *Bruguiera* (1 sp.), *Cassipourea* (1 sp.), and *Rhizophora* (1 sp.)

Economic plants and products: Tannin (from bark and foliage) and wood (for charcoal and underwater pilings) from *Rhizophora* and several other genera. *Anopyxis* and *Rhizophora* (red mangrove) sometimes cultivated.

Commentary: The taxonomic placement of the Rhizophoraceae has been problematical (see Cronquist 1981; Dahlgren 1988; and Juncosa and Tomlinson 1988, for summaries). The family has traditionally been placed in the Myrtales; affinities with the families of the Cornales (Thorne 1983) and Celastrales (Dahlgren 1988) also have been suggested. Although well known, the four mangrove genera (*Rhizophora*, *Bruguiera*, *Kandelia*, and *Ceriops*; about 17 species total) actually compose only a small portion of the family, with the majority of species being shrubs and climbers of (inland) tropical rain forests (Graham 1964). The term "mangrove" has also been applied to several genera of other families, notably *Avicennia* (Avicenniaceae, Verbenaceae) and *Laguncularia* and *Conocarpus* (both Combretaceae).

Rhizophora trees and shrubs exhibit a characteristic "mangrove habit" (Gill and Tomlinson 1969, 1972). A plant is supported by large downward-curving prop roots (or aerial roots) that arise from the main trunk or from large branches. These roots eventually interlace around the base of the trunk, forming a series of emergent loops. Conspicuous white lenticels on the aerial roots function in gas exchange to the absorptive underground root system.

The flowers of *Rhizophora mangle* are anemophilous, although visiting bees probably collect the pollen (Tomlinson et al. 1979; Tomlinson 1986). Each anther possesses numerous cross partitions and dehisces with a flap before anthesis (Fig. 59: g,h). The hairy inner surface of the corolla (Fig. 59: d,e) traps the pollen grains, which are dispersed as the petals recurve. The stigma becomes receptive after the pollen is shed, promoting cross-pollination.

The seeds of *Rhizophora* germinate while on the tree (vivipary) without undergoing a dormant period (Fig. 59: a,j–n). As the embryo develops, the endosperm grows out of the micropyle and forms an aril-like covering on the seed. The cotyledons (fused into a tube), and then the hypocotyl, emerge through the micropyle. When the hypocotyl is fully elongate, the seedling (hypocotyl plus plumule) detaches from the cotyledons (and the rest of the fruit) and falls from the tree into the water (Fig. 59: m,n). Thus, the seedling itself is the unit of dispersal (Rabinowitz 1978).

REFERENCES CITED

Cronquist, A. 1981. *An integrated system of classification of flowering plants*, pp. 655–659. Columbia University Press, New York.

Dahlgren, R. M. T. 1988. Rhizophoraceae and Anisophyllaceae: Summary statement, relationships. *Ann. Missouri Bot. Gard.* 75:1259–1277.

Gill, A. M., and P. B. Tomlinson. 1969. Studies on the growth of red mangrove (*Rhizophora mangle* L.). 1. Habit and general morphology. *Biotropica* 1:1–9.

——— and ———. 1972. Ibid., 2. Growth and differentiation of aerial roots. *Biotropica* 3:63–77.

Graham, S. A. 1964. The genera of Rhizophoraceae and Combretaceae in the southeastern United States. *J. Arnold Arbor.* 45:285–301.

Juncosa, A. M., and P. B. Tomlinson. 1988. A historical and taxonomic synopsis of Rhizophoraceae and Anisophyllaceae. *Ann. Missouri Bot. Gard.* 75:1278–1295.

Rabinowitz, D. 1978. Dispersal properties of mangrove propagules. *Biotropica* 10:47–57.

Thorne, R. F. 1983. Proposed new realignments in the angiosperms. *Nordic J. Bot.* 3:85–117.

Tomlinson, P. B. 1986. *The botany of mangroves*, pp. 317–360. Cambridge University Press, Cambridge.

———, R. B. Primack, and J. S. Bunt. 1979. Preliminary observations on floral biology in mangrove Rhizophoraceae. *Biotropica* 11:256–277.

OXALIDACEAE

Oxalis, Sheep-sorrel, or Wood-sorrel Family

Perennial or annual herbs, sometimes suffrutescent or shrubs, with acrid juice, often with fleshy rhizomes, bulb-like tubers, and contractile roots. **Leaves** pinnately or more often palmately compound (then usually trifoliolate) or sometimes simple due to suppression of leaflets, alternate, often forming basal rosettes or apical clusters, with long petioles and entire-margined leaflets often emarginate at apex, with pulvinus at petiole base and at each petiolule base, usually exstipulate. **Inflorescence** determinate, cymose and often appearing umbellate or sometimes racemose, axillary or seemingly terminal. **Flowers** actinomorphic, perfect, hypogynous, small and showy, often heterostylous (dimorphic or trimorphic), sometimes cleistogamous. **Calyx** of 5 sepals, distinct or basally connate, imbricate, persistent. **Corolla** of 5 petals, distinct or sometimes basally connate, often clawed, often yellow, white, or purple, convolute or sometimes imbricate. **Androecium** of 10 stamens, biseriate with the outer whorl shorter than the inner whorl, opposite the petals, and sometimes reduced to staminodes; filaments basally connate and forming a ring or tube (monadelphous), those of the outer whorl with basal nectariferous thickening or glands, persistent; anthers dorsifixed, versatile, dehiscing longitudinally, introrse. **Gynoecium** of 1 pistil, 5-carpellate with carpels often incompletely connate (fused only along adaxial sutures); ovary superior, 5-locular; ovules 1 or more in each locule, anatropous or sometimes hemitropous, pendulous, placentation axile; styles 5, distinct, persistent; stigmas 5, terminal on each style, usually capitate, punctate, or 2-lobed. **Fruit** usually a loculicidal capsule, often deeply 5-angled; seeds usually arillate and discharged from capsule by elastic separation of aril from testa; endosperm copious, oily, fleshy; embryo large, spatulate, straight, enveloped by endosperm.

Family characterization: Herbaceous plants with bulbous or tuberous stems and acrid juice; palmately compound leaves in basal or apical clusters; leaflets with pulvini responsible for "sleep movements"; 5-merous, often heterostylous flowers; monadelphous stamens; 5, distinct, persistent styles; 5-angled loculicidal capsule; and seeds with arils separating elastically from testa. Acrid sap composed of oxalic acid (in the form of dissolved potassium oxalate) secreted as calcium oxalate often appearing on plant parts as white, red, or brown deposits (Fig. 60: 1e). Anatomical features: scattered secretory cavities in the mesophyll; and tenuinucellate ovules.

Genera/species: *Oxalis* (500–800 spp.), *Biophytum* (50–70 spp.), *Sarcotheca* (11 spp.), *Dapania* (3 spp.), *Averrhoa* (2 spp.), and *Eichleria* (2 spp.) (Thorne pers. comm.)

Distribution: Primarily pantropical, but also widespread in temperate regions.

U.S./Canadian representatives: *Oxalis* (30 spp.)

Economic plants and products: Edible fruits from *Averrhoa* (carambola, bilimbi) and edible leaves and tubers from species of *Oxalis* (oca). Several weedy *Oxalis* spp. Ornamental plants: *Averrhoa*, *Biophytum* (life plant), and *Oxalis* (Irish shamrock, lady's-sorrel, sheep-sorrel, wood-sorrel).

Commentary: Most of the species in the family belong to the large genus *Oxalis* (600 species), which is divided into sections and species according to characters of the leaflets (size, shape, and number), inflorescence (type), and corolla (color). Some authors separate the woody members of the Oxalidaceae as a segregate family, the Averrhoaceae.

The wood-sorrels are identified easily in the field by their folded, "clover-like" leaves. As in the Fabaceae, the leaflets assume a "sleep position" (bend downwards) at night or in cold weather due to changes in the turgidity of the pulvini (Fig. 60: 1b,c; see also the commentary under Fabaceae).

Oxalis flowers attract various insects (such as bees and butterflies) that visit for the nectar secreted at the filament bases into the (more or less) tubular flower. In many species, nectar guides (lines) also occur on the petals (Fig. 60: 1d,f). Cleistogamous flowers, which resemble buds (see Violaceae), are prevalent in several species.

Heterostyly is common in the family, especially in *Oxalis* (see the commentaries under Rubiaceae and Pontederiaceae for detailed explanations and references to appropriate illustrations). Several *Oxalis* species are distylous (long- and short-length styles), and many are tristylous (long-, medium-, and short-length styles; Eiten 1959, 1963; Denton 1973; Robertson 1975). The derivation and operation of the complex outcrossing systems in these *Oxalis* species have received much attention in the literature (Ornduff 1972; Weller 1976). The varying style and stamen lengths of the two (or three) flower morphs often are correlated with differences in the pollen grain size, stigma and style morphology, and the pubescence of filaments and styles (Richards and Barrett 1992). Ideally, a legitimate cross may occur only between flowers with stamens and styles in a similar position (e.g., pollen from "long" stamen with long-styled stigma), although some tristylous *Oxalis* species also are self-compatible.

The loculicidal capsule of *Oxalis* appears septicidal because of the deep lobing between the carpels, which in many species are incompletely fused (Fig. 60: 2a). Each of the five carpels splits along the abaxial suture (Fig. 60: 2b) to expose the seeds. Any disturbance may then cause the explosive ejection of the seeds by means of their arils, which rapidly split abaxially and turn inside out (Fig. 60: 2c–e), a reaction caused by the turgid cells along the inner surface of the aril.

REFERENCES CITED

Denton, M. E. 1973. A monograph of *Oxalis*, section *Ionoxalis* (Oxalidaceae) in North America. *Publ. Mus. Michigan State Univ. Biol. Ser.* 4:455–615.

Eiten, G. 1959. Taxonomy and regional variation of *Oxalis* section *Corniculatae*. Ph.D. dissertation, Columbia University, New York.

———. 1963. Taxonomy and regional variation of *Oxalis* section *Corniculatae*. 1. Introduction, keys and synopsis of the species. *Amer. Midl. Naturalist* 69:257–309.

Ornduff, R. 1972. The breakdown of trimorphic incompatibility in *Oxalis* section *Corniculatae*. *Evolution* 26:52–65.

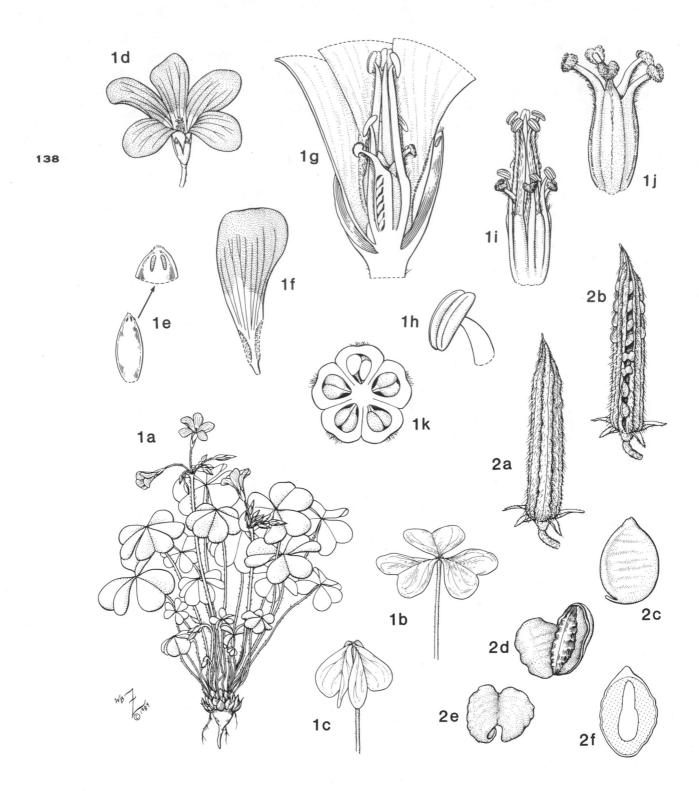

Figure 60. Oxalidaceae. 1, *Oxalis corymbosa*: **a,** habit, ×½; **b,** leaf with leaflets expanded, ×⅗; **c,** leaf with leaflets in "sleep position," ×⅗; **d,** flower, ×1¾; **e,** sepal, ×3½, with detail of apex showing dark calcium oxalate deposits, ×14; **f,** petal, ×3½; **g,** longitudinal section of flower, ×8½; **h,** anther, ×30; **i,** androecium and gynoecium, ×7; **j,** pistil, ×12; **k,** cross section of ovary, ×18. **2.** *Oxalis stricta*: **a,** capsule, ×3; **b,** dehiscing capsule, ×3; **c,** seed completely enclosed by aril, ×15; **d,** seed with dehiscing aril, ×15; **e,** aril after seed expelled, ×15; **f,** longitudinal section of seed, ×18.

Richards, J. H., and S. C. H. Barrett. 1992. The development of heterostyly. In *Evolution and function of heterostyly*, ed. S. C. H. Barrett, pp. 85–127. Monogr. Theor. & Appl. Genet. 15. Springer-Verlag, Berlin.

Robertson, K. R. 1975. The genera of Oxalidaceae in the southeastern United States. *J. Arnold Arbor.* 56:223–239.

Weller, S. G. 1976. The genetic control of tristyly in *Oxalis* section *Ionoxalis*. *Heredity* 37:387–393.

GERANIACEAE
Geranium Family

Annual or perennial herbs (sometimes suffrutescent) or shrubs, with pubescence of glandular hairs containing ethereal oils, often with rhizomes; stems jointed at nodes. **Leaves** palmately or pinnately compound or simple (then lobed or divided), entire or serrate, opposite or alternate, palmately veined, stipulate. **Inflorescence** determinate, cymose or often appearing umbellate, terminal or axillary, or sometimes flower solitary and axillary. **Flowers** actinomorphic to sometimes slightly zygomorphic, perfect, hypogynous, showy, with extrastaminal glands alternating with the petals. **Calyx** of usually 5 sepals, distinct or basally connate, imbricate or sometimes valvate. **Corolla** of usually 5 petals, distinct, often emarginate, variously colored (often pink, red, or purple), usually imbricate. **Androecium** of usually 10 stamens in 2 whorls (sometimes those of the outer whorl reduced to staminodes); filaments basally connate; anthers versatile, dehiscing longitudinally, introrse. **Gynoecium** of 1 pistil, usually 5-carpellate with carpels adnate to an elongate central column; ovary superior, 5-locular, lobed or grooved; ovules 2 in each locule, anatropous to campylotropous, pendulous, placentation axile; styles 5, usually basally connate/coherent, persistent; stigmas 5, distinct, ligulate, papillose. **Fruit** usually a schizocarp with 5 segments dehiscing elastically from a persistent central column; seeds 1 in each segment; endosperm scanty or usually absent; embryo curved or sometimes straight, with plicate to convolute cotyledons, often green.

Family characterization: Aromatic herbaceous to shrubby plants with glandular hairs containing ethereal oils; palmately or pinnately lobed or compound, stipulate leaves; 5-merous flowers; biseriate stamens with basally connate filaments; a beaked and lobed schizocarp elastically dehiscing into 1-seeded segments curling up on the beak; and nonendospermous seeds. Tannins and ellagic and/or gallic acids present. Tissues with calcium oxalate crystals. Anatomical feature: multicellular and capitate glandular hairs containing ethereal oils (Fig. 61: b).

Genera/species: 14/775

Distribution: Widely distributed in temperate to subtropical regions; relatively few representatives in tropical areas (where species usually restricted to high altitudes).

Major genera: *Geranium* (300–400 spp.), *Pelargonium* (250–280 spp.), and *Erodium* (60–90 spp.)

U.S./Canadian representatives: *Geranium* (42 spp.), *Erodium* (11 spp.), and *Pelargonium* (9 spp.)

Economic plants and products: Geranium oil (essential oils) used in perfumes extracted from *Pelargonium* spp. Several weedy plants such as species of *Erodium* (filaree) and *Geranium* (wild geranium). Ornamental plants (species of about 6 genera), including *Erodium* (heron's-bill, stork's-bill), *Geranium* (crane's-bill), *Monsonia*, and *Pelargonium* (geranium).

Commentary: The Geraniaceae traditionally are broadly defined, including five tribes (or five subfamilies) primarily distinguished by the number of ovules per locule and differences in fruit morphology. Most of the species in the family belong to the well-defined and monophyletic tribe Geranieae/subfamily Geranioideae (5 or 6 genera/750 spp.), characterized by two ovules per carpel and an elastic schizocarp with awned segments attached to a beak. The four remaining tribes/subfamilies sometimes are treated as segregate families (e.g., the Vivianiaceae). The affinities of these taxa need further critical study (Robertson 1972; Price and Palmer 1993). Taxonomists generally agree that the geranium family is closely related to the Oxalidaceae (Fig. 60): for example, the androecium in both families is characterized by ten (or fifteen) basally connate stamens.

The characteristic gynoecium of the family consists of a five-carpellate and -lobed ovary terminated by an elongated, branched style with five distinct stigmas (Fig. 61: i,k). Each carpel (ovary plus style branch) is adnate or adherent to an elongated central column (columella). The style-column structure persists in the fruit as a conspicuous beak (Fig. 61: l–n), hence the common names crane's-bill (for *Geranium*) and stork's-bill or heron's-bill (for *Erodium*). At maturity, the schizocarp splits elastically into five one-seeded segments (mericarps) that separate from the persistent column and either roll up and outwards (Fig. 61: n) or become spirally twisted. Each style section (or style branch) remains attached to its corresponding fruit segment (carpel), forming an apical awn. In many *Geranium* species, each seed is expelled when a segment itself dehisces as it simultaneously separates from the column. In the fruits of *Erodium*, *Pelargonium*, and certain

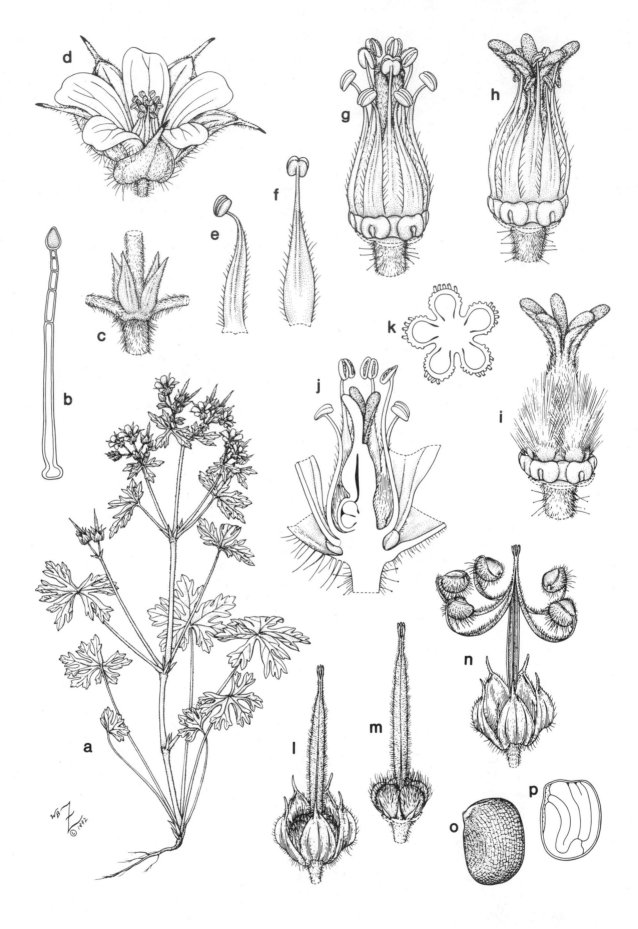

Geranium species, the seeds are retained inside closed segments whose hygroscopic awns twist into a cork-screw. As the awn uncoils (wet) and contracts (dry) in response to moisture, the pointed fruit segment is driven into the ground. Retrorse hairs (functioning as barbs) secure the dispersal unit in the soil (van der Pijl 1982; Yeo 1984, 1990).

The flowers of the Geraniaceae usually are brightly colored (shades of red or blue) with darker, conspicuous veins functioning as nectar guides. Nectar is produced by the five extrastaminal glands (Fig. 61: g–j) between the stamens and sepals; in *Pelargonium*, the adaxial sepal is modified into a nectar-secreting spur (Link 1990). The flowers of some species also produce an odor. Bees and other insects (butterflies, flies, beetles) are the principal pollinators, and cross-pollination is reinforced by protandry. In species with biseriate stamens (e.g., *Geranium* spp.), the inner stamens (Fig. 61: g), outer stamens (Fig. 61: h), and stigmas (Fig. 61: i) mature successively (Willson et al. 1979). The filaments of each whorl straighten as the anthers dehisce and recurve as the anthers wither and fall off. After the outer, shorter whorl of stamens sheds pollen, the branches of the style expand and expose the receptive stigmatic surfaces (Fig. 61: i).

REFERENCES CITED

Link, D. A. 1990. The nectaries of Geraniaceae. In *Proceedings of the international Geraniaceae symposium*, ed. P. Vorster, pp. 216–233. University of Stellenbosch, South Africa.

van der Pijl, L. 1982. *Principles of dispersal in higher plants*, pp. 18, 116–117. Springer-Verlag, Berlin.

Price, R. A., and J. D. Palmer. 1993. Phylogenetic relationships of Geraniaceae and Geraniales from *rbc*L sequence comparisons. *Ann. Missouri Bot. Gard.* 80:661–671.

Robertson, K. R. 1972. The genera of Geraniaceae in the southeastern United States. *J. Arnold Arbor.* 53:182–201.

Willson, M. F., L. J. Miller, and B. J. Rathcke. 1979. Floral display in *Phlox* and *Geranium*: Adaptive aspects. *Evolution* 33:52–63.

Yeo, P. F. 1984. Fruit-discharge type in *Geranium* (Geraniaceae): Its use in classification and its evolutionary implications. *Bot. J. Linn. Soc.* 89:1–36.

———. 1990. The classification of Geraniaceae. In *Proceedings of the international Geraniaceae symposium*, ed. P. Vorster, pp. 1–22. University of Stellenbosch, South Africa.

MALPIGHIACEAE
Barbados-cherry or Malpighia Family

Woody vines, shrubs, or small trees, with pubescence of unicellular 2-armed hairs ("malpighian hairs"). **Leaves** simple, entire, usually opposite, often with petiolar glands, with jointed petioles, stipulate. **Inflorescence** determinate, cymose and often appearing racemose or paniculate, frequently terminal. **Flowers** usually zygomorphic, perfect, hypogynous, with convex receptacle, on jointed pedicel, showy. **Calyx** of 5 sepals, distinct or basally connate, typically with abaxial oil glands, imbricate. **Corolla** of 5 petals, distinct, usually unequal, clawed, with fringed or toothed margin, variously colored, convolute-imbricate. **Androecium** of typically 10 stamens in 2 whorls, with outer whorl opposite the petals and often reduced to staminodes; filaments basally connate into a short tube; anthers basifixed, 2- or 1-locular, often with enlarged connective, usually dehiscing longitudinally, introrse. **Gynoecium** of 1 pistil, usually 3-carpellate; ovary superior, 3-locular; ovules 1 in each locule, pendulous, anatropous or hemitropous, placentation axile; styles 3, usually distinct, unequal; stigmas 3, terminal. **Fruit** usually a samara, schizocarp (splitting into 3 samaras), drupe, or capsule; endosperm scanty or absent; embryo large, straight to circinate, oily.

Family characterization: Woody plants with opposite, stipulate leaves; sepals with conspicuous abaxial oil glands; clawed, hooded, unequal petals with fringed margins; 10 stamens united basally into a short tube; distinct unequal styles; and a samara, samaroid schizocarp, or drupe as the fruit type. Anatomical features: vestiture of unicellular, stiff, 2-armed hairs (malpighian hairs, discussed below); water-storage cells in the mesophyll; mucilaginous epidermal cells (leaves); and various types of anomalous secondary growth.

Genera/species: 66/1,200

Distribution: Pantropical, with a few representatives in subtropical areas; especially diverse in South America.

Major genera: *Byrsonima* (120–150 spp.), *Heteropterys*

Figure 61. Geraniaceae. *Geranium carolinianum*: **a,** habit, ×½; **b,** capitate glandular hair from stem, ×120; **c,** node showing stipules, ×4; **d,** flower, ×5; **e,** stamen from outer whorl, ×12; **f,** stamen from inner whorl, ×12; **g,** androecium and gynoecium at anthesis (inner whorl of anthers dehiscing, stigmas unexpanded), ×12; **h,** androecium and gynoecium from older flower (anthers of both whorls dehisced, stigmas expanded), ×12; **i,** pistil and disc, ×12; **j,** longitudinal section of flower, ×12; **k,** cross section of ovary, ×20; **l,** schizocarp before dehiscence with persistent calyx, ×2½; **m,** schizocarp before dehiscence (persistent calyx removed), ×2½; **n,** dehisced schizocarp (seeds expelled), ×2½; **o,** seed, ×10; **p,** longitudinal section of seed, ×10.

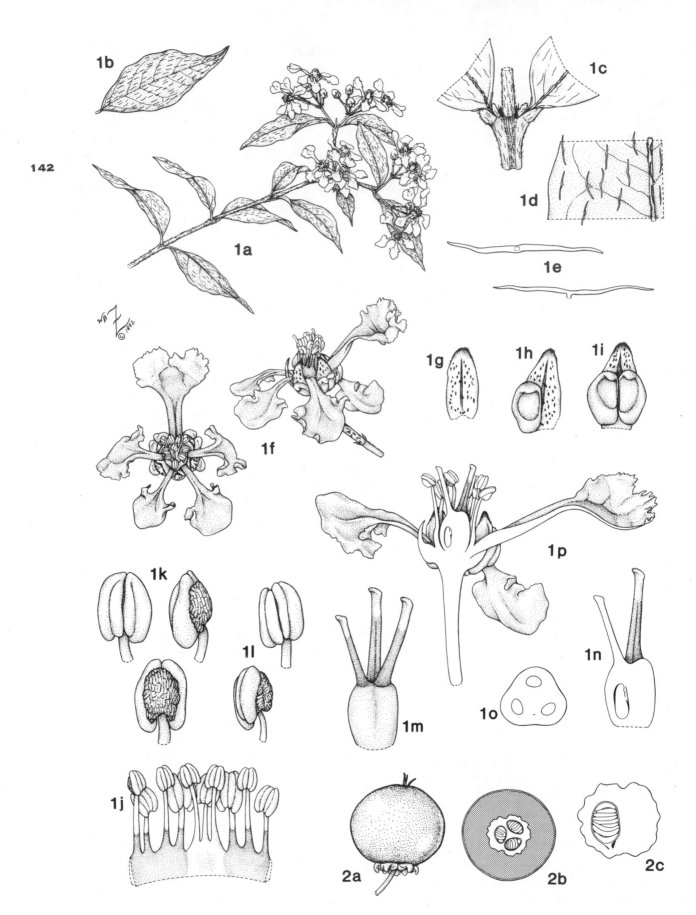

(100–120 spp.), *Banisteriopsis* (92–100 spp.), *Stigmaphyllon* (60–100 spp.), and *Tetrapterys* (80–90 spp.)

U.S./Canadian representatives: 11 genera/38 spp.; largest genera: *Malpighia* and *Stigmaphyllon*

Economic plants and products: Edible fruit from species of *Bunchosia*, *Byrsonima*, and *Malpighia* (Barbados-cherry, West Indian–cherry). Cordage from *Banisteria* spp. Hallucinogenic compounds (alkaloids) used by Amazonian Indians from species of several genera, such as *Banisteriopsis*. Ornamental plants (species of about 11 genera), including *Bunchosia*, *Galphimia*, *Malpighia*, and *Stigmaphyllon* (Brazilian golden vine).

Commentary: The Malpighiaceae have been divided into two or three subfamilies (see Robertson 1972; Anderson 1977) and several tribes based on fruit morphology. The family is not particularly well represented in our area: a few genera of shrubs to trees with usually drupaceous fruit. Worldwide, the species of this pantropical family usually are lianas (with anomalous secondary growth) and have conspicuous, paired petiolar glands (extrafloral nectaries). The fruit type also varies: schizocarp, capsule, samara, or drupe.

The family is characterized by the vesture of unusual stiff unicellular hairs, composed of two horizontal arms and attached to the plant by a long to short, central stalk (Fig. 62: 1d,e). These so-called malpighian hairs are not restricted to the Malpighiaceae, however (e.g., see Sapotaceae, Fig. 35: c, where these hairs are flexible). Although typically T- or Y-shaped with equal to subequal arms, the hairs in some species appear simple due to the reduction of one branch. In some species of *Malpighia*, the stiff hairs become embedded in the skin when touched and may be quite irritating.

The attractive flowers (Fig. 62: 1f) with fringed, clawed petals are mostly "bee flowers" (Anderson 1979, 1990). The sepals of most New World species are characterized by prominent oil glands on the abaxial surface that are conspicuous even on buds. The glands usually are paired (Fig. 62: 1i), although some sepals within a calyx may have only one gland (Fig. 62: 1h) or none (Fig. 62: 1g; Vogel 1990). Oil-gathering bees pollinate the flowers as they reach for the glands between the long claws of the petals. In the Old World, these pollinators are lacking, and the malpighian flowers do not have oil glands. Bees probably visit to collect pollen. Some of these species, however, produce sugary nectar in nectaries evidently homologous to the oil glands of the New World flowers.

REFERENCES CITED

Anderson, W. R. 1977. Byrsonimoideae, a new subfamily of the Malpighiaceae. *Leandra* 7:5–18.
——. 1979. Floral conservation in neotropical Malpighiaceae. *Biotropica* 11:219–223.
——. 1990. The origin of the Malpighiaceae—The evidence from morphology. *Mem. New York Bot. Gard.* 64:210–224.
Robertson, K. R. 1972. The Malpighiaceae in the southeastern United States. *J. Arnold Arbor.* 53:101–112.
Vogel, S. 1990. History of the Malpighiaceae in the light of pollination ecology. *Mem. New York Bot. Gard.* 55:130–142.

POLYGALACEAE
Milkwort Family

Annual or perennial herbs, vines, or shrubs, often with taproots. **Leaves** simple, entire, usually alternate, sometimes reduced to scales, exstipulate or stipules represented by glands. **Inflorescence** indeterminate, capitate, spicate, racemose, or paniculate, or sometimes flower solitary, terminal or axillary. **Flowers** zygomorphic, perfect, hypogynous, showy, each subtended by a bract and 2 bracteoles, sometimes with intrastaminal disc or nectariferous gland. **Calyx** of 5 sepals, distinct or the 2 lower (abaxial) connate, often with the 2 inner (lateral) sepals (wings) enlarged and petaloid, imbricate, persistent or caducous. **Corolla** of basically 5 petals but usually reduced to 3 (2 upper and 1 lower), variously connate and/or basally adnate to the androecium and forming a tube, with the abaxial (lower) petal (keel) often concave and crested with fringe, white, pink, yellow, or orange, imbricate. **Androecium** of usually 8 stamens; filaments connate into a tube-like sheath with an adaxial split above; anthers basifixed, often confluently 1-locular, dehiscing by an apical or subapical pore or by a V-shaped slit, introrse. **Gynoecium** of 1 pistil, usually 2-carpellate; ovary superior, 2-locular; ovules 1 in each locule, anatropous to

Figure 62. Malpighiaceae. 1, *Malpighia glabra*: **a,** flowering branch, ×¾; **b,** adaxial side of leaf (covered with malpighian hairs), ×⅕; **c,** node showing stipules, ×5; **d,** detail of adaxial leaf surface with malpighian hairs, ×6; **e,** two views of malpighian hair, ×35; **f,** two views of flower, ×3; **g,** abaxial sepal (no oil gland), ×7; **h,** one of two lateral sepals (one abaxial oil gland), ×7; **i,** one of two adaxial sepals (paired abaxial oil glands), ×7; **j,** expanded androecium, ×8; **k,** three views of anther from outer whorl of stamens, ×14; **l,** two views of anther from inner whorl of stamens, ×14; **m,** pistil, ×10; **n,** longitudinal section of pistil, ×10; **o,** cross section of ovary, ×10; **p,** longitudinal section of flower, ×6. **2,** *Byrsonima lucida*. **a,** drupe, ×2; **b,** cross section of drupe (note three-seeded pyrene), ×2; **c,** longitudinal section of pyrene, ×6.

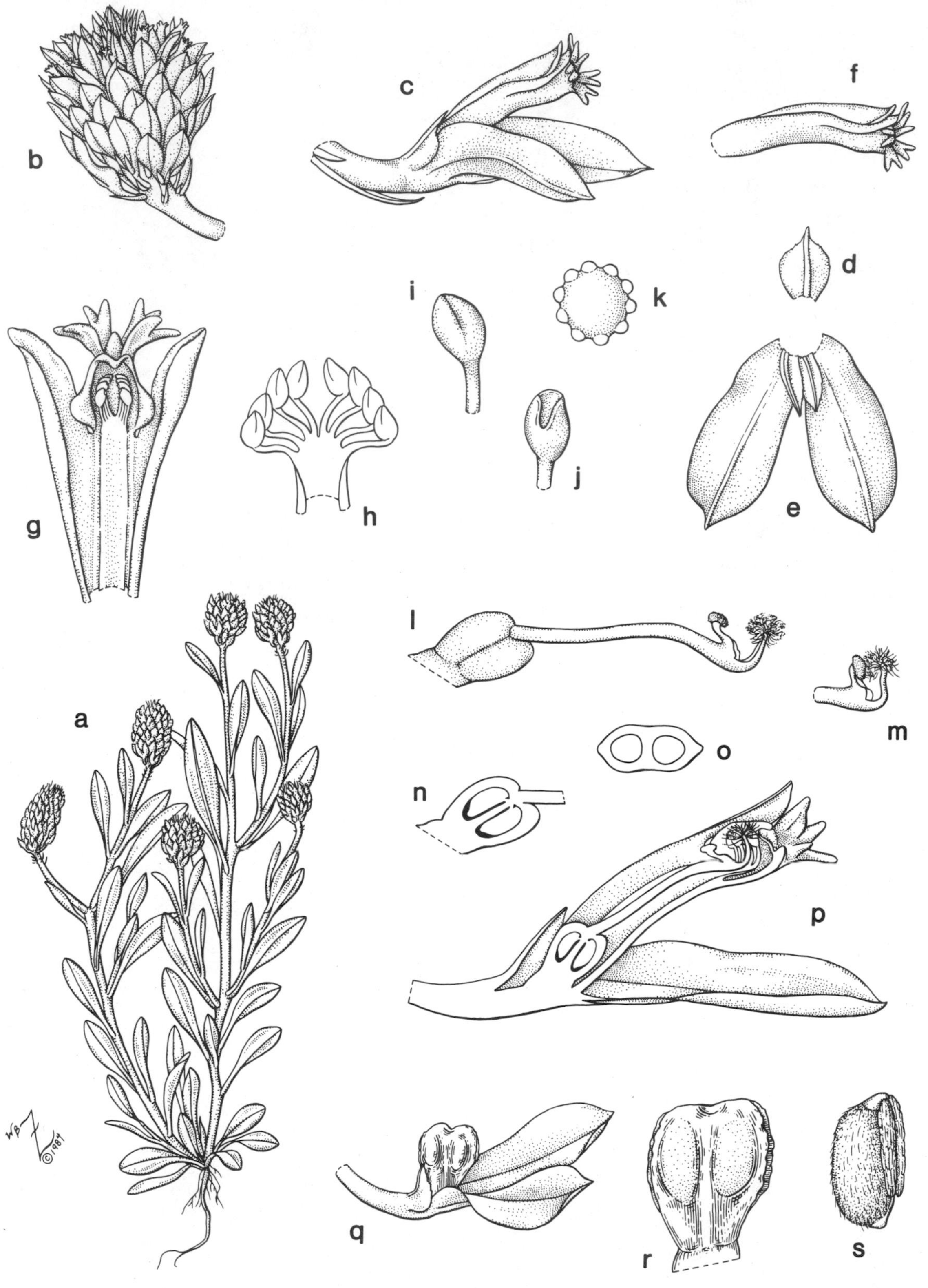

144

Figure 63. Polygalaceae. *Polygala lutea*: **a,** habit, ×³/₅; **b,** inflorescence, ×2⅓; **c,** flower, ×7; **d,** adaxial sepal, ×7; **e,** abaxial and lateral ("wing") sepals, ×7; **f,** corolla, ×7; **g,** expanded corolla and androecium, ×12; **h,** expanded androecium, ×18; **i,** anther, ×45; **j,** dehisced anther, ×45; **k,** pollen grain (polycolporate), ×270; **l,** pistil showing orientation of receptive and sterile lobes of style before dehiscence of anthers, ×15; **m,** apex of style from older flower, ×15; **n,** longitudinal section of ovary, ×15; **o,** cross section of ovary, ×15; **p,** longitudinal section of flower, ×12; **q,** capsule with persistent calyx, ×5½; **r,** capsule (calyx removed), ×11; **s,** seed, ×20.

hemitropous, pendulous, placentation axile; style 1, often apically bilobed with one lobe receptive and the other sterile and tufted with hairs; stigma capitate. **Fruit** usually a loculicidal capsule splitting into 2 1-seeded valves; seeds often with stiff and rigid hairs, with conspicuous aril-like outgrowth at the micropyle; endosperm usually copious, soft, fleshy, oily and proteinaceous; embryo straight.

Family characterization: Herbaceous plants; brightly colored flowers with modified perianths superficially resembling those of the Faboideae ("papilionaceous"); calyx of 5 sepals with the 2 inner petaloid and wing-like; reduced corolla of 2 + 1 petals with the lower (keel) boat-shaped and fringed; 8 stamens with filaments connate into a split sheath; confluently 1-locular anthers with poricidal dehiscence; bilobed style with a receptive lobe and a hairy, sterile lobe; and hairy seeds with micropylar outgrowths. Pollen grains distinctively polycolporate (Fig. 63: k). Saponins commonly present. Tissues commonly with calcium oxalate crystals. Anatomical features: closed ring of xylem in young stems; and lysigenous secretory cavities or oil ducts (*Polygala*).

Genera/species: 15/800

Distribution: Cosmopolitan; absent from New Zealand and the Arctic (and with only a few introduced weeds in Polynesia).

Major genera: *Polygala* (500–600 spp.) and *Monnina* (150 spp.)

U.S./Canadian representatives: *Polygala* (61 spp.), *Securidaca* (2 spp.), and *Monnina* (1 sp.)

Economic plants and products: Medicinal roots (due to saponins) from several *Polygala* species, such as *P. senega* (snakeroot—with seregin). Ornamental plants: species of *Monnina*, *Polygala* (milkwort), and *Securidaca*.

Commentary: The 600 or so species of *Polygala* constitute about two-thirds of the species in the Polygalaceae, a well-defined and presumably monophyletic family. Tribes (and genera) are based mainly upon fruit morphology, with *Polygala* having capsular fruits (Miller 1971).

A novice in the field could casually mistake a milkwort for an orchid or a legume. Actually, a *Polygala* flower (Fig. 63: c) does superficially resemble the papilionaceous flower of the Fabaceae, but the similar parts are not homologous. The conspicuous "wings" are the enlarged and petaloid lateral sepals (Fig. 63: e). The corolla (Fig. 63: f), reduced to three petals, usually is adnate to the staminal "tube" or sheath (slit on adaxial side; Fig. 63: g). The lower median petal, often

bearing a fringed crest, forms a "keel" closely surrounding the stigma and anthers (Fig. 63: p; Holm 1929).

The pollination of *Polygala* species is complex and little-studied. Self-pollination may occur in species in which the sterile apical lobe of the stigma consists of a tuft of hairs that catch the pollen when the anthers dehisce. As the flower develops, the sterile and the receptive lobes of the stigma may touch each other (see Fig. 63: l,m), resulting in the transfer of pollen. In other species where insect pollination has been observed closely, the pollen accumulates in a trough-like or horizontal extension of the style, which lacks the hairy sterile tip. Basically, insects (mainly bees) seek nectar at the base of the flower and land upon the keel, exposing the stigmas and anthers. The insect may then pick up pollen from the trough as it leaves and/or enters the flower.

The seeds of *Polygala* (Fig. 63: s) have a two- or three-lobed aril-like outgrowth that develops from the tissues of the outer integuments at the micropylar end of the seed and is, thus, not properly termed an aril (which develops from the funicle at the hilum) or a caruncle (which develops from integuments at the hilum). The seeds are distributed by ants, which utilize the outgrowths as a food source (Verkerke 1985).

REFERENCES CITED

Holm, T. 1929. Morphology of North American species of *Polygala*. *Bot. Gaz.* 88:167–185.
Miller, N. G. 1971. The genera of Polygalaceae in the southeastern United States. *J. Arnold Arbor.* 52:267–284.
Verkerke, W. 1985. Ovules and seeds of Polygalaceae. *J. Arnold Arbor.* 66:353–394.

RUTACEAE

Citrus or Rue Family

Usually shrubs or trees, with aromatic parts (due to ethereal oils). **Leaves** simple or palmately or pinnately compound, entire, alternate or sometimes opposite, coriaceous, punctate, sometimes reduced to spines, exstipulate. **Inflorescence** determinate, cymose and often appearing corymbose or paniculate, axillary or terminal, or sometimes flower solitary and axillary. **Flowers** usually actinomorphic, perfect or imperfect (then plants dioecious, monoecious, or polygamous), hypogynous, generally showy, with intrastaminal nectariferous disc. **Calyx** of typically 4 or 5 sepals, distinct or basally connate, often glandular, imbricate. **Corolla** of typically 4 or 5 petals, distinct, usually white, greenish yellow, or yellow, glandular, imbricate or valvate. **Androecium** of 4 or 5 stamens (uniseriate and alter-

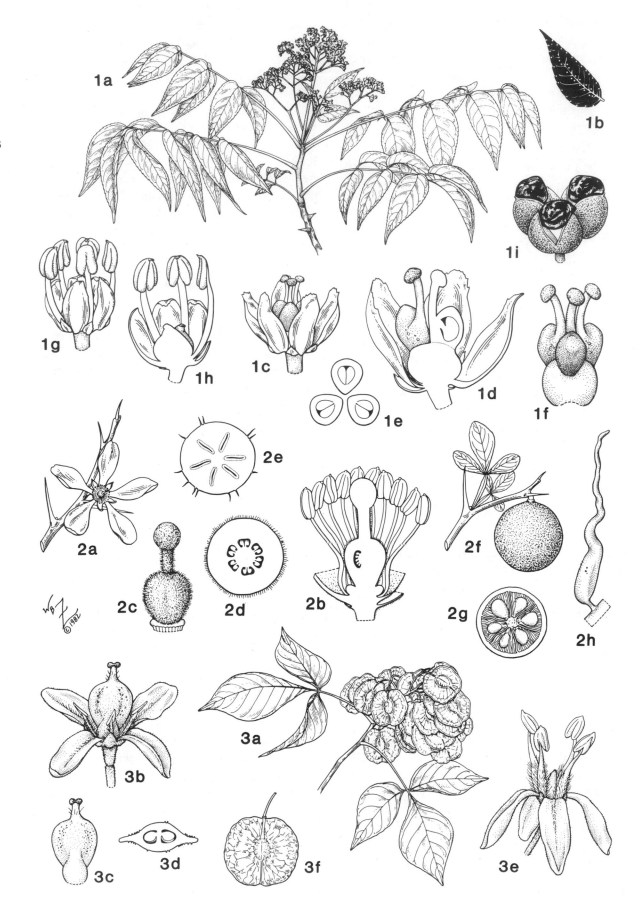

nate with petals) or 8, 10, to numerous stamens (biseriate, with the outer whorl opposite the petals and longer than the inner whorl); filaments distinct or basally connate, often dilated at base; anthers dorsifixed, versatile, often with glandular apex, dehiscing longitudinally, introrse; inner whorl sometimes reduced to staminodes. **Gynoecium** apocarpous to more commonly syncarpous, usually 2- or 5-carpellate; ovary superior, 4- or 5-locular, deeply lobed, sometimes with gynophore; ovules 1 to many in each locule, anatropous, placentation axile; style(s) 1 or as many as carpels, distinct, connivent, or connate, basal, lateral, or terminal; stigma simple or lobed. **Fruit** a leathery-skinned berry (hesperidium), capsule, drupe, samara, or schizocarp, or an aggregate of 2 to 5 drupes or follicles, often with pitted and glandular pericarp; endosperm fleshy or absent; embryo large, straight or bent.

Family characterization: Shrubs or trees with aromatic parts; simple to palmately or pinnately compound (glandular) punctate leaves; 4- or 5-merous perianth; intrastaminal nectariferous disc; and deeply lobed ovary with 2 to 5 carpels. Triterpenes (limonoids: bitter-tasting substances; Taylor 1983) and often also various alkaloids present. Tissues with calcium oxalate crystals. Anatomical features: tissues with lysigenous secretory cavities (from oil cells) containing ethereal oils (appearing on the leaves as pellucid dots; Fig. 64: 1b); and stylar canals (one associated with each carpel; Fig. 64: 2e).

Genera/species: 154/925

Distribution: In tropical and subtropical regions; also many representatives in warm temperate zones. Especially well represented in southern Africa, Australia, and South America.

Major genera: *Zanthoxylum* (20–200 spp.), *Agathosma* (135–180 spp.), *Pelea* (70–75 spp.), and *Glycosmis* (50–60 spp.)

U.S./Canadian representatives: 20 genera/130 spp.; largest genera: *Pelea* and *Citrus*

Economic plants and products: Fruit from *Citrus* (oranges, grapefruits, tangerines, limes, lemons, citron) and *Fortunella* (kumquat). Medicinal plants (due to oils and/or alkaloids): *Citrus* (bergamot oil), *Pilocarpus* (jaborandi leaves), *Ruta* (oil of rue), and *Zanthoxylum* (toothache bark). Yellow-colored silkwoods from *Zanthoxylum flavum*. Ornamental trees and shrubs (species of 50 genera), including *Murraya* (orange-jasmine), *Phellodendron* (cork tree), *Poncirus* (trifoliate-orange), *Ptelea* (hop tree), *Severinia* (boxthorn), and *Triphasia* (limeberry).

Commentary: The Rutaceae are allied with families in the Rutales-Sapindales complex, such as the Sapindaceae, Meliaceae, and Anacardiaceae (see Hegnauer 1983 for a historical summary). The diverse family is divided generally into five to seven subfamilies (e.g., da Silva et al. 1988) that often are treated as distinct families (e.g., Flindersiaceae).

Subfamilies are separated primarily on the basis of fruit type and carpel number and connation (see Saunders 1934). For example, *Zanthoxylum clava-herculis* (Rutoideae) has three distinct carpels (Fig. 64: 1f) that become follicles (Fig. 64: 1i). The syncarpous two-carpellate gynoecium (Fig. 64: 3c,d) of *Ptelea trifoliata* (Toddalioideae/Rutoideae) yields a samara (Fig. 64: 3f). *Poncirus trifoliata* and *Citrus* species (Aurantoideae), also with syncarpous gynoecia (carpels six or more; Fig. 64: 2c,d), have a specialized leathery-skinned berry called a hesperidium (Fig. 64: 2f,g; Swingle 1967). The "peel" comprises a tough exocarp dotted with numerous oil glands plus a thinner, more or less fleshy, mesocarp. The fleshy pulp is composed of juice sacs, club-shaped structures (glandular hairs) that project from the inner wall of the ovary (Fig. 64: 2h). In *Citrus*, the edible segments (carpels) are each surrounded by a transparent membrane representing the endocarp.

With strong scent, abundant nectar (from the discs), and often showy corollas, rutaceous flowers generally attract flies and bees as principal pollinators. Many in the family are characterized by imperfect flowers, such

Figure 64. Rutaceae. 1, *Zanthoxylum clava-herculis*: **a,** carpellate flowering branch, ×²⁄₅; **b,** leaf graphically depicting punctate surface, ×½; **c,** carpellate flower, ×6; **d,** longitudinal section of carpellate flower, ×12; **e,** cross section of ovaries, ×12; **f,** gynoecium and nectariferous disc, ×12; **g,** staminate flower, ×4½; **h,** longitudinal section of staminate flower, ×5; **i,** fruit (aggregate of three follicles), ×2⅓. **2,** *Poncirus trifoliata*: **a,** flowering branch, ×²⁄₃; **b,** longitudinal section of flower, ×3; **c,** pistil, ×3; **d,** cross section of ovary, ×5; **e,** cross section of style showing stylar canals, ×20; **f,** fruiting branch with mature leaves and one hesperidium, ×½; **g,** cross section of hesperidium, ×½; **h,** juice sac growing out of inner hesperidium wall, ×4. **3,** *Ptelea trifoliata*: **a,** fruiting branch (with mature leaves), ×²⁄₅; **b,** carpellate flower (note: perianth and androecium also often 5-merous), ×6; **c,** pistil and disc (from carpellate flower), ×6; **d,** cross section of ovary, ×10; **e,** staminate flower (note: perianth and androecium also often 5-merous), ×4; **f,** samara, ×¾.

as *Zanthoxylum clava-herculis* (Fig. 64: 1c,g), which is dioecious. Cross-pollination of perfect flowers generally is reinforced by marked protandry, although some species (e.g., *Citrus*, *Poncirus*) are self-compatible (Brizicky 1962).

REFERENCES CITED

Brizicky, G. K. 1962. The genera of Rutaceae in the southeastern United States. *J. Arnold Arbor.* 43:1–22.

Hegnauer, R. 1983. Chemical characters and classification of the Rutales. In *Chemistry and chemical taxonomy of the Rutales*, ed. P. G. Waterman and M. F. Grundon, pp. 401–440. Annual Proc. Phytochem. Soc. Eur. No. 22. Academic Press, London.

Saunders, E. R. 1934. On carpel polymorphism. IV. *Ann. Bot.* (Oxford) 48:643–692.

da Silva, M. F. das G. F., O. R. Gottlieb, and F. Ehrendorfer. 1988. Chemosystematics of the Rutales: Suggestions for a more natural taxonomy and evolutionary interpretation of the family. *Pl. Syst. Evol.* 161:97–134.

Swingle, W. T. 1967. The botany of *Citrus* and its wild relatives (rev. P. C. Reece). In *The citrus industry*, vol. 1, *History, world distribution, botany and varieties*, ed. W. Reuther, H. J. Webber, and L. D. Batchelor, pp. 190–430. University of California Division of Agricultural Science, Berkeley.

Taylor, D. A. H. 1983. Biogenesis, distribution, and systematic significance of limonoids in the Meliaceae, Cneoraceae, and allied taxa. In *Chemistry and chemical taxonomy of the Rutales*, ed. P. G. Waterman and M. F. Grundon, pp. 353–375. Annual Proc. Phytochem. Soc. Eur. No. 22. Academic Press, London.

MELIACEAE
Mahogany Family

Shrubs to trees, with aromatic parts. **Leaves** pinnately or bipinnately compound, entire or serrate, usually alternate (often clustered at branch tips), exstipulate. **Inflorescence** determinate, cymose and often appearing paniculate, usually axillary. **Flowers** actinomorphic, perfect or often functionally imperfect (then plants usually polygamous or monoecious), hypogynous, usually small, usually with intrastaminal nectariferous disc. **Calyx** of typically 4 or 5 sepals, usually basally connate, usually imbricate. **Corolla** of typically 4 or 5 petals, distinct or sometimes basally connate, usually reflexed, sometimes adnate to staminal column, variously colored (white, purple, greenish, yellowish), contorted/convolute, imbricate, or valvate. **Androecium** of usually 8 or 10 stamens; filaments occasionally distinct or typically united into a tube (monadelphous) and usually with apical appendages; anthers basifixed or dorsifixed (versatile), dehiscing longitudinally, introrse. **Gynoecium** of 1 pistil, 2- to 5-carpellate; ovary superior, with as many locules as carpels (septa often incomplete at apex); ovules 1 to numerous (but usually 2) in each locule, anatropous, campylotropous, or orthotropous, pendulous, placentation axile; style 1 or absent; stigma capitate or discoid, sometimes lobed. **Fruit** a septifragal or loculicidal capsule, a drupe, or sometimes a berry; seeds thin, often winged or with an aril and/or a sarcotesta; endosperm copious, oily, fleshy, or sometimes absent; embryo spatulate.

Family characterization: Aromatic trees or shrubs with pinnately compound leaves; imperfect flowers with 4- or 5-merous perianth and androecium; monadelphous stamens with apical appendages; and a woody or leathery capsule (with winged or occasionally arillate seeds) or a drupe (with a multi-seeded stone) as the fruit types. Triterpenes (limonoids: bitter-tasting substances) present (da Silva et al. 1984). Tissues with calcium oxalate crystals. Anatomical feature: secretory cells and/or cavities (containing resins) in leaves and stems.

Genera/species: 51/550 (Pennington and Styles 1975)

Distribution: Widespread in tropical to subtropical regions, with relatively few representatives in temperate areas; particularly common as understory trees in rain forests.

Major genera: *Aglaia* (100 spp.), *Trichilia* (66 spp.), *Turraea* (65 spp.), *Dysoxylum* (61 spp.), and *Guarea* (35 spp.) (Pennington and Styles 1975)

U.S./Canadian representatives: *Trichilia* (3 spp.), *Guarea* (2 spp.), *Cedrela* (1 sp.), *Melia* (1 sp.), and *Swietenia* (1 sp.)

Economic plants and products: High-quality timber from *Cedrela* (West Indian–cedar), *Khaya* (African-mahogany), and *Swietenia* (mahogany). Edible fruits from *Lansium* (langsat) and *Sandoricum* (santol). Insec-

Figure 65. Meliaceae. 1, *Melia azedarach*: **a,** flowering branch, ×¼; **b,** flower, ×3; **c,** androecium (monadelphous), ×6; **d,** expanded androecium, ×6; **e,** longitudinal section of flower, ×5; **f,** cross section of ovary below middle, ×12; **g,** cross section of ovary above middle, ×12; **h,** drupe, ×1½; **i,** longitudinal section of drupe, ×1½. **2,** *Swietenia mahagoni*: **a,** staminate flower, ×5; **b,** staminate flower with petals and half of androecium removed, ×10; **c,** carpellate flower, ×5; **d,** carpellate flower with petals and half of androecium removed, ×10; **e,** capsule (starting to dehisce), ×⅖; **f,** dehiscing capsule with one valve removed, exposing the seeds, ×⅖; **g,** seed, ×⅔.

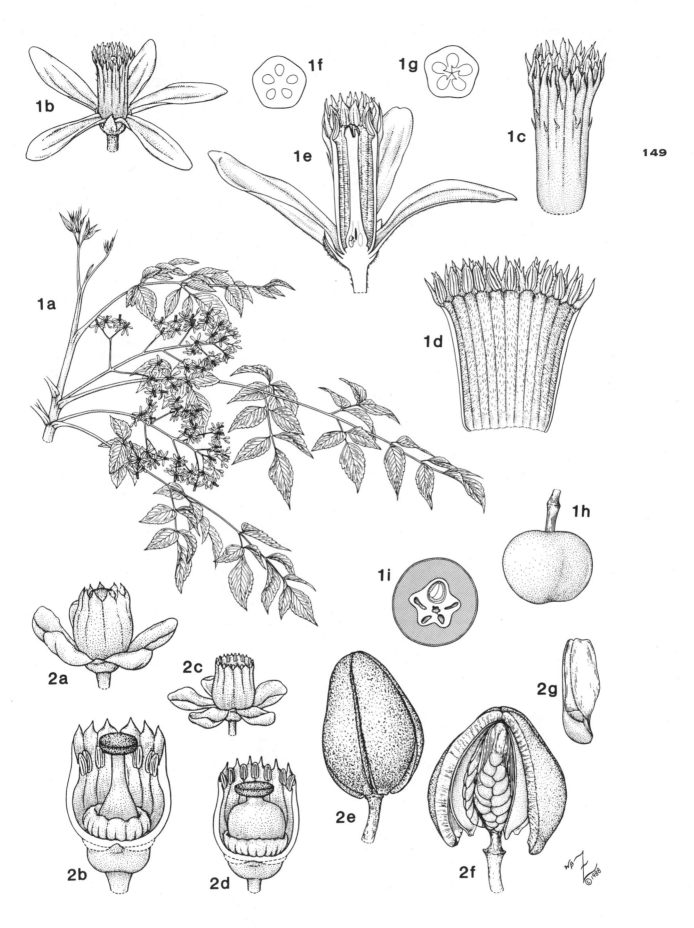

ticides (triterpenoid substances) extracted from fruits and seeds of species of *Azadirachta* (neem tree) and *Melia*. Ornamental plants (species of at least 11 genera), including *Cedrela*, *Melia* (china-berry, umbrella tree), *Swietenia*, and *Turraea* (South African–honeysuckle).

Commentary: The Meliaceae are divided into four subfamilies primarily based on stamen morphology, fruit and seed type, and wood anatomy (Pennington and Styles 1975). Because two of these taxa are monotypic and restricted to Madagascar, the Melioideae and the Swietenioideae constitute most of the family. The detailed monograph by Pennington and Styles (1975) estimates the number of species in the family to be about half the often-cited and inflated figure of 1,000 or more. Taxonomists generally agree on a close relationship to the Rutaceae and other families of the Rutales-Sapindales complex (Miller 1990).

Important generic and specific characters include those of the androecium, as well as the shape of the disc and stigma. The stamens typically are monadelphous, although the filaments are distinct in *Cedrela* and allies and vary from distinct to monadelphous in species of *Trichilia*. The anthers are inserted within the variously shaped tube near its apex, which often is fringed with various kinds of appendages; for example, in *Melia*, 20 teeth border the cylindrical staminal tube (Fig. 65: 1c,d), whereas the urceolate tube of *Swietenia* (Fig. 65: 2a,c) has ten teeth. Depending on the species, the apex of the cylindrical tube of *Guarea* varies from entire to appendaged.

The flowers usually are small, attracting pollinators with nectar and usually also fragrance. The shape of the nectariferous disc, situated between the base of the ovary and the staminal tube, varies from annular and inconspicuous (*Melia*; Fig. 65: 1e) to well developed and lobed (*Swietenia*; Fig. 65: 2b,d). Insects must insert their proboscises between the anthers (within the staminal tube) to reach the nectar at the base. The most important pollinators are bees and moths (Styles and Khosla 1976; Bawa et al. 1985).

Imperfect flowers are much more common in the family than previously recorded; many so-called perfect flowers actually are functionally staminate or carpellate, although the differences between the flower types of a species may not be morphologically obvious (Pennington 1981). For example, the flowers of *Melia azedarach* (Fig. 65: 1b) actually are perfect and staminate, differing in position on the three-flowered cymose units of the inflorescences (Styles 1972). The few perfect flowers occur as the central (terminal) element of the cymule and open first; the staminate flowers are situated as the two lateral elements of the cyme (and sometimes also as the central flower). The two flower types appear superficially identical, although the staminate form, presumably with abortive ovules, falls off soon after flowering, while the perfect flowers usually persist, maturing into the fruit. In other members of the family, the different flower forms are clearly distinguishable, as in *Swietenia*, with clearly dimorphic staminate and carpellate flowers (Lee 1967). The larger staminate flower (Fig. 65: 2a,b) has a long pistil with a stigma reaching above the insertion of the anthers on the tube; the smaller carpellate flowers (Fig. 65: 2c,d) are characterized by a short pistil with a globose ovary and a style at or below the anthers (with abortive pollen).

The various fruit and seed types of the family include drupes with multi-seeded stones (*Melia*; Fig. 65: 1h,i) and woody or leathery capsules (e.g., Fig. 65: 2e–g) with winged (*Swietenia*, *Cedrela*) or arillate (certain *Trichilia* species) seeds. The seeds of some capsules are characterized by a sarcotesta (e.g., *Guarea*). The erect, septicidal valves of *Swietenia* capsules dehisce from the base to the apex (Fig. 65: 2f), leaving a persistent central column.

REFERENCES CITED

Bawa, K. S., S. H. Bullock, D. R. Perry, R. E. Corville, and M. H. Grayum. 1985. Reproductive biology of tropical lowland rain forest trees. II. Pollination systems. *Amer. J. Bot.* 72:346–356.

Lee, H.-Y. 1967. Studies in *Swietenia* (Meliaceae): Observations on the sexuality of the flowers. *J. Arnold Arbor.* 48:101–104.

Miller, N. G. 1990. The genera of Meliaceae in the southeastern United States. *J. Arnold Arbor.* 71:453–486.

Pennington, T. D. 1981. Meliaceae. *Fl. Neotrop. Monogr.* 28:1–470.

—— and B. T. Styles. 1975. A generic monograph of the Meliaceae. *Blumea* 22:419–540.

da Silva, M. F. das G. F., O. R. Gottlieb, and D. L. Dreyer. 1984. Evolution of limonoids in the Meliaceae. *Biochem. Syst. & Ecol.* 12:299–310.

Styles, B. T. 1972. The flower biology of the Meliaceae and its bearing on tree breeding. *Silvae Genet.* 21:175–181.

—— and P. K. Khosla. 1976. Cytology and reproductive biology of Meliaceae. In *Tropical trees: Variation, breeding and conservation*, ed. J. Burley and B. T. Styles, pp. 61–67. Linn. Soc. Symp. Ser. No. 2. Academic Press, London.

ANACARDIACEAE
Cashew or Sumac Family

Trees, shrubs, or occasionally woody vines, with resinous bark, branches, leaves, flowers, and fruits, sometimes with milky sap. **Leaves** trifoliolate, pinnately compound, or sometimes simple, entire to serrate, alternate, deciduous or persistent, usually exstipulate.

Inflorescence determinate, cymose and appearing paniculate, terminal or axillary. **Flowers** actinomorphic, perfect or more often imperfect (then plants dioecious, polygamous, or polygamodioecious), hypogynous or rarely perigynous or epigynous, small, with annular intrastaminal nectariferous disc. **Calyx** of typically 5 sepals, usually basally connate, usually imbricate. **Corolla** of typically 5 petals, distinct or basally connate, white, green, or yellow, usually imbricate. **Androecium** of usually 5 or 10 (biseriate) stamens, arising upon or outside the disc; filaments distinct or rarely basally connate; anthers dorsifixed, versatile, dehiscing longitudinally, introrse; staminodes often present in carpellate flowers. **Gynoecium** of 1 pistil, basically 3-carpellate (2 carpels usually aborting); ovary typically superior, usually 1-locular; ovule solitary, anatropous, often with thickened funicle, placentation basal, parietal, or apical; style(s) 1 to 3, often widely separated; stigma(s) 1 to 3; rudimentary pistil often present in staminate flowers. **Fruit** usually a drupe with resinous mesocarp; endosperm scanty or absent; embryo oily, usually curved.

Family characterization: Trees and shrubs with resinous parts; trifoliolate or pinnately compound leaves; often imperfect flowers with intrastaminal nectariferous disc and 5-merous perianth and androecium; 3-carpellate and 1-locular ovary with a solitary ovule; and drupe with resinous mesocarp. Tissues characteristically with calcium oxalate crystals and a high tannin content. Anatomical features: well-developed schizogenous or lysigenous resin canals (exuding material that turns black on drying and that often contains irritant substances) or sometimes latex channels in the bark, leaves, flower, and/or fruits; and trilacunar nodes (Fig. 66: 1f).

Genera/species: 70/600

Distribution: Mainly pantropical, with a few representatives extending into temperate areas of Eurasia and North America.

Major genera: *Rhus* (200–250 spp.), *Lannea* (40–70 spp.), *Semecarpus* (50–60 spp.), and *Trichoscypha* (50 spp.)

U.S./Canadian representatives: 11 genera/34 spp.; largest genera: *Rhus* and *Toxicodendron*

Economic plants and products: Edible seeds or fruits from *Anacardium* (cashew), *Harpephyllum* (Kaffir-plum), *Mangifera* (mango), *Pistacia* (pistachio), and *Spondias* (mombin, Jamaica-plum, hog-plum). Timber from *Astronium* (kingwood, zebrawood) and *Schinopsis* (quebracho). Resins, oils, and lacquers from several, such as species of *Pistacia* (mastic tree) and *Toxico-*

dendron (varnish tree). Tannic acid from *Cotinus*, *Pistacia*, *Rhus*, and *Schinopsis*. Many toxic plants (due to irritant phenolic compounds; see the discussion below): *Metopium* (poisonwood) and *Toxicodendron* (poison-ivy, -oak, and -sumac). Ornamental plants (species of 15 genera), including *Cotinus* (smoke tree), *Mangifera*, *Rhus* (sumac), and *Schinus* (Brazilian pepper tree, Florida-holly; but see Morton 1979).

151

Commentary: The Anacardiaceae traditionally have been divided into five tribes based mainly upon the number and degree of fusion of the carpels; recent studies, however, indicate that these taxa may not be monophyletic (Wannan and Quinn 1991). A notable taxonomic problem within the family is the delimitation of the heterogeneous *Rhus-Toxicodendron* complex (100 to 250 spp.), which has a confusing history (see Brizicky 1962, 1963; Gillis 1971). The family generally has been allied with families of the Sapindales-Rutales complex (see Wannan and Quinn 1991 for a historical summary).

Certain members of this family are the most common and best known plants that cause contact dermatitis (Baer 1983). Direct or indirect contact as well as volatile emanation (from burning the plants) may cause an allergic reaction in sensitive individuals. The severe skin irritation is caused by certain derivative phenolic compounds (alkenyl phenols), called urushiols, in the exudate produced in the characteristic resin or latex canals. These toxic principles may be distributed throughout various parts of the plant or concentrated in particular organs. For example, all parts of several *Toxicodendron* species (poison-ivy, -oak, and -sumac) and all species of *Metopium* (poisonwood) are allergenic. The lacquer produced from *Toxicodendron vernicifluum* and the black ink from *Semecarpus anacardium* (marking-nut tree) may cause allergic reactions. Although the flesh of the mango fruit (*Mangifera*) is edible, the skin may be an irritant to some people. The potentially irritating oil of *Anacardium occidentale* (cashew) is rendered harmless by heat.

The flowers of the Anacardiaceae are usually entomophilous. Various insects are attracted to the small flowers (in large inflorescences) with exposed nectar secreted by the fleshy disc (Tomlinson 1980). The dioecious and polygamodioecious nature of the plants promotes cross-pollination.

Baer, H. 1983. Allergic contact dermatitis from plants. In *Handbook of natural toxins*, vol. 1, *Plant and fungal toxins*, ed. R. F. Keeler and A. T. Tu, pp. 421–442. Marcel Dekker, New York.

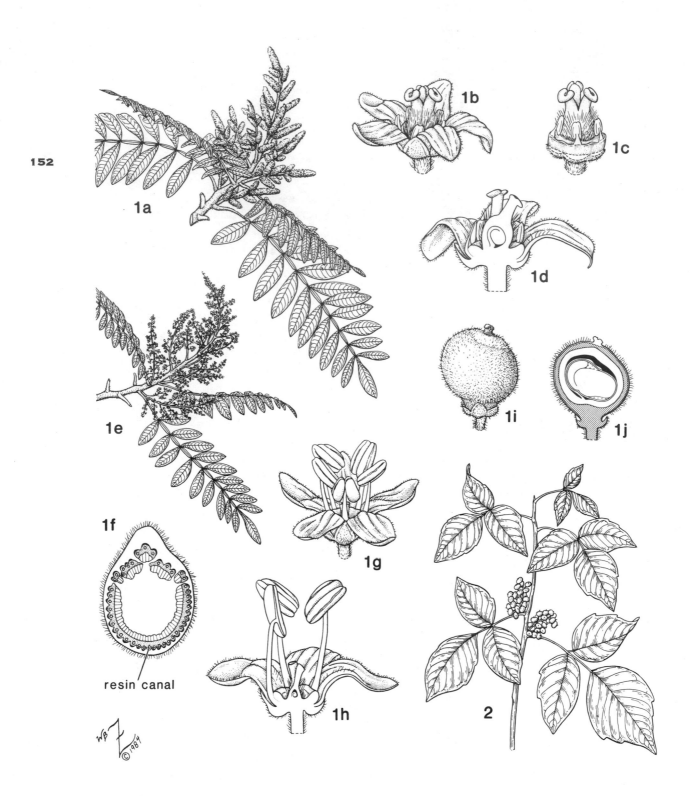

Figure 66. Anacardiaceae. 1, *Rhus copallina*: **a,** carpellate flowering branch, ×⅓; **b,** carpellate flower, ×12; **c,** androecium and gynoecium of carpellate flower, ×14; **d,** longitudinal section of carpellate flower, ×15; **e,** staminate flowering branch, ×⅓; **f,** cross section of node (three-trace trilacunar), ×3; **g,** staminate flower, ×9; **h,** longitudinal section of staminate flower, ×12; **i,** drupe, ×5; **j,** longitudinal section of drupe, ×5. **2,** *Toxicodendron radicans* (*Rhus radicans*): fruiting branch, ×⅓.

Brizicky, G. K. 1962. The genera of Anacardiaceae in the southeastern United States. *J. Arnold Arbor.* 43:359–375.

———. 1963. Taxonomic and nomenclatural notes on the genus *Rhus* (Anacardiaceae). *J. Arnold Arbor.* 44:60–80.

Gillis, W. T. 1971. The systematics and ecology of poison-ivy and the poison-oaks (*Toxicodendron*, Anacardiaceae). *Rhodora* 73:72–159, 161–237, 370–443, 465–540.

Morton, J. F. 1979. Brazilian pepper: Its impact on people, animals, and the environment. *Econ. Bot.* 32:353–359.

Tomlinson, P. B. 1980. *The biology of trees native to tropical Florida*, pp. 79–91. "Publ. by the author," Petersham, Mass.

Wannan, B. S., and C. J. Quinn. 1991. Floral structure and evolution in the Anacardiaceae. *Bot. J. Linn. Soc.* 107:349–385.

SAPINDACEAE

Soapberry Family

Including the Aceraceae: Maple Family,
    and the Hippocastanaceae:
        Horse-chestnut Family

Trees, shrubs, or sometimes woody vines with tendrils. **Leaves** usually pinnately compound or sometimes simple (then often with palmate venation) or palmately compound, usually serrate, alternate or sometimes opposite, persistent or deciduous, exstipulate or occasionally stipulate (in woody vines). **Inflorescence** determinate, cymose and appearing paniculate, racemose, corymbose, or umbellate (fasciculate), terminal or axillary. **Flowers** actinomorphic to zygomorphic, perfect and/or usually imperfect (then plants monoecious, dioecious, polygamous, or most often polygamodioecious), usually hypogynous, minute to occasionally showy, with well-developed extrastaminal or sometimes intrastaminal disc. **Calyx** of 4 or usually 5 sepals, distinct to sometimes basally connate, often unequal, occasionally reddish, usually imbricate. **Corolla** of 4 or usually 5 petals or absent, distinct, sometimes unequal, often clawed, frequently with scaly or hair-tufted basal appendages, greenish, white, yellow, or red, imbricate. **Androecium** of basically 10 stamens in 2 whorls of 5 but often reduced to 4 to 8 stamens, variously inserted within, around, or upon the disc; filaments distinct, often hairy; anthers dorsifixed and versatile or basifixed, dehiscing longitudinally, introrse; staminodes sometimes present in carpellate flowers. **Gynoecium** of 1 pistil, 2- or usually 3-carpellate; ovary usually superior, with as many locules as carpels; ovule(s) 1 or sometimes 2 in each locule, anatropous to campylotropous or orthotropous, typically lacking a well-developed funicle and broadly attached to placental protuberance (obturator), placentation axile or sometimes parietal; style(s) 1 (then sometimes bifid or trifid) to 3; stigma(s) simple, lobed, or along the inner surface of style; rudimentary pistil usually present in staminate flowers. **Fruit** a loculicidal or septifragal capsule, or a nut, arilloid berry, or schizocarp (splitting into drupe-like, nut-like, or samaroid segments), often 1-seeded and -locular (by abortion); seeds often arillate or with sarcotesta; endosperm usually absent; embryo curved, plicate, and/or twisted, oily and/or starchy, with radicle enclosed in infolding of testa (radicular pocket).

Family characterization: Trees, shrubs, or woody vines; perfect and/or imperfect, often small flowers with well-developed nectariferous disc; 4- or 5-merous perianth; androecium reduced to 8 or fewer stamens from basic number of 10; 2- or 3-carpellate, superior ovary with 1 or 2 ovule(s) per locule; "sessile" ovules lacking a distinct funicle and thus broadly attached to placental protuberance (obturator); various fruit types often 1-locular and -seeded by abortion; and nonendospermous seeds with curved to folded embryo and a radicular pocket. Saponins (especially in tissues of stem and/or fruit), various cyclopropane amino acids (Umadevi and Daniel 1991), and tannins present. Tissues with calcium oxalate crystals. Anatomical features: parenchymatous tissues with scattered secretory cells containing resins or laticiferous compounds (appearing on dried leaves as transparent dots or streaks); epidermis of leaves often with mucilage cells (appearing on leaves as transparent dots); and trilacunar nodes.

Genera/species: 147/2,215

Distribution: Primarily tropical to subtropical and extending into the north temperate region; especially abundant in Asia and America.

Major genera: *Serjania* (215 spp.), *Acer* (111–200 spp.), *Paullinia* (180–194 spp.), and *Allophylus* (1–190 spp.)

U.S./Canadian representatives: 19 genera/64 spp.; largest genera: *Acer, Aesculus,* and *Sapindus*

Economic plants and products: Edible seeds and/or arils from *Blighia* (akee; deadly poisonous if unripe), *Euphoria* (longan), *Litchi* (lychee or litchi), and *Melicoccus* (genip, Spanish-lime). Sugar and syrup from the sap of several *Acer* species, especially *Acer saccharum* (sugar maple). Caffeine-rich beverage and herbal medicines from seeds of *Paullinia cupana* (guarana). Timber from *Acer* (maple, sycamore) and *Aesculus* (buckeye, horsechestnut). Ornamental plants (species of 24 genera), including *Acer, Aesculus, Cardiospermum* (balloon vine, heartseed), *Dodonaea, Euphoria, Koelreuteria* (golden-

154

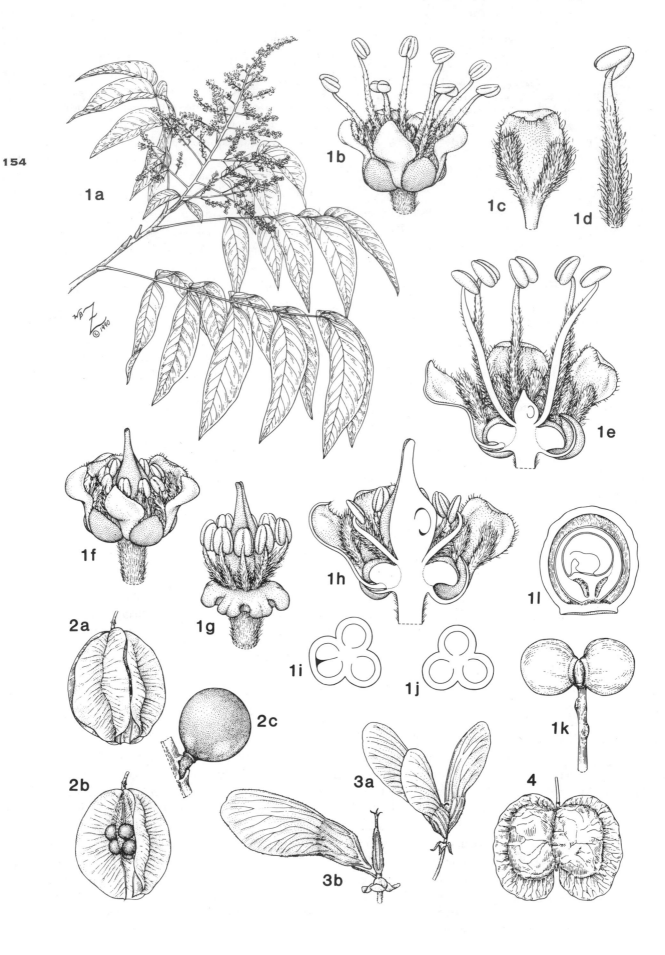

rain tree), *Litchi*, *Melicoccus*, *Sapindus* (soapberry), and *Xanthoceras*.

Commentary: The Sapindaceae are defined broadly here (Thorne 1992) to include the two small subfamilies Aceroideae and Hippocastanoideae, which usually are treated as the familiar segregate families Aceraceae and Hippocastanaceae. Both of these subfamilies are predominantly temperate and have long been recognized as closely related to the tropical Sapindaceae *s.s.* Table 8 and Figs. 67, 68, and 69 represent an attempt to summarize the features historically used to distinguish these three taxa, which actually overlap in most of the character states.

Although this is probably disconcerting to most traditionalists, an expanded Sapindaceae has been suggested by several authors (for various reasons), including Hutchinson (1926), who submerged the Hippocastanaceae in his initial delimitation of Sapindaceae, and Bentham and Hooker (1862), who included the Aceraceae. Heimsch (1942) noted that features of wood anatomy (highly specialized rays) supported the inclusion of both taxa in the Sapindaceae *s.l.* More recent authors also have advocated the retention of the Aceraceae and Hippocastanaceae within the Sapindaceae, based on palynological (Muller and Leenhouts 1976) and phytochemical (Umadevi et al. 1986; Umadevi and Daniel 1991) studies of these taxa; in their monograph on fossil *Acer*, Wolfe and Tanai (1987) proposed Aceraceae as a tribe within the Sapindaceae (because they considered both taxa as evolved from a common ancestor). Aceraceae and Hippocastanaceae are substantiated as derived groups within the Sapindaceae *s.s.* by Judd et al. (1994) in their preliminary cladistic study incorporating nineteen representative genera of the three taxa. Figure 70 is a simplified version of their consensus tree. Of note are that the Hippocastanaceae (*Aesculus* and *Billia*) plus the sapindaceous *Handeliodendron* constitute a group (clade), and that the traditional Aceraceae (*Acer* and *Dipteronia*) plus certain members of the Sapindaceae *s.s.* form a "samaroid clade" (characterized by winged schizocarps).

Also, these results do not support Aceraceae as "primitive" within the Sapindaceae. The authors conclude that the Sapindaceae *s.s.* clearly are paraphyletic and that "Aceraceae" and "Hippocastanaceae" may be monophyletic or paraphyletic. The numerous tribes (up to fourteen; Brizicky 1963) of the Sapindaceae *s.s.* need much further revision due to potential polyphyly.

The flowers of the Sapindaceae vary from actinomorphic and small to zygomorphic and somewhat showy (see Table 8). The petals are variously shaped but are usually clawed. In *Aesculus*, for example, the claws are long and wide (Fig. 69: e,f) and clasp the stamen filaments. Frequently petaloid appendages (often hairy) occur on the inner surface above the claw, as in *Sapindus* (appendage bifid; Fig. 67: 1c). Sometimes the petals are unequal, as in *Aesculus pavia*, where the upper pair (Fig. 69: e) are longer and much narrower than the laterals (Fig. 69: f). In *Acer*, the petals are reduced and sepal-like (Fig. 68: b,g).

The conspicuous nectariferous disc typically is extrastaminal, annular, and lobed, with the stamens inserted on the inner margin of the disc, as in *Sapindus* (Fig. 67: 1e). Sometimes the nectary is unilateral, as in *Cardiospermum* (one gland) or *Aesculus* (incomplete ring; Fig. 69: c,g). The disc of *Acer* (Fig. 68: c,d,h) may be misinterpreted as "intrastaminal" in the literature due to displacement of the stamens (Judd et al. 1994).

Little has been reported for the pollination of many of the Sapindaceae, particularly the tropical species. Generally, the small flowers are aggregated into conspicuous inflorescences, and bees (as well as other insects and hummingbirds) visit for the copious nectar produced by the disc. However, a few species (e.g., in *Acer* and *Dodonaea*) are anemophilous. Cross-pollination is reinforced by the strong tendency in the family for various forms of dioecism, coupled with the production of both perfect and imperfect flowers. Even in the monoecious *Sapindus*, one sex tends to predominate on a particular plant (Tomlinson 1980). Trees frequently are "functionally dioecious" with the perfect, staminate, and/or carpellate flowers in different locations/inflorescences on the tree and/or maturing

Figure 67. Sapindaceae. 1, *Sapindus marginatus*: **a,** flowering branch, ×⅓; **b,** staminate flower, ×7; **c,** adaxial side of petal showing paired appendages, ×12; **d,** stamen, ×12; **e,** longitudinal section of staminate flower, ×12; **f,** carpellate flower, ×7; **g,** carpellate flower (perianth removed) showing disc, ×10; **h,** longitudinal section of carpellate flower, ×12; **i,** cross section of ovary (one of the two ovules per carpel aborted in two locules), ×12; **j,** cross section of ovary (one of the two ovules per carpel aborted in all locules), ×12; **k,** schizocarp (two-carpellate, with third aborted carpel in center; fruits also occasionally one- or three-carpellate), ×⅚; **l,** longitudinal section of one drupe-like schizocarp segment, ×2. **2,** *Koelreuteria elegans*: **a,** capsule, ×¾; **b,** capsule with one valve removed, exposing the seeds, ×¾; **c,** seed (note obturator), ×3¾. **3,** *Thouinia discolor*: **a,** schizocarp (compare with Fig. 68: j), ×3; **b,** dehiscing schizocarp with one attached segment (compare with Fig. 68: k), ×4½. **4,** *Dodonaea viscosa*: capsule (two-carpellate; fruits also sometimes three-carpellate), ×1½.

156

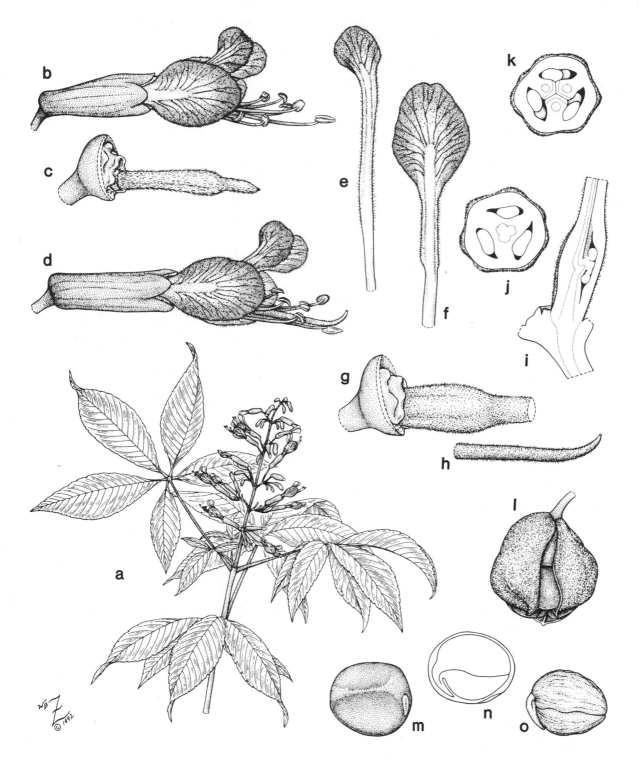

Figure 68 [facing page]. Sapindaceae (continued). *Acer rubrum*: **a,** carpellate flowering branch, ×1; **b,** carpellate flower, ×7; **c,** androecium, gynoecium, and disc of carpellate flower, ×7; **d,** longitudinal section of carpellate flower, ×11; **e,** cross section of ovary, ×11; **f,** staminate flowering branch, ×1; **g,** staminate flower, ×5½; **h,** longitudinal section of staminate flower, ×7; **i,** fruiting branch (with immature leaves), ×½; **j,** schizocarp, ×1¼; **k,** dehiscing schizocarp with one attached segment, ×1¼; **l,** longitudinal section of schizocarp segment, ×2½; **m,** mature leaf, ×⅗.

Figure 69 [above]. Sapindaceae (continued). *Aesculus pavia*: **a,** flowering branch, ×⅓; **b,** staminate flower, ×2¼; **c,** rudimentary pistil and disc from staminate flower, ×5; **d,** perfect flower, ×2¼; **e,** one of two adaxial petals, ×2½; **f,** lateral petal, ×2½; **g,** ovary and disc, ×5; **h,** apex of style with stigma (at tip), ×5; **i,** longitudinal section of ovary, ×5; **j,** cross section of ovary near base, ×10; **k,** cross section of ovary near apex, ×10; **l,** capsule, ×½; **m,** seed (note conspicuous hilum), ×¾; **n,** cross section of seed (note "radicular pocket"), ×¾; **o,** embryo, ×¾.

TABLE 8. Major morphological differences of three major traditional groups of the Sapindaceae *s.l.* (Thorne 1992): the Sapindaceae *s.s.*, Aceroideae ("Aceraceae"), and Hippocastanoideae ("Hippocastanaceae").

CHARACTER	SAPINDACEAE S.S. (5 SUBFAMILIES)	ACEROIDEAE ("ACERACEAE")	HIPPOCASTANOIDEAE ("HIPPOCASTANACEAE")
GENERA/SPECIES	143/2,000	2/200	2/15
DISTRIBUTION	primarily tropical to subtropical	temperate northern hemisphere	temperate northern hemisphere and tropical America
HABIT	trees, shrubs, or sometimes vines	trees, or sometimes shrubs	trees or shrubs
LEAVES	usually pinnately compound, or occasionally simple, palmately compound, or trifoliate usually alternate persistent or deciduous exstipulate or stipulate	simple and palmately lobed/veined, or sometimes pinnately compound or trifoliate opposite deciduous exstipulate	palmately compound opposite deciduous or persistent exstipulate
INFLORESCENCE	paniculate or racemose	corymbose, umbellate (fasciculate), racemose or paniculate	paniculate or racemose
FLOWER	actinomorphic to slightly zygomorphic small	actinomorphic small	zygomorphic relatively large and showy
NECTARIES	extrastaminal annular, or sometimes unilateral	presumably intrastaminal annular	extrastaminal unilateral
SEPALS	distinct to (sometimes) basally connate	distinct to (sometimes) basally connate	connate into a tube
PETALS	equal to (sometimes) unequal clawed frequently appendaged	equal unclawed unappendaged	unequal clawed "unappendaged" (appendages highly modified)
STAMEN NUMBER	10, often reduced to 4, 5, or 8	often 8, sometimes 4, 5, or 10	usually 5 to 8
ANTHERS	dorsifixed, versatile	basifixed	± dorsifixed, versatile
CARPEL NUMBER	3, or occasionally 2	2	usually 3
OVULES/CARPEL	1	2	2
STYLE NUMBER	1, sometimes trifid or bifid	2, or 1 and bifid	1 (unbranched)
STIGMA	simple or lobed	along inner surface of style	simple
FRUIT TYPE	capsule, nut, berry, or winged schizocarp	winged schizocarp	capsule

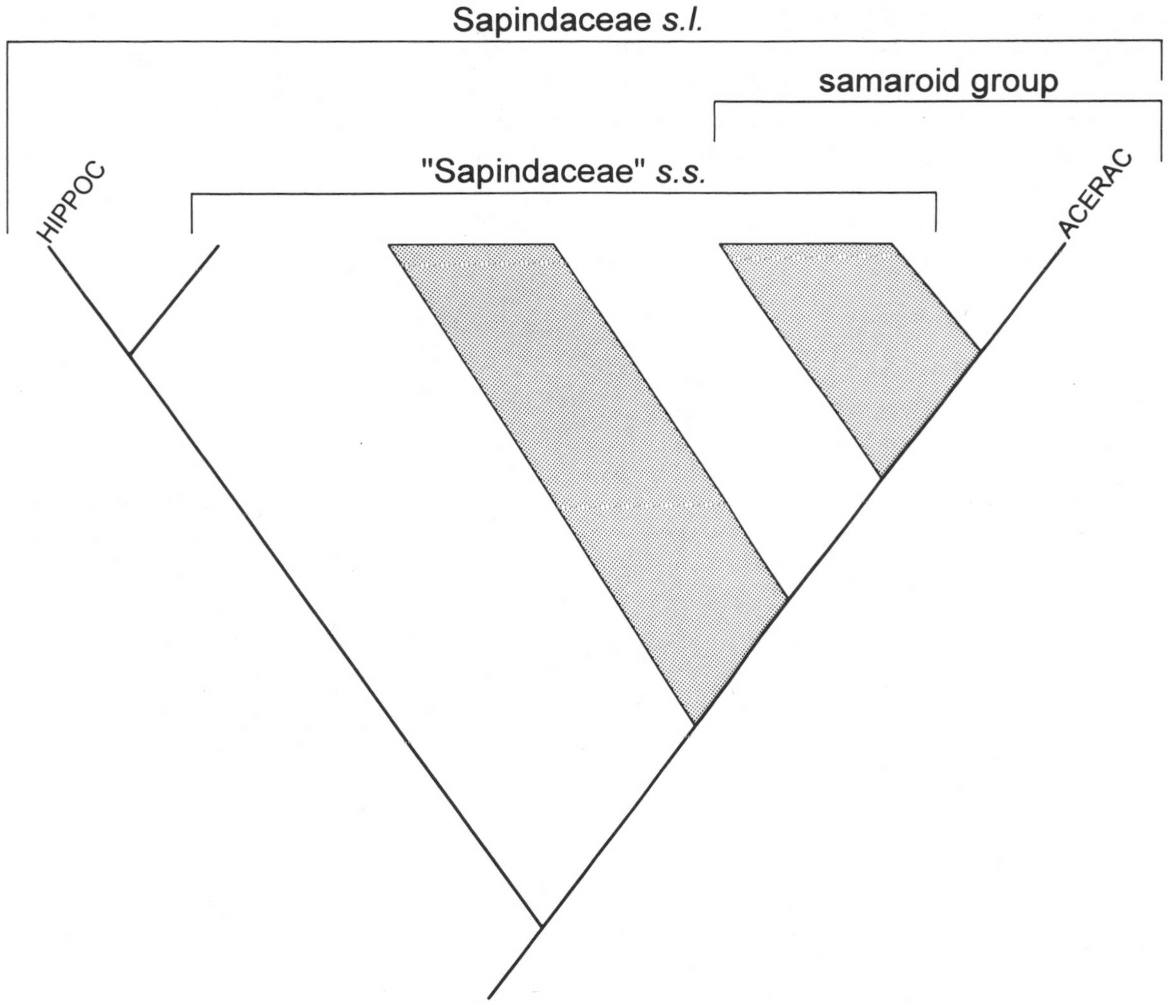

Figure 70. Relationships of the Sapindaceae-Aceraceae-Hippocastanaceae complex. Simplification of consensus tree from Judd et al. (1994), showing the relationships between the "Aceraceae," "Hippocastanaceae," and some major groupings of the Sapindaceae *s.s.* The monophyletic Sapindaceae *s.l.*, *sensu* Thorne (1992), comprise all groups shown here. Abbreviations: ACERAC (Aceraceae), HIPPOC (Hippocastanaceae).

at different times (e.g., *Acer, Aesculus, Hypelate*). The distinction between perfect and imperfect flowers may be unclear, as when apparently well-developed stamens in carpellate flowers do not dehisce. The numerous staminate flowers (Fig. 69: b) in an *Aesculus* inflorescence are readily identifiable by the rudimentary pistil (Fig. 69: c), but some of the few "perfect" flowers may be functionally carpellate because the anthers fall off before maturing.

The various fruit types of the family often are one-locular and one-seeded by abortion. In *Aesculus*, for example, the loculicidal capsule (Fig. 69: l) consists of a thick and leathery husk enclosing basically three, but sometimes two or usually one, large chestnut-like seed(s). Often the fruit is a schizocarp splitting into segments of various kinds, as in *Sapindus*, with drupe-like portions (Fig. 67: 1k,l). However, this fruit is quite

leathery (rather than fleshy), with the basic three segments often reduced to two or one. The schizocarp segments may be winged, as in *Acer* (two parts; Fig. 68: j,k) or *Thouinia* (three parts; Fig. 67: 3a,b). Bladdery capsules characterize *Cardiospermum* (usually septifragal), *Dodonaea* (septifragal), and *Koelreuteria* (loculicidal; Fig. 67: 2a). *Dodonaea* capsules are conspicuously winged (Fig. 67: 4). Although this is not true in most of our representatives, the family in general is characterized by arillate seeds, as exemplified by *Cupania* (loculicidal capsules with seeds subtended by cup-shaped arils). The variously described baccate fruit type of genera such as *Melicoccus* and *Litchi* consists of a single arillate seed surrounded by a leathery pericarp ("ariloid berry").

A derived feature of the family is the "radicular pocket" formed by a deep fold in the seed coat that encloses the radicle (embryonic root), separating it from the rest of the embryo (Figs. 67: 1l; 69: n; van der Pijl 1957). Another characteristic of the seed is a conspicuous, often relatively large scar in the micropylar region, which is often whitish as in *Aesculus* (Fig. 69: m) or *Cardiospermum*. This pseudo-hilum represents the broad attachment point of the developing ovule directly to the placental obturator (as evident in Fig. 67: 2c).

REFERENCES CITED

Bentham, G., and J. D. Hooker. 1862. *Genera plantarum*, vol. 1(1), pp. 388– 413. A. Black, London.

Brizicky, G. K. 1963. The genera of Sapindales in the southeastern United States. *J. Arnold Arbor.* 44:462–501.

Heimsch, C. 1942. Comparative anatomy of the secondary xylem of the "Gruinales" and "Terebinthales" of Wettstein with reference to taxonomic groupings. *Lilloa* 8:83–198.

Hutchinson, J. 1926. *The families of flowering plants.* Vol. 1, *Dicotyledons*, p. 242. Macmillan, London.

Judd, W. S., R. W. Sanders, and M. J. Donoghue. 1994. Angiosperm family pairs: Preliminary cladistic analyses. *Harvard Pap. Bot.* 5:1–51.

Muller, J., and P. W. Leenhouts. 1976. A general survey of pollen types in Sapindaceae in relation to taxonomy. In *The evolutionary significance of the exine*, ed. I. K. Ferguson and J. Muller, pp. 407–445. Linn. Soc. Symp. Ser. No. 1. Academic Press, London.

van der Pijl, L. 1957. On the arilloids of *Nephelium, Euphoria, Litchi* and *Aesculus* and the seeds of Sapindaceae in general. *Acta Bot. Neerl.* 6:618– 641.

Thorne, R. F. 1992. Classification and geography of the flowering plants. *Bot. Rev.* (Lancaster) 58:225–348.

Tomlinson, P. B. 1980. *The biology of trees native to tropical Florida*, pp. 379–381. "Publ. by the author," Petersham, Mass.

Umadevi, I., and M. Daniel. 1991. Chemosystematics of the Sapindaceae. *Feddes Repert.* 102:607–612.

——, ——, and S. D. Sabnis. 1986. Interrelationships among the families Aceraceae, Hippocastanaceae, Melianthaceae and Staphyleaceae. *J. Pl. Anat. Morphol.* 3:169–172.

Wolfe, J. A., and T. Tanai. 1987. Systematics, phylogeny, and distribution of *Acer* (maples) in the Cenozoic of North America. *J. Fac. Sci. Hokkaido Imp. Univ.*, Ser. 4, *Geol.* 22:1–246.

FABACEAE OR LEGUMINOSAE
Legume or Pea Family

Herbs, shrubs, or trees; roots often with nitrogen-fixing bacterial nodules. **Leaves** pinnately, bipinnately, or sometimes palmately compound or reduced to one leaflet, entire, usually alternate, sometimes with tendrils, with a prominent pulvinus at the base of the petiole and each petiolule, stipulate. **Inflorescence** indeterminate, racemose, paniculate, spicate, umbellate, or capitate, or sometimes flower solitary, terminal or axillary. **Flowers** zygomorphic or actinomorphic, usually perfect, slightly perigynous with a cupular short hypanthium, typically showy, with a nectariferous disc. **Calyx** synsepalous with 5 lobes, tubular, imbricate or valvate. **Corolla** of 5 petals, distinct or the 2 anterior petals variously connate, unequal or sometimes equal, variously colored, valvate or imbricate. **Androecium** of often 10 stamens (but sometimes fewer or more); filaments distinct, all united into a tube (monadelphous), or 9 united into a strap with 1 free (diadelphous); anthers dorsifixed or basifixed, dehiscing longitudinally or sometimes by apical pores, introrse. **Gynoecium** of 1 carpel; ovary superior, usually 1-locular, sometimes stipitate (with gynophore); ovules 2 to many in 2 alternating rows, anatropous or campylotropous, pendulous or obliquely ascending, placentation axile along the adaxial suture; style 1, arching; stigma 1. **Fruit** a legume dehiscing along both abaxial and adaxial sutures, or sometimes a loment breaking transversely into 1-seeded segments, or indehiscent, typically flattened; seeds usually with hard seed coat; endosperm absent or sometimes scanty (then glassy and hard); embryo with large, flat, and fleshy cotyledons.

Family characterization: Herbs to trees with bacterial nodules in the roots; alternate, stipulate, compound leaves with entire margins; leaflets with pulvini responsible for "sleep movements"; zygomorphic or actinomorphic, perigynous flowers with short hypanthium; distinct, monadelphous, or diadelphous stamens; unicarpellate gynoecium with long arching style; parietal placentation with ovules in two alternating rows along the adaxial suture; a legume or a loment as the fruit type; and nonendospermous seeds. Various alkaloids and cyanogenic glycosides often present. Tissues with calcium oxalate crystals. Anatomical feature: parenchymatous root nodules formed by nitrogen-fixing bacteria (see the discussion below).

Genera/species: 630/18,000

Distribution: Cosmopolitan.

Figure 71. Fabaceae. 1, *Centrosema virginianum*: **a,** habit, ×⅓; **b,** longitudinal section of flower, ×3; **c,** expanded androecium (diadelphous), ×4; **d,** nectary at base of ovary, ×6; **e,** portion of longitudinal section of ovary, ×12; **f,** cross section of ovary, ×12; **g,** androecium and gynoecium, ×3. **2,** *Crotalaria spectabilis*: **a,** upper portion of plant, ×⅓; **b,** root nodules, ×3; **c,** two views of flower, ×1¼; **d,** standard, ×1¼; **e,** wings, ×1¼; **f,** expanded keel, ×1¼; **g,** expanded androecium (monadelphous), ×2 ⅓; **h,** cross section of bud, ×5; **i,** androecium and gynoecium ×3 ½. For floral formula of *Crotalaria spectabilis*, see Fig. 4: 3d.

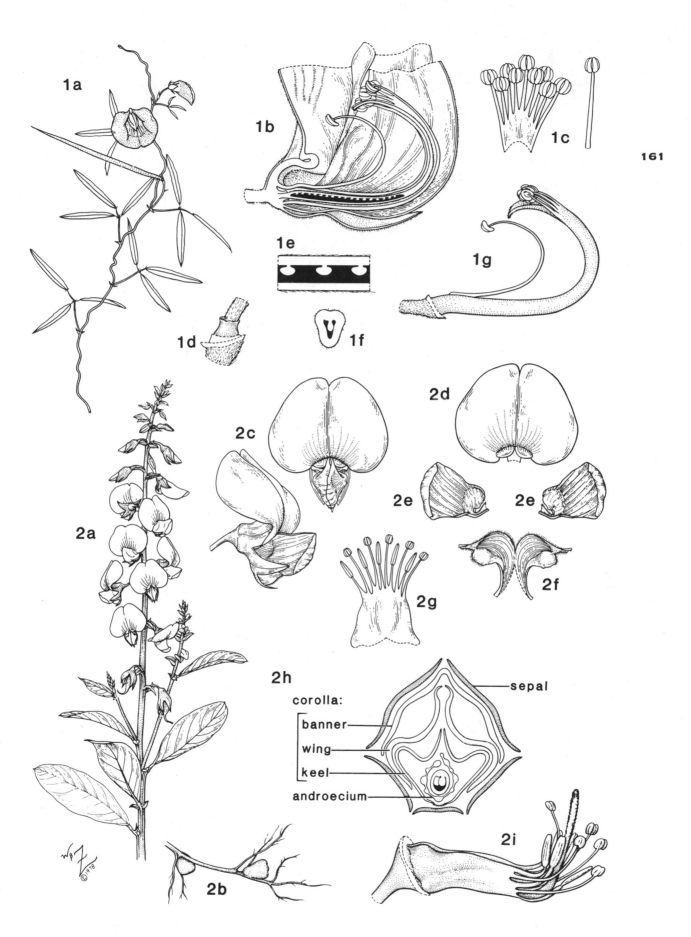

161

1a

1b

1c

1e

1d

1f

1g

2a

2b

2c

2d

2e

2e

2f

2g

2h

sepal

corolla:

banner

wing

keel

androecium

2i

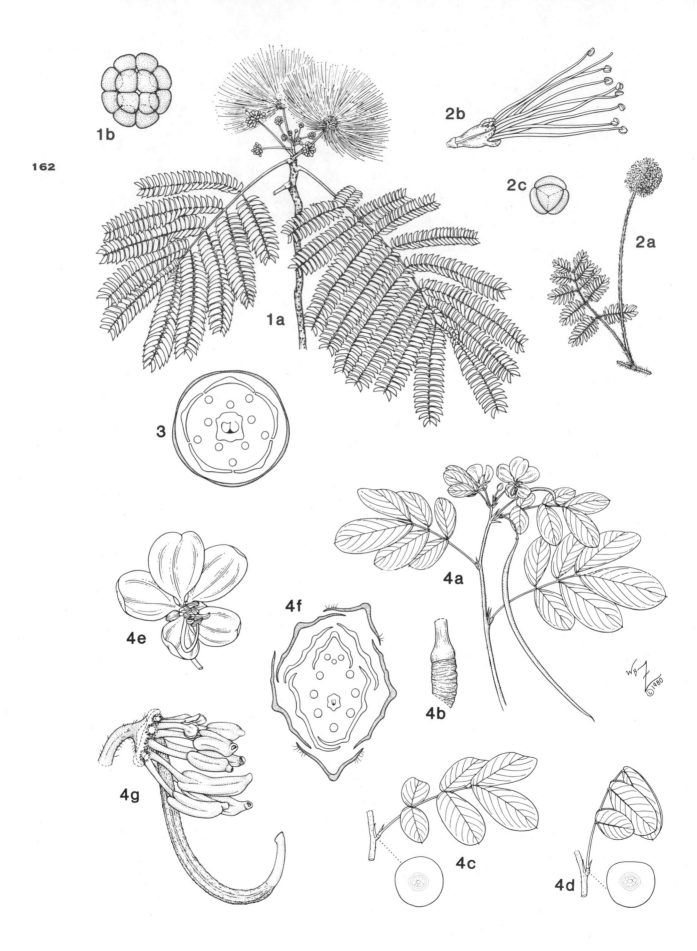

162

1b

1a

2b

2c

2a

3

4e

4f

4a

4b

4g

4c

4d

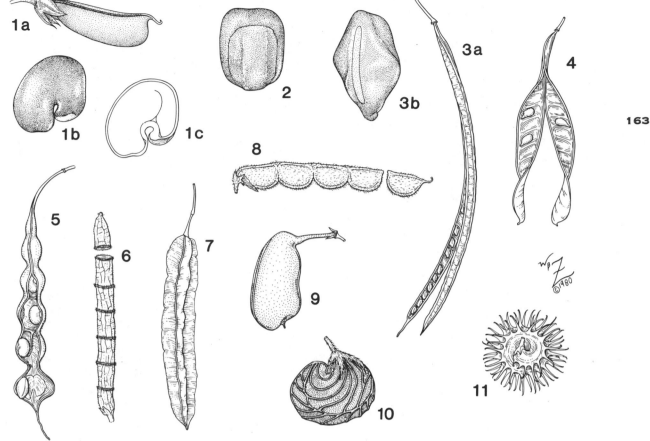

Figure 73. Fabaceae (continued). 1, *Crotalaria spectabilis:* **a,** legume, ×⁵⁄₆; **b,** seed, ×5; **c,** longitudinal section of seed, ×5. **2,** *Mimosa quadrivalvis (Schrankia microphylla):* seed (note U-shaped pleurogram), ×6. **3,** *Senna obtusifolia (Cassia obtusifolia):* **a,** legume, ×½; **b,** seed (note closed, elliptical pleurogram), ×5. **4,** *Chamaecrista fasciculata (Cassia fasciculata):* legume, ×¾. **5,** *Erythrina herbacea:* legume, ×²⁄₅. **6,** *Alysicarpus vaginalis (A. ovalifolius):* dehiscing loment, ×2²⁄₃. **7,** *Sesbania punicea:* legume, ×½. **8,** *Desmodium incanum:* dehiscing loment, ×2¼. **9,** *Baptisia alba:* legume, ×¾. **10,** *Medicago lupulina:* legume, ×12. **11,** *Medicago polymorpha (M. hispida):* legume, ×2²⁄₃.

Major genera: *Astragalus* (2,000 spp.), *Acacia* (750–1,200 spp.), *Indigofera* (700 spp.), *Crotalaria* (550–650 spp.), *Mimosa* (450–500 spp.), *Desmodium* (300–450 spp.), *Tephrosia* (300–400 spp.), *Trifolium* (238–300 spp.), *Dalbergia* (100–300 spp.), *Chamaecrista* (*Cassia* segregate, 260 spp.), and *Senna* (*Cassia* segregate, 250 spp.)

U.S./Canadian representatives: 142 genera/1,521 spp.; largest genera: *Astragalus, Lupinus, Trifolium, Dalea,* and *Desmodium*

Economic plants and products: Food crops: *Arachis* (peanuts), *Cicer* (chick peas), *Glycine* (soybeans), *Lens* (lentils), *Phaseolus* (various beans), *Pisum* (peas), *Vicia* (faba beans), and *Vigna* (cowpeas). Fodder and forage: *Lupinus* (lupine), *Medicago* (alfalfa), *Melilotus* (sweet-clover), *Trifolium* (clover), and *Vicia* (vetch). Fertilizer ("green manure" plowed into the ground in crop rotation to enrich the soil with nitrogen) from several, including *Lupinus* and *Trifolium.* Insecticide (rotenone) from *Derris* and *Lonchocarpus.* Several toxic plants, including *Astragalus* (toxic selenium levels), *Cassia* (al-

Figure 72 [facing page]. Fabaceae (continued). 1, *Albizia julibrissin:* **a,** flowering branch, ×½; **b,** pollen polyad, ×210. **2,** *Mimosa strigillosa:* **a,** inflorescence and leaf, ×²⁄₅; **b,** flower, ×5; **c,** pollen tetrad, ×1,250. **3,** *Mimosa quadrivalvis (Schrankia microphylla):* cross section of bud (somewhat diagrammatic), ×25. **4,** *Senna obtusifolia (Cassia obtusifolia):* **a,** upper portion of plant, ×½; **b,** pulvinus of leaf, ×2²⁄₃; **c,** "open" leaf (daylight), ×²⁄₅, with cross section of pulvinus, ×5; **d,** "closed" leaf (night), ×²⁄₅, with cross section of pulvinus, ×5; **e,** flower, ×1½; **f,** cross section of bud, ×5; **g,** androecium and gynoecium, ×4½.

TABLE 9. Morphological characters used to distinguish the three subfamilies of the Fabaceae. Features compiled from Cronquist (1981; taxa as segregate families), Elias (1974), and Robertson and Lee (1976).

CHARACTER	MIMOSOIDEAE ("MIMOSACEAE")	CAESALPINIOIDEAE ("CAESALPINIACEAE")	FABOIDEAE (PAPILIONOIDEAE) ("FABACEAE" s.s.)
GENERA/SPECIES	40/2,500	150/2,700	429/12,615
REPRESENTATIVE GENERA	*Acacia, Albizia, Calliandra, Mimosa, Pithecellobium*	*Bauhinia, Caesalpinia, Cassia, Cercis, Delonix, Gleditsia, Parkinsonia, Senna*	*Astragalus, Baptisia, Crotalaria, Desmodium, Glycine, Lupinus, Phaseolus, Pisum, Trifolium, Wisteria*
DISTRIBUTION	primarily tropical, with a few in temperate areas	primarily tropical, with a few in temperate areas	cosmopolitan
HABIT	trees or shrubs, occasionally herbs	trees or shrubs, occasionally herbs	herbs, shrubs, or trees
LEAVES	usually bipinnately compound; sometimes reduced to phyllodes	pinnately, or sometimes bipinnately, compound	pinnately compound, trifoliate, or occasionally reduced to one leaflet
FLORAL SYMMETRY	actinomorphic	usually ± zygomorphic	zygomorphic
COROLLA AESTIVATION	valvate	imbricate, with upper petal innermost	imbricate, with upper (or odd) petal outermost
COROLLA MORPHOLOGY	lobes equal; often connate into a tube	petals unequal, with uppermost often smaller than the laterals; distinct	petals unequal, with uppermost (or odd) the largest, and 2 basals connate/coherent at apex ("keel")
ANDROECIUM	10 to many; distinct, or sometimes basally connate	10 or fewer; usually distinct	10 distinct or fused (monadelphous), or 9 connate + 1 distinct (diadelphous)
POLLEN	often united into masses (tetrads to polyads)	single (monads)	single (monads)
U-SHAPED PLEUROGRAM (LINE) ON SEED	present on both seed faces	lacking (but closed pleurograms in some, such as *Senna* spp.)	lacking
OTHER	flowers small, usually in dense heads filaments long and brightly colored		

164

kaloids), and *Senna* (alkaloids). Valuable wood from *Dalbergia* (rosewood) and *Robinia* (locust). Various commercial products from *Acacia* (gums), *Copaifera* (resins), and *Indigofera* (indigo dyes). Ornamental plants (species of over 140 genera), including *Acacia*, *Albizia* (mimosa), *Bauhinia* (orchid tree), *Calliandra* (powder-puff), *Cassia* (various "shower trees"), *Cercis* (redbud), *Delonix* (poinciana), *Erythrina* (cockspur coral tree, cherokee-bean), *Lathyrus* (sweet-pea), *Parkinsonia* (Jerusalem-thorn), *Senna*, and *Wisteria*.

Commentary: The Fabaceae have been treated traditionally as a large, somewhat heterogeneous taxon (Polhill et al. 1981). The three conventional subfamilies (Faboideae, Caesalpinioideae, and Mimosoideae), sometimes treated as segregate families (as in Cronquist 1981), are distinguished by the characters listed in Table 9 and illustrated in Figs. 71, 72, and 73.

Notable among these features is the papilionaceous (butterfly-like) flower type (Fig. 71: 2c) of the subfamily Faboideae (or Papilionoideae). This specialized subfamily includes more than two-thirds of the species in the family and most of the common members. The zygomorphic corolla consists of a large posterior and upright standard (or banner; Fig. 71: 2d), a lateral pair of long-clawed wings (Fig. 71: 2e), and an innermost boat-shaped keel (Fig. 71: 2f) formed by the connation or coherence of two petals. The large standard encloses the other petals in bud, and flaps on the wings interlock with the keel projections or depressions (Fig. 71: 2h). The petals of flowers of the Caesalpinioideae (Fig. 72: 4e) also are unequal, but the uppermost (frequently the smallest) is situated inside the two lateral ones (Fig. 72: 4f). The actinomorphic flowers of the Mimosoideae (Fig. 72: 1a, 2a) have equal, valvate petals (Fig. 72: 3) often basally connate into a tube. See Table 9 for the particular androecial morphology associated with each of these three flower types.

Much research (e.g., Vincent 1982) has been devoted to the root nodules of the legumes, the site of nitrogen fixation. In this process, N_2 (atmospheric nitrogen) is reduced to NH_4+ (ammonium ions)—thus making nitrogen available in a form that may be incorporated into organic compounds (amino acids; Bergersen 1980, 1982). Nitrogen fixation is carried out most commonly by symbiotic bacteria, and the Fabaceae are the largest group of plants associated with these microorganisms (Allen and Allen 1981; see also the commentary under Myricaceae). The bacteria are supplied by the plant with organic compounds (carbohydrates) as an energy source; the plant obtains nitrogen in a form that may be incorporated in the production of plant proteins. Legumes are associated with *Rhizobium* species, aerobic bacteria that live as saprophytes in the soil until penetrating the root hair cells of legume seedlings (Hubbell 1981). The tumor-like, pinkish outgrowths (Fig. 71: 2b) are the result of the proliferation and enlargement of infected cortical cells. The shape of the nodules varies from spherical or cylindrical to digitate or coral-like. Generally, each *Rhizobium* species is effective with only one legume species.

Another outstanding vegetative character of the Fabaceae is the pulvinus, a conspicuous joint-like thickening at each petiole and petiolule base (Fig. 72: 4b). This structure consists of a large amount of cortical parenchyma tissue, which is associated with leaf movement. In specific areas, the cells undergo a decrease in turgor when water moves into intercellular spaces: the leaflets "close" when adaxial parenchyma cells lose turgidity (Fig. 72: 4d) and spread out when abaxial cells are involved (Fig. 72: 4c). The leaflets of many legumes open in the day and close at night, but the "sleep movement" is actually a circadian phenomenon that persists even when the plants are in continuous light or darkness (Herbert 1989). In some legumes, such as *Mimosa*, the leaflets fold up in response to touch.

The specialized papilionaceous corolla type has a significant role in various pollination mechanisms (Arroyo 1981). Basically, the conspicuous standard acts as a visual attractant, the wings serve as a landing platform, and the keel encloses and protects the stamens and stigma. Nectar is secreted by a disc at the ovary base (Fig. 71: 1d), and often the flowers are fragrant. As a bee probes for the concealed nectar, it pushes against the standard and depresses the wings, and subsequently also the keel; the exposed stigma and anthers then touch the underside of the bee. In many species the keel springs back over the anthers and stigma when the bee leaves, but in some species (e.g., *Medicago*) only one visit is possible since the exposure is explosive, and parts do not return to their original positions.

Pollination in the other two subfamilies, with flowers of a more open type, is more straightforward (Schrire 1989). The tiny flowers of the Mimosoideae often are clustered into heads (Fig. 72: 2a,b) and have brightly colored exerted stamens and styles. Several of the Caesalpinioideae (e.g., *Chamaecrista*, *Senna*) are visited by bees that collect the pollen. Cross-pollination often is promoted by protandry.

The unique and unifying character of the family is the unicarpellate gynoecium (with a double row of ovules; Fig. 71: 1e,f), which develops into a legume (Heywood 1971). This fruit type dehisces elastically along both sutures. Legumes generally are flattened (Fig. 73: 4) but can be terete and rod-like (*Senna*; Fig. 73: 3a), spirally coiled (*Medicago*; Fig. 73: 1l), or inflated with freely shaking seeds (*Crotalaria*; Fig. 73: 1a). The surface can vary from dry to fleshy and from smooth to spiny. In some genera, such as *Desmodium* (Fig. 73: 8) or *Alysicarpus* (Fig. 73: 6), the legume is modified into a loment that breaks transversely into indehiscent one-seeded segments.

REFERENCES CITED

Allen, O. N. and E. K. Allen. 1981. *The Leguminosae: A source book of characteristics, uses, and nodulation*, pp. xix–xx, lii–lxi, 707–727. University of Wisconsin Press, Madison.

Arroyo, M. T. K. 1981. Breeding systems and pollination biology in Leguminosae. In *Advances in legume systematics*, part 2, ed. R. M. Polhill and P. H. Raven, pp. 723–769. Royal Botanic Gardens, Kew.

Bergersen, F. J. 1980. Mechanisms associated with the fixation of nitrogen in the legume root nodule. In *Advances in legume science*, ed. R. J. Summerfield and A. H. Bunting, pp. 61–67. Proc. Int. Legume Conf., Vol. 1. Royal Botanic Gardens, Kew.

———. 1982. *Root nodules of legumes: Structure and functions*, pp. 9–21, 61–79, 81–95. John Wiley and Sons, Chichester, England.

Cronquist, A. 1981. *An integrated system of flowering plants*, pp. 592–601. Columbia University Press, New York.

Elias, T. E. 1974. The genera of Mimosoideae (Leguminosae) in the southeastern United States. *J. Arnold Arbor.* 55:67–118.

Herbert, T. J. 1989. A model of daily leaf movement in relation to the radiation regime. In *Advances in legume biology*, ed. C. H. Stirton and J. L. Zarucchi, pp. 629–643. Monogr. Syst. Bot. No. 29. Missouri Botanical Garden, St. Louis.

Heywood, V. H. 1971. The Leguminosae—A systematic purview. In *Chemotaxonomy of the Leguminosae*, ed. J. B. Harborne, D. Boulter, and B. L. Turner, pp. 1–29. Academic Press, London.

Hubbell, D. H. 1981. Legume infection by *Rhizobium*: A conceptual approach. *BioScience* 31:832–837.

Polhill, R. M., P. H. Raven, and C. H. Stirton. 1981. Evolution and systematics of the Leguminosae. In *Advances in legume systematics*, part 1, ed. R. M. Polhill and P. H. Raven, pp. 1–26. Royal Botanic Gardens, Kew.

Robertson, K. R., and Y.-T. Lee. 1976. The genera of Caesalpinioideae (Leguminosae) in the southeastern United States. *J. Arnold Arbor.* 57:1–53.

Schrire, B. D. 1989. A multidisciplinary approach to pollination biology in the Leguminosae. In *Advances in legume biology*, ed. C. H. Stirton and J. L. Zarucchi, pp. 181–242. Monogr. Syst. Bot. No. 29. Missouri Botanical Garden, St. Louis.

Vincent, J. M., ed. 1982. *Nitrogen fixation in legumes*. Academic Press, Sydney.

PLATANACEAE
Plane Tree or Sycamore Family

Trees, with pubescence of branched, stellate-like hairs; bark peeling off in large pieces (leaving smooth surface), forming furrows in old trunks. **Leaves** simple, broadly palmately lobed, serrate, alternate, palmately veined, with petiole swollen basally and enclosing the axillary bud, fragrant, deciduous, stipulate; stipules large and conspicuous, encircling the stem above leaf insertion, caducous. **Inflorescence** indeterminate (Boothroyd 1930), capitate/globose, unisexual, pendulous, axillary. **Flowers** actinomorphic, imperfect (plants monoecious), hypogynous, very reduced, minute. **Calyx** of 3 to 7 sepals or lobes, distinct or basally connate, minute. **Corolla** present in staminate flowers: of 3 to 7 petals, distinct, fleshy, minute; absent in carpellate flowers. **Androecium** of 3 to 7 stamens, opposite the sepals; filaments distinct, very short to absent; anthers basifixed or sessile, with pronounced peltate connective, dehiscing longitudinally, latrorse; staminodes usually present in carpellate flowers. **Gynoecium** apocarpous and of 5 to 9 carpels in 2 or 3 whorls, not sealed distally; ovaries superior, 1-locular; ovules usually solitary, anatropous to slightly hemitropous, pendulous, placentation parietal; style 1, linear, recurved, persistent; stigma 1, decurrent on inner margin of style; rudimentary carpels sometimes present in staminate flowers. **Fruit** aggregates of achenes closely packed into a globose head; achenes linear, 4-angled, subtended by long bristles; endosperm scanty, oily, proteinaceous; embryo straight, slender.

Family characterization: Monoecious trees with peeling bark; fragrant, broad, palmately lobed/veined leaves; large stipules encircling the stem above leaf insertion; long petioles with dilated bases enclosing the axillary buds; numerous, imperfect, reduced flowers packed into dangling, spherical, unisexual heads; anthers with conspicuous, peltate connectives; distinct, incompletely sealed carpels with hooked style and decurrent stigma; and globose head of achene aggregates, with each achene 4-angled, top-shaped, and subtended by bristles. Cyanogenic glycosides present. Anatomical features: vesture of stellate branched hairs ("candelabra hairs," discussed below); and multilacunar nodes (Fig. 75: b).

Genus/species: *Platanus* (6–9 spp.); a monotypic family

Distribution: Temperate to subtropical zones of the Northern Hemisphere: North America, Eurasia, and Indochina.

U.S./Canadian representatives: *Platanus* (4 spp.)

Economic plants and products: Wood for veneers (lacewood) and attractive ornamental trees (several species and horticultural hybrids of *Platanus*: buttonball tree, buttonwood tree, plane tree, sycamore).

Commentary: The position of this distinctive monotypic family in relation to other taxa has been problematic for taxonomists. Previously the family was included in the Urticales, but a close relationship to the Hamamelidaceae, as in Thorne (1992) and Cronquist (1988), is more likely (Hufford and Crane 1989; Hufford 1992; and see the interesting results of Schwarzwalder and Dilcher 1991).

A distinctive character of *Platanus* is the vesture of unusual stellate hairs ("candelabra hairs"; Fig. 74: c), each with a long, multicellular stalk or column bearing several whorls of three to five elongate unicellular branches. The deciduous hairs, which form a felt-like

167

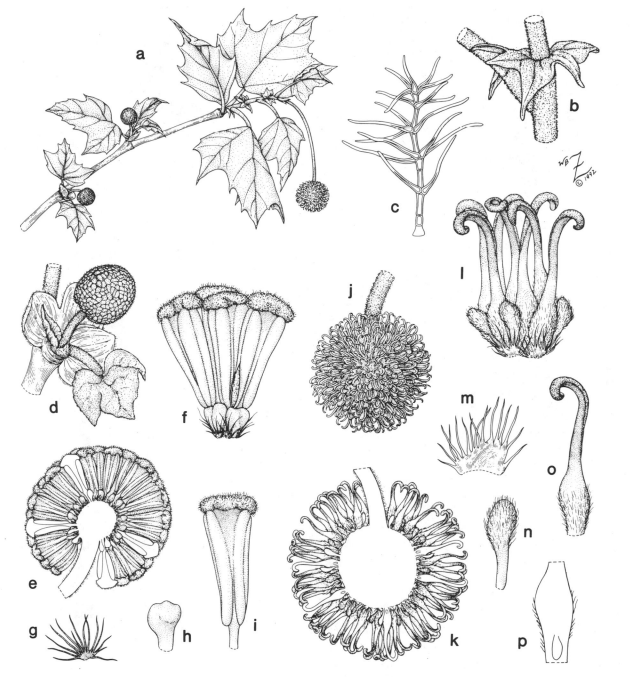

Figure 74. Platanaceae. *Platanus occidentalis*: **a,** flowering branch (with immature leaves), ×½; **b,** node showing large stipule united around stem, ×1¾; **c,** branched hair from young leaf, ×50; **d,** staminate inflorescence, ×2; **e,** longitudinal section of staminate inflorescence, ×5; **f,** staminate flower, ×12; **g,** sepal from staminate flower, ×20; **h,** petal from staminate flower, ×20; **i,** stamen, ×20; **j,** carpellate inflorescence, ×2½; **k,** longitudinal section of carpellate inflorescence, ×4; **l,** carpellate flower, ×12; **m,** sepal from carpellate flower, ×20; **n,** staminode, ×20; **o,** carpel, ×12; **p,** longitudinal section of ovary, ×15.

layer on young vegetative parts, may cause allergic reactions in sensitive individuals as the hairs drop from the growing leaves and stems. Similar hairs (with reduced branches) occur on the calyx and ovary (persisting on the mature fruit).

The reduced, imperfect flowers are packed into uni-sexual globose heads (Fig. 74: a,d,j), so individual flowers and their constituent parts are difficult to discern (Fig. 74: e,k). The rudimentary and nonvascularized perianth-like structures have been variously interpreted (and may actually represent scales/bracts of involucral origin; Ernst 1963). The calyx generally

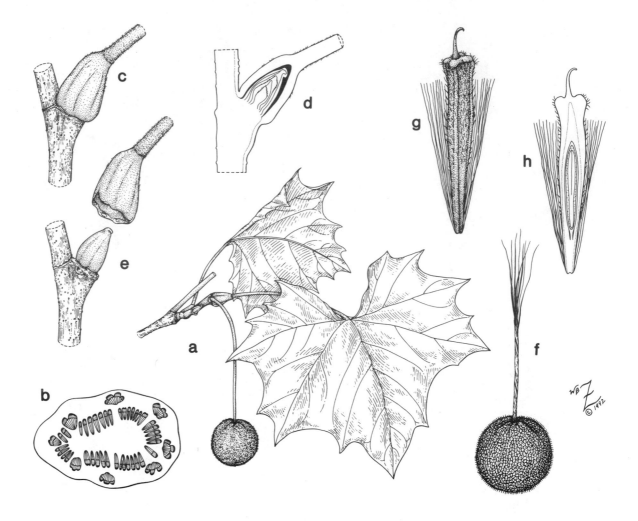

Figure 75. Platanaceae (continued). *Platanus occidentalis*: **a,** fruiting branch (with mature leaves), ×²⁄₅; **b,** cross section of node (multilacunar), ×6; **c,** node showing enlarged petiole base, ×2; **d,** longitudinal section of node showing hollow petiole base enclosing the axillary bud, ×2; **e,** node with cap-like petiole base removed, exposing axillary bud, ×2; **f,** infructescence (aggregates of achenes), ×½; **g,** achene, ×5; **h,** longitudinal section of achene, ×5.

comprises three to seven minute and hairy sepals or lobes (Fig. 74: g,m); the larger petals, occurring only in staminate flowers, are fleshy, glabrous, and spatulate (Fig. 74: h). The hairy staminodes of the carpellate flowers (Fig. 74: n) somewhat resemble the petals of the staminate flowers in size and shape (Boothroyd 1930).

Each carpel (Fig. 74: o) of the apocarpous gynoecium has a characteristic hook-like style and is incompletely fused at the distal end. As the carpels mature, they remain distinct but become jammed together, forming a 2–3 cm wide, globose fruit ball ("button ball") that dangles from a long fibrous peduncle (Fig. 75: a,f). Due to the mechanical stress, the elongate achenes are characteristically four-angled and top-shaped (Fig. 75: g,h). Each achene is surrounded by pappus-like bristles (derived from the accrescent perianth). The achenes are individually wind or water

dispersed as the infructescence falls apart (Endress 1977).

The minute flowers are anemophilous and protogynous. As the closely packed anthers dehisce, pollen becomes trapped under the appressed peltate connectives and is released when the individual stamens fall from the receptacle (Knuth 1909).

REFERENCES CITED

Boothroyd, L. E. 1930. The morphology and anatomy of the inflorescence and flower of the Platanaceae. *Amer. J. Bot.* 17:678–693.

Cronquist, A. 1988. *The evolution and classification of flowering plants*, 2d ed., pp. 297–299. New York Botanical Garden, Bronx.

Endress, P. K. 1977. Evolutionary trends in the Hamamelidales-Fagales-group. *Pl. Syst. Evol.*, Suppl. 1:321–347.

Ernst, W. R. 1963. The genera of Hamamelidaceae and Platanaceae in the southeastern United States. *J. Arnold Arbor.* 44:193–210.

Hufford, L. 1992. Rosidae and their relationships to other nonmagnoliid dicotyledons: A phylogenetic analysis using morphological and chemical data. *Ann. Missouri Bot. Gard.* 79:218–248.

—— and P. R. Crane. 1989. A preliminary phylogenetic analysis of the "lower" Hamamelidae. In *Evolution, systematics, and fossil history of the Hamamelidae,* ed. P. R. Crane and S. Blackmore, vol. 1, *Introduction and "lower" Hamamelidae,* pp. 175–192. Syst. Assoc. Special Vol. 40A. Clarendon Press, Oxford.

Knuth, P. 1909. *Handbook of flower pollination.* Trans. H. R. A. Davis. Vol. 3, *Goodenovieae to Cyadeae,* pp. 374–375. Oxford University Press, Oxford.

Schwarzwalder, R. N., and D. L. Dilcher. 1991. Systematic placement of the Platanaceae in the Hamamelidae. *Ann. Missouri Bot. Gard.* 78:962–969.

Thorne, R. F. 1992. Classification and geography of the flowering plants. *Bot. Rev.* (Lancaster) 58:225–348.

HAMAMELIDACEAE
Witch-hazel Family

Trees or shrubs; pubescence often of stellate or tufted hairs. **Leaves** simple, serrate and/or palmately lobed, alternate, deciduous or persistent, stipulate. **Inflorescence** indeterminate, usually spicate or capitate, occasionally subtended by brightly colored bracts, terminal or axillary. **Flowers** actinomorphic or less often zygomorphic, perfect or imperfect (then plants usually monoecious), hypogynous to usually epigynous, inconspicuous. **Calyx** of usually 4 or 5 sepals or sometimes reduced or absent, basally connate, persistent, usually imbricate. **Corolla** of 4 or 5 petals or absent, distinct, small, greenish, white, yellow, or reddish, imbricate or valvate. **Androecium** of 4 stamens and 4 staminodes or of numerous stamens; filaments distinct; anthers with 2 or 4 pollen sacs, 2-locular at anthesis, often with prolonged connective, dehiscing longitudinally or by upturning flap-like valves, introrse or latrorse. **Gynoecium** of 1 pistil, 2-carpellate, syncarpous but often carpels diverging and separating apically; ovary superior to inferior (usually half-inferior), 2-locular; ovules 1 to (less often) many in each locule (then often all but 1 in each locule aborting), anatropous, apical, pendulous, placentation axile; styles 2, distinct, spreading and recurved, hard, persistent; stigmas 2, decurrent. **Fruit** usually a loculicidal capsule (sometimes also with septicidal dehiscence) or septicidal capsule, often with woody or leathery exocarp and bony endocarp; seeds sometimes winged, with shiny and hard seed coat; endosperm thin, oily, fleshy; embryo large, spatulate, straight, embedded in endosperm.

Family characterization: Trees or shrubs with pubescence often of stellate or tufted hairs; small, inconspicuous flowers with reduced to absent perianth; 2-carpellate, half-inferior ovary with diverging, hardened styles; and woody or leathery capsule. Tissues with tannins and calcium oxalate crystals. Anatomical features: sclereids commonly in the leaf mesophyll; secretory cells or canals (containing tannins and/or mucilage) often present in parenchymatous tissues; and trilacunar nodes.

Genera/species: 30/120

Distribution: Primarily temperate to subtropical areas, but very discontinuous: mainly eastern Asia, with some representatives in eastern North America, Australia, southern Africa, and Madagascar.

Major genera: *Corylopsis* (7–20 spp.), *Dicoryphe* (15 spp.), and *Distylium* (12–15 spp.)

U.S./Canadian representatives: *Fothergilla* (2 spp.), *Hamamelis* (2 spp.), and *Liquidambar* (1 sp.)

Economic plants and products: Gum-resin (storax or styrax, used in perfumery) and timber from species of *Altingia* (rasamala) and *Liquidambar.* Witch-hazel liniment, an extract of bark and leaves of *Hamamelis virginiana.* Ornamental plants (species of 14 genera), including *Corylopsis* (winter-hazel), *Fothergilla* (witch-alder), *Hamamelis* (witch-hazel), *Liquidambar* (sweetgum), and *Loropetalum.*

Commentary: The Hamamelidaceae have been variously subdivided into subfamilies and tribes, primarily based on the inflorescence type, sexual condition of the flowers, extent of perianth reduction, androecial constituents, and the number of ovules per locule (Endress 1989b). About half of the 30 or so genera of the family are monotypic. *Liquidambar* (4 spp.), *Altingia* (7 spp.), and *Semiliquidambar* (3 spp.), constituting the subfamily Altingioideae, sometimes have been considered as a segregate family (Altingiaceae; see Ferguson 1989). The cladistic analysis of Hufford and Crane (1989) supports the Hamamelidaceae as a monophyletic sister group to the Platanaceae (*Platanus*).

As noted above, the members of the family exhibit diversity in several aspects of floral morphology, such as in the reduction of the perianth and androecium (see the summaries in Ernst 1963; Endress 1989a; and Ferguson 1989). Because suppression of one sex is common in the family, functionally unisexual and/or perfect flowers characterize some species (e.g., those of *Hamamelis*; Fig. 76: 2b), while only imperfect flowers occur in others (e.g., *Liquidambar*; Fig. 76: 1c,f). Worldwide, the flowers of the family also vary considerably in ovary position (superior to inferior), although

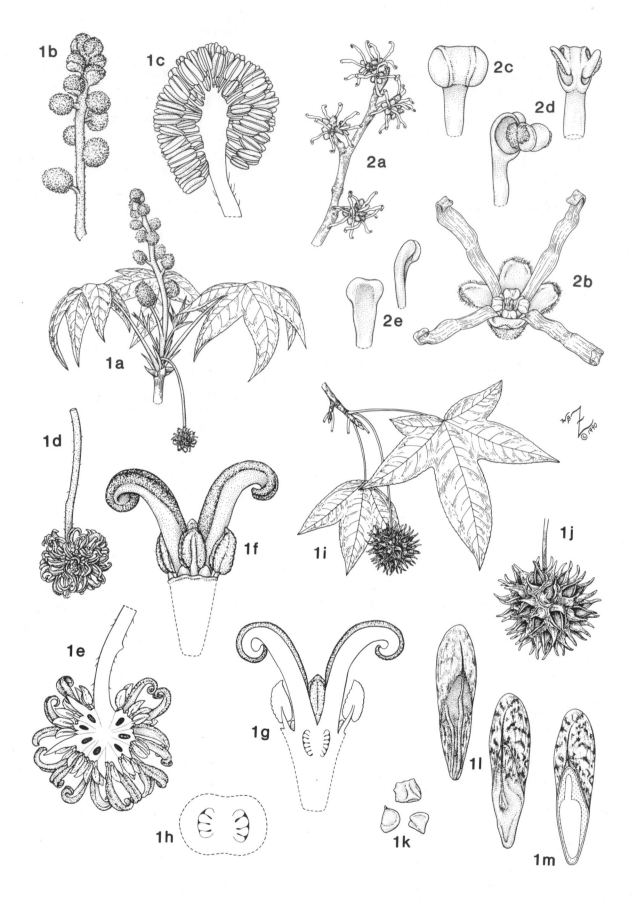

in our representatives the ovary is barely (*Fothergilla*, *Hamamelis*) to partly (*Liquidambar*; Fig. 76: 1g) inferior. Four or five strap-like petals with an equal number of stamens (plus alternating staminodes) characterize species such as those of *Hamamelis* (Fig. 76: 2b). Many representatives of the family (e.g., *Fothergilla* and *Liquidambar*) are apetalous. The androecium may comprise numerous stamens (*Fothergilla*, staminate flowers of *Liquidambar*; Fig. 76: 1c); four staminodes are present in the carpellate flowers of *Liquidambar* (Fig. 76: 1f). Anthers dehisce longitudinally (*Liquidambar*; Fig. 76: 1c) or by up-turning flaps (as in *Hamamelis*; Fig. 76: 2c,d).

Both entomophily and anemophily occur in the family (Endress 1977). The flowers of *Hamamelis*, for example, with strap-like, coiled, yellow or reddish petals, nectariferous staminodes, and unpleasant odor, are adapted to attract flies. Bees collect the copious pollen from the conspicuous, thickened stamens of *Fothergilla* flowers, which are arranged in showy, bottle-brush inflorescences. Many genera, characterized by reduced flowers packed into somewhat catkin-like inflorescences, are wind-pollinated. In *Liquidambar*, for example, the staminate flowers, individually indistinguishable (Fig. 76: 1c), are aggregated into globular masses arranged in erect racemes (Fig. 76: 1b); the numerous carpellate flowers form a spherical head that dangles from a long peduncle (similar to those of *Platanus*; Fig. 76: 1d).

As in the Fagaceae and Juglandaceae, fertilization of the ovules frequently is delayed until long after pollination. Each ovary locule often is ultimately uniovulate, since in multiovulate genera (such as *Liquidambar*) all ovules but one abort (see Fig. 76: 1k). The two-carpellate gynoecium (Fig. 76: 1h), fused at the base with two recurved and divergent styles (Fig. 76: 1g), ripens into a woody, two- or four-valved capsule. The capsules of *Hamamelis* forcibly eject the seeds, a dispersal mechanism common in the family (Tiffney 1986). The winged seeds of *Liquidambar* (Fig. 76: 1l) are easily shaken out of the pendulous, spherical infructescence (syncarp-like ball of capsules; Fig. 76: 1j).

REFERENCES CITED

Endress, P. K. 1977. Evolutionary trends in the Hamamelidales-Fagales-group. *Pl. Syst. Evol.*, Suppl. 1:321–347.

——. 1989a. Phylogenetic relationships in the Hamamelidoideae. In *Evolution, systematics, and fossil history of the Hamamelidae*, ed. P. R. Crane and S. Blackmore, vol. 1, *Introduction and "lower" Hamamelidae*, pp. 227–248. Syst. Assoc. Special Vol. 40A. Clarendon Press, Oxford.

——. 1989b. A suprageneric taxonomic classification of the Hamamelidaceae. *Taxon* 38:371–376.

Ernst, W. R. 1963. The genera of Hamamelidaceae and Platanaceae in the southeastern United States. *J. Arnold Arbor.* 44:193–210.

Ferguson, D. K. 1989. A survey of the Liquidambaroideae (Hamamelidaceae) with a view to elucidating its fossil record. In *Evolution, systematics, and fossil history of the Hamamelidae*, ed. P. R. Crane and S. Blackmore, vol. 1, *Introduction and "lower" Hamamelidae*, pp. 249–272. Syst. Assoc. Special Vol. 40A. Clarendon Press, Oxford.

Hufford, L. D., and P. R. Crane. 1989. A preliminary phylogenetic analysis of the "lower" Hamamelidae. In *Evolution, systematics, and fossil history of the Hamamelidae*, ed. P. R. Crane and S. Blackmore, vol. 1, *Introduction and "lower" Hamamelidae*, pp. 175–192. Syst. Assoc. Special Vol. 40A. Clarendon Press, Oxford.

Tiffney, B. H. 1986. Fruit and seed dispersal and the evolution of the Hamamelidae. *Ann. Missouri Bot. Gard.* 73:394–416.

JUGLANDACEAE
Walnut Family

Usually trees, aromatic and resinous. **Leaves** pinnately compound, serrate, usually alternate, deciduous, glandular-dotted beneath (lepidote) and aromatic, exstipulate. **Inflorescence** basically determinate, generally spicate (erect or pendulous), with spikes sometimes grouped into panicles, few-flowered (carpellate inflorescences) to many-flowered (staminate inflorescences), terminal or axillary. **Flowers** actinomorphic, imperfect (plants usually monoecious), epigynous, small and inconspicuous, typically associated with bracts. **Perianth** of typically 4 tepals, sometimes modified into a disc or absent, distinct, minute, scale-like, free or adnate to subtending bracts, persistent. **An-**

Figure 76. Hamamelidaceae. 1, *Liquidambar styraciflua*: **a,** flowering branch (with immature leaves), ×⅔; **b,** staminate inflorescence, ×¾; **c,** longitudinal section of staminate inflorescence, ×2; **d,** carpellate inflorescence, ×2; **e,** longitudinal section of carpellate inflorescence, ×4; **f,** carpellate flower, ×12; **g,** longitudinal section of carpellate flower, ×12; **h,** cross section of ovary, ×22; **i,** fruiting branch (with mature leaves), ×⅖; **j,** infructescence (syncarp-like ball of capsules), ×⅘; **k,** aborted ovules, ×6; **l,** two views of seed, ×6; **m,** longitudinal section of seed, ×6. **2,** *Hamamelis virginiana*: **a,** flowering branch, ×1¼; **b,** flower, ×6; **c,** stamen before dehiscence, ×25; **d,** two views of dehiscing stamen, ×25; **e,** two views of staminode, ×32.

droecium of 3 to many stamens; filaments distinct, short; anthers basifixed, dehiscing longitudinally. **Gynoecium** of 1 pistil, typically 2-carpellate; ovary inferior, 1-locular above and usually 2-locular at base (or apparently 4- to 8-locular due to "false" partitions), adnate (at least at base) to involucre; ovule solitary, orthotropous, erect, appearing basal (especially in young flowers) but actually at the apex of the incomplete septum; styles 2, distinct or often basally connate, short, fleshy; stigmas 2, along inner sides of style branches, often plumose or papillose; rudimentary pistil sometimes present in staminate flowers. **Fruit** a drupe-like nut with dehiscent or indehiscent, leathery to fibrous husk (fused involucre and perianth); seed solitary, large; endosperm absent; embryo oily, often massive, often with variously lobed and corrugated cotyledons.

Family characterization: Aromatic and resinous trees with pinnately compound glandular-dotted leaves; usually spicate, erect (carpellate) or pendulous (staminate) inflorescences; reduced imperfect flowers with uniseriate or absent perianth and associated bracts; inferior ovary with solitary "basal" ovule; drupe-like nut with leathery to fibrous husk; and large nonendospermous seeds with lobed and corrugated cotyledons. Tissues commonly with tannins and calcium oxalate crystals. Anatomical features: peltate glands (lepidote) on leaves (Fig. 77: 1c, 2) that secrete resinous substances; trilacunar nodes; and unitegmic ovules.

Genera/species: 8/59

Distribution: Primarily in temperate to subtropical regions of the Northern Hemisphere; also extending into montane areas of tropical America and tropical Asia. Most abundant in eastern Asia and eastern North America.

Major genera: *Carya* (17–25 spp.) and *Juglans* (15–21 spp.)

U.S./Canadian representatives: *Carya* (13 spp.) and *Juglans* (8 spp.)

Economic plants and products: Edible seeds from *Carya* (pecans, hickory nuts) and *Juglans* (walnuts, butternuts). Valuable wood from *Carya, Englehardtia,* and *Juglans.* Ornamental plants (species of 5 genera), including *Carya, Juglans,* and *Pterocarya* (wingnut).

Commentary: About two-thirds of this small family belong to the genera *Carya* and *Juglans.* Although numerous studies in morphology (e.g., Manning 1938, 1940, 1948) have attempted to discern "trends" within the Juglandaceae (e.g., Stone 1989), the phylogenetic position of the family in relation to other groups has remained controversial, with authors citing either the Fagales (Fagaceae, Betulaceae, and allies) or Rutales (Rutaceae and allies) as the most closely related taxa (see Elias 1972; Manning 1978; and Thorne 1989, for summaries). More critical review (based on shared derived characters) is needed concerning the phylogenetic position of the family.

The imperfect flowers of *Juglans* and *Carya,* each often subtended by three bracts (e.g., Fig. 77: 1k), form unisexual inflorescences, and the trees usually are monoecious. The anemophilous flowers open in the late spring after the leaves unfold. As in the Fagaceae, the ovules are not developed at the time of pollination.

The staminate aments (Fig. 77: 1j), sometimes grouped into panicles, occur on the branches of the previous year (Fig. 77: 1a). The staminate flowers vary in the number of stamens and the development of the perianth (e.g., no perianth in the flowers of *Carya*; Fig. 77: 1k). The carpellate flowers form a few-flowered and erect inflorescence (Fig. 77: 1a,d) that is terminal on the current year's leafy shoot (Abbe 1974). The bracts, which are variously united with the ovary (and perianth, when present), form a cupulate involucre (Fig. 77: 1e) that develops into a husk (Fig. 77: 1l,m) in the mature fruit (Manning 1940; Hjelmqvist 1948).

The drupe-like fruits of *Carya* and *Juglans* are dispersed by animals (Stone 1973; Tiffney 1986). In *Carya,* the husk usually splits regularly into four valves (Fig. 77: 1l), revealing a smooth nut (Fig. 77: 1m). The husk of the fruit of *Juglans,* containing a sculptured nut, is more fleshy and dehisces irregularly by decaying. The split of the walnut "shell" corresponds to the midline of the two carpels, and not to the suture of the carpels.

REFERENCES CITED

Abbe, E. C. 1974. Flowers and inflorescences of the "Amentiferae." *Bot. Rev.* (Lancaster) 40:159–261.

Elias, T. S. 1972. The genera of Juglandaceae in the southeastern United States. *J. Arnold Arbor.* 53:26–51.

Hjelmqvist, H. 1948. Studies on the floral morphology and phylogeny of the Amentiferae. *Bot. Not.,* Suppl. 2:1–171.

Manning, W. E. 1938. The morphology of the flower of the Juglandaceae. I. The inflorescence. *Amer. J. Bot.* 25:407–419.

——. 1940. Ibid., II. The pistillate flowers and fruit. *Amer. J. Bot.* 27:839–852.

——. 1948. Ibid., III. The staminate flowers. *Amer. J. Bot.* 35:606–621.

——. 1978. The classification within the Juglandaceae. *Ann. Missouri Bot. Gard.* 65:1058–1087.

Stone, D. E. 1973. Patterns in the evolution of amentiferous fruits. *Brittonia* 25:371–384.

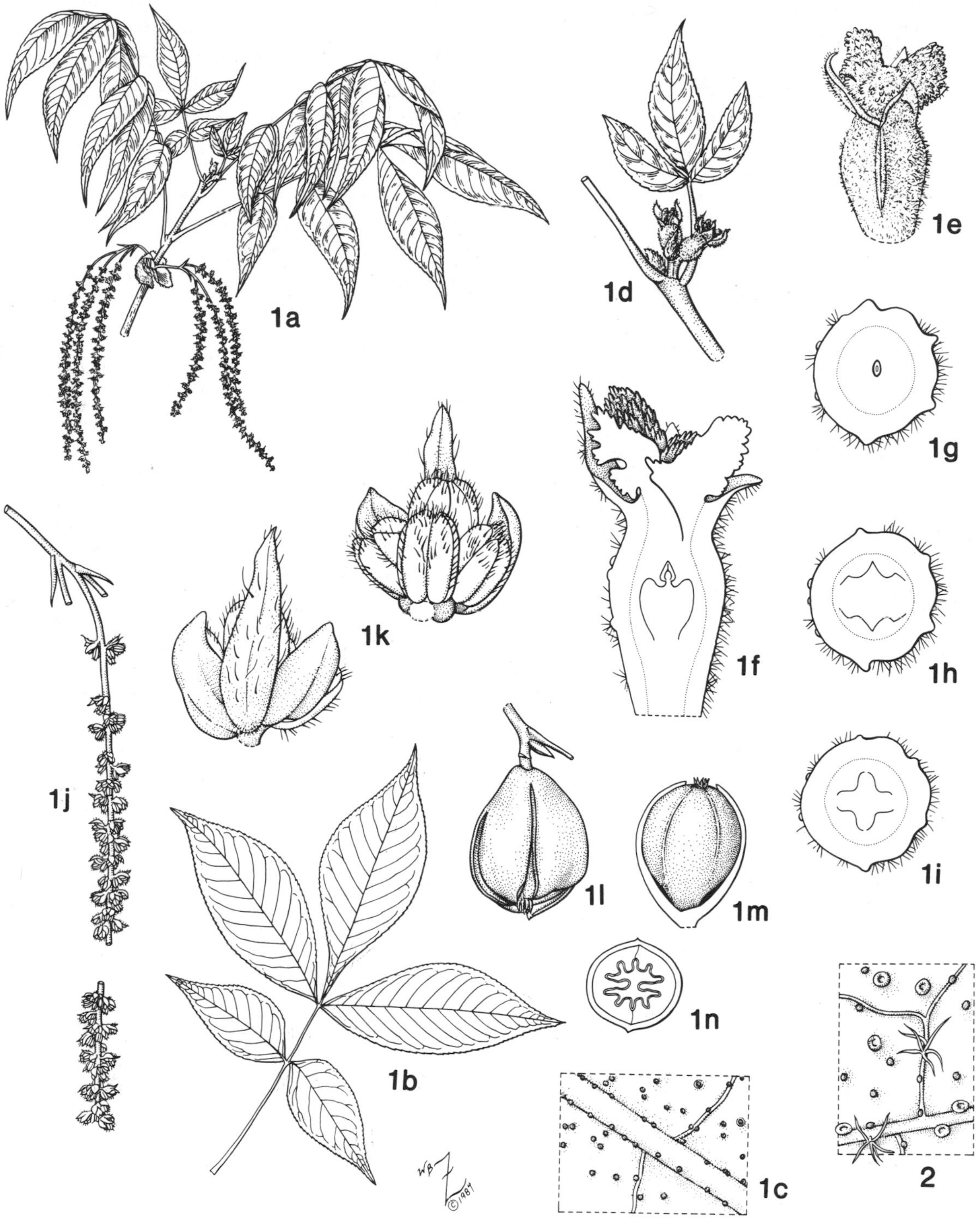

Figure 77. Juglandaceae. **1,** *Carya glabra*: **a,** flowering branch (with immature leaves), ×⅓; **b,** mature leaf, ×⅓; **c,** detail of abaxial surface of leaf showing peltate glands, ×25; **d,** carpellate inflorescence, ×1⅓; **e,** carpellate flower, ×3; **f,** longitudinal section of carpellate flower, ×5; **g,** cross section of ovary (one-locular) at level of ovule, ×5; **h,** cross section of ovary (two-locular) near middle, ×5; **i,** cross section of ovary (four-locular) near base, ×5; **j,** staminate inflorescence, ×1⅓; **k,** two views of staminate flower, ×10; **l,** fruit (nut subtended by dehiscing husk), ×½; **m,** fruit (half of husk removed), ×½; **n,** cross section of nut (note corrugated cotyledons), ×½. **2,** *Carya tomentosa*: detail of abaxial surface of leaf showing peltate glands (and stellate hairs), ×25.

———. 1989. Biology and evolution of temperate and tropical Juglandaceae. In *Evolution, systematics, and fossil history of the Hamamelidae*, ed. P. R. Crane and S. Blackmore, vol. 2, *"Higher" Hamamelidae*, pp. 117–145. Syst. Assoc. Special Vol. 40B. Clarendon Press, Oxford.

Thorne, R. F. 1989. "Hamamelididae": A commentary. In *Evolution, systematics, and fossil history of the Hamamelidae*, ed. P. R. Crane and S. Blackmore, vol. 1, *Introduction and "lower" Hamamelidae*, pp. 9–16. Syst. Assoc. Special Vol. 40A. Clarendon Press, Oxford.

Tiffney, B. H. 1986. Fruit and seed dispersal and the evolution of the Hamamelidae. *Ann. Missouri Bot. Gard.* 73:394–416.

MYRICACEAE

Bayberry or Wax-myrtle Family

Shrubs to small trees, generally growing in swampy acidic soils, aromatic; roots with nitrogen-fixing bacterial nodules. **Leaves** simple, entire to irregularly serrate, alternate, persistent or deciduous, often leathery, yellow glandular-dotted beneath, usually exstipulate. **Inflorescence** determinate, cymose and often appearing spicate (usually erect), unisexual, axillary. **Flowers** actinomorphic, usually imperfect (plants dioecious or occasionally monoecious), hypogynous to usually epigynous, very reduced, subtended by scalelike bract(s). **Perianth** absent. **Androecium** of 2 to 8 stamens; filaments distinct or basally connate; anthers basifixed or dorsifixed, dehiscing longitudinally, extrorse or latrorse. **Gynoecium** of 1 pistil, 2-carpellate; ovary superior to usually inferior, 1-locular; ovule solitary, orthotropous, erect, placentation basal; styles 2, distinct or basally connate; stigmas 2. **Fruit** a drupe or occasionally a nutlet, often tuberculate and coated with whitish wax, with persistent bracts; seed with thin seed coat; endosperm scanty to absent; embryo straight.

Family characterization: Shrubs to small trees growing in swampy acidic soils or xeric habitats; roots with bacterial nodules; aromatic leaves with yellow punctate dots; reduced imperfect flowers lacking a perianth but with sepal-like bract(s); unisexual aments; 2-carpellate, 1-locular ovary with a solitary, erect, basal ovule; and tuberculate drupelet with waxy coating.

Tissues commonly with tannins and calcium oxalate crystals. Anatomical features: peltate yellow glands on leaves (which secrete resins or aromatic oils, discussed below); and unitegmic ovules.

Genera/species: *Myrica* (38–50 spp.), *Canacomyrica* (1 sp.), and *Comptonia* (1 sp.)

Distribution: Widespread, but mostly in temperate to subtropical regions (excluding Australia); often in xeric or swampy acidic soils.

U.S./Canadian representatives: *Myrica* (10 spp.) and *Comptonia* (1 sp.)

Economic plants and products: Aromatic wax (palmitins, palmitic acid; from fruit), tannins, and medicinal substances (from bark) from several *Myrica* species (tallow shrub, candle-berry, bayberry). Ornamental plants: *Comptonia peregrina* (sweet-fern) and several species of *Myrica* (sweet gale, wax-myrtle).

Commentary: Depending on the authority, the family has been considered to comprise one (*Myrica s.l.*) to four genera (Elias 1971). The two distinct subgenera (or sections) of *Myrica* (38 spp.) sometimes have been elevated to generic level, *Morella/"Gale"* (monotypic) and *Myrica*; the genera *Comptonia* and *Canacomyrica* are monotypic. The inclusion of *Canacomyrica* (restricted to New Caledonia), characterized by features such as perfect flowers, is especially problematical (Hjelmqvist 1948; MacDonald 1989).

A characteristic vegetative feature of the Myricaceae is the glandular hair type (Fig. 78: c–e) occurring on the leaves, bracts, and young stems. These peltate hairs consist of a multicellular head plus a multi- or unicellular stalk that is partially embedded in the epidermis. The cells have yellow walls and resinous contents and appear macroscopically as golden yellow or brownish dots.

Another vegetative feature reported for most species of the family (*Myrica* spp., *Comptonia peregrina*) is the formation of root nodules (Fig. 78: b), the result of harboring nitrogen-fixing microorganisms. This phenomenon is discussed under the commentary for the Fabaceae. Besides the legumes, species of at least fifteen genera (e.g., of *Alnus*, *Ceanothus*, and *Myrica*) of

Figure 78. Myricaceae. *Myrica cerifera*: **a,** carpellate flowering branch, ×½; **b,** root nodules, ×10; **c,** abaxial surface of leaf showing glandular hairs (sunken into pits), ×20; **d,** top view of glandular hair, ×210; **e,** cross section of leaf surface showing lateral view of glandular hair, ×210; **f,** carpellate inflorescence, ×9; **g,** carpellate flower and subtending bracts (abaxial view), ×20; **h,** carpellate flower and subtending bracts (adaxial view), ×20; **i,** carpellate flower with secondary bracts (primary bract removed), ×20; **j,** longitudinal section of ovary, ×20; **k,** staminate flowering branch, ×½; **l,** staminate inflorescence, ×4; **m,** staminate flower and subtending bract (abaxial view), ×10; **n,** staminate flower and subtending bract (adaxial view), ×10; **o,** fruiting branch, ×½; **p,** drupe, ×7; **q,** longitudinal section of drupe, ×9.

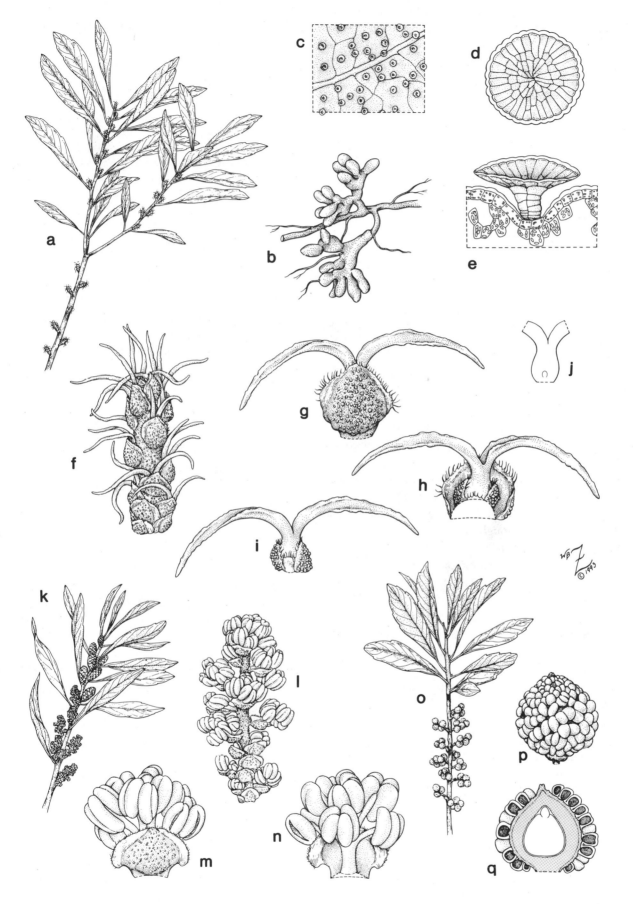

175

more or less related woody plants also have a symbiotic association with soil bacteria (Torrey 1978). The symbionts in these cases, however, are actinomycetes (filamentous bacteria), rather than the *Rhizobium* species associated with the legumes, and the root nodules induced by these bacteria differ morphologically and anatomically from the legume nodules. Most of these shrubs and trees are pioneers on nitrogen-deficient soils. The infection initially is manifested by tuberous, modified lateral roots that repeatedly branch, eventually producing a more or less spherical cluster that is often knobbed and coral-like (Fig. 78: b).

The Myricaceae traditionally have been allied to other amentiferous families, such as the Juglandaceae, Betulaceae, and Fagaceae, and generally are considered to be closely related to the Juglandaceae (e.g., Thorne 1992). The reduced flowers of these families are anemophilous and are packed into dense unisexual spikes. In Myricaceae, the plants usually are dioecious (Fig. 78: a,k), but in at least some species, the sex of the imperfect flowers on an individual plant or even on particular branches may vary from season to season (Davey and Gibson 1917).

The morphology of the reduced flowers has been the subject of numerous detailed studies (e.g., Hjelmqvist 1948; Abbe 1974; and others summarized in MacDonald 1989). The staminate flowers (Fig. 78: m,n) consist of usually one to five stamens subtended by one or two bracts; the carpellate flowers (Fig. 78: g–i) consist of one pistil (usually with two styles) plus two or more bracts. Although "nude" (i.e., lacking a perianth), the flower actually is epigynous in most members of the family, depending on the pattern of ovary and fruit wall maturation (see the summary in MacDonald 1989). In *Comptonia peregrina* the ovary is superior and remains so during fruit maturation (MacDonald 1974); in *Myrica*, however, intercalary growth beneath the base of the gynoecium results in a "circumlocular wall," composed of inflorescence axis (plus appendages) at the base and gynoecial tissue (and sometimes also receptacular tissue) toward the apex (MacDonald 1979)—thus, in this case, the ovary is inferior. In *Myrica gale*, the development of this wall is delayed until after pollination, so the flower initially is hypogynous, but the fruit (mature ovary) is inferior (MacDonald and Sattler 1973; MacDonald 1989).

The subtending sepal-like bract(s) may persist as accrescent structures in the fruit (e.g., as a cupule in *Comptonia*; a pair of "floaters" in *Myrica gale*). The fruit of *Myrica* (Fig. 78: p,q), however, most often is a druplet with a hard mesocarp and tuberculate parenchymatous exocarp, covered with a white layer of waxy material that aids in preventing desiccation in arid habitats and may help to keep a water-borne fruit afloat. These drupes are eaten and dispersed by birds (Tiffney 1986).

REFERENCES CITED

Abbe, E. C. 1974. Flowers and inflorescences of the "Amentiferae." *Bot. Rev.* (Lancaster) 40:159–261.

Davey, A. J., and C. M. Gibson. 1917. Note on the distribution of sexes in *Myrica Gale*. *New Phytol.* 16:147–151.

Elias, T. S. 1971. The genera of Myricaceae in the southeastern United States. *J. Arnold Arbor.* 52:305–318.

Hjelmqvist, H. 1948. Studies on the floral morphology and phylogeny of the Amentiferae. *Bot. Not.*, Suppl. 2:1–171.

MacDonald, A. D. 1974. Floral development of *Comptonia peregrina* (Myricaceae). *Canad. J. Bot.* 52:2165–2169.

——. 1979. Development of the female flower and gynecandrous partial inflorescence of *Myrica californica*. *Canad. J. Bot.* 57:141–151.

——. 1989. The morphology and relationships of the Myricaceae. In *Evolution, systematics, and fossil history of the Hamamelidae*, ed. P. R. Crane and S. Blackmore, vol. 2, *"Higher" Hamamelidae*, pp. 147–165. Syst. Assoc. Special Vol. 40B. Clarendon Press, Oxford.

—— and R. Sattler. 1973. Floral development of *Myrica gale* L. and the controversy over floral concepts. *Canad. J. Bot.* 51:1965–1975.

Thorne, R. F. 1992. Classification and geography of the flowering plants. *Bot. Rev.* (Lancaster) 58:225–348.

Tiffney, B. H. 1986. Fruit and seed dispersal and the evolution of the Hamamelidae. *Ann. Missouri Bot. Gard.* 73:394–416.

Torrey, J. G. 1978. Nitrogen fixation by actinomycete-nodulated angiosperms. *BioScience* 28:586–592.

BETULACEAE

Birch Family

Trees or shrubs; roots commonly mycorrhizal. **Leaves** simple, serrate, alternate, deciduous, often ovate, plicate along lateral veins, stipulate. **Inflorescence** determinate, cymose and appearing spicate (erect or pendulous) or capitate, unisexual, terminal or lateral. **Flowers** actinomorphic, imperfect (plants monoecious), epigynous, very reduced, associated with bracts. **Perianth** of 1 to 6 tepals or absent, distinct to connate, minute, scale-like, slightly imbricate. **Androecium** of 1 to 4 stamens; filaments distinct or basally connate, sometimes split, short; anthers dorsifixed, 2-locular with locules sometimes split, dehiscing longitudinally, extrorse. **Gynoecium** of 1 pistil, 2-carpellate; ovary inferior, usually 2-locular at base and 1-locular at apex; ovules 2 in each locule (all but 1 aborting), anatropous, pendulous, placentation axile; styles 2, persistent; stigmas slightly decurrent; rudimentary pistil sometimes present in staminate flowers. **Fruit** a nut, nutlet, or samara, subtended or enclosed by bracts; seed solitary; endosperm essentially absent; embryo large, straight, oily.

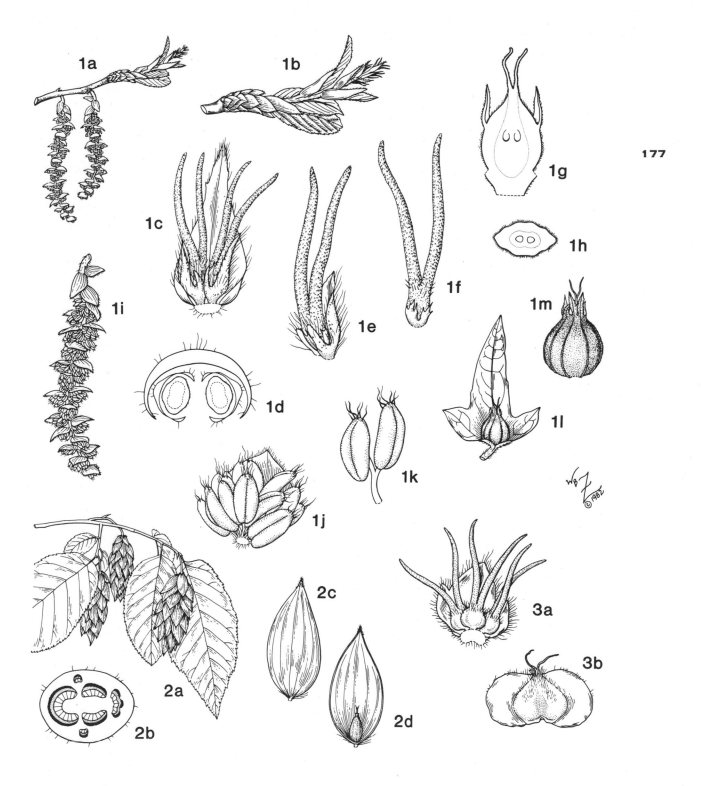

Figure 79. Betulaceae. 1, *Carpinus caroliniana*: **a,** flowering branch (with immature leaves), ×1¼; **b,** carpellate inflorescence, ×3; **c,** carpellate cymule, ×12; **d,** cross section of carpellate cymule showing subtending bracts (outer primary bract with each flower immediately enclosed by a central secondary bract fused to two lateral tertiary bracts), ×25; **e,** carpellate flower and subtending fused (secondary and tertiary) bracts, ×25; **f,** carpellate flower, ×25; **g,** longitudinal section of carpellate flower after pollination (one month after anthesis), ×12; **h,** cross section of ovary, ×12; **i,** staminate inflorescence, ×3; **j,** staminate cymule, ×12; **k,** stamen, ×17; **l,** nut with subtending bracts, ×1⅔; **m,** nut (with persistent perianth), ×4. **2,** *Ostrya virginiana*: **a,** fruiting branch, ×⅓; **b,** cross section of node (three-trace trilacunar), ×12; **c,** fruit (nutlet subtended by fused bracts), ×1½; **d,** fruit with half of subtending bracts removed, ×1½. **3,** *Betula nigra*: **a,** carpellate cymule, ×9; **b,** samara, ×5.

Family characterization: Monoecious trees or shrubs with stipulate, serrate leaves; reduced imperfect flowers with minute or absent perianth (the staminate in pendent catkins and the carpellate in pendent or erect catkins); inferior 2-carpellate ovary; and a nutlet or samara subtended by bracts as the fruit type. Tissues with calcium oxalate crystals and a high tannin content. Anatomical features: unitegmic ovules; and trilacunar nodes (Fig. 79: 2b).

Genera/species: *Betula* (60 spp.), *Alnus* (35 spp.), *Carpinus* (35 spp), *Corylus* (10–15 spp.), *Ostrya* (10 spp.), and *Ostryopsis* (2 spp.)

Distribution: In northern and southern temperate regions, and also in the alpine tropics.

U.S./Canadian representatives: *Betula* (18 spp.), *Alnus* (8 spp.), *Corylus* (3 spp.), *Ostrya* (3 spp.), and *Carpinus* (1 sp.)

Economic plants and products: Edible nuts from *Corylus* (hazelnuts, filberts). Lumber from *Alnus* (alder) and *Betula* (birch). Ornamental trees (species of 5 genera), including *Alnus*, *Betula*, and *Carpinus* (hornbeam).

Commentary: The Betulaceae usually are divided into two (or occasionally three) distinct subfamilies, the Betuloideae (staminate flowers with perianth in three-flowered groups, carpellate flowers naked) and Coryloideae (solitary, naked staminate flowers, carpellate flowers with perianth). European taxonomists frequently treat the latter subfamily as a segregate, the Corylaceae (Furlow 1990). The Betulaceae generally are considered closely related to the Fagaceae (see Crane 1989 for a summary). The family (along with the related Fagaceae) was included in the traditional "Amentiferae," a polyphyletic assemblage of woody, anemophilous families with very reduced flowers arranged in "aments" or catkins (Stern 1973; Thorne 1973, 1989; Dilcher and Zavada 1986).

The minute and highly modified flowers and inflorescences have a very complex morphology (see Abbe 1935, 1938, 1974; and Hjelmqvist 1948). By examining the comparative morphology, floral vasculature, and ontogeny, Abbe determined that the basic inflorescence unit of the Betulaceae is a cyme composed of three florets associated with primary, secondary, and tertiary axes and bract systems (Fig. 79: 1d). The cymules have been greatly modified by reduction of parts, compression, and shortening of the internodes. In all genera, the cymules of the flexuous staminate inflorescences are three-flowered. In comparison, the carpellate inflorescences, which are more compact and often erect, have either three- (*Betula*) or two-flowered (by reduction; *Carpinus*) cymules (see Fig. 79: 3a vs. 1c). Due to the lack of a perianth in some genera, the reduced staminate (*Carpinus*) or carpellate (*Betula*) florets can be difficult to interpret: clustered staminate florets appear to be one flower (Fig. 79: 1j), while the atepalous carpellate florets are not visibly epigynous.

The specialized flowers of the Betulaceae are wind-pollinated, and produce large quantities of pollen. In many species (e.g., those of *Alnus*, *Betula*, *Ostrya*), the staminate inflorescences are produced the season before flowering, while the carpellate catkins develop later on new growth, opening as the young leaves unfold. When the staminate flowers are shedding pollen, the pistil of each carpellate flower consists of only two long styles and an undeveloped ovary. The pollen tubes remain dormant in the style tissue (often for several months) until the ovary and ovules mature. The growing pollen tube does not penetrate the micropyle (the usual route) but enters the nucellus via the placenta and funicle (Hall 1952).

The infructescences are somewhat cone-like (e.g., Fig. 79: 2a). The one-seeded nutlets usually are subtended or enclosed by bracts, which act as wings, aiding in dispersal by wind (Stone 1973), but animal dispersal may be important in certain genera (Tiffney 1986). The nuts of *Carpinus* (Fig. 79: 1l), for example, are subtended by a central secondary and two tertiary bracts; *Ostrya* (Fig. 79: 2c,d) has a nutlet enclosed in a membranous bladder formed by fused secondary and two tertiary bracts. Some fruit pericarps, as in *Betula* (Fig. 79: 3b), also are winged (samaras). In certain species of *Alnus*, the fruits are dispersed by water.

REFERENCES CITED

Abbe, E. C. 1935. Studies in the phylogeny of the Betulaceae. I. Floral and inflorescence anatomy and morphology. *Bot. Gaz.* 97:1–67.

——. 1938. Ibid., II. Extremes in the range of variation of floral and inflorescence anatomy. *Bot. Gaz.* 99:431–469.

——. 1974. Flowers and inflorescences of the "Amentiferae." *Bot. Rev.* (Lancaster) 40:159–261.

Crane, P. R. 1989. Early fossil history and evolution of the Betulaceae. In *Evolution, systematics, and fossil history of the Hamamelidae*, ed. P. R. Crane and S. Blackmore, vol. 2, *"Higher" Hamamelidae*, pp. 87–116. Syst. Assoc. Special Vol. 40B. Clarendon Press, Oxford.

Dilcher, D. L., and M. S. Zavada. 1986. Phylogeny of the Hamamelidae: An introduction. *Ann. Missouri Bot. Gard.* 73:225–226.

Furlow, J. J. 1990. The genera of Betulaceae in the southeastern United States. *J. Arnold Arbor.* 71:1–67.

Hall, J. W. 1952. The comparative anatomy and phylogeny of the Betulaceae. *Bot. Gaz.* 113:235–270.

Hjelmqvist, H. 1948. Studies on the floral morphology of the Amentiferae. *Bot. Not.*, Suppl. 2:1–171.

Stern, W. L. 1973. Development of the amentiferous concept. *Brittonia* 25:316–333.

Stone, D. E. 1973. Patterns in the evolution of amentiferous fruits. *Brittonia* 25:371–384.

Thorne, R. F. 1973. The "Amentiferae" or Hamamelidae as an artificial group: A summary statement. *Brittonia* 25: 395–405.

——. 1989. "Hamamelididae": A commentary. In *Evolution, systematics, and fossil history of the Hamamelidae*, ed. P. R. Crane and S. Blackmore, vol. 1, *Introduction and "lower" Hamamelidae*, pp. 9–16. Syst. Assoc. Special Vol. 40A. Clarendon Press, Oxford.

Tiffney, B. H. 1986. Fruit and seed dispersal and the evolution of the Hamamelidae. *Ann. Missouri Bot. Gard.* 73:394–416.

FAGACEAE

Beech or Oak Family

Trees or shrubs; roots mycorrhizal. **Leaves** simple, entire or serrate to variously lobed, alternate, coriaceous, deciduous or persistent, stipulate (stipules minute). **Inflorescence** determinate, cymose and often appearing spicate (erect or pendulous) or capitate, or reduced to a small cyme or a solitary flower, unisexual. **Flowers** actinomorphic, imperfect (then plants usually monoecious), epigynous, very reduced; carpellate flowers with involucre of many adnate and imbricate bracts. **Perianth** of staminate flowers: 4- to 8-lobed, scale-like, imbricate; of carpellate flowers: 3- to 8-lobed, inconspicuous. **Androecium** of 4 to numerous stamens; filaments distinct; anthers basifixed or dorsifixed, dehiscing longitudinally, extrorse or introrse. **Gynoecium** of 1 pistil, 3-carpellate; ovary inferior, 3-locular; ovules 2 in each locule (all but 1 aborting), anatropous, pendulous, placentation axile; styles 3; stigma along upper side of style or a terminal pore. **Fruit** a nut, 1 to 3 subtended or completely enclosed by a muricate to spiny involucre (cupule); seed solitary; endosperm absent; embryo large, straight.

Family characterization: Monoecious trees or shrubs; reduced imperfect flowers with uniseriate perianth (the staminate in erect to pendent spicate or capitate inflorescences, and the carpellate in reduced cymes); inferior 3-carpellate ovary; and nut(s) subtended by an involucre. Tissues with calcium oxalate crystals and a high tannin content. Anatomical feature: trilacunar nodes.

Genera/species: 9/700–800

Distribution: Predominantly in broad-leaved forests of northern temperate regions and mountainous tropical areas (excluding tropical and southern Africa).

Major genera: *Quercus* (450–600 spp.), *Lithocarpus* (300 spp.), and *Castanopsis* (110–120 spp.)

U.S./Canadian representatives: *Quercus* (79 spp.), *Castanea* (4 spp.), *Castanopsis* (2 spp.), *Fagus* (2 spp.), and *Lithocarpus* (1 sp.)

Economic plants and products: Tannic acid extracted from fruits and insect galls of certain *Quercus* spp. (oak). Edible nuts from *Castanea* (chestnuts), *Fagus* (beechnuts for livestock), and *Quercus* (acorns for livestock). Cork from *Quercus suber* (cork oak). Timber from *Fagus* (beech) and *Quercus*. Ornamental shade trees (species of 6 genera), including *Castanea* (chestnut), *Fagus* (beech), and *Quercus* (oak).

Commentary: Taxonomists have not agreed upon the circumscription and relationships of the genera of the Fagaceae and traditionally have grouped them into three or four subfamilies (Elias 1971; Abbe 1974). Current cladistic studies (Nixon 1989), however, support the recognition of two subfamilies (the Fagoideae and Castaneoideae), with *Nothofagus* (southern beech) segregated as a monotypic family (as in Thorne 1992). The entomophilous Castaneoideae and anemophilous Fagoideae are differentiated on the basis of inflorescence, floral, and pollen features that probably reflect the different pollination strategies (discussed below).

The interpretation of the cupule, a characteristic feature of the family, has received much attention (see the summaries in Nixon 1989 and Jenkins 1993). This structure varies in morphology; for example, in *Castanea* (chestnut) four spine-bearing valves enclose one to three variously compressed nuts (Fig. 80: 2b), and in *Fagus* four bristly valves contain two triangular nuts. The acorn of *Quercus* (Fig. 80: 1j) comprises a smooth or scaly cupule enveloping one rounded or turbinate nut. Some recent studies, such as those by Fey and Endress (1983), indicate that the cupule probably represents part of a complex cymose structure in which the outer inflorescence axes are variously modified into valves (fused in *Quercus*). However, other botanists, such as Jenkins (1993), advocate derivation of the cupule from the outer perianth whorl.

The flowers of the Fagaceae generally are anemophilous and, like the Betulaceae, the family was included within the polyphyletic group "Amentiferae," characterized by reduced flowers arranged in "aments" or catkins (see the references under Betulaceae); for example, *Quercus* and *Fagus* both have pendulous staminate inflorescences (Fig. 80: 1h) with inconspicuous flowers. Other genera are more adapted for insect pol-

179

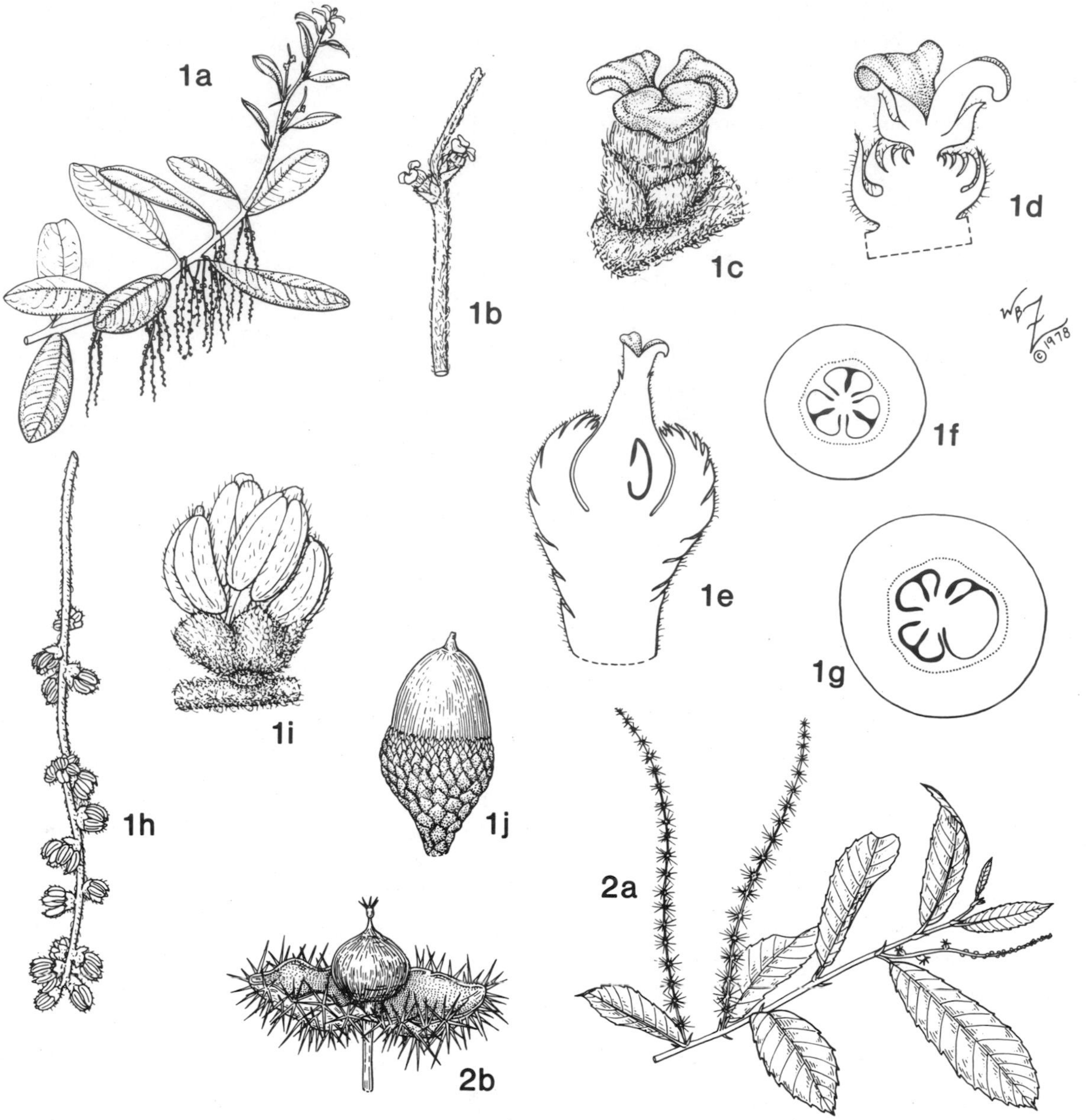

Figure 80. Fagaceae. 1, *Quercus geminata*: **a,** flowering branch, ×²⁄₅; **b,** carpellate inflorescence, ×3; **c,** carpellate flower, ×12; **d,** longitudinal section of carpellate flower at anthesis, ×12; **e,** longitudinal section of carpellate flower after pollination (two months after anthesis), ×6; **f,** cross section of ovary, ×12; **g,** cross section of ovary later in development, all but one ovule aborted, ×12; **h,** staminate inflorescence, ×3; **i,** staminate flower, ×12; **j,** fruit (nut subtended by scaly cupule), ×1¼. **2,** *Castanea pumila*: **a,** flowering branch, ×²⁄₅; **b,** fruit (nut subtended by spiny cupule), ×¾.

lination. For example, the staminate inflorescence of *Castanea* is somewhat erect (Fig. 80: 2a), and the staminate flowers have conspicuous pale yellow perianths and anthers, a strong scent, and sticky pollen; the carpellate flowers generally have a stiff style and sticky stigma. These flowers are visited and pollinated by various insects.

REFERENCES CITED

Abbe, E. C. 1974. Flowers and inflorescences of the "Amentiferae." *Bot. Rev.* (Lancaster) 40:159–261.

Elias, T. S. 1971. The genera of Fagaceae in the southeastern United States. *J. Arnold Arbor.* 52:159–195.

Fey, B. S., and P. K. Endress. 1983. Development and morphological interpretation of the cupule in Fagaceae. *Flora, Morphol. Geobot. Oekophysiol.* 173:451–468.

Jenkins, R. 1993. The origin of the fagaceous cupule. *Bot. Rev.* (Lancaster) 59:81–111.

Nixon, K. C. 1989. Origins of Fagaceae. In *Evolution, systematics, and fossil history of the Hamamelidae*, ed. P. R. Crane and S. Blackmore, vol. 2, *"Higher" Hamamelidae*, pp. 23–43. Syst. Assoc. Special Vol. 40B. Clarendon Press, Oxford.

Thorne, R. F. 1992. Classification and geography of the flowering plants. *Bot. Rev.* (Lancaster) 58:225–348.

ROSACEAE
Rose Family

Trees, shrubs, or perennial herbs, often armed with thorns or prickles. **Leaves** simple to palmately or pinnately compound, serrate, alternate, cauline or forming basal rosettes, stipulate. **Inflorescence** indeterminate or sometimes determinate, racemose, cymose, or sometimes reduced to a single flower, terminal or axillary. **Flowers** actinomorphic, perfect, perigynous to sometimes epigynous, with a cup-like hypanthium bearing a nectariferous disc, generally showy, sometimes with an epicalyx. **Calyx** of 5 sepals, basally connate, valvate. **Corolla** of 5 petals, distinct, short-clawed with horizontally spreading limbs, caducous, white, yellow, pink, purple, or orange, imbricate. **Androecium** of numerous stamens in several whorls (inner shorter than outer); filaments distinct or basally connate; anthers dorsifixed, versatile, dehiscing longitudinally, introrse. **Gynoecium** apocarpous and of 1 to many carpels (spirally or cyclically arranged on the receptacle) or sometimes syncarpous and of 1 compound pistil of 2 to 5 carpels; ovary (ovaries) superior (within free hypanthium) to sometimes completely inferior (adnate to hypanthium), with as many locules as carpels; ovules few to several in each carpel, anatropous, basal or pendulous, placentation axile; styles the same number as the carpels, distinct or sometimes basally connate, deciduous or persistent; stigmas terminal and discoid or in decurrent bands along the styles. **Fruit** a drupe or pome, or an aggregate of achenes (exposed or enclosed by hypanthium), follicles, or drupelets, often with enlarged receptacle and/or hypanthium; endosperm absent or sometimes scanty; embryo small, spatulate.

Family characterization: Herbaceous to woody plants often armed with thorns or prickles; alternate, stipulate, simple to compound leaves with serrate margins; actinomorphic, perigynous to epigynous flowers with a well-developed hypanthium; 5 distinct, short-clawed petals; numerous stamens; and seeds with little or no endosperm. Saponins and cyanogenic glycosides commonly present.

Genera/species: 100/2,000

Distribution: Generally cosmopolitan; most abundant in temperate regions of eastern Asia, North America, and Europe.

Major genera: *Rubus* (250–3,000+ spp.), *Potentilla* (500 spp.), *Prunus* (36–430 spp.), *Crataegus* (200–280+ spp.), *Alchemilla* (250 spp.), *Rosa* (100–250 spp.), *Cliffortia* (80–115 spp.), and *Acaena* (100 spp.). Species delimitations in *Rosa*, *Crataegus*, and *Rubus* are particularly difficult due to extensive hybridization and apomixis (Grant 1971; Robertson 1974).

U.S./Canadian representatives: 62 genera/870 spp.; largest genera: *Rubus*, *Crataegus*, and *Potentilla*

Economic plants and products: Fruit crops: *Cydonia* (quince), *Eriobotrya* (loquat), *Fragaria* (strawberry), *Malus* (apple), *Mespilus* (medlar), *Prunus* (almond, apricot, cherry, nectarine, peach, plum, prune), *Pyrus* (pear), and *Rubus* (blackberry, raspberry). Several toxic plants (with seeds and other plant parts containing amygdalin and prunasin, which are hydrolized by enzymes into poisonous cyanic acid), including *Prunus* species (cherry, apricot). Medicinal plants: *Prunus* (laetrile from apricot seeds) and *Rosa* ("rose hips," the mature fruits and hypanthium). Perfumes (essential oils) from *Rosa* (attar of roses). Ornamental trees and shrubs (species of 70 genera), including *Crataegus* (hawthorne), *Photinia*, *Pyracantha* (firethorn), *Rhaphiolepis* (India-hawthorne), *Rosa* (rose), and *Spiraea*.

Commentary: The Rosaceae generally are accepted as a monophyletic group and usually are divided into four subfamilies based primarily on gynoecial and fruit characters, as shown in Table 10 and Figs. 81 and 82. The Chrysobalanaceae, formerly sometimes included as an additional subfamily, are now excluded (see Prance 1970). The taxonomic rank given the subfamilies varies in the literature, and some have been elevated to familial status (e.g., Malaceae) by different authors. Recent critical work (e.g., Kalkman 1988; Phipps et al. 1991) has attempted to clarify the circumscription and relationships of the subfamilies.

The habit in the rose family varies from low-growing herbs (*Fragaria*), to scrambling bushes (*Rubus*; Fig. 81: 2a), to trees or shrubs (*Crataegus*, *Prunus*). Often the plants are armed with thorns (*Crataegus*; Fig. 82: 3a) or prickles (*Rosa*; Fig. 81: 1a).

A unifying feature of the family is the actinomorphic, perfect flower with a prominent cup-like hypanthium and many exserted stamens (e.g., Figs. 81: 1c, 2b; 82: 1c, 2d, 3b). The shallow flowers have easily

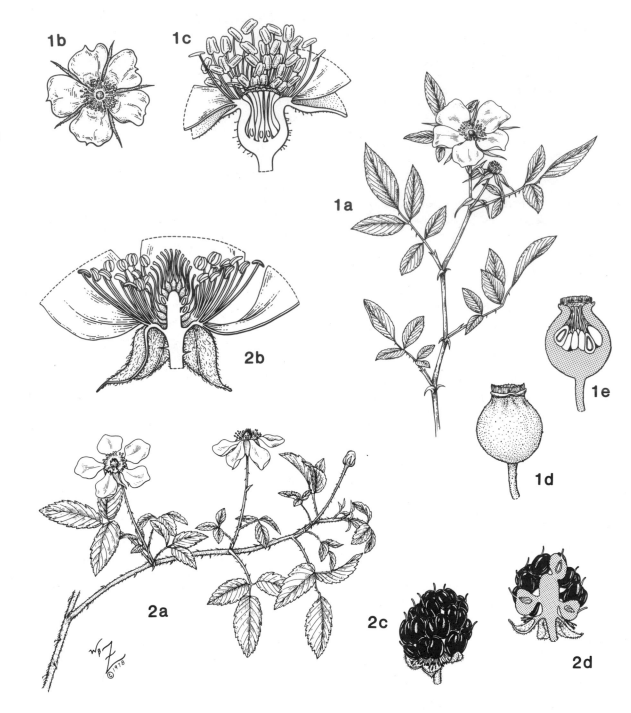

Figure 81 [above]. Rosaceae. 1, *Rosa palustris*: **a,** flowering branch, ×½; **b,** flower, ×⅔; **c,** longitudinal section of flower, ×3⅔; **d,** fruit (aggregate of achenes surrounded by fleshy hypanthium), ×2; **e,** longitudinal section of fruit, ×2. **2,** *Rubus trivialis*: **a,** flowering branch, ×½; **b,** longitudinal section of flower, ×3⅔; **c,** fruit (aggregate of drupelets), ×2; **d,** longitudinal section of fruit, ×2.

Figure 82 [facing page]. Rosaceae (continued). 1, *Spiraea thunbergii*: **a,** portion of flowering branch (with immature leaves), ×2½; **b,** flower, ×5; **c,** longitudinal section of flower, ×10; **d,** fruit (aggregate of follicles), ×10. **2,** *Prunus serotina*: **a,** flowering branch, ×½; **b,** top view of flower, ×4; **c,** lateral view of flower, ×6; **d,** longitudinal section of flower, ×6; **e,** drupe, ×2¼; **f,** longitudinal section of drupe, ×2¼. **3,** *Crataegus uniflora*: **a,** flowering branch, ×½; **b,** longitudinal section of flower, ×3⅔; **c,** cross section of ovary, ×7; **d,** pome, ×1½; **e,** longitudinal section of pome, ×1½.

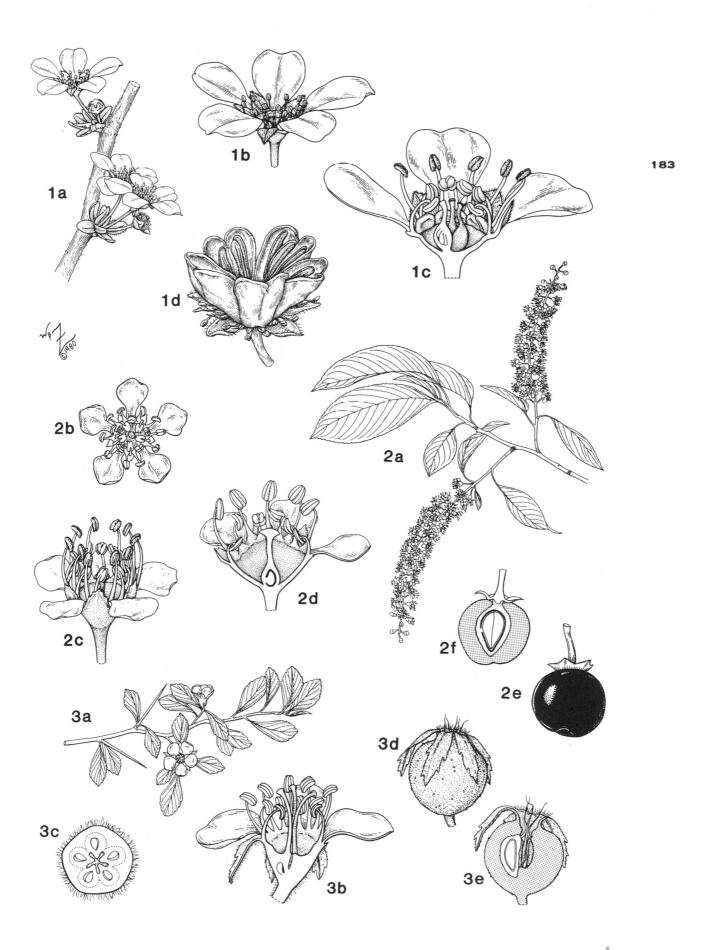

TABLE 10. Morphological characters used to distinguish the traditional four subfamilies of the Rosaceae (Cronquist 1981; Phipps et al. 1991). Numbers of genera and species are rough estimates.

CHARACTER	ROSOIDEAE	SPIRAEOIDEAE	AMYGDALOIDEAE (PRUNOIDEAE)	MALOIDEAE (POMOIDEAE)
GENERA/SPECIES	40–50/300–500	17/140	1/400–430	28/940–1,110
REPRESENTATIVE GENERA	*Cercocarpus, Fragaria, Geum, Potentilla, Rosa, Rubus*	*Physocarpus, Spiraea*	*Prunus*	*Amelanchier, Crataegus, Cydonia, Eriobotrya, Malus, Photinia, Pyracantha, Pyrus, Sorbus*
HABIT	herbs, shrubs, or small trees	shrubs	shrubs or trees	shrubs or trees
LEAVES	usually pinnately or palmately compound; stipulate	simple, or sometimes pinnately compound; exstipulate	simple; stipulate	simple, or sometimes pinnately compound; stipulate
OVARY POSITION	superior (flowers perigynous; carpels on elongated receptacle, or ± enclosed by urceolate floral cup + receptacle)	superior (flowers perigynous; carpels at base of floral cup)	superior (flowers perigynous; carpel at base of floral cup)	inferior to ± half-inferior (flowers fully to partially epigynous; carpels ± adnate to floral cup)
CARPEL NUMBER	usually many	2 to usually 5	usually 1	2 to usually 5
CARPEL FUSION	usually distinct	usually distinct	——	± connate
FRUIT TYPE	aggregate of achenes or drupelets	usually aggregate of follicles	drupe	pome

accessible nectar produced by a nectariferous disc or ring that lines the hypanthium apex, and insects also collect the abundant pollen. Protogyny is prevalent in the family. Smaller flowers (such as in *Prunus*) mostly are visited by flies and short-tongued bees. Larger bees and other Hymenoptera, Coleoptera, and Lepidoptera pollinate larger flowers (such as those of *Rosa* and *Rubus*).

The fruit morphology in the family is particularly diverse, and several aggregate fruit types are derived from perigynous apocarpous flowers (Table 10). The rose hip (*Rosa*) consists of a swollen hypanthium surrounding numerous small achenes (Fig. 81: 1d,e). The strawberry (*Fragaria*) fruit is an enlarged fleshy receptacle covered with achenes, and the aggregate fruit of *Rubus* (blackberry, raspberry) has an elongate receptacle bearing drupelets (Fig. 81: 2c,d). The fruit of *Spiraea* is an aggregate of follicles (Fig. 82: 1d). The unicarpellate flower of *Prunus* (Fig. 82: 2d) produces one drupe (Fig. 82: 2e,f). Genera such as *Crataegus* (hawthorne),

Malus (apple), and *Pyrus* (pear) have a two- to five-carpellate compound pistil (e.g., Fig. 82: 3b,c), which is adnate to the hypanthium; the unusual fruit of these flowers is a "pome" (Fig. 82: 3d,e). The outer part of a pome is fleshy and the center has papery (*Malus*) or cartilaginous (*Crataegus*) structures surrounding the seeds. The epigynous nature of the flower is exemplified by the apple, which, even at maturity, has persistent floral parts on the "bottom" (actually the apex of the floral axis). Most of the fleshy tissue of an apple is the swollen hypanthium; only the core and some surrounding tissue are derived from the ovary (Rohrer et al. 1991).

REFERENCES CITED

Cronquist, A. 1981. *An integrated system of classification of flowering plants*, p. 576. Columbia University Press, New York.

Grant, V. 1971. *Plant speciation*, pp. 269–292. Columbia University Press, New York.

Kalkman, C. 1988. The phylogeny of the Rosaceae. *Bot. J. Linn. Soc.* 98:37–59.

Phipps, J. B., K. R. Robertson, J. R. Rohrer, and P. G. Smith. 1991. Origins and evolution of subfam. Maloideae (Rosaceae). *Syst. Bot.* 16:303–332.

Prance, G. T. 1970. The genera of Chrysobalanaceae in the southeastern United States. *J. Arnold Arbor.* 51:521–528.

Robertson, K. R. 1974. The genera of Rosaceae in the southeastern United States. *J. Arnold Arbor.* 55:303–332.

Rohrer, J. R., K. R. Robertson, and J. E. Phipps. 1991. Variation in structure among fruits of Maloideae (Rosaceae). *Amer. J. Bot.* 78:1617–1635.

HYDRANGEACEAE

Hydrangea Family

Usually shrubs, small trees, or woody vines. **Leaves** simple, serrate, sometimes lobed, opposite, exstipulate. **Inflorescence** determinate, cymose and often appearing corymbose or paniculate, terminal or axillary. **Flowers** actinomorphic, usually perfect, sometimes sterile, epigynous to half-epigynous, small or large and showy, with short hypanthium, with a nectariferous disc at ovary apex. **Calyx** synsepalous with 4 or 5 lobes, enlarged and petaloid in sterile flowers, valvate or imbricate, persistent. **Corolla** of 4 or 5 petals, distinct, usually white, pink, or blue, valvate, imbricate, or convolute. **Androecium** of 8 to numerous stamens in 2 to several whorls; filaments distinct or basally connate; anthers basifixed, dehiscing longitudinally, introrse or latrorse. **Gynoecium** of 1 pistil, usually 2- to 5-carpellate; ovary half-inferior to inferior, with as many locules as carpels or sometimes 1-locular (due to incomplete septa), capped by nectariferous disc; ovules usually several to numerous, anatropous, placentation axile, parietal (in 1-locular ovaries) or axile above and parietal below (in incompletely septate ovaries); styles 2 to 5, distinct or basally connate, short, stout, persistent; stigmas 2 to 5, capitate or somewhat decurrent on styles. **Fruit** usually a septicidal or loculicidal capsule; seeds numerous, small, often winged; endosperm copious, fleshy; embryo straight, linear, enveloped by endosperm.

Family characterization: Woody plants with simple, opposite, exstipulate leaves; semi-epigynous to epigynous flowers with short hypanthium; numerous stamens in 2 to several whorls; ovary capped by nectariferous disc and distinct styles; and capsular fruit. Tannins, saponins, and iridoid compounds commonly present. Anatomical features: raphide sacs (which often also contain mucilage) in leaves and stems; and unitegmic and tenuinucellate ovules.

Genera/species: 17/250

Distribution: Primarily in temperate to subtropical areas of the Northern Hemisphere, with a few representatives in western South America.

Major genera: *Hydrangea* (23–80 spp.), *Philadelphus* (65–75 spp.), and *Deutzia* (40–50 spp.)

U.S./Canadian representatives: 11 genera/49 spp.; largest genera: *Philadelphus*, *Fendlera*, and *Hydrangea*

Economic plants and products: Roots and leaves of several species of *Hydrangea* with poisonous or medicinal properties (due to substances such as alkaloids—e.g., hydrangin). Ornamental plants (species of at least 13 genera), including *Decumaria* (climbing-hydrangea), *Deutzia*, *Hydrangea*, and *Philadelphus* (mock-orange, syringa).

Commentary: The Hydrangeaceae formerly have been included as a subfamily (Hydrangeoideae) within a broadly interpreted Saxifragaceae, a traditionally large, heterogeneous, and polyphyletic family divided into as many as 17 subfamilies (e.g., Thorne 1976) and numerous tribes (see Soltis et al. 1990 and Morgan and Soltis 1993 for historical summaries). These subfamilies (and groups within them) are now often treated as segregate families (e.g., Escalloniaceae, Grossulariaceae, Parnassiaceae), leaving a much more narrowly circumscribed Saxifragaceae of generally 30 genera (550 herbaceous spp.). The Hydrangeaceae are allied usually with other woody segregates formerly associated with the Saxifragaceae (e.g., Escalloniaceae; see Table 11 and Fig. 84). Several morphological and molecular cladistic studies, summarized by Morgan and Soltis (1993), demonstrate that some of these "woody saxifrages" (e.g., *Itea*) are closely allied with the herbaceous Saxifragaceae *s.s.*, whereas other taxa (e.g., the Hydrangeaceae) are only distantly related. The Hydrangeaceae actually are closely related to the Cornaceae complex of families ("Corniflorae"; Xiang et al. 1993).

Some species (e.g., those of *Philadelphus*) have large and showy, fragrant flowers; in contrast, species of *Hydrangea* (and some allied genera) are characterized by inconspicuous, fertile flowers (Fig. 83: 1b,d) and showy, sterile flowers (Fig. 83: 1b,c). This "neutral" floral form has an enlarged petaloid calyx (white, pink, or blue) and abortive gynoecium and androecium, although the stamens may appear to be fertile. The sterile flowers terminate the branches of mixed inflorescences (Fig. 83: 1b) or may compose the entire inflorescence (e.g., cultivated forms of *Hydrangea*).

The flowers of the Hydrangeaceae attract a wide variety of insect pollinators (bees, flies, lepidopterans) that probe between the stamens for the nectar secreted

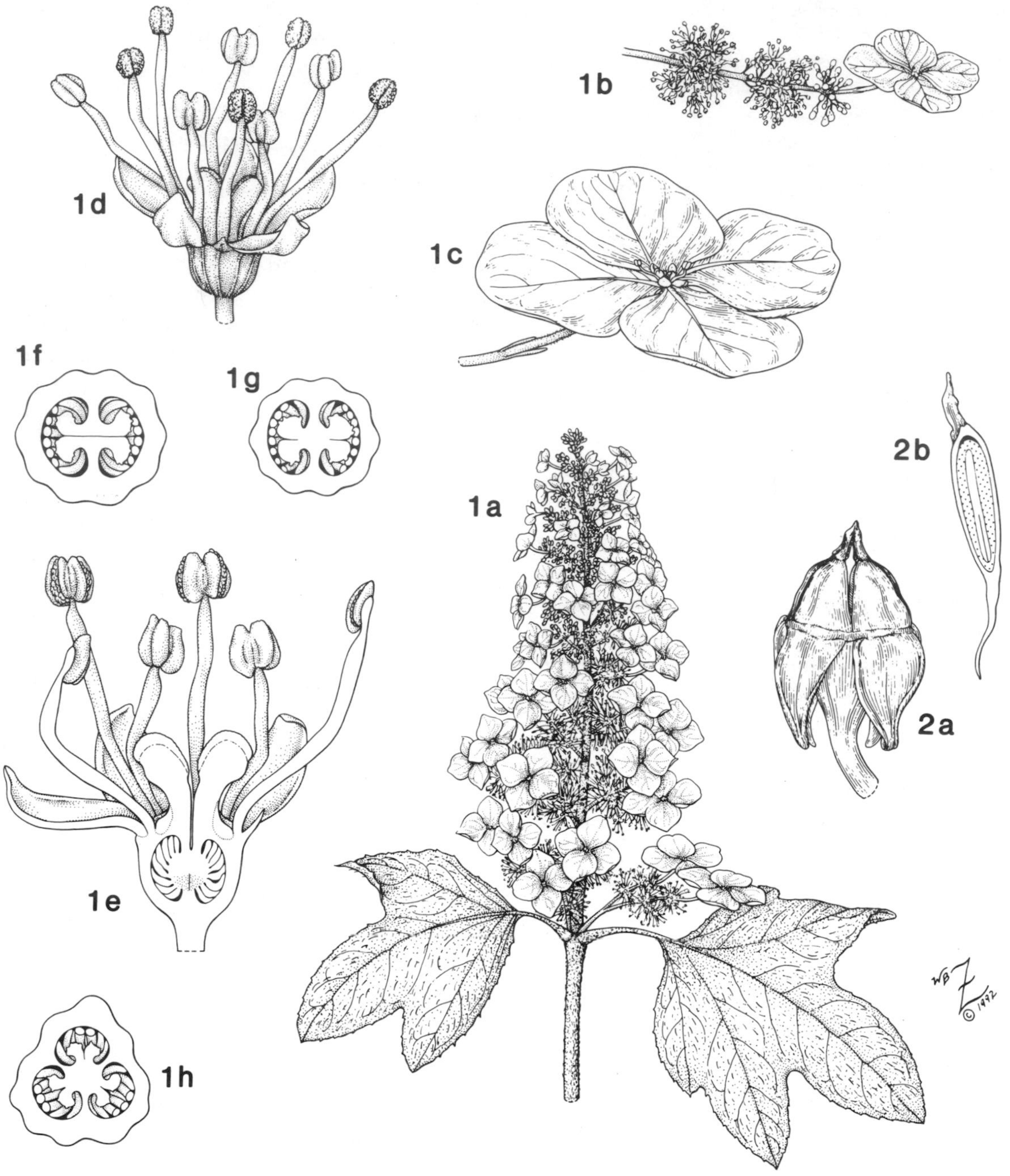

186

Figure 83. Hydrangeaceae. 1, *Hydrangea quercifolia*: **a,** flowering branch, ×²⁄₅; **b,** branch of inflorescence with terminal sterile flower, ×½; **c,** sterile flower, ×2¼; **d,** flower, ×8; **e,** longitudinal section of flower, ×10; **f,** cross section of two-carpellate ovary near base, ×15; **g,** cross section of two-carpellate ovary above middle, ×15; **h,** cross section of three-carpellate ovary above middle, ×15. 2, *Philadelphus inodorus*: **a,** capsule, ×2½; **b,** longitudinal section of seed, ×12.

by the disc topping the ovary. The outer (or outermost) whorl of stamens dehisces first. Although the flowers are protogynous, self-pollination is also possible (Spongberg 1972).

REFERENCES CITED

Morgan, D. R., and D. E. Soltis. 1993. Phylogenetic relationships among members of Saxifragaceae sensu lato based on *rbc*L sequence data. *Ann. Missouri Bot. Gard.* 80:631–660.

Soltis, D. E., P. S. Soltis, M. T. Clegg, and M. Durbin. 1990. *rbc*L sequence divergence and phylogenetic relationships in Saxifragaceae sensu lato. *Proc. Natl. Acad. Sci. U.S.A.* 87:4640–4644.

Spongberg, S. A. 1972. The genera of Saxifragaceae in the southeastern United States. *J. Arnold Arbor.* 53:409–498.

Thorne, R. F. 1976. A phylogenetic classification of the Angiospermae. *Evol. Biol.* 9:35–106.

TABLE 11. Major morphological differences between the Hydrangeaceae and Escalloniaceae, two woody segregate families formerly included within the Saxifragaceae *s.l.* The variable Escalloniaceae, as circumscribed here (as in Thorne 1992), may be polyphyletic; in particular, *Itea* is most likely not closely related to other members of the Escalloniaceae and probably should be considered as a segregate family (Morgan and Soltis 1993).

CHARACTER	HYDRANGEACEAE	ESCALLONIACEAE
GENERA/SPECIES	17/250	15/200
REPRESENTATIVE GENERA	*Decumaria, Fendlera, Hydrangea, Philadelphus*	*Escallonia, Itea*
DISTRIBUTION	primarily temperate to subtropical Northern Hemisphere, with a few in western South America	primarily montane-temperate Southern Hemisphere, with a few in North America
HABIT	shrubs, small trees, or woody vines	usually shrubs or trees
LEAVES	usually opposite serrate to lobed exstipulate	usually alternate serrate (often glandular toothed) exstipulate, or sometimes stipulate (stipules small)
INFLORESCENCE	cymose, often corymbose or paniculate	usually racemose
STAMEN NUMBER	8 to numerous (in 2 to several whorls)	usually 5
OVARY POSITION	half-inferior to inferior with short hypanthium	hypogynous to epigynous
CARPEL NUMBER	usually 2 to 5	1 to 6
FRUIT TYPE	usually a capsule	capsule or berry
ENDOSPERM	copious	scanty to copious

———. 1992. Classification and geography of the flowering plants. *Bot. Rev.* (Lancaster) 58:225–348.

Xiang, Q.-Y., D. E. Soltis, D. R. Morgan, and P. S. Soltis. 1993. Phylogenetic relationships of *Cornus* L. sensu lato and putative relatives inferred from *rbc*L sequence data. *Ann. Missouri Bot. Gard.* 80:723–734.

VITACEAE
Grape Family

Climbing woody vines, usually with tendrils opposite the leaves; stems sympodial, with swollen or jointed nodes. **Leaves** simple or pinnately or palmately compound, often palmately lobed and/or veined, generally alternate (lower ones sometimes opposite), distichous, pellucid punctate, stipulate; stipules small, caducous. **Inflorescence** determinate, cymose and sometimes appearing racemose or paniculate, opposite a leaf or sometimes also terminal. **Flowers** actinomorphic, perfect or imperfect (then plants usually monoecious or polygamodioecious), hypogynous (or slightly perigynous), minute, with well-developed intrastaminal disc or sometimes 5 distinct glands. **Calyx** of 4 or 5 often indistinct teeth or lobes, sometimes reduced to a rim around the ovary, cup-like. **Corolla** typically of 4 or 5 petals, distinct or sometimes apically connate and caducous as a calyptra (cap), usually greenish, valvate. **Androecium** of 4 or 5 stamens, opposite the petals; filaments distinct; anthers dorsifixed, distinct or connate, dehiscing longitudinally, introrse; staminodes present in carpellate flowers. **Gynoecium** of 1 pistil, 2-carpellate; ovary superior, 2-locular (septa often incomplete), variously adnate to the disc; ovules 1 or 2 on each placenta, anatropous, erect, placentation axile; style 1, short; stigma 1, small, discoid or capitate; rudimentary pistil present in staminate flowers. **Fruit** a berry; seeds 1 to 4, with hard, bony, or crustaceous testa, with conspicuous "chalazal knot" and 2 deep adaxial grooves; endosperm copious, usually 3-lobed, hard-fleshy, oily, proteinaceous; embryo small, straight, spatulate, surrounded by endosperm.

Family characterization: Sympodial woody vines with tendrils and swollen nodes; pellucid punctate, simple to compound, often palmately veined or lobed leaves; inflorescences arising opposite the leaves; small flowers with 4- or 5-merous perianth and 4 or 5 sta-

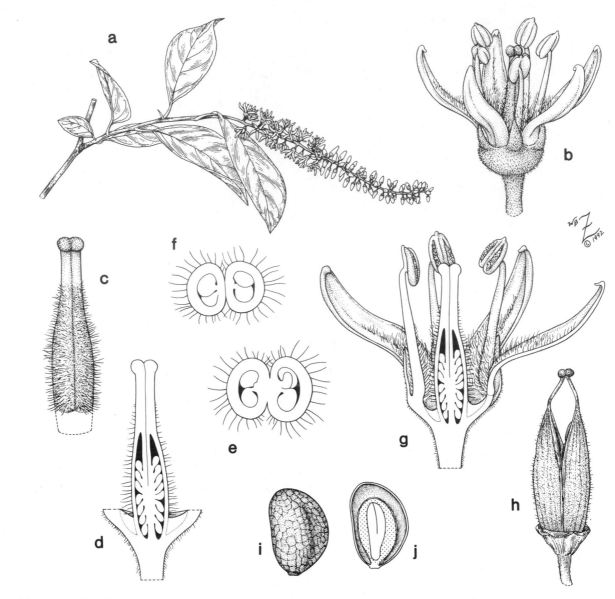

Figure 84. Escalloniaceae. *Itea virginica*: **a,** flowering branch, ×⅔; **b,** flower, ×7; **c,** pistil, ×10; **d,** longitudinal section of pistil, ×10; **e,** cross section of ovary below middle (above base), ×22; **f,** cross section of ovary near apex, ×22; **g,** longitudinal section of flower, ×9; **h,** capsule, ×7; **i,** seed, ×22; **j,** longitudinal section of seed, ×22. See Table 11 for family characteristics.

mens opposite the petals; and a berry with seeds having characteristic morphology (see the discussion below). Tannins present. Anatomical features: raphide sacs and mucilage cells (often also containing raphides) in stems and leaves.

Genera/species: 13/735

Distribution: Primarily pantropical, with a few representatives in temperate regions.

Major genera: *Cissus* (350 spp.) and *Cyphostemma* (150 spp.)

U.S./Canadian representatives: *Vitis* (24 spp.), *Cissus* (8 spp.), *Ampelopsis* (4 spp.), and *Parthenocissus* (4 spp.)

Economic plants and products: Wine, juice, and table grapes, and muscatel, raisins, and "currants" from fruits of several *Vitis* species. Ornamental vines (species of 7 genera), including *Cissus* (grape-ivy), *Parthenocissus* (Boston-ivy, Virginia-creeper, woodbine), and *Vitis*.

Commentary: The Vitaceae constitute a presumably monophyletic family. *Leea*, a genus of the Old World tropics, sometimes is excluded as a monotypic family, the Leeaceae (as in Ridsdale 1974 and Cronquist 1981). Placement of the Vitaceae varies: the family usually is allied with the Rhamnaceae (Cronquist 1988), whereas Thorne (1992) includes them in the Cornales (Cornaceae and allies). The genera are sepa-

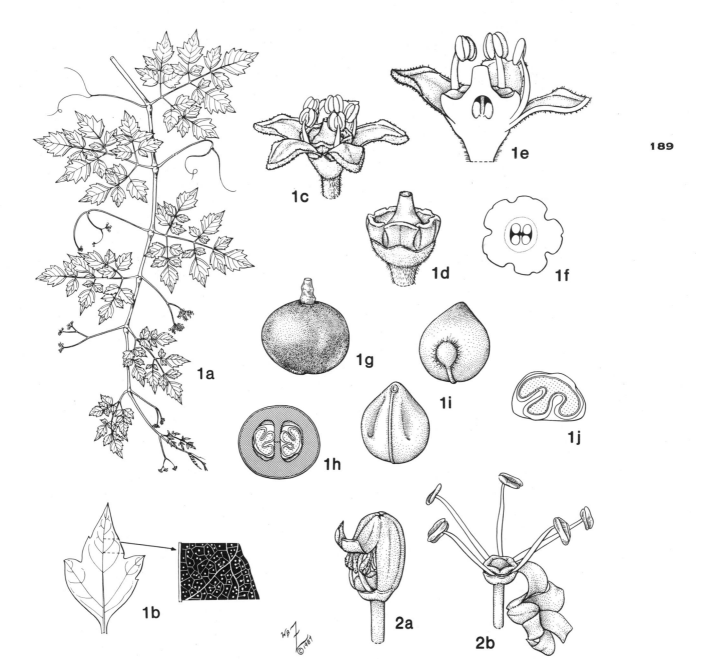

Figure 85. Vitaceae. 1, *Ampelopsis arborea*: **a,** habit, ×⅓; **b,** leaflet, ×1½, with detail showing punctate surface, ×6; **c,** flower, ×9; **d,** pistil and nectariferous disc, ×12; **e,** longitudinal section of flower, ×12; **f,** cross section of ovary, ×12; **g,** berry, ×2; **h,** cross section of berry, ×2; **i,** two views of seed, ×4½; **j,** cross section of seed, ×4½. **2,** *Vitis rotundifolia*: **a,** staminate flower bud immediately before anthesis (corolla starting to detach), ×12; **b,** staminate flower with corolla detached as a calyptra, ×12.

rated primarily on the basis of the morphology of the style (relative length), nectariferous disc, and endosperm (lobing), as well as characters of the corolla (free or apically connate), inflorescence, and tendrils (attach by coiling or with disc-like structures).

Most of the Vitaceae are vines with tendrils (and inflorescences) that occur on the stem opposite the leaves (Fig. 85: 1a), an unusual situation that suggests that the stem is sympodial (Brizicky 1965). The tendril generally has been regarded as representing a main axis that has been subordinated (or pushed aside) by the more vigorous growth of the branch in the opposing leaf axil. However, the situation in *Cissus* and *Vitis* is more complex, with the leaves having axillary buds as well as opposing tendrils (Cronquist 1981). Since flower-bearing tendrils may occur in some species (e.g., *Ampelopsis arborea*; Fig. 85: 1a), the tendrils may represent modified inflorescences.

The small flowers secrete nectar from the disc (Fig. 85: 1d) and often are fragrant and usually entomophilous. However, a few species (e.g., *Vitis* spp.) apparently are anemophilous. Cross-pollination of perfect flowers generally is promoted by protandry (e.g., *Ampelopsis*), but homogamy and self-pollination are not uncommon, as in some species of *Vitis* with perfect flowers.

The seeds of the Vitaceae have a distinctive morphology (Brizicky 1965; Corner 1976). The seed coat consists of a thin outer membrane and a hard (often bony) inner layer (sclerotesta) that forms two deep grooves on the adaxial surface (Fig. 85: 1i, lower left). This infolding conforms the endosperm into a characteristic three-lobed configuration (Fig. 85: 1j). A conspicuous ridge (raphe) occurs between the grooves and extends from the hilum to the seed apex (on the adaxial side). It terminates on the abaxial surface in a conspicuous, depressed or elevated area called a "chalazal knot" (Fig. 85: 1i, upper right).

REFERENCES CITED

Brizicky, G. K. 1965. The genera of Vitaceae in the southeastern United States. *J. Arnold Arbor.* 46:48–67.

Corner, E. J. H. 1976. *The seeds of dicotyledons*, vol. 1, pp. 277–280. Cambridge University Press, Cambridge.

Cronquist, A. 1981. *An integrated system of classification of flowering plants*, pp. 746–750. Columbia University Press, New York.

———. 1988. *The evolution and classification of flowering plants*, 2d ed., pp. 396–397. New York Botanical Garden, Bronx.

Ridsdale, C. E. 1974. A revision of the family Leeaceae. *Blumea* 22:57–100.

Thorne, R. F. 1992. Classification and geography of the flowering plants. *Bot. Rev.* (Lancaster) 58:225–348.

CORNACEAE
Dogwood Family
Including the Nyssaceae: Tupelo Family

Usually trees or shrubs, or sometimes suffrutescent subshrubs. **Leaves** simple, entire to obscurely serrate, opposite or sometimes alternate, often with arching secondary veins, usually deciduous, exstipulate. **Inflorescence** determinate, cymose and appearing corymbose, umbellate, capitate, paniculate, or sometimes reduced to a single flower, sometimes subtended by large (usually white) petaloid bracts, terminal or axillary. **Flowers** actinomorphic, perfect or imperfect (then plants monoecious, dioecious, or polygamodioecious), epigynous, small, with a nectariferous disc at ovary apex, sessile or nearly so, subtended by bracts. **Calyx** of 4 or sometimes 5 sepals, distinct, small and inconspicuous, often reduced to teeth around upper edge of ovary, valvate. **Corolla** of 4 or sometimes 5 petals, distinct, white, purple, or yellow, valvate or sometimes imbricate. **Androecium** of 4 or sometimes 5 stamens (uniseriate), or sometimes of 10 stamens (biseriate with outer whorl opposite the petals); filaments distinct; anthers usually dorsifixed and versatile or sometimes basifixed, dehiscing longitudinally, introrse. **Gynoecium** of 1 pistil, 2-carpellate or sometimes appearing 1-carpellate (pseudomonomerous); ovary inferior, 2- or sometimes 1-locular, capped by nectariferous disc at apex; ovules 1 in each locule, anatropous or hemitropous, pendulous, placentation axile; style 1; stigmas capitate or lobed; rudimentary pistil occasionally present in staminate flowers. **Fruit** usually a drupe with 1 furrowed pyrene (stone); endosperm copious, oily; embryo small to large, straight, spatulate.

Family characterization: Trees or shrubs; small, epigynous flowers with 4- or 5-merous perianth and androecium; very reduced sepals; 2-carpellate inferior ovary capped by a nectariferous disc; a solitary, pendulous ovule in each locule; and a drupe with a single furrowed stone. Iridoid compounds commonly present. Tissues with calcium oxalate crystals. Anatomical features: unitegmic ovules; and trilacunar nodes.

Genera/species: *Cornus* (4–46 spp.), *Mastixia* (13–25 spp.), *Nyssa* (5–10 spp.), *Camptotheca* (1 sp.), *Davidia* (1 sp.), and *Diplopanax* (1 sp.)

Distribution: Widespread; most common in north temperate regions.

U.S./Canadian representative: *Cornus* (14 spp.) and *Nyssa* (3 spp.)

Economic plants and products: Edible fruits from *Cornus mas* (Cornelian-cherry). Wood from several species of *Cornus* and *Nyssa*. Ornamental trees and shrubs: species of *Camptotheca*, *Cornus* (dogwood), *Davidia* (dove tree), and *Nyssa* (tupelo).

Commentary: In this treatment, the Cornaceae comprise *Camptotheca*, *Cornus*, *Davidia*, *Diplopanax*, *Mastixia*, and *Nyssa* (Thorne 1992). The Cornaceae have been variously delimited in the literature, including from four (Ferguson 1977) to sixteen (Ferguson 1966) genera. The various sections (or subgenera) of the polymorphic *Cornus s.l.* have been recognized as segregate genera by some taxonomists (Eyde 1988; Murrell 1993); depending on the authority, several genera (including some of these *Cornus* segregates) are either included within the Cornaceae *s.l.* or separated into small (often monotypic) families (e.g., Curtisiaceae, Davidiaceae, Mastixiaceae). For example, *Nyssa* and *Camptotheca* (plus sometimes also *Davidia*) frequently are placed in the segregate family Nyssaceae, distin-

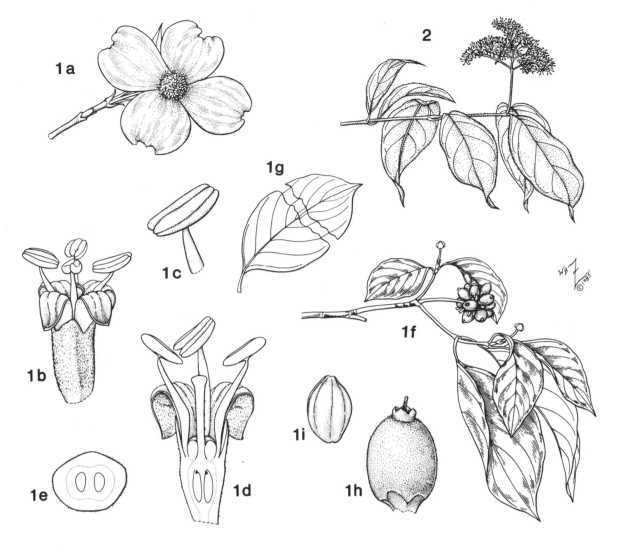

Figure 86. Cornaceae. 1, *Cornus florida*: **a,** flowering branch, ×½; **b,** flower, ×6; **c,** anther, ×12; **d,** longitudinal section of flower, ×8; **e,** cross section of ovary, ×10; **f,** fruiting branch (with mature leaves), ×⅓; **g,** torn leaf connected by unraveled spiral thickenings of the vessel elements ("*Cornus* test"), ×⅓; **h,** drupe, ×2½; **i,** pyrene, ×2½. **2,** *Cornus asperifolia*: flowering branch, ×½.

guished from the rest of the Cornaceae primarily by the five-merous perianth and androecium and the triangular germination valve on the pyrene (Eyde 1966). The close relationship of the Nyssaceae and Cornaceae *s.s.* has been supported by cladistic analysis of morphological (Hufford 1992) and molecular (Xiang et al. 1993) data.

As circumscribed by Thorne (1992), the family is represented by *Cornus* and *Nyssa* in the United States. Of note is that sterile *Cornus* plants are readily recognizable by the arcuate venation of the leaves, and the "*Cornus* test" (see Fig. 86: 1g) may verify the identification of the genus: if the leaf blade is carefully torn in two, the main veins remain connected by delicate threads (unraveled spiral thickenings in the vessel elements).

The small flowers of the family often are clustered into conspicuous inflorescences and attract various insect pollinators (bees, flies, beetles). Nectar is secreted by the disc topping the ovary and surrounding the style (Figs. 86: 1d; 87: c), and the flowers of many species are also fragrant. The species of *Nyssa* in North America (e.g., Fig. 87) have imperfect flowers: staminate plus carpellate (or functionally carpellate), and/ or perfect forms; our representatives of *Cornus* (e.g., Fig. 86) are characterized by perfect flowers. *Cornus* flowers are generally homogamous or occasionally protandrous. The capitate inflorescences of several *Cornus* species, such as *C. florida*, are encircled by large, petaloid, usually white bracts (Fig. 86: 1a). These form a conspicuous pseudanthium, with the bracts functioning as "petals," and the small flowers in the center,

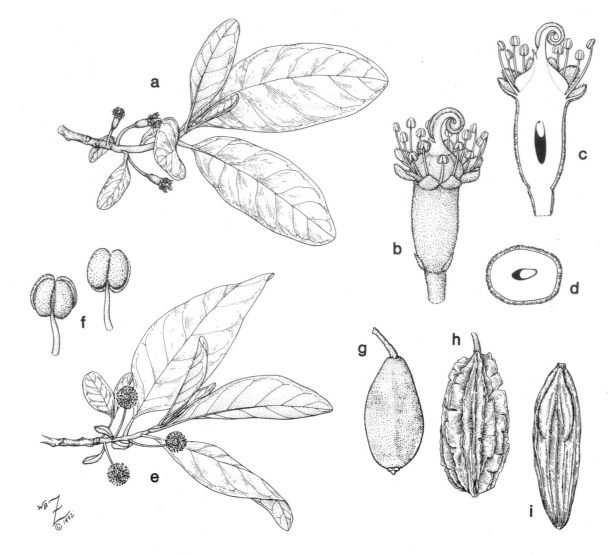

Figure 87. Cornaceae (continued). *Nyssa ogeche*: **a,** carpellate flowering branch, ×½; **b,** carpellate flower, ×3; **c,** longitudinal section of carpellate flower, ×3; **d,** cross section of ovary, ×5; **e,** staminate flowering branch, ×½; **f,** two views of anther, ×12; **g,** drupe, ×⅔; **h,** pyrene, ×2; **i,** pyrene with wings removed showing germination valve, ×2¼.

as "stamens and pistils." Cross- and self-pollination may occur as the insects walk over the surface of the inflorescence, although cross-pollination is somewhat favored by the different filament and style lengths.

The blue-black, purple, or red drupes are globose to ellipsoid and crowned by the persistent disc. The endocarp of the single pyrene is ridged (Figs. 86: 1i; 87: i), and in *Nyssa ogeche* (Fig. 87: h) the stone also has cartilaginous wings. In *Nyssa*, each locule of the pyrene opens apically by a triangular germination valve (Fig. 87: i).

REFERENCES CITED

Eyde, R. H. 1966. The Nyssaceae in the southeastern United States. *J. Arnold Arbor.* 47:117–125.

——. 1988. Comprehending *Cornus*: Puzzles and progress in the systematics of the dogwoods. *Bot. Rev.* (Lancaster) 54:233–351.

Ferguson, I. K. 1966. The Cornaceae in the southeastern United States. *J. Arnold Arbor.* 47:106–116.

——. 1977. Cornaceae. *World Pollen Spore Fl.* 6:1–34.

Hufford, L. 1992. Rosidae and their relationships to other nonmagnoliid dicotyledons: A phylogenetic analysis using morphological and chemical data. *Ann. Missouri Bot. Gard.* 79:218–248.

Murrell, Z. E. 1993. Phylogenetic relationships in *Cornus* (Cornaceae). *Syst. Bot.* 18:469–495.

Thorne, R. F. 1992. Classification and geography of the flowering plants. *Bot. Rev.* (Lancaster) 58:225–348.

Xiang, Q.-Y., D. E. Soltis, D. R. Morgan, and P. S. Soltis. 1993. Phylogenetic relationships of *Cornus* L. sensu lato and putative relatives inferred from *rbc*L sequence data. *Ann. Missouri Bot. Gard.* 80:723–734.

APIACEAE OR UMBELLIFERAE

Carrot or Parsley Family

Including the Araliaceae: Ginseng Family

Annual, biennial, or perennial herbs, or shrubs to trees, aromatic; stems stout, furrowed, with large pith and often hollow internodes. **Leaves** pinnately or palmately compound or decompound, sometimes simple, alternate, often in basal rosettes, sometimes heteromorphic, with petioles broad and sheathing at the base, sometimes stipulate; stipules often modified. **Inflorescence** indeterminate, basically of simple umbels, these often in umbellate, racemose, or paniculate arrangements, sometimes reduced to a head, subtended by an involucre of bracts, terminal. **Flowers** actinomorphic, perfect or sometimes imperfect (then plants polygamodioecious, dioecious, or monoecious), epigynous, small, with a nectariferous disc at ovary apex. **Calyx** absent or of 5 sepals, distinct, small, inconspicuous, often reduced to teeth or a rim around the upper edge of the ovary, persistent. **Corolla** of 5 petals, distinct, apically pointed and inflexed, caducous, white, green, yellow, or pink to purple, valvate. **Androecium** of 5 stamens; filaments distinct; anthers basifixed or dorsifixed, versatile, dehiscing longitudinally, introrse. **Gynoecium** of 1 pistil, 2- or 5-carpellate; ovary inferior, with as many locules as carpels, capped by nectariferous disc at apex; ovules 1 in each locule, anatropous, pendulous, placentation axile; styles 2 or 5 or absent, usually basally connate and adnate to the disc (forming the stylopodium); stigmas not well differentiated, terminal and/or along adaxial surface of each style. **Fruit** a schizocarp composed of 2 1-seeded segments (that remain suspended by the split central column) or a drupe with 5 pyrenes (stones); seeds often united to pericarp; endosperm copious, oily, usually firm; embryo minute.

Family characterization: Herbs to trees that are often aromatic; pithy, furrowed stem that often becomes hollow at the internodes; compound leaves with sheathing petioles; small, numerous, epigynous flowers in umbellate inflorescences; 5-merous perianth and androecium; inconspicuous calyx; epigynous nectariferous disc that is often adnate to the style bases; single pendulous ovule in each locule; and a schizocarp or a drupe as the fruit type. Saponins present. Anatomical feature: tissues with numerous schizogenous secretory canals (Figs. 88: 1g; 89: d) containing ethereal oils, resins, or gums (Crowden et al. 1969).

Genera/species: 460/4,250

Distribution: Widespread; most common in north temperate regions.

Major genera: *Eryngium* (230 spp.), *Schefflera* (200 spp.), *Ferula* (133–172 spp.), *Pimpinella* (150 spp.), *Bupleurum* (70–150 spp.), and *Hydrocotyle* (75–100 spp.)

U.S./Canadian representatives: 94 genera/440 spp.; largest genera: *Eryngium*, *Cymopterus*, and *Angelica*

Economic plants and products: Food and spice plants: *Anethum* (dill), *Anthriscus* (chervil), *Apium* (celery), *Carum* (caraway), *Coriandrum* (coriander), *Daucus* (carrot), *Foeniculum* (fennel), *Levisticum* (lovage), *Pastinaca* (parsnip), *Petroselinum* (parsley), and *Pimpinella* (anise). Several toxic plants (with poisonous resins or alkaloids), including *Cicuta* (water-hemlock) and *Conium* (poison-hemlock). Herbal plants: *Aralia* (wild sarsaparilla) and *Panax* (ginseng). Chinese rice paper from the pith of *Tetrapanax papyrifera*. Ornamental plants (species of 70 genera), including *Aralia*, *Eryngium*, *Fatsia*, *Hedera* (English-ivy), *Schefflera* (umbrella tree), and *Trevesia*.

Commentary: The circumscription of the Apiaceae *s.l.* presented here, combining the Apiaceae *s.s.* and the Araliaceae, follows the previous delineation of Thorne (1973, 1983). Thorne (1992), however, recently dismantled the broadly defined family for reasons unrelated to plant phylogeny (Thorne pers. comm.): although systematists have long recognized a close relationship between the two taxa (Constance 1971), the concept of an expanded Apiaceae has met with considerable opposition. The two taxa are closely linked by vegetative, anatomical, floral, and chemical features (Jurica 1922; Mittal 1961; and see the family characterization below), and nearly all the features of the apiads can be found within the araliads (Rodriguez 1957, 1971). Generally, the apiads have derived expressions of the characters already present in the more diverse araliads.

Table 12 summarizes the major morphological differences traditionally used to distinguish between these two groupings. Basically, the "Araliaceae," somewhat difficult to define, are characterized as tropical, woody plants with stipulate leaves, usually five-carpellate and -locular ovaries, and usually baccate fruits (Philipson 1970). In contrast, the Apiaceae *s.s.* (Apioideae, Saniculoideae, and Hydrocotyloideae) are characterized as temperate, herbaceous plants with sheathing exstipulate leaves, two-carpellate and -locular ovaries, and schizocarps with two one-seeded segments (Graham 1966). However, some araliads are temperate and herbaceous; some apiads are woody, and a few are tropical. In addition, some araliad genera have bicarpellate and biovulate ovaries, and a few, such as *Myodocarpus*, produce a schizocarp with two one-

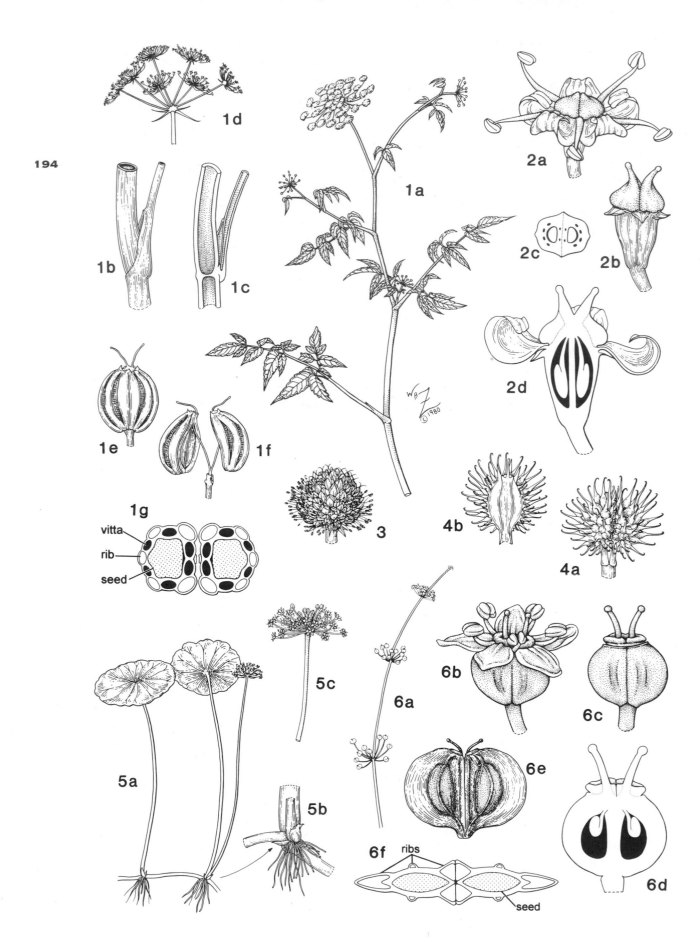

194

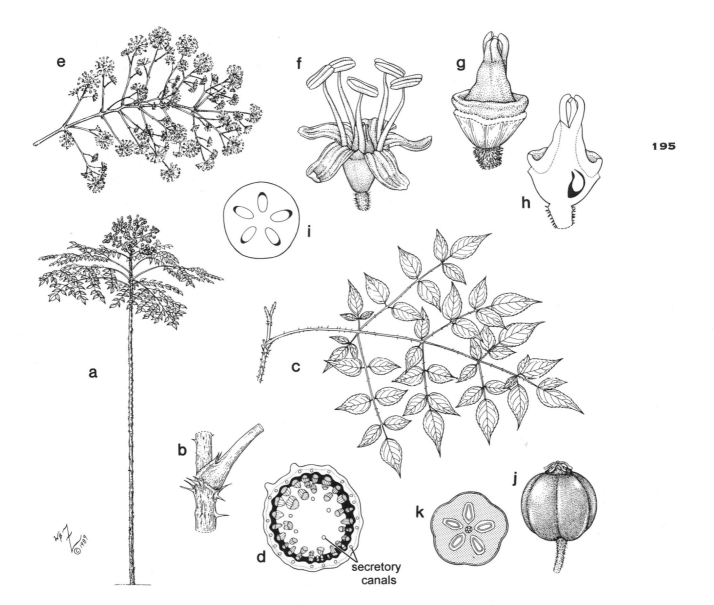

Figure 89 [above]. Apiaceae (continued). *Aralia spinosa*: **a,** habit, ×1/30; **b,** node showing stipules, ×1/3; **c,** leaf, ×1/6; **d,** cross section of petiole showing secretory canals, ×12; **e,** portion of inflorescence, ×1/6; **f,** flower, ×5; **g,** pistil and disc, ×9; **h,** longitudinal section of pistil, ×9; **i,** cross section of ovary ×9; **j,** drupe, ×3; **k,** cross section of drupe, ×3.

Figure 88 [facing page]. Apiaceae. 1, *Cicuta mexicana*: **a,** upper portion of plant, ×1/6; **b,** node showing sheathing leaf base, ×1/2; **c,** longitudinal section of node showing hollow internode, ×1/2; **d,** longitudinal section of inflorescence (compound umbel), ×1/3; **e,** schizocarp before dehiscence, ×6; **f,** dehiscing schizocarp, ×6; **g,** cross section of schizocarp, ×12. **2,** *Oxypolis filiformis*: **a,** flower at anthesis, ×8; **b,** mature gynoecium with disc (stamens fallen off, petals removed), ×9; **c,** cross section of ovary, ×12; **d,** longitudinal section of flower (stamens fallen off), ×12. **3,** *Eryngium yuccifolium*: inflorescence (congested head), ×1 1/4. **4,** *Sanicula canadensis*: **a,** schizocarp starting to dehisce, ×6; **b,** one segment of schizocarp, ×6. **5,** *Hydrocotyle umbellata*: **a,** habit, ×1/2; **b,** node with stipules, ×2; **c,** inflorescence (simple umbel), ×1 1/4. **6,** *Hydrocotyle verticillata*: **a,** inflorescence, ×1/2; **b,** flower at anthesis, ×15; **c,** mature gynoecium with disc (stamens fallen off, petals removed), ×15; **d,** longitudinal section of gynoecium, ×20; **e,** schizocarp (before dehiscence), ×10; **f,** cross section of schizocarp (note lack of vittae), ×20.

TABLE 12. Major morphological differences traditionally used to distinguish the Apiaceae *s.s.* (subfamilies Apioideae, Hydrocotyloideae, and Saniculoideae of the Apiaceae *s.l.*) and the "Araliaceae" (Aralioideae). These taxa are here combined as the Apiaceae *s.l.* (*sensu* Thorne 1973, 1983); see the text for discussion.

CHARACTER	APIOIDEAE, HYDROCOTYLOIDEAE, AND SANICULOIDEAE ("APIACEAE" *S.S.*)	ARALIOIDEAE ("ARALIACEAE")
GENERA/SPECIES	410/3,100	50/1,150
DISTRIBUTION	nearly cosmopolitan, but most diverse in north temperate regions	widespread in tropical to subtropical regions, with a few in temperate areas
HABIT	perennial or annual herbs, occasionally shrubs to trees	trees, shrubs, woody vines, or sometimes perennial herbs
STIPULES	usually lacking (present in Hydrocotyloideae)	usually present (often modified)
INFLORESCENCE	compound or sometimes simple umbels, or occasionally capitate	simple umbels, usually in racemose or paniculate arrangements
CARPEL NUMBER	2	usually 2 to 5
FRUIT TYPE	schizocarp splitting into 2 segments, usually attached apically to persistent, free carpophore vittae usually present (absent in Hydrocotyloideae)	drupe, berry, or rarely a schizocarp with persistent carpophore vittae usually absent

seeded segments having ribs and oil tubes. The subfamily Hydrocotyloideae, usually considered within the Apiaceae *s.s.* or treated as a segregate family (Hydrocotylaceae, as in Thorne 1992), are characterized by stipules and a schizocarp that usually lacks a free carpophore.

Recent cladistic analyses of molecular (Plunkett et al. 1992) and morphological (Judd et al. 1994) data support the original broad circumscription proposed by Thorne (1973, 1983): the traditional Apiaceae *s.s.* probably are polyphyletic, and the Araliaceae, paraphyletic (see Fig. 90, a graphic representation of this hypothesis). Thorne's subsequent (1992) acceptance of three families, the Araliaceae, Hydrocotylaceae, and more restricted Apiaceae *s.s.*, may resolve the polyphyly of the traditional Apiaceae, but it does not solve the problem of the paraphyly of the Araliaceae. A combined Araliaceae-Apiaceae (including the Hydrocotylaceae), however, is clearly monophyletic in Fig. 90.

Important taxonomic features of the Apiaceae *s.l.* include inflorescence morphology, the nature of the epigynous disc, and characters of the fruit. The umbellate inflorescence is a unifying familial feature, and the arrangement of umbels provides generic and specific characters. Although some species of a few genera (*Hydrocotyle*; Fig. 88: 5a) have simple umbels, the inflorescence of most members, such as *Cicuta* (Fig. 88:

1d), consists of simple umbels arranged into compound umbels. The pedicels of the simple umbel are called primary rays. In a compound umbel, each primary ray is terminated by secondary rays (the pedicels of the flowers of each simple or secondary umbel). Subtending bracts often form a whorl at the base of the compound umbel (involucre) and around each simple umbel (involucel). The umbels of some genera, such as *Aralia* (Fig. 89: e), are arranged in panicles or racemes, and in *Eryngium* (Fig. 88: 3) the flowers with reduced pedicels are crowded into dense heads.

The size and shape of the nectariferous tissue that caps the ovary (and surrounds the styles) also vary considerably throughout the family (e.g., Figs. 88: 2b,d; 89: g,h). In the literature, the term "disc" usually refers to this tissue when it is free from the styles; "stylopodium" is used for the structure formed by the adnation of the style bases and associated nectariferous tissue (Rodgers 1950). In the case of the stylopodium, the style bases often are expanded and the apices of the styles project from the top of the structure (Fig. 88: 2b).

Characters of the fruit provide criteria for separating subfamilies, tribes, genera, and, to a lesser extent, species (Jackson 1933). The mature fruit of the subfamily Aralioideae usually is a drupe with five pyrenes (Fig. 89: j,k). The remaining three subfamilies (Hydro-

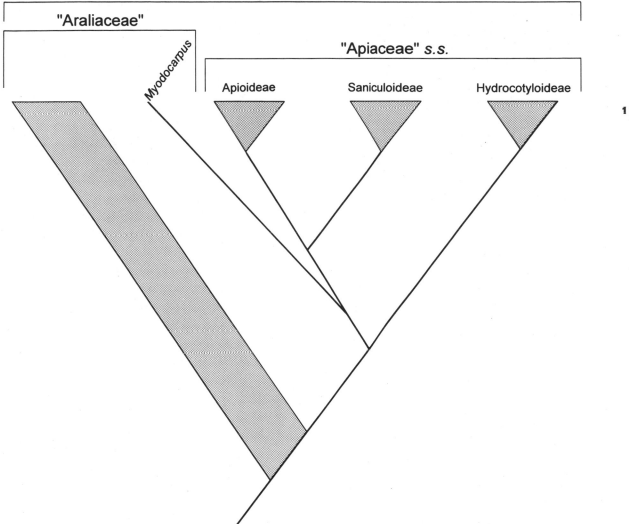

Apiaceae *s.l.*

"Araliaceae"

Myodocarpus

"Apiaceae" *s.s.*

Apioideae   Saniculoideae   Hydrocotyloideae

197

Figure 90. Relationships between the subfamilies of the Apiaceae *s.l.* Shown are the "Araliaceae" and Apiaceae *s.s.*, as traditionally circumscribed: the Araliaceae are polyphyletic, and the Apiaceae *s.s.*, paraphyletic. Cladogram modified and simplified from Judd et al. (1994).

cotyloideae, Saniculoideae, and Apioideae) are characterized by a schizocarp that splits into two one-seeded segments (Fig. 88: 1e,f, 4a,b, 6e); the interior face of each segment, along which the pair are initially joined, is termed the "commissure." The schizocarp varies in shape from terete to flattened (parallel or perpendicular to the plane of the commissure). The wall structure of the segments also is important taxonomically. Each segment often has five projecting longitudinal "primary ribs" or "ridges" corresponding to five vascular bundles (see Fig. 88: 1g, 6f); the ribs sometimes are obscure. Schizogenous oil ducts (vittae), which appear as cavities in cross section, typically occur between the ribs. The fruit is capped by the rim of the persistent calyx and the remains of the style. After dehiscence,

the segments often remain suspended at their apices by a thin, wiry central column (carpophore; Fig. 88: 1f). Wing-like and spine-covered ribs may aid in dispersal (e.g., Fig. 88: 4a,b, 6e).

The small flowers, aggregated into dense conspicuous inflorescences, attract small flies, bees, beetles, moths, and various other pollinators (Bell 1971). The widely open flowers have easily accessible nectar secreted by the disc at the ovary apex. Cross-pollination is promoted by protandry or protogyny.

REFERENCES CITED

Bell, C. R. 1971. Breeding systems and floral biology of the Umbelliferae, or evidence for specialization in unspecialized flowers. In *The biology and chemistry of the Umbelliferae,* ed. V. H. Heywood, pp. 93–107. Bot. J. Linn. Soc. vol. 64, Suppl. 1. Academic Press, London.

Constance, L. 1971. History of the classification of Umbelliferae (Apiaceae). In *The biology and chemistry of the Umbelliferae,* ed. V. H. Heywood, pp. 1–11. Bot. J. Linn. Soc. vol. 64, Suppl. 1. Academic Press, London.

Crowden, R. K., J. B. Harborne, and V. H. Heywood. 1969. Chemosystematics of the Umbelliferae—A general survey. *Phytochemistry* 8:1963–1984.

Graham, S. A. 1966. The genera of Araliaceae in the southeastern United States. *J. Arnold Arbor.* 47:126–136.

Jackson, G. A. 1933. A study of the carpophore of the Umbelliferae. *Amer. J. Bot.* 20:121–144.

Judd, W. S., R. W. Sanders, and M. J. Donoghue. 1994. Angiosperm family pairs: Preliminary cladistic analyses. *Harvard Pap. Bot.* 5:1–51.

Jurica, H. S. 1922. A morphological study of the Umbelliferae. *Bot. Gaz.* 74:292–307.

Mittal, S. P. 1961. Studies in the Umbellales. II. The vegetative anatomy. *J. Indian Bot. Soc.* 40:424–443.

Philipson, W. R. 1970. Constant and variable features of the Araliaceae. In *New research in plant anatomy,* ed. N. K. B. Robson, D. F. Cutler, and M. Gregory, pp. 87–100. Bot. J. Linn. Soc. vol. 63, Suppl. 1. Academic Press, London.

Plunkett, G. M., D. E. Soltis, and P. S. Soltis. 1992. Molecular phylogenetic study of Apiales (Apiaceae, Araliaceae, and Pittosporaceae). *Amer. J. Bot.* 79(b, part 2): 158. [Abstract].

Rodgers, C. L. 1950. The Umbelliferae of North Carolina and their distribution in the Southeast. *J. Elisha Mitchell Sci. Soc.* 66:195–266.

Rodriguez, R. L. 1957. Systematic anatomical studies of *Myrrhidendron* and other woody Umbellales. *Univ. Calif. Publ. Bot.* 29:145–318.

———. 1971. The relationships of the Umbellales. In *The biology and chemistry of the Umbelliferae,* ed. V. H. Heywood, pp. 63–91. Bot. J. Linn. Soc. vol. 64, Suppl. 1. Academic Press, London.

Thorne, R. F. 1973. Inclusion of the Apiaceae (Umbelliferae) in the Araliaceae. *Notes Roy. Bot. Gard. Edinburgh* 32:161–165.

———. 1983. Proposed new realignments in the angiosperms. *Nordic J. Bot.* 3:85–117.

———. 1992. Classification and geography of the flowering plants. *Bot. Rev.* (Lancaster) 58:225–348.

CAPRIFOLIACEAE

Honeysuckle Family

Shrubs, woody vines, or occasionally perennial herbs. **Leaves** simple, entire to serrate, opposite (paired leaves sometimes united at base), decussate, persistent or deciduous, usually exstipulate. **Inflorescence** determinate, cymose and appearing racemose or spicate, or flowers paired (then joined at the ovaries) or solitary, terminal or axillary. **Flowers** zygomorphic, perfect, epigynous, showy, with floral tube distinctly constricted below the calyx lobes. **Calyx** synsepalous with 4 or usually 5 lobes or teeth, campanulate to urceolate,

persistent, usually imbricate. **Corolla** sympetalous with 4 or usually 5 lobes, campanulate, tubular, funnelform, to often bilabiate, sometimes basally spurred/gibbose, with glandular hairs on inner surface, variously colored, usually imbricate. **Androecium** of 4 or usually 5 stamens, epipetalous; filaments distinct; anthers dorsifixed, versatile, dehiscing longitudinally, introrse. **Gynoecium** of 1 pistil, 2- to 5-carpellate; ovary inferior, basically with as many locules as carpels (but several carpels often aborting) or sometimes 1-locular at apex (due to incomplete septa); ovules 1 to numerous in each locule, anatropous, pendulous, placentation axile or axile at base and parietal at apex of ovary (when septa incomplete); style 1, elongate; stigma capitate. **Fruit** a berry, drupe, capsule, or achene; endosperm copious, oily, fleshy; embryo straight, small.

Family characterization: Shrubs or woody vines with simple, opposite leaves; sympetalous, zygomorphic flowers conspicuously constricted below the calyx lobes; showy, tubular (campanulate, funnelform, to bilabiate) corolla; dorsifixed, versatile anthers; and 2- to 5-carpellate, inferior ovary with several carpels often aborting. Iridoid compounds and saponins present. Tissues with calcium oxalate crystals. Anatomical feature: unitegmic and tenuinucellate ovules.

Genera/species: 12/450

Distribution: Primarily in the temperate to subtropical areas of the Northern Hemisphere; most diverse in eastern North America and eastern Asia.

Major genera: *Lonicera* (180–200 spp.) and *Abelia* (30 spp.)

U.S./Canadian representatives: *Lonicera* (29 spp.), *Symphoricarpos* (13 spp.), *Diervilla* (3 spp.), *Triosteum* (3 spp.), *Linnaea* (1 sp.), and *Weigelia* (1 sp.)

Economic plants and products: *Lonicera japonica,* naturalized in the eastern United States, a serious invasive weed. Ornamental plants (species of about 10 genera), including *Abelia, Diervilla* (bush-honeysuckle), *Kolkwitzia* (beautybush), *Linnaea* (twinflower), *Lonicera* (honeysuckle), and *Symphoricarpos* (coral-berry, snowberry).

Commentary: As suggested in a series of papers by Donoghue (1983; Donoghue et al. 1992) and subsequently adopted by Thorne (1983, 1992), the Caprifoliaceae are here rather strictly defined, including only the ten core genera (plus a few segregates) and excluding *Sambucus* and *Viburnum* (which have been transferred to the Adoxaceae). Table 13 and Figs. 91 and 92 compare the main morphological differences between the Caprifoliaceae *s.s.* and the Adoxaceae *s.l.* thus de-

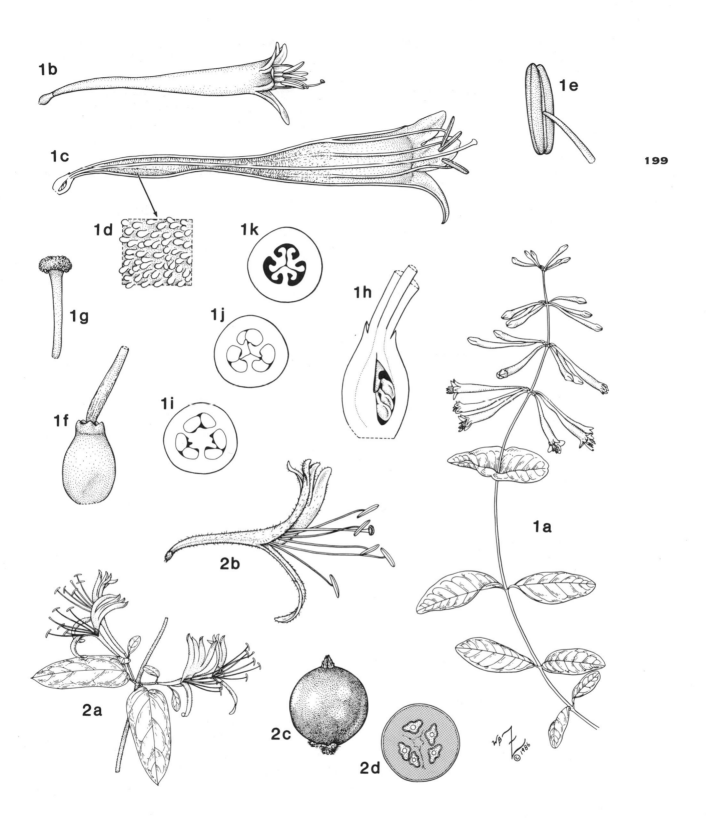

Figure 91. Caprifoliaceae. 1, *Lonicera sempervirens*: **a,** habit, ×³⁄₈; **b,** flower, ×1¼; **c,** longitudinal section of flower, ×2¼; **d,** nectariferous hairs on inside surface of saccate corolla tube base, ×25; **e,** anther, ×5; **f,** ovary, ×9; **g,** stigma, ×9; **h,** longitudinal section of ovary, ×10; **i,** cross section of ovary near base, ×12; **j,** cross section of ovary near middle, ×12; **k,** cross section of ovary near apex, ×12. **2,** *Lonicera japonica*: **a,** habit, ×½; **b,** flower, ×1; **c,** berry, ×3½; **d,** cross section of berry, ×3½.

TABLE 13. Major morphological differences between the Adoxaceae (including *Sambucus* and *Viburnum*) and the Caprifoliaceae *s.s.*

CHARACTER	CAPRIFOLIACEAE S.S.	ADOXACEAE (INCL. *VIBURNUM* AND *SAMBUCUS*)
GENERA/SPECIES	12/450	5/243
DISTRIBUTION	primarily northern hemisphere	primarily northern hemisphere, with a few representatives in Australia and montane South America
HABIT	usually shrubs or woody vines	herbs, shrubs, or small trees
LEAVES	simple usually exstipulate	simple to compound often stipulate (stipules reduced)
INFLORESCENCE	various (cymose, racemose, spicate, reduced to 1 or 2 flowers), but not umbellate	umbellate
FLOWER	zygomorphic large	actinomorphic small
NECTARIES	simple glandular hairs lining corolla tube near base	disc topping the ovary, or multicellular hairs at corolla base
CALYX LOBES	moderately sized to small	very small
COROLLA	tubular: campanulate, funnelform, or bilabiate variously colored	rotate white, yellowish, or pinkish
ANTHERS	introrse	introrse or extrorse
POLLEN	relatively large often spinose	relatively small reticulate
OVARY POSITION	fully inferior	partially inferior
STYLE	elongate	short to absent
STIGMA	capitate	lobed
FRUIT TYPE	berry, or sometimes capsule, achene, or dry drupe	fleshy drupe with 1 or 3 to 5 stones

fined. The Caprifoliaceae *s.s.* have been divided into three or four tribes, primarily based on the number of carpels (and fertile locules) and the fruit type.

Taxonomists have long grappled with the problem of circumscribing the family and ascertaining its phylogenetic position (see the summaries in Ferguson 1966 and Donoghue 1983). The group has been allied with such families as the Rubiaceae and Cornaceae. The placement of *Sambucus*, *Viburnum*, and *Adoxa*—all distinct in many characters—has been the subject of much speculation. *Viburnum*, usually placed in the Caprifoliaceae *s.l.*, has occupied a pivotal position, seemingly linking *Sambucus* (and *Adoxa*) with the core genera of the Caprifoliaceae. The genus *Adoxa*, also sometimes included in the Caprifoliaceae, generally is segregated as the Adoxaceae, a little-known family comprising one to a few poorly known herbaceous species. The placement of *Sambucus* has been particularly perplexing, due to unusual characters such as compound leaves and extrorse anthers. This genus,

Figure 92. Adoxaceae. 1, *Viburnum nudum*: **a,** flowering branch, ×½; **b,** flower, ×8; **c,** portion of corolla with epipetalous stamen, ×11; **d,** two views of anther, ×12; **e,** pistil, ×15; **f,** cross section of ovary, ×15; **g,** longitudinal section of flower, ×11. **2,** *Viburnum rufidulum*: **a,** drupe, ×2¼; **b,** cross section of drupe, ×2¼. **3,** *Sambucus canadensis*: **a,** flowering branch, ×⅓; **b,** leaf, ×⅜; **c,** flower, ×6; **d,** portion of corolla with epipetalous stamen, ×9; **e,** pistil, ×12; **f,** longitudinal section of pistil, ×12; **g,** cross section of ovary, ×12; **h,** drupe, ×4; **i,** cross section of drupe, ×4. See Table 13 for family characteristics.

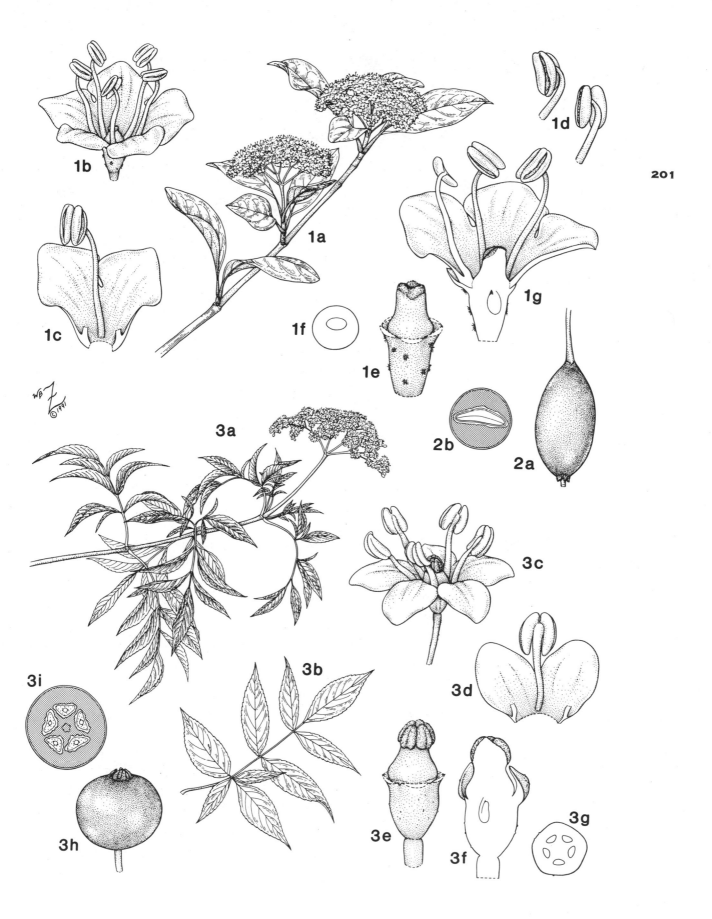

201

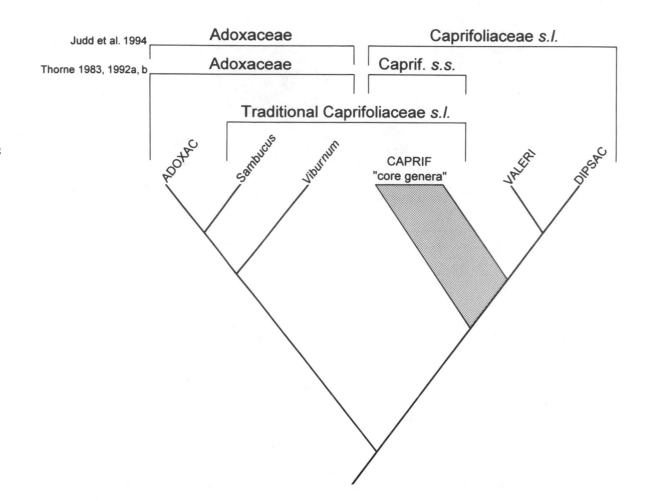

Figure 93. Relationships between groups of the Adoxaceae-Caprifoliaceae complex. Cladogram representing the results of analyses (Judd et al. 1994) of the Caprifoliaceae-Valerianaceae-Dipsacaceae-Adoxaceae complex, showing the primary division between the *Viburnum-Sambucus*-Adoxaceae clade and the Caprifoliaceae *s.s.*-Valerianaceae-Dipsacaceae clade. Simplified from Judd et al. (1994). Abbreviations: ADOXAC (Adoxaceae), CAPRIF (Caprifoliaceae), DIPSAC (Dipsacaceae), VALERI (Valerianaceae).

generally considered closely related to *Viburnum* but not to the rest of the Caprifoliaceae *s.l.*, sometimes has been segregated into its own monotypic family (Sambucaceae).

The preliminary cladistic study by Donoghue (1983) suggested the transfer of *Sambucus* and *Viburnum* to the Adoxaceae. This hypothesis has been supported by more detailed morphological (Judd et al. 1994) and molecular (Donoghue et al. 1992; Olmstead et al. 1993) analyses. Figure 93 summarizes the conclusion presented in Judd et al. (1994): the Caprifoliaceae *s.l.* indeed are a paraphyletic group, and *Viburnum* is more closely related to *Sambucus* and the Adoxaceae (*Adoxa*) than it is to the Caprifoliaceae *s.s.* However, their results also demonstrate that the Caprifoliaceae *s.s.* as defined by Thorne (1992) also are paraphyletic, and this core group probably should be merged with the Valerianaceae and Dipsacaceae (resulting in a very broadly defined Caprifoliaceae). This is supported by

strong evidence concerning floral anatomy, particularly the abortion of ovary locules (carpels) during development (see the discussion below).

The floral symmetry of the Caprifoliaceae basically is zygomorphic, a condition even noticeable in the campanulate corollas of *Symphoricarpos* and *Linnaea*. It is most pronounced in genera with a bilabiate corolla, such as *Lonicera* (Fig. 91: 1b, 2b). The long and slender tube often is basally saccate, and the two-lipped limb comprises a four-lobed upper lip with the fifth lobe as the lower lip.

The Caprifoliaceae are characterized by unusual carpel and ovule development in which several locules (carpels) often abort (Wilkinson 1949). The number of carpels (and fertile locules) composing the compound ovary varies considerably in the family. In *Leycesteria*, for example, all five carpels are fertile. The four-carpellate ovary of *Symphoricarpos* usually has two uniovulate fertile locules plus two several-ovulate sterile (abortive) locules. Three locules of *Triosteum* are uniovulate and the fourth is empty. Many genera in the family are characterized by three-carpellate ovaries, such as *Abelia* (only one locule with ovules) and *Linnaea* (one uniovulate fertile locule plus two several-ovulate sterile locules). The two carpels of the *Diervilla* gynoecium are both fertile with numerous ovules. Species of *Lonicera* (Fig. 91: 1i) usually are two- or three-carpellate with a few to several ovules per locule, and, as occurs occasionally in the family, the ovary is apically unilocular due to incomplete septa (placentation parietal at apex; Fig. 91: 1j,k). Also, ovaries of paired flowers are fused in various species of this genus.

The white or conspicuously colored flowers attract insects and hummingbirds as pollinators. The lower part of the corolla tube has nectariferous hairs inside and is sometimes gibbose to spurred (Fig. 91: 1c,d). Flowers have been variously described as protogynous and protandrous, and the corollas of a few species (e.g., in *Lonicera* and *Diervilla*) change color as they pass from one sexual stage to the other. In some taxa (e.g., *Linnaea*, *Lonicera*) the prolongation of the mature stigma on the long style far beyond the anthers helps to prevent self-pollination.

REFERENCES CITED

Donoghue, M. J. 1983. The phylogenetic relationships of *Viburnum*. In *Advances in cladistics*, vol. 2, ed. N. I. Platnick and V. A. Funk, pp. 143–166. Columbia University Press, New York.

——, R. G. Olmstead, J. F. Smith, and J. D. Palmer. 1992. Phylogenetic relationships of Dipsacales based on *rbc*L sequences. *Ann. Missouri Bot. Gard.* 79:333–345.

Ferguson, I. K. 1966. The genera of Caprifoliaceae in the southeastern United States. *J. Arnold Arbor.* 47:33–59.

Judd, W. S., R. W. Sanders, and M. J. Donoghue. 1994. Angiosperm family pairs: Preliminary cladistic analyses. *Harvard Pap. Bot.* 5:1–51.

Olmstead, R. G., B. Bremer, K. M. Scott, and J. D. Palmer. 1993. A parsimony analysis of the Asteridae sensu lato based on *rbc*L sequences. *Ann. Missouri Bot. Gard.* 80:700–722.

Thorne, R. F. 1983. Proposed new realignments in the angiosperms. *Nordic J. Bot.* 3:85–117.

——. 1992. Classification and geography of the flowering plants. *Bot. Rev.* (Lancaster) 58:225–348.

Wilkinson, A. M. 1949. Floral anatomy and morphology of *Triosteum* and of the Caprifoliaceae in general. *Amer. J. Bot.* 36:481–489.

ASTERACEAE OR COMPOSITAE
Aster, Composite, or Sunflower Family

Annual to perennial herbs or sometimes shrubs or trees, with resinous or milky sap, often with taproots or tubers. **Leaves** simple or pinnately or palmately lobed (sometimes deeply so), alternate or sometimes opposite, in basal rosettes and/or cauline, petiolate or sessile, often decurrent, exstipulate. **Inflorescence** indeterminate, capitate, with heads often in paniculate, racemose, cymose, or corymbose arrangements, with each head subtended by an involucre of bracts (phyllaries), terminal or axillary. **Flowers** actinomorphic or zygomorphic, perfect or imperfect (then plants monoecious or sometimes dioecious), epigynous, small, sessile, sometimes subtended by bracts (pales). **Calyx** absent or represented by a pappus of hairs, bristles, scales, or a ring, persistent. **Corolla** sympetalous with 5 lobes, usually tubular (disc flower) or ligulate (ray flower), variously colored, valvate. **Androecium** of 5 stamens, epipetalous; filaments distinct; anthers basifixed, connate (syngenesious) and forming a cylinder around the style, often with apical or basal appendages, dehiscing longitudinally, introrse. **Gynoecium** of 1 pistil, 2-carpellate; ovary inferior, 1-locular; ovule solitary, anatropous, ascending, placentation basal; style bifid (undivided in staminate flowers), with pollen-collecting hairs on outside of branches; stigmas with very restricted receptive surface (often along 2 marginal lines of each style branch). **Fruit** an achene, compressed, crowned by persistent pappus, sometimes winged or spiny; endosperm absent; embryo large, straight.

Family characterization: Herbs or shrubs, with leaves often in basal rosettes; inflorescence a head subtended by an involucre of bracts (phyllaries); small, epigynous flowers (florets); tubular or strap-shaped sympetalous corolla; reduced calyx (pappus); 5 syngenesious, appendaged anthers; bicarpellate ovary with a single, basal ovule; bifid styles with pollen-collecting hairs and restricted stigmatic surfaces; achene with persistent pappus; and nonendospermous seeds. Many with inulins (unusual storage polysaccharides) in the roots and tubers (Hegnauer 1977). Anatomical feature: unitegmic and tenuinucellate ovules.

Genera/species: 1,160/19,085; one of the largest angiosperm families.

Distribution: Cosmopolitan; many especially well adapted to temperate, montane, or dry regions.

Major genera: *Senecio* (1,500 spp.), *Hieracium* (1,000 spp.), *Vernonia* (1,000 spp.), *Cousinia* (600 spp.), *Centaurea* (500–600 spp.), *Helichrysum* (500 spp.), *Artemisia* (400 spp.), *Baccharis* (400 spp.), *Saussurea* (300+ spp.), *Mikania* (300 spp.), *Cirsium* (250–300 spp.), *Aster* (250 spp.), and *Jurinea* (250 spp.); largest genera/species estimates from various listings in many detailed papers

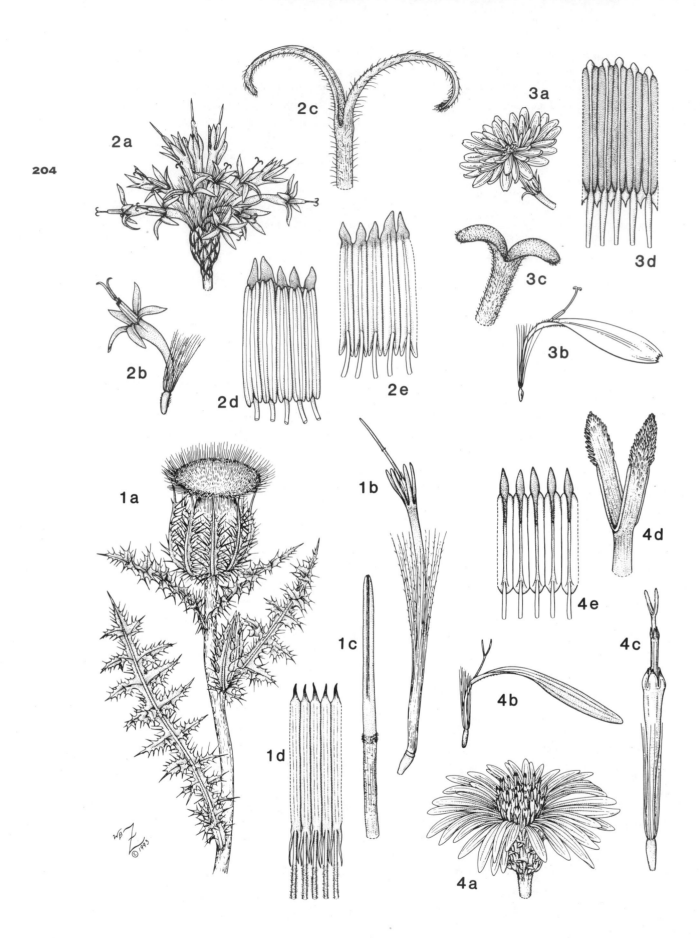

(on tribes) in symposium volumes edited by Heywood et al. (1977).

U.S./Canadian representatives: 346 genera/2,687 spp.; largest genera: *Erigeron, Aster, Senecio, Cirsium, Solidago, Eupatorium,* and *Artemisia*

Economic plants and products: Food plants: *Carthamus* (safflower oil, dye), *Cichorium* (chicory, endive), *Cynara* (artichoke), *Helianthus* (Jerusalem-artichoke, sunflower seeds, oil), and *Lactuca* (lettuce). Pyrethrins (ester-based contact insecticides) from *Chrysanthemum cinerariifolium*. Medicinal plants (contain resins and oils): *Anthemis* (chamomile), *Artemisia* (wormwood), and *Tussilago* (colt's-foot). Many weedy species, including *Ambrosia* (ragweed), *Bidens* (beggarticks), *Pyrrhopappus* (false dandelion), and *Taraxacum* (dandelion). Ornamental plants (species of more than 200 genera), including *Aster, Calendula, Chrysanthemum, Dahlia, Helianthus* (sunflower), *Senecio* (Mexican flame vine), *Tagetes* (French marigold), *Wedelia,* and *Zinnia*.

Commentary: The Asteraceae have been divided historically into two subfamilies, the Cichorioideae (Lactucoideae) and Asteroideae, primarily based on the types of florets (defined below) and the presence of laticifers (Heywood et al. 1977). The family has been customarily further divided into 13 or 14 tribes: traditionally, the Cichorioideae (ligulate heads, latex) comprised only the tribe Cichorieae, with the remaining tribes placed in the Asteroideae (discoid or radiate heads; e.g., Cronquist 1955, 1977). Now at least 17 tribes are recognized (Thorne 1992), with several transferred from a redefined Asteroideae (radiate heads, resin canals, robust stigma with marginal stigmatic lines) to an expanded Cichorioideae (discoid or ligulate heads, laticifers or resin canals, slender style branches with stigmatic inner surface; Carlquist 1976; Wagenitz 1976). In addition, a new third subfamily, the Barnadesioideae, characterized by unique axillary spines and long, unicellular hairs, has been proposed (Bremer 1987; Bremer and Jansen 1992).

Table 14 lists some of the important morphological characters (illustrated in Figs. 94, 95, and 96) often used to distinguish major tribal groupings. Important features include the style morphology and the development of the anther appendages (Bremer 1987). However, although some tribes probably are monophyletic, much of the intrafamilial classification of the Asteraceae currently is in flux as the result of recent analyses of molecular and morphological data (see the summary in Bremer et al. 1992). For example, although the Asteroideae and Barnadesioideae are monophyletic, the Cichorioideae may be paraphyletic (Karis et al. 1992). Much further critical study is especially needed to better evaluate tribal relationships (e.g., Kim et al. 1992).

The unifying feature of this complex and diverse family is the inflorescence consisting of a head (capitulum) with many tiny flowers (florets) crowded onto a receptacle (e.g., Fig. 96: 1c; Burtt 1977; Cronquist 1980). Each head is subtended by numerous imbricate bracts (phyllaries), which protect the buds and also close over the mature flower head in cold or wet weather. The receptacle may vary in shape from conic to flattened, and in texture, from smooth to pitted at flower and fruit insertion. The three basic types of florets on the heads are classified according to corolla shape. A zygomorphic corolla with a short tube and a strap-shaped, apically toothed limb characterizes a ligulate or ray floret. Often these two terms are used interchangeably, but some authors restrict the term "ligulate" to perfect florets; then the strap-like limb is typically 5-toothed (Fig. 94: 3b). "Ray" floret then refers to flowers lacking an androecium and/or style (therefore, sterile or carpellate); the limb of this corolla generally has two or three terminal points (Fig. 95: 2b). The actinomorphic disc floret (e.g., Fig. 95: 2c) has a conspicuous corolla tube and a short limb with five lobes. The bilabiate floret, the third type, is not common in wild composites in the United States and is restricted primarily to the largely ornamental tribe Mutisieae. In many tribes, each floret is subtended by a chaffy bract (pale; Fig. 95: 3c).

The homogamous head is the simplest case of floret arrangement, in which all the florets are the same type

Figure 94. Asteraceae. 1, *Cirsium horridulum*: **a,** upper portion of plant with inflorescence (discoid head), ×½; **b,** disc floret, ×2; **c,** apex of style (branches fused; receptive on inner surface), ×9; **d,** expanded androecium (abaxial surface), ×11. **2,** *Vernonia gigantea*: **a,** inflorescence (discoid head), ×2½; **b,** disc floret, ×3¾; **c,** apex of style (branches receptive on inner surface near base), ×20; **d,** expanded androecium (adaxial surface), ×12; **e,** expanded androecium (abaxial surface), ×12. **3,** *Pyrrhopappus carolinianus*: **a,** inflorescence (ligulate head), ×¾; **b,** ligulate floret, ×2⅓; **c,** apex of style, ×30; **d,** expanded androecium (abaxial surface), ×15. **4,** *Aster carolinianus*: **a,** inflorescence (radiate head), ×2; **b,** ray floret, ×3½; **c,** disc floret, ×5; **d,** apex of style from disc floret (note marginal stigmatic lines and sterile hairy appendages), ×23; **e,** expanded androecium (abaxial surface), ×15. See Table 14 for descriptions of these general characters (representing features of various tribes).

TABLE 14. Major morphological characters used to distinguish the major tribes of the Asteraceae. General characters from Solbrig (1963), Wagenitz (1976), and Bremer (1987); those for the Barnadesieae, from Cabrera (1977), and for the Liabeae,

CHARACTER	BARNADE- SIEAE	MUTISIEAE	CARDUEAE (CYNAREAE)	VERNONIEAE	LIABEAE	LACTUCEAE (CICHORIEAE)	ARCTOTEAE
GENERA/ SPECIES	9/95	81/800	80/2,610	70/1,500	15/155–160	70/2,300	13/180
REPRESEN- TATIVE GENERA	Barnadesia, Chuquiraga, Dasyphyllum	Chaptalia, Gerbera, Gochnatia, Mutisia	Carduus, Centaurea, Cirsium, Cynara	Elephantopus, Erlangea, Stokesia, Vernonia	Liabum, Munnozia	Cichorium, Crepis, Hieracium, Lactuca, Pyrrhopappus, Taraxacum	Arctotis, Berkheya
DISTRIBU- TION	South America	mainly tropical to subtropical America	cosmopolitan; es- pecially diverse in Mediterra- nean region	mostly tropical America	tropical America	cosmopolitan; es- pecially diverse in Mediterra- nean region	mainly Africa
SUBFAMILY	Barnedesioideae	Lactucoideae	Lactucoideae	Lactucoideae	Lactucoideae	Lactucoideae	Lactucoideae
HABIT	shrubs, trees, or sometimes herbs; glabrous to variously pubescent; with axillary spines	shrubs, or some- times herbs; pubescent	robust herbs; spiny, woolly	herbs, shrubs, sometimes trees; glabrous to pubescent	herbs, shrubs, or sometimes trees, often with milky sap; pubescent	usually herbs with milky sap; glabrous to pubescent	usually herbs; woolly or glabrous, some- times spiny
LEAVES	alternate; entire; often spine- tipped	alternate or basal; entire to pin- nately divided	alternate; usually pinnately di- vided; spiny	usually alternate; entire or toothed	opposite or basal; entire, toothed, or lobed	alternate or basal; entire to pin- nately divided	alternate or basal; lobed or pin- nately divided
INVOLUCRAL BRACTS	several series scarious to ± her- baceous	several series herbaceous	several series herbaceous, ± membranous margin, sharp- pointed	several series herbaceous	3 to several series ± herbaceous	1 or several series membranous or herbaceous	several series ± scarious, often sharp-pointed
RECEP- TACLE	usually pubes- cent/ bristly	naked	bristly or naked	usually naked	usually naked	usually naked	naked or bristly
FLORET TYPE(S)	disc (deeply lobed), or disc + bilabiate (4+1 lobes)	bilabiate (2+3 lobes), or sometimes bilabiate + ray	usually disc (deeply lobed)	disc (deeply lobed)	disc (deeply lobed), or ray + disc	ligulate	usually disc (deeply lobed) + ray
PAPPUS	scales/bristles with long hairs	bristles or capil- lary hairs	bristles or capil- lary hairs	bristles	usually capillary bristles	usually capillary bristles	absent, short cup, or scales
ANTHER BASE	blunt to long- tailed	long-tailed	long-tailed	pointed	pointed	pointed to short- tailed	pointed to short- tailed
STYLE BRANCH AND STIGMA MORPHOL- OGY	short, blunt; stig- matic on inner surface	short, blunt, tipped with brush-like hairs; stigmatic on inner sur- face	short (often fused) with ring of hairs below bifurca- tion; stigmatic on inner sur- face	long, slender, tapering; stig- matic on inner surface near base	short to long, blunt; stigmatic on inner sur- face	slender, cylindri- cal; stigmatic on inner sur- face	linear to oblong, blunt, with ring of hairs below bifurcation; stigmatic on in- ner surface

from Robinson (1983). As numerous studies further reevaluate the phylogeny of the Asteraceae, the circumscriptions of various tribes are changing.

ASTEREAE	ANTHEMIDAE	INULEAE (INCL. GNAPHALIEAE)	SENECIONEAE	CALENDULEAE	EUPATORIEAE	HELIANTHEAE (INCL. HELENIEAE, COREOPSIDEAE, TAGETEAE, AMBROSIEAE)
135/2,500	107/1,650	185/2,050	100/2,000	7/110	60/2,000	226/2,405
Aster, Conyza, Erigeron, Haplopappus, Olearia, Solidago	*Achillea, Anthemis, Artemisia, Chrysanthemum*	*Antennaria, Gnaphalium, Inula, Pluchea*	*Arnica, Erechtites, Senecio*	*Calendula, Osteospermum*	*Carphephorus, Eupatorium, Liatris, Mikania*	*Ambrosia, Berlandiera, Bidens, Coreopsis, Cosmos, Dahlia, Helianthus, Rudbeckia, Wedelia, Zinnia*
cosmopolitan; especially diverse in temperate and montane regions	mainly Old World, especially Mediterranean region	cosmopolitan; especially diverse in South Africa and Australia	cosmopolitan	mainly Africa and Mediterranean region	cosmopolitan; mostly tropical America	mainly America
Asteroideae	Asteroideae	Asteroideae	Asteroideae	Asteroideae	Asteroideae	Asteroideae
herbs or shrubs, often resinous; glabrous to pubescent	herbs or shrubs, aromatic; woolly or hispid	herbs or shrubs; woolly or glandular	herbs, shrubs, or trees; with soft hairs	herbs or shrubs; glabrous to variously hairy	usually herbs or shrubs; often pubescent	herbs or shrubs; usually scabrous or hirsute
alternate; usually entire or toothed	alternate; pinnately divided	usually alternate; entire, or sometimes toothed-lobed	usually alternate or basal; entire to pinnately divided	usually alternate; entire, toothed, or lobed	usually opposite; entire to pinnately divided	usually opposite; entire to pinnately divided
several series ± herbaceous	several series membranous or herbaceous ± membranous margin	several series white or scarious, often hairy	1 or occasionally 2 series herbaceous or membranous	1 to 3 series herbaceous ± membranous margin	2 to several series ± herbaceous	2 to several series herbaceous
naked	naked or chaffy	naked or chaffy	naked	usually naked	usually naked	naked or chaffy
usually disc + ray, or disc (lobes short)	usually disc + ray, or disc (lobes short)	usually disc (lobes short)	disc (lobes short), or disc + ray	disc (lobes short) + ray	disc (lobes short)	usually disc + ray, or disc (lobes short)
usually bristles	absent or short cup	usually capillary hairs	capillary bristles	none	bristles	scales, awns, bristles, or absent
usually blunt	blunt to pointed	long-tailed	blunt to pointed	blunt to short-tailed	blunt	blunt to pointed
flat, tipped with sterile hairy appendage; marginal stigmatic lines	blunt, flat, with short hairs at tips; marginal stigmatic lines	usually blunt, flat; marginal stigmatic lines	flat, blunt, tipped with hairs; marginal stigmatic lines (not meeting at base)	usually flat, blunt; marginal stigmatic lines	semicylindrical, ± clavate, blunt, tipped with long sterile appendage; marginal stigmatic lines only near base	± blunt, ± appendaged; marginal stigmatic lines

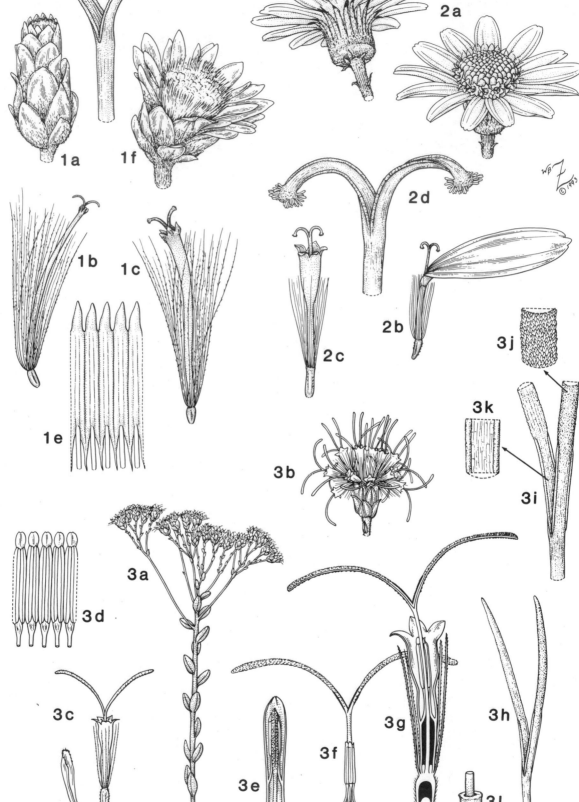

and perfect: a ligulate head (Fig. 94: 3a) or a discoid head (Fig. 94: 2a). More often, the heads are heterogamous with ray florets along the head margin and disc florets in the center, forming a radiate head (Fig. 95: 2a). In this case, the disc florets are either perfect or functionally staminate; the ray florets, carpellate or sterile. The heads of tiny flowers are biologically analogous to a single flower (a "pseudanthium," as in the Euphorbiaceae). This is especially apparent in the radiate head, comprised of a "calyx" (protective involucre of bracts), "corolla" (ray florets around the head margin), and "pistils and stamens" (disc florets in the center).

The aggregated tiny flowers form conspicuous heads, and pollinators (usually Lepidoptera, flies, or beetles) may visit many flowers in a short time. The florets on a head develop from the outside toward the center (i.e., peripheral florets open first; as in Figs. 95: 2a; 96: 1c). A ring-shaped nectary (Fig. 95: 3l) at the style base secretes nectar, which accumulates and rises to the upper part of the corolla. The appendaged anthers (e.g., Fig. 94: 1d) probably protect the nectar from dilution by rain.

The Asteraceae have an unusual pollen presentation mechanism (Leins and Erbar 1990; see also the discussion under Campanulaceae). Before anthesis, the syngenesious introrse anthers, which form a cylinder around the style, dehisce and fill the tube with pollen. At this stage, the unexpanded style is very short. As the style elongates, the collecting hairs on the back of the unexpanded style arms act as a bottle brush or plunger (Fig. 95: 3e) and push the pollen through the cylinder, thereby exposing the pollen to pollinators. Much later, the style arms separate and expose the receptive stigmatic surface (Fig. 95: 3f). A few of the Asteraceae, such as *Ambrosia* (Fig. 96: 2), are completely anemophilous. The carpellate heads (Fig. 96: 2e,f) are composed of florets with reduced corollas and large, prominently exposed stigmas. The pendulous staminate heads easily disseminate airborne pollen from the distinct stamens (Fig. 96: 2b–d).

Some composites have effective means of animal dispersal, with spines, hooks, or hairs on the pappus (*Bidens*; Fig. 96: 5), phyllaries (*Xanthium*; Fig. 96: 4), or fruits (*Berlandiera*; Fig. 96: 1d). In others, such as *Pyrrhopappus* (Fig. 96: 3a), a pappus of fine hairs functions as a wind-borne parachute. Persistent chaffy bracts (pales) on a fruiting receptacle forming a shaker-type structure, or hygroscopic phyllaries, may also aid in dispersal.

REFERENCES CITED

Bremer, K. 1987. Tribal interrelationships of the Asteraceae. *Cladistics* 3:210–253.

—— and R. K. Jansen. 1992. A new subfamily of the Asteraceae. *Ann. Missouri Bot. Gard.* 79:414–415.

——, ——, P. O. Karis, M. Källersjö, S. C. Keely, et al. 1992. A review of the phylogeny and classification of the Asteraceae. *Nordic J. Bot.* 141–148.

Burtt, B. L. 1977. Aspects of diversification of the capitulum. In *The biology and chemistry of the Compositae*, vol. 1, ed. V. H. Heywood, J. B. Harborne, and B. L. Turner, pp. 41–59. Academic Press, London.

Cabrera, A. L. 1977. Mutisieae—Systematic review. In *The biology and chemistry of the Compositae*, vol. 2, ed. V. H. Heywood, J. B. Harborne, and B. L. Turner, pp. 1039–1066. Academic Press, London.

Carlquist, S. 1976. Tribal interrelationships and phylogeny of the Asteraceae. *Aliso* 8:465–492.

Cronquist, A. 1955. Phylogeny and taxonomy of the Compositae. *Amer. Midl. Naturalist* 53:478–511.

——. 1977. The Compositae revisited. *Brittonia* 19:137–153.

——. 1980. *Vascular flora of the southeastern United States*. Vol. 1, *Asteraceae*, pp. 3–4. University of North Carolina Press, Chapel Hill.

Hegnauer, R. 1977. The chemistry of the Compositae. In *The biology and chemistry of the Compositae*, vol. 1, ed. V. H. Heywood, J. B. Harborne, and B. L. Turner, pp. 283–335. Academic Press, London.

Heywood, V. H., J. B. Harborne, and B. L. Turner. 1977. An overture to the Compositae. In *The biology and chemistry of*

Figure 95. Asteraceae (continued). 1, *Gnaphalium obtusifolium*: **a,** inflorescence (discoid head), ×6; **b,** carpellate discoid floret (floret type composing most of head), ×14; **c,** perfect discoid floret (a few of this floret type occurring in center of head), ×14; **d,** apex of style from perfect floret (note marginal stigmatic lines), ×65; **e,** expanded androecium (abaxial surface), ×35; **f,** infructescence, ×6. **2,** *Senecio glabellus*: **a,** two views of inflorescence (radiate head), ×2½; **b,** ray floret, ×6; **c,** disc floret, ×6; **d,** apex of style from disc floret (note marginal stigmatic lines), ×35. **3,** *Carphephorus corymbosus*: **a,** upper portion of plant, ×½; **b,** inflorescence (discoid head), ×1⅓; **c,** disc floret and subtending chaffy bract ("pale"), ×2¼; **d,** expanded androecium (adaxial surface), ×8; **e,** longitudinal section of bud showing position of style branches before anthesis, ×5; **f,** androecium and expanded style branches, ×5; **g,** longitudinal section of floret, ×5; **h,** apex of style (note marginal stigmatic lines), ×6; **i,** base of style branches, ×10; **j,** "pollen brush" hairs on back of style branches, ×25; **k,** detail of receptive marginal stigma, ×25; **l,** nectary at apex of ovary, ×12. See Table 14 for descriptions of these general characters (representing features of various tribes).

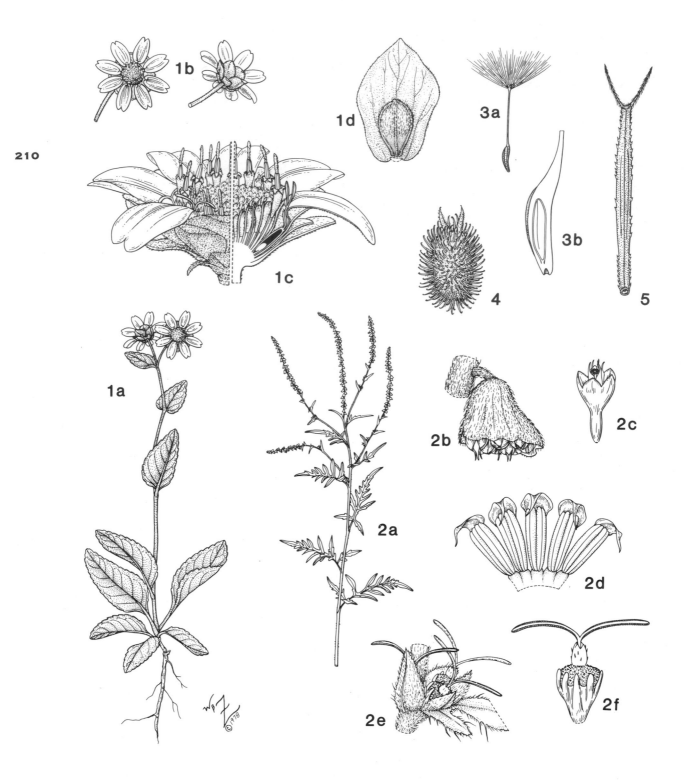

Figure 96. Asteraceae (continued). 1, *Berlandiera pumila*: **a,** habit, ×⅜; **b,** two views of inflorescence (radiate head), ×½; **c,** detail of head and longitudinal section of head, ×2⅔; **d,** achene with adnate involucral bract, ×3. **2,** *Ambrosia artemisiifolia*: **a,** habit, ×⅖; **b,** staminate inflorescence, ×12; **c,** staminate floret, ×15; **d,** expanded androecium (adaxial surface), ×35; **e,** cluster of carpellate involucres, ×12; **f,** involucre with a single carpellate floret, ×15. **3,** *Pyrrhopappus carolinianus*: **a,** achene (with persistent pappus), ×1½; **b,** longitudinal section of achene (pappus removed), ×6. **4,** *Xanthium strumarium*: fruiting head, ×1½. **5,** *Bidens alba*: achene, ×6.

the Compositae, vol. 1, ed. V. H. Heywood, J. B. Harborne, and B. L. Turner, pp. 1–20. Academic Press, London.

Karis, P. O., M. Källersjö, and K. Bremer. 1992. Phylogenetic analysis of the Cichorioideae (Asteraceae), with emphasis on the Mutisieae. *Ann. Missouri Bot. Gard.* 79:416–427.

Kim, K.-J., R. K. Jansen, R. S. Wallace, H. J. Michaels, and J. D. Palmer. 1992. Phylogenetic implications of *rbc*L sequence variation in the Asteraceae. *Ann. Missouri Bot. Gard.* 79:428–445.

Leins, P., and C. Erbar. 1990. On the mechanisms of secondary pollen presentation in the Campanulales-Asterales-complex. *Bot. Acta* 103:87–92.

Robinson, H. 1983. A generic review of the tribe Liabeae. *Smithsonian Contr. Bot.* 54:1–69.

Solbrig, O. T. 1963. The tribes of Compositae in the southeastern United States. *J. Arnold Arbor.* 44:436–461.

Thorne, R. F. 1992. Classification and geography of the flowering plants. *Bot. Rev.* (Lancaster) 58:225–348.

Wagenitz, G. 1976. Systematics and phylogeny of the Compositae (Asteraceae). *Pl. Syst. Evol.* 125:29–46.

CAMPANULACEAE
Bellflower Family

Annual or perennial herbs or sometimes shrubs, with milky or watery sap. **Leaves** simple, entire to pinnately divided, alternate, exstipulate. **Inflorescence** determinate, cymose and appearing racemose, paniculate, or sometimes capitate, terminal, or flowers solitary in leaf axils. **Flowers** actinomorphic or zygomorphic, usually perfect, epigynous or sometimes half-epigynous, generally showy, with nectariferous disc, sometimes resupinate (pedicel twisted 180° during development), sometimes cleistogamous. **Calyx** synsepalous with typically 5 lobes, imbricate or valvate, often persistent. **Corolla** sympetalous with typically 5 lobes, campanulate, tubular, to bilabiate (then often split down one side), blue, red, violet, or sometimes white, valvate. **Androecium** of usually 5 stamens, epipetalous or attached to disc, sometimes unequal in length; filaments distinct to connate, often expanded basally (and forming a chamber over the disc; anthers basifixed, distinct, coherent, or connate (syngenesious), forming a cylinder around the style (even when anthers distinct), dehiscing longitudinally, introrse. **Gynoecium** of 1 pistil, typically 2-, 3-, or 5-carpellate; ovary usually inferior or sometimes half-inferior, usually with as many locules as carpels, lobed, commonly capped by nectariferous disc; ovules numerous, anatropous, placentation axile or occasionally parietal; style 1, sometimes 2-, 3-, or 5-branched, with pollen-collecting hairs below the stigmas; stigmas 2, 3, or 5, globose to cylindrical. **Fruit** a poricidal capsule with apical or basal pores and opening with flaps or slits or sometimes a circumscissile capsule or a berry; seeds numerous, small; endosperm copious, fleshy, oily; embryo small, straight, short to spatulate.

Family characterization: Herbs to subshrubs, often with milky sap; epigynous, zygomorphic or actinomorphic flowers with 5-merous perianth and androecium; coherent to connate anthers forming a cylinder around the style; numerous ovules on axile placentae; styles with pollen-collecting hairs; and a poricidal or circumscissile capsule as the fruit type. Tissues commonly with calcium oxalate crystals. Inulin (an unusual storage polysaccharide) and cyanogenic glycosides present. Anatomical features: articulated laticifers in phloem of stem and leaves; often cystoliths in leaf epidermal cells; frequently calcification or silicification of leaf epidermal cell walls; unitegmic and tenuinucellate ovules; and unilacunar nodes.

Genera/species: 65/2,000

Distribution: Generally widely distributed, especially in temperate and subtropical regions and the montane tropics.

Major genera: *Lobelia* (200–365 spp.), *Campanula* (300 spp.), *Centropogon* (230 spp.), and *Siphocampylus* (215 spp.)

U.S./Canadian representatives: 23 genera/290 spp.; largest genera: *Cyanea, Lobelia, Clermontia,* and *Campanula*

Economic plants and products: Several poisonous and medicinal plants (due to alkaloids), such as species of *Laurentia* (isotomin, a heart poison) and *Lobelia* (lobeline, a narcotic). Ornamental plants (species of 25 genera), including *Campanula* (bellflower, bluebell, harebell), *Codonopsis* (bonnet-bellflower), *Edraianthus* (glassy-bells), *Jasione* (sheep's-bit), *Lobelia* (cardinal-flower), *Phyteuma* (horned rampion), *Triodanis* (Venus'-looking-glass), and *Wahlenbergia.*

Commentary: The Campanulaceae often are divided into three or four subfamilies, and some authors treat these groups as segregate families. The Lobelioideae, for example, sometimes are separated into their own family (Lobeliaceae) due to the zygomorphic and resupinate flowers, connate anthers, and distinctive alkaloids (Lammers 1992). Genera within the Campanulaceae as a whole often are distinguished by fruit morphology, especially the mode of dehiscence (e.g., capsules with apical or basal pores; Rosatti 1986).

The colorful flowers of the Campanulaceae attract various insects (especially bees), and in some cases birds, as the pollinators. Although the flowers of the family vary from open (*Triodanis;* Fig. 97: 1c) to tubular and bilabiate (*Lobelia;* Fig. 97: 3a), they share a pro-

212

tandrous pollen presentation mechanism similar to that of the Asteraceae (Leins and Erbar 1990). The expanded filament bases fit closely together around the nectariferous disc at the base of the style, allowing only the insertion of a proboscis between them. The stamens, with filaments and/or anthers coherent to connate, form a tube around the style and the appressed, unexposed stigmas (Fig. 97: 1d). Before or shortly after anthesis, the introrse anthers dehisce and fill the tube with pollen, which clings to the copious hairs often present on the style. As the "bottle-brush" style elongates, the pollen is pushed out and exposed to pollinators (Fig. 97: 1e,f). In some genera (e.g., *Campanula*), the pollen-collecting hairs retract into the stylar tissue, thereby releasing the pollen grains (Shetler 1979; Erbar and Leins 1989). Later the stigmas separate to expose the receptive upper surfaces (Fig. 97: 1g). In some species, self-pollination may result when the stigmas recurve far enough to pick up some pollen still adhering to their own style.

REFERENCES CITED

Erbar, C., and P. Leins. 1989. On the early floral development and mechanisms of secondary pollen presentation in *Campanula, Jasione,* and *Lobelia. Bot. Jahrb. Syst.* 111: 29–55.

Lammers, T. G. 1992. Circumscription and phylogeny of the Campanulales. *Ann. Missouri Bot. Gard.* 79:388–413.

Leins, P., and C. Erbar. 1990. On the mechanisms of secondary pollen presentation in the Campanulales-Asterales-complex. *Bot. Acta* 103:87–92.

Rosatti, T. J. 1986. The genera of Sphenocleaceae and Campanulaceae in the southeastern United States. *J. Arnold Arbor.* 67:1–64.

Shetler, S. G. 1979. Pollen-collecting hairs of *Campanula* (Campanulaceae), I: Historical review. *Taxon* 28:205–215.

SOLANACEAE
Nightshade Family

Annual, biennial, or perennial herbs, or sometimes shrubs or trees, often with prickles. **Leaves** simple or sometimes pinnately divided, alternate (often becoming opposite in the inflorescence), exstipulate. **Inflorescence** determinate, cymose, or sometimes flower solitary, axillary. **Flowers** usually actinomorphic, perfect, hypogynous, generally showy. **Calyx** synsepalous with 5 lobes, persistent and often enlarging in fruit. **Corolla** sympetalous with 5 lobes, rotate, funnelform, or salverform, variously colored, plicate or convolute. **Androecium** of 5 stamens, epipetalous; filaments distinct; anthers basifixed, sometimes connivent, sometimes with enlarged connective, equal or unequal, dehiscing longitudinally or by apical pores. **Gynoecium** of 1 pistil, 2-carpellate; ovary superior, 2-locular; ovules numerous, anatropous to campylotropous, placentation axile on protruding placentae; style single; stigma 2-lobed. **Fruit** a berry (sometimes enclosed by inflated persistent calyx) or a septicidal capsule; seeds numerous, flattened and discoid; endosperm copious, fleshy; embryo straight or bent.

Family characterization: Herbaceous to shrubby plants with alternate leaves; actinomorphic, hypogynous flowers with 5-merous perianth; sympetalous, plicate or convolute corolla; 5 epipetalous stamens; 2-carpellate ovary with axile placentation; and a berry or a capsule as the fruit type. Various alkaloids present. Anatomical features: intraxylary phloem in the petiole (bicollateral bundles) and stem; and unitegmic and tenuinucellate ovules.

Genera/species: 76/2,900

Distribution: Cosmopolitan; most abundant in Central and South America.

Major genera: *Solanum* (1,000 spp.), *Lycianthes* (200 spp.), and *Cestrum* (175 spp.); see D'Arcy (1991)

U.S./Canadian representatives: 33 genera/199 spp.; largest genera: *Solanum, Physalis,* and *Lycium*

Economic plants and products: Food crops: *Capsicum* (cayenne pepper, red and green peppers), *Lycopersicon* (tomato), and *Solanum* (eggplant, potato). Numerous medicinal and toxic plants (discussed below). Ornamental plants (species of 30 genera), including *Brunfelsia* (yesterday-today-and-tomorrow), *Cestrum* (night-blooming-jessamine), *Datura* (thorn-apple, angel's-trumpet), *Petunia, Physalis* (ground-cherry, Chinese-lantern), *Solandra* (chalice vine), and *Solanum* (nightshade, potato vine).

213

Figure 97. Campanulaceae. 1, *Triodanis perfoliata*: **a,** habit, ×³⁄₅; **b,** cleistogamous flower, ×5½; **c,** flower, ×3²⁄₃; **d,** androecium and gynoecium from a bud, ×5½; **e,** androecium and gynoecium at anthesis (anthers dehisced), ×5½; **f,** androecium and gynoecium of older flower (pollen deposited on pollen-collecting hairs of style branches), ×5½; **g,** mature, fully expanded stigma, ×5½; **h,** longitudinal section of flower, ×5; **i,** cross section of ovary at base, ×15; **j,** cross section of ovary near apex, ×15; **k,** capsule, ×7. **2,** *Wahlenbergia marginata*: flower, ×5½. **3,** *Lobelia glandulosa*: **a,** flower, ×2½; **b,** cross section of ovary, ×6; **c,** androecium and gynoecium, ×3½; **d,** detail of anthers (syngenesious) and stigma, ×6.

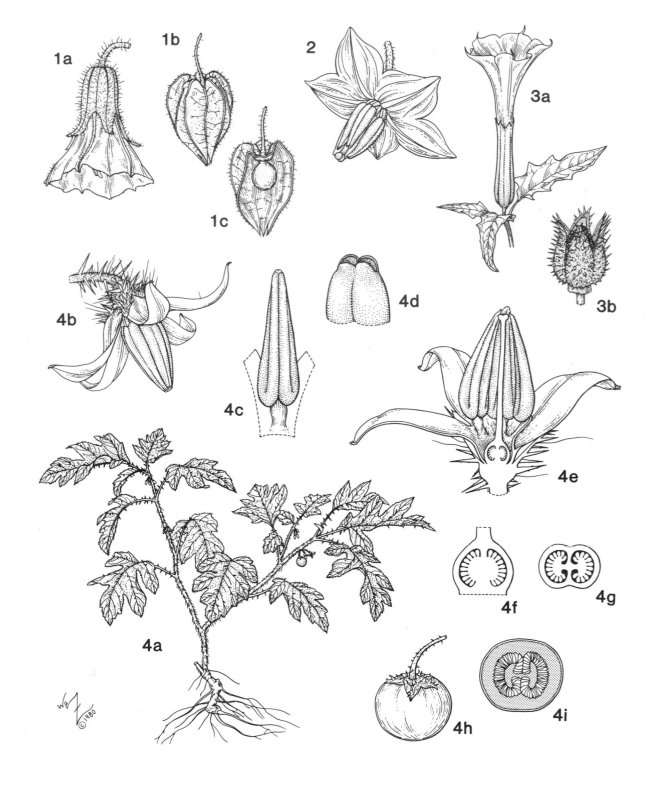

Figure 98. Solanaceae. 1, *Physalis heterophylla*: **a,** flower, ×2; **b,** fruit (berry subtended by inflated calyx), ×⅔; **c,** fruit with half of calyx removed, ×⅔. **2,** *Solanum carolinense*: flower, ×2. **3,** *Datura stramonium*: **a,** flowering branch, ×⅜; **b,** capsule, ×½. **4,** *Solanum capsicoides* (*S. ciliatum*): **a,** habit, ×⅕; **b,** flower, ×3; **c,** stamen, ×6; **d,** detail of abaxial side of anther apex showing pores, ×25; **e,** longitudinal section of flower, ×4½; **f,** longitudinal section of ovary, ×12; **g,** cross section of ovary, ×12; **h,** berry, ×⅔; **i,** cross section of berry, ×⅔.

Commentary: The Solanaceae usually are divided into five subfamilies with ten tribes, but many taxonomists (e.g., Thorne 1992) consider the Nolanoideae (2 genera/22 spp.) as a segregate family, the Nolanaceae. Cladistic analyses by Olmstead and Palmer (1992) support the inclusion of this group within the Solanaceae. Depending on the authority, one-third to almost one-half of the family is composed of the 1,000 to 3,700 species of *Solanum*, one of the largest genera of angiosperms. According to D'Arcy (1991), the actual number is close to 1,000: the inflated estimates reflect overdescribed variants of species by botanists and horticulturists emphasizing the economic importance of the group.

The family has been a popular subject for ethnobotanists, and the various alkaloids occurring in the family have been studied extensively (see Evans 1979; Tétényi 1987; Roddick 1986). In both the Old and New Worlds, many species have long been utilized for their medicinal, hallucinogenic, and poisonous properties; today the family is the source of several commercial pharmaceutical drugs (Roddick 1991). Presumably, the family name (from the genus *Solanum*) is derived from the Latin "solamen" meaning "quieting," which alludes to the narcotic properties of certain plants (Heiser 1969); sometimes the quieting effect is permanent. Similarly, the English name "nightshade" probably refers to the alkaloids. The "magical" roots of *Mandragora* (mandrake, with two alkaloids), for example, have been used widely in herbal medicine as a powerful narcotic. Other drug plants (and drugs) of the family include *Nicotiana* (tobacco: nicotine), *Hyoscyamus* (henbane: hyocyanine), *Atropa* (belladona: atropine), and *Datura* (jimson-weed: stramonium). Because even the berries of *Solanum* species can be very poisonous, Europeans did not readily accept certain solanaceous food plants (such as tomatoes) when these were first introduced.

The conspicuous flowers of the family are entomophilous. A characteristic *Solanum* flower has a rotate corolla (often with recurved lobes; Fig. 98: 2, 4b) and large, prominent anthers, which dehisce by apical pores (Fig. 98: 4c,d). Most *Solanum* species have no nectar, and pollinators (Hymenoptera and Diptera) visit the flowers to collect pollen (Symon 1979; Buchmann 1986). The insects manipulate the anthers (by vibrations or "milking" action) in order to shake out the pollen. Some members of the family, such as *Physalis* (Fig. 98: 1a), have salverform to funnelform corollas and produce nectar that is stored at the base of the corolla tube.

REFERENCES CITED

Buchmann, S. L. 1986. Vibratile pollination in *Solanum* and *Lycopersicon*: A look at pollen chemistry. In *Solanaceae: Biology and systematics*, ed. W. G. D'Arcy, pp. 237–252. Columbia University Press, New York.

D'Arcy, W. G. 1991. The Solanaceae since 1976, with a review of its biogeography. In *Solanaceae III: Taxonomy, chemistry, evolution*, ed. J. G. Hawkes, R. N. Lester, M. Nee, and N. Estrada, pp. 75–137. Royal Botanic Gardens, Kew.

Evans, W. C. 1979. Tropane alkaloids of the Solanaceae. In *The biology and taxonomy of the Solanaceae*, ed. J. G. Hawkes, R. N. Lester, and A. D. Shelding, pp. 241–254. Linn. Soc. Symp. Ser. No. 7. Academic Press, London.

Heiser, C. B. 1969. *Nightshades—The paradoxical plants*, pp. 1–5. W. H. Freeman, San Francisco.

Olmstead, R. G., and J. D. Palmer. 1992. A chloroplast DNA phylogeny of the Solanaceae: Subfamilial relationships and character evolution. *Ann. Missouri Bot. Gard.* 79:346–360.

Roddick, J. G. 1986. Steroidal alkaloids of the Solanaceae. In *Solanaceae: Biology and systematics*, ed. W. G. D'Arcy, pp. 201–222. Columbia University Press, New York.

———. 1991. The importance of the Solanaceae in medicine and drug therapy. In *Solanaceae III: Taxonomy, chemistry, evolution*, ed. J. G. Hawkes, R. N. Lester, M. Nee, and N. Estrada, pp. 7–23. Royal Botanic Gardens, Kew.

Symon, D. E. 1979. Sex forms in *Solanum* (Solanaceae) and the role of pollen collecting insects. In *The biology and taxonomy of the Solanaceae*, ed. J. G. Hawkes, R. N. Lester, and A. D. Shelding, pp. 385–397. Linn. Soc. Symp. Ser. No. 7. Academic Press, London.

Tétényi, P. 1987. A chemotaxonomic classification of the Solanaceae. *Ann. Missouri Bot. Gard.* 74:600–608.

Thorne, R. F. 1992. Classification and geography of the flowering plants. *Bot. Rev.* (Lancaster) 58:225–348.

CONVOLVULACEAE
Morning-glory Family

Annual or perennial vines, sometimes shrubs or small trees, commonly with milky sap, with rhizomes or tuberous roots or stems. **Leaves** simple, entire, lobed, or pinnately divided to pectinate, alternate, exstipulate. **Inflorescence** determinate, cymose, or flowers solitary, axillary, with jointed peduncles. **Flowers** actinomorphic, perfect, hypogynous, often large and showy, ephemeral, usually with intrastaminal disc, generally subtended by a pair of bracts (sometimes enlarged and forming an involucre). **Calyx** of 5 sepals, distinct or sometimes basally connate, sometimes unequal, imbricate, persistent. **Corolla** sympetalous, entire to slightly 5-lobed, funnelform or salverform, plicate, brightly colored (commonly red, violet, blue, or white), induplicate-valvate and/or convolute (twisted) in bud. **Androecium** of 5 stamens, epipetalous at corolla base;

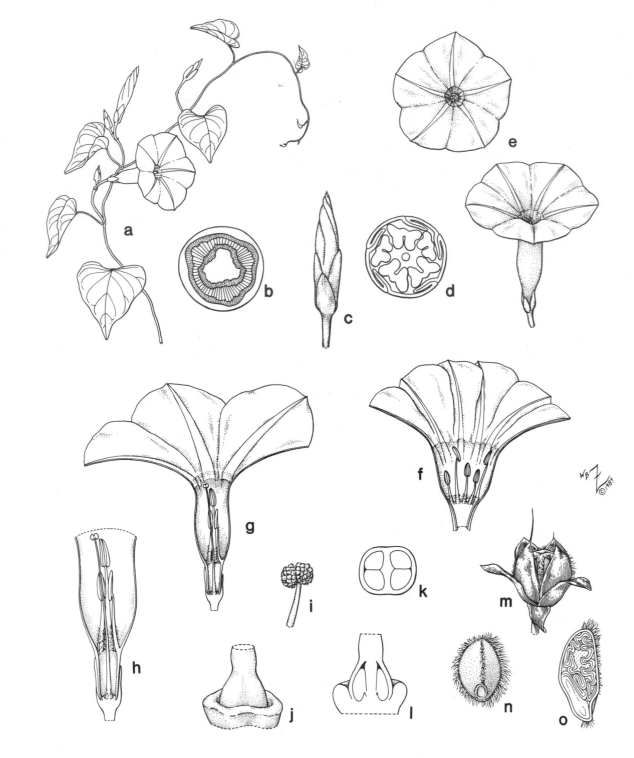

Figure 99. Convolvulaceae. *Ipomoea pandurata*: **a,** habit, ×⅓; **b,** cross section of stem showing intraxylary phloem, ×15; **c,** bud, ×1; **d,** cross section of bud showing induplicate-valvate and convolute aestivation of corolla, ×6; **e,** two views of flower, ×½; **f,** expanded corolla and epipetalous androecium, ×½; **g,** longitudinal section of flower, ×¾; **h,** longitudinal section of flower (apical portion of corolla removed), ×1; **i,** stigma, ×6; **j,** ovary with nectariferous disc, ×6; **k,** cross section of ovary, ×9; **l,** longitudinal section of ovary, ×6; **m,** capsule, ×1½; **n,** seed, ×3; **o,** longitudinal section of seed, ×4½.

filaments distinct, often unequal; anthers dorsifixed, dehiscing longitudinally, usually introrse. **Gynoecium** of 1 pistil, 2-carpellate; ovary superior, 2-locular or sometimes appearing 4-locular due to false septa, sometimes with dense covering of hairs; ovules 2 in each locule, anatropous, sessile, placentation basal or basal-axile; style simple and filiform or forked; stigma(s) 1 or 2, linear, lobed or capitate. **Fruit** usually a 4-valved septifragal capsule; seeds smooth or hairy; endosperm scanty, hard, cartilaginous; embryo large, straight or curved, with folded or coiled, emarginate to bifid cotyledons, surrounded by endosperm.

Family characterization: Vines with milky sap; showy, actinomorphic, funnelform to salverform, plicate corolla with induplicate-valvate and/or convolute aestivation; 5 epipetalous stamens; 2-carpellate ovary with axile placentation; septifragal capsule; and large embryo with folded, often bifid cotyledons. Various alkaloids and cyanogenic glycosides present. Tissues commonly with calcium oxalate crystals. Anatomical features: articulated latex canals or latex cells; intraxylary phloem in the petiole (bicollateral bundles) and stem (Fig. 99: b); and unitegmic, generally tenuinucellate ovules.

Genera/species: 59/1,830

Distribution: Primarily in the tropics and subtropics, with representatives having ranges extending into north and south temperate regions; particularly abundant in tropical America and tropical Asia.

Major genera: *Ipomoea* (500 spp.), *Convolvulus* (250 spp.), *Cuscuta* (145–170 spp.), and *Jacquemontia* (120 spp.)

U.S./Canadian representatives: 18 genera/198 spp.; largest genera: *Ipomoea*, *Cuscuta*, and *Calystegia*

Economic plants and products: Edible tubers from *Ipomoea batatas* (sweet-potatoes, "yams"). Powerful drugs from several, such as species of *Convolvulus* (scammony, a purgative from the tubers) and *Ipomoea* (jalap, a purgative from the tubers, and lysergic acid, a hallucinogen from the seeds). Several weedy plants, such as *Convolvulus* (bindweed) and *Cuscuta* (dodder). Ornamental plants (species of 13 genera), including *Calystegia* (bindweed), *Convolvulus*, *Dichondra*, *Ipomoea* (morning-glory, cypress vine), *Porana* (Christmas vine), and *Stylisma*.

Commentary: The Convolvulaceae have been divided into three or four subfamilies (sometimes segregated as distinct families) and/or three to ten tribes. Although the relationships between these groups have been generally agreed upon, the taxonomic rank (family, subfamily, or tribe) is a matter of controversy (see Wilson 1960). A notable segregate group, the Cuscutoideae or Cuscutaceae (a monotypic taxon), has been separated from the rest of the Convolvulaceae by some botanists on the basis of the parasitic habit with related specializations of the corolla and embryo (Momin 1977).

Authors also disagree on the delimitation of the various genera within the family, such as *Ipomoea* (Sengupta 1972). The generic lines depend upon characters of the bracts, sepals, corolla, pollen, stigma(s), and fruit. For example, the sepals vary in size, shape, and pubescence, and the stigmas may be simple, lobed, or globose. In addition, seed characters (e.g., type of pubescence) are important for species delimitation.

Morning-glories are easy to spot in the field with their twining habit and generally large, white or brightly colored, and funnel-shaped corolla. The corollas are twisted clockwise in bud and strongly plicate (Fig. 99: c,d; Allard 1947). Usually a flower is open for only one day (for a few hours); the corolla then incurves as it wilts. The corolla is characteristically divided longitudinally by five obvious demarcations that occur along the middle of the five lobes of the limb (see Fig. 99: e). These markings taper toward the apex and usually twist in the clockwise direction.

The flowers attract various insects (and in species of *Ipomoea*, birds), which visit for the nectar secreted by the hypogynous disc (Fig. 99: j; Govil 1975). The stamens closely surround the style by forming a short column in the center of the flower (Fig. 99: g,h), and five narrow passages between the filament bases lead to the nectar. The insect may touch the protruding stigma as it enters the flower, and then it becomes dusted with pollen from the introrse anthers as it reaches for the nectar near the base. Self-pollination may occur when the flower wilts.

REFERENCES CITED

Allard, H. A. 1947. The direction of twist of the corolla in the bud, and twining of the stems in Convolvulaceae and Dioscoreaceae. *Castanea* 12:88–94.

Govil, C. M. 1975. Phylogeny of floral nectary in Convolvulaceae. *Curr. Sci.* 44:518–519.

Momin, A. R. 1977. Bearing of embryological data on taxonomy of Convolvulaceae. *J. Univ. Bombay* 44:50–65.

Sengupta, S. 1972. On the pollen morphology of the Convolvulaceae with special reference to taxonomy. *Rev. Palaeobot. Palynol.* 13:157–212.

Wilson, K. A. 1960. The genera of Convolvulaceae in the southeastern United States. *J. Arnold Arbor.* 41:298–317.

BORAGINACEAE

Borage Family

Annual or perennial herbs commonly with rhizomes or taproots, or sometimes shrubs or trees, usually scabrous or hispid. **Leaves** simple, entire, alternate (lower ones sometimes opposite), exstipulate. **Inflorescence** determinate, basically a coiled cyme (helicoid or scorpioid cyme) and often appearing racemose or spicate, with the coiled axis bearing flowers along the "upper" side and straightening as the flowers mature, terminal. **Flowers** actinomorphic to slightly zygomorphic, usually perfect, hypogynous, occasionally heterostylous (dimorphic), usually showy, generally with nectariferous disc. **Calyx** of 5 sepals, distinct or basally connate, generally campanulate, usually imbricate, often persistent. **Corolla** sympetalous with 5 lobes, rotate, salverform, funnelform, or campanulate, often with projecting infoldings or appendages (scales) in the throat, commonly blue, white, pink, or yellow, imbricate or contorted. **Androecium** of 5 stamens, epipetalous; filaments distinct; anthers basifixed or basally dorsifixed, dehiscing longitudinally, introrse. **Gynoecium** of 1 pistil, 2-carpellate; ovary superior, 2-locular but becoming 4-locular due to false septa, often deeply 4-lobed; ovules 4 (1 in each ovary section), anatropous to hemitropous, placentation axile but often appearing basal; style(s) 1 or 2, gynobasic or terminal; stigma(s) 1 (2-lobed) or 2, usually capitate, papillose. **Fruit** a schizocarp splitting into basically 4 nutlets, or sometimes a 1- to 4-seeded drupe; endosperm scanty and fleshy or absent; embryo spatulate, erect, straight or sometimes curved.

Family characterization: Often scabrous or hispid herbs; circinate and one-sided cymose inflorescence with the axis uncoiling as the flowers mature; corolla tube often with projecting infoldings or scales (a "corona"); deeply 4-lobed to unlobed, 2-carpellate ovary with gynobasic or terminal style; a schizocarp (splitting into 4 nutlets) or drupe as the fruit type; and seeds with little or no endosperm. Iridoid compounds and various alkaloids commonly present. Tissues frequently with calcium oxalate crystals. Anatomical features: firm, unicellular hairs with thick, often calcified and silicified walls and basal cystoliths; xylem (even in herbs) forming a continuous ring with narrow rays; and unitegmic and tenuinucellate ovules.

Genera/species: 117/2,400

Distribution: Cosmopolitan, but particularly well represented in temperate and subtropical regions; centers of diversity in the Mediterranean region and western North America.

Major genera: *Cordia* (250 spp.), *Heliotropium* (250 spp.), *Tournefortia* (150 spp.), *Onosma* (150 spp.), and *Cryptantha* (100 spp.)

U.S./Canadian representatives: 34 genera/384 spp.; largest genera: *Cryptantha*, *Plagiobothrys*, and *Hackelia*

Economic plants and products: Timber from *Cordia* spp. Alkanet (a red dye) from roots of *Alkanna*. Medicinal herbs: *Borago* (borage), *Lithospermum* (puccoon—used as a contraceptive by certain Indians of North America), and *Symphytum* (comfrey). Ornamental plants (species of 32 genera), including *Borago*, *Cerinthe* (honeywort), *Cordia* (geiger tree), *Cynoglossum* (hound's-tongue), *Echium* (viper's-bugloss), *Heliotropium* (heliotrope), *Mertensia* (Virginia-bluebell), *Myosotis* (forget-me-not), and *Pulmonaria* (lungwort).

Commentary: The Boraginaceae are generally divided into two to five subfamilies, primarily based upon characters of the style (simple or bilobed, terminal or gynobasic) and fruit (schizocarp or drupe); nutlet morphology is important for the recognition of the four to thirteen tribes (see Al-Shehbaz 1991 for a complete summary). Mature fruits also are essential in most treatments for the identification of genera. The subfamilies have been variously recognized as segregate families (e.g., Ehretiaceae); the phylogenetic position of the Boraginaceae also has been controversial. Although many taxonomists have allied the family with the Lamiaceae-Verbenaceae (Lamiales), others (e.g., Thorne 1992) consider the Boraginaceae to be related to the Solanales complex of families (e.g., Convolvulaceae, Solanaceae, and Polemoniaceae; see Cantino 1982 and Olmstead et al. 1993).

The characteristic dorsiventral and coiled inflorescences are complex variations on a basically cymose arrangement (Fig. 100: 1a). The major axes are sympodial, producing circinate cymes that uncoil progressively as the flowers open. The borage inflorescence type has been generally termed "helicoid" and/or "scorpioid" by various authors (see Rickett 1955). The coiling is caused by the development of only one flower (or branch) of each original lateral pair of the cyme. Generally, in a helicoid cyme the lateral branches develop from the same side of the axis, and in a scorpioid cyme, on alternating sides in a stepwise manner (see Fig. 100: 1a–c; Prior 1960). A seemingly spicate or racemose arrangement results.

Various insects (bees, Lepidoptera, flies) are the primary pollinators that visit the flowers for the nectar secreted at the base. Birds and bats also have been reported as pollinators of several species (e.g., those of

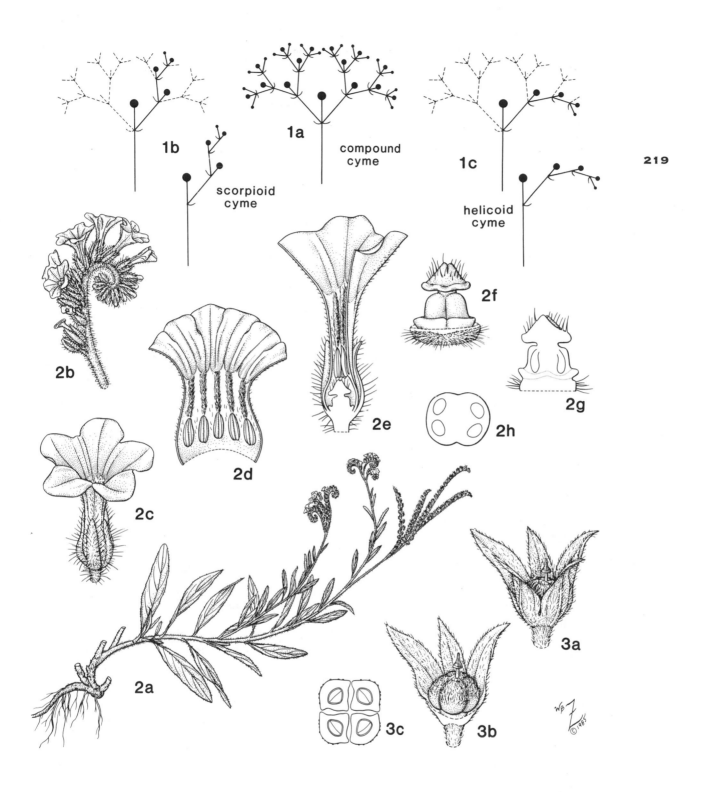

Figure 100. Boraginaceae. 1, Inflorescence diagrams: **a,** compound cyme; **b,** scorpioid cyme; **c,** helicoid cyme. **2,** *Heliotropium amplexicaule*: **a,** habit, ×⅓; **b,** inflorescence, ×2; **c,** flower, ×4½; **d,** expanded corolla and epipetalous androecium, ×6; **e,** longitudinal section of flower, ×8; **f,** pistil and nectariferous disc, ×15; **g,** longitudinal section of pistil, ×15; **h,** cross section of ovary, ×25. **3,** *Heliotropium polyphyllum*: **a,** schizocarp with persistent calyx, ×6; **b,** schizocarp (portion of calyx removed), ×6; **c,** cross section of schizocarp, ×10.

Cordia; Al-Shehbaz 1991). Infoldings, appendages, and/or hairs of the corolla function as nectar guides (Fig. 100: 2d,e) and aid in the protection and partial concealment of the nectar. The corollas of several members (e.g., *Myosotis* and *Cryptantha*) undergo striking color changes, which in at least some cases signal the pollinators about nectar availability (Casper and La Pine 1984). Self-pollination evidently is prevalent in species with relatively inconspicuous flowers (Knuth 1909). Species of several genera (e.g., *Cordia* and *Cryptantha*) have distylous flowers (see the discussions on heterostyly in the commentaries under Rubiaceae and Lythraceae; Ganders 1979).

REFERENCES CITED

Al-Shehbaz, I. A. 1991. The genera of Boraginaceae in the southeastern United States. *J. Arnold Arbor.*, Suppl. Ser. 1:1–169.

Cantino, P. D. 1982. Affinities of the Lamiales: A cladistic analysis. *Syst. Bot.* 7:237–248.

Casper, B. B., and T. R. La Pine. 1984. Changes in corolla color and other floral characteristics in *Cryptantha humilis* (Boraginaceae): Cues to discourage pollinators? *Evolution* 38:128–141.

Ganders, F. R. 1979. The biology of heterostyly. *New Zealand J. Bot.* 17:607–635.

Knuth, P. 1909. *Handbook of flower pollination*. Trans. J. R. A. Davis. Vol. 3, *Goodenovieae to Cyadeae*, pp. 115–142. Oxford University Press, Oxford.

Olmstead, R. G., B. Bremer, K. M. Scott, and J. D. Palmer. 1993. A parsimony analysis of the Asteridae sensu lato based on *rbc*L sequences. *Ann. Missouri Bot. Gard.* 80:700–722.

Prior, P. V. 1960. Development of the helicoid and scorpioid cymes in *Myosotis laxa* Lehm. and *Mertensia virginica* L. *Proc. Iowa Acad. Sci.* 67:76–81.

Rickett, H. W. 1955. Materials for a dictionary of botanical terms. III. Inflorescences. *Bull. Torrey Bot. Club* 82:419–445.

Thorne, R. F. 1992. Classification and geography of the flowering plants. *Bot. Rev.* (Lancaster) 58:225–348.

POLEMONIACEAE
Phlox Family

Annual, biennial, or perennial herbs, occasionally shrubs, small trees, or vines, often with noxious odor (due to substances in gland-tipped hairs). **Leaves** simple or occasionally compound, entire, or pinnately or palmately divided, alternate to sometimes opposite, exstipulate. **Inflorescence** determinate, generally cymose and often appearing corymbose or capitate, terminal, or flowers occasionally solitary and axillary. **Flowers** usually actinomorphic, perfect, hypogynous, generally showy, usually with intrastaminal nectariferous disc. **Calyx** synsepalous with 5 lobes, campanulate or salverform, imbricate or valvate, persistent. **Corolla** sympetalous with 5 lobes, rotate, campanulate, funnelform, or salverform, plicate, variously colored, convolute/contorted in bud. **Androecium** of 5 stamens, epipetalous, sometimes inserted at unequal levels within the corolla tube; filaments distinct, sometimes unequal; anthers dorsifixed or basifixed, dehiscing longitudinally, introrse. **Gynoecium** of 1 pistil, 3-carpellate; ovary superior, 3-locular; ovules 1 to many in each locule, anatropous or hemitropous, sessile, placentation axile; style 1, trifid at apex; stigmas 3, along upper surface of style branches, linear. **Fruit** usually a loculicidal capsule with 3 valves separating from a central column; seeds with seed coat often sticky or mucilaginous when moistened; endosperm usually copious, fleshy, oily; embryo spatulate, straight or slightly curved, embedded in endosperm.

Family characterization: Herbaceous plants with noxious odor; showy, salverform to funnelform, plicate corolla with convolute/contorted aestivation; 5 epipetalous stamens; 3-carpellate gynoecium with apically trifid style; and loculicidal capsules with numerous seeds becoming mucilaginous when moistened. Inulin (an unusual storage polysaccharide), saponins, and various flavonoids usually present. Anatomical features: vascular tissue (even in herbs) forming a continuous ring (Fig. 101: b); and unitegmic and tenuinucellate ovules.

Genera/species: 16/320

Distribution: Mainly in temperate North America, but extending into the Andes Mountains of South America and also into temperate Eurasia; most diverse in western North America.

Major genera: *Gilia* (25–120 spp.), *Phlox* (67 spp.), *Linanthus* (35–50 spp.), *Polemonium* (25–50 spp.), and *Navarretia* (30 spp.)

U.S./Canadian representatives: 14 genera/283 spp.; largest genera: *Phlox*, *Gilia*, *Linanthus*, *Ipomopsis*, and *Polemonium*

Economic plants and products: Ornamental plants (species of at least 12 genera), including *Cobaea* (cup-and-saucer vine, Mexican-ivy), *Gilia*, *Linanthus*, *Phlox*, and *Polemonium* (Greek-valerian, Jacob's-ladder).

Commentary: The Polemoniaceae often are divided into four or five tribes primarily based on habit type, leaf morphology, inflorescence form, and the mode of stamen insertion on the corolla. Although previously allied with such diverse families as the Caryophyllaceae and Geraniaceae, the family now is generally placed in the Solanales (Solanaceae, Convolvulaceae, Boragina-

220

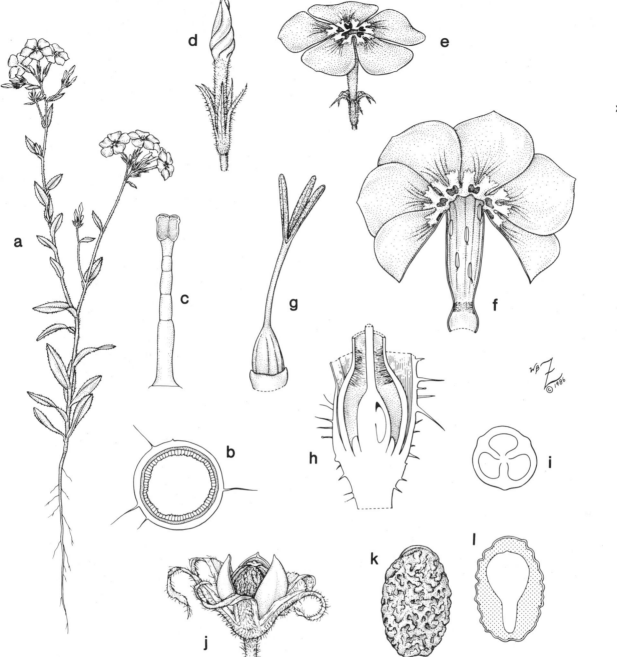

Figure 101. Polemoniaceae. *Phlox drummondii*: **a,** habit, ×²⁄₅; **b,** cross section of stem showing vascular tissue in a continuous ring, ×8; **c,** capitate glandular hair from stem, ×170; **d,** bud, ×2; **e,** flower, ×1½; **f,** expanded corolla and epipetalous androecium, ×2½; **g,** pistil, ×12; **h,** longitudinal section of ovary (and flower base), ×15; **i,** cross section of ovary, ×25; **j,** capsule with persistent calyx, ×5; **k,** seed, ×10; **l,** longitudinal section of seed, ×10.

ceae; e.g., Thorne 1992). However, this placement merits further investigation: recent cladistic analyses of molecular data suggest that the Polemoniaceae are not closely related to these families and may instead be allied to other groups such as the Ericaceae and Diapensiaceae (Olmstead et al. 1992, 1993). Generic de-limitation within the family, especially of temperate groups, often is variously interpreted, such as with *Gilia* and segregates (Smith et al. 1977). Interpretations of numerous polymorphic species, in particular, have resulted in different taxonomic treatments for such genera as *Polemonium* and *Phlox* (Wilson 1960).

The white or brightly colored, sympetalous corolla is characterized by a well-developed long tube with a spreading limb, and it varies in overall shape from campanulate (*Cobaea*), salverform (*Phlox*; Fig. 101: e), funnelform (*Gilia, Polemonium*), to almost rotate (*Polemonium*; Grant 1959). In some species of *Gilia*, the tube is very narrow and filiform. The pollination mechanisms of the family have been studied extensively by Grant and Grant (1965). The various corolla shapes, correlated with particular groups of pollinators, evidently have diverged within the family under the selective pressure of different pollinators, which include various insects (bees, flies, beetles, butterflies, moths), hummingbirds, and bats.

The attractive flowers, often fragrant or with an unpleasant odor, are arranged in conspicuous, clustered inflorescences. Nectar, secreted by the annular disc (Fig. 101: g), accumulates at the base of the internally hairy tube. Often an eye and/or streaks of contrasting color on the limb (Fig. 101: e,f) serve as nectar guides. Flowers usually are protandrous, with the style elongating beyond the anthers after the pollen has been shed.

REFERENCES CITED

Grant, V. 1959. *Natural history of the phlox family*. Vol. 1, *Systematic botany*, pp. 23–25, 35–51. Martinus Nijhoff, The Hague.

—— and K. A. Grant. 1965. *Flower pollination in the phlox family*, pp. 143–165. Columbia University Press, New York.

Olmstead, R. G., B. Bremer, K. M. Scott, and J. D. Palmer. 1993. A parsimony analysis of the Asteridae sensu lato based on *rbc*L sequences. *Ann. Missouri Bot. Gard.* 80:700–722.

——, H. J. Michaels, K. M. Scott, and J. D. Palmer. 1992. Monophyly of the Asteridae and identification of their major lineages inferred from DNA sequences of *rbc*L. *Ann. Missouri Bot. Gard.* 79:249–265.

Smith, D. M., C. W. Glennie, J. B. Harborne, and C. A. Williams. 1977. Flavonoid diversification in the Polemoniaceae. *Biochem. Syst. & Ecol.* 5:107–115.

Thorne, R. F. 1992. Classification and geography of the flowering plants. *Bot. Rev.* (Lancaster) 58:225–348.

Wilson, K. A. 1960. The genera of Hydrophyllaceae and Polemoniaceae in the southeastern United States. *J. Arnold Arbor.* 41:197–212.

LYTHRACEAE
Loosestrife Family

Annual or perennial herbs, sometimes shrubs, or rarely trees. **Leaves** simple, entire, opposite or whorled, exstipulate or stipulate; stipules minute. **Inflorescence** determinate, cymose and often appearing fasciculate, racemose, or paniculate, or flowers solitary, terminal or axillary. **Flowers** actinomorphic to occasionally zygomorphic, perfect, perigynous, often showy, sometimes heterostylous (di- or trimorphic) with well-developed campanulate to tubular hypanthium, often with nectariferous disc, frequently subtended by an epicalyx. **Calyx** of 4 to 8 sepals, distinct, valvate. **Corolla** of 4 to 8 petals or (seldom) absent, distinct, arising from rim or upper, inner surface of hypanthium, clawed, with erose margins, typically fugacious, often white, pink, red, or purple, imbricate and crumpled in bud. **Androecium** of often twice as many stamens as petals (biseriate), sometimes of as many stamens as petals (uniseriate), or seldom of numerous stamens, typically inserted within the hypanthium below the petals or occasionally on the rim or at the base; filaments distinct, usually inflexed in bud, unequal; anthers dorsifixed, versatile, dehiscing longitudinally, introrse. **Gynoecium** of 1 pistil, typically 2- to 6- to sometimes many-carpellate; ovary usually superior, with as many locules as carpels; ovules several to numerous in each locule, anatropous, placentation axile or sometimes axile at base and parietal at apex (due to incomplete septa); style 1; stigma usually capitate. **Fruit** usually a variously dehiscent capsule enclosed by persistent hypanthium, or rarely a berry; seeds sometimes winged, seldom with sarcotesta; endosperm absent; embryo straight, oily.

Family characterization: Herbs to shrubs with opposite, entire-margined leaves; perigynous flowers

Figure 102. Lythraceae. 1, *Lythrum alatum* var. *lanceolatum*: **a,** flowering branch, ×²⁄₃; **b,** cross section of stem (internode) showing intraxylary phloem, ×20; **c,** cross section of node (one-trace unilacunar), ×20; **d,** bud, ×8; **e,** petal, ×5½; **f,** three views of anther, ×15; **g–j,** short-styled flower: **g,** short-styled flower, ×4; **h,** expanded perianth, hypanthium, and androecium (long stamens), ×5½; **i,** pollen grain from long stamen, ×800; **j,** pistil (short-styled), ×5½; **k–q,** long-styled flower: **k,** long-styled flower, ×4; **l,** expanded perianth, hypanthium, and androecium (short stamens), ×5½; **m,** pollen grain from short stamen, ×800; **n,** pistil (long-styled), ×5½; **o,** abaxial view of expanded perianth and hypanthium, ×5½; **p,** cross section of ovary, ×25; **q,** longitudinal section of ovary and hypanthium base, ×15; **r,** fruit (capsule subtended by persistent hypanthium), ×10; **s,** capsule (persistent hypanthium removed), ×10; **t,** seed, ×25; **u,** longitudinal section of seed, ×25. **2,** *Cuphea hyssopifolia*: **a,** expanded perianth, hypanthium, and androecium (stamens biseriate), ×5; **b,** cross section of ovary, ×25.

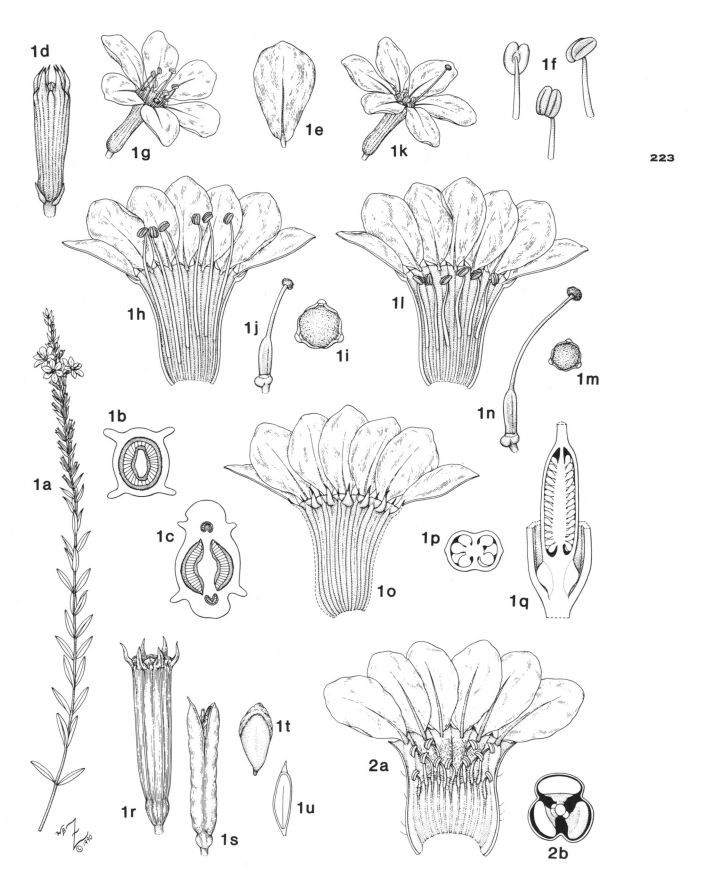

223

with well-developed campanulate to tubular hypanthium; clawed petals with erose margins crumpled in bud; stamens with unequal filaments inserted on the inner surface of the hypanthium below petals; and a variously dehiscent capsule enclosed by the persistent hypanthium as the typical fruit type. Various alkaloids and tannins present. Tissues with calcium oxalate crystals. Anatomical features: mucilage cells often in leaf epidermis or mesophyll; intraxylary phloem in the stems (a continuous internal ring or in bicollateral vascular bundles; Fig. 102: 1b,c); and unilacunar nodes (Fig. 102: 1c).

Genera/species: 25/460

Distribution: Mainly subtropical to tropical areas, with some representatives in temperate regions; especially diverse in the American tropics. The temperate species are mainly herbaceous to suffrutescent and often grow in damp habitats; the tropical species typically are shrubs or trees.

Major genera: *Cuphea* (250 spp.), *Nesaea* (50–56 spp.), *Lagerstroemia* (53 spp.), *Diplusodon* (50 spp.), and *Rotala* (44–50 spp.)

U.S./Canadian representatives: 12 genera/36 spp.; largest genera: *Lythrum* and *Cuphea*

Economic plants and products: Several dye plants, such as *Lawsonia inermis* (henna, an orange-red dye extracted from dried leaves). Pomegranates, the fruits of *Punica granatum*. Ornamental plants (species of 12 genera), including *Cuphea* (cigar-flower), *Lagerstroemia* (crape- or crepe-myrtle), *Lawsonia* (mignonette tree), *Lythrum* (loosestrife), *Punica*, and *Woodfordia*.

Commentary: The concept of Lythraceae presented here, with four subfamilies, is rather broad (Dahlgren and Thorne 1984; Thorne 1992), including the monotypic Punicaceae (*Punica*, the pomegranate; 2 spp.) and the Sonneratiaceae, two genera of mangrove (*Sonneratia*, 4 spp.) and rain forest (*Duabanga*, 3 spp.) trees of the Old World tropics. Both groups differ from the Lythraceae *s.s.* by the more numerous carpels and stamens and partly to fully inferior ovary, as well as a berry fruit type in *Punica* and *Sonneratia* (Graham et al. 1990). Several similarities in anatomical and embryological features, summarized in detail in Dahlgren and Thorne (1984), closely link the Lythraceae *s.s.* with these two groups; the Lythraceae *s.l.* also are supported by the cladistic analyses of Johnson and Briggs (1984). The subfamily Lythroideae (Lythraceae *s.s.*) sometimes has been divided into two tribes, based on whether the septa are complete in the apex of the ovary locule (Graham 1964).

A characteristic feature of the family (and of other families in the order Myrtales) is the prominent hypanthium ("calyx tube"), which varies in shape from campanulate (*Decodon*) to tubular (*Lythrum*; Fig. 102: 1d,g). Sometimes, as in *Cuphea*, the hypanthium is spurred. The calyx lobes appear marginal or on the rim, and the petals also arise on the rim or along the upper, inner surface (Fig. 102: 1o). The stamens usually are inserted below the petals on the inner surface within the cup (Fig. 102: 1h,l; 2a). The hypanthium is conspicuously ribbed or nerved, often with appendages (bracts composing an epicalyx) alternating with the nerves and calyx lobes (Fig. 102: 1d,o). Each appendage apparently represents two fused bracts (or stipules) associated with each calyx lobe.

The hypanthium persists in fruit (Fig. 102: 1r) and encloses the various capsule types: loculicidal (e.g., *Decodon*), septicidal (*Rotala*), septifragal (several *Lythrum* species), and irregularly dehiscent (*Ammannia*). In *Punica*, the hypanthium forms a crown at the apex of the berry, characterized by a thick and leathery rind. The thin walls of the carpels divide the berry into chambers filled with seeds surrounded by pulpy sarcotestae.

The attractively colored and conspicuous flowers of the family attract various pollinators (insects, birds; bats in *Sonneratia*). Often nectar, secreted by the disc, accumulates at the base of the hypanthium cup; for example, in the slightly zygomorphic flowers of *Lythrum*, pollinators (e.g., bees and flies) reach for the nectar through an open space created between the anthers and style, which are curved to one side. Long-tongued insects and hummingbirds pollinate the spurred and long-tubular flowers of *Cuphea*. The flowers of *Punica* do not produce nectar, and insects (such as beetles) collect pollen from the numerous, conspicuous stamens with orange-red filaments. The several apetalous species in the family (such as those of *Ammannia* and *Peplis*) are presumed to be self-pollinating (cleistogamous or pseudocleistogamous).

A large number of species in the family exhibit flower heteromorphism, either distyly (e.g., *Rotala*, *Lythrum* species) or tristyly (*Decodon*, *Lagerstroemia*, several *Lythrum* species); the discussions under Rubiaceae and Pontederiaceae, respectively, detail the aspects of these two outcrossing mechanisms. *Lythrum salicaria*, a European species (escaped in the United States), has been studied extensively as one of the classic examples of trimorphic heterostyly (see Weller 1992 for a summary). Evidently, dimorphic heterostyly in the family, such as in other species of *Lythrum* (*L. alatum*; Fig. 102: 1g–n), has been derived from the tristylous condition.

REFERENCES CITED

Dahlgren, R., and R. F. Thorne. 1984. The order Myrtales: Circumscription, variation, and relationships. *Ann. Missouri Bot. Gard.* 71:633–699.

Graham, A., S. A. Graham, J. W. Nowicke, V. Patel, and S. Lee. 1990. Palynology and systematics of the Lythraceae. III. Genera *Physocalymma* through *Woodfordia*, addenda and conclusions. *Amer. J. Bot.* 77:159–177.

Graham, S. A. 1964. The genera of Lythraceae in the southeastern United States. *J. Arnold Arbor.* 45:235–250.

Johnson, L. A. S., and B. G. Briggs. 1984. Myrtales and Myrtaceae—A phylogenetic analysis. *Ann. Missouri Bot. Gard.* 71:700–756.

Thorne, R. F. 1992. Classification and geography of the flowering plants. *Bot. Rev.* (Lancaster) 58:225–348.

Weller, S. G. 1992. Evolutionary modifications of tristylous breeding systems. In *Evolution and function of heterostyly*, ed. S. C. H. Barrett, pp. 247–272. Monogr. Theor. & Appl. Genet. 15. Springer-Verlag, Berlin.

MELASTOMATACEAE
Meadow-beauty or Melastome Family

Annual or perennial herbs, shrubs, or trees, erect or climbing; stems square or round in cross section. **Leaves** simple, entire or serrate, opposite (one of pair sometimes smaller than the other), decussate, sessile to petiolate, with usually 3 to 5 prominent and subparallel veins diverging from the base and converging at the apex (connected by transverse cross-veins), exstipulate. **Inflorescence** determinate, cymose, terminal or axillary. **Flowers** slightly zygomorphic (due to orientation of androecium), perfect, epigynous or sometimes half-epigynous or perigynous, large and showy to minute, sometimes subtended by showy bracts, with well-developed and often urceolate (urn-shaped) hypanthium. **Calyx** of usually 4 or 5 sepals, often distinct, imbricate, persistent. **Corolla** of usually 4 or 5 petals, distinct, usually spreading, often red, blue, white, pink, or purple, usually imbricate and/or convolute. **Androecium** of twice as many stamens as petals, biseriate, arising from the hypanthium; filaments distinct, inflexed in bud, geniculate; anthers 2- or sometimes 1-locular, basifixed, often with various appendages at base of connectives, each dehiscing by an apical pore. **Gynoecium** of 1 pistil, usually 3- to 5-carpellate; ovary inferior (adnate to hypanthium) or sometimes half-inferior or superior (within free hypanthium), with as many locules as carpels; ovules numerous, anatropous or sometimes campylotropous, placentation usually axile; style 1, simple; stigma 1, capitate or punctate. **Fruit** a loculicidal capsule or a berry; seeds numerous, small; endosperm absent; embryo minute, straight to curved, usually with unequal cotyledons.

Family characterization: Herbaceous or woody plants with opposite, simple, exstipulate leaves sometimes with prominent subparallel venation (connected by transverse cross-veins); 4- or 5-merous flowers with urceolate hypanthium; stamens with geniculate filaments and often appendaged anthers opening with apical pores; and a loculicidal capsule or berry as the fruit type. Tannins and calcium oxalate crystals commonly present. Anatomical features: intraxylary phloem (internal phloem) in the stem (Fig. 103: 1b); often vascular bundles in the cortex and pith; and unilacunar nodes.

Genera/species: 244/3,360

Distribution: Pantropical; most diverse in South America, especially in tropical montane habitats.

Major genera: *Miconia* (700–1,000 spp.), *Medinilla* (150–400 spp.), *Tibouchina* (200–350 spp.), *Memecylon* (150–300 spp.), *Leandra* (200 spp.), and *Clidemia* (145–165 spp.)

U.S./Canadian representatives: 19 genera/65 spp.; largest genera: *Miconia* and *Rhexia*

Economic plants and products: Lumber (used locally for furniture and construction) from *Astronia* and *Memecylon*. Edible fruits from *Medinilla*, *Melastoma*, and *Mouriri*. A few dye plants, such as *Memecylon*. Ornamental plants (species of 19 genera), including *Dissotis*, *Heterocentron*, *Medinilla*, *Melastoma*, *Rhexia* (meadow-beauty), and *Tibouchina* (princess-flower).

Commentary: The Melastomataceae are a well-defined and apparently monophyletic family (Dahlgren and Thorne 1984). In the continental United States, the family is represented by *Rhexia* (all species occurring in the southeast; e.g., Fig. 103: 1) and *Tetrazygia bicolor* (Fig. 103: 2; Wurdack and Kral 1982).

The plants are easily recognized in the field by the distinctive leaf venation (not obvious in *Rhexia*) and stamens. The opposite leaves (e.g., *Tetrazygia*; Fig. 103: 2a) often are characterized by prominent secondary veins that diverge from the base and converge at the apex; these major veins and the midvein are connected by transverse tertiary veins. The stamens typically are geniculate (bent), dehisce by apical pores, and have modified appendages of the connective (Fig. 103: 1e,f; Wilson 1950). The morphology of these appendages (e.g., awl-shaped, spiny, curved) has been used to distinguish genera and species.

The showy flowers generally do not produce nectar, and insects (often bumblebees) visit to collect pollen

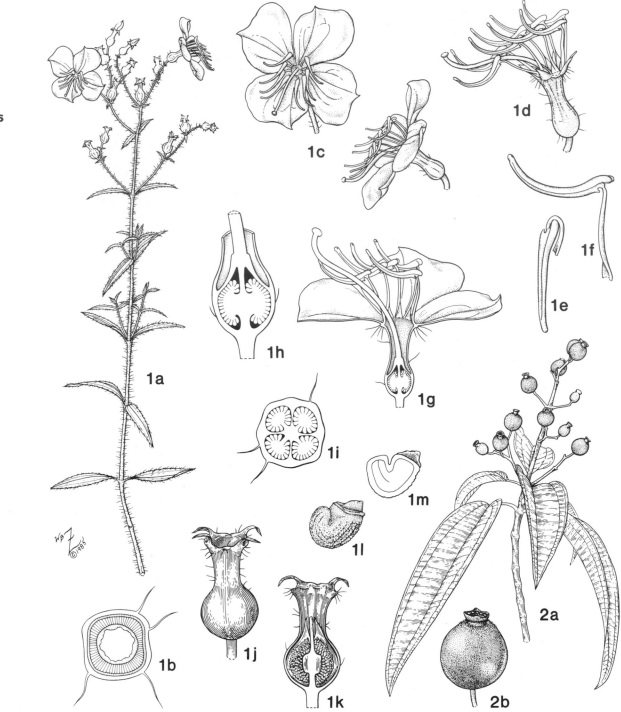

Figure 103. Melastomataceae. 1, *Rhexia mariana*: **a,** habit, ×½; **b,** cross section of stem showing intraxylary phloem, ×9; **c,** two views of flower, ×1; **d,** flower with petals removed, ×2; **e,** stamen from bud, ×4½; **f,** stamen from mature flower, ×4½; **g,** longitudinal section of flower, ×2; **h,** longitudinal section of ovary, ×6; **i,** cross section of ovary, ×6; **j,** fruit (capsule subtended by hypanthium), ×3; **k,** longitudinal section of fruit, ×3; **l,** seed, ×30; **m,** longitudinal section of seed, ×30. **2,** *Tetrazygia bicolor*: **a,** fruiting branch, ×½; **b,** berry, ×2.

by vibrating or otherwise manipulating the terminally pored anthers (Renner 1989). Self-pollination generally is prevented by the effective spatial separation of anthers and stigma: the long style (with small stigma) is curved and exserted beyond the shorter stamens. In *Tetrazygia*, the style protrudes before the flower opens (Tomlinson 1980). A few tropical species (e.g., *Medinilla* and *Tibouchina* spp.) are nectariferous and are pollinated by bees, hummingbirds, bats, and rodents (Stein and Tobe 1989).

The distinctive fruit of *Rhexia* (Fig. 103: 1j,k) is a capsule enclosed within the lower globose portion of the persistent flask-shaped hypanthium. The fruit of *Tetrazygia* is a berry (Fig. 103: 2b).

REFERENCES CITED

Dahlgren, R. M. T., and R. F. Thorne. 1984. The order Myrtales: Circumscription, variation, and relationships. *Ann. Missouri Bot. Gard.* 71:633–699.

Renner, S. S. 1989. A survey of reproductive biology in neotropical Melastomataceae and Memecylaceae. *Ann. Missouri Bot. Gard.* 76:496–518.

Stein, B. A., and H. Tobe. 1989. Floral nectaries in Melastomataceae and their systematic and evolutionary implications. *Ann. Missouri Bot. Gard.* 76:519–531.

Tomlinson, P. B. 1980. *The biology of trees native to tropical Florida*, pp. 238–240. "Publ. by the author," Petersham, Mass.

Wilson, C. L. 1950. Vasculation of the stamen in the Melastomataceae, with some phyletic implications. *Amer. J. Bot.* 37:431–444.

Wurdack, J. J., and R. Kral. 1982. The genera of Melastomataceae in the southeastern United States. *J. Arnold Arbor.* 63:429–439.

COMBRETACEAE
Indian-almond or White Mangrove Family

Trees, shrubs, or woody vines. **Leaves** simple, entire, alternate, opposite, or sometimes whorled, persistent or deciduous, often with 2 glandular cavities (extrafloral nectaries) at blade base and several pits (domatia) on abaxial surface, generally exstipulate. **Inflorescence** determinate, spicate, racemose, capitate, or paniculate, terminal or axillary. **Flowers** actinomorphic or sometimes slightly zygomorphic, perfect or sometimes imperfect (then plants dioecious or polygamous), epigynous, usually small, sessile, with hypanthium shortly to conspicuously prolonged beyond ovary, usually with a pubescent nectariferous disc at ovary apex. **Calyx** of usually 4 or (more often) 5 sepals, distinct, valvate or sometimes imbricate, persistent. **Corolla** of 4 or (more often) 5 petals or absent, distinct, small, greenish white, pink, red, or orange, imbricate or valvate. **Androecium** of 8 or (more often) 10

(biseriate) or sometimes of 4 or 5 (uniseriate) stamens; filaments distinct, inflexed in bud; anthers versatile, dehiscing longitudinally, introrse. **Gynoecium** of 1 pistil, 2- to 5-carpellate; ovary inferior, 1-locular, often ribbed, capped by nectariferous disc at apex; ovules 2 to 6, anatropous, pendulous, with long and slender funicles, placentation apical; style 1; stigma capitate or punctate. **Fruit** a drupe, commonly with leathery exocarp and spongy mesocarp, often somewhat flattened, ribbed to winged; seed 1; endosperm absent; embryo large, straight, often green, oily, with convolute, plicate, or contorted cotyledons.

Family characterization: Woody plants often growing in coastal mangrove swamps and hammocks; simple, entire leaves with paired glands at each blade base and cavities (domatia) on the abaxial surface; small, epigynous flowers with hypanthium prolonged beyond the ovary; 1-locular ovary; pendulous ovules suspended from ovary apex by long, slender funicles; and leathery, ribbed to winged, drupaceous fruit with spongy mesocarp. Tannins and often saponins present. Anatomical features: stellate idioblasts containing calcium oxalate crystals in the leaf mesophyll (appearing as translucent dots if leaf is not too thick); compartmented, thick-walled, unicellular hairs that appear bicellular at base (due to included membrane; "combretaceous hairs," discussed below); and intraxylary phloem (internal phloem) in the leaf veins (bicollateral bundles) and stem (Fig. 104: 1f).

Genera/species: 20/600

Distribution: Mainly tropical, and extending into subtropical areas; especially well represented in Africa, where the species often are shrubby and occur in savannas.

Major genera: *Combretum* (250 spp.) and *Terminalia* (150–250 spp.)

U.S./Canadian representatives: *Bucida* (2 spp.), *Terminalia* (1 or 2 spp.), *Buchenavia* (1 sp.), *Conocarpus* (1 sp.), and *Laguncularia* (1 sp.)

Economic plants and products: Timber, tannins, and edible seeds from *Terminalia* species (Indian-almond, tropical-almond). Ornamental plants (species of at least 6 genera), including *Bucida* (black-olive), *Combretum*, *Conocarpus* (button mangrove, buttonwood, silver buttonbush), *Pteleopsis*, *Quisqualis* (Rangoon-creeper), and *Terminalia*.

Commentary: The Combretaceae generally have been divided into two subfamilies and five tribes based upon the habit type, presence of a corolla, and fruit morphology (Graham 1964). More than two-thirds of the family belong to the large genera *Combretum* and

228

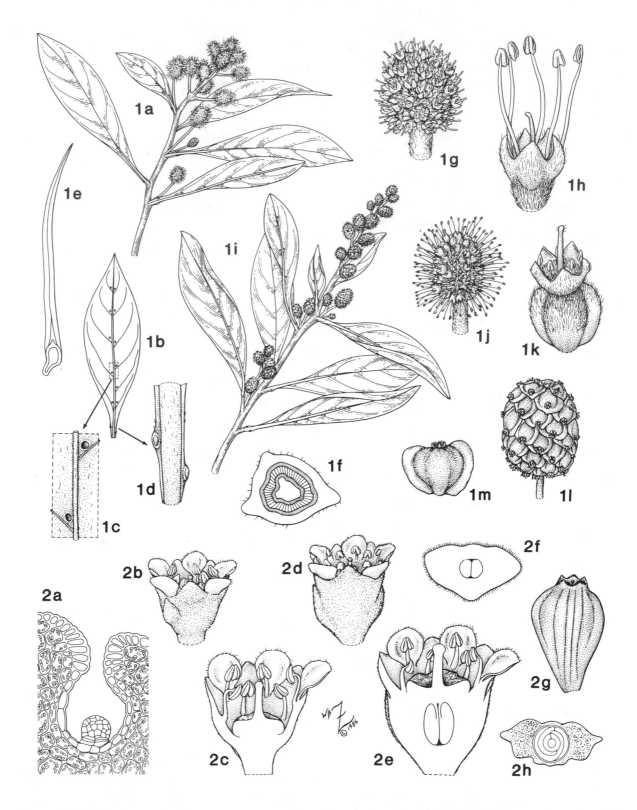

Figure 104. Combretaceae. 1, *Conocarpus erectus*: **a,** staminate flowering branch, ×¾; **b,** leaf (abaxial view), ×¾; **c,** detail of abaxial leaf surface showing midvein and domatia, ×5; **d,** detail of petiole showing extrafloral nectaries, ×5; **e,** combretaceous hair from leaf, ×225; **f,** cross section of stem showing intraxylary phloem, ×12; **g,** staminate inflorescence, ×4; **h,** staminate flower, ×12; **i,** carpellate flowering branch, ×¾; **j,** carpellate inflorescence, ×4½; **k,** carpellate flower, ×12; **l,** infructescence, ×3; **m,** drupe, ×6. **2,** *Laguncularia racemosa*: **a,** cross section of abaxial side of leaf through domatium (note gland at base), ×215; **b,** staminate flower, ×7½; **c,** longitudinal section of staminate flower, ×10; **d,** carpellate flower, ×7½; **e,** longitudinal section of carpellate flower, ×10; **f,** cross section of ovary, ×10; **g,** drupe, ×2; **h,** cross section of drupe, ×2½.

Terminalia. A few are mangrove genera: *Conocarpus*, *Laguncularia*, and *Lumnitzera*.

The family is characterized by several unusual vegetative characters. For example, peculiar unicellular hairs ("combretaceous hairs"; Fig. 104: 1e) with a sharply pointed apex and a thickened, bulbous base are distributed widely in the family (Stace 1965b). These hairs appear to be basally bicellular, as though one smaller hair were enclosed by another (similar to the tufted hair components of the Cistaceae)—an illusion caused by a conical or concave cellulose membrane within the cell wall. Each hair is inserted between the epidermal cells by means of a peg-like foot or stalk at the base. At the base of the leaf blade is a pair of conspicuous glands (extrafloral nectaries; Fig. 104: 1d). On the abaxial (*Conocarpus*; Fig. 104: 1b,c) and sometimes also adaxial (*Laguncularia*) leaf surfaces are several to numerous pits called domatia (Stace 1965a). These small cavities of varying morphology, generally restricted to certain woody dicotyledons, are found on the vegetative organs. Domatia usually develop at the midvein–secondary vein junctions and sometimes elsewhere, such as along the outer leaf margins at the ends of the secondary veins (*Laguncularia*). These cavities frequently are inhabited by mites (and other minute arthropods). In *Laguncularia racemosa*, each flask-shaped domatium contains a capitate, multicellular gland (Fig. 104: 2a).

A well-developed hypanthium, commonly tubular or flattened laterally, characterizes the epigynous flowers of the Combretaceae (Fig. 104: 2c,e). The lower portion of the hypanthium is fused to the ovary, with the upper part prolonged above, forming a shallow to deep cup. The inside base of the cup is lined by a hairy, nectariferous disc that caps the ovary.

Although pollinators (including primates and small mammals for *Combretum*) have been noted for tropical species, little is known about the pollination biology of our members of the family. The small flowers usually are protogynous and probably are pollinated by insects that are attracted to the nectar secreted by the disc. In *Bucida*, the flowers are perfect; the receptive stigma protrudes before anthesis (when the anthers expand). The distinctly unisexual flowers of *Conocarpus erectus* (Fig. 104: 1h,k), packed into globose heads (Fig. 104: 1g,j), occur on separate plants. *Laguncularia racemosa* is functionally dioecious, with very similar staminate and perfect flowers (Fig. 104: 2b,c vs. 2d,e). The staminate form, with a disc and style but lacking an ovary, is somewhat smaller. The pollen from the perfect ("carpellate") flowers is apparently functional. The flowers of *Terminalia catappa* also are staminate and perfect, but both types occur on the same tree (Tomlinson 1980).

The fruits (drupes) of the Combretaceae often are well adapted for water dispersal: a thick exocarp encloses a spongy, aerenchymatous mesocarp that serves as "flotation tissue." The drupes often are flattened, winged, and/or ribbed, and are crowned by the persistent hypanthium. The ribbed fruit of *Laguncularia*, for example, is flattened laterally with two thick and spongy wings (Fig. 104: 2g,h). The seeds sprout while the fruit is still attached to the tree (vivipary; see Rhizophoraceae). The leathery, scale-like fruits of *Conocarpus* also are flattened and two-winged (Fig. 104: 1m) but are aggregated into a cone-like infructescence that eventually breaks apart (Fig. 104: 1l). The almond-shaped fruits of *Terminalia* are slightly winged. Although hard and somewhat nut-like, the "drupes" of *Bucida* have a mesocarp with several aerenchymatous layers.

REFERENCES CITED

Graham, S. A. 1964. The genera of Rhizophoraceae and Combretaceae in the southeastern United States. *J. Arnold Arbor.* 45:285–301.
Stace, C. A. 1965a. Cuticular studies as an aid to plant taxonomy. *Bull. Brit. Mus. (Nat. Hist.), Bot.* 4:1–78.
——. 1965b. The significance of the leaf epidermis in the taxonomy of the Combretaceae. I. A general review of tribal, generic and specific characters. *J. Linn. Soc., Bot.* 59:229–252.
Tomlinson, P. B. 1980. *The biology of trees native to tropical Florida*, pp. 143–153. "Publ. by the author," Petersham, Mass.

ONAGRACEAE
Evening-primrose Family

Annual, biennial, or usually perennial herbs or sometimes shrubs or trees, occasionally aquatic. **Leaves** simple, entire, serrate, or lobed, alternate, opposite, or whorled, stipulate (stipules minute and caducous) or exstipulate. **Inflorescence** indeterminate, spicate, racemose, paniculate, or flowers solitary and axillary. **Flowers** actinomorphic or sometimes zygomorphic, usually perfect, epigynous, with well-developed tubular to campanulate hypanthium (adnate to and often produced beyond the ovary), often showy, with nectariferous disc (at base of style) or nectaries (at lower part of hypanthium). **Calyx** of usually 4 sepals, distinct, deciduous or persistent, valvate. **Corolla** of usually 4 petals, distinct, often clawed, yellow, white, red, or orange, convolute, imbricate or valvate. **Androecium** of 8 (biseriate) or sometimes 4 (uniseriate) stamens, arising from or near hypanthium rim; filaments distinct; anthers versatile, basifixed, sometimes cross-partitioned, dehiscing longitudinally; pollen in tetrads

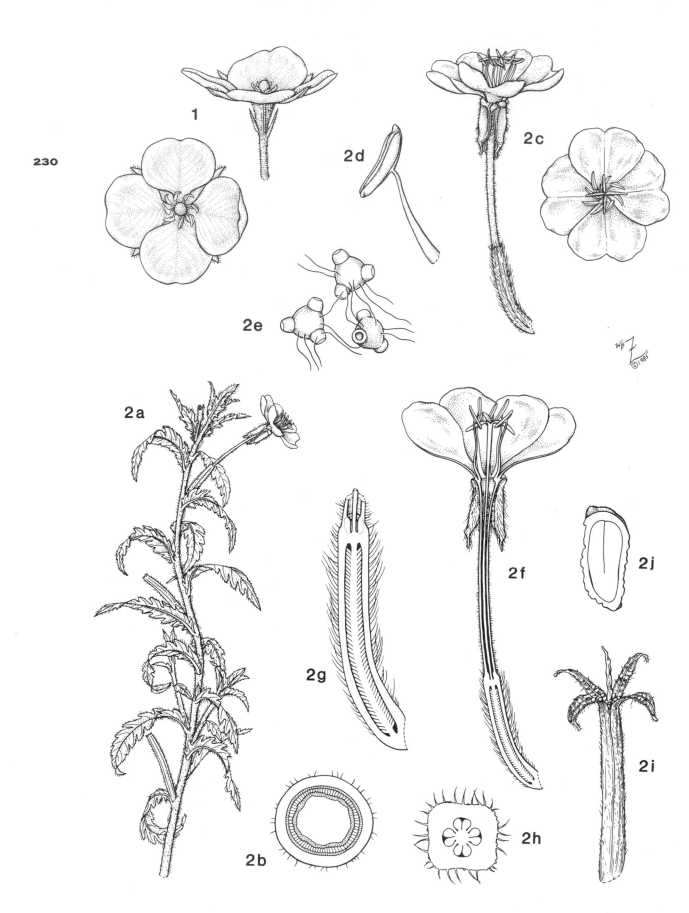

or monads, connected by viscin strands, each grain with 3 pores and protruding "stoppers" (apertures). **Gynoecium** of 1 pistil, 4-carpellate; ovary inferior, 4-locular but sometimes with septa incomplete at base; ovules several to numerous in each locule, anatropous, placentation axile; style 1, slender; stigma 1 and capitate, notched, or 4-lobed, or of 4 radiate branches. **Fruit** usually a loculicidal capsule or sometimes a berry or nutlet; seeds usually numerous, comose or glabrous; embryo oily, generally straight; endosperm absent.

Family characterization: Perennial herbs or shrubs; 4-merous, epigynous flowers with short to elongate hypanthium; distinctive pollen grains (with 3 protruding apertures) connected by viscin strands (see the discussion below); and inferior ovary (often terminated by prolonged hypanthium). Tissues with raphides (bundles of needle-like calcium oxalate crystals) and tannins. Anatomical features: intraxylary phloem (internal phloem) in the stems (Fig. 105: 2b); and unilacunar nodes.

Genera/species: 18/650

Distribution: Temperate and subtropical regions (especially in the New World); most diverse in the western United States and Mexico.

Major genera: *Epilobium* (200–215 spp.), *Fuchsia* (100 spp.), *Oenothera* (80 spp.), and *Ludwigia* (75 spp.)

U.S./Canadian representatives: 13 genera/252 spp.; largest genera: *Camissonia*, *Oenothera*, and *Clarkia*

Economic plants and products: Ornamental plants (species of 12 genera), including *Clarkia* (farewell-to-spring, godetia), *Fuchsia* (lady's-eardrops), *Gaura*, and *Oenothera* (evening-primrose).

Commentary: The Onagraceae are a highly distinctive and monophyletic group characterized by several features (Hoch et al. 1993), including the abundant raphides (bundles of needle-like calcium oxalate crystals) and the unusual pollen morphology (discussed below). The family is included in the order Myrtales (with such families as the Myrtaceae, Melastomataceae, and Lythraceae) in modern classification schemes, sharing such characters as the distinctive hypanthium and internal phloem (Dahlgren and Thorne 1984). The literature on the family is extensive, especially concerning embryology, cytology, and chemistry (see the summary of literature in Conti et al. 1993). Modern revisions are available for most, if not all, genera; *Oenothera*, in particular, has been the subject of many detailed studies (e.g., Cleland 1972). The genera are grouped into several small tribes (most monophyletic) that are separated primarily on the basis of fruit characters (Hoch et al. 1993).

The distinctive pollen of the Onagraceae has long intrigued botanists (see Skvarla et al. 1975, 1976). These relatively large grains often are more or less triangular in outline due to three protruding stopper-like apertures that have a complex morphology. The pollen grains may be shed in monads or, in many species (e.g., many *Ludwigia* spp.), in tetrads. Viscin threads, composed of an elastic and somewhat viscid material, connect the pollen grains (Fig. 105: 2e). These slender strands, which vary in number and structure, also occur on the pollen of certain Ericaceae (i.e., subfamily Rhododendroideae).

The attractive flowers usually are entomophilous (or sometimes bird-pollinated), and typically open in the evening for pollinators (usually bees and Lepidoptera). Nectar, which accumulates in the floral tube, is secreted by a disc at the base of the style or by nectaries within the lower part of the hypanthium. The pollen adheres to the insects by means of the sticky substance (viscin) that forms strands between the grains (or tetrads). Although cross-pollination generally is promoted by protandry, self-pollination also is very prevalent in the family (Raven 1979).

REFERENCES CITED

Cleland, R. E. 1972. *Oenothera: Cytogenetics and evolution.* Academic Press, London.

Conti, E., A. Fischback, and K. J. Sytsma. 1993. Tribal relationships in Onagraceae: Implications from *rbc*L data. *Ann. Missouri Bot. Gard.* 80:672–685.

Dahlgren, R. M. T., and R. F. Thorne. 1984. The order Myrtales: Circumscription, variation, and relationships. *Ann. Missouri Bot. Gard.* 71:633–699.

Hoch, P. C., J. V. Crisci, H. Tobe, and P. E. Berry. 1993. A cladistic analysis of the plant family Onagraceae. *Syst. Bot.* 18:31–47.

Raven, P. H. 1979. A survey of reproductive biology in Onagraceae. *New Zealand J. Bot.* 17:575–593.

Figure 105. Onagraceae. 1, *Ludwigia peruviana*: two views of flower, ×1. **2,** *Oenothera laciniata*: **a,** habit, ×3/5; **b,** cross section of stem showing intraxylary phloem, ×11; **c,** two views of flower, ×1½; **d,** anther, ×11; **e,** pollen grains connected by viscin strands, ×96; **f,** longitudinal section of flower, ×1⅘; **g,** longitudinal section of ovary, ×3½; **h,** cross section of ovary, ×11; **i,** capsule, ×2⅓; **j,** longitudinal section of seed, ×22.

Skvarla, J. J., P. H. Raven, and J. Praglowski. 1975. The evolution of pollen tetrads in Onagraceae. *Amer. J. Bot.* 62:6–35.

——, ——, and ——. 1976. Ultrastructural survey of Onagraceae pollen. In *The evolutionary significance of the exine*, ed. I. K. Ferguson and J. Muller, pp. 447–479. Linn. Soc. Symp. Ser. No. 1. Academic Press, London.

MYRTACEAE
Myrtle Family

Trees or shrubs, aromatic (due to ethereal oils). **Leaves** simple, entire, opposite or alternate, coriaceous, persistent, glandular-punctate, exstipulate. **Inflorescence** determinate, cymose and often appearing racemose, umbellate, or paniculate, or sometimes flower solitary and axillary. **Flowers** actinomorphic, perfect, epigynous or sometimes half-epigynous, generally showy, with well-developed hypanthium (often prolonged beyond the ovary), with nectariferous disc (on summit of ovary or lining the hypanthium), usually subtended by 2 bracts. **Calyx** of 4 or 5 sepals (sometimes much reduced or absent), distinct or basally connate, imbricate or sometimes undivided in bud (then splitting irregularly at anthesis or deciduous as a calyptra). **Corolla** of 4 or 5 petals, distinct (or sometimes connivent and forming a calyptra or absent), usually white or red, imbricate. **Androecium** of numerous stamens; filaments distinct or sometimes basally connate into 4 or 5 fascicles opposite the petals, inflexed in bud; anthers dorsifixed, versatile, with connective apex often glandular, usually dehiscing longitudinally, introrse. **Gynoecium** of 1 pistil, usually 2- to 5-carpellate; ovary inferior or sometimes half-inferior, 1-locular or usually with as many locules as carpels; ovules 2 to many in each locule, anatropous or campylotropous, pendulous, placentation basically parietal (1-locular ovary) but often appearing axile due to coalescent intruded placentae; style 1, elongate; stigma 1, capitate. **Fruit** usually a berry or loculicidal capsule; endosperm scanty and starchy or absent; embryo more or less bent or spirally rolled, variously shaped.

Family characterization: Trees or shrubs with aromatic parts; simple, coriaceous, glandular-punctate leaves; 4- or 5-merous perianth and numerous stamens; well-developed hypanthium; inferior ovary with axile or deeply intruding parietal placentae; and a berry or loculicidal capsule as the fruit type. Tissues with calcium oxalate crystals and a high tannin content. Anatomical features: schizogenous secretory cavities and lysigenous glands containing ethereal oils (appearing on the leaves as translucent dots; Fig. 106: 1b);

intraxylary phloem in the petiole (bicollateral bundles) and stem; and unilacunar nodes.

Genera/species: 144/3,000

Distribution: Tropical and subtropical regions; particularly diverse in Australia and tropical America.

Major genera: *Eugenia* (1,000 spp.), *Syzygium* (500 spp.), *Eucalyptus* (450–500 spp.), and *Myrcia* (250–500 spp.)

U.S./Canadian representatives: 17 genera/80 spp.; largest genera: *Eugenia* and *Calyptranthes*

Economic plants and products: Edible fruits from *Eugenia* (Surinam-cherry), *Myrciaria* (jaboticaba), *Psidium* (guava), and *Syzygium* (rose-apple). Spices and ethereal oils from *Eucalyptus* (oil), *Melaleuca* (cajeput oil), *Pimenta* (allspice, the unripe berries; pimento; oil-of-bay-rum), and *Syzygium* (cloves, the dried flower buds). Timbers from several, such as species of *Eucalyptus* and *Eugenia*. Weedy species in south Florida: *Melaleuca quinquenervia* and *Rhodomyrtus tomentosus*. Ornamental trees and shrubs (species of 38 genera), including *Callistemon* (bottlebrush), *Eucalyptus* (gum tree), *Eugenia* (Australian bush-cherry, Surinam-cherry), *Feijoa* (pineapple-guava), *Melaleuca* (bottlebrush, cajeput tree, punk tree), *Myrtus* (myrtle), and *Rhodomyrtus* (downy-myrtle).

Commentary: Depending on the circumscription, the Myrtaceae are divided into two to four subfamilies primarily based upon the fruit type (berry, capsule, or dry and indehiscent fruit; Schmid 1980; Dahlgren and Thorne 1984). The generic delimitations (e.g., *Eugenia–Syzygium*) and relationships vary considerably with different classifications (see Wilson 1960; Johnson and Briggs 1984). A problem with establishing generic limits is the relatively uniform flower structure (discussed below) in this apparently monophyletic family (Tomlinson 1980).

The prominent field characters are the glandular-punctate leaves with a spicy odor, epigynous flowers, and numerous stamens on a cup-shaped receptacle. Sometimes, as in *Psidium*, the sepals are fused into a cap that falls off when the flower expands. A well-developed nectariferous disc lines the hypanthium or tops the ovary (e.g., Fig. 106: 1e). The description of myrtaceous flowers as "hypogynous to epigynous" by some authors reflects the variation in the degree of fusion of the hypanthium to the top or sides of the ovary. According to Briggs and Johnson (1979), the ovary is never completely inferior or superior, and the general flower type of the family is, thus, probably best described as "perigynous."

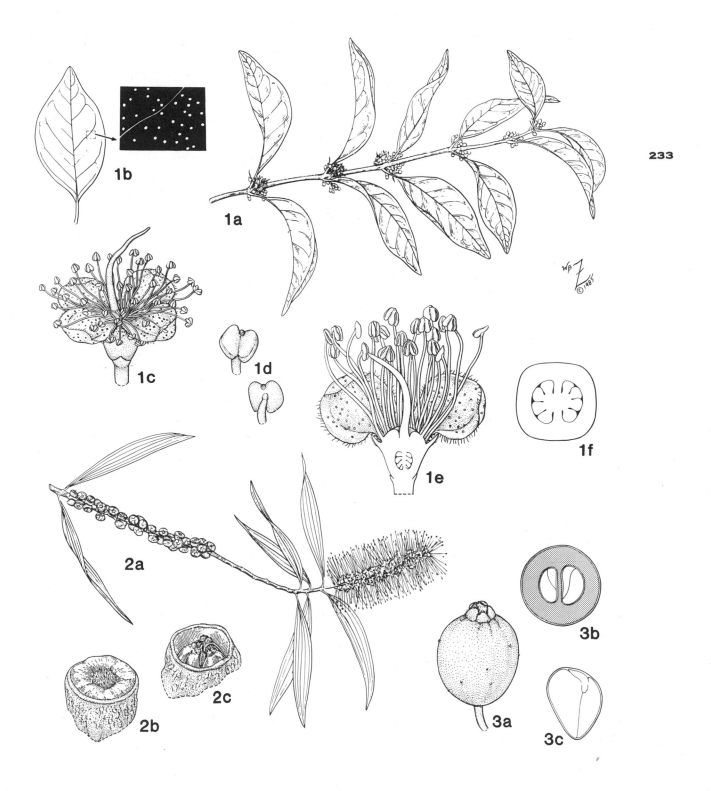

Figure 106. Myrtaceae. 1, *Eugenia axillaris*: **a,** flowering branch, ×½; **b,** leaf, ×½, with detail showing punctate surface, ×6; **c,** flower, ×8; **d,** two views of anther, ×25; **e,** longitudinal section of flower, ×9; **f,** cross section of ovary, ×20. **2,** *Melaleuca quinquenervia*: **a,** flowering and fruiting branch, ×½; **b,** fruit (capsule enclosed by hypanthium), ×4½; **c,** dehisced fruit, ×4½. **3,** *Myrcianthes fragrans*: **a,** berry, ×2; **b,** cross section of berry, ×2; **c,** longitudinal section of seed, ×3½.

Insects (and sometimes birds) seeking nectar are attracted to the flowers, which usually have very conspicuous stamens (e.g., *Melaleuca*, Fig. 106: 2a; *Callistemon*) creating a "bottle-brush" effect, or showy petals with tufts of stamens as in *Myrtus* or *Myrcianthes* (Briggs and Johnson 1979). Myrtaceous flowers often are sweet scented.

234

REFERENCES CITED

Briggs, B. G., and L. A. S. Johnson. 1979. Evolution in the Myrtaceae—Evidence from inflorescence structure. *Proc. Linn. Soc. New South Wales* 102:157–256.

Dahlgren, R. M. T., and R. F. Thorne. 1984. The order Myrtales: Circumscription, variation, and relationships. *Ann. Missouri Bot. Gard.* 71:633–699.

Johnson, L. A. S., and B. G. Briggs. 1984. Myrtales and Myrtaceae—A phylogenetic analysis. *Ann. Missouri Bot. Gard.* 71:700–756.

Schmid, R. 1980. Comparative anatomy and morphology of *Psiloxylon* and *Heteropyxis*, and the subfamilial and tribal classification of the Myrtaceae. *Taxon* 29:559–595.

Tomlinson, P. B. 1980. *The biology of trees native to tropical Florida*, pp. 265–285. "Publ. by the author," Petersham, Mass.

Wilson, K. A. 1960. The genera of Myrtaceae in the southeastern United States. *J. Arnold Arbor.* 41:270–278.

LOGANIACEAE
Logania Family

Woody vines, shrubs, trees, or occasionally herbs. **Leaves** simple, entire, opposite, stipulate; stipules usually interpetiolar, membranous or often reduced to a ridge, typically with adaxial glands (colleters) at base (and on adjoining stem surface). **Inflorescence** determinate, cymose, or sometimes flowers solitary, terminal or axillary. **Flowers** actinomorphic to slightly zygomorphic, usually perfect, hypogynous, showy, sometimes heterostylous (dimorphic), sometimes with small nectariferous disc. **Calyx** usually synsepalous with 4 or 5 lobes, variously basally connate (to almost distinct), tubular, often with glands (colleters) on inner surface, imbricate. **Corolla** sympetalous with 4 or 5 lobes, tubular, funnelform, or urceolate, variously colored, imbricate, convolute, or valvate. **Androecium** of 4 or 5 stamens (as many as corolla lobes), epipetalous; filaments distinct; anthers basifixed, often with sagittate base, dehiscing longitudinally, usually introrse. **Gynoecium** of 1 pistil, usually 2-carpellate; ovary superior, 2-locular; ovules few to numerous, anatropous to hemitropous, placentation axile; style 1, simple or 2- or 4-branched; stigma capitate or lorate. **Fruit** usually a septicidal capsule (with also loculicidal dehiscence) or sometimes a berry; seeds flattened or ellipsoidal, often winged; endosperm usually copious, fleshy to cartilaginous; embryo straight, embedded in endosperm.

Family characterization: Woody plants (lianas, shrubs, or trees); opposite, simple, stipulate leaves with entire margins; interpetiolar stipules (or ridge) with glands (colleters) on adaxial surface; sympetalous flowers with 4- or 5-merous perianth and androecium; 2-carpellate, superior ovary with axile placentation; and often a septicidal capsule that also dehisces along the locules as the fruit type. Iridoid compounds and various alkaloids present. Tissues with calcium oxalate crystals. Anatomical features: intraxylary phloem in the petiole (bicollateral bundles) and stem (Fig. 107: c); unitegmic and tenuinucellate ovules; and unilacunar nodes (Fig. 107: c). Interxylary phloem (included phloem) present in the wood of several genera (e.g., *Strychnos*).

Genera/species: 17/480

Distribution: Widespread in tropical and subtropical regions, with a few representatives in temperate areas; often in dry, lowland habitats.

Major genera: *Strychnos* (190–200 spp.), *Geniostoma* (52–60 spp.), and *Spigelia* (50 spp.)

U.S./Canadian representatives: *Geniostoma* (25 spp.), *Spigelia* (6 spp.), *Mitreola* (= *Cynoctonum*; 3 spp.), *Gelsemium* (2 spp.), and *Strychnos* (1 sp.)

Economic plants and products: Many poisonous/medicinal plants (due to alkaloids, iridoids, and/or saponins), such as *Gelsemium* (yellow-jessamine root; alkaloids such as sempervirine and gelsemine) and *Strychnos* (curare component/medications; alkaloids such as strychnine and brucine extracted from bark and seeds). Species of at least six genera cultivated as ornamentals, including *Desfontainia*, *Gelsemium* (Carolina-jessamine, yellow-jessamine), *Geniostoma*, *Logania*, *Spigelia* (Indian-pink, pinkroot, worm-grass), and *Strychnos* (Natal-orange).

Commentary: The circumscription of the Loganiaceae, a "garbage can" family, has been variously interpreted (see Rogers 1986 and Bremer and Struwe 1992 for summaries). As many as five segregate families (e.g., Strychnaceae, Spigeliaceae) have been recognized, and several genera (e.g., *Polypremum*) sometimes have been transferred to other related families. The Buddlejaceae (*Buddleja* plus several genera of uncertain placement), previously often included within the Loganiaceae (Tiagi and Kshetrapal 1972), now are generally treated as a distinct family in a different order (Scrophulariales; Cronquist 1988; Thorne 1992). The Loganiaceae here are rather strictly defined (Thorne unpublished update [1992]), including basically only

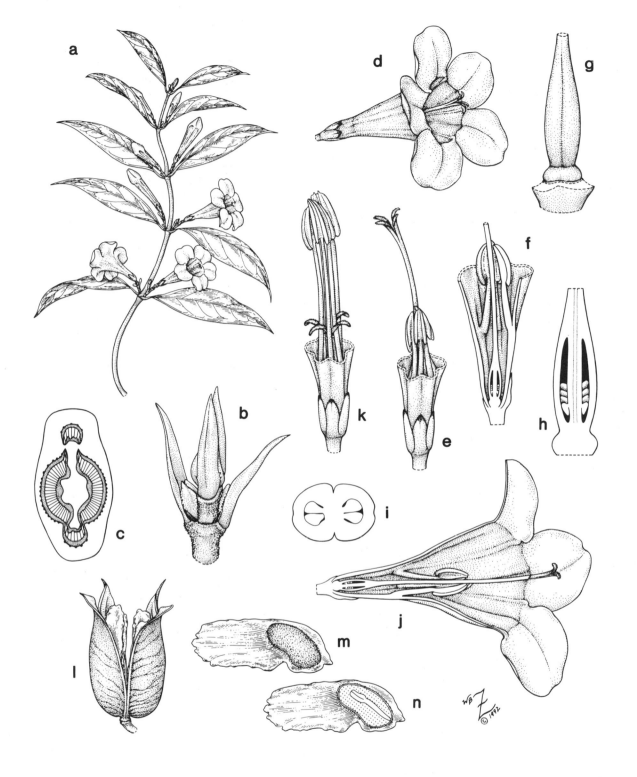

Figure 107. Loganiaceae. *Gelsemium sempervirens*: **a,** habit, ×½; **b,** node showing stipules and colleters on shoot tip, ×10; **c,** cross section of node (one-trace unilacunar) showing intraxylary phloem, ×10; **d–j,** long-styled flower: **d,** long-styled flower, ×1½; **e,** long-styled flower (corolla cut near base; stamens short), ×2½; **f,** longitudinal section of long-styled flower base, ×3; **g,** ovary and disc, ×8; **h,** longitudinal section of ovary, ×10; **i,** cross section of ovary, ×20; **j,** longitudinal section of long-styled flower, ×2½; **k,** short-styled flower (corolla cut near base; stamens long), ×2½; **l,** capsule, ×2; **m,** seed, ×4; **n,** longitudinal section of seed, ×4.

the former subfamily Loganioideae, a conclusion reached by Jensen (1992) in a detailed examination of the morphology and chemistry of the Loganiaceae *s.l.* Cladistic analyses of morphological data by Bremer and Struwe (1992) also indicate that the Loganiaceae *s.l.* are paraphyletic, and that further analyses, including all genera, are needed. Thorne suggests the transfer of *Polypremum* to the Scrophulariaceae, and of the Potalioideae (*Anthocleista*, *Fagraea*, and *Potalia*) to the Gentianaceae, as well as the elevation of the Plocospermatioideae (*Plocosperma*) to a segregate family. The Loganiaceae *s.s.* (Loganioideae) generally are divided into a varying number of tribes depending on the corolla aestivation and fruit morphology. The close relationship of certain taxa (*Gelsemium* and allies) to the Rubiaceae is supported by cladistic analyses of morphological characters (Bremer and Struwe 1992).

Vegetatively, the woody plants are characterized by the opposite, stipulate leaves with entire margins. The stipules are situated between the petioles (interpetiolar, as in many of the Rubiaceae) and vary from membranous structures (*Spigelia*) to a line or ridge between the petioles (*Mitreola* species). As in the Rubiaceae, colleters (glandular hairs) often occur on the inner surface of the stipules, and sometimes also on the leaf bases, the bracts, or inside the calyx (e.g., *Spigelia*). For example, in *Gelsemium*, the petiole bases are connected by a stipular flange and colleters occur in the leaf axils and between the petioles (Fig. 107: b).

The tubular to urceolate corolla of the family generally attracts various insects, as well as hummingbirds (e.g., for *Spigelia*) as pollinators. Sometimes the flowers are fragrant and have a nectar disc, as in *Gelsemium* (Fig. 107: g). Notable also is the distylous outcrossing mechanism, discussed in detail under the Rubiaceae and Pontederiaceae, found in *Gelsemium* (see Fig. 107: e,f,k; Ornduff 1970).

The fruit type of the family usually is a more or less compressed, septicidal capsule that also dehisces along the locules in varying degrees (Fig. 107: l). The two or four valves break away from the central axis to release the seeds, which are often winged (Fig. 107: m). The fruit of *Strychnos* is a berry.

REFERENCES CITED

Bremer, B., and L. Struwe. 1992. Phylogeny of the Rubiaceae and the Loganiaceae: Congruence or conflict between morphological and molecular data? *Amer. J. Bot.* 79:1171–1184.

Cronquist, A. 1988. *The evolution and classification of flowering plants*, 2d ed., pp. 431–436. New York Botanical Garden, Bronx.

Jensen, S. R. 1992. Systematic implications of the distribution of iridoids and other chemical compounds in the Loganiaceae and other families of the Asteridae. *Ann. Missouri Bot. Gard.* 79:284–302.

Ornduff, R. 1970. The systematics and breeding system of *Gelsemium* (Loganiaceae). *J. Arnold Arbor.* 51:1–17.

Rogers, G. K. 1986. The genera of Loganiaceae in the southeastern United States. *J. Arnold Arbor.* 67:143–185.

Thorne, R. F. 1992. Classification and geography of the flowering plants. *Bot. Rev.* (Lancaster) 58:225–348.

Tiagi, Y. D., and S. Kshetrapal. 1972. Studies on the floral anatomy, evolution of the gynoecium and relationships of the family Loganiaceae. In *Advances in plant morphology*, ed. Y. S. Murty, B. M. Johri, H. Y. Mohan Ram, and T. M. Varghese, pp. 408–416. Sarita Prakashan, Meerut, India.

RUBIACEAE
Coffee or Madder Family

Most frequently trees or shrubs (in the tropics), sometimes lianas, or herbs (in temperate regions). **Leaves** simple, usually entire, opposite or apparently whorled, decussate, stipulate; stipules interpetiolar or sometimes intrapetiolar, with glands (colleters) on adaxial surface or margins. **Inflorescence** determinate, basically cymose and appearing paniculate or sometimes capitate, or reduced to a single flower, terminal or axillary. **Flowers** actinomorphic, perfect, epigynous, often heterostylous (dimorphic), with nectariferous disc. **Calyx** synsepalous with 4 or 5 lobes, sometimes unequal with one or more lobes enlarged, sometimes with glands (colleters) on inner surface, lobes not overlapping in bud, persistent and sometimes becoming enlarged in fruit. **Corolla** sympetalous with 4 or 5 lobes, variously shaped but usually rotate, salverform, or funnelform, often hairy inside throat, white or brightly colored, valvate, imbricate, or contorted. **Androecium** of 4 or 5 stamens (as many as corolla lobes), epipetalous; filaments distinct; anthers dorsifixed, dehiscing longitudinally, introrse. **Gynoecium** of 1 pistil, usually 2-carpellate; ovary inferior, usually 2-locular, capped by nectariferous disc; ovules 1 to many in each locule, anatropous, erect, pendulous, or horizontal, placentation axile or appearing basal or apical; style bifid or sometimes simple; stigma(s) linear (on bifid styles) or capitate or 2-lobed (on simple style). **Fruit** a loculicidal or septicidal capsule, a schizocarp splitting into 2 segments, or sometimes a berry or drupe; seeds sometimes winged; endosperm usually copious, fleshy; embryo small, usually straight.

Family characterization: Herbs to shrubs; opposite, simple, stipulate leaves with entire margins; interpetiolar stipules with glands (colleters) on adaxial surface or margins; epigynous, sympetalous, actinomorphic flowers (which are often heterostylous); 4- or 5-mer-

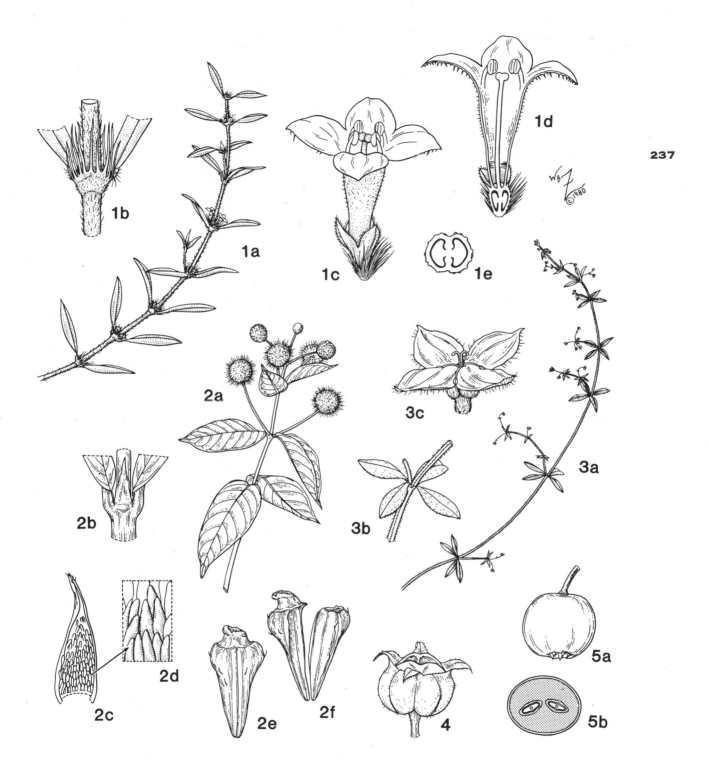

Figure 108. Rubiaceae. 1, *Diodia teres*: **a,** habit, ×½; **b,** detail of node showing stipules, ×2½; **c,** flower, ×6; **d,** longitudinal section of flower, ×6; **e,** cross section of ovary, ×12. **2,** *Cephalanthus occidentalis*: **a,** flowering branch, ×¼; **b,** detail of young node showing stipules, ×2¼; **c,** adaxial side of stipule with colleters, ×6; **d,** colleters from stipule, ×25; **e,** schizocarp before dehiscence, ×4½; **f,** dehiscing schizocarp, ×4½. **3,** *Galium hispidulum*: **a,** habit, ×½; **b,** node showing foliaceous stipules, ×2¼; **c,** flower, ×12. **4,** *Hedyotis corymbosa*: capsule, ×12. **5,** *Chiococca alba*: **a,** drupe, ×3; **b,** cross section of drupe, ×3.

ous androecium and perianth; and 2-carpellate ovary with bifid styles. Various alkaloids and iridoid compounds commonly present. Tissues with calcium oxalate crystals (commonly raphides). Anatomical feature: unitegmic and tenuinucellate ovules.

Genera/species: 500–600/9,000

Distribution: Predominantly tropical to subtropical (trees and shrubs) and most abundant in northern South America; several herbaceous representatives predominate in temperate zones.

Major genera: *Psychotria* (700–1,400 spp.), *Galium* (400 spp.), *Ixora* (400 spp.), *Pavetta* (350–400 spp.), *Tarenna* (370 spp.), *Oldenlandia* (300 spp.), *Randia* (200–300 spp.), *Gardenia* (200–250 spp.), and *Mussaenda* (200 spp.)

U.S./Canadian representatives: 60 genera/317 spp.; largest genera: *Galium, Hedyotis,* and *Psychotria*

Economic plants and products: Medicinal plants (contain alkaloids): *Cephaelis* (ipecac), *Cinchona* (quinine), and *Coffea* (coffee). Dyes from *Rubia* (madder) and *Uncaria* (gambier). Ornamental plants (species of 66 genera), including *Gardenia, Hamelia* (scarletbush), *Ixora, Pentas* (starcluster), and *Serissa.*

Commentary: The delimitation of the intrafamilial taxa of the Rubiaceae is controversial, and the family historically has been divided into an array of various subfamilies and tribes (Verdcourt 1958; Bremekamp 1966). In addition, many of the large genera are taxonomically difficult, as for example *Psychotria* and the *Hedyotis–Oldenlandia* complex. Opinion also differs greatly on the relationships of the Rubiaceae to other families (see Lee and Fairbrothers 1978 for a summary). Recent cladistic analyses of molecular data (Bremer and Jansen 1991) have evaluated and supported the redefinition of certain tribes; phylogenetic analyses of morphological characters (Bremer and Struwe 1992) support the close relationship of the family to certain taxa (*Gelsemium* and allies) of the Loganiaceae.

Rubiaceous plants have prominent stipules, which occur usually between the petioles (interpetiolar) or sometimes between the petiole and stem (intrapetiolar). The stipules often subtend and protect the apex of the growing shoot. Often a stipule of one leaf is connate with the adjacent stipule of the opposite leaf, forming a sheath (*Hedyotis*) or even a foliaceous structure, as in *Galium* (Fig. 108: 3b). The number of whorled "leaves" at each node of *Galium* varies depending on the amount of fusion and splitting of the stipules. Generally, the true leaves can be discerned by the presence of axillary buds. Sometimes the stipules

in herbaceous species are split into bristle-like structures (as in *Diodia*; Fig. 108: 1b), which may have apical glands (*Pentas*). Rubiaceous plants consistently have glandular hairs called colleters on the adaxial surface or margins of the stipules, and sometimes inside the calyx (Fig. 108: 2c,d; Lersten 1975). In most of the Rubiaceae, the colleters are cylindrical to more or less conical. The waxy or mucilaginous secretory products of these glands may aid in the protection of buds (Robbrecht 1988).

Another prominent feature of the family is the occurrence of heterostylous flowers in a number of species. Heterostyly ("different styles") is a genetically controlled type of flower polymorphism in which flowers of the same species have mature styles of two (distylous flowers) or three (tristylous flowers) lengths (Barrett 1992). In the Rubiaceae, heterostyly involves two style lengths, "long" and "short," as shown for the Turneraceae (Fig. 52: e–k), Lythraceae (Fig. 102: 1g–n), and Loganiaceae (Fig. 107: e,f,k); for a discussion of trimorphic heterostyly, see the commentary under Pontederiaceae (and Fig. 146). Each of the style lengths usually is associated with a different corolla tube length, filament length, stigmatic papilla size, and pollen grain size and ornamentation (Faegri and van der Pijl 1980). The genetic mechanism of heterostyly, which safeguards cross-pollination, has been studied extensively (see Vuilleumier 1967; Ganders 1979; Lewis and Jones 1992). A much higher degree of fertility results from the pollination of a long-styled flower with pollen from a short-styled flower (or vice versa) than from the pollination of a stigma by pollen from a flower of the same type. The Rubiaceae probably have more species with dimorphic heterostyly than any other angiosperm family (Barrett and Richards 1990), and perhaps even more than those of all other angiosperms combined (Anderson 1973). Although the phenomenon occurs in numerous genera throughout the family, heterostyly is most common in the herbaceous species.

The white or brightly colored rubiaceous flowers easily attract pollinators to the nectar secreted by the epigynous disc. Often small flowers are aggregated into conspicuous dense inflorescences, as in *Cephalanthus* (Fig. 108: 2a). The long-tubed flowers of many species often are visited by Hymenoptera and Lepidoptera. The more open flowers of several plants, such as *Galium* (Fig. 108: 3c), are visited chiefly by small insects, such as flies. The flowers of many woody species, such as those of *Gardenia* and *Ixora*, are characterized by a stylar pollen presentation mechanism similar to that of the Asteraceae and Campanulaceae (Robbrecht 1988): pollen is deposited (often in bud) onto the upper portion of the style and/or outside the (nonreceptive) stigma; the receptive stigma expands later.

REFERENCES CITED

Anderson, W. R. 1973. A morphological hypothesis for the origin of heterostyly in the Rubiaceae. *Taxon* 22:537–542.

Barrett, S. C. H. 1992. Heterostylous genetic polymorphisms: Model systems for evolutionary analysis. In *Evolution and function of heterostyly*, ed. S. C. H. Barrett, pp. 1–29. Monogr. Theor. & Appl. Genet. 15. Springer-Verlag, Berlin.

—— and J. H. Richards. 1990. Heterostyly in tropical plants. *Mem. New York Bot. Gard.* 55:35–61.

Bremekamp, C. E. B. 1966. Remarks on the position, the delimitation and subdivision of the Rubiaceae. *Acta Bot. Neerl.* 15:1–33.

Bremer, B., and R. K. Jansen. 1991. Comparative restriction site mapping of chloroplast DNA implies new phylogenetic relationships within the Rubiaceae. *Amer. J. Bot.* 78: 198–213.

—— and L. Struwe. 1992. Phylogeny of the Rubiaceae and the Loganiaceae: Congruence or conflict between morphological and molecular data? *Amer. J. Bot.* 79:1171–1184.

Faegri, K., and L. van der Pijl. 1980. *The principles of pollination ecology*, pp. 31–33. Pergamon Press, Oxford.

Ganders, F. R. 1979. The biology of heterostyly. *New Zealand J. Bot.* 17:607–635.

Lee, S. Y., and D. E. Fairbrothers. 1978. Serological approaches to the systematics of the Rubiaceae and related families. *Taxon* 27:159–185.

Lersten, N. R. 1975. Colleter types in Rubiaceae, especially in relation to the bacterial leaf nodule symbiosis. *Bot. J. Linn. Soc.* 71:311–319.

Lewis, D., and D. A. Jones. 1992. The genetics of heterostyly. In *Evolution and function of heterostyly*, ed. S. C. H. Barrett, pp. 129–150. Monogr. Theor. & Appl. Genet. 15. Springer-Verlag, Berlin.

Robbrecht, E. 1988. Tropical woody Rubiaceae. *Opera Bot. Belg.* 1:1–271.

Verdcourt, B. 1958. Remarks on the classification of the Rubiaceae. *Bull. Jard. Bot. État* 28:209–290.

Vuilleumier, B. S. 1967. The origin and evolutionary development of heterostyly in the angiosperms. *Evolution* 21:210–226.

APOCYNACEAE
Dogbane Family
Including the Asclepiadaceae:
Milkweed Family

Perennial herbs, lianas, shrubs, or sometimes small trees, sometimes succulent, with milky sap and often fleshy or woody tubers. **Leaves** simple, entire, opposite or whorled, decussate, with colleters frequently at petiole base (and on adjoining stem surface), exstipulate or stipulate (stipules minute). **Inflorescence** determinate, cymose and appearing racemose or umbellate, terminal or lateral and interpetiolar. **Flowers** actinomorphic, perfect, hypogynous to slightly perigynous, showy. **Calyx** of 5 sepals, distinct or basally connate, often with adaxial glands (colleters) or nonglandular projections at base, imbricate. **Corolla** sympetalous with 5 lobes, rotate or sometimes salverform to funnelform, often with appendages either inside the tube or terminating the tube, color extremely variable, contorted, imbricate, or valvate. **Androecium** of 5 stamens, adherent or adnate to the stigmatic head (forming a gynostegium), epipetalous; filaments short, connate into a tube, often with various appendages; anthers 2- or 4-locular, basifixed, connivent to connate, often apically and sometimes basally appendaged, dehiscing longitudinally or by an apical pore, introrse; pollen grains in tetrads, granular or grouped into waxy masses (pollinia). **Gynoecium** of 1 pistil with only styles and/or stigmas connate, 2-carpellate; ovaries superior to slightly inferior (Woodson 1930), distinct, 1-locular; ovules numerous in each carpel, anatropous, pendulous, placentation parietal; style(s) connate and 1, or distinct and 2; stigma 1, variously shaped, bilobed to 5-angled, greatly enlarged, with various receptive areas (entire surface to a few very restricted areas on the stigmatic head). **Fruit** 2 (aggregate) or 1 (by abortion) follicle(s) dehiscing along adaxial suture, linear to ovoid, fleshy to woody, smooth or variously armed, or occasionally indehiscent and a berry or drupe; seeds ovate to oblong, compressed, usually with an apical tuft of long silky hairs (coma); endosperm scanty to copious, fleshy to cartilaginous; embryo usually large, straight or bent.

Family characterization: Herbs to shrubs with laticiferous parts; opposite or whorled, entire leaves with colleters at petiole base; calyx often with adaxial glandular (colleters) or nonglandular projections at base; epipetalous stamens and sympetalous corolla with various appendages; tetradenous pollen often united into pollinia; 2-carpellate pistil with distinct ovaries (connate only at the styles and/or stigmas) and a large, specialized stigmatic head; and follicular fruit with comose seeds. Gynostegium (compound structure formed by fusion of androecium and gynoecium) present in more derived apocynads. Various alkaloids and glycosides present. Tissues with calcium oxalate crystals. Anatomical features: unsegmented laticiferous tubes; intraxylary phloem in the petiole (bicollateral bundles) and stem (Figs. 109: 1b; 110: 1b); unitegmic and tenuinucellate ovules; and unilacunar nodes (Fig. 109: 1b).

Genera/species: 355/3,700

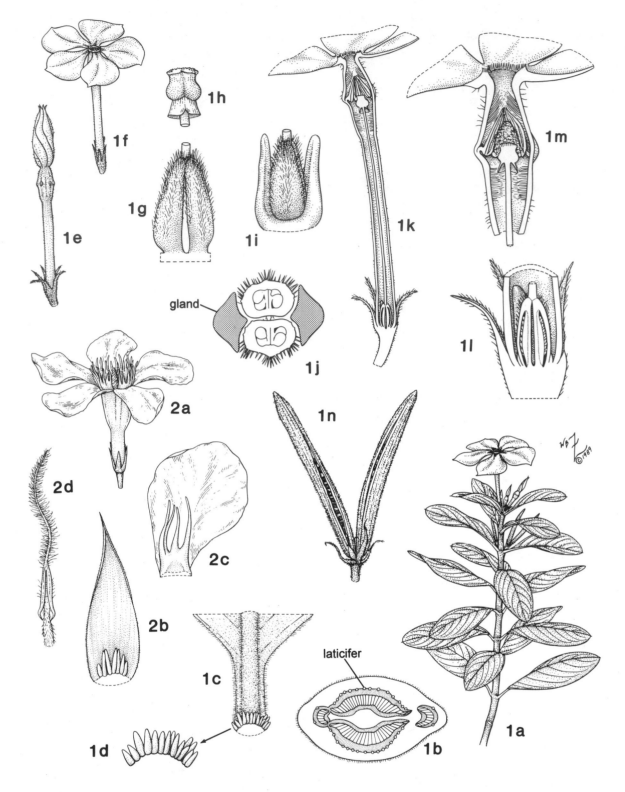

240

gland

laticifer

Figure 109. Apocynaceae. 1, *Catharanthus roseus*: **a,** habit, ×½; **b,** cross section of node (one-trace unilacunar) showing intraxylary phloem and laticifers, ×5; **c,** detail of adaxial side of petiole base showing colleters, ×6; **d,** colleters, ×12; **e,** bud, ×1½; **f,** flower, ×¾; **g,** ovaries, ×12; **h,** stigmatic head, ×12; **i,** ovaries and glands, ×12; **j,** cross section of ovaries and glands, ×20; **k,** longitudinal section of flower, ×3; **l,** longitudinal section of basal portion of flower, ×7; **m,** longitudinal section of upper portion of flower (dehisced anthers depositing pollen on sterile apex of stigmatic head), ×5; **n,** follicles (with persistent calyx and glands at base), ×2¼. **2,** *Nerium oleander*: **a,** flower, ×⅔; **b,** adaxial surface of sepal showing colleters at base, ×4; **c,** adaxial side of corolla lobe with appendage, ×1¾; **d,** stamen, ×3½.

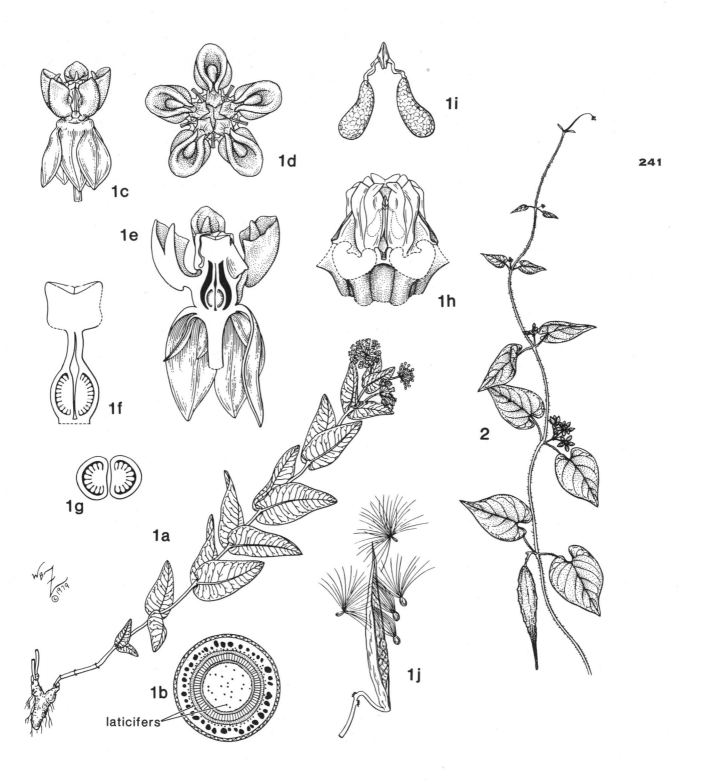

Figure 110. Apocynaceae (continued). 1, *Asclepias humistrata*: **a,** habit, ×⅒; **b,** cross section of stem showing laticifers and intraxylary phloem, ×5; **c,** lateral view of flower, ×3; **d,** top view of flower, ×5; **e,** longitudinal section of flower, ×5; **f,** longitudinal section of gynoecium, ×10; **g,** cross section of ovaries, ×12; **h,** gynostegium ("hoods" and "horns" removed) showing the position of one pair of pollinia, ×10; **i,** pair of pollinia plus gland and translator arms, ×20; **j,** follicle with comose seeds, ×⅓. **2,** *Matelea floridana*: habit, ×⅓.

TABLE 15. Major morphological differences traditionally used to separate the "Apocynaceae" *s.s.* and the "Asclepiadaceae," which are combined here as one family, the Apocynaceae *s.l.* (Thorne 1992). Note that many of these features actually overlap (see the family commentary). For a detailed comparison of these two taxa, see the discussion in Rosatti (1989).

CHARACTER	"APOCYNACEAE" *s.s.* (Apocynoideae and Plumerioideae)	"ASCLEPIADACEAE" (Asclepiadoideae, Periplocoideae, and Secamonoideae)
GENERA/SPECIES	188/1,500–2,900	167/2,000–2,100
DISTRIBUTION	mainly tropical to subtropical, with a few in temperate areas	tropical to subtropical, with relatively few in temperate areas
HABIT	trees, shrubs, woody vines, or sometimes herbs	herbs, woody or herbaceous vines, or seldom shrubs or trees
COROLLA SHAPE	funnelform or salverform	rotate, or often tubular with reflexed lobes, sometimes salverform
"CORONA"	often with distinct appendages or ring within corolla tube no filament appendages (i.e., no true "corona")	sometimes with distinct appendages or ring within corolla tube usually with well-developed filament appendages ("hood" + "horn"; true "corona")
NECTARIES	conspicuous disc or glands around gynoecium base, or absent	secretory cells lining stigmatic chamber
FILAMENT FUSION	distinct	distinct to (usually) connate into a tube
ANTHER FUSION	distinct or connivent free or adherent to stylar head	coherent to connate adherent to usually adnate to stylar head (forming the gynostegium)
POLLEN	monads, or sometimes tetrads (then ± cohering)	tetrads; granular, or usually connate into loose to firm pollinia
FRUIT TYPE	1 or 2 follicles, or sometimes a berry, drupe, or capsule	1 or 2 follicles
SEEDS	comose, winged, or lacking such structures	usually comose

Distribution: Almost exclusively pantropical; relatively few representatives in temperate regions.

Major genera: *Hoya* (90–200 spp.), *Cynanchum* (100–150 spp.), *Matelea* (130 spp.), and *Asclepias* (120 spp.)

U.S./Canadian representatives: 42 genera/223 spp.; largest genera: *Asclepias, Matelea, Cynanchum,* and *Amsonia*

Economic plants and products: Many toxic plants (with alkaloids and glucosides that commonly poison livestock), including species of *Apocynum* (dogbane) and *Asclepias* (milkweed). Medicinal plants: *Catharanthus* (vincristine and vinblastine, antileukemia drugs), *Rauvolfia* (reserpine, a hypertension drug), and *Strophanthus* (strophanthin, a cardiac glycoside). Ornamental plants (species of 60 genera), including *Allamanda*, *Amsonia* (bluestar), *Asclepias* (butterfly-weed, milkweed), *Carissa* (Natal-plum), *Catharanthus* (Madagas-car-periwinkle), *Hoya* (wax plant), *Nerium* (oleander), *Plumeria* (frangipani), *Stapelia* (carrion-flower), *Thevetia* (luckynut), and *Vinca* (periwinkle).

Commentary: In accordance with Safwat (1962) and others (such as Demeter 1922), Thorne (1992) combines the Apocynaceae *s.s.* with the Asclepiadaceae into a single family (with five subfamilies). A close relationship between the two groups has long been acknowledged (Rosatti 1989), and they probably have been maintained as separate families based on tradition when only the temperate representatives were well known. The Apocynaceae *s.s.* clearly are paraphyletic and the "Asclepiadaceae," monophyletic, as shown by the preliminary cladistic analyses by Judd et al. (1994); the Apocynaceae *s.l.*, as delimited by Thorne (1992), are monophyletic (see also Downie and Palmer 1992).

Table 15 summarizes the character states, mainly

those concerning pollination-apparatus specialization, conventionally used to separate these two groups. However, the taxa share many morphological features—the most notable being the peculiar two-carpellate gynoecium with distinct ovaries and united styles and/or stigmas. Even specialized features historically associated with the "Asclepiadaceae" also intergrade with those of the Apocynaceae *s.s.* (Safwat 1962). For example, the Asclepiadaceae are distinguished by the tetradenous pollen in pollinia and the specialized stigma that secretes structures to connect them. Although these characters hold for the most derived asclepiads, a few have granular pollen (still in tetrads). In comparison, apocynaceous pollen typically is granular, but in certain genera it remains in tetrads, which are united into a mass resembling a pollinium. In most specialized asclepiads, the pollinia from adjacent anthers are connected in pairs by "translator arms," which themselves are joined by a two-parted structure or "gland" (corpusculum; see Fig. 110: 1i) secreted by restricted glandular regions of the stigma. Homologous to this specialized translator arm–gland apparatus is the adhesive material (discussed below) secreted by the median zone of the stigmatic head of apocynads (Schick 1980). In the most specialized condition (*Apocynum*), the secretion becomes organized into five amorphous bodies (plates, a type of translating apparatus) that may facilitate the transference of pollen.

Other interesting features of the family as a whole are the appendages of the corolla and androecium (see Bookman 1981). The term "corona," variously used in the literature, actually refers to three structures according to the monographer Woodson (1941): a "faucal annulus" derived and arising from the corolla tube; various fleshy radial elaborations or enations of the filaments (true corona, or "hood and horn"; Fig. 110: 1d,e); and sterile apical anther appendages. Asclepiads have one or more of these structures. The corolla tubes of apocynads often are appendaged with laminar scales, hairs, or other outgrowths (e.g., Fig. 109: 1f,m; 2a,c), which actually represent coronas of the faucal annulus type (Safwat 1962). Often the anthers become apically elongated into various processes (Fig. 109: 2d) comparable to the appendages of the asclepiads.

As in the Orchidaceae, the members of the family exhibit numerous fascinating and complex pollination mechanisms. Various insects (especially Hymenoptera and Lepidoptera) are attracted to the showy, richly nectariferous flowers. In less specialized species (Apocynoideae and Plumerioideae), glands (or a disc) at the base of the ovary secrete nectar, which is stored at the base of the corolla tube (Woodson and Moore 1938). The enlarged stigmatic head of these flowers comprises three distinct functional and morphological regions (Fig. 109: 1h,m; Schick 1980, 1982). The receptive stigmatic tissue is restricted to the base of the head under a collar-like extension (pollen scraper) that forms a chamber. Sticky fluid is produced by the middle portion of the head, an area Schick (1982) calls the "glue zone." At the apex is the sterile pollen depository, where pollen accumulates from the introrsely dehiscing anthers (Fig. 109: 1m). Guided by the sterile basal tails of the anthers, a visiting insect's proboscis may easily probe between the connivent anthers for the concealed nectar. As the proboscis is withdrawn, it first contacts the pollen scraper, thereby depositing pollen from a previous flower into the stigmatic chamber underneath. The proboscis becomes sticky as it next passes the adhesive zone, so that new pollen situated at the apex of the stylar head may be taken up and transferred to the next flower.

Of the more derived pollination apparatuses, the specialized features of *Asclepias*, in particular, have received much attention in the literature (Robertson 1886; Woodson 1954). The gynostegium, situated at the center of the flower, consists of five coherent stamens surrounding the enlarged stigmatic head (Fig. 110: 1h). This structure is comparable to the column in orchids. Also as in the Orchidaceae, the pollen is massed into pollinia (Fig. 110: 1i). The receptive portion of the stigma consists of five narrow slits or chambers, one of which occurs behind each hood between adjacent anthers. Secretory cells lining the stigmatic chamber produce nectar, which accumulates in the corona (hoods; Galil and Zeroni 1965; Schnepf and Christ 1980). The broad stigma apex (Fig. 110: 1f) is nonreceptive. When an insect's leg slips on the slippery stigma apex down into the space between the anthers, a spur of the leg becomes snagged on the pollinia gland; the entire apparatus (gland, translators, and two pollinia) is then pulled out as the leg is dislodged. Cross-pollination occurs when the insect visits another flower and the burdened leg slips into a stigmatic slit and deposits the pollinia. The pollinium shape matches the contour of the stigmatic chamber, so that its convex side, from which the growing pollen tubes emerge, is adjacent to the receptive surface (Galil and Zeroni 1969). If the pollinia are not inserted at this correct orientation, the pollen tubes grow away from the stigmatic surface.

REFERENCES CITED

Bookman, S. S. 1981. The floral morphology of *Asclepias speciosa* (Asclepiadaceae) in relation to pollination and a clarification in terminology for the genus. *Amer. J. Bot.* 68:675–679.

Demeter, K. 1922. Vergleichende Asclepiadeenstudien. *Flora* 115:130–176.

Downie, S. R., and J. D. Palmer. 1992. Restriction site mapping of the chloroplast DNA inverted repeat: A molecular phylogeny of the Asteridae. *Ann. Missouri Bot. Gard.* 79:266–283.

Galil, J., and M. Zeroni. 1965. Nectar system of *Asclepias curassavica*. *Bot. Gaz.* 126:144–148.

—— and ——. 1969. On the organization of the pollinium in *Asclepias curassavica*. *Bot. Gaz.* 130:1–4.

Judd, W. S., R. W. Sanders, and M. J. Donoghue. 1994. Angiosperm family pairs: Preliminary cladistic analyses. *Harvard Pap. Bot.* 5:1–51.

Robertson, C. 1886. Notes on the mode of pollination of *Asclepias*. *Bot. Gaz.* 11:262–269.

Rosatti, T. J. 1989. The genera of suborder Apocynineae (Apocynaceae and Asclepiadaceae) in the southeastern United States. *J. Arnold Arbor.* 70:307–401, 443–514.

Safwat, F. M. 1962. The floral morphology of *Secamone* and the evolution of the pollinating apparatus in Asclepiadaceae. *Ann. Missouri Bot. Gard.* 49:95–129.

Schick, B. 1980. Untersuchungen über Biotechnik der Apocynaceenblüte. I. Morphologie und Funktion des Narbenkopfes. *Flora, Morphol. Geobot. Oekophysiol.* 170:394–432.

——. 1982. Ibid., II. Bau und Funktion des Bestäubungsapparates. *Flora, Morphol. Geobot. Oekophysiol.* 172:347–371.

Schnepf, E., and P. Christ. 1980. Unusual transfer cells in the epithelium of the nectaries of *Asclepias curassavica* L. *Protoplasma* 105:135–148.

Thorne, R. F. 1992. Classification and geography of the flowering plants. *Bot. Rev.* (Lancaster) 58:225–348.

Woodson, R. E. 1930. Studies in the Apocynaceae. I. A critical study of the Apocynoideae (with special reference to the genus *Apocynum*). *Ann. Missouri Bot. Gard.* 17:1–212.

——. 1941. The North American Asclepiadaceae. I. Perspective of the genera. *Ann. Missouri Bot. Gard.* 28:193–244.

——. 1954. The North American species of *Asclepias* L. *Ann. Missouri Bot. Gard.* 41:1–211.

—— and J. A. Moore. 1938. The vascular anatomy and comparative morphology of apocynaceous flowers. *Bull. Torrey Bot. Club* 65:135–166.

GENTIANACEAE
Gentian Family

Annual, biennial, or perennial herbs, or sometimes shrubs to small trees; stems winged; roots often mycorrhizal. **Leaves** simple, entire, opposite, decussate, sessile, basally connate or connected by transverse line on the stem, occasionally reduced to scales, frequently with glands (colleters) at petiole base (and on adjoining stem surface), exstipulate. **Inflorescence** determinate, cymose, axillary or terminal, or sometimes flowers solitary. **Flowers** actinomorphic, usually perfect, hypogynous, showy, with nectariferous disc or glands around ovary base. **Calyx** synsepalous with 4 or 5 lobes, tubular or sometimes campanulate, frequently with glands (colleters) on inner surface, persistent, imbricate or valvate. **Corolla** sympetalous with 4 or 5 lobes, rotate, salverform, funnelform, or campanulate, sometimes plicate at sinuses, often with nectary pits and/or fimbriate/scaly in the tube, marcescent, usually brightly colored (commonly white, blue, or pink), convolute. **Androecium** of 4 or 5 stamens, epipetalous; filaments distinct; anthers versatile or sometimes basifixed, often twisted, dehiscing longitudinally, introrse. **Gynoecium** of 1 pistil, 2-carpellate; ovary superior, 1- or occasionally 2-locular, frequently with glands at base; ovules numerous, anatropous, placentation usually parietal on deeply intruded placentae (1-locular ovary) or sometimes axile (2-locular ovary); style 1; stigmas generally 2, linear, sometimes spirally twisted. **Fruit** usually a septicidal capsule; seeds numerous, small; endosperm usually copious, oily, fleshy; embryo small, straight, cylindrical to spatulate; embedded in endosperm.

Family characterization: Herbaceous plants with winged stems and mycorrhizal roots; entire, opposite, exstipulate leaves with sessile and connate bases (or connected by line on the stem); attractively colored corolla of various types (rotate, salverform, funnelform, campanulate) often with hairs/scales and/or nectary pits within the tube; epipetalous stamens; 2-carpellate, superior, often 1-locular ovary; usually parietal placentation; and septicidal capsule with many small seeds. Iridoid compounds (e.g., gentiopicroside) present. Tissues commonly with calcium oxalate crystals. Anatomical features: xylem forming a continuous ring; intraxylary phloem in the petiole (bicollateral bundles) and stem (Fig. 111: 1b); medullary bundles or isolated strands of phloem often in the pith; and unitegmic ovules.

Genera/species: 83/965

Distribution: Cosmopolitan; most common in temperate to subtropical regions and in the montane tropics.

Major genera: *Gentiana* (300–400 spp.), *Gentianella* (125 spp.), *Sebaea* (60–100 spp.), *Swertia* (50–100 spp.), and *Halenia* (70 spp.)

U.S./Canadian representatives: 17 genera/117 spp.; largest genera: *Gentiana*, *Centaurium*, *Sabatia*, and *Frasera*

Economic plants and products: Several medicinal plants (due to bitter glucosides in roots and rhizomes), such as several *Gentiana* species (gentian root). Ornamental plants (species of at least 17 genera), including *Centaurium* (centaury), *Eustoma* (prairie-gentian), *Ex-*

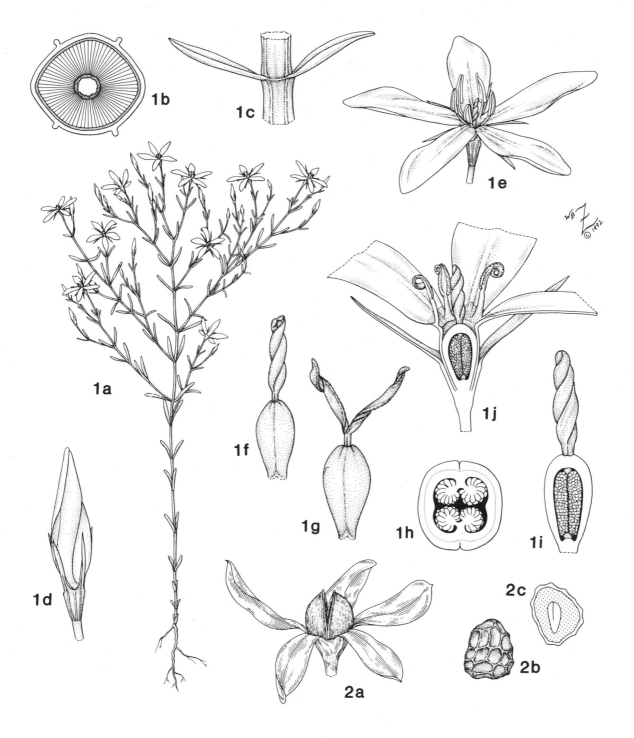

Figure 111. Gentianaceae. 1, *Sabatia brevifolia*: **a,** habit, ×²⁄₅; **b,** cross section of stem showing "wings" (ridges) and intraxylary phloem, ×10; **c,** node showing sessile leaves, ×2½; **d,** bud, ×3; **e,** flower at anthesis, ×2½; **f,** pistil (at anthesis, styles coiled), ×7; **g,** pistil (styles expanded), ×7; **h,** cross section of ovary, ×20; **i,** longitudinal section of pistil, ×10; **j,** longitudinal section of flower, ×6. **2,** *Sabatia calycina*: **a,** capsule with persistent calyx, ×2½; **b,** seed, ×35; **c,** longitudinal section of seed, ×35.

acum (German-violet), *Gentiana* (gentian), and *Sabatia* (rose-pink).

Commentary: The Gentianaceae have been divided traditionally into two subfamilies, Gentianoideae and Menyanthoideae; the latter is now generally consid-

ered a segregate family (Menyanthaceae) allied with the Asterales, not even very closely related to the Gentianaceae *s.s.* (e.g., Cronquist 1988; Thorne 1992; Michaels et al. 1993). The tribes and particularly the genera of the Gentianaceae have been difficult for taxonomists to define (see Wood and Weaver 1982 for a summary). Tribes have been delimited primarily on characters of the style (relative length), ovary (locule number), and especially pollen (size, shape, exine morphology, germ-pore number).

Plants of the family are commonly mycorrhizal, and a few species (e.g., *Bartonia*, *Obolaria virginica*) presumably are saprophytic or parasitic. These small, low-growing herbs have coral-like mycorrhizae, and the leaves are reduced to scales (Gillett 1959). Chlorophyll evidently is lacking in *Obolaria virginica*.

The brightly colored flowers (often blue, purple, or pink) are some of the world's most beautiful wildflowers. Fringe or scales on the corolla or an eye of contrasting color may serve as nectar guides for various insect pollinators (e.g., bees and lepidopterans) that collect the nectar secreted by glandular tissue at the ovary base (*Gentiana*, *Eustoma*), epipetalous glands (*Gentianella*), or a well-developed disc (many woody species of the American tropics). The showy, nectariferous flowers mostly are protandrous; for example, in some species (e.g., of *Sabatia* and *Centaurium*), the stigmatic branches are tightly twisted at anthesis (Fig. 111: 1e,f), uncurling and exposing the receptive surface after the anthers have dehisced (Fig. 111: 1g).

REFERENCES CITED

Cronquist, A. 1988. *The evolution and classification of flowering plants*, 2d ed., pp. 417–425. New York Botanical Garden, Bronx.

Gillett, J. M. 1959. A revision of *Bartonia* and *Obolaria* (Gentianaceae). *Rhodora* 61:43–62.

Michaels, H. J., K. M. Scott, R. G. Olmstead, T. Szaro, R. K. Jansen, et al. 1993. Interfamilial relationships of the Asteraceae: Insights from the *rbc*L sequence variation. *Ann. Missouri Bot. Gard.* 80:742–751.

Thorne, R. F. 1992. Classification and geography of the flowering plants. *Bot. Rev.* (Lancaster) 58:225–348.

Wood, C. E., and R. E. Weaver. 1982. The genera of Gentianaceae in the southeastern United States. *J. Arnold Arbor.* 63:441–487.

OLEACEAE
Olive Family

Trees, shrubs, or sometimes lianas. **Leaves** simple or pinnately compound, usually entire, opposite, deciduous or persistent, often abaxially punctate, exstipulate. **Inflorescence** determinate, cymose and appearing racemose, paniculate, or fasciculate, terminal or axillary. **Flowers** actinomorphic, perfect or sometimes imperfect (then plants dioecious, polygamous, or polygamodioecious), hypogynous, small, with nectariferous disc often present around base of ovary. **Calyx** synsepalous with usually 4 lobes, valvate. **Corolla** usually sympetalous with 4 lobes or sometimes petals distinct or absent, variously colored, usually imbricate or valvate. **Androecium** of typically 2 stamens, epipetalous; filaments distinct, short; anthers basifixed, often apiculate, dehiscing longitudinally. **Gynoecium** of 1 pistil, 2-carpellate; ovary superior, 2-locular; ovules usually 2 in each locule, anatropous or amphitropous, placentation axile; style 1 or absent; stigma 2-lobed or simple, capitate. **Fruit** a drupe, loculicidal capsule, circumscissile capsule (pyxis), or samara; endosperm oily or absent; embryo straight, spatulate.

Family characterization: Usually trees or shrubs with opposite, exstipulate and often abaxially punctate leaves; usually 4-merous perianth; 2 epipetalous stamens; and 2-locular superior ovary. Tissues commonly with calcium oxalate crystals. Iridoid compounds and glycosides commonly present. Anatomical features: peltate trichomes (sometimes secretory) often on leaves and twigs (appearing on leaves as greenish or sunken dots; Fig. 112: 1c); unitegmic and tenuinucellate ovules; and 1-trace unilacunar nodes (Fig. 112: 1d).

Genera/species: 29/600

Distribution: Nearly cosmopolitan; particularly diverse in Asia and Australasia.

Major genera: *Jasminum* (300 spp.), *Chionanthus* (80–100 spp.), and *Fraxinus* (70 spp.)

U.S./Canadian representatives: 12 genera/65 spp.; largest genera: *Fraxinus* and *Forestiera*

Economic plants and products: Fruit and oil from *Olea europaea* (olive). Perfume from several species of *Jasminum* (jasmine). Lumber from *Fraxinus* (ash). Ornamental trees and shrubs (species of 16 genera), including *Chionanthus* (fringe tree), *Forsythia* (goldenbells), *Jasminum*, *Ligustrum* (privet), *Osmanthus* (fragrant-olive), and *Syringa* (lilac).

Commentary: The Oleaceae generally are considered to be a monophyletic family, but the affinities to other groups are uncertain: the family has been allied with the Gentianales, the Scrophulariales, or placed in its own order (Oleales or Ligustrales; see Wilson and Wood 1959; Cronquist 1981; Baas et al. 1988). Preliminary molecular data support the relationship of the family with those of the Scrophulariales-Lamiales complex (Olmstead et al. 1993). The Oleaceae usually are divided into two subfamilies (and several tribes) based upon the fruit type and ovule/seed position. For example, the fruit of *Fraxinus* is a dry, one-seeded samara (Fig. 112: 3b); *Syringa* has a loculicidal capsule; and the fruits of *Chionanthus* (Fig. 112: 2d,e), *Osmanthus*, and *Ligustrum* are drupaceous.

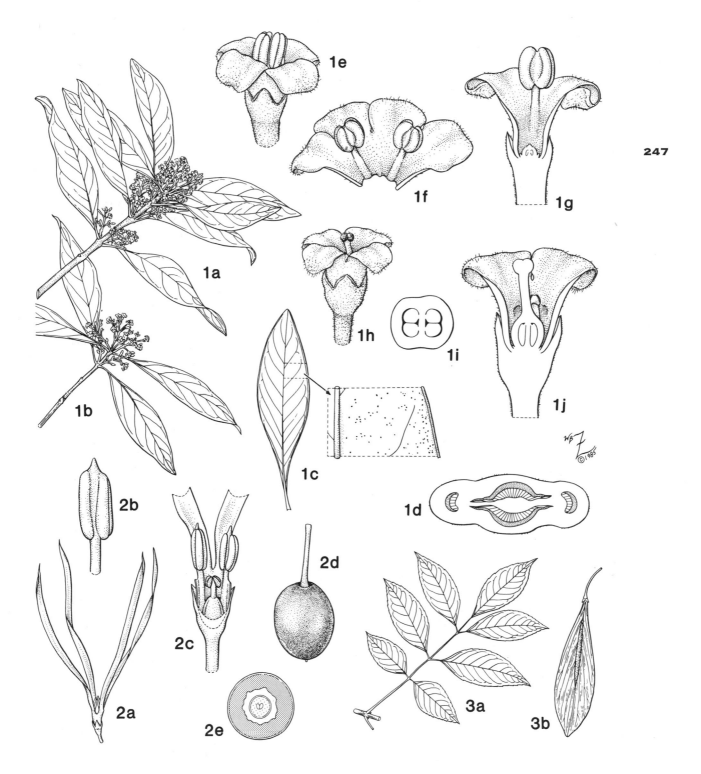

Figure 112. Oleaceae. 1, *Osmanthus americana*: **a,** staminate flowering branch, ×⅓; **b,** carpellate flowering branch, ×⅓; **c,** abaxial side of leaf, ×⅓, with detail showing punctate surface (peltate trichomes), ×2; **d,** cross section of node (one-trace unilacunar), ×4½; **e,** staminate flower, ×6; **f,** expanded corolla and epipetalous androecium of staminate flower, ×6; **g,** longitudinal section of staminate flower, ×7½; **h,** carpellate flower, ×6; **i,** cross section of ovary, ×18; **j,** longitudinal section of carpellate flower, ×9. **2,** *Chionanthus virginica*: **a,** perfect flower, ×2; **b,** anther, ×10; **c,** detail of perfect flower (portion of perianth removed), ×7; **d,** drupe, ×1½; **e,** cross section of drupe, ×1½. **3,** *Fraxinus caroliniana*: **a,** leaf, ×⅓; **b,** samara, ×1¼.

The small flowers, often clustered into dense inflorescences, generally are more or less tubular with spreading limbs. In some species, the flowers develop before the leaves expand (e.g., *Forsythia*). Various insects are attracted to the nectar (concealed at the corolla tube base) and often also to the strong sweet odor of the flowers. Flowers of the family are perfect to sometimes imperfect. Several species, such as those of *Syringa* and *Ligustrum*, are homogamous and evidently self-compatible; others (e.g., *Jasminum* and *Forsythia* spp.) are heterostylous. The often imperfect *Fraxinus* flowers, which have reduced (or absent) perianths, are anemophilous.

REFERENCES CITED

Baas, P., P. M. Esser, M. E. T. van der Westen, and M. Zandee. 1988. Wood anatomy of the Oleaceae. *I.A.W.A. Bull.*, n.s. 9:103–182.

Cronquist, A. 1981. *An integrated system of classification of flowering plants*, pp. 948–950. Columbia University Press, New York.

Olmstead, R. G., B. Bremer, K. M. Scott, and J. D. Palmer. 1993. A parsimony analysis of the Asteridae sensu lato based on *rbc*L sequences. *Ann. Missouri Bot. Gard.* 80:700–722.

Wilson, K. A., and C. E. Wood. 1959. The genera of Oleaceae in the southeastern United States. *J. Arnold Arbor.* 40:369–384.

BIGNONIACEAE
Bignonia or Trumpet Vine Family

Trees, shrubs, or most often woody vines (then with adventitious roots or tendrils). **Leaves** simple or more often palmately or pinnately compound with terminal leaflet sometimes tendril-like, entire to serrate, opposite or occasionally whorled, decussate, often with glands at the petiole bases or in the leaf-vein axils, exstipulate. **Inflorescence** determinate, cymose and often appearing racemose, or flower solitary, terminal or axillary. **Flowers** zygomorphic, perfect, hypogynous, usually large and showy, with nectariferous disc. **Calyx** synsepalous with 5 lobes or teeth, sometimes truncate, campanulate or bilabiate. **Corolla** sympetalous with usually 5 lobes, campanulate, funnelform, to bilabiate, variously colored, imbricate. **Androecium** of typically 4 stamens and 1 staminode, epipetalous; filaments distinct; anthers basifixed, coherent in pairs or sometimes free, with divergent and often unequal locules (one seemingly above the other), dehiscing longitudinally, introrse. **Gynoecium** of 1 pistil, 2-carpellate; ovary superior, 2-locular or occasionally 1-locular; ovules numerous, anatropous, erect, placentation axile (in 2-locular ovary) or parietal (in 1-locular ovary); style simple, filiform; stigma 1, 2-lobed (lobes flap-like). **Fruit** usually a 2-valved septicidal, loculicidal, or sometimes septifragal capsule, often woody, often elongate, or sometimes a berry or indehiscent pod; seeds numerous, compressed, winged, sometimes comose; endosperm typically absent; embryo straight.

Family characterization: Trees or woody vines with opposite, usually compound leaves; zygomorphic, campanulate to bilabiate, showy corolla; androecium of 4 didynamous, epipetalous stamens and 1 staminode; divergent anther locules; 2-valved, woody, capsular fruit; and winged nonendospermous seeds. Tissues with calcium oxalate crystals. Iridoid compounds and glycosides commonly present. Anatomical features: anomalous secondary growth (see the discussion below); unitegmic and tenuinucellate ovules; and 3- to several-trace unilacunar nodes (Fig. 113: 1b).

Genera/species: 113/800

Distribution: Primarily tropical and subtropical, with a few representatives in the temperate zone; particularly diverse in northern South America.

Major genera: *Tabebuia* (100 spp.), *Arrabidaea* (50–70 spp.), *Adenocalymna* (40–50 spp.), and *Jacaranda* (30–50 spp.)

U.S./Canadian representatives: 18 genera/28 spp.; largest genera: *Tabebuia* and *Catalpa*

Figure 113. Bignoniaceae. 1, *Campsis radicans*: **a,** habit, ×⅓; **b,** cross section of node (several-trace unilacunar), ×6; **c,** two views of flower, ×½; **d,** expanded corolla and epipetalous androecium (note staminode), ×⅔; **e,** two views of anther, ×2; **f,** ovary and nectary, ×3; **g,** stigma, ×2; **h,** longitudinal section of ovary, ×3; **i,** cross section of ovary, ×6; **j,** longitudinal section of flower, ×¾; **k,** capsule, ×½; **l,** seed, ×2½; **m,** longitudinal section of seed, ×2½. 2, *Bignonia capreolata*: node showing one leaf and branched tendril, ×⅔. 3, *Macfadyena unguis-cati*: **a,** node showing one leaf and claw-like tendril, ×½; **b,** cross section of young stem with some secondary growth, ×20; **c,** cross section of older stem with four-lobed xylem and phloem wedges, ×15; **d,** cross section of stem (older than **c**) with additional phloem wedges, ×10; **e,** cross section of older stem with phloem wedges more developed than in **d**, ×7½. 4, *Clytostoma callistegioides*: **a,** cross section of stem with four-lobed xylem and phloem wedges, ×4; **b,** cross section of older stem with vascular tissue divided by wide vascular rays, ×5. 5, *Pithecoctenium crucigerum* (*P. echinatum*): **a,** cross section of stem with four-lobed xylem and phloem wedges, ×4 ½; **b,** cross section of older stem with included phloem, ×2.

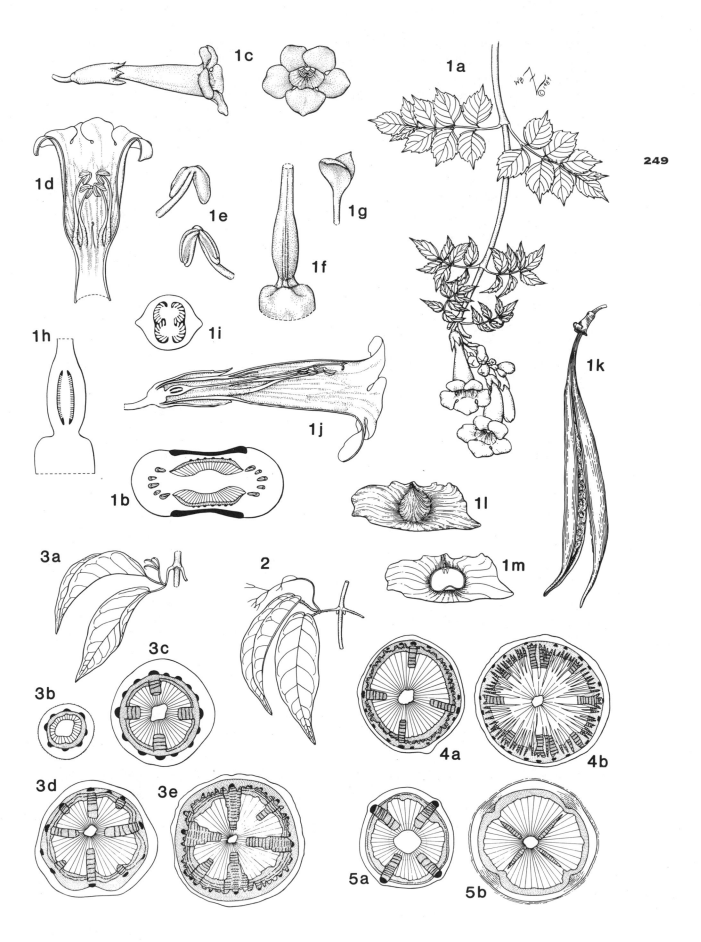

249

Economic plants and products: Lumber from *Catalpa* and *Tabebuia* (West Indian–boxwood). Ornamental trees and vines (species of 45 genera), including *Bignonia* (cross vine, trumpet-flower), *Campsis* (trumpet-creeper, trumpet vine), *Clytostoma* (painted-trumpet), *Crescentia* (calabash tree), *Jacaranda*, *Kigelia* (sausage tree), *Macfadyena* (catclaw), *Pandorea*, *Paulownia* (empress tree), *Pyrostegia* (flame vine), *Spathodea* (flame tree, African tulip tree), *Tabebuia* (gold tree, roble blanco), *Tecoma* (yellow-elder), and *Tecomaria* (Cape-honeysuckle).

Commentary: The Bignoniaceae are divided into four to eight tribes (Armstrong 1985) based on characters of the ovary (2- or 1-locular), fruit (various capsules or berry), and seeds (presence of wings). Specimens with both flowers and fruits are essential for critical determinations of genera and species (Gentry 1973, 1980; Gentry and Tomb 1979). The family is compared to the related Scrophulariaceae, Acanthaceae, Verbenaceae, and Lamiaceae in Table 16.

The majority of species are woody vines that are root climbers (*Campsis radicans*; Fig. 113: 1a) or tendril climbers (most species, e.g., *Bignonia*). The tendrils, which represent modified leaflets, terminate the compound leaves. They may be branched (*Bignonia capreolata*; Fig. 113: 2) and sometimes also end in hooks (*Macfadyena unguis-cati*; Fig. 113: 3a).

The stems of many lianas in this family undergo characteristic anomalous secondary growth that results in a lobed or furrowed xylem (Metcalfe 1985). After the usual type of secondary growth is produced, four strips of cambium become unidirectional, ceasing to produce xylem and producing only phloem (Dobbins 1981); these anomalously developed phloem panels or wedges then become flanked by xylem (which is produced by the normal cambium), and the xylem eventually develops a four-lobed pattern (e.g., Fig. 113: 3c). The common name "cross vine" for *Bignonia capreolata*, for example, refers to this configuration, which is more or less visible macroscopically in cut stems. In some species, this four-lobed pattern remains constant as the stem increases in diameter, but many (e.g., *Macfadyena unguis-cati*) develop additional cambia (and the corresponding phloem wedges) between the original four xylem segments, resulting in further dissection of the woody tissue (Fig. 113: 3d,e; Dobbins 1971). The xylem may be additionally furrowed or fissured by the development of wide rays (*Clytostoma callistegioides*; Fig. 113: 4b) or by proliferation of pith and wood parenchyma. Included phloem may occur when the xylem outgrows and surrounds the phloem wedges, isolating them from the rest of the secondary phloem (Fig. 113: 5b).

Insects, hummingbirds, and bats visit the colorful, bell- or funnel-shaped flowers for the nectar secreted and concealed at the base (Gentry 1990). The chances of self-pollination are reduced by means of the "sensitive stigma" (Fig. 113: 1g), which consists of two flap-like lobes (Gentry 1974). As a pollinator enters the flower, it first contacts the stigma and deposits pollen (from a previous flower) onto the inner surfaces of the spreading lobes; the sensitive lobes then close together, and when the pollinator (covered with new pollen) backs away from that flower, it touches only the nonreceptive outer surface of the stigmatic lobes.

REFERENCES CITED

Armstrong, J. E. 1985. The delimitation of Bignoniaceae and Scrophulariaceae based on floral anatomy, and the placement of problem genera. *Amer. J. Bot.* 72:755–766.

Dobbins, D. R. 1971. Studies on the anomalous cambial activity in *Doxantha unguis-cati* (Bignoniaceae). II. A case of differential production of secondary tissues. *Amer. J. Bot.* 58:697–705.

———. 1981. Anomalous secondary growth in lianas of the Bignoniaceae is correlated with the vascular pattern. *Amer. J. Bot.* 68:142–144.

Gentry, A. H. 1973. Generic delimitations of Central American Bignoniaceae. *Brittonia* 25:226–242.

———. 1974. Coevolutionary patterns in Central American Bignoniaceae. *Ann. Missouri Bot. Gard.* 61:728–759.

———. 1980. Bignoniaceae—Part 1 (Crescentieae and Tourrettieae). *Fl. Neotrop. Monogr.* 25:1–130.

———. 1990. Evolutionary patterns in neotropical Bignoniaceae. *Mem. New York Bot. Gard.* 55:118–129.

——— and A. S. Tomb. 1979. Taxonomic implications of Bignoniaceae palynology. *Ann. Missouri Bot. Gard.* 66:756–777.

Metcalfe, C. R. 1985. Anomalous structure. In *Anatomy of the dicotyledons*, 2d ed., vol. 2, *Wood structure and conclusion of the general introduction*, ed. C. R. Metcalfe and C. Chalk, pp. 58–61. Oxford University Press, Oxford.

SCROPHULARIACEAE
Figwort or Snapdragon Family
Including the Orobanchaceae:
   Broomrape Family

Annual, biennial, or perennial herbs or sometimes undershrubs, hemiparasitic, parasitic, or myco-parasitic, occasionally achlorophyllous. **Leaves** simple, entire to deeply pinnately lobed, opposite or alternate, sometimes scale-like, exstipulate. **Inflorescence** indeterminate or determinate, racemose, spicate, cymose, paniculate, or flower solitary, terminal or axillary. **Flowers** zygomorphic, perfect, hypogynous, typically showy, usually with nectariferous disc. **Calyx** of 4 or 5 sepals, usually basally connate, equal or unequal, imbricate or valvate, persistent. **Corolla** sympetalous with 4 or 5

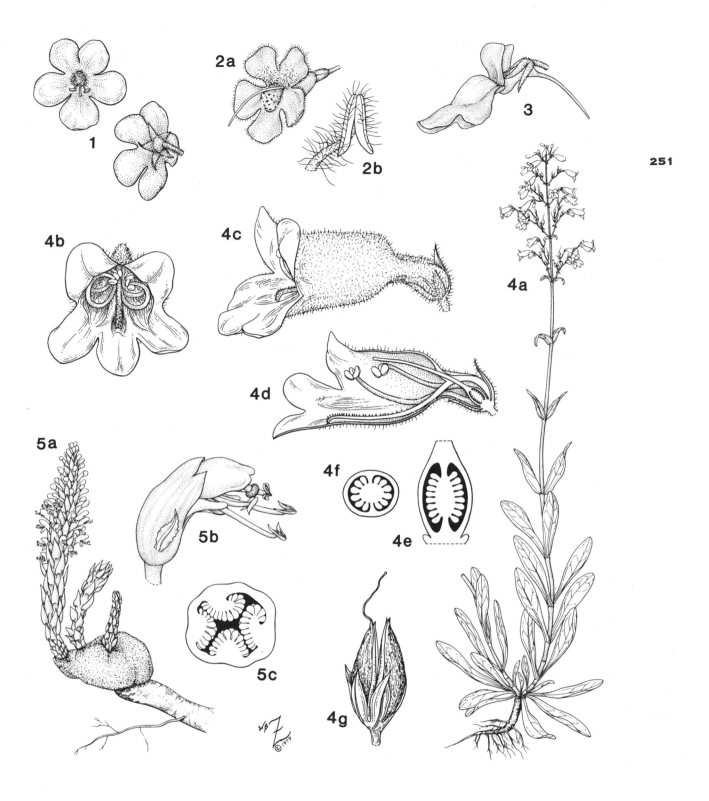

Figure 114. Scrophulariaceae. 1, *Verbascum blattaria*: two views of flower, ×¾. **2,** *Agalinis fasciculata*: **a, flower,** ×¾; **b,** anther, ×6. **3,** *Linaria canadensis*: flower (note spur), ×3. **4,** *Penstemon multiflorus*: **a,** habit, ×¹⁄₁₀; **b,** front view of flower, ×3; **c,** lateral view of flower, ×3; **d,** longitudinal section of flower, ×3; **e,** longitudinal section of ovary, ×12; **f,** cross section of ovary, ×12; **g,** capsule, ×3. **5,** *Conopholis americana*: **a,** flowering plant growing on *Quercus laurifolia* root, ×²⁄₅; **b,** flower, ×3½; **c,** cross section of ovary, ×10.

lobes, rotate, salverform or campanulate to bilabiate (then sometimes personate) and sometimes spurred at base, color extremely variable, imbricate. **Andro-ecium** of typically 4 stamens with one staminode sometimes present, didynamous, epipetalous; filaments distinct or sometimes coherent in pairs; anthers dorsifixed or basifixed, sometimes sagittate, dehiscing longitudinally, introrse. **Gynoecium** of 1 pistil, 2-carpellate; ovary superior, 1- or 2-locular; ovules numerous, anatropous, placentation axile or parietal on large placentae; style simple, filiform; stigma(s) 1 and bilobed or 2 and slender or capitate. **Fruit** usually a septicidal capsule enveloped by persistent calyx; seeds small, numerous; endosperm fleshy; embryo straight or slightly curved, rarely undifferentiated.

Family characterization: Herbaceous to shrubby plants; sympetalous zygomorphic flowers; attractively colored corolla of various shapes (rotate, salverform, campanulate, bilabiate); 4 didynamous epipetalous stamens; 2-carpellate superior ovary with axile or parietal placentation (see Armstrong 1985); and septicidal capsule with many small seeds. Iridoid compounds and orobanchin (a glycoside) present. Anatomical feature: unitegmic and tenuinucellate ovules.

Genera/species: 220/3,000

Distribution: Cosmopolitan; most species in north temperate regions.

Major genera: *Pedicularis* (350–500 spp.), *Calceolaria* (300–400 spp.), *Verbascum* (360 spp.), *Veronica* (250–300 spp.), *Scrophularia* (200–300 spp.), *Penstemon* (250 spp.), and *Linaria* (100–150 spp.)

U.S./Canadian representatives: 73 genera/861 spp.; largest genera: *Penstemon, Castilleja,* and *Mimulus*

Economic plants and products: Cardiac glycosides from *Digitalis* (digitoxin, digoxin, and digitalin). Ornamental plants (species of 64 genera), including *Antirrhinum* (snapdragon), *Calceolaria* (slipper-flower), *Digitalis* (foxglove), *Penstemon* (beardtongue), *Russelia* (firecracker plant), and *Veronica* (speedwell).

Commentary: The Scrophulariaceae *s.s.* have been divided traditionally into three subfamilies based on leaf arrangement and corolla aestivation (Thieret 1967); these taxa are occasionally considered to be segregate families (e.g., Globulariaceae as in Thorne 1992). Thorne (1992) includes the Orobanchaceae as an additional subfamily of the Scrophulariaceae. The Orobanchoideae are a small group (13 genera, 150 spp.) of achlorophyllous parasitic herbs most common in north temperate Eurasia. The subfamily is represented in the United States by species of *Boschniakia, Conopholis* (Fig. 114: 5), *Epifagus,* and *Orobanche.*

Traditionally, certain gynoecial characters have been used to separate the Scrophulariaceae *s.s.* (bilocular ovary, axile placentation) and Orobanchaceae (unilocular ovary, parietal placentation), but the Orobanchoideae are similar to certain members of the Scrophulariaceae in numerous aspects of vegetative and floral morphology. For example, according to anatomical studies of Y. D. Tiagi (1962; cited in Thieret 1971), the placentation in both groups is parietal, with a clear transition from the two-locular ovary of most Scrophulariaceae to the one-locular ovary of the Orobanchaceae (Arekal 1963).

The Orobanchoideae also have been separated on the basis of the very specialized parasitic habit. However, the Scrophulariaceae subfamily Rhinanthoideae exhibit a continuous progression of parasitism, from slight root parasites (*Agalinis* spp.) to highly parasitic genera (*Harveya* and *Hyobanche*); the Orobanchoideae complete this line of parasitism (Boeshore 1920; Cassera 1935). Additional characters of affinity include pollen morphology (Minkin and Eshbaugh 1989), reduced or modified anther locules, and the mode of development of gametophytes, endosperm, embryo, and seeds (B. Tiagi 1963).

The plants of the Scrophulariaceae sometimes are confused in the field with those of other related families with bilabiate corollas (see Table 16). According to the monographer Pennell (1920, 1935), the range of variation in corolla structure and color in the Scrophulariaceae probably is exceeded only by the orchids. Insects and hummingbirds are attracted to the brightly colored flowers and to the nectar secreted by the disc around the base of the ovary. The corolla throat often has darker spots or lines, which serve as nectar guides. Various corolla morphologies can be correlated with particular pollinators. For example, the rotate *Verbascum*-type flower (Fig. 114: 1) has a short tube, easily accessible nectar, and projecting stamens and pistil; the pollinators are bees and flies. In contrast, the bilabiate and personate corolla of *Linaria* (Fig. 114: 3) has a long, wide throat and an anterior spur, which stores the nectar; the lower lip (with rounded protrusions) serves as a landing platform, and the stamens and stigmas touch the backs of large bees as they enter the corolla tube to gather the nectar in the spur.

REFERENCES CITED

Arekal, G. D. 1963. Embryological studies in Canadian representatives of the tribe Rhinantheae, Scrophulariaceae. *Canad. J. Bot.* 41:267–302.

Armstrong, J. E. 1985. The delimitation of Bignoniaceae and Scrophulariaceae based on floral anatomy, and the placement of problem genera. *Amer. J. Bot.* 72:755–766.

TABLE 16. Major morphological differences between several "bilabiate-corolla" families of Thorne's (1992) "Lamiales" (Bignoniales, Scrophulariales), sometimes confused in the field. For a more detailed comparison between the Verbenaceae and Lamiaceae, as defined in this text, see Table 17.

CHARACTER	BIGNONIACEAE	SCROPHULARIACEAE	ACANTHACEAE	VERBENACEAE	LAMIACEAE
GENERA/SPECIES	113/800	220/3,000	256/2,770	36/1,035	258/6,970
DISTRIBUTION	Mainly tropical to subtropical; a few in temperate regions	Cosmopolitan; most in north temperate regions	Pantropical; a few in temperate regions	Tropical to subtropical; a few in temperate regions	Cosmopolitan; mainly in temperate regions
HABIT	woody vines, trees, or shrubs	herbs, or sometimes shrubs	herbs, vines, or sometimes shrubs	herbs or shrubs, seldom trees or vines	herbs, shrubs, or trees
LEAVES	simple, or more often pinnately or palmately compound / opposite, or sometimes whorled / entire to serrate	simple / opposite or alternate / entire to deeply lobed	simple / opposite / usually entire	simple / opposite, or sometimes whorled / entire, serrate, to lobed	simple, or pinnately or palmately compound / opposite, or sometimes whorled / usually serrate to dissected
INFLORESCENCE	determinate	determinate or indeterminate	determinate	indeterminate	determinate
COROLLA SHAPE	5-lobed / campanulate, funnelform, to bilabiate	4- or 5-lobed / rotate, salverform, campanulate, to bilabiate; sometimes spurred	5-lobed / usually bilabiate	5-lobed / salverform and ± bilabiate	5-lobed / usually tubular and bilabiate
ANDROECIUM	2 + 2 stamens + 1 staminode	2 + 2 stamens ± 1 staminode	2 + 2 stamens, or 2 stamens ± 2 staminodes	2 + 2 stamens	2 + 2, 4, or 2 stamens
OVARY LOCULE NUMBER	2, or sometimes 1	1 or 2	2	2 (appearing as 4 due to false septa)	2 (appearing as 4 due to false septa)
PLACENTATION	axile or parietal	axile or parietal	axile	axile (often appearing basal)	axile
OVULE NUMBER	numerous	numerous	2 to 10	usually 4	4
STYLE	terminal	terminal	terminal	terminal	gynobasic to terminal
STIGMA	2 flap-like lobes	2-lobed or capitate	funnelform or 2-lobed	2-lobed	at tips of style branches
FRUIT TYPE	septicidal, loculicidal, or septifragal capsule, or sometimes a berry or indehiscent pod	septicidal capsule	loculicidal capsule	schizocarp or drupe	schizocarp or drupe
ENDOSPERM	usually absent	copious	usually absent	absent	usually absent
OTHER	terminal leaflet sometimes tendril-like / seeds winged	some achlorophyllous parasites	showy bracts / seed with modified funiculus	plants often aromatic / stems square	plants often aromatic / stems square

Boeshore, I. 1920. The morphological continuity of Scrophulariaceae and Orobanchaceae. *Contr. Bot. Lab. Morris Arbor. Univ. Pennsylvania* 5:139–177.

Cassera, J. D. 1935. Origin and development of the female gametophyte, endosperm, and embryo in *Orobanche uniflora*. *Bull. Torrey Bot. Club* 62:455–466.

Minkin, J. P., and W. H. Eshbaugh. 1989. Pollen morphology of the Orobanchaceae and rhinanthoid Scrophulariaceae. *Grana* 28:1–18.

Pennell, F. W. 1920. Scrophulariaceae of the southeastern United States. *Contr. New York Bot. Gard.* 221:224–291.

———. 1935. *The Scrophulariaceae of eastern temperate North America.* Academy of Natural Sciences of Philadelphia, Monogr. No. 1.

Thieret, W. 1967. Supraspecific classification in the Scrophulariaceae; A review. *Sida* 3:87–106.

———. 1971. The genera of Orobanchaceae in the southeastern United States. *J. Arnold Arbor.* 52:404–434.

Thorne, R. F. 1992. Classification and geography of the flowering plants. *Bot. Rev.* (Lancaster) 58:225–348.

Tiagi, B. 1963. Studies in the family Orobanchaceae. IV. Embryology of *Boschniakia himalaica* Hook. and *B. tuberosa* (Hook.) Jepson, with remarks on the evolution of the family. *Bot. Not.* 116:81–93.

Tiagi, Y. D. 1962. Anatomical studies of the vascular equipment of the flower of some species of the families Orobanchaceae and Scrophulariaceae. (In Russian.) *Vestn. Moskovsk. Univ., Ser. 6, Biol.* 2:29–52.

PLANTAGINACEAE
Plantain Family

Usually annual or perennial herbs. **Leaves** simple, entire to serrate, alternate, in basal rosettes, parallel-veined, linear to lanceolate, not clearly differentiated into blade and petiole, sheathing at base, exstipulate. **Inflorescence** indeterminate, spicate or capitate, terminal on a stout or wiry scape, axillary. **Flowers** usually actinomorphic to slightly zygomorphic, usually perfect, hypogynous, small and inconspicuous, each subtended by 1 bract. **Calyx** of basically 4 sepals (with the 2 abaxial sometimes connate), distinct or basally connate, membranous, persistent, imbricate. **Corolla** sympetalous with 4 lobes, salverform, scarious, persistent, imbricate. **Androecium** of 4 stamens, epipetalous near apex of corolla tube, long-exserted; filaments distinct, long, slender; anthers dorsifixed, versatile, relatively large, dehiscing longitudinally, latrorse or introrse. **Gynoecium** of 1 pistil, 2-carpellate; ovary superior, usually 2-locular (occasionally appearing 4-locular due to false septa); ovules 1 to numerous in each locule, anatropous to hemitropous, placentation usually axile; style 1, slender; stigma simple or 2-lobed, somewhat plumose. **Fruit** a circumscissile capsule, membranous; seeds concave, peltately attached, often with mucilaginous seed coat when moistened; endosperm copious, firm, translucent; embryo usually straight, spatulate, enveloped by endosperm.

Family characterization: Weedy, scapose herbs with basal rosettes of simple leaves; sheathing, parallel-veined leaves not clearly differentiated into petiole and blade; small flowers crowded into dense spicate or capitate inflorescences; 4-merous, membranous perianth; sympetalous, salverform corolla; 4 long-exserted stamens with wiry filaments and large, versatile anthers; 2-locular ovary; and membranous circumscissile capsule with peltately attached, concave seeds becoming mucilaginous when moistened. Iridoid compounds present. Anatomical feature: unitegmic and tenuinucellate ovules.

Genera/species: *Plantago* (216–265 spp.), *Littorella* (3 spp.), and *Bougueria* (1 sp.)

Distribution: Widely distributed in temperate regions and in the montane tropics; especially abundant in weedy, xeric, or saline areas.

U.S./Canadian representatives: *Plantago* (42 spp.) and *Littorella* (1 sp.)

Economic plants and products: Several noxious lawn weeds, such as *Plantago lanceolata* (English plantain) and *P. major* (common plantain). Mucilaginous seeds of several *Plantago* species used medicinally as laxatives (psyllium, ispaghula, or spogel seeds). A few species of *Plantago* rarely cultivated as rock garden plants; *Littorella uniflora* (shoreweed, shore-grass) sometimes grown in aquaria.

Commentary: The Plantaginaceae (220–270 species) comprise three genera of unequal size: *Plantago* (216–265 spp., cosmopolitan), *Littorella* (3 spp., Europe and the Antarctic), and *Bougueria* (1 sp., Andes of South America). Taxonomists currently concur that the family is related to the Scrophulariaceae, and plantains are viewed as wind-pollinated, scarious, small scrophs (Rosatti 1984). Like the four-merous flower type of *Veronica* (Scrophulariaceae), the four-merous plantain flowers also may have been derived from a basically five-merous plan (due to fusion of two lobes).

The cosmopolitan, weedy plants often grow in xeric waste places, as well as maritime or alpine locales. Plantains are easily recognized in the field by the scapose habit (Fig. 115: a) with basal rosettes of simple, often narrow, leaves with sheathing bases and parallel venation. The leaf blade, often not clearly differentiated from the petiole, apparently represents an expanded and flattened petiole ("phyllode").

The numerous, small flowers are crowded into a dense head or spike (Fig. 115: b) on a stout or wiry scape. Flowers are protogynous and generally anemophilous (Soekarjo 1992). Long, feathery, receptive stigmas first protrude from the unopened flower buds (Fig. 115: d). When the stigma of a flower withers, the

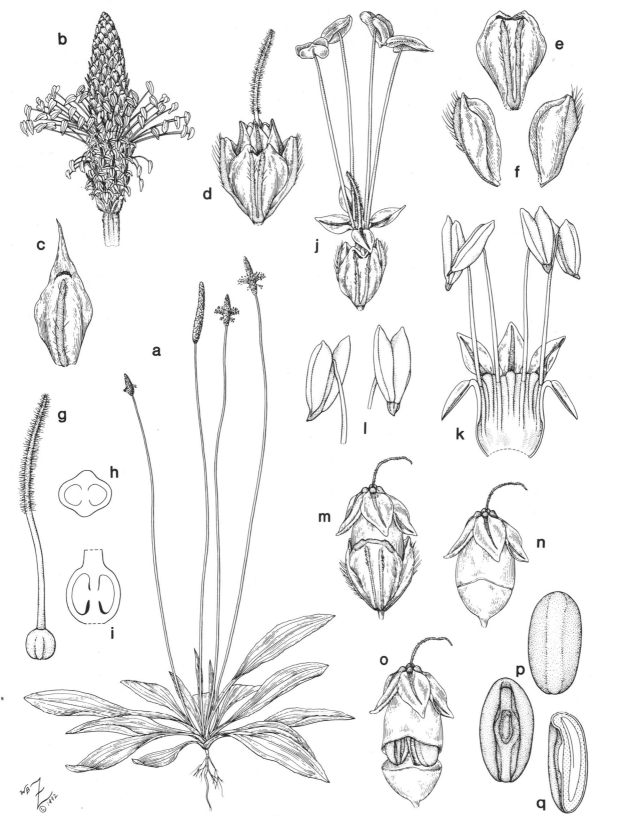

Figure 115. Plantaginaceae. *Plantago lanceolata*: **a,** habit, ×⅓; **b,** inflorescence, ×2; **c,** bract, ×12; **d,** flower ("female stage," corolla lobes and stamens unexpanded), ×10; **e,** fused abaxial sepals, ×12; **f,** lateral sepals, ×12; **g,** pistil, ×15; **h,** cross section of ovary, ×30; **i,** longitudinal section of ovary, ×30; **j,** older flower ("male stage," corolla lobes and stamens fully expanded), ×8; **k,** expanded corolla and epipetalous androecium, ×10; **l,** two views of anther, ×12; **m,** pyxis with persistent perianth, ×10; **n,** pyxis (persistent calyx removed), ×10; **o,** dehiscing pyxis (persistent calyx removed), ×10; **p,** two views of seed, ×15; **q,** longitudinal section of seed, ×15.

corolla lobes reflex, and the filaments elongate and project from the corolla tube (Fig. 115: j). The large and versatile anthers are long-exserted on a flexible filament and produce copious pollen.

Starting at the base of the inflorescence, each whorl of flowers matures sequentially from the female to the male stage (see Fig. 115: b). However, self-pollination is possible, because the stigmas may remain receptive while stamens of the same flower mature; in addition, upper flowers in the female stage may be pollinated by pollen from the lower, more advanced (male) flowers on the same inflorescence. In some species (e.g., *Plantago lanceolata*), some plants may be functionally carpellate (van Damme 1992a,b).

Although predominantly anemophilous, the flowers of several species attract insects (bees and flies) that visit to collect the copious pollen (Clifford 1962; Stelleman 1978). The long filaments are quite conspicuous and also may be colorful, as in *Plantago media* (purple) or *P. major* (brown); some species (*Plantago media*) also are fragrant.

REFERENCES CITED

Clifford, H. T. 1962. Insect pollinators of *Plantago lanceolata* L. *Nature* 193:196.

van Damme, J. M. M. 1992a. Breeding systems in *Plantago*. In *Plantago: A multidisciplinary study*, ed. P. J. C. Kuiper and M. Bos, pp. 12–18. Ecol. Stud. vol. 89. Springer-Verlag, Berlin.

———. 1992b. Selection for the maintenance of sex polymorphism in *Plantago*. In *Plantago: A multidisciplinary study*, ed. P. J. C. Kuiper and M. Bos, pp. 204–222. Ecol. Stud. vol. 89. Springer-Verlag, Berlin.

Rosatti, T. J. 1984. The Plantaginaceae in the southeastern United States. *J. Arnold Arbor.* 65:533–562.

Soekarjo, R. 1992. General morphology in *Plantago*. In *Plantago: A multidisciplinary study*, ed. P. J. C. Kuiper and M. Bos, pp. 6–12. Ecol. Stud. vol. 89. Springer-Verlag, Berlin.

Stelleman, P. 1978. The possible role of insect visits in pollination of reputedly anemophilous plants, exemplified by *Plantago lanceolata*, and syrphid flies. In *The pollination of flowers by insects*, ed. A. J. Richards, pp. 41–46. Linn. Soc. Symp. Ser. No. 6. Academic Press, London.

LENTIBULARIACEAE
Bladderwort Family

Annual or perennial herbs growing in moist to aquatic habitats, insectivorous, scapose, sometimes rootless and free-floating, occasionally epiphytic. **Leaves** simple, alternate, sometimes in basal rosettes, entire or finely divided (when submersed), highly modified (with inrolling margins or bearing insectivorous bladders), with stalked and/or sessile glands, exstipulate. **Flowers** solitary or in racemose arrangement on terminal scape, zygomorphic, perfect, hypogynous, showy. **Calyx** of 4 or 5 sepals, distinct to basally connate, often bilabiate, usually imbricate, persistent. **Corolla** sympetalous with 5 lobes, bilabiate with lip basally saccate or spurred and often personate, white, blue, purple, or yellow, imbricate. **Androecium** of 2 stamens and occasionally 2 staminodes, epipetalous at corolla base; filaments distinct; anthers 1-locular, basifixed, dehiscing longitudinally, introrse or extrorse. **Gynoecium** of 1 pistil, 2-carpellate; ovary superior, 1-locular; ovules numerous, anatropous, often somewhat sunken into the globose placenta, placentation free-central; style very short to absent; stigma unequally 2-lobed (upper lobe reduced), papillose. **Fruit** a capsule, circumscissile (pyxis) or dehiscing by 2 or 4 valves or irregularly dehiscent; seeds minute; embryo poorly differentiated; endosperm absent.

Family characterization: Insectivorous, scapose herbs growing in wet to aquatic areas; specialized, glandular leaves with inrolling margins or insect-trapping bladders; bilabiate corolla with basally saccate/spurred and often personate lip; 2 stamens with 1-locular anthers; 1-locular, 2-carpellate, superior ovary with free-central placentation; unequally 2-lobed stigma; and a capsule containing small, nonendospermous seeds with poorly differentiated embryo. Iridoid compounds often present. Anatomical features: specialized leaf morphology for ensnaring and digesting insects and other small prey (see Lloyd 1942, and the discussion below); and unitegmic and tenuinucellate ovules.

Genera/species: *Utricularia* (120–180 spp.), *Pinguicula* (46 spp.), *Genlisea* (15–16 spp.), and *Polypompholyx* (2–3 spp.)

Distribution: Cosmopolitan in wetlands, aquatic areas, and moist forests.

U.S./Canadian representatives: *Utricularia* (22 spp.) and *Pinguicula* (9 spp.)

Economic plants and products: Species of *Genlisea*, *Pinguicula* (butterwort), and *Utricularia* (bladderwort) grown as novelty house plants.

Commentary: The Lentibulariaceae are well known for their insectivory and the remarkable leaf specializations for ensnaring and digesting minute organisms (insects, crustaceans). Due to the similarity in floral morphology (zygomorphic flowers with reduced androecium, capsular fruit), the family is closely allied with the Scrophulariaceae in classification schemes. The circumscription of *Utricularia*, the largest genus, has varied with different authorities: it sometimes has been divided into several smaller genera (e.g., *Biovula-*

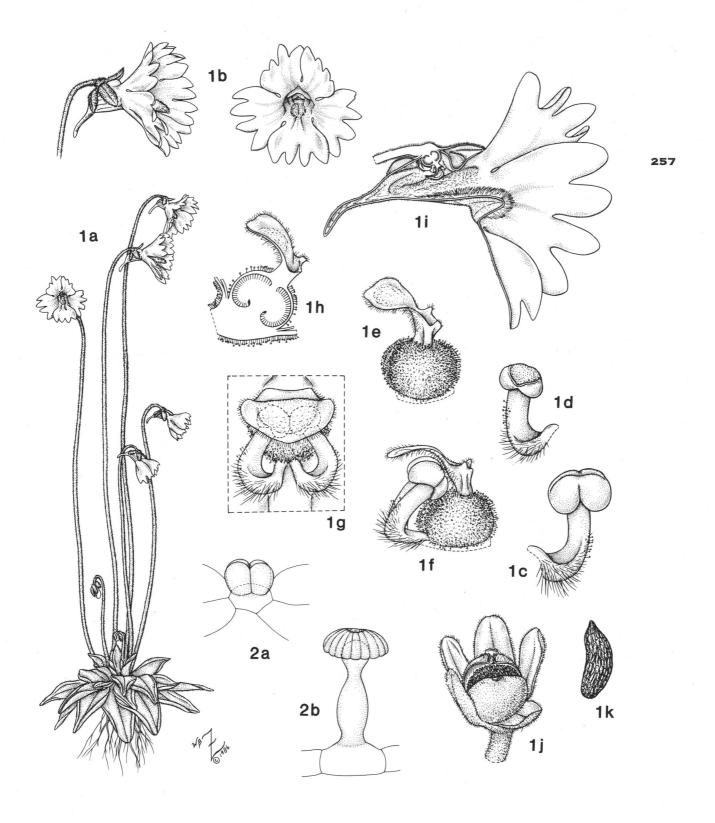

Figure 116. Lentibulariaceae. 1, *Pinguicula lutea*: **a,** habit, ×⅜; **b,** two views of flower, ×1; **c,** stamen (from bud), ×9; **d,** dehiscing stamen, ×9; **e,** pistil, ×10; **f,** androecium and pistil, ×10; **g,** top view of androecium and pistil, ×10; **h,** longitudinal section of pistil, ×10; **i,** longitudinal section of flower, ×2¼; **j,** capsule with persistent calyx, ×3; **k,** seed, ×32. **2,** *Pinguicula primuliflora*: **a,** digestive gland from adaxial leaf surface, ×265; **b,** adhesive gland from leaf, ×225.

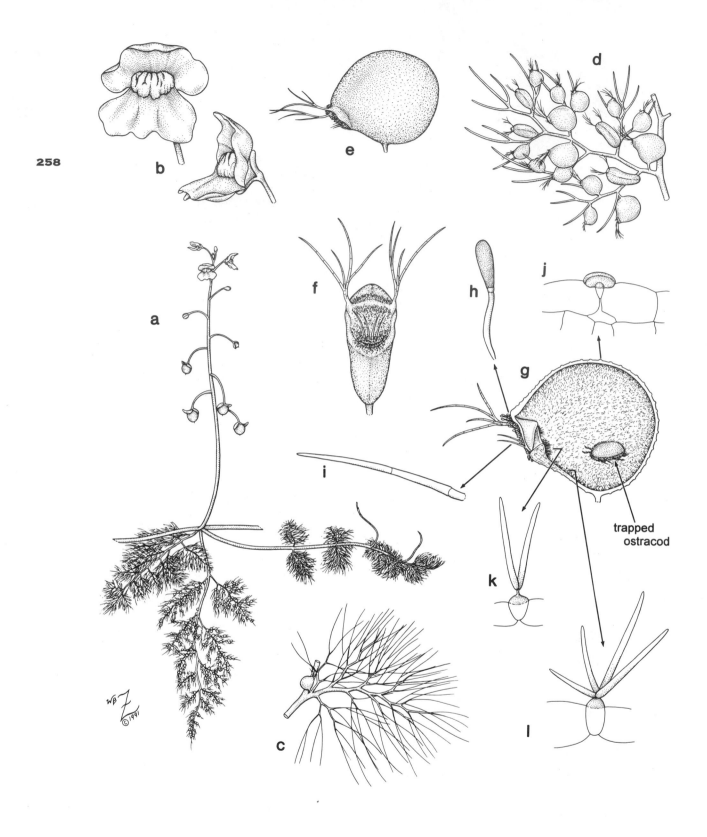

258

Figure 117. Lentibulariaceae (continued). *Utricularia foliosa*: **a,** habit, ×²⁄₅; **b,** two views of flower, ×2½; **c,** portion of "leafy" branch (with only one bladder), ×3; **d,** portion of branch with bladders (insectivorous traps), ×6; **e,** lateral view of bladder, ×18; **f,** front view of bladder showing trap door and position of the four "trigger" bristles, ×25; **g,** longitudinal section of bladder with trapped animal (ostracod), ×22; **h,** mucilage gland, ×250; **i,** "trigger" bristle, ×120; **j,** cross section of bladder surface showing capitate mucilage gland, ×250; **k,** two-armed digestive hair, ×250; **l,** four-armed digestive hair, ×250.

ria, a segregate) or expanded by the inclusion of *Genlisea* and/or *Polypompholyx* (see Taylor 1989).

The classic work on carnivorous plants by Lloyd (1942) details the morphology of the unusual family, summarized in Juniper et al. (1989). The plants grow generally in moist to aquatic, nitrogen-poor areas and vary from terrestrial (*Pinguicula*, *Genlisea*, some *Utricularia* species), epiphytic (*Utricularia*, some *Pinguicula* species), to free-floating (*Utricularia*). The family is represented in our area by *Pinguicula* and *Utricularia*. In *Pinguicula* (Fig. 116: 1a), the plants are attached to the moist soil with true roots. A flat rosette of simple, entire leaves with uprolled margins surrounds the scape(s). The leaves feel soft and greasy, and the surface glistens due to two kinds of glandular hairs. Stalked glands ("adhesive glands"; Fig. 116: 2b), with a multicellular head of radiating cells, contain mucilage that entraps prey (aphids, ants, small flies). After an insect is ensnared, the margins of the leaf roll inward, as enzymes secreted by sessile "digestive glands" (Fig. 116: 2a) break down the prey (the leaf margins later uncurl to their original position). These glands are similar in structure to the stalked form but have a smaller cap of four to eight cells and are sunken somewhat into the leaf surface. A few glands of this type also may occur on the lower surface of the leaf; both sessile and stalked glands also are found on the scape.

Although some species of *Utricularia* are terrestrial or epiphytic, many are submersed aquatic plants lacking roots (Fig. 117: a). In these species, the vascular system is much reduced, and the distinction between stems and leaves is unclear. The habit comprises a scape plus dissected, photosynthetic "branches" or appendages that are interpreted as either stems or leaves. The branches are dimorphic: either composed of mostly "leaf" material (highly dissected axes; Fig. 117: c), or primarily consisting of tiny (ca. 0.3 mm wide) bladder-like traps (Fig. 117: d,e), each attached to the axis by a short stalk. Small, capitate, mucilage glands (Fig. 117: j) are scattered on the outside surface of the bladder, and in many species, antennae-like bristles arch over the opening of the trap, giving the pear-shaped bladder a crustacean-like appearance (Fig. 117: e–g). The mouth of the bladder is hermetically sealed by a complex system of two sensitive valves (hinged trapdoors) associated with glands and bristle-like hairs (see Fig. 117: f,g). The outside surface of the door itself is covered with conspicuous, stalked mucilage glands (Fig. 117: h) that probably are attractive to prey (aquatic insects, larvae, small crustaceans). Four stiff, tapering bristles on the door near the entrance act as the trigger (Fig. 117: f,i): when touched, the slightly displaced valve opens the bladder to a sudden rush of inflowing water that sucks the prey inside. Glandular hairs that secrete digestive enzymes line the inside of the bladder proper (four-armed hairs; Fig. 117: l), as well as the inner threshold of the door (two-armed hairs; Fig. 117: k). The trap resets itself when water is actively withdrawn from the bladder lumen.

The broad lower lip of the bilabiate flowers of *Utricularia* (Fig. 117: b) and *Pinguicula* (Fig. 116: 1b) serves as a landing platform for the pollinators (usually bees) that visit for the nectar secreted in the spur. The reddish lines or streaks on the flowers of many *Utricularia* species are nectar guides. Due to the proximity of the two corolla lips, the entrance to the *Utricularia* flower is essentially closed until the bee depresses the lower lip. As the bee enters and probes for the nectar, it first contacts the bilobed stigma, which is sensitive (i.e., closing when touched) in many species of *Utricularia*. Later, the pollen-dusted bee retreats past the closed stigma, thus preventing self-pollination of the flower. In *Pinguicula*, the large upper lobe of the stigma completely covers the anthers (Fig. 116: 1f,g,i). A pollinator first encounters the papillose lower lobe, covering it with pollen from the previous flower; as the bee (covered with new pollen) leaves the flower, it forces the stigmatic lobe upward, thereby avoiding contact (and self-pollination of the flower).

REFERENCES CITED

Juniper, B. E., R. J. Robbins, and D. M. Joel. 1989. *The carnivorous plants*, pp. 43–44, 51, 64–71, 108, 117–125. Academic Press, London.

Lloyd, F. E. 1942. *The carnivorous plants*, pp. 90–94, 106–114, 213–270. Chronica Botanica, Waltham, Mass.

Taylor, P. 1989. The genus *Utricularia*—A taxonomic monograph. *Kew Bull.*, Addit. Ser. 14:1–724.

ACANTHACEAE
Acanthus Family

Perennial herbs, vines, or occasionally shrubs, sometimes armed. **Leaves** simple, usually entire, opposite, decussate, exstipulate. **Inflorescence** determinate, cymose and often appearing racemose, paniculate, or spicate (flowers congested in leaf axils), or sometimes flower solitary, axillary. **Flowers** zygomorphic, perfect, hypogynous, showy, with nectariferous disc, often subtended by conspicuous bracts. **Calyx** synsepalous with 5 or sometimes 4 lobes, convolute or imbricate, persistent. **Corolla** sympetalous with 5 lobes, usually bilabiate, variously colored, imbricate or convolute. **Androecium** of usually 4 stamens and didynamous (then 1 staminode sometimes also present) or sometimes reduced to 2 stamens (then 2 staminodes

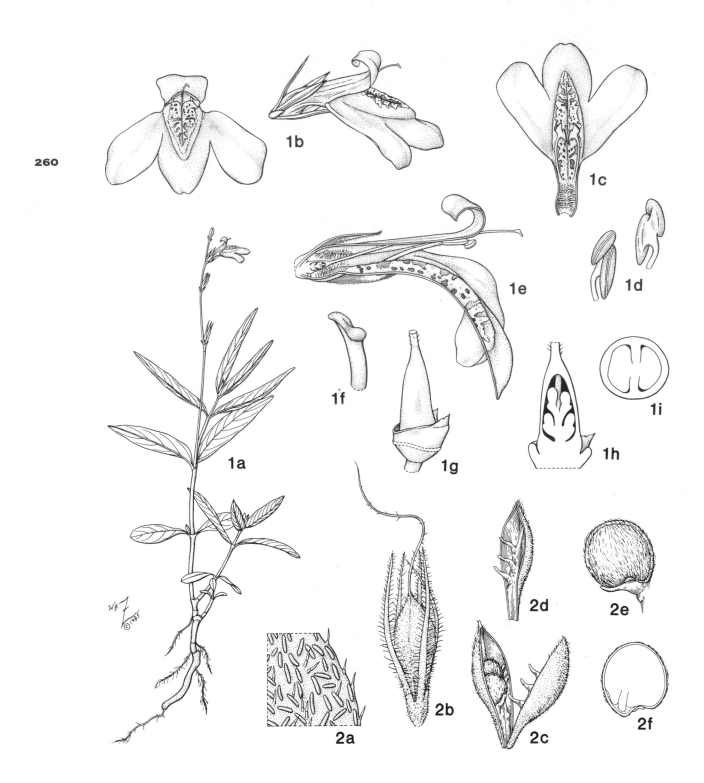

Figure 118. Acanthaceae. 1, *Justicia angusta* (*J. ovata* var. *angusta*): **a,** habit, ×½; **b,** two views of flower, ×2; **c,** lower corolla lip and epipetalous androecium, ×2; **d,** two views of anther, ×6; **e,** longitudinal section of flower, ×3½; **f,** stigma, ×25; **g,** ovary and nectary, ×10; **h,** longitudinal section of ovary, ×12; **i,** cross section of ovary, ×25. **2,** *Ruellia caroliniensis*: **a,** portion of dried leaf showing cystoliths, ×25; **b,** capsule before dehiscence, ×3; **c,** dehiscing capsule (persistent calyx removed), ×3; **d,** one valve of capsule showing retinacula (seeds ejected), ×3; **e,** seed (dry) with retinaculum, ×6; **f,** longitudinal section of seed, ×6.

frequently present), epipetalous; filaments distinct or coherent in pairs; anthers more or less basifixed, 2- or 1-locular, sometimes with locules unequal and at different levels, sometimes spurred and/or hairy, sometimes with prominent connective, dehiscing longitudinally. **Gynoecium** of 1 pistil, 2-carpellate; ovary superior, 2-locular; ovules 2 to 10 in 2 rows in each locule, anatropous, amphitropous, or campylotropous, with modified funicle (developed in fruit into a hook-shaped jaculator or retinaculum), placentation axile; style 1, slender, filiform; stigma(s) 1 or 2, funnelform or 2-lobed, often with reduced posterior lobe. **Fruit** an elastic loculicidal capsule with 2 recurving valves; seeds usually flat, supported by small hook-like projections (retinacula or jaculators), with very thin, often mucilaginous (when moistened) seed coat; endosperm usually absent; embryo large, curved, bent, or sometimes straight.

Family characterization: Herbs with opposite, simple leaves; bracteate flowers with usually bilabiate corollas; androecium of 4 (didynamous) or 2 stamens; anthers with unusual morphology (with hairs, spurs, expanded connective, and/or reduced to 1 locule); 2-valved, elastically dehiscing loculicidal capsule; and mucilaginous seeds with specialized hook-like funicles (retinacula or jaculators). Tissues often with calcium oxalate crystals. Anatomical features: cystoliths in the parenchyma and epidermal cells in the leaves and stems (often appearing as raised streaks or lines; Fig. 118: 2a); phloem with bundles of acicular fibers; and unitegmic and tenuinucellate ovules.

Genera/species: 256/2,770

Distribution: Pantropical, with only a few representatives in temperate areas; centers of distribution in Indo-Malaya, Africa, Brazil, and Central America.

Major genera: *Justicia* (300–420 spp.), *Strobilanthes* (250 spp.), *Barleria* (230–250 spp.), *Aphelandra* (170–200 spp.), *Thunbergia* (100–200 spp.), *Dicliptera* (150 spp.), and *Ruellia* (5–150 spp.)

U.S./Canadian representatives: 24 genera/98 spp.; largest genera: *Ruellia* and *Justicia*

Economic plants and products: A few dye plants, such as *Adhatoda* (yellow dye) and *Strobilanthes* (blue dye). Ornamental plants (species of 36 genera), including *Acanthus* (bear's-breech), *Aphelandra* (zebra plant), *Asystasia* (coromandel), *Beloperone* (shrimp plant), *Crossandra*, *Eranthemum* (blue-sage), *Fittonia* (mosaic plant), *Graptophyllum* (caricature plant), *Justicia*, *Odontonema* (firespike), *Pachystachys* (cardinal's-guard, yellow shrimp plant), *Pseuderanthemum*, *Sanchezia*, and *Strobilanthes*.

Commentary: The Acanthaceae have been divided into three or four subfamilies, with the majority of species placed in the Acanthoideae, a well-defined and presumably monophyletic group. The other subfamilies, sometimes segregated as families, have less-modified funicles and lack other characters, such as cystoliths (Bremekamp 1965). The phylogenetic relationships of the Acanthaceae also are controversial, with affinities to such groups as the Scrophulariaceae, Gesneriaceae, Verbenaceae, and Lamiaceae cited in the literature (see Table 16; Mohan Ram and Wadhi 1965; and Long 1970, for summaries).

Genera are distinguished by characters of the bracts (size and form), corolla (shape), stamens (number and morphology), and ovules (number). Pollen morphology and sculpturing also are often used to delimit genera (Scotland 1992). Cystoliths, which often are evident as raised streaks on the leaf blades (Fig. 118: 2a), are characteristic of certain genera and tribes. In addition, numerous embryological studies have clarified the relationships of genera (and subfamilies).

The anthers, in particular, show a great variety of number, form, and position in different genera. For example, the flowers of *Thunbergia* have four stamens, while those of *Justicia* (Fig. 118: 1c) have only two. The anther locules, often at different levels, also may be unequal in size (or sometimes one is absent) and often are separated by a well-developed connective. In addition, the anthers frequently are spurred and/or hairy.

With brightly colored corollas, often showy bracts, and nectar produced by the discs, the flowers of the Acanthaceae attract insects (or sometimes birds). The flowers are often protogynous. The corolla generally is either tubular with a five-lobed limb (*Ruellia*) or, more often, more or less bilabiate (*Justicia*; Fig. 118: 1b) with an erect upper lip (bifid at apex) and a lower 3-lobed horizontal lip. The lower lip of the bilabiate flower serves as a landing platform for insect visitors. An insect becomes dusted with pollen as it enters and touches the anther locule tips (or spurs); it then transfers the pollen to the stigma of the next flower visited.

Most genera of the Acanthaceae are characterized by a bilocular capsule (Fig. 118: 2c) that splits elastically, leaving a persistent central column. The seeds are situated upon hook-shaped projections that represent modified funicles (called retinacula or jaculators; Fig. 118: 2d,e); as the valves of the capsule dehydrate and recurve, the jaculators (usually slightly twisted to one side) help to direct the seeds laterally.

261

REFERENCES CITED

Bremekamp, C. E. B. 1965. Delimitation and subdivision of the Acanthaceae. *Bull. Bot. Surv. India* 7:21–30.

Long, R. W. 1970. The genera of Acanthaceae in the southeastern United States. *J. Arnold Arbor.* 51:257–309.

Mohan Ram, H. Y., and M. Wadhi. 1965. Embryology and the delimitation of the Acanthaceae. *Phytomorphology* 15: 201–205.

Scotland, R. W. 1992. Commentary. Pollen morphology and taxonomic characters in Acanthaceae. *Syst. Bot.* 17:337–340.

VERBENACEAE
Verbena Family

Annual or perennial herbs, shrubs, or rarely trees or vines, often aromatic, often armed with prickles and/or thorns; stems frequently square in cross section. **Leaves** simple, entire or serrate to lobed, opposite or sometimes whorled, decussate, exstipulate. **Inflorescence** indeterminate, often appearing racemose, spicate, or capitate, sometimes subtended by an involucre, terminal or axillary. **Flowers** zygomorphic, perfect, hypogynous, showy, usually with inconspicuous nectariferous disc. **Calyx** synsepalous with 5 lobes or teeth, campanulate or tubular, aestivation open, persistent, sometimes expanded or enlarged in fruit. **Corolla** sympetalous with 5 lobes, salverform (with very narrow tube and abruptly spreading limb), often somewhat bilabiate (with 2-lobed upper lip and 3-lobed lower lip), variously colored, imbricate. **Androecium** of 4 stamens, didynamous, epipetalous; filaments distinct; anthers dorsifixed, dehiscing longitudinally, introrse. **Gynoecium** of 1 pistil, 2-carpellate; ovary superior, basically 2-locular but typically appearing 4-locular due to ovary wall intrusions (false septa), usually slightly to moderately 4-lobed; ovules 1 in each apparent locule, usually anatropous, erect, placentation axile with ovules marginally attached (directly to carpel margins); style 1, terminal; stigma 1, lobed. **Fruit** a drupe with 2 or 4 pyrenes, or a schizocarp splitting into 2 or 4 nutlets, enclosed or subtended by persistent calyx; endosperm absent; embryo straight, oily.

Family characterization: Aromatic herbs or shrubs often with square stems and opposite leaves; indeterminate (racemose) inflorescences; sympetalous, salverform, and bilabiate corolla; 4 epipetalous and didynamous stamens; 2-carpellate ovary with false partitions forming 4 locules; 1 marginally attached ovule in each apparent locule; a drupe (with 2 or 4 stones) or a schizocarp (splitting into 2 or 4 nutlets) as the fruit types; and seeds with no endosperm. Iridoid compounds or essential oils present. Tissues often with calcium oxalate crystals. Anatomical features: exclusively unicellular nonglandular trichomes; xylem (even in herbs) forming a continuous ring with narrow rays (Fig. 119: 1d, 2b); unitegmic and tenuinucellate ovules; and unilacunar nodes (Fig. 119: 1c).

Genera/species: 36/1,035

Distribution: Tropical and subtropical, with only a limited number of representatives (usually herbs) in temperate regions.

Major genera: *Verbena* (250 spp.), *Lippia* (200–220 spp.), and *Lantana* (150 spp.)

U.S./Canadian representatives: 14 genera/108 spp.; largest genera: *Verbena*, *Lantana*, and *Glandularia*

Economic plants and products: Timber from *Citharexylum* (fiddlewood). Essential oils from *Lippia* (verbena oil), and toxic substances (triterpenes) in *Lantana* (shrub-verbena). Herbal remedies from *Verbena* (vervain) and tea-leaves and edible tubers from *Priva*. Ornamental plants (species of 13 genera), including *Citharexylum*, *Congea* (woolly congea), *Duranta* (golden-dewdrop), *Lantana*, *Petrea* (queen's-wreath), *Stachytarpeta*, and *Verbena* (vervain).

Commentary: The circumscription of the Verbenaceae is restricted substantially here, including only the 36 or so genera (1,035 spp.) of the traditional subfamily Verbenoideae (excluding tribe Monochileae), according to the studies of the Lamiaceae-Verbenaceae by Cantino (1992a,b), adopted by Thorne (1992) for his phylogenetic scheme. Examples of familiar genera still included in this Verbenaceae *s.s.* include *Citharexylum*, *Duranta*, *Glandularia*, *Lantana*, *Lippia*, *Phyla*, and *Verbena*. The remaining two-thirds of the

Figure 119. Verbenaceae. 1, *Lantana camara*: **a,** habit, ×¼; **b,** node, ×3; **c,** cross section of node (one-trace unilacunar), ×5; **d,** cross section of young stem (internode) showing continuous ring of vascular tissue, ×11; **e,** inflorescence (indeterminate), ×1½; **f,** two views of flower, ×3; **g,** longitudinal section of flower, ×6; **h,** pistil, ×10; **i,** longitudinal section of ovary, ×18; **j,** cross section of ovary near base, ×18; **k,** cross section of ovary near middle (note: ovary reduced to one locule), ×18; **l,** expanded corolla and epipetalous androecium, ×4½; **m,** drupe, ×3; **n,** cross section of drupe (note two-seeded pyrene), ×3. For floral diagram and floral formula of *Lantana camara*, see Fig. 4: 4a–c. **2,** *Verbena bonariensis*: **a,** habit, ×⅓; **b,** cross section of stem showing continuous ring of vascular tissue, ×5; **c,** two views of flower, ×10; **d,** pistil, ×22; **e,** longitudinal section of ovary, ×25; **f,** cross section of ovary near base, ×40; **g,** cross section of ovary through middle, ×40; **h,** schizocarp (half of persistent calyx removed), ×12; **i,** longitudinal section of schizocarp segment, ×12.

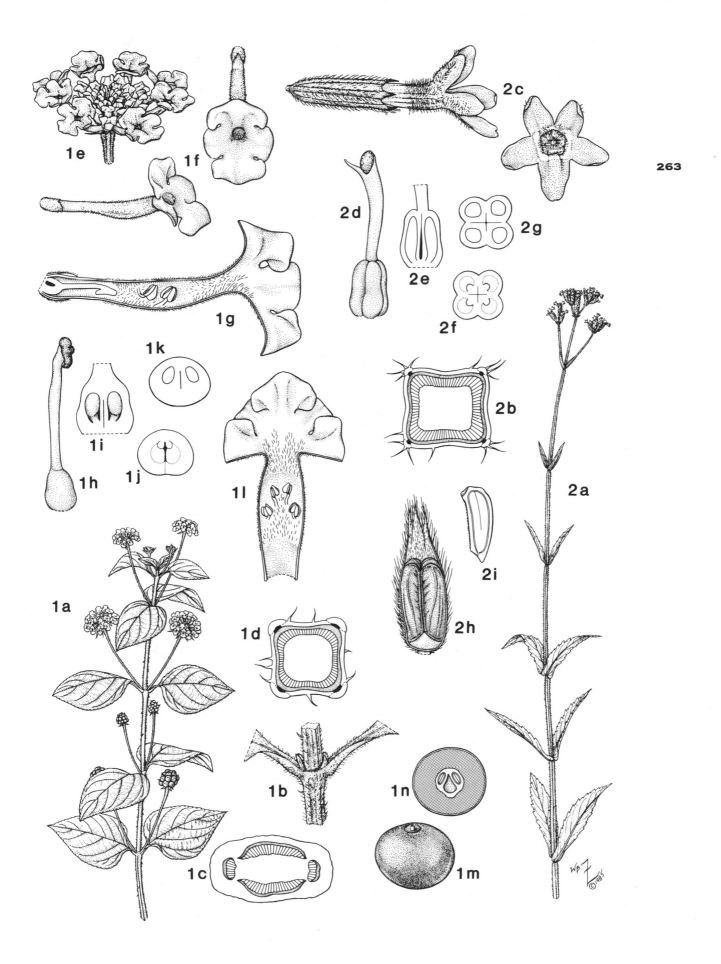

263

TABLE 17. Comparison of the Verbenaceae *s.s.* and the Lamiaceae *s.l.* as circumscribed in this text (*sensu* Cantino 1992a,b, and Thorne 1992). The most important features are in *italic*. See the commentary under both families for explanation and for a discussion of the characters shared by the two taxa.

CHARACTER	VERBENACEAE S.S.	LAMIACEAE S.L.
GENERA/SPECIES	36/1,035	258/6,970
SUBFAMILIES OF FORMER VERBENACEAE S.L. INCLUDED	only Verbenoideae (excluding tribe Monochileae)	Viticoideae, Caryopteridoideae, and Chloanthoideae
EXAMPLES OF GENERA TRADITIONALLY INCLUDED IN THE VERBENACEAE S.S.	*Citharexylem, Duranta, Glandularia, Lantana, Lippia, Petrea, Phyla, Verbena*	*Callicarpa, Clerodendrum, Cornutia, Holmskioldia, Petitia, Tectona, Vitex*
DISTRIBUTION	tropical to subtropical, with a few temperate representatives	cosmopolitan; majority of species in temperate areas
HABIT	herbs or shrubs, seldom trees or vines	herbs, shrubs, or trees
LEAVES	simple (but may be deeply dissected)	simple to pinnately or palmately compound
INFLORESCENCE	*indeterminate ("racemose")*	*determinate ("cymose")*
FLORAL SYMMETRY	zygomorphic	zygomorphic to actinomorphic
CALYX	tubular to campanulate	tubular, campanulate, to spreading
COROLLA	salverform and somewhat bilabiate	tubular and typically bilabiate, or seldom unilabiate
STAMENS	2 + 2 (didynamous)	2 + 2 (didynamous), 4 (of equal length), or sometimes 2
OVARY LOBING	slight to moderate	slight to deep
OVULE ATTACHMENT	*marginal (directly on carpel margin)*	*lateral on false septa*
STYLE	simple terminal	often bifid gynobasic to terminal
STIGMA	conspicuous lobed	usually inconspicuous at tips of stylar branches
FRUIT TYPE	schizocarp (2 or 4 nutlets) or drupe (2 or 4 stones)	schizocarp (4 nutlets) or drupe (1 to 5 stones)
OTHER	*pollen exine thickened near apertures (pores) nonglandular trichomes exclusively unicellular*	

conventional Verbenaceae *s.l.* have been transferred to the Lamiaceae *s.l.* (Cantino et al. 1992), which now include the former subfamilies Caryopteridoideae, Chloanthoideae, and the well-known Viticoideae (*Callicarpa, Clerodendrum, Vitex*). The detailed commentary under the Lamiaceae summarizes the reasons for these new delimitations, and the two families, as redefined, are compared and contrasted in Table 17. Basically, the Lamiaceae *s.s.* and Verbenaceae *s.l.*, as traditionally recognized, are polyphyletic and paraphyletic, respec-

tively. The close affinity of the two taxa generally has been agreed upon by taxonomists (El-Gazzar and Watson 1970), but the boundary between the two traditional families, which form the basis of the "Lamiales," has always been unclear due to transitional groupings (see the commentary under Lamiaceae and Tables 16 and 17 for some shared character states).

As emphasized in Table 17, the Verbenaceae *s.s.* are distinguished from the Lamiaceae *s.l.* by the indeterminate ("racemose") inflorescence (Fig. 119: 1e) and the marginally attached ovules that arise directly on the carpel margin where it curves in toward the interior of the ovary (Fig. 119: 1j, 2f). The styles are exclusively terminal and unbranched, with a well-developed, lobed stigma (Fig. 119: 1h, 2d). Cantino (1992a,b) also emphasizes the thickening of the pollen exine near the apertures (pores), as well as the exclusively unicellular, nonglandular trichomes (in those species that are pubescent). Although Cantino's (1992a,b) cladistic analyses (summarized in Fig. 122) did not include the Verbenoideae (equivalent to the Verbenaceae as here defined), the molecular study by Olmstead et al. (1993) indicates that the Verbenaceae *s.s.* are basal, or nearly so, within the Verbenaceae-Lamiaceae group.

To date, Cantino's analyses also have not directly addressed the placement of *Avicennia* (black mangrove) and *Phryma* (lopseed), both of which have been either included within the Verbenaceae *s.l.* as monotypic subfamilies or treated as segregate families (Avicenniaceae and Phrymaceae, as in Moldenke 1971). Thieret (1972) and Chadwell et al. (1992) suggest that *Phryma* is closely related to but not part of the Verbenaceae. Previously Thorne (1983) included *Avicennia*, but now (1992) he excludes it. More complete future investigations of the Lamiaceae-Verbenaceae should include these two taxa, as well as representatives of the Verbenaceae *s.s.* (Verbenoideae).

The corolla of the verbenaceous flower typically is salverform, with a long and slender tube plus an abruptly spreading limb that usually is somewhat bilabiate (e.g., Fig. 119: 1f, 2c). The colorful flowers are congested into various kinds of conspicuous, indeterminate inflorescences. Bees and flies seek the nectar secreted at the base of the ovary by the disc. The flowers generally are protandrous, and pollination probably occurs in a pattern similar to that in the mints.

The fruit often is drupaceous, although a schizocarp occurs in some genera, such as *Verbena* (Fig. 119: 2h) and *Glandularia* (both with four nutlets), and *Lippia* and *Phyla* (two nutlets). Genera with drupes generally are characterized by the arrangement of the pyrenes: for example, the drupe of *Duranta* has four two-chambered stones, and two two-chambered stones occur in *Citharexylum*. Most *Lantana* species are characterized by drupes with two one-locular stones, although *L. camara* has one two-chambered stone (Fig. 119: 1m,n).

REFERENCES CITED

Cantino, P. D. 1992a. Evidence for a polyphyletic origin of the Labiatae. *Ann. Missouri Bot. Gard.* 79:361–379.

———. 1992b. Toward a phylogenetic classification of the Labiatae. In *Advances in labiate science*, ed. R. M. Harley and T. Reynolds, pp. 27–37. Royal Botanic Gardens, Kew.

———, R. M. Harley, and S. J. Wagstaff. 1992. Genera of Labiatae: Status and classification. In *Advances in labiate science*, ed. R. M. Harley and T. Reynolds, pp. 511–522. Royal Botanic Gardens, Kew.

Chadwell, T. B., S. J. Wagstaff, and P. D. Cantino. 1992. Pollen morphology of *Phryma* and some putative relatives. *Syst. Bot.* 17:210–219.

El-Gazzar, A., and L. Watson. 1970. A taxonomic study of Labiatae and related genera. *New Phytol.* 69:451–486.

Moldenke, H. N. 1971. *A fifth summary of the Verbenaceae, Avicenniaceae, Stilbaceae, Dicrastylidaceae, Symphoremaceae, Nyctanthaceae, and Eriocaulaceae of the world as to valid taxa, geographic distribution, and synonymy.* 2 vols. "Publ. by the author," Wayne, N.J.

Olmstead, R. G., B. Bremer, K. M. Scott, and J. D. Palmer. 1993. A parsimony analysis of the Asteridae sensu lato based on *rbc*L sequences. *Ann. Missouri Bot. Gard.* 80:700–722.

Thieret, J. W. 1972. The Phrymaceae in the southeastern United States. *J. Arnold Arbor.* 53:227–233.

Thorne, R. F. 1983. Proposed new realignments in the angiosperms. *Nordic J. Bot.* 3:85–117.

———. 1992. Classification and geography of the flowering plants. *Bot. Rev.* (Lancaster) 58:225–348.

LAMIACEAE OR LABIATAE
Mint Family

Annual or perennial herbs, shrubs, or trees, aromatic, usually with pubescence of glandular hairs containing ethereal oils; stems frequently square in cross section. **Leaves** simple to pinnately or palmately dissected or compound, usually serrate, opposite or sometimes whorled, decussate, exstipulate. **Inflorescence** determinate, often of paired cymes congested into whorls but frequently appearing racemose, spicate, paniculate, or capitate, sometimes subtended by bracts, terminal or axillary. **Flowers** zygomorphic to occasionally actinomorphic, perfect, hypogynous, showy, often with fleshy nectariferous disc. **Calyx** synsepalous with 5 lobes, actinomorphic to bilabiate, tubular to broadly campanulate or spreading, often oblique at the throat, imbricate, persistent and sometimes enlarged in fruit. **Corolla** sympetalous with 5 lobes, tubular and typically bilabiate with 2-lobed upper lip and 3-lobed lower lip, variously colored, imbricate. **Androecium** of 4 stamens and didynamous (anterior pair longer) or of equal length, or sometimes reduced to 2, epipetalous; anthers basifixed, often with prominent connective, dehiscing longitudinally, introrse. **Gynoecium** of 1 pistil, 2-carpellate; ovary superior, 2-locular but appearing 4-locular due to ovary wall intrusions (false septa), slightly to deeply 4-lobed; ovules 1 in each apparent locule, anatropous, erect, placentation

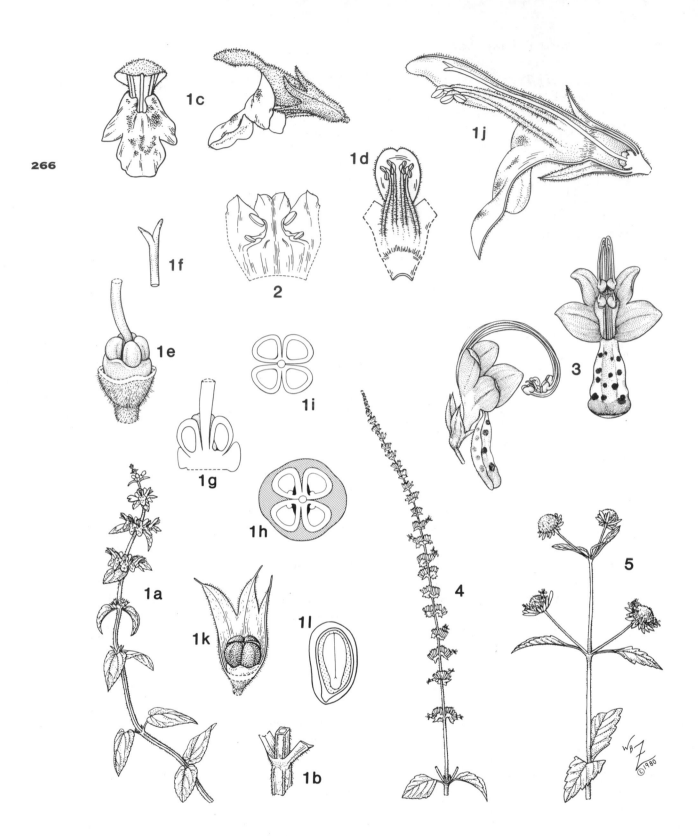

Figure 120. Lamiaceae. 1, *Stachys floridana*: **a,** habit, ×²/₅; **b,** node, ×3; **c,** two views of flower, ×3; **d,** epipetalous androecium with portion of corolla, ×3; **e,** ovary and disc, ×12; **f,** style apex (stigmas terminal), ×12; **g,** longitudinal section of ovary, ×18; **h,** cross section of ovary (and disc) at base, ×20; **i,** cross section of ovary near middle, ×20; **j,** longitudinal section of flower, ×6; **k,** schizocarp (four "nutlets") with half of persistent calyx removed, ×5; **l,** longitudinal section of schizocarp segment ("nutlet"), ×12. **2,** *Salvia lyrata*: epipetalous androecium with portion of corolla, ×3. **3,** *Trichostema dichotomum*: two views of flower, ×2³/₄. **4,** *Hyptis mutabilis*: inflorescence, ×²/₅. **5,** *Hyptis alata*: upper portion of plant showing capitate inflorescences, ×¹/₂.

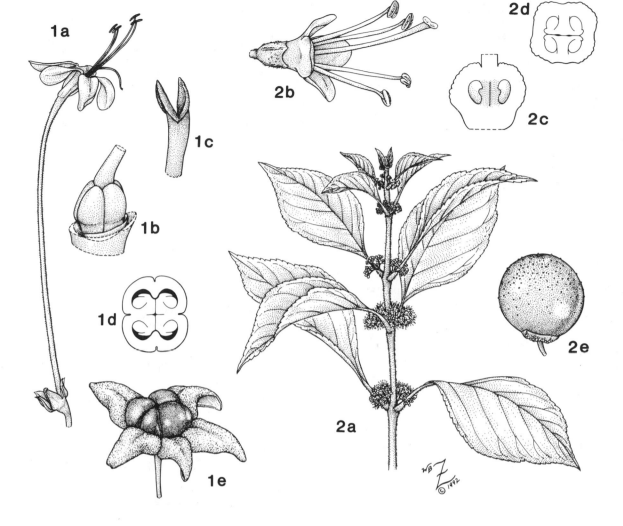

Figure 121. Lamiaceae (continued). 1, *Clerodendrum indicum*: **a,** flower, ×³⁄₄; **b,** ovary and disc, ×6; **c,** stigma, ×12; **d,** cross section of ovary, ×10; **e,** drupe with accrescent calyx, ×1. **2,** *Callicarpa americana*: **a,** habit, ×²⁄₅; **b,** flower, ×6; **c,** longitudinal section of ovary, ×30; **d,** cross section of ovary, ×30; **e,** drupe, ×4½.

axile (often appearing basal) with ovules laterally attached on the false septa just short of inrolled carpel margins; style 1, gynobasic (arising from central depression of ovary lobes) to terminal, bifid at apex; stigmas minute at stylar branch tips. **Fruit** a schizocarp splitting into 4 nutlets or a drupe with 1 to 5 pyrenes, subtended by or enclosed within persistent calyx; endosperm absent or scanty and fleshy; embryo generally straight.

Family characterization: Aromatic, herbaceous to woody plants with square stems and opposite leaves; cymose inflorescence often congested into false whorls at the nodes; bilabiate sympetalous corolla; usually 4 epipetalous and didynamous stamens; 4-lobed, 2-carpellate ovary with false partitions forming 4 locules; each apparent locule with 1 ovule laterally

attached on the false septa (just short of the carpel margin); terminal to gynobasic style; a schizocarp (splitting into 4 nutlets) or a drupe (with 1 to 5 pyrenes) as the fruit types; and seeds with little or no endosperm. Essential oils or iridoid compounds present. Anatomical features: unitegmic and tenuinucellate ovules; and unilacunar nodes.

Genera/species: 258/6,970

Distribution: Cosmopolitan, with the majority of species in temperate zones; particularly diverse in the Mediterranean region.

Major genera: *Salvia* (700–900 spp.), *Hyptis* (400 spp.), *Clerodendrum* (400 spp.), *Thymus* (300–400 spp.), *Scutellaria* (300 spp.), *Stachys* (300 spp.), *Teucrium* (100–300 spp.), *Vitex* (250 spp.), and *Premna* (200 spp.)

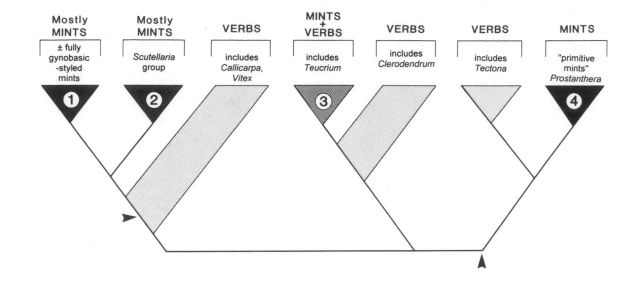

Figure 122. Relationships of major groupings of the Lamiaceae *s.s.* and Verbenaceae *s.l.* Simplification of detailed consensus tree (106 genera; 85 characters) from Cantino (1992a) showing the traditional "mints" (Lamiaceae *s.s.*) derived at least four times within the Verbenaceae *s.l.* (i.e., the four clades that are numbered here). The Lamiaceae *s.l.* (*sensu* Thorne 1992 and Cantino 1992a,b) comprise all groups represented in this cladogram. The two arrows indicate possible positions of the most likely roots for this tree. Abbreviations: MINTS = genera traditionally included in the Lamiaceae *s.s.*; VERBS = genera traditionally included in the Verbenaceae *s.l.*

U.S./Canadian representatives: 77 genera/552 spp.; largest genera: *Salvia, Scutellaria, Stachys, Stenogyne,* and *Phyllostegia*

Economic plants and products: Condiments and perfumes from *Hedeoma* (pennyroyal), *Hyssopus* (hyssop), *Lavandula* (lavender), *Mentha* (peppermint, spearmint), *Ocimum* (basil), *Origanum* (oregano, pot marjoram), *Pogostemon* (patchouly), *Rosmarinus* (rosemary), *Salvia* (sage), *Satureja* (savory), and *Thymus* (thyme). Edible tubers from *Stachys*. Timber from *Tectona* (teak) and *Vitex*. Ornamental plants (species of more than 80 genera), including *Ajuga* (bugleweed), *Callicarpa* (beauty-berry), *Clerodendrum* (bleedingheart, glorybower), *Coleus, Holmskioldia* (Chinese-hat plant), *Pycnanthemum* (mountain-mint), *Salvia, Scutellaria* (skullcap), and *Vitex* (chaste tree, monk's-pepper).

Commentary: The Lamiaceae are here very broadly circumscribed (*sensu* Thorne 1992, Cantino 1992a,b), including approximately two-thirds of the genera and species formerly included in the traditional Verbenaceae *s.l.* (e.g., *Callicarpa, Clerodendrum, Vitex*; see Cantino et al. 1992 for the complete list of transfers). The "mints," as conventionally defined, have been accepted widely as a "natural" and monotypic family, most diverse in temperate areas. The family presumably evolved from the predominantly tropical Verbenaceae *s.l.* These two families form the core of the Lamiales (as defined, e.g., in Thorne 1983), sharing numerous characters, including square stems; opposite, exstipulate leaves; basically zygomorphic, bilabiate corollas; two-carpellate ovaries with false partitions; and two ovules per carpel (Cantino 1982; Table 16). Historically, the Lamiaceae *s.s.* have been primarily distinguished by the deeply four-lobed ovary with gynobasic style, as compared to the unlobed ovary with terminal style in the customary Verbenaceae *s.l.* However, the variation of these gynoecial characters for these two taxa actually exhibits a continuum rather than a clear-cut distinction, with intermediate character states found in two mint tribes plus several verbenaceous genera. Thus, the boundary between the two families, as conventionally circumscribed, is somewhat arbitrary, and other characters (discussed below and emphasized in Table 17) may more accurately reflect the phylogenetic relationships.

As first indicated by pollen studies (Abu-Asab and Cantino 1989) and subsequently supported by detailed morphological and anatomical cladistic analyses (Cantino 1992a), the Lamiaceae *s.s.* as traditionally circumscribed probably are polyphyletic, and the Verbenaceae *s.l.* are paraphyletic (see the summaries in Cantino 1992b and Judd et al. 1994). Figure 122, a simplified rendering of Cantino's (1992a,b) cladogram, represents his hypothesis resulting from these preliminary investigations of the relationships of all mint lineages.

The diagram clearly depicts the Lamiaceae *s.s.* ("MINTS") evolving at least four times within the Verbenaceae *s.l.* ("VERBS") in three widely divergent parts of the tree. Of note is that the "gynobasic-styled mints" with four-nutlet fruits (group or clade #1) would include most genera (about 90%) of the Lamiaceae *s.s.* The results of this anatomical and morphological survey are also somewhat supported by molecular data (Olmstead et al. 1992; Olmstead et al. 1993), which incorporated only a small sample set of this complex (10 genera).

The problem of the polyphyly of the Lamiaceae *s.s.* and the paraphyly of the Verbenaceae *s.l.* apparently may be rectified in several ways (Cantino 1992a,b). One solution—combining the two taxa—would only move the level of the issue, since dividing the resultant family into monophyletic subgroups (subfamilies, tribes) still would be problematical. Another obvious resolution would be the redefinition of the Lamiaceae as group #1 on the cladogram (gynobasic-styled mints), but the remaining Verbenaceae *s.l.* would still be a paraphyletic taxon. Therefore, Cantino (1992a,b) recommends an expanded Lamiaceae and a restricted Verbenaceae primarily based on inflorescence and ovule characters (summarized in Table 17). Although Cantino (1992a,b) stresses the preliminary nature of his work at this point, Thorne (1992) has adopted this hypothesis and resolution in his phylogenetic scheme.

The Lamiaceae *s.l.*, thus defined, include the verbenaceous genera with cymose (determinate) inflorescences and laterally attached ovules on false septa (i.e., the ovule is attached just short of the carpel margin where it curves into the interior of the ovary; Figs. 120: 1h; 121: 1d, 2d; see also Junell 1934). These genera formerly constituted the subfamilies Caryopteridoideae, Chloanthoideae, and Viticoideae of the Verbenaceae *s.l.* (Cantino et al. 1992). As discussed in the commentary under that family, the Verbenaceae *s.s.* now are restricted to the original subfamily Verbenoideae (excluding the tribe Monochileae), characterized by indeterminate ("racemose") inflorescences, and ovules attached directly on the incurved carpel margins. Further analyses are needed on the mint-verb complex—especially concerning the circumscription of the numerous tribes/subfamilies of the traditional Lamiaceae *s.s.* (El-Gazzar and Watson 1970; Cantino and Sanders 1986), many of which are polyphyletic (Cantino 1992a,b).

As mentioned above, the inflorescence of the Lamiaceae is composed basically of axillary cymes bearing three to many flowers. These cymose units often surround the stem, forming a congested axillary cluster (as in *Callicarpa americana*; Fig. 121: 2a) or more typically a "false whorl" (verticil) at each node (as in *Stachys floridana*; Fig. 120: 1a). A racemose or spicate inflorescence results when the leaves (bracts) become very reduced (*Hyptis mutabilis*, Fig. 120: 4; *Clerodendrum indicum*). Further condensation of the axis results in a terminal head (*Hyptis alata*; Fig. 120: 5). A reduction in the number of the verticils or the number of flowers per cyme may produce an inflorescence consisting of a single cyme or a solitary flower.

In most temperate genera, the lobes of the bilabiate corolla are in a "two-three" arrangement, with the lips separated by relatively deep sinuses (Fig. 120: 1c). The two lobes of the arching upper lip are equal in size, but the median lobe on the lower lip commonly differs from the two lateral lobes in size and shape. Occasionally, as in *Teucrium* or *Trichostema* (Fig. 120: 3), the original upper lip is deeply cleft between the two widely divergent lobes, forming a limb with apparently a single, five-lobed ("lower") lip. In other genera, such as *Callicarpa* (Fig. 121: 2b), the funnelform (to salverform) corolla is actinomorphic.

Pollinators of the Lamiaceae usually are bees, although other insects and birds also have been frequently observed (Huck 1992), and some species of *Monarda* are visited by butterflies. The general bilabiate corolla type of the family is well adapted for pollination by Hymenoptera (van der Pijl 1972): the arching upper lip protects the stigma and stamens, while the lower lip acts as a visible attractant and landing place for the pollinators. The length of the corolla tube varies depending on the kinds of visitors. The pollen lands on the insect's back as it stands on the lower lip and probes for the nectar secreted by the disc near the base of the flower. Flowers commonly are protandrous.

A schizocarp splitting into four nutlets (derived from a deeply lobed ovary with gynobasic style) is the fruit type defining the traditional Lamiaceae *s.s.* (e.g., Fig. 120: 1g,k). However, the expanded circumscription of the family adopted here includes many species characterized by an ovary with terminal style that develops into a drupe (typically with four pyrenes), as in *Clerodendrum* (deeply four-lobed drupe; Fig. 121: 1e) and *Callicarpa* (Fig. 121: 2e). The drupe of *Vitex* has a single four-locular stone.

REFERENCES CITED

Abu-Asab, M. S., and P. D. Cantino. 1989. Pollen morphology of *Trichostema* (Labiatae) and its systematic implications. *Syst. Bot.* 14:359–369.

Cantino, P. D. 1982. Affinities of the Lamiales: A cladistic analysis. *Syst. Bot.* 7:237–248.

———. 1992a. Evidence for a polyphyletic origin of the Labiatae. *Ann. Missouri Bot. Gard.* 79:361–379.

———. 1992b. Toward a phylogenetic classification of the Labiatae. In *Advances in labiate science*, ed. R. M. Harley and T. Reynolds, pp. 27–37. Royal Botanic Gardens, Kew.

———, R. M. Harley, and S. J. Wagstaff. 1992. Genera of Labiatae: Status and classification. In *Advances in labiate science*, ed. R. M. Harley and T. Reynolds, pp. 511–522. Royal Botanic Gardens, Kew.

—— and R. W. Sanders. 1986. Subfamilial classification of Labiatae. *Syst. Bot.* 11:163–185.

El-Gazzar, A., and L. Watson. 1970. A taxonomic study of Labiatae and related genera. *New Phytol.* 69:451–486.

Huck, R. B. 1992. Overview of pollination biology in the Lamiaceae. In *Advances in labiate science*, ed. R. M. Harley and T. Reynolds, pp. 167–181. Royal Botanic Gardens, Kew.

Judd, W. S., R. W. Sanders, and M. J. Donoghue. 1994. Angiosperm family pairs: Preliminary cladistic analyses. *Harvard Pap. Bot.* 5:1–51.

Junell, S. 1934. Zur Gynäceummorphologie und Systematik der Verbenaceen und Labiaten. *Symb. Bot. Upsal.* 4:1–219.

Olmstead, R. G., B. Bremer, K. M. Scott, and J. D. Palmer. 1993. A parsimony analysis of the Asteridae sensu lato based on *rbc*L sequences. *Ann. Missouri Bot. Gard.* 80:700–722.

———, H. J. Michaels, K. M. Scott, and J. D. Palmer. 1992. Monophyly of the Asteridae and identification of its major lineages inferred from *rbc*L sequences. *Ann. Missouri Bot. Gard.* 79:249–265.

van der Pijl, L. 1972. Functional considerations and observations on the flowers of some Labiatae. *Blumea* 20:93–103.

Thorne, R. F. 1983. Proposed new realignments in the angiosperms. *Nordic J. Bot.* 3:85–117.

———. 1992. Classification and geography of the flowering plants. *Bot. Rev.* (Lancaster) 58:225–348.

LILIACEAE

Lily Family

Perennial herbs, with bulbs and contractile roots or occasionally with rhizomes. **Leaves** simple, entire, alternate, cauline or in basal rosettes, parallel-veined, sometimes dark-spotted, often sheathing at base. **Inflorescence** usually determinate, often racemose or umbellate or reduced to a single flower, terminal. **Flowers** actinomorphic to slightly zygomorphic, perfect, hypogynous, showy. **Perianth** of 6 tepals in 2 whorls of 3, petaloid, distinct, variously colored, frequently with striations or spots, imbricate or contorted. **Androecium** of 6 stamens, biseriate; filaments distinct; anthers basifixed or dorsifixed and versatile, dehiscing longitudinally, introrse or extrorse. **Gynoecium** of 1 pistil, 3-carpellate; ovary superior, 3-locular; ovules several to many in each locule, anatropous, placentation axile; style simple or sometimes branched; stigma(s) on 3 stigmatic crests or 3-lobed. **Fruit** usually a loculicidal capsule or (seldom) a berry; seeds typically flat, disc-shaped; endosperm copious, oily; embryo generally undifferentiated, ovoid or ellipsoid.

Family characterization: Perennial herbs with bulbs or rhizomes; showy flowers on terminal inflorescences; 6-merous perianth of petaloid, often spotted tepals; 6 distinct and free stamens; 3-carpellate superior ovary with axile placentation; a loculicidal capsule or berry as the fruit type; and flat seeds with copious, oily endosperm and undifferentiated embryo. Saponins often present. Anatomical feature: unitegmic and tenuinucellate ovules.

Genera/species: 22/485

Distribution: Predominantly temperate regions of the Northern Hemisphere; especially diverse in southwestern and Himalayan Asia and China.

Major genera: *Tulipa* (100 spp.), *Fritillaria* (85–100 spp.), *Lilium* (75–100 spp.), and *Gagea* (70–90 spp.)

U.S./Canadian representatives: 13 genera/141 spp.; largest genera: *Calochortus*, *Lilium*, *Fritillaria*, and *Erythronium*

Economic plants and products: Ornamental plants (species of at least 15 genera), including *Calochortus* (mariposa-lily, globe-tulip), *Erythronium* (adder's-tongue, dog-tooth-violet, trout-lily), *Fritillaria* (fritillary), *Lilium* (lily), and *Tulipa* (tulip).

Commentary: The Liliaceae are rather narrowly circumscribed here, including only 22 genera (Thorne 1992). Seven of the eight genera in Thorne's subfamily Tricyrtidoideae (e.g., *Uvularia*) are considered as a closely related segregate family, the Uvulariaceae, by Dahlgren et al. (1985), who place the eighth genus, *Scoliopus*, in the Trilliaceae. They also segregate *Calochortus* as the monotypic Calochortaceae, leaving an even more restricted Liliaceae *s.s.* of 13 genera. In an unpublished update on the classification of the group (see Zomlefer 1989), Thorne (pers. comm.) had previously included *Trillium* (and relatives; Fig. 124), which now are (Thorne 1992) established as the Trilliaceae, still maintained within the Liliales. Some authors (e.g., Dahlgren and Clifford 1982; Dahlgren et al. 1985), however, consider the Trilliaceae as only distantly related to the Liliaceae *s.s.* (see Table 19). Obviously, more study is needed on the Liliaceae *s.s.*, especially concerning the relationships of the genera mentioned above, plus the problematic placement of *Medeola*, *Clintonia*, *Diospora*, and *Streptopus*.

The traditional Liliaceae *s.l.* included a heterogeneous array of variously delimited subfamilies and tribes (see Sato 1943 and Conran 1989 for historical sum-

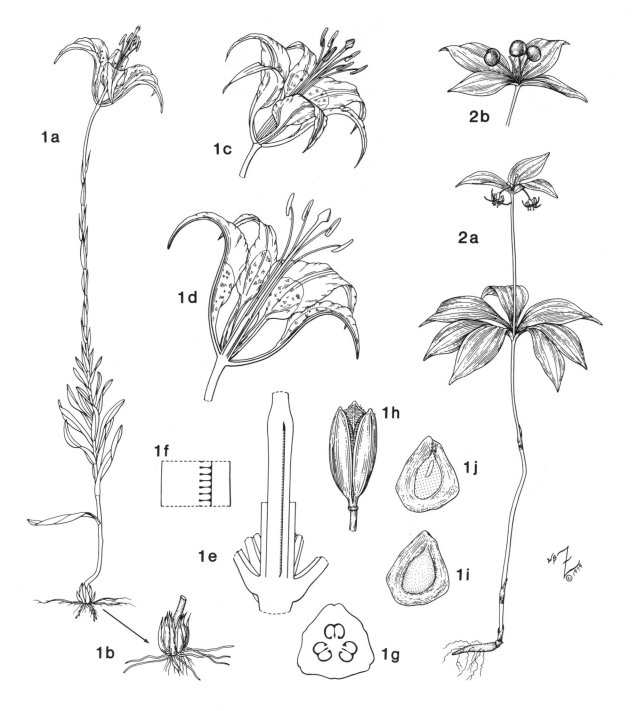

Figure 123. Liliaceae. 1, *Lilium catesbaei*: **a,** habit, ×⅓; **b,** bulb, ×⅔; **c,** flower, ×⅖; **d,** longitudinal section of flower, ×½; **e,** longitudinal section of ovary, ×2; **f,** detail of longitudinal section of ovary, ×6; **g,** cross section of ovary, ×6; **h,** capsule, ×½; **i,** seed, ×4; **j,** longitudinal section of seed, ×4. **2,** *Medeola virginiana*: **a,** habit, ×⅓; **b,** infructescence (berries), ×⅜.

maries). The recent critical studies by Dahlgren and associates (Dahlgren and Clifford 1982; Dahlgren et al. 1985) have dismantled this polyphyletic complex into numerous segregates placed into at least two orders. Many of these families (e.g., the Asparagaceae, Agavaceae, Amaryllidaceae, and Hemerocallidaceae) have been transferred to the Asparagales, while a few

(such as the Alstroemeriaceae) still are considered to be allied with the Liliaceae *s.s.* in the Liliales. Although the rank of order is not emphasized in this book, Table 18 (comparing these two orders) is included since so many of these segregates, formerly constituting familiar groupings within the Liliaceae *s.l.* of Cronquist (1981, 1988) and Thorne (1983), now are in the As-

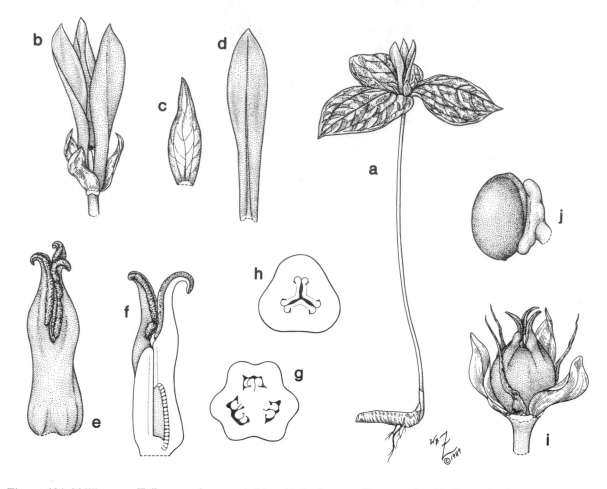

Figure 124. Trilliaceae. *Trillium maculatum*: **a,** habit, ×¼; **b,** flower, ×⅔; **c,** sepal, ×¾; **d,** petal, ×¾; **e,** pistil, ×2½; **f,** longitudinal section of pistil, ×2½; **g,** cross section of ovary near base, ×5; **h,** cross section of ovary above base, ×5; **i,** berry ("arilloid berry") with persistent perianth, ×⅘; **j,** seed (note aril), ×5. See Table 19 for family characteristics.

paragales. The monophyly of the Asparagales is well supported by recent preliminary investigations of molecular data (Duvall et al. 1993).

The primary characters separating the two orders involve features of the seed coat found in families of the Asparagales (but lacking in those of the Liliales). In particular, the inner integument deteriorates (forming a membrane), and the outer layers of the seed coat often are encrusted with phytomelan—a dark, brittle, charcoal-like substance (Dahlgren and Clifford 1982). Although blackish seeds are found throughout the angiosperms, the substance phytomelan probably is restricted to the seeds of the Asparagales (Dahlgren et al. 1985). Phytomelan is present in seeds in most species with capsules, but in the seeds of baccate fruits, the outer epidermis usually becomes obliterated and the black crust is not formed (but see the discussion under Asparagaceae and the accompanying Fig. 129: 1j vs. 2f).

Table 19 summarizes the major morphological differences between the Liliaceae *s.s.* and several familiar segregate families often formerly included within the Liliaceae *s.l.* Of note is that Thorne (1992) sometimes combines some of the segregates of Dahlgren et al. (1985), as in Thorne's more broadly circumscribed Liliaceae (mentioned above), Asparagaceae, and Dracaenaceae. Also, the placement of the Trilliaceae (as mentioned above) and the Melanthiaceae differs as shown in the table.

Liliaceous flowers (Fig. 123: 1c), typically large with brightly colored perianths and conspicuous stamens, usually attract bees and various Lepidoptera. Nectar often is secreted by nectaries at the base of the tepals (as in *Lilium*) and accumulates at the base of the flower. Some flowers, such as those of certain *Tulipa* species, may be devoid of nectar, and pollinators visit to collect pollen.

REFERENCES CITED

Conran, J. G. 1989. Cladistic analysis of some net-veined Liliiflorae. *Pl. Syst. Evol.* 168:123–141.

TABLE 18. Comparison between the orders Liliales and Asparagales, modified from Dahlgren et al. (1985). The most important characters are those of the seed coat (see text for discussion). See Table 19 for the somewhat more restricted circumscriptions of some of these families by Dahlgren et al. (1985) in comparison to those of Thorne (1992), as well as the differing placements of the Melanthiaceae and Trilliaceae (not listed here).

CHARACTER	LILIALES	ASPARAGALES
REPRESENTATIVE FAMILIES	Liliaceae (incl. Uvulariaceae and Calochortaceae), Alstroemeriaceae, Iridaceae	Asparagaceae, Agavaceae, Amaryllidaceae, Alliaceae, Hemerocallidaceae, Dracaenaceae
HABIT	herbs / not succulent	herbs to woody plants / sometimes succulent
TEPALS	often spotted	generally unspotted
NECTARY POSITION	usually at the base of tepals or filaments	usually septal
STYLE(S)	1 (then often trifid) or 3	usually 1 and simple
STARCH IN IMMATURE SEED COAT	probably always present (but disappearing at maturity)	usually none
INNER INTEGUMENT OF SEED COAT	cellular structure usually retained	usually collapsed, forming a reddish-brown or colorless membrane
OUTER EPIDERMIS OF SEED COAT	always present and well-developed	present in seeds from dry fruits, but obliterated in most baccate fruits
PHYTOMELAN CRUST ON SEED COAT	lacking (*seeds not black*)	present on seeds from dry fruits (*seeds black*), but seldom present in seeds of baccate fruits (since outer epidermis is obliterated)

Cronquist, A. 1981. *An integrated system of classification of flowering plants*, pp. 1208–1211. Columbia University Press, New York.

———. 1988. *The evolution and classification of flowering plants*, 2d ed., p. 516. New York Botanical Garden, Bronx.

Dahlgren, R. M. T., and H. T. Clifford. 1982. *The monocotyledons: A comparative study*, pp. 26–30, 228–231, 328–329. Academic Press, London.

———, ———, and P. F. Yeo. 1985. *The families of the monocotyledons*, pp. 92–96, 129–214, 220–238. Springer-Verlag, Berlin.

Duvall, M. R., M. T. Clegg, M. W. Chase, W. D. Clark, W. J. Kress, et al. 1993. Phylogenetic hypotheses for the monocotyledons constructed from *rbc*L data. *Ann. Missouri Bot. Gard.* 80:607–619.

Sato, D. 1943. Karyotype alteration and phylogeny in Liliaceae and allied families. *Jap. J. Bot.* 12:58–161.

Thorne, R. F. 1983. Proposed new realignments in the angiosperms. *Nordic J. Bot.* 3:85–117.

———. 1992. Classification and geography of the flowering plants. *Bot. Rev.* (Lancaster) 58:225–348.

Zomlefer, W. B. 1989. *Flowering plants of Florida: A guide to common families*, pp. 139–144. Biological Illustrations, Gainesville.

IRIDACEAE
Iris Family

Perennial herbs, often scapose, with rhizomes, bulbs, or corms; roots fibrous, mycorrhizal. **Leaves** simple, entire, alternate and basal, distichous, equitant, parallel-veined, numerous, usually narrow, conduplicate, with open basal sheath. **Inflorescence** determinate, cymose and often appearing racemose or paniculate, or sometimes flowers solitary, terminal. **Flowers** actinomorphic or zygomorphic, perfect, epigynous, usually large and showy, individually or collectively subtended by 1 or 2 expanded bract(s) (forming a spathe). **Perianth** of 6 tepals, biseriate, similar or differentiated, generally all petaloid, distinct or usually basally connate and forming a tube, variously colored, overlapping and twisted in bud. **Androecium** of 3 stamens, opposite and often adnate to the outer tepals; filaments distinct or sometimes basally connate and forming a tube; anthers basifixed, dehiscing longitudi-

TABLE 19. Major morphological differences between the Liliaceae *s.s.* (*sensu* Thorne 1992) and several segregate families with "lily-like" flowers often included within the traditional Liliaceae *s.l.* (Cronquist 1981, Thorne 1983). Note that the circumscription of the Liliaceae, Asparagaceae, and Dracaenaceae of Dahlgren et al. (1985) is more restricted than that of Thorne (1992); also, the

CHARACTER	LILIACEAE	TRILLIACEAE	ALSTROEMERIACEAE	MELANTHIACEAE
GENERA/SPP. (THORNE 1992)	22/485	2/50	4/160	25/155
DAHLGREN ET AL. (1985) SEGREGATES	Calochortaceae, Uvulariaceae, *Scoliopus* in Trilliaceae			
REPRESENTATIVE GENERA	*Calochortus, Erythronium, Fritillaria, Lilium, Tulipa*	*Paris, Trillium*	*Alstroemeria*	*Aletris, Schoenocaulon, Tofieldia, Veratrum, Zigadenus*
ORDER (THORNE 1992)	Liliales	Liliales	Liliales	Liliales
ORDER (DAHLGREN ET AL. 1985)	Liliales	Dioscoreales	Liliales	Melanthiales
DISTRIBUTION	mostly temperate Northern Hemisphere	mostly temperate Northern Hemisphere	restricted to Central and South America	mostly temperate to boreal Northern Hemisphere, with a few in South America
HABIT	perennial herbs; bulbs, or seldom rhizomes	perennial herbs; rhizomes	perennial herbs, often twining; rhizomes with swollen, fleshy roots	perennial herbs; rhizomes
LEAF POSITION	cauline or basal	cauline	cauline	cauline and/or ± basal
LEAF MODIFICATIONS		whorled; net-veined	inverted (twisted)	
INFLORESCENCE	racemose, umbellate, or flower solitary	flowers solitary	umbellate helicoid cymes; flowers subtended by leaf-like bracts	racemose or spicate
FLORAL SYMMETRY	actinomorphic, or occasionally zygomorphic	actinomorphic	actinomorphic to slightly zygomorphic	usually actinomorphic
NECTARY POSITION	base of tepals	base of petals, or septal	base of tepals	base of tepals, or sometimes septal
PERIANTH DIFFERENTIATION	tepals; showy, often spotted	often sepals and petals; showy	tepals; showy, often spotted	tepals; inconspicuous, occasionally spotted
PERIANTH FUSION	distinct	distinct	usually distinct	distinct, or sometimes basally connate
OVARY POSITION	superior	superior	inferior	superior to half-inferior
FRUIT TYPE	capsule, occasionally berry	berry or capsule	usually capsule	capsule, or aggregate of follicles
SEED SHAPE	flat or rounded	globose to ellipsoid	globose or ellipsoid	rounded and often winged
PHYTOMELAN CRUST	none (outer epidermis present)	none (outer epidermis present)	none (outer epidermis present)	none (outer epidermis present)
ENDOSPERM DEVELOPMENT	nuclear	helobial or nuclear	nuclear	helobial
OTHER	sometimes saponins	saponins raphides	saponins	saponins raphides

placement of the Trilliaceae and Melanthiaceae differs. Recent molecular analysis (Duvall et al. 1993) indicates that the Melanthiaceae, a problematic family, should be divided into several families that are probably not closely related to each other. Character states are summarized from Dahlgren and Clifford (1982) and Dahlgren et al. (1985).

ASPARAGACEAE	HEMEROCALLIDACEAE	AGAVACEAE	DRACAENACEAE	AMARYLLIDACEAE	ALLIACEAE
26/440	1/16	8/300	9/230	50/860	30/720
Convallariaceae, Ruscaceae, Herreriaceae			Nolinaceae, Asteliaceae		
Asparagus, Convallaria, Maianthemum (incl. *Smilacina*), *Polygonatum*	*Hemerocallis*	*Agave, Yucca*	*Cordyline, Dracaena, Nolina, Sansevieria*	*Crinum, Hymenocallis, Narcissus, Zephyranthes*	*Agapanthus, Allium, Nothoscordum*
Asparagales	Asparagales	Asparagales	Asparagales	Asparagales	Asparagales
Asparagales	Asparagales	Asparagales	Asparagales	Asparagales	Asparagales
Northern Hemisphere and Old World tropics, with a few in South America	mostly temperate Asia, and extending into southern Europe	subtropical to tropical regions	tropical to subtropical regions	warm-temperate to tropical regions	widely distributed
perennial herbs, vines, or shrubs rhizomes	perennial herbs rhizomes with swollen, fleshy roots	large perennial herbs to tree-like, with secondary growth rhizomes	large perennial herbs to tree-like, with secondary growth rhizomes	perennial or biennial herbs bulbs	perennial herbs bulbs
cauline, or in basal rosettes	mostly basal rosettes	dense apical rosettes	basal or apical rosettes	basal rosettes	basal rosettes
often scale-like		often stiff and/or succulent			
umbellate, racemose, or spicate	cymose (flowers few), scapose	racemose or paniculate; large	racemose, paniculate, or umbellate	umbellate, subtended by bracts, scapose	umbellate, subtended by bracts, scapose
actinomorphic	zygomorphic	actinomorphic to slightly zygomorphic	actinomorphic	actinomorphic to slightly zygomorphic	actinomorphic or zygomorphic
septal	septal	septal	septal	usually septal	septal
tepals usually inconspicuous	tepals showy, ± streaked	tepals showy	tepals generally showy	tepals, sometimes with corona showy	tepals ± showy
distinct or basally connate	basally connate (tubular)	distinct or basally connate	distinct, or often basally connate (tubular)	distinct or basally connate	distinct to often connate (tubular or campanulate)
superior	superior	superior or inferior	superior	inferior	superior
usually berry	capsule	capsule, or sometimes berry	berry or nut-like	capsule	capsule
globose	subglobose or prismatic	often flat	globose to ellipsoid	often rounded	tetrahedral, triangular, or rounded
often present (but sometimes outer epidermis obliterated)	present	present	usually none (outer epidermis obliterated)	often present (but sometimes outer epidermis obliterated)	usually present
nuclear	nuclear	usually helobial	nuclear	nuclear or helobial	nuclear or helobial
saponins raphides	[chemistry unknown] raphides	saponins dimorphic chromosomes	saponins raphides	"amaryllis" alkaloids raphides	saponins, sulfuric compounds

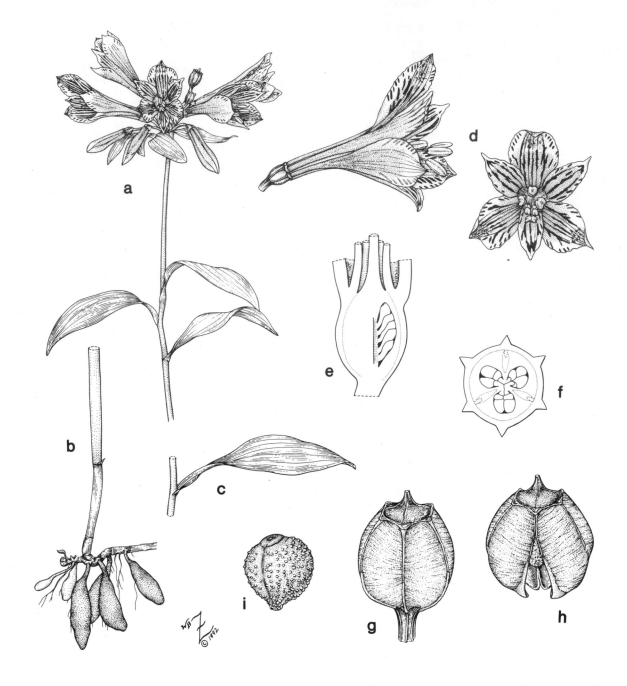

Figure 125. Alstroemeriaceae. *Alstroemeria pulchella*: **a,** upper portion of flowering plant, ×½; **b,** base of plant showing swollen roots, ×⅖; **c,** leaf (note twisted base), ×½; **d,** two views of flower, ×1; **e,** longitudinal section of ovary, ×4½; **f,** cross section of ovary, ×5; **g,** capsule before dehiscence, ×2; **h,** dehiscing capsule, ×2; **i,** seed, ×7. See Table 19 for family characteristics.

nally, extrorse. **Gynoecium** of 1 pistil, 3-carpellate; ovary inferior, 3-locular; ovules few to many in each locule, anatropous, placentation axile; style 1, often trifid with branches simple, flattened, or enlarged and petaloid; stigmas 3, terminal (style branches simple) or on abaxial surface of petaloid style branches, papillose. **Fruit** a 3-valved loculicidal capsule; seeds numerous, sometimes arillate or with sarcotesta; endosperm copious, fleshy; embryo small, linear, straight.

Family characterization: Perennial herbs with rhizomes, bulbs, or corms; equitant, basal, linear to ensiform leaves; 3-merous flowers with petaloid perianth and 3 stamens; inferior ovary; and a loculicidal capsule

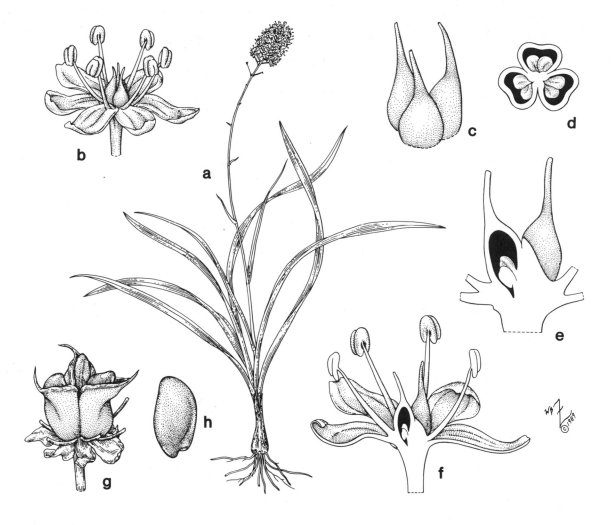

Figure 126. Melanthiaceae. *Zigadenus muscaetoxicus*: **a,** habit, ×¼; **b,** flower, ×4; **c,** gynoecium, ×8; **d,** cross section of ovary, ×10; **e,** longitudinal section of gynoecium, ×10; **f,** longitudinal section of flower, ×6; **g,** capsule, ×4; **h,** seed, ×6. See Table 19 for family characteristics.

as the fruit type. Saponins and various terpenoid compounds commonly present. Tissues with tannins and calcium oxalate crystals (long and prismatic). Anatomical feature: mucilage ducts.

Genera/species: 77/1,655

Distribution: Cosmopolitan; centers of diversity in South Africa, the eastern Mediterranean region, and tropical America.

Major genera: *Iris* (300 spp.), *Gladiolus* (180–300 spp.), *Moraea* (100–111 spp.), and *Sisyrinchium* (100 spp.)

U.S./Canadian representatives: 18 genera/109 spp.; largest genera: *Iris* and *Sisyrinchium*

Economic plants and products: Saffron (a dye and a spice) from the stigmas of *Crocus sativus*. Orris root (a fragrant substance used in perfumes and cosmetics) from rhizomes of *Iris* spp. Ornamental plants (species of 46 genera), including *Crocus*, *Dietes*, *Eustylis*, *Freesia*, *Gladiolus*, *Iris* (flag, fleur-de-lis), *Ixia* (corn-lily), *Moraea* (butterfly-iris, Natal-lily), *Neomarica* (fan-iris), *Sisyrinchium* (blue-eyed-grass), and *Tigridia* (tiger-flower).

Commentary: Botanists generally concur that the Iridaceae are related to the Liliaceae *s.s.* and allied families of the Liliales (see the commentary under Liliaceae; Dahlgren and Clifford 1982; Dahlgren et al. 1985). The family is divided into subfamilies, tribes, and genera based on the rootstock (rhizomes, bulbs, or corms), inflorescence type, floral symmetry, and style morphology (e.g., three-branched with terminal stigmas, or petaloid with sub-terminal stigmas; Goldblatt 1990).

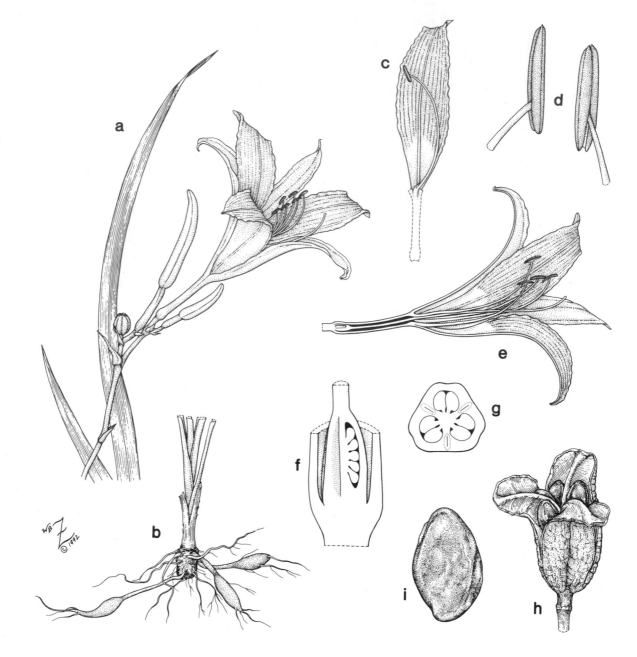

Figure 127. Hemerocallidaceae. *Hemerocallis fulva*: **a,** upper portion of flowering plant, ×²⁄₅; **b,** base of plant showing swollen roots, ×¹⁄₃; **c,** tepal and epitepalous stamen, ×¹⁄₂; **d,** two views of anther, ×4¹⁄₂; **e,** longitudinal section of flower, ×4¹⁄₂; **f,** longitudinal section of ovary and perianth tube base, ×5; **g,** cross section of ovary, ×8; **h,** capsule, ×1¹⁄₃; **i,** seed (note seed coat with phytomelan), ×4¹⁄₂. See Table 19 for family characteristics.

The brightly colored flowers usually are pollinated by insects (such as bees and flies) that seek the nectar accumulated at the base of the floral tube. Nectaries may occur in the form of septal glands (*Gladiolus*) or at the base of the tepals or stamens (*Iris*). The pollination mechanism depends upon the particular floral morphology. For example, the perianth of *Sisyrinchium* (Fig. 128: 1b) consists of six similar tepals forming a short tube with spreading lobes. Often a yellow or white "eye" targets the center of the flower. The monadel-

phous stamens surround the style, which has filiform branches. The extrorse anthers typically release the pollen before the style branches expand.

In comparison, a highly modified perianth and style characterize the complex *Iris* flower (Fig. 128: 2a). The large tepals (called "falls") of the outer whorl (Fig. 128: 2b) are spreading or deflexed, while those of the inner whorl ("standards"; Fig. 128: 2c) are smaller and erect. Each outer tepal often is bearded with a crest of hairs and marked with nectar guides (lines). The flat and

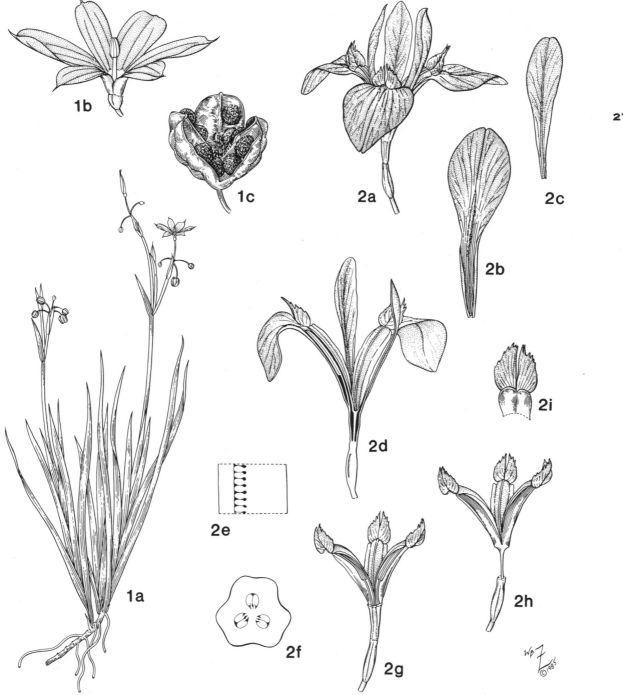

Figure 128. Iridaceae. 1, *Sisyrinchium atlanticum*: **a,** habit, ×½; **b,** flower, ×3; **c,** capsule, ×4½. **2,** *Iris hexagona* var. *savannarum*: **a,** flower, ×½; **b,** outer tepal (one of the three "falls"), ×½; **c,** inner tepal ("standard"), ×½; **d,** longitudinal section of flower, ×⅔; **e,** detail of longitudinal section of ovary, ×3; **f,** cross section of ovary, ×3; **g,** androecium and gynoecium, ×½; **h,** pistil, ×½; **i,** underside of style branch apex showing receptive area (flap), ×¾.

petaloid style branches curve along the outer tepals, forming a protective covering over the stamens (Fig. 128: 2g,h). A flap-like stigma (Fig. 128: 2i) is situated on the underside of each style branch near the distal end. Only the adaxial surfaces of the stigmas are receptive.

An insect uses a large tepal of an *Iris* flower as a landing platform. It may then transfer pollen from a previous flower to the receptive stigmatic surface while it probes for the nectar through the tube (or passageway) formed by the tepal and the corresponding style branch. Further down, it becomes dusted with new pollen from the anther. As the insect retreats from the flower, it encounters only the nonreceptive abaxial surface of the stigma.

REFERENCES CITED

Dahlgren, R. M. T., and H. T. Clifford. 1982. *The monocotyledons: A comparative study*, pp. 26–30. Academic Press, London.
——, ——, and P. F. Yeo. 1985. *The families of the monocotyledons*, pp. 238–249. Springer-Verlag, Berlin.
Goldblatt, P. 1990. Phylogeny and classification of Iridaceae. *Ann. Missouri Bot. Gard.* 77:607–627.

ASPARAGACEAE
Asparagus Family
Including the Convallariaceae, Ruscaceae,
    and Herreriaceae

Perennial herbs, vines, or shrubs, frequently with flat and leaf-like branchlets (phylloclades), with sympodial or monopodial rhizome. **Leaves** simple, entire, alternate or sometimes opposite or whorled, cauline or in basal rosettes, parallel-veined, frequently scale-like, sessile or sometimes petiolate. **Inflorescence** determinate, umbellate, racemose, or spicate, axillary or sometimes terminal on a scape, or flowers few to solitary in leaf axils. **Flowers** actinomorphic, perfect or occasionally imperfect, hypogynous, generally inconspicuous, often with septal nectaries. **Perianth** of 6 tepals, biseriate, petaloid, distinct or basally connate, spreading, campanulate, or urceolate, usually white, yellow, green, or purple, imbricate. **Androecium** of 6 stamens, epitepalous; filaments usually distinct; anthers dorsifixed or basifixed, dehiscing longitudinally, introrse; staminodes present in carpellate flowers. **Gynoecium** of 1 pistil, 3-carpellate; ovary superior, 3-locular; ovules 2 to 12 in each locule, hemitropous to orthotropous, placentation usually axile; style 1; stigma 1, capitate or lobed. **Fruit** usually a berry; seeds globose, frequently with black crust (or sometimes outer epidermis obliterated); endosperm copious, fleshy; embryo straight to slightly curved, embedded in endosperm.

Family characterization: Perennial herbs to shrubs with rhizomes; inconspicuous flowers with 6 petaloid tepals; 6 epitepalous stamens; 3-carpellate, superior ovary; and a berry (often with black seeds) as the fruit type. Leaves sometimes photosynthetic and parallel-veined, but often reduced and scale-like (then plants with phylloclades; see the discussion below). Saponins (e.g., diosgenin and gentrogenin) and chelidonic acid present. Tissues with raphides (bundles of needle-like calcium oxalate crystals). Anatomical feature: roots of several genera with multiseriate epidermis (velamen; discussed under Orchidaceae).

Genera/species: 26/240

Distribution: Widely distributed in the Northern Hemisphere and the Old World tropics; a few representatives in South America.

Major genera: *Asparagus* (100–312 spp.), *Polygonatum* (50–55 spp.), *Ophiopogon* (20–50 spp.), *Maianthemum* (incl. *Smilacina*; 33 spp.), and *Tupistra* (6–25 spp.)

U.S./Canadian representatives: *Polygonatum* (5 spp.), *Maianthemum* (incl. *Smilacina*; 5 spp.), *Asparagus* (4 spp.), and *Convallaria* (2 spp.)

Economic plants and products: Edible young shoots from *Asparagus officinalis* (asparagus), as well as edible roots or tubers from species of *Asparagus* and *Polygonatum*. Poisons and medicines (glucosides) derived from rhizomes of *Convallaria* and *Polygonatum* species. Ornamental plants (species of 11 genera), including *Asparagus* (asparagus-fern), *Aspidistra*, *Convallaria* (lily-of-the-valley), *Liriope* (lily-turf), *Maianthemum* (incl. *Smilacina*; false lily-of-the-valley, false Solomon's-seal), *Ophiopogon* (mondo-grass), *Polygonatum* (Solomon's-seal), and *Ruscus* (butcher's-broom).

Commentary: Due to lack of critical revision, the Asparagaceae (and associated segregate families) have been conveniently submerged as tribes or subfamilies within a very broadly defined Liliaceae (e.g., Cronquist 1981; Thorne 1983). Recent work (see Dahlgren and Clifford 1982; Dahlgren et al. 1985) has established the order Asparagales, comprising many segregate taxa from the Liliaceae *s.l.* (see the commentary in Liliaceae and Tables 18 and 19). As presented here, Thorne's (1992) concept of the Asparagaceae includes four subfamilies (Asparagoideae, Convallarioideae, Ruscoideae, Herrerioideae), treated as closely related segregate families by some taxonomists (Dahlgren et al. 1985). More work is needed, for the placement of a few genera (e.g., *Clintonia*) in this family in some treatments is still questionable.

Most of the species in the family belong to the genus *Asparagus*. Many species of *Asparagus* (and also *Ruscus*) that grow in warm and dry climates are characterized by modified, needle-like or flattened branchlets (phylloclades, cladodes; Fig. 129: 1d) that function as the primary photosynthetic surface (Hirsch 1977). The real leaves are reduced to scale-like structures and sometimes also are modified into short, recurved spines (Fig. 129: 1b,c).

The fruit type of the family usually is a berry (Fig.

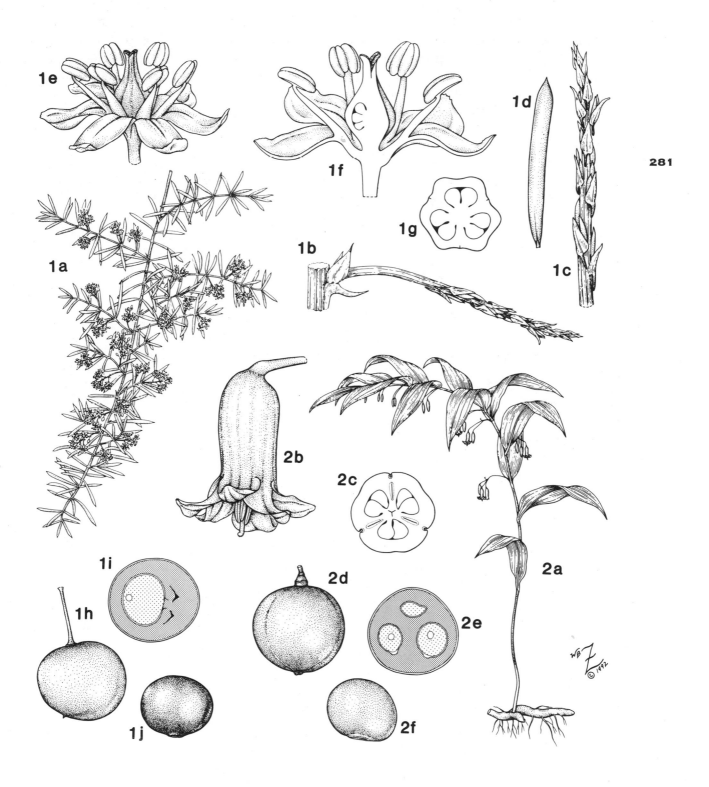

Figure 129. Asparagaceae. 1, *Asparagus densiflorus*: **a,** habit, ×½; **b,** node showing scale-leaf (with spine) subtending branchlet, ×2; **c,** axillary branchlet with scale-leaves, ×3; **d,** phylloclade, ×3; **e,** flower, ×10; **f,** longitudinal section of flower, ×14; **g,** cross section of ovary, ×20; **h,** berry, ×3; **i,** cross section of berry, ×3; **j,** seed (note seed coat with phytomelan), ×4. **2,** *Polygonatum biflorum*: **a,** habit, ×¼; **b,** flower, ×3; **c,** cross section of ovary, ×8; **d,** berry, ×2; **e,** cross section of berry, ×2; **f,** seed (note seed coat lacking phytomelan), ×4.

129: 1h, 2d), although a few (e.g., *Liriope*) have capsules. In general, the outer epidermis of the seeds in baccate fruits of the order Asparagales becomes obliterated so that the characteristic black crust (of phytomelan) is not present at maturity (e.g., *Polygonatum*; Fig. 129: 2f). However, in many of the Asparagaceae with berries (*Asparagus*, Fig. 129: 1j; *Ruscus*), this layer is intact in the mature, shiny, black seeds (see the commentary under Liliaceae).

The flowers in the family are somewhat small and inconspicuous (white, green, yellow, or sometimes purple) and usually have septal nectaries (Fig. 129: 1g, 2c). Many also are fragrant (e.g., species of *Asparagus* and *Convallaria*). Insects (usually bees) visit to collect nectar (which accumulates at the base of the flower) or sometimes pollen (in flowers lacking nectaries, such as those of *Convallaria*).

REFERENCES CITED

Cronquist, A. 1981. *An integrated system of classification of flowering plants*, pp. 1208–1211. Columbia University Press, New York.

Dahlgren, R. M. T., and H. T. Clifford. 1982. *The monocotyledons: A comparative study*, pp. 27–29. Academic Press, London.

——, ——, and P. F. Yeo. 1985. *The families of the monocotyledons*, pp. 129–144. Springer-Verlag, Berlin.

Hirsch, A. M. 1977. A developmental study of the phylloclades of *Ruscus aculeatus* L. *Bot. J. Linn. Soc.* 74:355–365.

Thorne, R. F. 1983. Proposed new realignments in the angiosperms. *Nordic. J. Bot.* 3:85–117.

——. 1992. Classification and geography of the flowering plants. *Bot. Rev.* (Lancaster) 58:225–348.

AGAVACEAE
Century Plant Family

Perennial herbs to arborescent shrubs, often growing in arid habitats, large and robust, with rhizomes or bundles of roots; stems often stout, simple or sparingly branched, with secondary thickening growth. **Leaves** simple, often spinose-serrate, alternate and spirally arranged, often crowded into dense rosettes (terminating the stem), parallel-veined, typically lanceolate or linear, sessile, stiff (very fibrous) and/or succulent, often sharp-pointed at apex, persistent. **Inflorescence** cymose and often appearing racemose or paniculate, large, terminal. **Flowers** actinomorphic to slightly zygomorphic, perfect, hypogynous or epigynous, showy, usually with septal nectaries, subtended by conspicuous bracts. **Perianth** of 6 tepals, biseriate, distinct or basally connate (forming a short tube), petaloid, thick and fleshy, greenish, white, or yellow, imbricate. **Androecium** of 6 stamens, biseriate; filaments distinct, basally adnate to perianth, sometimes dilated at base; anthers dorsifixed, versatile, dehiscing longitudinally, introrse. **Gynoecium** of 1 pistil, 3-carpellate; ovary superior or inferior, 3-locular; ovules several to many in each locule, usually anatropous, placentation axile; style 1, short to long; stigma capitate or lobed. **Fruit** usually a loculicidal capsule or (seldom) a berry; seeds flattened, with black crust; endosperm copious, very hard; perisperm sometimes present (around endosperm); embryo straight, linear, embedded in endosperm.

Family characterization: Large herbs to shrubs with secondary thickening growth; linear to lanceolate, fleshy and/or fibrous, prickly-margined and sharp-pointed leaves clustered into rosettes; very large, terminal, racemose or paniculate inflorescence; 3-merous, fleshy and petaloid perianth; 6 stamens; superior or inferior ovary; and a capsule with black seeds as the typical fruit type. Steroidal saponins (sapogenins) and chelidonic acid present. Tissues with calcium oxalate crystals. Polyfructans (unusual storage polysaccharides) in vegetative organs. Anatomical features: secondary growth (produced by a special cambium); and well-developed fibrous cap on the vascular bundles of leaves.

Genera/species: 8/300

Distribution: Primarily in subtropical and tropical regions (especially arid and semiarid habitats) with a few representatives in temperate areas; native to the Americas (especially diverse in Mexico).

Major genera: *Agave* (200–300 spp.) and *Yucca* (40 spp.)

U.S./Canadian representatives: *Yucca* (31 spp.), *Agave* (27 spp.), *Manfreda* (5 spp.), and *Furcraea* (3 spp.)

Economic plants and products: Tequila and mescal from fermented sugary sap of certain species of *Agave*. Steroidal saponins (an active ingredient of oral contraceptives) from *Agave* and *Yucca*. Fiber for rope and cordage from leaves of *Agave* (sisal-hemp), *Furcraea* (Cuban-hemp), and *Yucca*. Ornamental plants (species of at least 7 genera), including *Agave* (century plant), *Furcraea*, *Manfreda*, *Polianthes*, and *Yucca* (Spanish-bayonet).

Commentary: The Agavaceae are here rather narrowly circumscribed (Dahlgren et al. 1985; Thorne 1992), including only the Agaveae (Agavoideae: *Agave* and allies) and the Yucceae (Yuccoideae: *Yucca* and allies). Many other genera commonly included in the Agavaceae *s.l.* have now been transferred to several other families (see Table 19), as for example *Nolina*

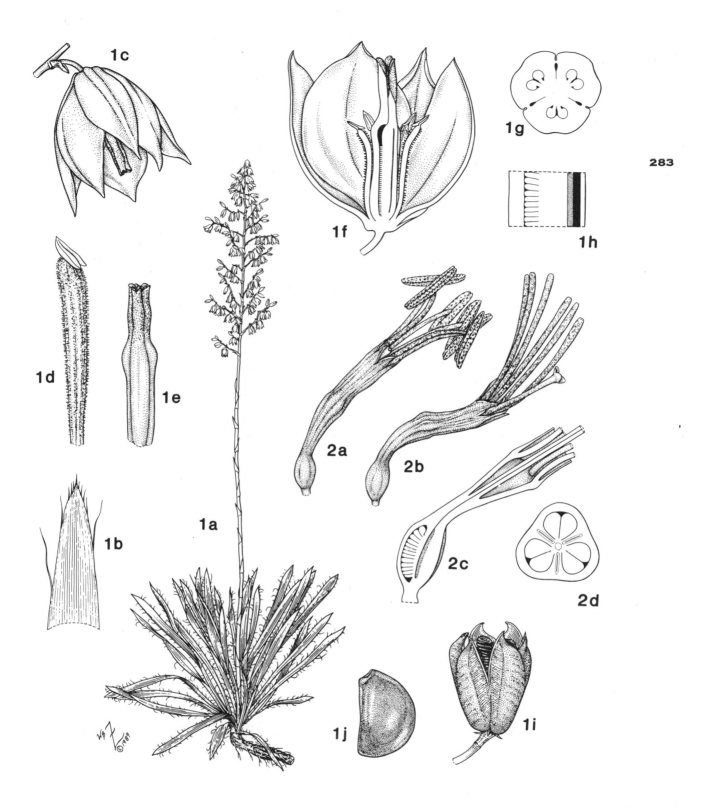

Figure 130. Agavaceae. 1, *Yucca flaccida*: **a,** habit, ×¹⁄₁₆; **b,** leaf apex, ×¹⁄₂; **c,** flower, ×³⁄₄; **d,** stamen, ×2¹⁄₂; **e,** pistil, ×1¹⁄₄; **f,** longitudinal section of flower, ×1¹⁄₄; **g,** cross section of ovary, ×4¹⁄₂; **h,** detail of longitudinal section of ovary, ×4¹⁄₂; **i,** capsule, ×³⁄₄; **j,** seed (note seed coat with phytomelan), ×4. **2,** *Manfreda virginica* (*Polianthes virginica*): **a,** flower at anthesis (anthers mature, stigma unexpanded), ×1³⁄₄; **b,** older flower (anthers fallen off, stigma receptive), ×1³⁄₄; **c,** longitudinal section through base of flower, ×2¹⁄₂; **d,** cross section of ovary, ×5.

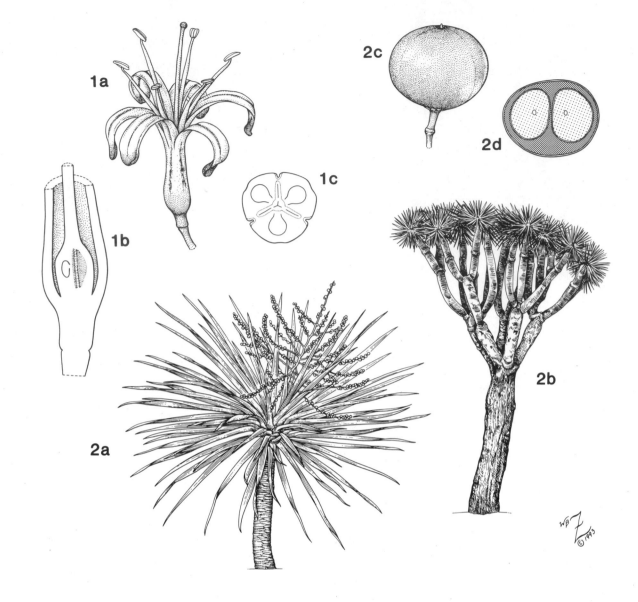

Figure 131. Dracaenaceae. 1, *Sansevieria aff. zeylanica*: **a,** flower, ×2 ½; **b,** longitudinal section of ovary and perianth tube base, ×6; **c,** cross section of ovary, ×10. **2,** *Dracaena draco*: **a,** relatively young shrubby plant with infructescence, ×1/20; **b,** much older plant (well-developed tree), much reduced; **c,** berry, ×1½; **d,** cross section of berry with two seeds, ×1½. See Table 19 for family characteristics.

(Nolinaceae or Dracaenaceae *s.l.*), *Sansevieria* (Dracaenaceae; Fig. 131), *Cordyline* (Asteliaceae or Dracaenaceae *s.l.*), and *Phormium* (Phormiaceae). Most of the species in the family belong to the large genus *Agave*. The literature particularly emphasizes the unusual karyotype of 5 large plus 25 small chromosomes as a unifying feature (McKelvey and Sax 1933). Formerly the Agavaceae were submerged in the convenient Liliaceae *s.l.* (e.g., Thorne 1983), or the Agavoideae were placed in the Amaryllidaceae (constituting an "inferior ovary group"), and the Yuccoideae, in the Liliaceae ("superior ovary group"). Lawrence (1951) includes a detailed historical summary of these various transfers.

The Agavaceae are xerophytes characterized by a shrubby to tree-like habit with a large, terminal inflorescence. Secondary thickening of the stem consists of the centrifugal formation of concentric vascular bundles and ground tissue produced by a special cambium that arises in the parenchyma outside the vascular bundles; this "secondary growth" is, thus, not similar to that in the "dicotyledons" (Tomlinson and Zimmerman 1969).

The thick and fibrous leaves are clustered in a dense

basal rosette on the stout stem (e.g., Fig. 130: 1a). In the Yuccoideae, the leaves are stiff, tough, and somewhat succulent. The plants of the Agavoideae, often succulent, are characterized by fleshy leaves with tough fibers and sunken stomata. Leaves of both subfamilies tend to have spiny margins and terminate in a sharp point (Fig. 130: 1b).

The large inflorescences of fragrant, often yellow or white flowers attract various pollinators: insects (e.g., moths in *Yucca*, see below), birds (in *Beschorneria*), or even bats (in *Agave* species). Nectar is produced by septal nectaries (Fig. 130: 1g, 2d). Most flowers are protandrous (e.g., *Manfreda*; Fig. 130: 2a), with the inner and outer stamen whorls dehiscing successively before the stigma matures (Fig. 130: 2b). Members of the Agavoideae are monocarpic (flowering once and then dying) but may reproduce vegetatively by stolons.

A classic case of pollinator-plant interdependency is exemplified by several *Yucca* species that are associated with the life cycle of a particular moth (several species of *Tegeticula* and one species of *Parategeticula*). The stigmas of the *Yucca* flower are elevated too far above the anthers for self-pollination to occur (Fig. 130: 1f). The moth gathers the putty-like pollen into a ball, then carries the mass under its head to another flower where it carefully presses the pollen into the stigmatic tube with its proboscis. The moth lays one egg in each ovary locule. As the larvae and ovules develop together, a few of the seeds are consumed. At maturity, the larvae eat through the fruit wall and drop to the ground, overwintering as cocoons and emerging (as moths) when the *Yucca* is blooming again. The moth-*Yucca* relationship actually is much more complex, since, for example, seed set has been noted in the absence of moth visits, indicating self-pollination or visits by other pollinators (Baker 1986).

REFERENCES CITED

Baker, H. G. 1986. Yuccas and yucca moths—A historical commentary. *Ann. Missouri Bot. Gard.* 73:556–564.

Dahlgren, R. M. T., H. T. Clifford, and P. F. Yeo. 1985. *The families of the monocotyledons*, pp. 157–161. Springer-Verlag, Berlin.

Lawrence, G. H. M. 1951. *Taxonomy of vascular plants*, pp. 413–415, 419. Macmillan, New York.

McKelvey, S. D., and K. Sax. 1933. Taxonomic and cytological relationships of *Yucca* and *Agave*. *J. Arnold Arbor.* 14: 76–81.

Thorne, R. F. 1983. Proposed new realignments in the angiosperms. *Nordic J. Bot.* 3:85–117.

——. 1992. Classification and geography of the flowering plants. *Bot. Rev.* (Lancaster) 58:225–348.

Tomlinson, P. B., and M. H. Zimmerman. 1969. Vascular anatomy of monocotyledons with secondary growth—An introduction. *J. Arnold Arbor.* 50:159–179.

AMARYLLIDACEAE
Amaryllis or Daffodil Family

Perennial or biennial herbs, scapose, with bulbs and contractile roots. **Leaves** simple, entire, alternate, distichous, basally clustered, usually linear, parallel-veined, sheathing at base. **Inflorescence** determinate, cymose and often appearing umbellate or sometimes reduced to a single flower, subtended by several bracts (spathes), terminal on a scape. **Flowers** actinomorphic to slightly zygomorphic, perfect, epigynous, showy, with septal nectaries. **Perianth** of 6 tepals, biseriate, distinct to basally connate (into a short to long tube), petaloid, sometimes with corona, white, yellow, purple, or red, imbricate. **Androecium** of 6 stamens, biseriate; filaments distinct or connate into a petaloid cup, sometimes appendaged (either apically on cup or basally on distinct filaments), adnate to base of tepals/perianth tube; anthers basifixed or versatile, dehiscing longitudinally, introrse. **Gynoecium** of 1 pistil, 3-carpellate; ovary inferior, 3-locular; ovules several to many in each locule, anatropous, placentation axile; style simple; stigma punctate, capitate, or 3-lobed. **Fruit** usually a loculicidal capsule or (rarely) a berry; seeds globose or flattened, typically with a black crust (lost in some seeds); endosperm copious, fleshy, oily or starchy; embryo small, straight or slightly curved, cylindrical or compressed.

Family characterization: Scapose herbs with bulbs and contractile roots; linear, basally clustered leaves; umbellate inflorescence subtended by spathaceous bracts; epigynous, showy flowers with 6 petaloid tepals and 6 stamens; and a loculicidal capsule with black seeds as the typical fruit type. Several "amaryllis" alkaloids (unique to the family) and chelidonic acid present. Tissues with calcium oxalate crystals (raphides within mucilage-filled cells or elongate sacs).

Genera/species: 50/860

Distribution: Predominantly in warm temperate to tropical regions; especially diverse in South Africa, Andean South America, and the Mediterranean region.

Major genera: *Crinum* (100–130 spp.), *Hippeastrum* (75 spp.), *Zephyranthes* (35–71 spp.), *Hymenocallis* (40–50 spp.), and *Haemanthus* (21–50 spp.)

U.S./Canadian representatives: 10 genera/49 spp.; largest genera: *Hymenocallis, Zephyranthes, Crinum,* and *Narcissus*

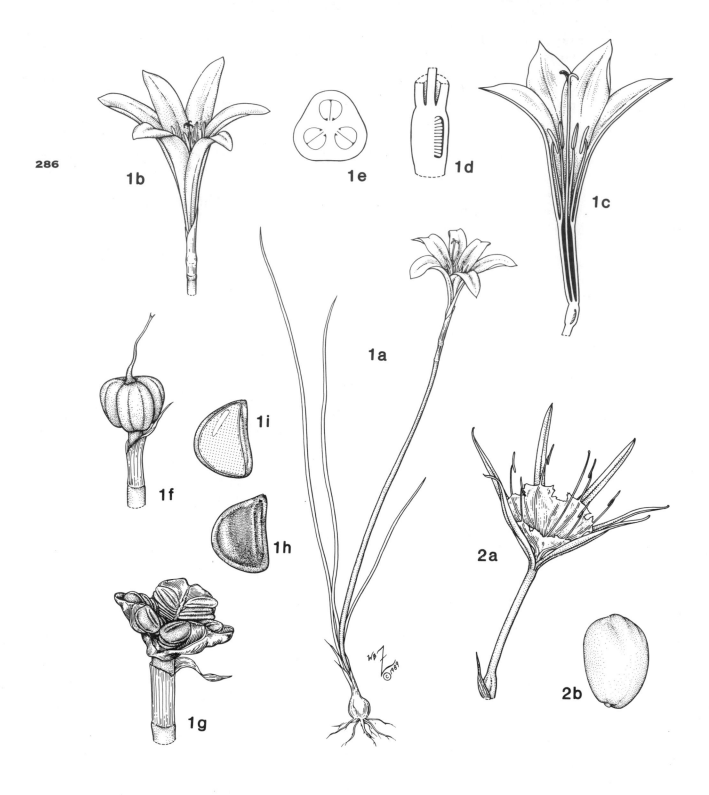

Figure 132. Amaryllidaceae. 1, *Zephyranthes treatiae*: **a,** habit, ×⅓; **b,** flower, ×⅔; **c,** longitudinal section of flower, ×¾; **d,** longitudinal section of ovary, ×3; **e,** cross section of ovary, ×5; **f,** immature capsule, ×1¼; **g,** capsule, ×1¾; **h,** seed (note seed coat with phytomelan), ×4; **i,** longitudinal section of seed, ×4. **2,** *Hymenocallis rotata*: **a,** flower, ×¼; **b,** seed ("water-rich" and green), ×1.

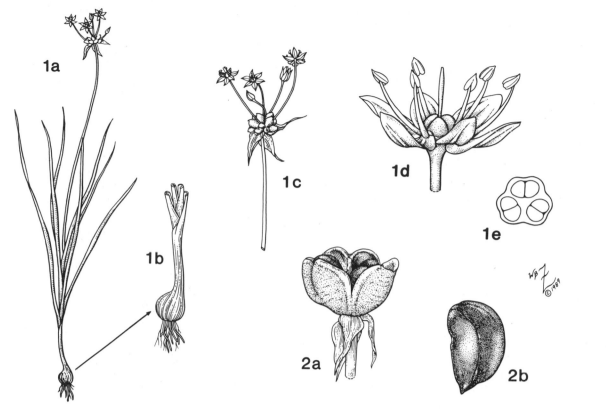

Figure 133. Alliaceae. 1, *Allium canadense*: **a,** habit, ×⅓; **b,** bulb, ×½; **c,** inflorescence (note bulbils), ×⅖; **d,** flower, ×3; **e,** cross section of ovary, ×6. **2,** *Allium bivalve* (*Nothoscordum bivalve*): **a,** capsule, ×5; **b,** seed (note seed coat with phytomelan), ×10. See Table 19 for family characteristics.

Economic plants and products: Ornamental plants (species of at least 30 genera), including *Clivia* (Kaffir-lily), *Crinum* (crinum-lily, spider-lily), *Eucharis*, *Galanthus* (snowdrops), *Haemanthus* (blood-lily), *Hippeastrum* (amaryllis, Barbados-lily), *Hymenocallis* (basket-flower, spider-lily), *Lycoris*, *Narcissus* (daffodil, jonquil), and *Zephyranthes* (rain-lily, zephyr-lily).

Commentary: The Amaryllidaceae are here rather narrowly defined (Dahlgren et al. 1985; Thorne 1992), including only 50 genera with approximately 860 species. The family has been divided into a varying number of tribes that are more or less restricted to certain geographical areas and are based on numerous characters, including floral symmetry, inflorescence composition, stamen and pollen morphology, and fruit and seed types (Traub 1957; also summarized in Dahlgren et al. 1985).

As with the Agavaceae *s.s.* (and many other families), the Amaryllidaceae have been included in the Liliaceae *s.l.* (e.g., Cronquist 1981; Thorne 1983). Except for the inferior ovary, the flowers superficially resemble those of the Liliaceae. On the basis of an inferior ovary, several segregate families (such as the

Agavaceae, Hypoxidaceae, Haemodoraceae, and Alstroemeriaceae) sometimes have been included in a broadly defined Amaryllidaceae. However, according to Dahlgren et al. (1985), the family probably is most closely related to segregates such as the Hyacinthaceae and Alliaceae—both with superior ovaries. The Alliaceae (see Fig. 133) have been associated with the Amaryllidaceae ever since Hutchinson (1934) deemphasized ovary position and stressed the umbellate inflorescence (with spathaceous bracts) as an important taxonomic character when he transferred many of the "Liliaceae" to the Amaryllidaceae (see the summary in Traub 1970). The choice of the umbellate inflorescence as a single distinctive character has been questioned, since developmental studies (such as Mann 1959) have demonstrated that these "umbellate" types actually are much more complex. In addition to the inflorescence type, the Alliaceae and Amaryllidaceae *s.s.* (see Table 19) both are characterized as scapose plants with bulbs, and the seeds have a phytomelan crust (lost in some of the Amaryllidaceae). Besides the ovary position, the two families differ chemically (sulfur compounds in the Alliaceae; unique alkaloids in the Amaryllidaceae) and anatomically (simple perforated vessels in roots of Alliaceae; scalariform perforation plates in Amaryllidaceae).

The brightly colored and often fragrant flowers of the Amaryllidaceae attract various pollinators (bees, butterflies, birds) that seek nectar (secreted by more or less distinct septal grooves on the ovary) or pollen. Both protandry and protogyny have been reported in the family. A showy petaloid corona, present in many species, may conceal the nectar at the tubular base of the flower. Depending on the genus, this structure is either androecial in origin or derived from the perianth (Arber 1937). In *Hymenocallis*, for example, the dilated filaments are fused into a large, fringed cup from which the stamens appear to arise (Fig. 132: 2a). In other cases (such as in a few *Eucharis* species), scale-like basal appendages on the filaments represent a reduced staminal corona. In genera such as *Narcissus* (daffodil), however, the cup-like to tubular corona is an actual outgrowth of the perianth tube itself.

A unifying feature (synapomorphy) of the Asparagales is the black crust (of phytomelan) on the seeds in nonbaccate (and sometimes baccate) fruits (e.g., Fig. 132: 1h; see the commentary under Asparagaceae, and Table 18). However, the outer portion of the seed coat is lost in several of the Amaryllidaceae (e.g., *Hymenocallis, Crinum, Amaryllis*) with green, "water-rich" seeds (75–90% water) that do not dehydrate as would a typical seed (Dahlgren et al. 1985). These seeds germinate soon after maturity and are dispersed when the embryo is still quite small, or are sometimes viviparous. The regular amaryllidaceous seed (12–25% water) with phytomelan crust is flattened to globose and has oily endosperm (e.g., Fig. 132: 1h). In contrast, the globose water-rich seeds are relatively large (e.g., Fig. 132: 2b); the main food reserve is starch, which is found in the seed coat, as well as in the endosperm. In *Crinum*, the fleshy seeds, which lack a true seed coat, resemble bulbils (van der Pijl 1982).

REFERENCES CITED

Arber, A. 1937. Studies in flower structure III. On the corona and androecium in certain Amaryllidaceae. *Ann. Bot.* (Oxford) 1:293–304.

Cronquist, A. 1981. *An integrated system of classification of flowering plants*, pp. 1208–1211. Columbia University Press, New York.

Dahlgren, R. M. T., H. T. Clifford, and P. F. Yeo. 1985. *The families of the monocotyledons*, pp. 199–206. Springer-Verlag, Berlin.

Hutchinson, J. 1934. *The families of flowering plants.* Vol. 2, *Monocotyledons*, pp. 128–135. Macmillan, London.

Mann, L. K. 1959. The *Allium* inflorescence: Some species of the section *Molium. Amer. J. Bot.* 46:730–739.

van der Pijl, L. 1982. *Principles of dispersal in higher plants*, pp. 117–119. Springer-Verlag, Berlin.

Thorne, R. F. 1983. Proposed new realignments in the angiosperms. *Nordic J. Bot.* 3:85–117.

———. 1992. Classification and geography of the flowering plants. *Bot. Rev.* (Lancaster) 58:225–348.

Traub, H. P. 1957. Classification of the Amaryllidaceae—Subfamilies, tribes and genera. *Pl. Life* 13:76–83.

———. 1970. An introduction to Herbert's "Amaryllidaceae, etc." 1837 and related works. In William Herbert, *Amaryllidaceae*, reprint ed., ed. J. Cramer and H. K. Swann, pp. 1–93. Historiae naturalis classica. J. Cramer, Lehre, Germany.

SMILACACEAE
Catbrier or Greenbrier Family

Herbaceous to partially woody vines, with thick rhizomes or tubers; stems armed with recurved prickles; roots mycorrhizal, lacking root hairs. **Leaves** simple, entire to spiny-serrate, often shallowly lobed, usually alternate, palmately veined with main veins connected by network of smaller veins, stiff and coriaceous, often armed with prickles, with paired tendrils (arising from near petiole base). **Inflorescence** determinate, of simple umbellate clusters sometimes in racemose or spicate arrangements, usually axillary. **Flowers** actinomorphic, usually imperfect (plants dioecious), hypogynous, small. **Perianth** of 6 tepals in 2 whorls of 3, distinct or basally connate, petaloid, inconspicuous, greenish, whitish, yellowish, or brownish, imbricate. **Androecium** of typically 6 stamens, biseriate; filaments distinct or basally connate, sometimes adnate to perianth; anthers basifixed, 2-locular but often appearing 1-locular (due to confluence), dehiscing longitudinally, introrse or latrorse; filiform staminodes commonly present in carpellate flowers. **Gynoecium** of 1 pistil, 3-carpellate; ovary superior, 3- or occasionally

Figure 134. Smilacaceae. 1, *Smilax tamnoides*: **a,** apex of new shoot (note scale-like leaves), ×³⁄₈; **b,** node of new shoot with scale-like leaf and tendrils, ×²⁄₅. **2,** *Smilax auriculata*: **a,** rhizome, ×¼; **b,** mature leaf with tendrils, ×⅝; **c,** carpellate flowering plant, ×½; **d,** carpellate flower, ×9; **e,** pistil, ×12; **f,** cross section of ovary, ×12; **g,** longitudinal section of carpellate flower, ×12; **h,** staminate flowering plant, ×½; **i,** staminate flower, ×6; **j,** stamen (from bud), ×10; **k,** stamen (anther dehisced), ×10; **l,** fruiting plant, ×½; **m,** berry, ×2½; **n,** cross section of berry with three seeds, ×2½; **o,** cross section of berry with two seeds, ×2½; **p,** cross section of berry with one seed, ×2½.

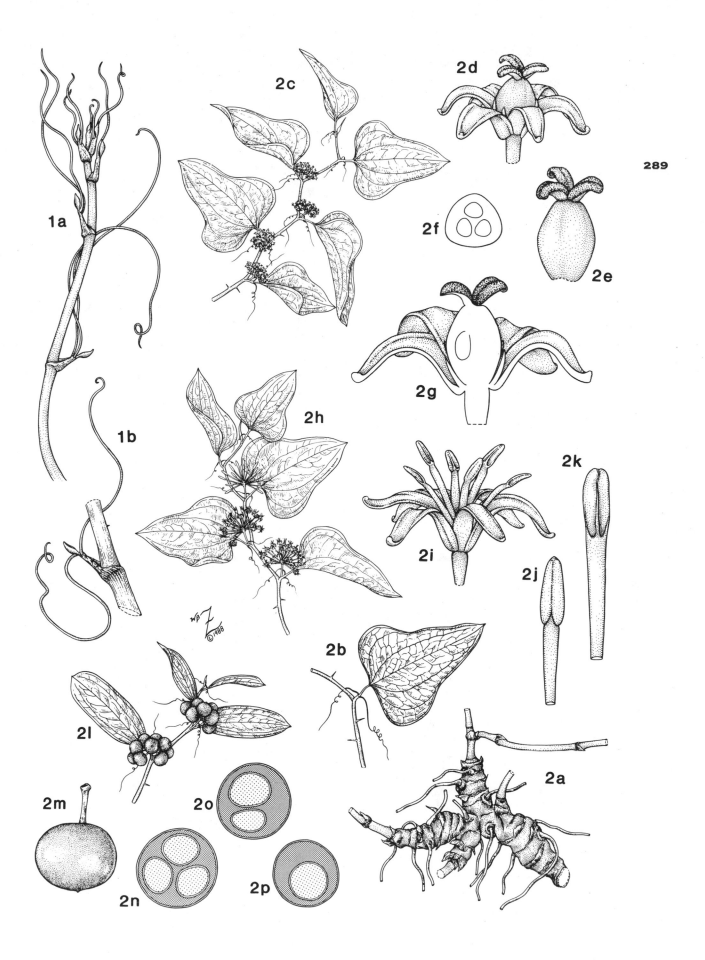

289

1-locular; ovules typically 1 or 2 in each locule, ortho-tropous or less often anatropous, hemitropous, to campylotropous, pendulous, placentation axile or occasionally parietal (in 1-locular ovary); styles 3, distinct or basally connate; stigmas 3, capitate or lobed. **Fruit** a berry; seeds 1 to 3, globose to ovoid; endosperm very hard; embryo small, straight, linear.

Family characterization: Dioecious vines armed with prickles; petiolate, net-veined leaves with paired tendrils near petiole base; small, imperfect, hypo-gynous flowers in axillary umbels; "1-locular" anthers (due to merging of pollen sacs); a few-seeded berry as the fruit type; and very hard endosperm. Saponins present. Tissues with calcium oxalate crystals (raph-ides in mucilage cells).

Genera/species: *Smilax* (200–350 spp.), *Heterosmilax* (15 spp.), and *Pseudosmilax* (2 spp.)

Distribution: Primarily tropical to subtropical and ex-tending into northern temperate regions; especially diverse in tropical America.

U.S./Canadian representatives: *Smilax* (25 spp.)

Economic plants and products: Extracts from tu-bers of various *Smilax* species used as medicinal tonics and stimulants (due to various saponins, e.g., smilasa-ponin and sarsasaponin) and as a flavoring (sarsapa-rilla); the berries, starchy tubers, and young stems of some species are edible. Several species of *Smilax* also cultivated as ornamentals, but many are tenacious weeds and hazards (due to well-developed prickles).

Commentary: Broadly circumscribed, the Smilaca-ceae may include as many as 12 (or more) genera. Some taxonomists variously segregate these into sev-eral closely related, small (Philesiaceae) or monotypic (Petermanniaceae, Ripogonaceae) families, leaving a very narrowly defined Smilacaceae *s.s.* According to Thorne's (1992) scheme, the family is restricted to the large genus *Smilax* (200–350 spp.) plus two small gen-era (*Heterosmilax* and *Pseudosmilax*). The Smilacaceae (and segregates) traditionally were included within the Liliaceae *s.l.* (Thorne 1976) or as a segregate family in the Liliales (e.g., Thorne 1983). Although the family is now included generally in the Dioscoreales (Dahlgren et al. 1985; Thorne 1992), recent molecular evidence does support its inclusion in the Liliales (Duvall et al. 1993).

Smilax vines are recognizable in the field by several vegetative characters, most obviously the recurved prickles (modified trichomes) on the scrambling and climbing stems (thus, the common names "green-brier" and "catbrier"). The plants arise from thickened rhizomes (Fig. 134: 2a) and tubers that are rich in starch, saponins, and calcium oxalate crystals. New succulent shoots from the rhizomes bear scale leaves and tendrils (Fig. 134: 1a,b). These rapidly growing stems become quite long before branching, eventually producing regular leaves with petiole plus well-developed, expanded blade (Godfrey and Wooten 1979). The stiff and leathery leaf blade typically is shallowly lobed and sometimes armed with spiny teeth along the margin (Fig. 134: 2b). The size and shape of *Smilax* leaves are extremely variable, even between plants of the same species (e.g., *S. auriculata*, Fig. 134: 2b vs. 2l). The "net-venation" consists of several (usu-ally three) palmate main veins connected by a network of smaller veins. The paired tendrils near the petiole base have been variously interpreted as modified stip-ules or as the midveins of two reduced lateral leaflets.

Little has been reported on the pollination biology of *Smilax* flowers, which are presumably pollinated by small insects (such as flies). The small, inconspicuous flowers are clustered into carpellate or staminate in-florescences on separate plants (Fig. 134: 2c,h). Nectar often is produced at the base of the stamens/stami-nodes, the inner basal surface of the tepals, and/or by septal nectaries. Sometimes the flowers have a fetid odor, as in *S. herbacea* ("carrion-flower").

REFERENCES CITED

Dahlgren, R. M. T., H. T. Clifford, and P. F. Yeo. 1985. *The families of the monocotyledons*, pp. 110–112. Springer-Verlag, Berlin.

Duvall, M. R., M. T. Clegg, M. W. Chase, W. D. Clark, W. J. Kress, et al. 1993. Phylogenetic hypotheses for the mono-cotyledons constructed from *rbc*L data. *Ann. Missouri Bot. Gard.* 80:607–619.

Godfrey, R. K., and J. W. Wooten. 1979. *Aquatic and wetland plants of southeastern United States: Monocotyledons*, pp. 571–572. University of Georgia Press, Athens.

Thorne, R. F. 1976. A phylogenetic classification of the Angiospermae. *Evol. Biol.* 9:35–106.

———. 1983. Proposed new realignments in the angiosperms. *Nordic J. Bot.* 3:85–117.

———. 1992. Classification and geography of the flowering plants. *Bot. Rev.* (Lancaster) 58:225–348.

DIOSCOREACEAE
Yam Family

Perennial herbs (mostly vines), with tubers or thick rhizomes; roots usually mycorrhizal, lacking root hairs. **Leaves** simple, entire, usually alternate, palmately veined with reticulate lateral veinlets, with broad blade and usually cordate base, sometimes with a bulbil in the axil; petioles commonly twisted and sometimes jointed at base, often with stipule-like flange, with a pulvinus at both ends. **Inflorescence** determinate, cymose and often appearing racemose, spicate, or paniculate, axillary. **Flowers** actinomorphic, usually imperfect (plants dioecious), epigynous, small and inconspicuous. **Perianth** of 6 tepals in 2 whorls, petaloid to somewhat sepaloid, distinct or basally connate into a short tube, spreading, rotate, or sometimes campanulate, white, cream, greenish, or brownish, imbricate. **Androecium** of 6 stamens, biseriate, inner whorl sometimes reduced to staminodes or absent, epitepalous at perianth base; filaments usually distinct; anthers basifixed, often with broad connective (then locules separated), dehiscing longitudinally, introrse or extrorse; staminodes often present in carpellate flowers. **Gynoecium** of 1 pistil, 3-carpellate; ovary inferior, 3-locular, with septal nectaries; ovules usually 2 in each locule, anatropous, placentation axile; styles 3, distinct or basally connate; stigmas 3, blunt; rudimentary pistil often present in staminate flowers. **Fruit** typically a loculicidal capsule, triangular and 3-winged, leathery or membranous; seeds flattened, often winged; endosperm copious, hard; embryo small, with broad cotyledon (and sometimes with vestigial second cotyledon), embedded in endosperm.

Family characterization: Rhizomatous or tuberous vines with mycorrhizal roots; broad, cordate-based leaves with reticulate venation; basally twisted petioles with conspicuous pulvini; small, imperfect, epigynous flowers; and 3-winged capsules with flat, winged seeds. Saponins (steroidal saponins, sapogenins; e.g., diosgenin), alkaloids (e.g., dioscorine), tannins, and chelidonic acid typically present. Seed coat with many cell layers containing red to yellowish-brown pigment (phlobaphene; Dahlgren et al. 1985). Anatomical features: raphide sacs (which also contain mucilage); relatively large sieve tubes; vascular bundles of stem in two rings; and very complex nodal vascularization (Ayensu 1972; see the discussion below).

Genera/species: *Dioscorea* (600+ spp.), *Rajania* (20–25 spp.), *Tamus* (4–5 spp.), *Stenomeris* (2 spp.), and *Avetra* (1 sp.)

Distribution: Widely distributed throughout tropical to subtropical regions, with relatively few representatives in the northern temperate areas; especially diverse in tropical America.

U.S./Canadian representatives: *Dioscorea* (12 spp.)

Economic plants and products: Yams (edible tubers), as well as poisonous tubers (due to alkaloids, sapogenins, and/or tannins) from several species of *Dioscorea*. Medicinal products (e.g., corticosteroids, used as an anti-inflammatory medication, and progesterone, used in contraceptives) prepared from the precursor diosgenin (a sapogenin—i.e., a compound derived from a steroidal saponin). Several species of *Dioscorea* (air-potato, cinnamon vine, elephant's-foot) also cultivated as ornamentals.

Commentary: The Dioscoreaceae have been divided into several subfamilies or tribes, some of which (e.g., Stenomeridaceae) have been treated as small, often monotypic families (see Al-Shehbaz and Schubert 1989 for a summary). The 600 (to 850 or more) species of *Dioscorea* make up most of the family (5 genera with about 600 spp.) and exhibit great variation in chemical constituents, vegetative features, and pollen morphology. The genus has been divided into as many as 75 sections. Not uncommonly, staminate and carpellate vines of the same species have been described as two species (even in different sections), contributing to overdescription and confusion.

The vines are easily recognizable in the field by the heart-shaped leaves with reticulate venation (Fig. 135: 1a,h; 2a): parallel and arcuate main veins (i.e., palmate primary veins) with a network of lateral veinlets (secondary and tertiary veins). The petioles typically are long with a pulvinus at both ends (Fig. 135: 1e, 2b). A stipule-like flange often occurs on the basal pulvinus, which is jointed and twisted (Fig. 135: 2b).

The vines develop from a well-developed tuber, a leafless, swollen storage organ derived from the seedling hypocotyl and/or the lowest internode(s) of the stem (Ayensu 1972). A thick layer of cork surrounds the ground tissue of parenchyma cells filled with starch grains. This perennial structure may increase in thickness year after year. In many areas of the world, the tubers of several *Dioscorea* species are a staple food (yams)—not to be confused with sweet-potatoes (tubers of *Ipomoea batatas* in the Convolvulaceae). The stems of numerous species of *Dioscorea* (e.g., *D. bulbifera*; Fig. 135: 2a) also produce an aerial storage organ (bulbil, "tuber") in the leaf axils. This vegetative propagule resembles a tuber anatomically and represents a modified branch.

Several anatomical features concerning stem vasculature of the Dioscoreaceae evidently are unique (Ayensu 1970, 1972). The vascular bundles of the stem are arranged in two distinct rings (Fig. 135: 1f): the bundles of the outer whorl ("common vascular bundles") are elliptical in shape; those of the inner ("cauline vascular bundles"), V- or U-shaped. The phloem elements occur in isolated groups ("phloem units") arranged in a particular pattern within the xylem. Metaxylem vessels plus tracheids compose the xylem. Scle-

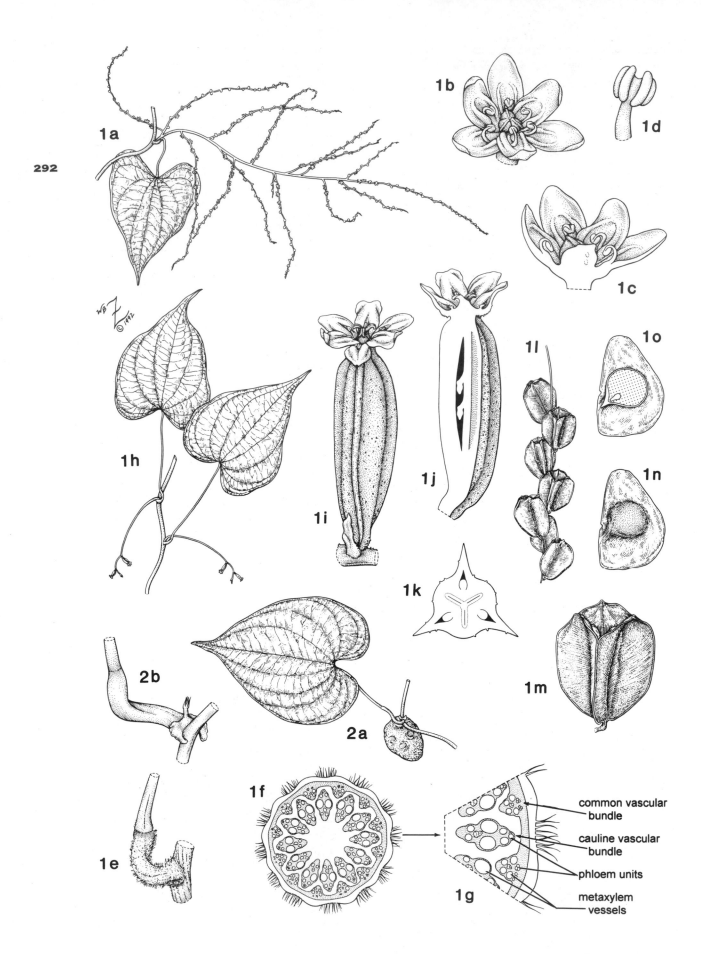

common vascular
bundle

cauline vascular
bundle

phloem units

metaxylem
vessels

renchyma or thick-walled parenchyma cells surround the vascular tissue (Fig. 135: 1g).

The nodal anatomy also is complex. The vascular elements are interlocked together, forming highly specialized xylem or phloem conglomerates ("glomeruli") that are closely associated in a knot-like structure at the nodes (Ayensu 1969). These highly connected systems may ensure the connection of wounded vascular bundles of these climbing plants, thereby guaranteeing the uninterrupted translocation of photosynthates or water (Behnke and Sukkri 1971).

Little is known about the pollination of the small and inconspicuous flowers of the family. Staminate plants usually greatly outnumber the carpellate ones. Several authors have assumed the prevalence of wind-pollination, but the flowers often have a scent and sticky pollen and probably are entomophilous.

REFERENCES CITED

Al-Shehbaz, I., and B. G. Schubert. 1989. The Dioscoreaceae of the southeastern United States. *J. Arnold Arbor.* 70:57–95.

Ayensu, E. S. 1969. Aspects of the complex nodal anatomy of the Dioscoreaceae. *J. Arnold Arbor.* 50:124–137.

——. 1970. Analysis of the complex vascularity in stems of *Dioscorea composita. J. Arnold Arbor.* 51:228–240.

——. 1972. *Dioscoreales*, vol. 6 of *Anatomy of the monocotyledons*, ed. C. R. Metcalfe, pp. 6–13, 41–51. Oxford University Press, Oxford.

Behnke, H.-D., and B. Sukkri. 1971. Anastomoses in the internode of *Dioscorea*: Their frequency, structure, and function. *Z. Pflanzenphysiol.* 66:82–92.

Dahlgren, R. M. T., H. T. Clifford, and P. F. Yeo. 1985. *The families of the monocotyledons*, pp. 115–119. Springer-Verlag, Berlin.

ORCHIDACEAE
Orchid Family

Perennial herbs, terrestrial, epiphytic, or sometimes saprophytic and without chlorophyll; stems leafy or plants scapose, often basally thickened and forming a pseudobulb in epiphytes; roots mycorrhizal, fleshy, sometimes tuberous (in terrestrials), or aerial and cord-like (in epiphytes). **Leaves** simple, entire, alternate, often distichous, parallel-veined, coriaceous or succulent, sessile but sometimes tapering to a petiole-like base, sheathing at base and encircling stem. **Inflorescence** indeterminate, spicate, racemose, paniculate, or sometimes a solitary flower, terminal or axillary. **Flowers** strikingly zygomorphic, perfect, epigynous, resupinate (pedicel and/or ovary twisted 180° during development), very small and inconspicuous to large and showy. **Perianth** of 6 tepals, biseriate, all petaloid or more or less differentiated into calyx and corolla, distinct, imbricate. **Calyx** sometimes unequal with median different from laterals, green or colored and petaloid. **Corolla** unequal with median petal larger, more differentiated than the 2 laterals, and often spurred, variously colored. **Androecium** of 1 or sometimes 2 stamen(s), adnate to style and stigma (forming a column); anthers cap-like, dehiscing longitudinally, introrse; pollen grains in tetrads, grouped into soft to bony masses (pollinia). **Gynoecium** of 1 pistil, 3-carpellate; ovary inferior, 1-locular; ovules very numerous, minute, anatropous, placentation parietal; style single, adnate to androecium; stigma basically 3-lobed, usually of 2 lateral receptive lobes and central sterile beak (rostellum). **Fruit** a capsule dehiscing by 3 or 6 median slits; seeds very abundant and minute, spindle-shaped; endosperm absent; embryo minute, undifferentiated.

Family characterization: Perennial herbs with mycorrhizal roots; simple, often xerophytic leaves with sheathing bases; epigynous, zygomorphic, resupinate flowers; 3-merous perianth with highly specialized median petal (lip); adnate androecium, style, and stigma forming a column; usually 1 stamen with cap-like anther; tetradenous pollen agglutinated into pollinia; ovary with numerous ovules and parietal placentation; capsule dehiscing by 3 or 6 median slits; and minute, extremely numerous seeds with undifferentiated embryos and no endosperm. Alkaloids commonly present. Tissues with raphides (bundles of needle-like calcium oxalate crystals). Anatomical feature: roots of most epiphytic and many terrestrial species with multiseriate epidermis (velamen; discussed below); and tenuinucellate ovules.

293

Figure 135. Dioscoreaceae. 1, *Dioscorea floridana*: **a,** staminate flowering plant, ×½; **b,** staminate flower, ×12; **c,** longitudinal section of staminate flower, ×15; **d,** stamen, ×32; **e,** node showing jointed, twisted pulvinus, ×2½; **f,** cross section of stem showing two rings of vascular bundles, ×7½; **g,** detail of **f** (stippled areas include fibers and/or tracheids and thick-walled parenchyma associated with the vascular bundles), ×14; **h,** carpellate flowering plant, ×½; **i,** carpellate flower, ×9; **j,** longitudinal section of carpellate flower, ×9; **k,** cross section of ovary, ×18; **l,** infructescence, ×½; **m,** capsule, ×1¾; **n,** seed, ×20; **o,** longitudinal section of seed, ×20. **2,** *Dioscorea bulbifera*: **a,** portion of plant with aerial bulbil, ×⅖; **b,** node showing jointed, twisted petiole with stipule-like flange, ×1½.

294

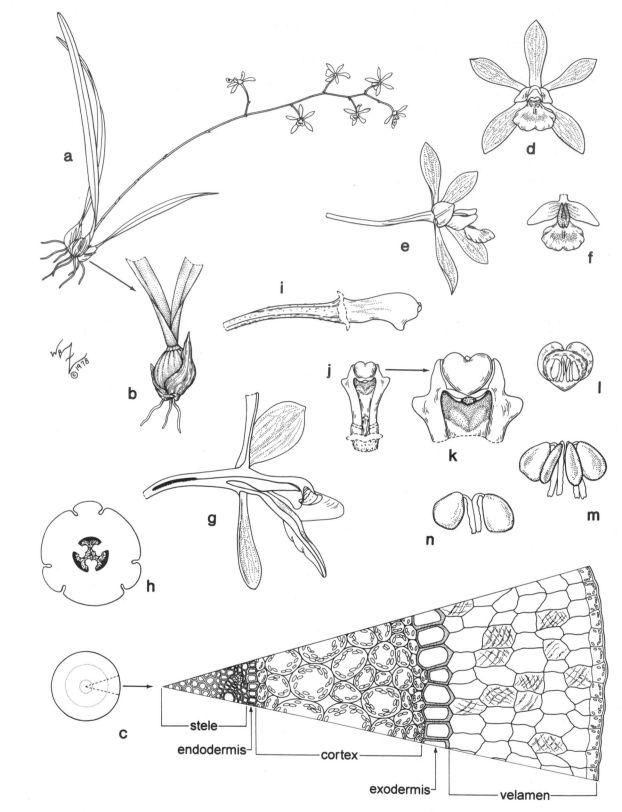

stele
endodermis
cortex
exodermis
velamen

Figure 136. Orchidaceae. *Encyclia tampensis*: **a,** habit, ×¼; **b,** base of plant showing pseudobulb and aerial roots, ×½; **c,** cross section of aerial root, ×10, with detail showing morphology (note velamen), ×135; **d,** front view of flower, ×1; **e,** lateral view of flower, ×1; **f,** lip, ×1; **g,** longitudinal section of flower, ×2⅔; **h,** cross section of ovary, ×18; **i,** column and ovary, ×2⅔; **j,** underside of column, ×2⅔; **k,** detail of column apex showing stigmatic surface and anther, ×6; **l,** anther with pollinia, ×6; **m,** pollinia from one anther, ×12; **n,** one pair of pollinia, ×12.

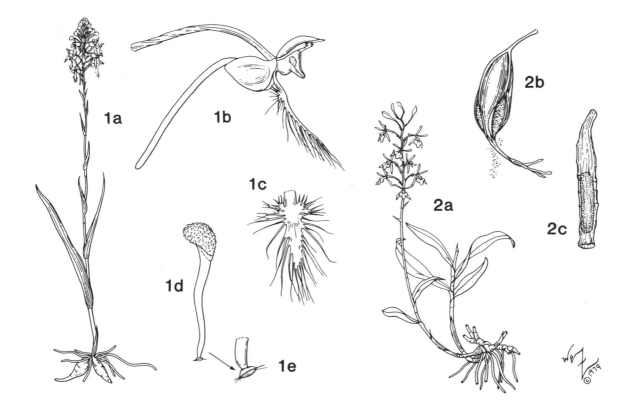

Figure 137. Orchidaceae (continued). 1, *Platanthera ciliaris:* **a,** habit, ×⅕; **b,** flower, ×2; **c,** fringed portion of lip (spur removed), ×2; **d,** pollinium, ×9; **e,** detail of caudicle base showing viscidium, ×25. **2,** *Epidendrum conopseum:* **a,** habit, ×⅓; **b,** capsule, ×1⅓; **c,** seed, ×70.

Genera/species: 775/19,500 (Dressler 1993); one of the largest angiosperm families

Distribution: Widely distributed; most diverse in the tropics (especially Indo-Malaya and America).

Major genera: *Pleurothallis* (1,120 spp.), *Bulbophyllum* (1,000 spp.), *Dendrobium* (900 spp.), *Epidendrum* (800 spp.), *Habenaria* (600 spp.), *Eria* (500 spp.), *Lepanthes* (460 spp.), *Maxillaria* (420 spp.), *Oncidium* (420 spp.), *Masdevallia* (380 spp.), *Stelis* (370 spp.), *Liparis* (350 spp.), *Malaxis* (300 spp.), and *Oberonia* (300 spp.) (Dressler 1993)

U.S./Canadian representatives: 88 genera/285 spp.; largest genera: *Spiranthes, Platanthera,* and *Epidendrum* (see also Luer 1972, 1975).

Economic plants and products: Natural flavoring extracted from immature capsules of *Vanilla planifolia.* Ornamental plants (species of more than 200 natural and hybridized genera), including *Cattleya* (corsage orchid), *Dendrobium, Epidendrum* (buttonhole orchid), *Paphiopedilum* (slipper orchid), *Phalaenopsis* (moth orchid), and *Vanda.*

Commentary: The taxonomic (and horticultural) literature on the Orchidaceae is voluminous, with many societies and journals devoted exclusively to the subject. Numerous new species from the tropics are described each year. Probably due to careless taxonomic work (Dressler pers. comm.), generic and specific lines are not always clear-cut. Many intrafamilial classification schemes have been devised (e.g., Dressler 1981, 1986, 1993; Burns-Balogh and Funk 1986), and subfamilies, tribes, and subtribes are based largely on technical features of the column, anthers, and pollinia. Some authorities consider the subfamilies as segregate families (e.g., Apostasiaceae and Cypripediaceae in Rasmussen 1985). Despite the diversity, the family clearly is a monophyletic group based on such characters as the reduction of the adaxial stamens and the numerous, small, nonendospermous seeds (Rasmussen 1985).

Two basic habit types occur in the family. In temperate regions, orchids typically are terrestrial with rhizomes or corms and fleshy to tuberous roots (Fig. 137: 1a). A few are saprophytic and achlorophyllous. Epiphytic orchids largely are limited to the tropics. Many of these have a basally thickened, bulbous stem (pseudobulb) and hanging, whitish, cord-like roots (Fig. 136: a,b); at maturity, these aerial roots have a well-developed multiseriate epidermis (velamen; Fig. 136: c), composed of nonliving cells, and also a cortex of chlorenchyma (i.e., photosynthetic tissue). A less obvious velamen also occurs in the roots of many terrestrial orchids (Pridgeon 1987). This spongy layer around the roots commonly harbors mycorrhizal fungi (Dressler 1981) and functions in water and nutrient absorption (Capesius and Barthlott 1975). Species of several other monocot families, such as the Amaryllidaceae, Asparagaceae, and Araceae, also have roots with a velamen.

The diversity and specialization in orchid floral morphology have fascinated botanists and collectors for centuries. The traditional view hypothesizes that the zygomorphic orchid flower arose from a three-merous liliaceous prototype (Garay 1960). The three outer tepals (or sepals) usually are smaller and less conspicuous than those of the inner whorl (petals). The petals of different species vary greatly in shape and color. The very specialized median petal (lip or labellum) may have various modifications, such as fleshy protuberances on the upper surface and/or a spur (Fig. 137: 1b). Although actually adaxial (uppermost), the lip typically appears abaxial due to a 180° twist (resupination) of the ovary and/or pedicel during floral development.

In most orchids, the stamens, styles, and stigmas are variously adnate into a single structure, the column (Fig. 136: i,j). The column is unique among monocots and is analogous to the gynostegium of the Apocynaceae. Although the ancestral orchids usually have two stamens, most orchids have only one stamen that is opposite the lip. The cap-like anther (Fig. 136: k,l) is situated at the apex of the column and contains tetradenous pollen grains agglutinated into pollinia (Fig. 136: m,n); very hard pollinia lack the substance (elastoviscin) that holds relatively soft pollinia together. One end of the pollinium may be attenuated into a sterile stalk (caudicle) composed of viscin and some pollen grains (Fig. 137: 1d,e). The three-lobed stigma occurs in a depressed area on the inner side of the column, immediately below the anther (Fig. 136: k). In most orchids, only two lobes are receptive. A portion of the central sterile lobe, called a rostellum, is modified into a sterile beak or flap. In more derived genera, a portion of the rostellum is specialized as sticky discs (viscidia) to which the caudicles are attached (Fig. 137: 1e). In the most derived genera, a nonviscid strap of column tissue (stipe) connects the pollinia caudicles to the viscidium.

Modifications of the perianth, androecium, and gynoecium represent the basis for a variety of floral adaptations for animal-mediated pollination. Numerous researchers from Darwin (1862) to van der Pijl and Dodson (1966) have examined the complex relationships between orchid species and their particular pollinators, which include bees (most frequently), moths, butterflies, flies, and hummingbirds. The pollinators are attracted by the color, shape, nectar, and odor of the flowers. Nectaries usually occur on the lip, and the most specialized orchids secrete nectar in the spur.

The lip functions additionally as a visual attractant, landing platform, and center guide. The introrse anthers on the column underside (opposite the lip) are ideally situated for pollinium deposition as the insect enters or leaves the flower via the lip. Numerous mechanisms enable the pollinia to adhere to the insect. Most orchids have a flap-like rostellum, which hangs down between stigma and anther; when the insect backs out of the flower, it brushes against the sticky rostellum and subsequently picks up the pollinia. In the most derived cases, pollinia with viscidia (sticky discs) adhere to the insect as a unit. Cross-pollination is completed when the pollinia are deposited on the stigma of the next flower.

The ovules of orchids do not develop until fertilization. The mature airborne seeds (Fig. 137: 2c) are exceedingly small and numerous and collectively look like a fine powder: several million seeds can fill a single capsule! Each seed comprises only a membranous testa surrounding a few undifferentiated cells (Knudson 1925). For germination, the seed (which lacks endosperm) requires nutrients supplied by a mycorrhizal relationship with a specific fungus.

REFERENCES CITED

Burns-Balogh, P., and V. A. Funk. 1986. A phylogenetic analysis of the Orchidaceae. *Smithsonian Contr. Bot.* 61: 1–79.

Capesius, I., and W. Barthlott. 1975. Isotopen-Markierungen und raster-elektronmikroskopische Untersuchungen des *Velamen radicum* der Orchideen. *Z. Pflanzenphysiol.* 75:436–448.

Darwin, C. R. 1862. *On the various contrivances by which British and foreign orchids are fertilized by insects, and on the good effects of intercrossing.* J. Murray, London.

Dressler, R. L. 1981. *The orchids: Natural history and classification,* pp. 25–26, 142–274. Harvard University Press, Cambridge, Mass.

——. 1986. Recent advances in orchid phylogeny. *Lindleyana* 1:5–20.

——. 1993. *Phylogeny and classification of the orchid family,* pp. 59–82, 267–278. Dioscorides Press, Portland, Ore.

Garay, L. A. 1960. On the origin of the Orchidaceae. *Bot. Mus. Leafl.* 19:57–96.

Knudson, L. 1925. Physiological study of the symbiotic germination of orchid seeds. *Bot. Gaz.* 79:345–380.

Luer, C. A. 1972. *The native orchids of Florida.* New York Botanical Gardens, Bronx.

———. 1975. *The native orchids of the United States and Canada excluding Florida.* New York Botanical Garden, Bronx.

van der Pijl, L., and C. H. Dodson. 1966. *Orchid flowers, their pollination and evolution.* University of Miami Press, Coral Gables.

Pridgeon, A. M. 1987. The velamen and exodermis of orchid roots. In *Orchid biology, reviews and perspectives,* vol. 4, ed. J. Arditti, pp. 139–192. Cornell University Press, Ithaca, N.Y.

Rasmussen, F. N. 1985. Orchids. In *The families of the monocotyledons,* ed. R. M. T. Dahlgren, H. T. Clifford, and P. F. Yeo, pp. 249–274. Springer-Verlag, Berlin.

ALISMATACEAE
Arrowhead or Water-plantain Family

Annual or usually perennial aquatic herbs, scapose, with milky sap; rhizome stout, often producing tubers. **Leaves** simple, entire, basal, parallel-veined with major veins converging at apex, erect, floating or submersed, long-petiolate with open basal sheath. **Inflorescence** determinate, cymose and appearing racemose or paniculate with flowers in whorls at the nodes, terminal. **Flowers** actinomorphic, perfect or imperfect (then plants monoecious, dioecious, or polygamous), hypogynous, showy to inconspicuous, with flat to globose receptacle. **Calyx** of 3 sepals, distinct, imbricate, persistent. **Corolla** of 3 petals, distinct, ephemeral, caducous, often showy, typically white or pink, imbricate and crumpled in bud. **Androecium** of 6 to many stamens; filaments distinct; anthers basifixed, dehiscing longitudinally, extrorse. **Gynoecium** apocarpous and of 6 to many carpels (spirally arranged on the receptacle); ovary superior, 1-locular; ovule solitary, anatropous, placentation basal; style short, persistent; stigma scarcely distinguishable. **Fruit** an aggregate of achenes; endosperm absent; embryo large, horseshoe-shaped.

Family characterization: Aquatic herbs with a milky latex; long-petiolate, emergent leaves with sheathing bases; 3 greenish sepals and 3 caducous, ephemeral, white or pink petals; gynoecium of many free carpels spirally arranged on the receptacle; achenes clustered into a head; and seeds with large, horseshoe-shaped embryos and no endosperm. Tissues often with calcium oxalate crystals. Anatomical features: secretory or excretory ducts in the leaves and stems; and schizogenous intercellular spaces in all plant parts (Stant 1964).

Genera/species: 16/100

Distribution: Generally cosmopolitan; primarily in freshwater swamps and streams in temperate and tropical regions of the Northern Hemisphere.

Major genera: *Echinodorus* (30–47 spp.) and *Sagittaria* (20–35 spp.)

U.S./Canadian representatives: *Sagittaria* (up to 27 spp.), *Echinodorus* (3–4 spp.), *Alisma* (3 spp.), and *Machaerocarpus* (1 sp.)

Economic plants and products: Edible tubers from *Sagittaria* spp. Ornamental plants for pools and aquaria: *Alisma* (water-plantain) and *Sagittaria* (arrowhead).

Commentary: The family boundaries and affiliations of the Alismataceae, as well as intrafamilial concepts, have been variously interpreted (Tomlinson 1982; Rogers 1983; Dahlgren et al. 1985). For example, several taxonomists (such as Thorne 1992) include the Limnocharitaceae (3 genera). In addition, the species of the family tend to be overdescribed due to environmental plasticity and variability (see Adams and Godfrey 1961). For example, as in most aquatic plants, the members of the Alismataceae exhibit great variation in leaf morphology depending on the locality and depth of water in which they grow. Submersed leaves tend to be linear, while erect leaves are long-petiolate with linear to elliptic, ovate, or sagittate blades. Sometimes several leaf forms with intermediates occur on the same plant.

The Alismataceae have been cited as an example of a "primitive" group of extant monocots due to the actinomorphic, hypogynous flowers with many free stamens and carpels (Fig. 138: 1d,f)—character states similar to those of the so-called Ranalian dicots, such as the Ranunculaceae and Nymphaeaceae. However, investigations (see Moore and Uhl 1973) of various morphological and anatomical aspects indicate that the family actually is not very similar to these dicot groups. In particular, the large carpel and stamen numbers actually are due to enlargement and multiplication of the originally few whorls of three.

The showy flowers attract flies, certain bees, and other small insects as principal pollinators. In most species, nectar is secreted at the base of the carpels (Dahlgren et al. 1985). In *Alisma* and *Echinodorus* the flowers are perfect, while in *Sagittaria* they usually are imperfect (plants typically monoecious).

REFERENCES CITED

Adams, P., and R. K. Godfrey. 1961. Observations on the *Sagittaria subulata* complex. *Rhodora* 63:247–266.

Dahlgren, R. M. T., H. T. Clifford, and P. F. Yeo. 1985. *The families of the monocotyledons,* pp. 95–98, 301–303. Springer-Verlag, Berlin.

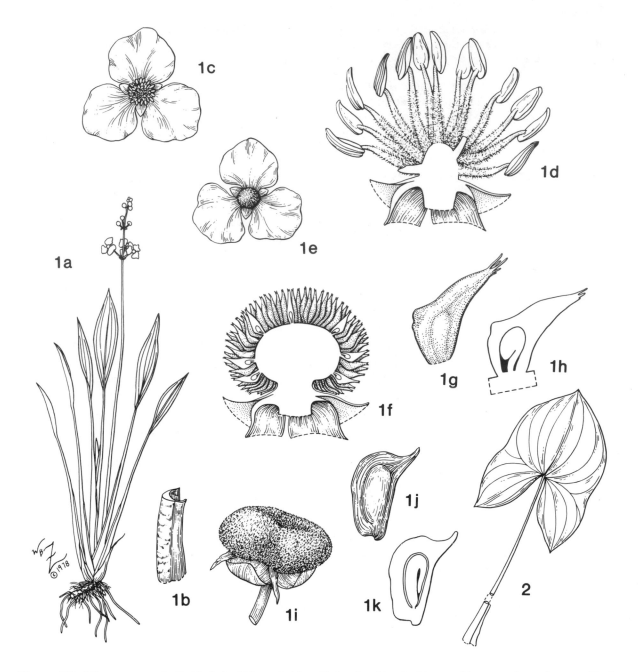

Figure 138. Alismataceae. 1, *Sagittaria lancifolia*: a, habit, ×⅕; b, sheathing leaf base, ×½; c, staminate flower, ×¾; d, longitudinal section of staminate flower, ×6; e, carpellate flower, ×¾; f, longitudinal section of carpellate flower, ×6; g, carpel, ×25; h, longitudinal section of carpel, ×25; i, fruiting head (aggregate of achenes), ×2; j, achene, ×12; k, longitudinal section of achene, ×12. 2, *Sagittaria latifolia*: leaf, ×⅛.

Moore, E. H., and N. W. Uhl. 1973. The monocotyledons: Their evolution and comparative biology. VI. Palms and the origin and evolution of monocotyledons. *Quart. Rev. Biol.* 48:414–436.

Rogers, G. K. 1983. The genera of Alismataceae in the southeastern United States. *J. Arnold Arbor.* 64:383–420.

Stant, M. Y. 1964. Anatomy of Alismataceae. *J. Linn. Soc., Bot.* 59:1–42.

Thorne, R. F. 1992. Classification and geography of the flowering plants. *Bot. Rev.* (Lancaster) 58:225–348.

Tomlinson, P. B. 1982. *Helobiae (Alismatidae)*, vol. 7 of *Anat-omy of the monocotyledons*, ed. C. R. Metcalfe, pp. 57–89. Oxford University Press, Oxford.

HYDROCHARITACEAE
Frog's-bit or Tape-grass Family

Perennial aquatic herbs, completely or partially submersed or floating, with creeping monopodial rhizome and/or erect main stem. **Leaves** simple, entire to serrate, crowded and basal or cauline (then alternate, opposite, or whorled), generally parallel-veined, extremely variable in shape and size, with sheathing bases. **Inflorescence** determinate, cymose and umbellate (staminate inflorescence) or flower solitary (carpellate inflorescence), subtended by 2 distinct or connate bracts (forming a spathe), axillary. **Flowers** generally actinomorphic, usually imperfect (then plants most often dioecious), epigynous, small and inconspicuous to fairly large and showy, sometimes with very long peduncles. **Calyx** of 3 sepals, distinct, valvate. **Corolla** of 3 petals or sometimes absent, distinct, delicate and fugacious, usually white, imbricate or convolute. **Androecium** of 2 or 3 to numerous stamens, often in 1 to 5 whorls of 3, inner or outer whorls sometimes reduced to staminodes; filaments distinct to sometimes connate (monadelphous); anthers dehiscing longitudinally; staminodes often present in carpellate flowers. **Gynoecium** of 1 pistil, usually 3- to 6-carpellate; ovary inferior, 1-locular; ovules numerous, anatropous, placentation parietal and often appearing axile due to 3 to many deeply intruding placentae, styles as many as carpels, often bifid; stigmas papillose; rudimentary pistil sometimes present in staminate flowers. **Fruit** berry-like, dry or fleshy, indehiscent or rupturing irregularly, globose to linear, submersed; seeds usually numerous; endosperm absent; embryo straight.

Family characterization: Submersed to floating aquatic herbs; inflorescence subtended by a spathe or a pair of bracts; imperfect flowers; syncarpous gynoecium with inferior, 1-locular ovary; parietal placentation with placentae deeply intruded into the ovary; and berry-like submersed fruit. Anatomical features: scattered tanniferous cells; and aerenchymatous stem with much-reduced vascular system.

Genera/species: 17/130

Distribution: Primarily in tropical marine and freshwater habitats, with a few representatives in temperate freshwaters.

Major genera: *Najas* (35–50 spp.) and *Ottelia* (40 spp.)

U.S./Canadian representatives: 10 genera/24 spp.; largest genera: *Elodea* and *Hydrocharis*

Economic plants and products: Several weedy plants, such as species of *Elodea* (waterweed, ditchmoss) and *Hydrilla*. Ornamental plants for aquaria (species of 9 genera), including *Egeria* (waterweed), *Elodea, Hydrilla, Limnobium* (American frog's-bit), and *Vallisneria* (eel-grass, tape-grass).

Commentary: The Hydrocharitaceae are divided into three to five subfamilies (see Dahlgren et al. 1985).

The two monotypic groups (Thalassioideae and Halophilioideae) are characterized by growing in salt water.

The floating to submersed vegetative habit varies considerably in the family (Ancibor 1979). For example, the linear and grass-like leaves of *Vallisneria* are clustered on short stems along the nodes of the rhizome. The leaves also are clustered in the marine genera, *Thalassia* (with ribbon-like leaves) and *Halophila* (leaves clustered at summits of stems). In *Hydrilla* and *Elodea* the small sessile leaves are in whorls along the submersed and free-floating stems. The leaves of *Limnobium* are clearly differentiated into a lamina and petiole. The juvenile form (Fig. 139: a) consists of floating rosettes of reniform leaves, each with a central area of spongy tissue. Eventually, more robust plants develop that bear the flowers and fruit (Fig. 139: b); the leaves of these mature plants are long-petiolate with rounded and leathery blades.

The flowers vary considerably in carpel and stamen number and typically are imperfect (Kaul 1968, 1970). Generally, the inflorescences are unisexual, with each subtended by a spathe. The carpellate inflorescences usually are one-flowered, and the staminate, one- to many-flowered (e.g., see Fig. 139: c,i).

The Hydrocharitaceae are notable for specialized pollination mechanisms (Sculthorpe 1967). Several with relatively large and showy flowers (e.g., *Egeria* and most *Limnobium* spp.) have nectaries and are pollinated by insects (such as flies and beetles). In others, such as *Hydrilla* and *Vallisneria*, the staminate flowers become detached and drift in the water to the carpellate inflorescences (Cook and Luond 1982; Lowden 1982). The long peduncles of the carpellate flowers of *Vallisneria* coil up after pollination, pulling the flowers beneath the surface of the water. The anthers of *Elodea* (staminate flowers sometimes detach) and *Hydrilla* explode and scatter pollen over the surface of the water. Pollination takes place underwater in the flowers of both *Thalassia* and *Halophila*, which release the pollen in connected chains (Tomlinson 1969).

REFERENCES CITED

Ancibor, E. 1979. Systematic anatomy of vegetative organs of the Hydrocharitaceae. *Bot. J. Linn. Soc.* 78:237–266.

Cook, C. D. K., and R. Luond. 1982. A revision of the genus *Hydrilla* (Hydrocharitaceae). *Aquatic Bot.* 13:485–504.

Dahlgren, R. M. T., H. T. Clifford, and P. F. Yeo. 1985. *The families of the monocotyledons*, pp. 303–307. Springer-Verlag, Berlin.

Kaul, R. B. 1968. Floral morphology and phylogeny in the Hydrocharitaceae. *Phytomorphology* 18:13–35.

——. 1970. Evolution and adaptation of the inflorescences in the Hydrocharitaceae. *Amer. J. Bot.* 57:708–715.

Lowden, R. M. 1982. An approach to the taxonomy of

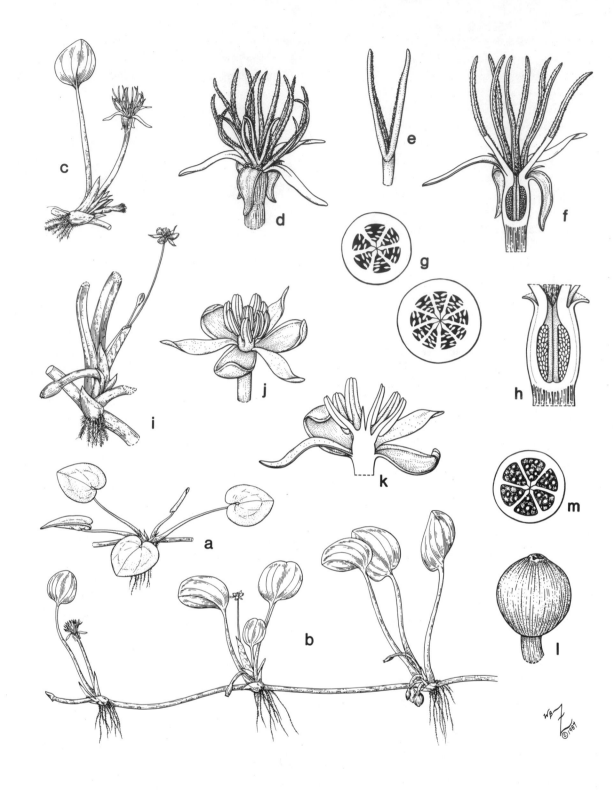

Figure 139. Hydrocharitaceae. *Limnobium spongia*: **a,** juvenile plant, ×¼; **b,** plant with flowers and fruit, ×¼; **c,** node with cluster of carpellate inflorescences, ×⅔; **d,** carpellate flower, ×2; **e,** stigma, ×3; **f,** longitudinal section of carpellate flower, ×2½; **g,** cross section of ovaries from two flowers showing varying carpel number among flowers, ×6; **h,** longitudinal section of ovary, ×4½; **i,** node with staminate inflorescence and young infructescence, ×⅔; **j,** staminate flower, ×3; **k,** longitudinal section of staminate flower, ×4; **l,** fruit, ×1½; **m,** cross section of fruit, ×1½.

Vallisneria L. (Hydrocharitaceae). *Aquatic Bot.* 13:269–298.

Sculthorpe, C. D. 1967. *The biology of aquatic vascular plants*, pp. 301–311. St. Martin's Press, New York.

Tomlinson, P. B. 1969. On the morphology and anatomy of turtle grass *Thalassia testudinum* (Hydrocharitaceae). III. Floral morphology and anatomy. *Bull. Mar. Sci.* 19:286–305.

ARACEAE
Arum Family

Perennial herbs or vines, terrestrial, aquatic, or sometimes epiphytic, with watery or milky pungent sap, with rhizomes or tubers; roots adventitious, mycorrhizal, lacking root hairs. **Leaves** simple to pinnately or palmately compound, entire, alternate, basally clustered or cauline, parallel-, pinnately-, or palmately veined, with membranous basal sheath. **Inflorescence** indeterminate, spicate and composed of numerous flowers packed onto a cylindrical fleshy axis (spadix), subtended by a large, foliose or petaloid bract (spathe). **Flowers** actinomorphic, perfect or imperfect (then plants usually monoecious), fundamentally hypogynous, minute, ebracteate. **Perianth** absent (in imperfect flowers) or present (in perfect flowers), of 4 to 6 tepals, distinct (scale-like) to completely connate (cup-like), fleshy, inconspicuous. **Androecium** of 1 to 6 stamens, opposite the tepals; filaments distinct to completely connate; anthers basifixed, distinct or connate, dehiscing by apical pores, extrorse. **Gynoecium** of 1 pistil, usually 1- to 3-carpellate; ovary basically superior (sometimes appearing inferior when embedded in spadix), typically 1-locular; ovules 1 to many, anatropous, amphitropous, or orthotropous, placentation basal, parietal, axile, or apical; stigma usually sessile. **Fruit** a berry; endosperm usually present, copious; embryo usually straight.

Family characterization: Rhizomatous or tuberous perennial herbs typically growing in wet areas; milky or watery, pungent sap; basal leaves each with sheathing petiole; reduced flowers crowded onto a fleshy axis (spadix) subtended by a showy bract (spathe); inconspicuous or absent perianth; anthers dehiscing by apical pores; and a berry as the fruit type. Tissues with raphides (bundles of needle-like calcium oxalate crystals), which can cause painful injury to the mouth and throat of an unwary consumer.

Genera/species: 104/2,500

Distribution: Mostly pantropical.

Major genera: *Anthurium* (700–1,000 spp.), *Philodendron* (350–600 spp.), *Arisaema* (150 spp.), *Homalomena* (140 spp.), *Amorphophallus* (100 spp.), and *Schismatoglottis* (100 spp.) (Grayum 1990)

U.S./Canadian representatives: 20 genera/40 spp.; largest genera: *Philodendron*, *Anthurium*, and *Xanthosoma*

Economic plants and products: Starchy corms (used as subsistence food sources) from *Alocasia*, *Colocasia* (taro), and *Xanthosoma* (yautia). Ornamental plants (species of more than 50 genera), including *Aglaonema* (Chinese evergreen, silverqueen), *Anthurium*, *Caladium*, *Colocasia* (elephant's-ear), *Dieffenbachia* (dumbcane), *Monstera* (windowleaf), *Philodendron*, *Pothos*, *Spathiphyllum* (spatheflower), *Syngonium*, and *Zantedeschia* (calla-lily).

Commentary: The Araceae exhibit much diversity in morphological and anatomical characters. The family apparently is monophyletic, based on characters such as the inflorescence type; of note is that the problematic, somewhat isolated genus *Acorus* has been removed from the family (Grayum 1990; Bogner and Nicolson 1991; Duvall et al. 1993; and under "*taxa incertae sedis*" in Thorne 1992). The family is thought to be derived from either a liliaceous or a palmaceous phylogenetic stock (Mookerjea 1955; Dahlgren and Clifford 1982; Dahlgren et al. 1985). Taxonomists have divided the Araceae into various subfamilies and tribes based primarily on habit and leaf morphology, details of floral structure, and anatomical characters such as the presence of latex and raphides (see Wilson 1960; Grayum 1990; and Bogner and Nicolson 1991, for summaries).

The habits of aroids are diverse: members of this family may be epiphytes (some *Anthurium* species), lianas (*Monstera*), terrestrials that often grow in moist habitats (*Arisaema*; Fig. 140: 1a), aquatics (*Orontium*; Fig. 140: 2a), or even floating aquatics (*Pistia*; Fig. 140: 3). The leaves are simple to compound, with the compound condition derived from a continuous blade that splits into leaflets. Blades may be laciniate, and in *Monstera* (windowleaf) and allies they may have large fenestrations. The shape of the blade varies from linear to ovate (with hastate or sagittate bases), and the venation is parallel, pinnate, or palmate. Often the juvenile and adult leaves differ greatly in size and shape (Bogner 1987).

Araceous flowers also have a wide range of morphology. The small flowers may be perfect (*Orontium*; Fig. 140: 2c,d) or imperfect (*Arisaema*; Fig. 140: 1e–g). The fleshy tepals, usually present only in perfect flowers, are distinct to completely connate. The stamens, which vary in number, also have different degrees of connation. The number of carpels and locules and the placentation type of the ovary are extremely variable.

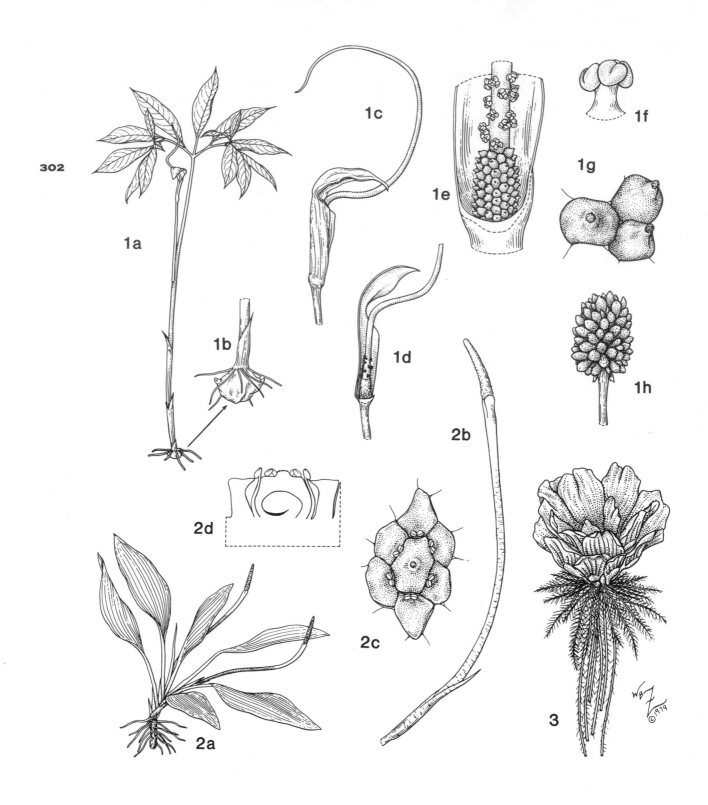

302

Figure 140. Araceae. 1, *Arisaema dracontium*: **a,** habit, ×⅙; **b,** corm, ×⅖; **c,** inflorescence (spadix subtended by spathe), ×½; **d,** inflorescence with half of spathe removed, ×½; **e,** detail of base of inflorescence (half of spathe removed), ×2⅓; **f,** staminate flower, ×12; **g,** carpellate flowers, ×12; **h,** infructescence (berries), ×⅖. **2,** *Orontium aquaticum*: **a,** habit, ×¼; **b,** inflorescence (spadix with spathe), ×⅖; **c,** flower, ×6; **d,** longitudinal section of flower, ×6. **3,** *Pistia stratiotes*: habit, ×⅙.

According to Grayum (1990), the ovary is always superior but appears inferior in some genera when the flowers are embedded in the inflorescence axis.

A unifying feature of the family is the inflorescence, a spadix comprising numerous small ebracteate flowers packed onto a fleshy axis (Fig. 140: 1c, 2b). Sometimes, as in *Arisaema dracontium*, the end of the spadix is naked and conspicuous. The spadix usually is subtended by a large bract (spathe). In aroids with imperfect flowers, the staminate and carpellate flowers usually occur on the same spadix with the carpellate flowers on the lower portion, as in Fig. 140: 1e.

The protogynous flowers are pollinated by insects (usually flies and beetles) that are attracted to the conspicuous spathe and spadix. The inflorescence also produces a strong disagreeable scent attractive to many carrion-visiting insects. Heat produced by the spadix apparently aids in diffusing the odor (Brown 1988). The perfect flowers are protogynous, and in most monoecious inflorescences, the carpellate flowers mature first.

REFERENCES CITED

Bogner, J. 1987. Morphological variation in aroids. *Aroideana* 10:4–16.

—— and D. H. Nicolson. 1991. A revised classification of Araceae with dichotomous keys. *Willdenowia* 21:35–50.

Brown, D. 1988. *Aroids*, pp. 37–54. Timber Press, Portland, Ore.

Dahlgren, R. M. T., and H. T. Clifford. 1982. *The monocotyledons: A comparative study*, pp. 292–295. Academic Press, London.

——, ——, and P. F. Yeo. 1985. *The families of the monocotyledons*, pp. 96, 277–278. Springer-Verlag, Berlin.

Duvall, M. R., M. T. Clegg, M. W. Chase, W. D. Clark, W. J. Kress, et al. 1993. Phylogenetic hypotheses for the monocotyledons constructed from *rbc*L data. *Ann. Missouri Bot. Gard.* 80:607–619.

Grayum, M. H. 1990. Evolution and phylogeny of the Araceae. *Ann. Missouri Bot. Gard.* 77:628–697.

Mookerjea, A. 1955. Cytology of different species of aroids with a view to trace the basis of their evolution. *Caryologia* 7:221–291.

Thorne, R. F. 1992. Classification and geography of the flowering plants. *Bot. Rev.* (Lancaster) 58:225–348.

Wilson, K. A. 1960. The genera of Arales in the southeastern United States. *J. Arnold Arbor.* 41:47–72.

LEMNACEAE
Duckweed Family

Perennial aquatic herbs, free-floating or submersed, undifferentiated into stem and leaves and reduced to a thallus ("frond"), small to minute, flat and leaf-like to globose, sometimes purplish beneath, often reproducing asexually (by budding); roots simple and thread-like or absent. **Flowers** produced only occasionally, solitary or paired in pouches (on margins and/or upper surface of thallus), imperfect (plants monoecious), hypogynous, extremely reduced and minute, often initially enclosed by a membranous sheath (spathe). **Perianth** absent. **Androecium** of usually 1 stamen; filaments filiform to fusiform or absent; anthers 2- or sometimes 1-locular, dehiscing longitudinally or transversely. **Gynoecium** of 1 pistil, 1-carpellate; ovary superior, 1-locular; ovule(s) usually 1 or 2, orthotropous, placentation basal; style 1, short; stigma 1, funnel-shaped. **Fruit** a utricle; seed(s) ovoid, usually 1 or 2, with cap-like operculum at micropylar end ("stopper" or inner integument); endosperm scanty, fleshy, sheathing the embryo, or absent; embryo relatively large, straight, consisting almost entirely of a large cotyledon.

Family characterization: Submersed or free-floating aquatic herbs occurring in still freshwaters; reduced plant body (without stem or leaves) of a small leaf-like thallus; frequent asexual reproduction (by budding) and only occasional flowering; imperfect flowers consisting of only one stamen (staminate flowers) or one carpel (carpellate flowers); and a utricle as the fruit type. Anatomical features: aerenchyma tissue (in thallus); and typically a reduced vascular system without xylem (although tracheids occur in the roots of *Spirodela*).

Genera/species: *Lemna* (9–15 spp.), *Wolffia* (7–10 spp.), *Wolffiella* (6–8 spp.), and *Spirodela* (4–6 spp.)

Distribution: Cosmopolitan; in quiet freshwater habitats.

U.S./Canadian representatives: *Lemna* (8 spp.), *Wolffia* (4 spp.), *Wolffiella* (3 spp.), and *Spirodela* (2 spp.)

Economic plants and products: Important food sources for waterfowl and fish, serious weeds of still waters, and ornamental plants for pools and aquaria (species of all 6 genera), including *Lemna* (duckweed, duck's-meat), *Spirodela* (duckweed), *Wolffia* (watermeal), and *Wolffiella* (mudmidget, bogmat).

Commentary: The Lemnaceae are the smallest of seed plants, with *Wolffia* species being the tiniest of all. They are closely related to and derived from the Araceae. Plants of *Spirodela*, the least reduced genus of the Lemnaceae, are somewhat similar morphologically to those of *Pistia* (Fig. 140: 3), the monotypic genus of free-floating aroids (see Wilson 1960; Hartog 1975; Dahlgren et al. 1985). Recent molecular studies (Duvall et al. 1993) ally *Pistia* with *Lemna*.

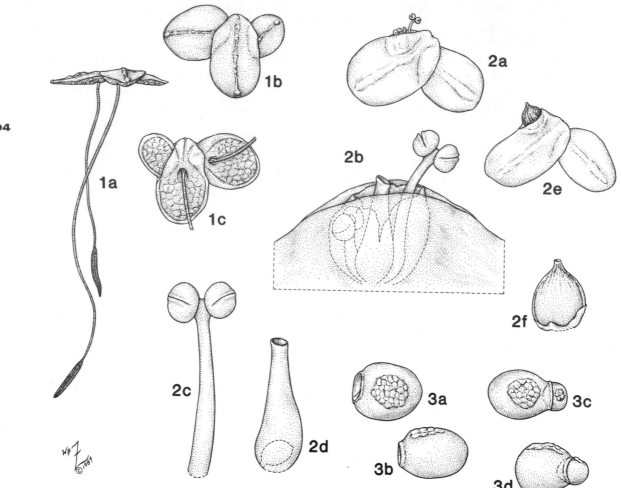

Figure 141. Lemnaceae. 1, *Lemna obscura*: **a,** lateral view of clone of three thalli, ×8; **b,** dorsal view of **a,** ×8; **c,** ventral view of **a,** ×8. 2, *Lemna minor*: **a,** dorsal view of flowering thallus, ×12; **b,** detail of thallus margin showing pouch with inflorescence, ×45; **c,** staminate flower, ×60; **d,** carpellate flower, ×60; **e,** dorsal view of fruiting thallus, ×12; **f,** utricle, ×25. 3, *Wolffia columbiana*: **a,** dorsal view of thallus, ×22; **b,** lateral view of thallus, ×22; **c,** dorsal view of thallus with vegetative bud, ×22; **d,** lateral view of **c,** ×22.

Duckweed plants form dense floating mats on ponds or pools. The vegetative structure consists of a specialized plant body (frond or thallus) of various shapes (circular, elongate, or ellipsoid) that represents a modified stem and/or leaf (see Hartog and van der Plas 1970; Landolt 1980, 1986). Vegetative buds (and in a few genera, flowers) develop in one or two reproductive pouches along the basal margins of the thallus. Most species have no roots (e.g., Fig. 141: 3a), although the thalli of *Spirodela* have several, and those of *Lemna* (Fig. 141: 1a,c), a single "root" with no vascular tissue.

Reproduction is primarily vegetative. New thalli, which emerge from the reproductive pockets, may separate or remain attached to the parent plant (Fig. 141: 1b; 3c,d). The main dispersers of the Lemnaceae are animals that live around the water. Bulblets—resting thalli with abundant starch reserves—are produced before winter or during conditions unfavorable for growth; depending on the species, the bulblets either sink to the bottom of the pond or remain in sheltered areas until favorable conditions return.

The flowers of the Lemnaceae, which develop in pockets on the thallus (Fig. 141: 2a,b), are rarely produced (Landolt and Kandeler 1987). The staminate flowers consist of a single stamen (Fig. 141: 2c), and the carpellate, of a single carpel (Fig. 141: 2d). The inflorescence of *Lemna* (Fig. 141: 2b) and of *Spirodela* consists of two staminate flowers and one carpellate flower subtended by a minute, rudimentary spathe. The spathe is absent in *Wolffia* and *Wolffiella*, which have only one staminate and one carpellate flower in each inflorescence. Flowers of most species are pro-togynous, and those in the same plant mature at different times. Insects and other aquatic animals, as well as direct contact of the flowers, may cause cross-pollination (Landolt 1986).

The seeds are released from the fruit by the disintegration of the pericarp.

REFERENCES CITED

Dahlgren, R. M. T., H. T. Clifford, and P. F. Yeo. 1985. *The families of the monocotyledons*, pp. 287–289. Springer-Verlag, Berlin.

Duvall, M. R., M. T. Clegg, M. W. Chase, W. D. Clark, W. J. Kress, et al. 1993. Phylogenetic hypotheses for the monocotyledons constructed from *rbc*L data. *Ann. Missouri Bot. Gard.* 80:607–619.

Hartog, C. den. 1975. Thoughts about the taxonomical relationships within the Lemnaceae. *Aquatic Bot.* 1:407–416.

—— and F. van der Plas. 1970. A synopsis of the Lemnaceae. *Blumea* 18:355–368.

Landolt, E. 1980. Biosystematic investigations in the family of duckweeds (Lemnaceae), vol. 1. Key to the determination of taxa within the family of Lemnaceae. *Veröff. Geobot. Inst. ETH Stiftung Rübel Zürich* 70:13–21.

——. 1986. Ibid., vol. 2. The family of Lemnaceae—A monographic study, vol. 1. *Veröff. Geobot. Inst. ETH Stiftung Rübel Zürich* 71:1–566.

—— and R. Kandeler. 1987. Ibid., vol. 4. The family of Lemnaceae—A monographic study, vol. 2. *Veröff. Geobot. Inst. ETH Stiftung Rübel Zürich* 95:1–638.

Wilson, K. A. 1960. The genera of Arales in the southeastern United States. *J. Arnold Arbor.* 41:47–72.

ARECACEAE OR PALMAE
Palm Family

Trees, shrubs, or lianas, with diffuse secondary thickening; stems erect or prostrate (then plants often appearing acaulescent), usually unbranched; roots mycorrhizal, lacking root hairs. **Leaves** large, simple to palmately or pinnately compound, entire, alternate, spirally arranged or distichous, usually in a terminal tuft, palmately or pinnately veined, folded (V- or Λ-shaped in cross section), sometimes with a woody ligule (hastula), coriaceous, persistent, with large basal sheath. **Inflorescence** determinate and often appearing paniculate to compound-spicate, very large and much branched, usually below or among the leaves, subtended by 1 or more bracts (prophylls). **Flowers** actinomorphic, perfect or (more often) imperfect (then plants dioecious, monoecious, or polygamous), hypogynous, sessile, small. **Calyx** of usually 3 sepals, distinct to connate, imbricate, often persistent. **Corolla** of 3 petals, distinct to connate, leathery or fleshy, inconspicuous, green and sepaloid to yellow, pink, brown, or white, imbricate or valvate. **Androecium** of generally 6 stamens, biseriate; filaments distinct to variously connate and/or adnate to petals; anthers basifixed or dorsifixed, dehiscing by longitudinal slits, introrse, latrorse, or extrorse; staminodes sometimes present in carpellate flowers. **Gynoecium** apocarpous and 1- to 3-carpellate, to syncarpous and 3-carpellate; ovary (ovaries) superior, 1- or 3 locular; ovules 1 in each locule, usually anatropous, placentation basal, apical, or axile; style(s) 1 (connate) or 3 (distinct), long to not clearly differentiated; stigmas erect or recurved; rudimentary pistil sometimes present in staminate flowers. **Fruit** usually a drupe or nut-like, with fleshy, fibrous, or leathery mesocarp; seed 1; endosperm copious, homogeneous or ruminate, oily or sometimes hard; embryo minute.

Family characterization: Trees or shrubs with well-developed, usually unbranched trunks; terminal cluster of extremely large, coriaceous, flabellate or pinnately compound, plicate leaves with expanded and persistent basal sheaths; numerous small flowers in spathaceous, highly branched inflorescences; and baccate fruit (usually a drupe). Pollen grains generally monocolpate (Sowunmi 1972). Tissues with tannins, calcium oxalate crystals (raphides), and silica. Anatomical feature: stems with an unusual apical meristem (which is primarily leaf-producing), a primary thickening meristem (which causes an increase in diameter), and diffuse secondary growth (Tomlinson 1961, 1964).

Genera/species: 200/2,780

Distribution: Primarily in tropical to subtropical regions, and also extending into warm temperate zones.

Major genera: *Calamus* (370 spp.), *Bactris* (239 spp.), *Pinanga* (120 spp.), *Daemonorops* (115 spp.), *Licuala* (108 spp.), and *Chamaedorea* (100 spp.) (Uhl and Dransfield 1987)

U.S./Canadian representatives: 16 genera/59 spp.; largest genera: *Pritchardia*, *Coccothrinax*, and *Sabal*

Economic plants and products: Food plants: *Areca* (betel-nut, palm heart), *Cocos* (coconut, copra, oil), *Elaeis* (oil), and *Phoenix* (dates). Commercial products from *Calamus* (rattan), *Copernicia* (carnauba wax), *Phytelepha* (vegetable-ivory), and *Raphia* (raffia). Ornamental trees (species of 120 genera), including *Caryota* (fishtail palm), *Chamaerops* (European fan palm), *Cocos* (coconut), *Livistona* (Chinese fan palm, fountain palm), *Phoenix* (date palm), *Roystonea* (royal palm), *Sabal* (cabbage palm), *Syagrus* (queen palm), and *Washingtonia* (California fan palm).

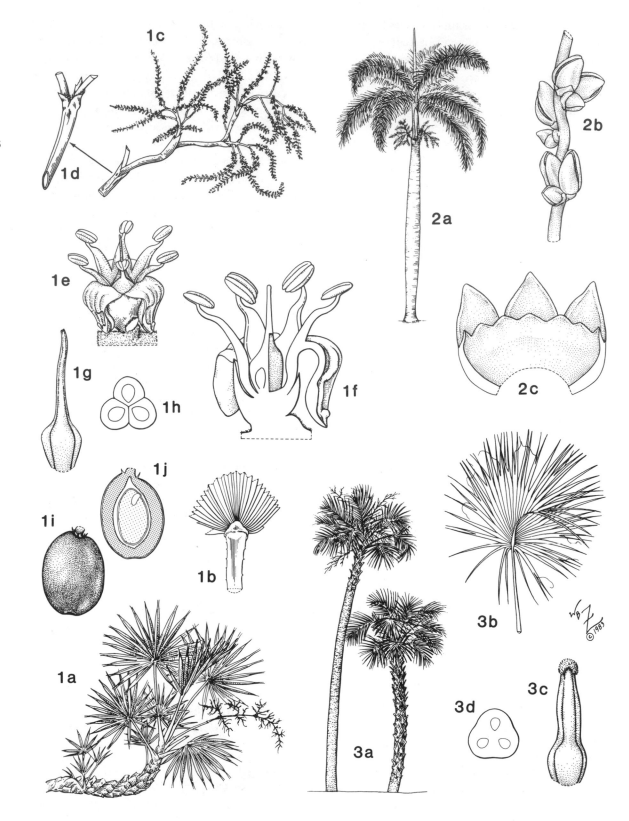

Figure 142. Arecaceae. 1, *Serenoa repens*: **a,** habit, much reduced; **b,** adaxial side of leaf blade base showing hastula, ×½; **c,** inflorescence, ×⅛; **d,** prophyll, ×⅕; **e,** flower, ×6; **f,** longitudinal section of flower, ×10; **g,** pistil, ×12; **h,** cross section of ovary, ×18; **i,** drupe, ×1¼; **j,** longitudinal section of drupe, ×1¼. **2,** *Roystonea regia*: **a,** habit, much reduced; **b,** portion of inflorescence showing clusters of two (larger) staminate flowers and one (smaller) carpellate flower, ×3; **c,** expanded corolla and androecium (staminodial) of carpellate flower, ×5. **3,** *Sabal palmetto*: **a,** habit, much reduced; **b,** abaxial view of leaf (costapalmate), ×1/60; **c,** pistil, ×9; **d,** cross section of ovary, ×12.

Commentary: Due to their large size and often inaccessible tropical distribution, palms often are poorly represented in herbarium collections (Dransfield 1986). For these reasons, the family has not been favored by much detailed taxonomic study in comparison to other important groups. Several schemes have been devised to divide the diverse palm family into tribes, subfamilies, and even families (see Moore 1961, 1973; Dransfield and Uhl 1986). The major groupings of palms are distinguished primarily on leaf characters, as well as on the presence of perfect versus imperfect flowers and features of the fruit (Moore and Uhl 1982; Uhl and Dransfield 1987).

Palm leaves exhibit a remarkable range of morphological diversity and also are characterized by a unique type of segmentation. Dissection of the palm leaf blade into segments is a very complicated process that involves differential growth in the lamina meristem (see Dengler et al. 1982; Tomlinson 1990). The leaflets or segments of dissected or compound leaves in other flowering plants ("dicots" and some monocots) are formed by the marginal lobing of the primordial leaf; in palms, leaflets result from the complex folding of the surface of a growing blade, followed by the separation of these plications. Palm leaves often are very large; for example, those of *Raphia* are over 25 m long—the largest in the plant kingdom. Blades are plicate (folded or corrugated) and thick-cuticled. The leaf segments often have pointed apices and spines along the margins, midribs, and petiole. Palm leaves usually are either palmate (fan palms; Fig. 142: 1a) or pinnate (feather palms; Fig. 142: 2a), although many are the intermediate costapalmate type (Fig. 142: 3b; Tomlinson 1990). In fan-leaved palms, the adaxial side of the petiole is prolonged into a scale-like projection (hastula, ligule) at the junction with the blade (Fig. 142: 1b). Principal classification features of the feather palms include the induplicate (V-shaped in cross section) or reduplicate (∧-shaped) folding of the pinnae, as well as the presence or absence of a terminal leaflet.

The "trunks" or stems of palms also are distinctive; they can vary in orientation (prostrate to erect) and in diameter (pencil-thin to 1 m). Each stem has a single growing point, and branching is rare. The "woody" and perennial trunk has no permanent cambium and actually is composed of many interconnected vascular bundles and fibers within softer ground tissues. The initial thickening of the shoot apex is caused by an increase in the amount of ground parenchyma produced by divisions of the primary thickening meristem, a mantle-like zone beneath the leaf primordia. Subsequent gradual increases in diameter are caused by enlargement and divisions of certain ground parenchyma cells, resulting in "diffuse secondary growth" (Tomlinson 1961, 1964, 1990). The stem may be further strengthened internally by fibers associated with the bundles, and externally by buttress roots and persistent fibrous leaf sheaths that do not decay readily.

Palm flowers are perfect or imperfect, and the fusion of the floral parts varies considerably among genera (Uhl and Dransfield 1987). For example, the perianth may be fused basally, forming a shallow cupule (as in *Serenoa*), or the sepals and petals may be distinct (*Cocos*). The stamens are variously connate and/or adnate to the corolla: in *Chamaerops*, for example, the fused filaments compose a thick-walled cup. When comprising more than one carpel, the gynoecium may be apocarpous (*Rhapidophyllum*, three distinct carpels) or completely syncarpous (*Sabal*; Fig. 142: 3c,d), or the carpels may be distinct at the base and joined at the styles (*Serenoa*; Fig. 142: 1g,h).

Although early monographers assumed that many palms were anemophilous, the flowers actually are predominantly entomophilous. Common insect vectors include beetles, Hymenoptera, and flies; bats and hummingbirds also have been noted (Henderson 1986). The small flowers are aggregated into conspicuous inflorescences. The flowers of many species have nectaries between the carpels (septal nectaries; Uhl and Moore 1971) and often have a sweet or musky odor. Some inflorescences emit heat (e.g., *Bactris* spp.).

Palm fruits characteristically are one-seeded berries or drupes (two carpels abort). The size may vary from pea-sized (*Euterpe*) to the large coconut (*Cocos*). Most palm fruits are dispersed by mammals and birds (Zona and Henderson 1989). The best-known palm fruit, the coconut (*Cocos*), is a "fibrous drupe" with a thin and smooth exocarp and a well-developed fibrous mesocarp. The thick and hard endocarp, the layer with the three "eyes," surrounds the thin seed coat, which is lined with a layer of solid endosperm (the edible part of the fruit). The central portion of the seed is filled with potable liquid.

REFERENCES CITED

Dengler, N. G., R. E. Dengler, and D. R. Kaplan. 1982. The mechanism of plication inception in palm leaves: Histogenic observations on the pinnate leaf of *Chrysalidocarpus lutescens. Canad. J. Bot.* 60:2976–2998.

Dransfield, J. 1986. A guide to collecting palms. *Ann. Missouri Bot. Gard.* 73:166–176.

—— and N. W. Uhl. 1986. An outline of the classification of palms. *Principes* 30:3–11.

Henderson, A. 1986. A review of pollination studies in the palms. *Bot. Rev.* (Lancaster) 52:221–259.

Moore, H. E. 1961. Botany and the classification of palms. *Amer. Hort. Mag.* 40:17–26.

——. 1973. The major groups of palms and their distribution. *Gentes Herb.* 11:27–141.

—— and N. W. Uhl. 1982. The major trends of evolution in palms. *Bot. Rev.* (Lancaster) 48:1–69.

Sowunmi, M. A. 1972. Pollen morphology of the Palmae and its bearing on taxonomy. *Rev. Palaeobotan. Palynol.* 13: 1–80.

Tomlinson, P. B. 1961. *Palmae*, vol. 2 of *Anatomy of the monocotyledons*, ed. C. R. Metcalfe, pp. 10–23. Oxford University Press, Oxford.

——. 1964. Stem structure in arborescent monocotyledons. In *The formation of wood in forest trees*, ed. M. H. Zimmerman, pp. 65–86. Academic Press, London.

——. 1990. *The structural biology of palms*, pp. 54–59, 219–230, 245–258. Clarendon Press, Oxford.

Uhl, N. W., and J. Dransfield. 1987. *Genera palmarum*, pp. 22–25, 66–73. L. H. Bailey Hortorium and International Palm Society, Ithaca, N.Y.

—— and H. E. Moore. 1971. The palm gynoecium. *Amer. J. Bot.* 58:945–992.

Zona, S., and A. Henderson. 1989. A review of animal-mediated seed dispersal in palms. *Selbyana* 11:6–21.

BROMELIACEAE
Bromeliad or Pineapple Family

Perennial herbs, epiphytic or sometimes terrestrial xerophytes, scapose; stems usually reduced or absent; roots adventitious. **Leaves** simple, entire or spinose-serrate, alternate and spirally arranged, clustered and forming basal rosettes, parallel-veined, sessile, strap-shaped and often trough-like, stiff, often reddish or purplish at base, with basal sheaths often overlapping and forming a cup. **Inflorescence** indeterminate, spicate, racemose, paniculate, or capitate, usually terminal. **Flowers** actinomorphic or slightly zygomorphic, usually perfect, hypogynous to (more often) epigynous, often showy, each usually in the axil of a brightly colored bract. **Calyx** of 3 sepals, distinct or basally connate, convolute and/or imbricate, persistent. **Corolla** of 3 sepals, distinct or basally connate, frequently with a pair of scale-like appendages at base, often brightly colored, convolute and/or imbricate. **Androecium** of 6 stamens, biseriate, often epipetalous; filaments distinct or basally connate; anthers basifixed or dorsifixed, versatile, dehiscing longitudinally, introrse. **Gynoecium** of 1 pistil, 3-carpellate; ovary inferior to sometimes superior, 3-locular; ovules usually numerous, anatropous, placentation axile; style 1; stigmas 3, often spirally twisted (at least in bud). **Fruit** a septicidal capsule or a berry (sometimes becoming a multiple fruit by connation of berries and axes of adjacent flowers); seeds often winged, caudate, or with plumose appendage; endosperm copious, mealy; embryo small, usually peripheral at base of endosperm.

Family characterization: Epiphytic scapose herbs with reduced stems and adventitious roots; strap-like, stiff, usually concave leaves often with colored and sheathing bases; 3-merous flowers subtended by bracts; spirally twisted stigmas; and seeds with plumose appendages. Anatomical features: mucilage cells and raphide sacs (in all organs), silica bodies (in epidermal cells), and tannin granules (in parenchymatous cells of shoot); and vesture of peltate scales (as well as numerous other adaptations of the leaves for water retention and absorption; see the discussion below).

Genera/species: 51/1,520

Distribution: Almost exclusively in tropical to warm temperate America (except for one species of *Pitcairnia*, which is African).

Major genera: *Tillandsia* (411–500 spp.), *Pitcairnia* (250–263 spp.), *Vriesea* (190–260 spp.), and *Aechmea* (150–172 spp.)

U.S./Canadian representatives: 10 genera/42 spp.; largest genera: *Tillandsia* and *Catopsis*

Economic plants and products: Edible fruits from *Ananas comosa* (pineapple). Cordage and fiber for fabric from several, such as species of *Aechmea* and *Ananas*. "Vegetable hair" (dried stems and leaves) used for upholstery stuffing from *Tillandsia* (Spanish-moss). Ornamental plants (species of 33 genera), including *Aechmea* (air-pine, living-vase), *Billbergia* (vase plant), *Bromelia* (pinquin), *Cryptanthus* (earthstar), *Guzmania*, *Nidularium*, *Pitcairnia*, *Tillandsia*, and *Vriesea*.

Commentary: The Bromeliaceae, a very distinctive and apparently monophyletic family, usually are divided into three or four subfamilies (see Dahlgren et al. 1985) primarily based upon features of the habit (terrestrial or epiphytic), ovary (superior to inferior), fruit (capsule or berry), and seeds (type of appendages). The family has been allied with various taxa, including the Zingiberales, Commelinales (Commelinaceae and allies), and Pontederiaceae and allies. Recently, molecular and morphological cladistic studies (e.g., Gilmartin and Brown 1987; Ranker et al. 1990) have examined the relationships of the subfamilies, as well as the phylogenetic position of the family, which still are not well resolved.

Bromeliads are xerophytes, and the majority are epiphytic. As with many epiphytes (as well as succulents of arid regions), species of several genera (e.g., *Billbergia*, *Tillandsia*) have the capacity for crassulacean acid metabolism (CAM), a specialized variant of the C_4 photosynthetic pathway (Medina 1974; Benzing 1980). The typical "tank-type" plant has a shortened axis tightly clasped by the well-developed leaf sheaths

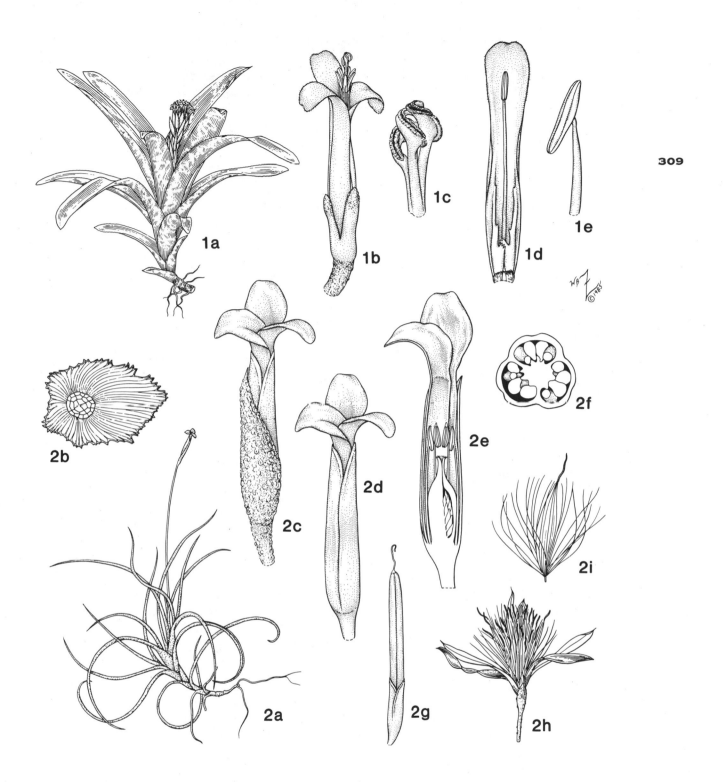

Figure 143. Bromeliaceae. 1, *Billbergia pyramidalis*: **a,** habit, ×¹⁄₁₀; **b,** flower, ×1¼; **c,** style apex showing twisted stigmas, ×6; **d,** adaxial view of petal showing adnate stamen and paired appendages, ×1½; **e,** anther, ×5. **2,** *Tillandsia recurvata*: **a,** habit, ×²⁄₃; **b,** top view of peltate scale from leaf, ×45; **c,** flower with subtending bract, ×6; **d,** flower, ×6; **e,** longitudinal section of flower, ×7; **f,** cross section of ovary, ×18; **g,** capsule before dehiscence, ×2; **h,** dehisced capsule, ×1¼; **i,** seed, ×1½.

(Fig. 143: 1a), thus forming a basin that collects water as well as decayed material. In most species, the leaf is the primary organ for absorbing water. The primary root usually is short-lived; the adventitious roots (developing at the leaf bases) function mainly to attach the plant to the substrate. Water and (often) dissolved organic compounds are absorbed by specialized peltate scales (Fig. 143: 2b) that occur within the leaf sheath or, in the most derived species (e.g., *Tillandsia* spp.), over the entire surface of the plant. Each trichome generally consists of a uniseriate stalk plus an expanded shield-like apex, which is composed of dead cells at maturity. When moistened, the empty "shield cells" expand, and water is drawn osmotically through the living stalk cells to the mesophyll (Tomlinson 1969; Benzing 1980). The leaves typically have water-storing parenchyma (water-storage tissue) between the assimilating mesophyll and the epidermis on the adaxial side. A thick cuticle helps reduce water loss, and in some bromeliads (*Tillandsia*) the dry and collapsed scales also provide some insulation (Smith and Wood 1975).

The flowers are protandrous, with the stigmas twisted spirally into a head at anthesis (Fig. 143: 1c). This "conduplicate-spiral" architecture probably is unique within the monocotyledons (Brown and Gilmartin 1989). After the pollen is shed, the stigmas expand to expose the receptive surfaces. Insects and birds (hummingbirds) are the principal pollinators, seeking nectar secreted by the septal nectaries (in the ovary) and sometimes by the appendages of the corolla. The flowers of many species also have an attractive and/or strong scent.

The fruit is either a capsule (*Tillandsia*; Fig. 143: 2g,h) or a berry (*Bromelia*). In *Ananas* (pineapple), the whole fleshy inflorescence (axis, bracts, and berries) forms a succulent multiple fruit. The apical crown of leaves represents the growth of the axis beyond the inflorescence.

REFERENCES CITED

Benzing, D. 1980. *The biology of the bromeliads*, pp. 58–77, 103–114. Mad River Press, Eureka, Calif.

Brown, G. K., and A. J. Gilmartin. 1989. Stigma types in the Bromeliaceae—A systematic study. *Syst. Bot.* 14:110–132.

Dahlgren, R. M. T., H. T. Clifford, and P. F. Yeo. 1985. *The families of the monocotyledons*, pp. 329–333. Springer-Verlag, Berlin.

Gilmartin, A. J., and G. K. Brown. 1987. Bromeliales, related monocots, and resolution of relationships among Bromeliaceae subfamilies. *Syst. Bot.* 12:493–500.

Medina, E. 1974. Dark CO_2 fixation, habitat preference and evolution within the Bromeliaceae. *Evolution* 28:677–686.

Ranker, T. A., D. E. Soltis, and A. J. Gilmartin. 1990. Subfamilial phylogenetic relationships of the Bromeliaceae: Evidence from chloroplast DNA restriction site variation. *Syst. Bot.* 15:425–434.

Smith, L. B., and C. E. Wood. 1975. The genera of Bromeliaceae in the southeastern United States. *J. Arnold Arbor.* 56:375–397.

Tomlinson, P. B. 1969. *Commelinales—Zingiberales*, vol. 3 of *Anatomy of the monocotyledons*, ed. C. R. Metcalfe, pp. 193–294. Oxford University Press, Oxford.

PONTEDERIACEAE
Pickerelweed or Water-hyacinth Family

Perennial or sometimes annual herbs, aquatic, floating or rooted in substrate and emergent, with rhizomes or stolons; stems sympodial, often spongy, short or erect, usually enveloped by sheathing leaf bases; roots fibrous. **Leaves** simple, entire, opposite or whorled, distichous, in basal rosettes or cauline, parallel-veined, lanceolate to broadly ovate, with fleshy and sheathing petiole (enveloping the stem). **Inflorescence** determinate, appearing racemose, spicate, or paniculate, usually subtended by a modified leaf (often reduced to a spathe-like sheath), terminal. **Flowers** actinomorphic to zygomorphic, perfect, hypogynous, showy, often heterostylous (trimorphic), frequently with septal nectaries. **Perianth** of 6 tepals in 2 whorls of 3, petaloid,

310

Figure 144. Pontederiaceae. *Pontederia cordata*: **a**, habit, ×⅙; **b**, inflorescence, ×⅓; **c–f**, long-styled flower: **c**, long-styled flower, ×2½; **d**, upper half of perianth and epitepalous androecium (short stamens), ×2½; **e**, lower half of perianth and epitepalous androecium (medium-length stamens), ×2½; **f**, pistil (long-styled), ×4; **g–j**, medium-styled flower: **g**, medium-styled flower, ×2½; **h**, upper half of perianth and epitepalous androecium (short stamens), ×2½; **i**, lower half of perianth and epitepalous androecium (long stamens), ×2½; **j**, pistil (medium-styled), ×4; **k–n**, short-styled flower: **k**, short-styled flower, ×2½; **l**, upper half of perianth and epitepalous androecium (medium-length stamens), ×2½; **m**, lower half of perianth and epitepalous androecium (long stamens), ×2½; **n**, pistil, (short-styled), ×4; **o**, longitudinal section of medium-styled flower, ×4½; **p**, cross section of ovary, ×25; **q**, short stamen, ×7; **r**, medium-length stamen, ×7; **s**, long stamen, ×7; **t**, pollen grain from short stamen, ×260; **u**, pollen grain from medium-length stamen, ×260; **v**, pollen grain from long stamen, ×260; **w**, fruit (nutlet enclosed by accrescent perianth tube), ×4½; **x**, longitudinal section of fruit, ×4½; **y**, nutlet (accrescent perianth removed), ×5. For diagrammatic representations of legitimate crosses among tristylous flowers of *Pontederia cordata*, see Fig. 146.

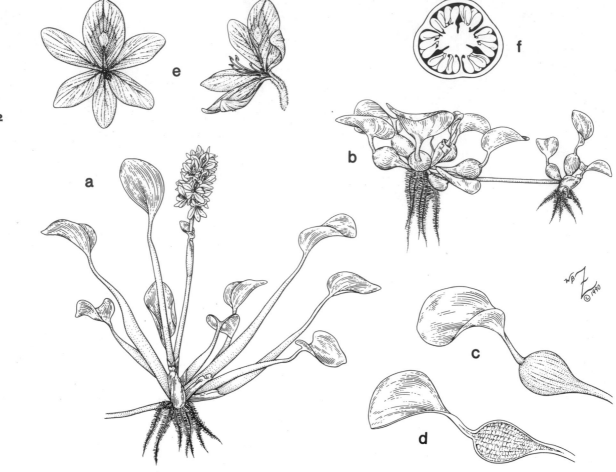

Figure 145. Pontederiaceae (continued). *Eichhornia crassipes*: **a,** flowering plant, ×1/10; **b,** vegetative plants connected by stolon, ×1/3; **c,** leaf from vegetative plant, ×1/2; **d,** longitudinal section of leaf from vegetative plant, ×1/2; **e,** two views of flower, ×1/2; **f,** cross section of ovary, ×10.

unequal with median inner tepal more differentiated (in zygomorphic flowers), distinct or usually basally connate (then forming a tube), funnelform to salverform, frequently blue, lilac, or white, sometimes persistent, imbricate. **Androecium** of typically 6 stamens, epitepalous and inserted at unequal levels within the perianth tube, usually unequal in length, size, and/or form; filaments distinct; anthers basifixed, versatile, dehiscing longitudinally, introrse. **Gynoecium** of 1 pistil, basically 3-carpellate with sometimes 2 carpels reduced; ovary superior, 3- or sometimes 1-locular; ovules numerous or sometimes solitary and pendulous (in 1-locular ovary), anatropous, placentation axile or sometimes parietal on intruded placentae (1-locular ovary); style 1, stigma 3-lobed or capitate. **Fruit** a loculicidal capsule or a nutlet (enclosed by warty, accrescent perianth tube); seeds small, ovoid, often ribbed longitudinally; endosperm copious, mealy,

starchy; embryo small, linear, straight, surrounded by endosperm.

Family characterization: Perennial, freshwater aquatic herbs with sheathing leaves enveloping the sympodial stems; terminal inflorescence subtended by a reduced leaf (often a spathe-like sheath); 6-merous, basally connate, petaloid perianth; 6 unequal, epitepalous stamens inserted at different levels within the perianth tube; and a nutlet enclosed by accrescent perianth tube or a capsule as the fruit type. Tissues with tannins and calcium oxalate crystals (commonly raphides).

Genera/species: 8/30 (Thorne pers. comm.)

Distribution: Widespread in the subtropics and tropics, with a few representatives extending into the northern temperate zone; most diverse in tropical America.

Legitimate Crosses – Trimorphic Heterostyly

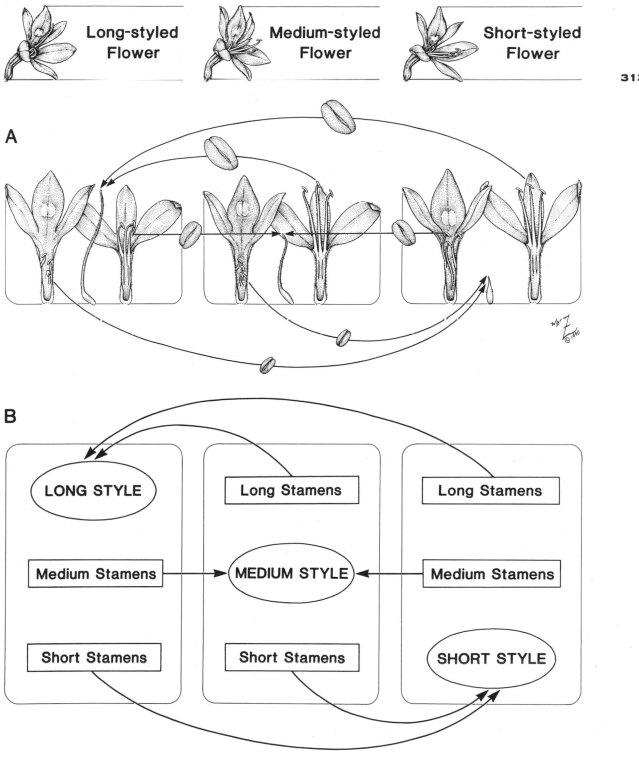

Long-styled Flower

Medium-styled Flower

Short-styled Flower

A

B

LONG STYLE

Long Stamens

Long Stamens

Medium Stamens

MEDIUM STYLE

Medium Stamens

Short Stamens

Short Stamens

SHORT STYLE

Figure 146. Legitimate crosses in tristylous flowers. Diagrammatic representations of legitimate crosses among tristylous flowers of *Pontederia cordata*: flowers, ×1¼; **A,** Illustrations of pertinent flower parts involved in legitimate crosses: expanded perianths and epipetalous androecia, ×2; pistils, ×2; pollen grains, ×150; **B,** Graphic representation of crosses illustrated in **A.**

Major genera: *Heteranthera* (10 spp.) and *Eichhornia* (7 spp.)

U.S./Canadian representatives: *Heteranthera* (5 spp.), *Eichhornia* (4 spp.), *Monochoria* (1 sp.), and *Pontederia* (1 sp.)

Economic plants and products: Several serious aquatic weeds, especially *Eichhornia crassipes*, as well as species of *Heteranthera* and *Pontederia*. Species of *Eichhornia* (water-hyacinth), *Heteranthera* (mud-plantain), *Monochoria*, and *Pontederia* (pickerelweed) cultivated as ornamentals for pools and ponds.

Commentary: The position of the Pontederiaceae in phylogenetic schemes has varied considerably: they have been allied with such diverse groups as the Bromeliaceae (Dahlgren et al. 1985; Thorne 1992) and the Liliaceae (Cronquist 1988). Two monotypic genera (*Eurystemon* and *Zosterella*) sometimes are segregated from *Heteranthera s.l.*, and *Reussia* (two or three species), from *Pontederia s.l.*

These freshwater aquatic herbs are characterized by spongy (aerenchymatous) sympodial stems. Leaf blades vary from linear to orbicular, sometimes with a sagittate or cordate base. The petioles are fleshy and inflated, and the sheathing bases more or less envelop the stem. In *Eichhornia*, for example, the leaf petioles of the vegetative clumps (Fig. 145: b) are bladder-like and filled with porous aerenchymatous tissue (Fig. 145: c,d). Each successive axis of flowering plants is terminated by an inflorescence subtended by a modified leaf, as in *Eichhornia* (Fig. 145: a). Often this bract lacks a blade and is reduced to a spathe-like leaf sheath (e.g., *Pontederia*, Fig. 144: b, and *Heteranthera*). Perennial species also reproduce vegetatively by fragmentation of rhizomes or stolons, as for example the plants of *Eichhornia crassipes* (a serious tropical aquatic weed), which form extensive radiating systems of stolons and vegetative rosettes (Fig. 145: b).

Several species of the family (*Eichhornia* and *Pontederia*) are heterostylous, an outcrossing mechanism discussed under the Rubiaceae. However, the flowers of these Pontederiaceae are trimorphic (three style lengths: three flower types), a condition evidently found in only three families: the Pontederiaceae, Lythraceae, and Oxalidaceae (Vuilleumier 1967; Ganders 1979; Weller 1992). Dimorphic heterostyly (two styles/flower types) is a much more common form of flower heteromorphism (see Rubiaceae, and also Figs. 52: e–k; 102: 1g–n; 107: e,f,k). In trimorphic heterostyly (three "different styles"), flowers of the same species have short, medium, or long mature style lengths (short-, medium-, and long-styled flowers, respectively; Richards and Barrett 1992). Each style length occurs in a flower with two particular sets of filament lengths (in the biseriate androecium): long styles with short + medium filaments, medium styles with short + long filaments, and short styles with medium + long filaments (see Figs. 144: c–n and 146; Ornduff 1966; Price and Barrett 1984). Other morphological differences between flower morphs in the Pontederiaceae include the pollen grain size, associated with a physiological self-incompatibility. Ideally, a legitimate cross resulting in optimum fertility (see Fig. 146) may occur only between flowers with stamens and styles in a similar position (e.g., pollen from a "long" stamen with long-styled stigma), although a few species with weakly developed pollen trimorphism (e.g., *Eichhornia crassipes*) also are self-compatible (Barrett 1977). In these species, one flower form frequently constitutes a population. The genetics of heterostyly in the Pontederiaceae has been studied extensively (Barrett et al. 1983; Lewis and Jones 1992) and probably is controlled by two diallelic loci. Cladistic studies indicate that tristyly probably arose only once within the family (Eckenwalder and Barrett 1986).

Various insect pollinators (bees, flies, Lepidoptera) visit for the nectar secreted by septal nectaries and/or to collect pollen. The enlarged upper (inner) tepal of *Eichhornia* (Fig. 145: e) and *Pontederia* (Fig. 144: c,g,k) flowers functions as a nectar guide, with streaks of violet-blue and/or a large yellow splotch. In the tristylous *Pontederia* and *Eichhornia* species, legitimate crosses may vary geographically due to the spread of the species to areas where the specialized pollinators are absent (Barrett 1980). The submerged flowers of some species, such as those of *Heteranthera dubia*, are cleistogamous.

The fruit of the family commonly is a capsule, but the pistil of *Pontederia*, with one fertile, uniovulate locule (Fig. 144: p), develops into a nutlet (sometimes referred to as a "utricle") that is enclosed by the accrescent, buoyant, and warty perianth tube (Fig. 144: w–y; Rosatti 1987).

REFERENCES CITED

Barrett, S. C. H. 1977. Tristyly in *Eichhornia crassipes* (Mart.) Solms (water hyacinth). *Biotropica* 9:230–238.

———. 1980. Sexual reproduction in *Eichhornia crassipes* (water hyacinth). II. Seed production in natural populations. *J. Appl. Ecol.* 17:113–124.

———, S. D. Price, and J. S. Shore. 1983. Male fertility and anisoplethic population structure in tristylous *Pontederia cordata* (Pontederiaceae). *Evolution* 37:745–759.

Cronquist, A. 1988. *The evolution and classification of flowering plants*, 2d ed., pp. 491–497. New York Botanical Garden, Bronx.

Dahlgren, R. M. T., H. T. Clifford, and P. F. Yeo. 1985. *The families of the monocotyledons*, pp. 323–325. Springer-Verlag, Berlin.

Eckenwalder, J. E., and S. C. H. Barrett. 1986. Phylogenetic systematics of Pontederiaceae. *Syst. Bot.* 11:373–391.

Ganders, F. R. 1979. The biology of heterostyly. *New Zealand J. Bot.* 17:607–635.

Lewis, D., and D. A. Jones. 1992. The genetics of heterostyly. In *Evolution and function of heterostyly*, ed. S. C. H. Barrett, pp. 129–159. Monogr. Theor. & Appl. Genet. 15. Springer-Verlag, Berlin.

Ornduff, R. 1966. The breeding system of *Pontederia cordata* L. *Bull. Torrey Bot. Club* 93:407–416.

Price, S. D., and S. C. H. Barrett. 1984. The function and adaptive significance of tristyly in *Pontederia cordata* L. (Pontederiaceae). *Bot. J. Linn. Soc.* 21:315–329.

Richards, J. H., and S. C. H. Barrett. 1992. The development of heterostyly. In *Evolution and function of heterostyly*, ed. S. C. H. Barrett, pp. 85–127. Monogr. Theor. & Appl. Genet. 15. Springer-Verlag, Berlin.

Rosatti, T. J. 1987. The genera of Pontederiaceae in the southeastern United States. *J. Arnold Arbor.* 68:35–71.

Thorne, R. F. 1992. Classification and geography of the flowering plants. *Bot. Rev.* (Lancaster) 58:225–348.

Vuilleumier, B. S. 1967. The origin and evolutionary development of heterostyly in the angiosperms. *Evolution* 21:210–226.

Weller, S. G. 1992. Evolutionary modifications of tristylous breeding systems. In *Evolution and function of heterostyly*, ed. S. C. H. Barrett, pp. 247–272. Monogr. Theor. & Appl. Genet. 15. Springer-Verlag, Berlin.

HAEMODORACEAE
Bloodwort or Redroot Family

Perennial herbs, with rhizomes and stolons, with reddish-orange sap; roots fibrous. **Leaves** simple, alternate and mostly basal (cauline leaves reduced), distichous, equitant, parallel-veined, narrow, conduplicate with edges becoming fused above, sheathing at base. **Inflorescence** determinate, cymose and appearing umbellate, corymbose, paniculate, or racemose, densely villose, terminal. **Flowers** actinomorphic to zygomorphic, perfect, hypogynous to epigynous, subtended by conspicuous bracts. **Perianth** of 6 tepals, uniseriate or biseriate, unequal in size with outer tepals shorter and narrower than inner tepals, distinct (erect-spreading) to connate into a tube, villose on abaxial surface, variously colored, valvate, persistent. **Androecium** of 3 or 6 stamens, epipetalous; filaments distinct, enclosed by inrolled margins of inner tepals; anthers basifixed (sagittate at base) or dorsifixed, dehiscing longitudinally, introrse. **Gynoecium** of 1 pistil, 3-carpellate; ovary superior to inferior, 3-locular, with septal nectaries; ovules few in each locule, orthotropous, placentation axile on margins of peltate placenta or in vertical rows; style 1, persistent; stigma 1, capitate

or trifid. **Fruit** a loculicidal capsule; seeds discoid or tetrahedral; endosperm copious, fleshy, starchy; embryo small, globose to ovoid, embedded in endosperm.

Family characterization: Perennial, rhizomatous herbs with reddish sap growing in acid soils in wet habitats; equitant, linear to ensiform, conduplicate leaves with folded portions becoming fused above; terminal, corymbose inflorescence with woolly tomentum; epigynous flowers with 6, fuzzy, pale yellow tepals; 3 epitepalous stamens opposite to and enclosed by the inner tepals; ovules inserted on margin of enlarged, peltate placenta; and globose, loculicidal capsule with discoid seeds. Chelidonic acid present. Red pigment (haemocorin, a glycoside) in roots and rhizomes. Anatomical feature: raphide sacs (which also contain mucilage).

Genera/species: 14/100

Distribution: Primarily in the Southern Hemisphere (Australasia [excluding New Zealand], South Africa, and tropical America), with one species in eastern North America; particularly diverse in Australia.

Major genera: *Conostylis* (23–35 spp.) and *Haemodorum* (20 spp.)

U.S./Canadian representative: *Lachnanthes caroliniana*

Economic plants and products: Red dyes and herbal remedies extracted from roots and rhizomes of *Lachnanthes caroliniana* (dyeroot, paintroot). Ornamental plants (species of about 7 genera), including *Anigozanthos* (kangaroo-paw), *Conostylis*, *Lachnanthes* (redroot), and *Xiphidium*.

Commentary: The circumscription of the Haemodoraceae has varied with different authors, with genera such as *Lophiola* and *Aletris* (both transferred to the Melanthiaceae) and *Sansevieria* (Dracaenaceae) having been included (see Simpson 1990 for a complete summary). *Lophiola* still is commonly associated with the Haemodoraceae in floras (e.g., Robertson 1976). As circumscribed here with 14 genera, the Haemodoraceae are monophyletic (Simpson 1990). Dahlgren et al. (1985) divide the family into two subfamilies (the Haemodoroideae and Conostyloideae); the latter group is restricted to western Australia. *Lachnanthes caroliniana* (the only representative of the family in our area; Fig. 147) is characterized by corymbose inflorescences, epigynous flowers, more or less distinct biseriate tepals, three stamens, and ovules on an enlarged peltate placenta. However, the characterization for all members of the family, which is diverse in Australia, South Africa, and South America, would expand those

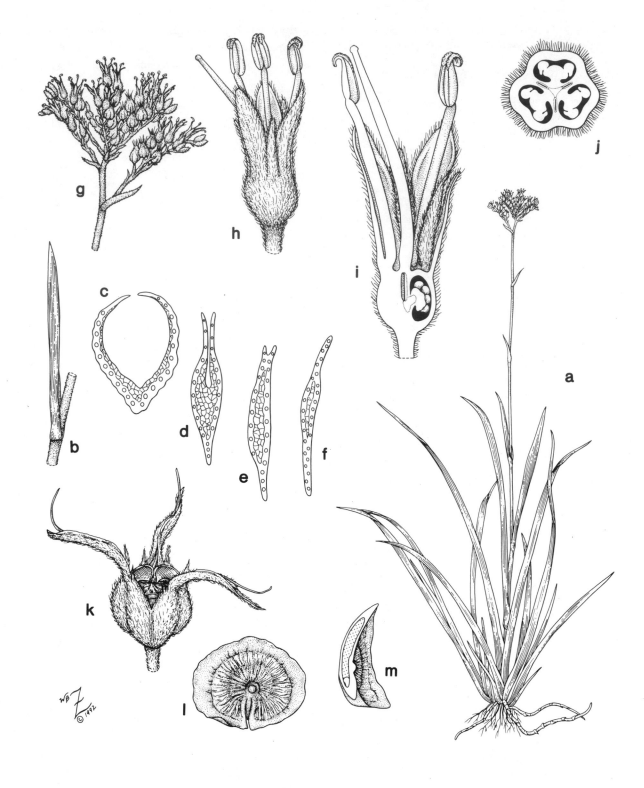

Figure 147. Haemodoraceae. *Lachnanthes caroliniana*: **a,** habit, ×⅙; **b,** node showing conduplicate leaf with adaxial surfaces fused at apex, ×½; **c,** cross section of leaf sheath (adaxial surfaces free), ×3 ½; **d,** cross section of leaf near middle (adaxial surfaces partially fused), ×3½; **e,** cross section of leaf above middle (adaxial surfaces almost totally fused), ×3½; **f,** cross section of leaf near apex (adaxial surfaces totally fused), ×3½; **g,** inflorescence, ×½; **h,** flower, ×3½; **i,** longitudinal section of flower, ×5; **j,** cross section of ovary, ×8; **k,** capsule with persistent tepals, ×5; **l,** seed, ×12; **m,** longitudinal section of seed, ×12.

character states to include racemose-paniculate inflorescences; hypogynous, zygomorphic flowers; uniseriate, fused tepals; six stamens; and ovules in vertical rows. The family sometimes has been included within the Liliaceae *s.l.* (e.g., Thorne 1976), but now it is generally allied with the Pontederiaceae and related families (Dahlgren et al. 1985; Thorne 1992), a placement supported by cladistic analysis of morphological and anatomical data (Simpson 1990).

The stout plants of *Lachnanthes caroliniana* grow in acid soils of wet habitats (swamps, ditches, bogs). The habit is somewhat scapose, but a few small leaves occur on the simple aerial stem that terminates in a corymbose inflorescence (Fig. 147: a,g). The folded parts of a cauline or a basal leaf (Fig. 147: b), sheathing below, become progressively fused together toward the apex of the blade (Fig. 147: c–f). The sap and rhizomes are reddish, and the outer tepals (Fig. 147: h,i), inflorescence axes (Fig. 147: g), and upper stems are covered with a woolly tomentum.

Little has been reported about the pollination of *Lachnanthes caroliniana* (although other members of the family are pollinated by insects and birds). Presumably, insects are attracted to the conspicuous hairy inflorescences with yellow flowers and collect the nectar produced by the septal glands.

The globose capsule with persistent tepals (Fig. 147: k) contains unusual discoid (coin-like) seeds (Fig. 147: l,m) that are peltately attached around the large placentae on the central column. The annular margin is winglike, and the seeds are wind dispersed.

REFERENCES CITED

Dahlgren, R. M. T., H. T. Clifford, and P. F. Yeo. 1985. *The families of the monocotyledons*, pp. 336–340. Springer-Verlag, Berlin.

Robertson, K. R. 1976. The genera of the Haemodoraceae in the southeastern United States. *J. Arnold Arbor.* 205–216.

Simpson, M. G. 1990. Phylogeny and classification of the Haemodoraceae. *Ann. Missouri Bot. Gard.* 77:722–784.

Thorne, R. F. 1976. A phylogenetic classification of the Angiospermae. *Pl. Syst. Evol.*, Suppl. 1:299–319.

———. 1992. Classification and geography of the flowering plants. *Bot. Rev.* (Lancaster) 58:225–348.

TYPHACEAE

Cat-tail Family

Including the Sparganiaceae: Burr-reed Family

Perennial herbs, aquatic and commonly emergent in shallow water, often forming large colonies; stems simple or branched, erect; rhizome sympodial, starchy. **Leaves** simple, entire, alternate and mostly basal, distichous, parallel-veined, linear, thick and somewhat spongy, sheathing at base. **Inflorescence** determinate, compound and complex, appearing spicate or of globose clusters in a racemose arrangement, unisexual with sexes superposed (staminate inflorescence[s] above carpellate), often subtended by a bract (spathe), axillary, terminal on stem. **Flowers** actinomorphic, imperfect (plants monoecious), hypogynous, reduced and inconspicuous, sessile, sometimes subtended by bracts; sterile carpellate flowers frequently present. **Perianth** of 3 or 4 sepaloid and bract-like tepals or of numerous long bristles or sometimes scales, distinct. **Androecium** of generally 1 to 6 stamens; filaments distinct or often variously basally connate; anthers basifixed, with pronounced connective separating the locules and extending beyond them, dehiscing longitudinally, extrorse. **Gynoecium** of 1 pistil, basically 2-carpellate with 1 carpel aborting, often appearing 1-carpellate (pseudomonomerous), commonly on gynophore; ovary superior, usually 1-locular; ovule 1, anatropous, pendulous, placentation apical; style usually 1, persistent, stigma linear to spatulate, decurrent on style. **Fruit** a follicle or a drupe; seed 1, with testa forming a lid; endosperm copious, in two parts ("chambers"; Dahlgren et al. 1985), mealy; embryo straight, cylindrical, enveloped by endosperm.

Family characterization: Perennial, rhizomatous herbs forming colonies in shallow water; distichous, mostly basal, linear leaves with sheathing bases; dense, unisexual, spicate or spherical inflorescences with the carpellate cluster(s) below the staminate; reduced perianth of scales or bristles; anthers with pronounced connective; 1-locular pistil with 1 ovule and decurrent stigma; and a follicle or drupe as the fruit type. Tissues commonly with tannins. Anatomical feature: raphide sacs (which also contain mucilage) in leaves and stems.

Genera/species: *Sparganium* (12–20 spp.) and *Typha* (10–20 spp.)

Distribution: Generally widespread, but absent from a few tropical regions (such as northern South America); typically growing along ditches or shores of freshwater lakes, streams, or swamps.

U.S./Canadian representatives: *Sparganium* (9 spp.) and *Typha* (4 spp.)

Economic plants and products: Matting and weaving material (leaves), food (edible pollen and rhizomes), kapok substitute (floss from seeds), and decorative components for dried floral arrangements (infructescence) from several species of *Typha*. Edible fruit (for wildfowl) and rhizomes from *Sparganium* species. Several species of *Sparganium* (burr-reed) and *Ty*-

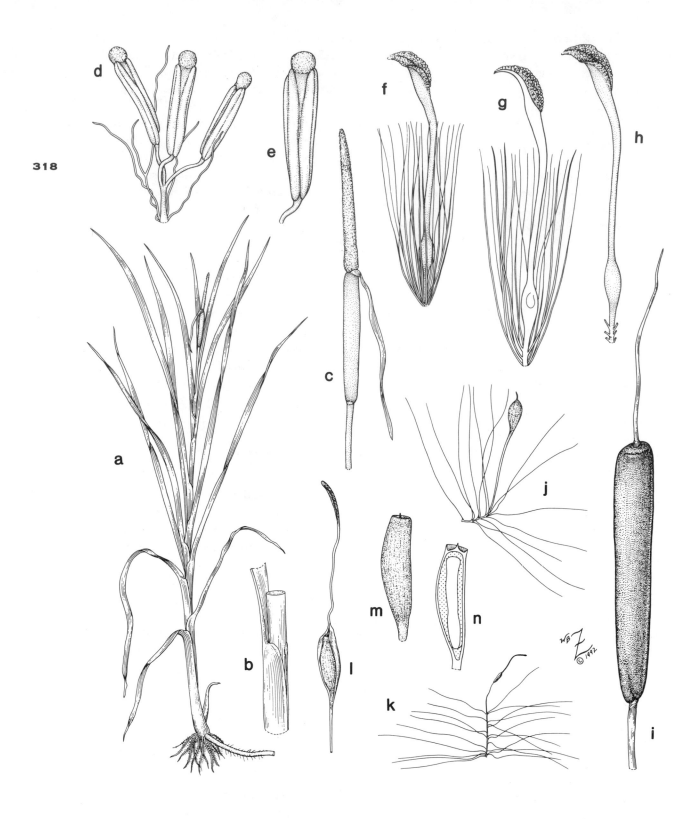

Figure 148. Typhaceae. *Typha latifolia*: **a,** habit, ×1/15; **b,** sheathing leaf base, ×1/2; **c,** inflorescence (staminate flowers above; carpellate, below), ×3/8; **d,** staminate flower, ×10; **e,** stamen, ×15; **f,** carpellate flower, ×20; **g,** longitudinal section of carpellate flower, ×25; **h,** pistil, ×25; **i,** infructescence, ×1/3; **j,** sterile, long-stipitate flower with aborted ovary (fruit) from young infructescence, ×4; **k,** fruit (follicle and persistent perianth), ×3; **l,** dehiscing follicle, ×15; **m,** seed, ×28; **n,** longitudinal section of seed, ×28.

pha (cat-tail, reedmace) also occasionally cultivated as ornamentals around pond margins.

Commentary: As defined by Thorne (1992), the Typhaceae include two monotypic subfamilies, Typhoideae (*Typhus*) and Sparganioideae (*Sparganium*). These taxa often are treated as two very closely related families constituting the order Typhales (Thieret 1982); however, several botanists besides Thorne (e.g., Bentham and Hooker 1883; Müller-Doblies and Müller-Doblies 1977) have submerged the Sparganiaceae within the Typhaceae *s.l.* Discerning the affinities of the Typhales (or Typhaceae *s.l.*) to other monocot groups has been problematic for taxonomists (see Bergner and Jensen 1989 for a summary). Recent analysis of molecular data, however, indicates a close relationship of this taxon to the Cyperales (Cyperaceae and Juncaceae; Duvall et al. 1993).

The characteristic inflorescences of the family terminate the erect stems and are unisexual, with the staminate cluster(s) above the carpellate. The reduced flowers are tightly packed into complex spicate (*Typha*; Fig. 148: c) or globose (*Sparganium*) units. In *Typha*, a complete inflorescence (spadix) consists of two superposed cylinders (male above female) that may be either contiguous or separated on the single stem. Each portion usually is subtended by a bract (spathe), although this is lacking for the carpellate inflorescences of a few species (e.g., *T. latifolia*; Fig. 148: c). The staminate flowers (Fig. 148: d) appear to arise directly on the axis, while the carpellate flowers (Fig. 148: f) are crowded on brief lateral outgrowths. The terminal flowers on these short branches (Fig. 148: j) usually are sterile (ovaries aborted). In *Sparganium*, the flowers are in spherical unisexual heads arranged in racemes, and the axis (stem) may be simple or branched. The carpellate clusters (and occasionally also the staminate heads) are subtended by bracts (small cauline leaves).

The mature infructescence of *Typha* is a large, brown, velvety cylinder that eventually breaks apart (Fig. 148: i). Each wind-dispersed unit comprises an achene-like fruit plus bristles (persistent perianth or "pappus"; Fig. 148: k). After dispersal, the "achene" (actually a membranous follicle) ultimately splits lengthwise (Fig. 148: l) to release the single seed (Fig. 148: m). In *Sparganium*, nut-like drupes (endocarp dry and spongy) form the muricate, globose fruiting heads. The fruit is dispersed by water and animals.

The reduced flowers are anemophilous and apparently protogynous.

REFERENCES CITED

Bentham, G., and J. D. Hooker. 1883. *Genera plantarum*, vol. 3(2). L. Reeve, London.

Bergner, I., and U. Jensen. 1989. Phytoserological contribution to the systematic placement of the Typhales. *Nordic J. Bot.* 8:447–456.

Dahlgren, R. M. T., H. T. Clifford, and P. F. Yeo. 1985. *The families of the monocotyledons*, pp. 344–346. Springer-Verlag, Berlin.

Duvall, M. R., M. T. Clegg, M. W. Chase, W. D. Clark, W. J. Kress, et al. 1993. Phylogenetic hypotheses for the monocotyledons constructed from *rbc*L data. *Ann. Missouri Bot. Gard.* 80:607–619.

Müller-Doblies, U., and D. Müller-Doblies. 1977. Ordnung Typhales. In *Illustrierte Flora von Mittel-Europa*, 3d ed., ed. G. Hegi, vol. 2, part 1(4), pp. 275–317.

Thieret, J. W. 1982. The Sparganiaceae in the southeastern United States. *J. Arnold Arbor.* 63:341–355.

Thorne, R. F. 1992. Classification and geography of the flowering plants. *Bot. Rev.* (Lancaster) 58:225–348.

ZINGIBERACEAE
Ginger Family

Perennial herbs, small to large, aromatic, with tuberous rhizomes; stems unbranched. **Leaves** simple, entire, alternate, distichous, large, oblong to broadly elliptic, pinnately veined (prominent midrib plus parallel and arching-convergent secondary veins), rolled in bud; basal sheaths open, overlapping and forming a pseudostem, with ligule at junction with petiole. **Inflorescence** determinate, cymose and often appearing spicate (cone-like) or racemose, with conspicuous spirally arranged primary bracts, terminal. **Flowers** usually zygomorphic, perfect, epigynous, showy, ephemeral, subtended by bracts. **Calyx** synsepalous with 3 equal or unequal lobes, tubular, sometimes split along one side, green or colorless, imbricate. **Corolla** sympetalous with 3 lobes, tubular, with median lobe larger than the laterals, variously colored, delicate, imbricate. **Androecium** of 1 stamen and usually 4 staminodes, with 2 staminodes fused into a large petaloid lip, basically biseriate, basally adnate to corolla; filament deeply grooved and enfolding the style; anther usually basifixed, with expanded laminar connective separating the locules (the style inserted between them), dehiscing longitudinally, introrse. **Gynoecium** of 1 pistil, 3-carpellate; ovary inferior, 3- or sometimes 1-locular, topped by 2 nectaries; ovules numerous, anatropous, placentation axile (3-locular ovaries) or sometimes parietal or basal (1-locular ovaries); style 1, very thin, inserted in the filament groove and between the anther locules; stigma 1, funnelform, ciliate, wet, protruding beyond the anther. **Fruit** usually a fleshy loculicidal capsule or sometimes a berry; seeds arillate, operculate; endosperm less abundant than and surrounded by perisperm, hard to mealy; perisperm copious, hard to mealy, starchy; embryo straight, cylindrical or clavate, embedded in endosperm.

320

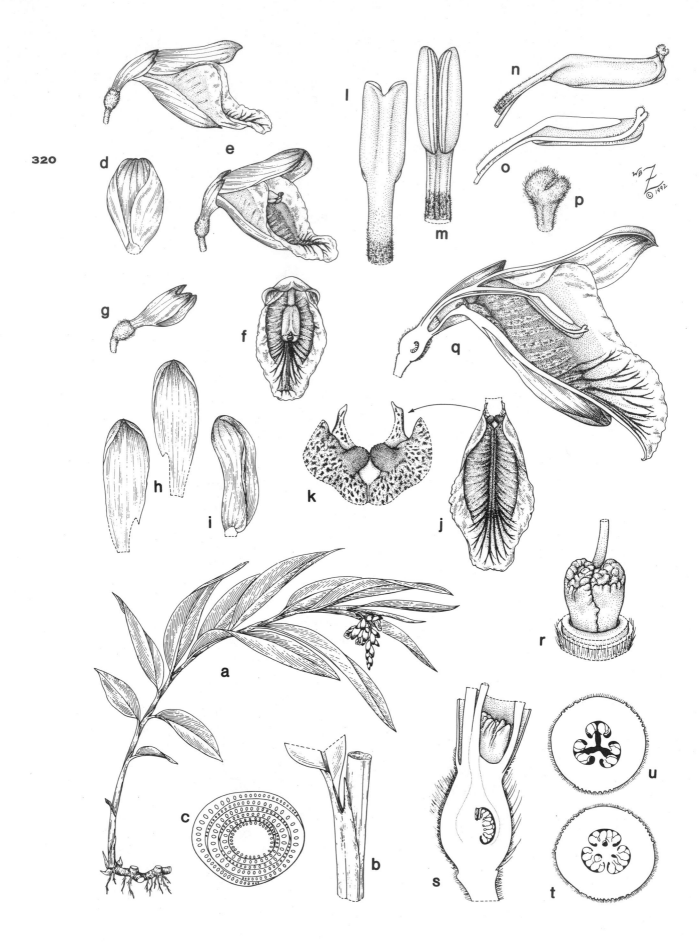

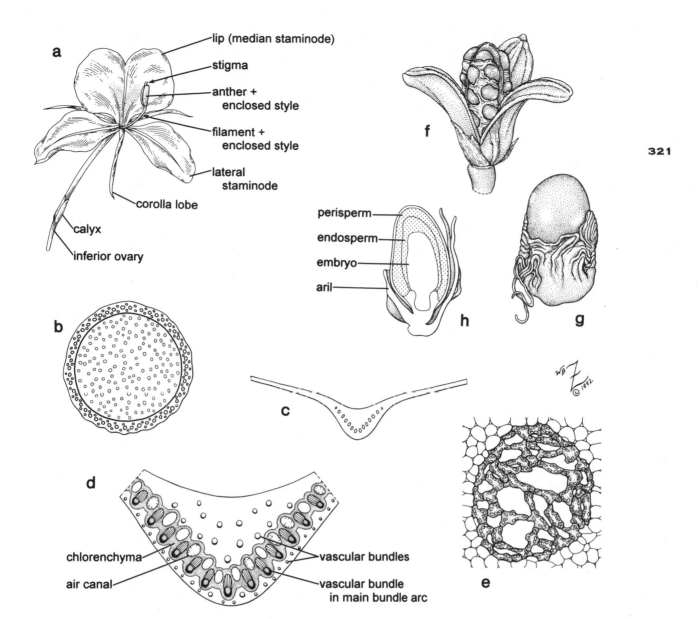

Figure 150 [above]. Zingiberaceae (continued). *Hedychium coronarium*: **a,** flower, ×½; **b,** cross section of stem showing fibrous cylinder, "pith," and "cortex," ×4; **c,** cross section of leaf showing air canals, ×3½; **d,** cross section of midvein, ×12; **e,** detail of air canal showing transverse diaphragm of lobed cells, ×140; **f,** capsule, ×1½; **g,** seed (with aril), ×6; **h,** longitudinal section of seed, ×6.

Figure 149 [facing page]. Zingiberaceae. *Alpinia speciosa*: **a,** habit, ×1/12; **b,** node showing sheathing leaf base and ligule, ×½; **c,** cross section of "pseudostem" (stem plus overlapping leaf sheaths), ×2½; **d,** bract, ×5/6; **e,** lateral views of flower, ×¾; **f,** front view of flower, ×¾; **g,** calyx and ovary, ×5/6; **h,** lateral petals, ×5/6; **i,** adaxial petal, ×5/6; **j,** staminodes (petaloid lip plus two scale-like lateral staminodes), ×5/6; **k,** detail of lip apex showing scale-like lateral staminodes, ×2¾; **l,** abaxial view of stamen, ×2¼; **m,** adaxial view of stamen, ×2¼; **n,** lateral view of stamen and protruding stigma (style inserted between anther locules), ×2¼; **o,** longitudinal section of stamen, style, and stigma, ×2¼; **p,** stigma, ×6; **q,** longitudinal section of flower, ×1¼; **r,** two nectaries at ovary apex, ×6; **s,** longitudinal section of ovary, ×4; **t,** cross section of ovary near middle, ×5; **u,** cross section of ovary at apex, ×5.

TABLE 20. Major morphological differences of seven families of the order Zingiberales. Characters from Dahlgren et al. (1985) and Kress (1990). See the commentary under the Zingiberaceae for a general discussion of characters of the order.

322

CHARACTERS	ZINGIBERACEAE	COSTACEAE	MUSACEAE
GENERA/SPECIES	50/1,000	4/150	2/42
DISTRIBUTION	pantropical	pantropical	Old World tropics
HABIT	small to large herbs glabrous	low to medium herbs hairy	large to tree-like herbs glabrous
LEAF ARRANGEMENT	distichous	spiral (1-ranked, spiral staircase-like)	spiral
LEAF BLADE	entire	entire	often torn
PETIOLE	short to long	short	long
BASAL SHEATH	open; overlapping to form a pseudostem	closed	open; overlapping to form a pseudostem
PRIMARY BRACTS	spiral conspicuous	spiral conspicuous (with nectariferous callus)	spiral conspicuous
FLORAL SYMMETRY	usually zygomorphic	zygomorphic	zygomorphic
NECTARIES	2 on top of the ovary	septal	septal
PERIANTH	3 fused sepals + 2+1 fused petals (median enlarged)	3 fused sepals + 2+1 fused petals (median enlarged)	5 fused + 1 (inner whorl; enlarged) tepals
FERTILE STAMEN(S)	1, embraces style	1, anther embraces upper style	usually 5
STAMINODES	2 fused (lip) ± 2 laterals	5 fused (lip)	1 minute
POLLEN EXINE	poorly developed	thin layer	poorly developed
OVARY LOCULE NUMBER	1 or 3	2 or 3	3
OVULES/LOCULE	numerous	numerous	numerous
STYLE	elongate; enclosed by filament + anther	elongate; apically enclosed by anther	elongate
STIGMA	funnelform	funnelform	trilobate
FRUIT TYPE	usually fleshy loculicidal capsule	loculicidal capsule	fleshy capsule to berry
ARIL	present	present	absent
EMBRYO	straight	straight	straight or curved
SILICA BODIES	in leaf epidermis	associated with vascular bundles	associated with vascular bundles
OTHER	aromatic (ethereal oils) flavonoids	pollen grains with thin exine air canals of leaf axis poorly developed	flowers imperfect laticifers with mucilaginous contents

STRELITZIACEAE	HELICONIACEAE	CANNACEAE	MARANTACEAE
3/7	1/250	1/50	30/450
tropical South America, southern Africa, and Madagascar	primarily American tropics	primarily tropical and subtropical America	pantropical
medium to tree-like herbs glabrous	small to large herbs usually glabrous	large herbs glabrous	low to medium herbs usually hairy
distichous	distichous	distichous (ranks spiral)	distichous to spiral
often torn	often torn	entire	entire
long	long	none (pseudopetiole)	long
open	open; overlapping to form a pseudostem	open	open; with pulvinus at petiole-blade junction
1 or a few conspicuous (boat-shaped)	distichous conspicuous (boat-shaped)	spiral inconspicuous	distichous to spiral conspicuous (spathe-like)
zygomorphic	zygomorphic	irregular	irregular (flowers in mirror-image pairs)
septal	septal	septal	septal
3 distinct sepals + 2 fused + 1 petals	5 fused + 1 (outer whorl) tepals	3 distinct sepals + 3 fused petals (1 smaller)	3 distinct sepals + 3 fused petals (1 enlarged)
usually 5	5	½, petaloid	½, petaloid
1 or none	1, ± petaloid	½ + 3 or 4 (1 enlarged into a lip)	½ + 3 or 4 (1 a hood over the style)
poorly developed	poorly developed	poorly developed	poorly developed
3	3	3	1 (2 abort)
numerous	1	numerous	1
elongate	elongate and apically thickened	petaloid, winged, fused to stamen and enclosing anther in bud	stout, curved, apically expanded and lobed
trilobate	capitate or trilobate	decurrent along edge of style	in depression between lobes
woody, loculicidal capsule	schizocarp with drupe-like segments	warty, irregularly dehiscing capsule	loculicidal capsule, berry, or nut-like
present	absent	absent (but seeds embedded in aril-like hairs)	present
straight	straight	straight	curved to plicate
associated with vascular bundles	associated with vascular bundles	associated with vascular bundles	associated with vascular bundles
woody stem (lost in some *Strelitzia* spp.)	inverted flowers (median sepal adaxial) heteropolar pollen grains	mucilage canals in stems and rhizomes secondary pollen presentation mechanism	complex "trigger" secondary pollination presentation mechanism

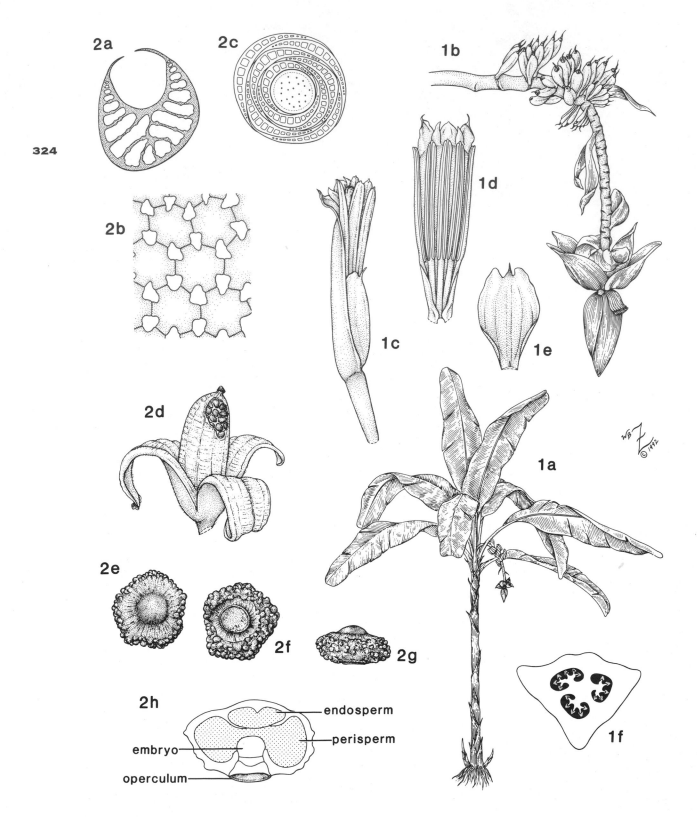

324

Figure 151. Musaceae. 1, *Musa balbisiana*: **a,** habit, greatly reduced; **b,** young infructescence (at base) and staminate inflorescence (at apex), ×⅛; **c,** staminate flower, ×1¼; **d,** adaxial corolla lobes and stamens, ×1½; **e,** abaxial petal and staminode (at base), ×1½; **f,** cross section of young fruit, ×2. **2,** *Musa velutina*: **a,** cross section of petiole showing air canals, ×3; **b,** stellate cells from transverse diaphragm of air canal, ×150; **c,** cross section of "pseudostem" (stem plus overlapping leaf sheaths), ×1; **d,** capsule, ×½; **e,** top view of seed, ×4; **f,** bottom view of seed, ×4; **g,** lateral view of seed, ×4; **h,** longitudinal section of seed, ×7. See Table 20 for family characteristics.

Family characterization: Aromatic rhizomatous herbs; large, distichous, pinnately veined leaves with ligulate, open basal sheaths overlapping to form a pseudostem; spicate or racemose inflorescence with conspicuous, spirally arranged, primary bracts; zygomorphic, epigynous, showy flowers with tubular calyx and corolla; androecium of 1 fertile stamen and several staminodes (2 fused into a petaloid lip); 2 nectaries topping the ovary; weak style encircled by grooved filament and anther locules; funnelform, wet stigma protruding from anther apex; fleshy capsule, and arillate seeds with perisperm. Pollen grains usually with very poorly developed exine. Flavonoids present. Tissues with tannins and calcium oxalate crystals. Anatomical features: secretory cells with ethereal oils (plus other substances, such as aromatic ketones) in ground tissues of all organs; silica bodies in epidermal cells of leaves; a smooth, fibrous cylinder dividing the stem into distinct cortex and pith (Fig. 150: b); and a single row of air canals (Fig. 150: c,d) with transverse diaphragms of stellate or lobed cells (Fig. 150: e) in the leaf axis (petiole and midvein; Tomlinson 1956, 1969).

Genera/species: 50/1,000

Distribution: Pantropical; most diverse in southeastern Asia. Typically in shady, lowland forests with rich soil.

Major genera: *Alpinia* (250 spp.), *Amomum* (90–150 spp.), *Zingiber* (85–90 spp.), *Kaempferia* (50–70 spp.), and *Hedychium* (50 spp.)

U.S./Canadian representatives: *Renealmia* (3 spp.), *Zingiber* (3 spp.), *Hedychium* (2 spp.), *Alpinia* (1 sp.), and *Curcuma* (1 sp.); none native to the continental U.S.

Economic plants and products: Numerous spices (rhizomes or seeds): ginger (*Zingiber*), galangal or Siamese-ginger (*Alpinia*), and other ginger-like spices (*Kaempferia*); turmeric (*Curcuma*), cardamom or cardamon (*Amomum, Elettaria*), and melagueta-pepper (*Aframomum*). "Arrowroot" starch from rhizomes of several *Curcuma* species. Perfumes (essential oils) extracted from species of several genera, such as *Alpinia* and *Hedychium*. Several (e.g., *Hedychium*) also sources of fiber and pulp. Ornamental plants (species of about 19 genera), including *Alpinia* (shell-ginger), *Curcuma* (hidden-lily), *Globba*, *Hedychium* (garland-lily, ginger-lily), *Kaempferia*, *Nicolaia* (torch-ginger), *Renealmia*, and *Zingiber* (ginger).

Commentary: The Zingiberaceae have been divided into three or four tribes (Burtt and Smith 1972) based primarily on lateral staminode development (conspicuous or reduced), ovary locule number (one or three),

and inflorescence type (cone-like or racemose). The family is here defined in a restricted sense: the Costaceae, often included as a subfamily (Costoideae; e.g., in Maas 1972), are treated as a segregate family (see Rogers 1984 for a historical summary). These two closely related, sister taxa differ in many morphological, embryological, palynological, and anatomical features (see the commentary under Costaceae, and Tomlinson 1956, 1962, 1969).

The Zingiberaceae and Costaceae are closely allied to the other families (Cannaceae, Marantaceae, Musaceae, Heliconiaceae, Strelitziaceae, and Lowiaceae) in the well-defined and clearly monophyletic order Zingiberales (the "Scitamineae" of earlier authors). Since some of these families have long been recognized as subfamilies, the Zingiberales complex may comprise four to eight families, depending on the author. For a complete overview of the order, see Dahlgren et al. (1985) and Kress (1990); the major morphological differences of seven of these families are summarized in Table 20 (see also Figs. 149 through 158).

Basically, all members of the Zingiberales are tropical to subtropical, rhizomatous herbs with large sheathing leaves that are rolled up in bud. The lateral veins are pinnate ("pinnate-parallel") and arching-convergent toward the entire blade margin. The showy, zygomorphic to irregular flowers are epigynous and are arranged in terminal inflorescences, often with conspicuous primary bracts. The most conspicuous component of the flower typically is androecial (various parts petaloid). The original six biseriate stamens are reduced into various numbers of stamen(s) plus staminode(s). Pollen grains typically have a poorly developed exine; stigmas are consistently wet. Perisperm is present in the seeds, which often are arillate.

As in the Musaceae and Heliconiaceae, the overlapping basal leaf sheaths in the Zingiberaceae form a "pseudostem" surrounding the true, somewhat flimsy, stem (Fig. 149: c). In addition, a membranous ligule is present at the upper margin of the sheath at the junction with the petiole (Fig. 149: b).

Portions of the androecium compose the most conspicuous parts of the highly specialized flowers, which are very diverse in appearance. Of the original outer whorl of three stamens, only the median (upper) stamen is fertile (Fig. 149: l,m). The other two are staminodial and fused into a single petaloid lip (labellum) opposite the stamen (Figs. 149: j; 150: a); this conspicuous structure, often two-lobed, varies greatly in coloration and shape and resembles an orchid lip (the median upper tepal). The median (lower) stamen of

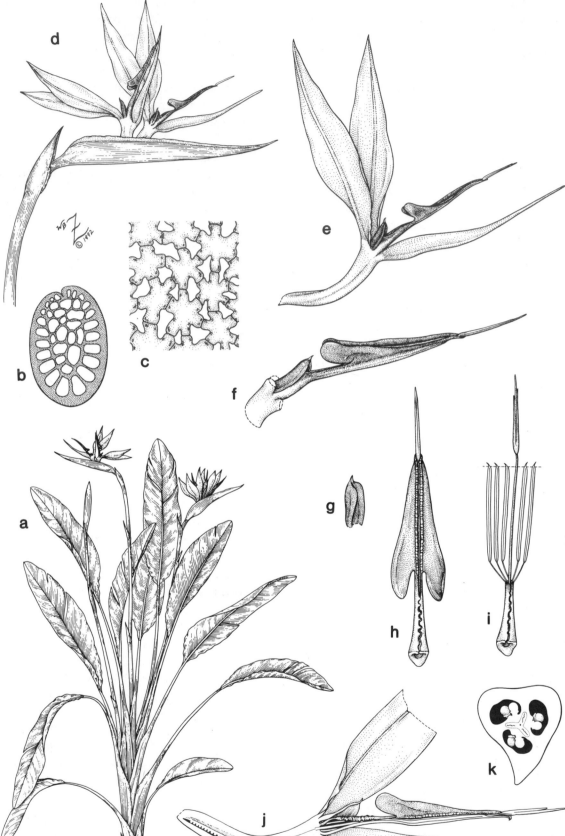

the original outer whorl is absent. The two lateral androecial elements of the outer whorl are modified into petaloid to scale-like staminodes (or sometimes these elements are absent); for example, in *Hedychium* (Fig. 150: a) the lateral staminodes are well developed, petaloid, and free from the lip, whereas in *Alpinia* they usually are represented by two small teeth at the base of the broad labellum (Fig. 149: j,k; Burtt 1972).

The conspicuous flowers are arranged in terminal inflorescences (Fig. 149: a), often subtended by colorful bracts (Fig. 149: d). Two nectaries occur at the apex of the ovary (Fig. 149: r), and the lip often has nectar guides (Fig. 149: e,f,j). In addition, flowers and plants are aromatic. The pollination mechanism is similar to that of the Costaceae. A common androecial-gynoecial structure is formed by a thin style threaded through a groove in the filament and enclosed by the anther locules (Fig. 149: n,o); the stigma protrudes through the anther apex. The various flower types are specialized for pollinators (bees, Lepidoptera, hummingbirds). For example, the long-exserted stamen-stigma structure of *Hedychium* (Fig. 150: a) touches the wings of hovering Lepidoptera, which carry the pollen to another flower.

The fruit is a colorful, fleshy capsule, and the seeds typically have red arils. For example, in *Hedychium* the bright orange capsule (Fig. 150: f) contains bright red seeds (plus red arils; Fig. 150: g,h). The attractive seeds of the family generally are bird dispersed.

REFERENCES CITED

Burtt, B. L. 1972. General introduction to papers on Zingiberaceae. *Notes Roy. Bot. Gard. Edinburgh* 31:155–165.

——— and R. M. Smith. 1972. Tentative keys to subfamilies, tribes and genera of Zingiberaceae. *Notes Roy. Bot. Gard. Edinburgh* 31:171–176.

Dahlgren, R. M. T., H. T. Clifford, and P. F. Yeo. 1985. *The families of the monocotyledons*, pp. 100, 102, 350–352. Springer-Verlag, Berlin.

Kress, W. J. 1990. The phylogeny and classification of the Zingiberales. *Ann. Missouri Bot. Gard.* 77:698–721.

Maas, P. J. M. 1972. Costoideae. *Fl. Neotrop. Monogr.* 8:1–139.

Rogers, G. K. 1984. The Zingiberales (Cannaceae, Marantaceae, and Zingiberaceae) in the southeastern United States. *J. Arnold Arbor.* 65:5–55.

Tomlinson, P. B. 1956. Studies in the systematic anatomy of the Zingiberaceae. *J. Linn. Soc., Bot.* 55:547–592.

———. 1962. Phylogeny of the Scitamineae—Morphological and anatomical considerations. *Evolution* 16:192–213.

———. 1969. *Commelinales—Zingiberales*, vol. 3 of *Anatomy of the monocotyledons*, ed. C. R. Metcalfe, pp. 293–302, 341–359. Oxford University Press, Oxford.

COSTACEAE
Costus Family

Perennial herbs, small to medium-sized, with tuberous rhizomes; stems usually unbranched, often spirally twisted. **Leaves** simple, entire, spirally arranged, large, oblong to broadly elliptic, with short petiole, pinnately veined (prominent midrib plus parallel and arching-convergent secondary veins), rolled in bud; basal sheaths closed, with short ligule at junction with petiole. **Inflorescence** determinate, spicate or capitate, terminal; primary bracts conspicuous, imbricate, spirally arranged, with nectariferous callus just below apex. **Flowers** zygomorphic, perfect, epigynous, showy, ephemeral, subtended by bracts. **Calyx** synsepalous with 3 lobes, tubular, variously colored, imbricate, persistent. **Corolla** sympetalous with 3 lobes, tubular, with upper medial lobe larger than the laterals, somewhat fleshy, variously colored, convolute. **Androecium** of 1 stamen and basically 5 staminodes, with staminodes fused into a large lip, petaloid, basically biseriate, basally adnate to corolla; filament expanded, flat, petaloid; anther basifixed, with expanded connective separating the locules (the style inserted between them), often with apical appendage, dehiscing longitudinally, introrse. **Gynoecium** of 1 pistil, 3-carpellate (occasionally 1 carpel aborting); ovary inferior, generally 3-locular, with 2 well-developed septal nectaries; ovules numerous, anatropous, placentation axile; style 1, slender, with apical portion inserted between the anther locules; stigma 1, funnelform, ciliate, wet, protruding beyond the anther. **Fruit** usually a loculicidal capsule crowned by the persistent calyx; seeds arillate, operculate; endosperm copious, starchy, mealy; perisperm copious, starchy, mealy; embryo straight, cylindrical, embedded in endosperm.

Figure 152. Strelitziaceae. *Strelitzia reginae:* **a,** habit, ×¹⁄₁₂; **b,** cross section of petiole showing air canals, ×2¼; **c,** stellate cells from transverse diaphragm of air canal, ×160; **d,** inflorescence, ×¼; **e,** flower, ×³⁄₈; **f,** flower with sepals removed, ×⁵⁄₈; **g,** adaxial petal, ×⁵⁄₈; **h,** fused lateral petals subtending androecium, style, and stigma, ×⁵⁄₈; **i,** androecium, style, and stigma (dotted line indicates fusion of stamens to lateral petals), ×⁵⁄₈; **j,** longitudinal section of flower, ×½; **k,** cross section of ovary, ×3. See Table 20 for family characteristics.

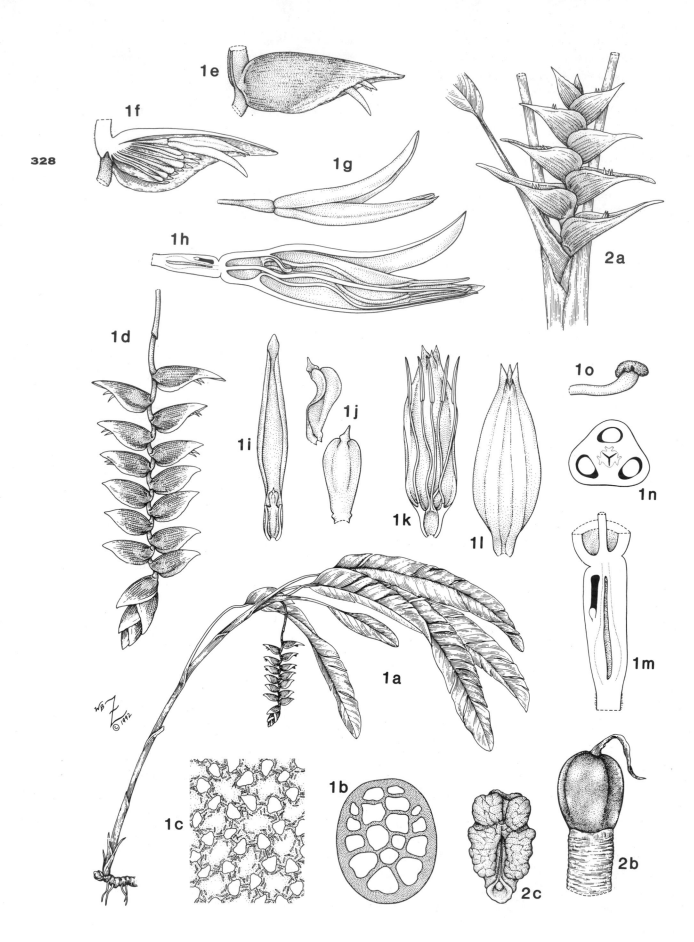

Family characterization: Rhizomatous herbs with spirally twisted stems; large, spirally arranged, pinnately veined leaves with ligulate, closed basal sheaths; colorful, nectariferous bracts spirally arranged in cone-like inflorescences; zygomorphic, epigynous, showy flowers with tubular calyx and corolla; petaloid androecium of 1 fertile stamen and a lobed lip (representing several fused staminodes); separated anther locules that enclose the apical portion of the style; funnelform, wet stigma protruding from the anther; capsule crowned by persistent calyx; and arillate seeds with perisperm. Pollen grains with distinct, but thin, exine. Tissues with calcium oxalate crystals. Anatomical features: a fluted, fibrous cylinder dividing the stem into distinct pith and cortex (Fig. 155: 1c,d); and a poorly developed system of air canals in the leaf axis (Tomlinson 1956, 1969).

Genera/species: *Costus* (90–150 spp.), *Tapeinocheilos* (15–20 spp.), *Dimerocostus* (2–8 spp.), and *Monocostus* (1 sp.)

Distribution: Pantropical, and most diverse in the American tropics; typically in rain forests.

U.S./Canadian representatives: *Costus* (2 spp.); not native to continental United States

Economic plants and products: Herbal remedies (used in the tropics) from various parts (root, stem, fruit) of several *Costus* spp. Several species of *Costus* (spiral flag, spiral-ginger), *Dimerocostus*, *Monocostus*, and *Tapeinocheilos* cultivated as ornamentals.

Commentary: The family consists of one large genus, *Costus* (with 90 to 150+ species), and three small genera (*Dimerocostus*, *Monocostus*, and *Tapeinocheilos*). The inclusion of the Costaceae as a subfamily (or tribe) within the Zingiberaceae is arbitrary: monographers of these particular groups tend to submerge the taxon (e.g., Maas 1972; Rogers 1984), while those who study family relationships on a large scale tend to treat the Costaceae as a segregate family (e.g., Cronquist 1981; Dahlgren et al. 1985; Thorne 1992). The recent cladistic studies of the ginger group by Kirchoff (1988) and of the Zingiberales by Kress (1990) support the two taxa as phylogenetically close (sister groups); however,

preliminary molecular data (Smith et al. 1993) conflict with these earlier studies based on morphology.

The Costaceae differ from the Zingiberaceae *s.s.* in several morphological, embryological, and anatomical features (refer to Table 20; compare the family characterizations of the two families; and see Tomlinson 1956, 1962, 1969 and Dahlgren et al. 1985 for complete summaries). The plants of Costaceae, which are not aromatic, have strongly developed main stems (i.e., no pseudostems) bearing spirally arranged leaves (Fig. 154: a) with closed leaf sheaths (Fig. 155: 1a). The ligule is short and annular (Fig. 155: 1b). Vegetative parts frequently are hairy, in comparison to the typically glabrous plants of the Zingiberaceae. Conspicuous, spirally arranged, primary bracts form compact heads or spikes (Fig. 155: 2a). Each bract bears a linear nectariferous gland (callus) just below the apex (Figs. 154: b; 155: 2a), which attracts ants. The androecium of the Costaceae differs from that of the Zingiberaceae (see below); most notable is that the lip of the Costaceae (Figs. 154: d; 155: 2c) probably represents five fused staminodes (compared to the two-staminode lip of the Zingiberaceae). Rather than capping the ovary apex (as in the Zingiberaceae), the septal nectaries are within cavities, sunken into the ovary (Fig. 154: l–n). Pollen grains have a thin, but distinct, exine, whereas this layer is very poorly developed on pollen grains in the other families of the Zingiberales.

The petaloid androecium comprises only one stamen (Fig. 154: i) opposite a large and conspicuous lip. The one fertile stamen represents the median (upper) element of the original inner whorl. Filament and connective are broad and flat and extend beyond the anther locules as an appendage (Fig. 154: h,i). The apical portion of the style is inserted between the separated anther sacs (Fig. 154: i,j). The three- or five-lobed lip (or labellum; Figs. 154: d; 155: 2c) generally is interpreted as a compound structure formed by the fusion of two lateral staminodes (inner whorl) with the three elements (staminodes) of the outer whorl.

The showy flowers with colorful bracts usually are pollinated by insects (bees) or hummingbirds (Maas 1972). *Costus* flowers have a pollination mechanism similar to that in *Iris*, with the large lip used as a landing

Figure 153. Heliconiaceae. 1, *Heliconia rostrata*: **a,** habit, ×¹⁄₂₀; **b,** cross section of petiole showing air canals, ×4; **c,** stellate cells from transverse diaphragm of air canal, ×140; **d,** inflorescence, ×¹⁄₆; **e,** bract (and flowers), ×¹⁄₂; **f,** longitudinal section of bract revealing subtended flowers, ×¹⁄₂; **g,** flower, ×1; **h,** longitudinal section of flower, ×1¾; **i,** adaxial tepal and staminode, ×1¼; **j,** two views of staminode, ×5; **k,** abaxial tepals and stamens (adaxial view), ×1¼; **l,** abaxial tepals (abaxial view), ×1¼; **m,** longitudinal section of ovary, ×4; **n,** cross section of ovary, ×6; **o,** stigma, ×10. **2,** *Heliconia caribaea*: **a,** inflorescence, ×¹⁄₅; **b,** fruit, ×1¾; **c,** seed, ×4. See Table 20 for family characteristics.

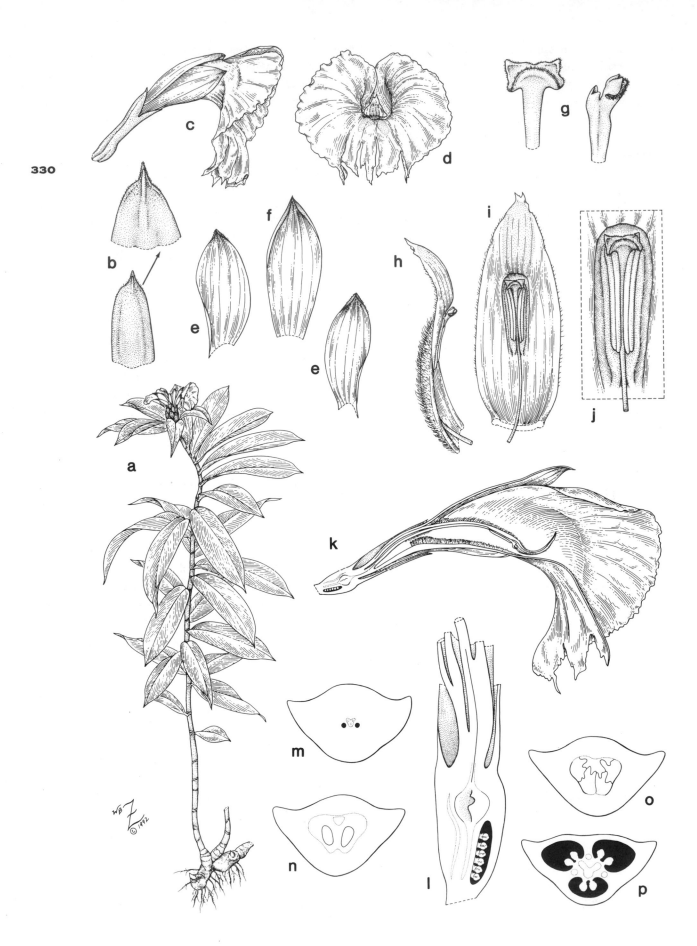

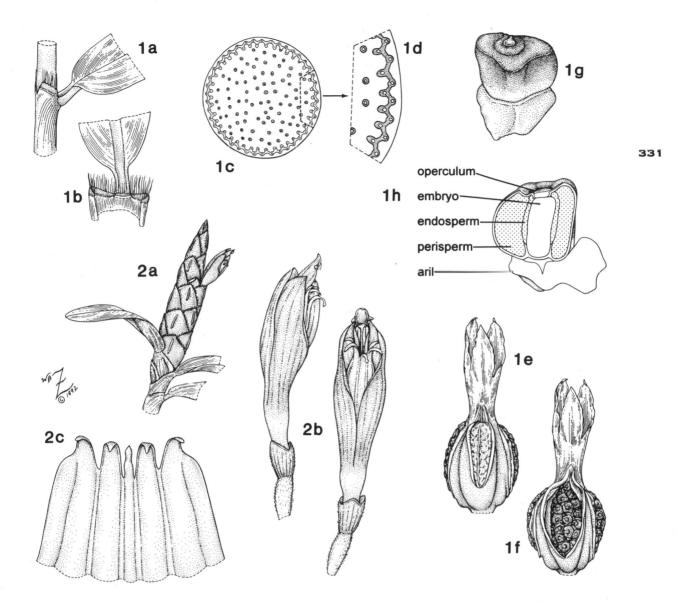

Figure 155 [above]. Costaceae (continued). 1, *Costus speciosus*: **a,** node showing sheathing leaf base and ligule, ×½; **b,** adaxial side of expanded leaf sheath base showing ligule, ×½; **c,** cross section of stem showing fluted, fibrous cylinder, ×4; **d,** detail of **c** showing fibrous cylinder and vascular bundles, ×9; **e,** dehiscing capsule (with persistent calyx), ×1¼; **f,** dehisced capsule (with persistent calyx), ×1¼; **g,** seed with aril, ×7½; **h,** longitudinal section of seed, ×8. **2,** *Costus pulverulentus*: **a,** inflorescence, ×½; **b,** two views of flower, ×1¼; **c,** expanded lip (staminode) apex (note five lobes), ×3.

Figure 154 [facing page]. Costaceae. *Costus speciosus*: **a,** habit, ×1/12; **b,** bract, ×5/6, with detail of apex showing nectariferous callus, ×3; **c,** lateral view of flower, ×¾; **d,** front view of flower (note five lobes of lip), ×¾; **e,** lateral petals, ×¾; **f,** adaxial petal, ×¾; **g,** two views of stigma, ×6; **h,** lateral view of petaloid stamen and subtended style and stigma, ×2; **i,** adaxial side of petaloid stamen and subtended style and stigma, ×2; **j,** detail of **i** showing style inserted between anther locules and projecting stigma, ×4; **k,** longitudinal section of flower, ×1; **l,** longitudinal section of ovary, ×3; **m,** cross section of ovary at apex showing cavities of the two well-developed septal nectaries, ×4; **n,** cross section of ovary through septal nectaries at apex (note two well-developed glands and one reduced gland), ×4; **o,** cross section of ovary through septal nectary, ×4; **p,** cross section of ovary through locule, ×4.

platform. Septal nectaries occur in a depressed cavity near the apex of the ovary (Fig. 154: m), and nectar accumulates in the tube formed by the fused labellum and androecium base (Fig. 154: k). The tube is bordered with hairs. Projecting in the center of the flower is the fertile stamen (enclosing the style) with the stigma protruding from the apex. Besides nectar, bees also visit to collect pollen (Sytsma and Pippen 1985). As the pollinator probes for the nectar, it touches the stigma, lifts the stamen, and contacts the anther. The thick walls of the floral tube around the nectar chamber prevent external puncturing by nonpollinators; also, the leathery bracts closely subtend the flower base.

REFERENCES CITED

Cronquist, A. 1981. *An integrated system of classification of flowering plants*, pp. 1180–1183. Columbia University Press, New York.

Dahlgren, R. M. T., H. T. Clifford, and P. F. Yeo. 1985. *The families of the monocotyledons*, pp. 365–367. Springer-Verlag, Berlin.

Kirchoff, B. K. 1988. Floral ontogeny and evolution in the ginger group of the Zingiberales. In *Aspects of floral development*, ed. P. Leins, S. C. Tucker, and P. K. Endress, pp. 45–56. J. Cramer, Berlin.

Kress, W. J. 1990. The phylogeny and classification of the Zingiberales. *Ann. Missouri Bot. Gard.* 77:698–721.

Maas, P. J. M. 1972. Costoideae. *Fl. Neotrop. Monogr.* 8:1–139.

Rogers, G. K. 1984. The Zingiberales (Cannaceae, Marantaceae, and Zingiberaceae) in the southeastern United States. *J. Arnold Arbor.* 65:5–55.

Smith, J. F., W. J. Kress, and E. A. Zimmer. 1993. Phylogenetic analysis of the Zingiberales based on *rbc*L sequences. *Ann. Missouri Bot. Gard.* 80:620–630.

Sytsma, K. J., and R. W. Pippen. 1985. Morphology and pollination biology of an intersectional hybrid of *Costus* (Costaceae). *Syst. Bot.* 10:353–362.

Thorne, R. F. 1992. Classification and geography of the flowering plants. *Bot. Rev.* (Lancaster) 58:225–348.

Tomlinson, P. B. 1956. Studies in the systematic anatomy of the Zingiberaceae. *J. Linn. Soc., Bot.* 55:547–592.

——. 1962. Phylogeny of the Scitamineae—Morphological and anatomical considerations. *Evolution* 16:192–213.

——. 1969. *Commelinales—Zingiberales*, vol. 3 of *Anatomy of the monocotyledons*, ed. C. R. Metcalfe, pp. 299–300, 360–364. Oxford University Press, Oxford.

CANNACEAE
Canna Family

Perennial herbs, large, coarse, and robust, with tuberous rhizomes; stems unbranched. **Leaves** simple, entire, spirally arranged, large, oblong to broadly elliptic, pinnately veined (prominent midrib plus parallel and arching-convergent secondary veins), rolled in bud, more or less sessile but tapering to a petiole-like base with open sheath. **Inflorescence** determinate, cymose and appearing racemose, spicate, or paniculate, terminal. **Flowers** irregular, perfect, epigynous, large and showy, ephemeral, each subtended by a bract. **Calyx** of 3 sepals, distinct, sometimes unequal, green or purplish, persistent, imbricate. **Corolla** of 3 petals, basally connate and adnate to the androecium and style (forming a tube), unequal with 1 smaller than other 2, often yellow, orange, or red, imbricate. **Androecium** of 1 stamen (partially fertile and partially staminodial) and usually 3 or 4 basally connate staminodes with one staminode more enlarged than the others and reflexed (forming a lip), basically biseriate, petaloid and conspicuous, basally adnate to corolla; anther lateral on margin of petaloid stamen near apex, 1-locular, dehiscing longitudinally, introrse. **Gynoecium** of 1 pistil, 3-carpellate; ovary inferior, 3-locular, tuberculate, with septal nectaries; ovules numerous, anatropous, placentation axile; style 1, petaloid, winged, adnate to floral tube, enclosing the anther in bud; stigma decurrent along an edge of the style, papillose, wet. **Fruit** a tuberculate capsule, usually splitting irregularly by disintegration of pericarp, crowned by persistent calyx; seeds spherical, black, operculate, embedded in aril-like hairs; endosperm thin, starchy; perisperm copious, starchy, very hard; embryo straight.

Family characterization: Rhizomatous robust herbs with unbranched stems; large, spirally arranged, pin-

Figure 156. Cannaceae. *Canna flaccida*: **a,** habit, ×⅛; **b,** cross section of upper petiole showing air canals, ×8; **c,** stellate cells from transverse diaphragm of air canal, ×80; **d,** two views of flower, ×⅖; **e,** adaxial corolla lobe, ×½; **f,** lateral corolla lobe, ×½; **g,** lower portion of flower (calyx, gynoecium, and androecial-corolla tube), ×½; **h,** longitudinal section of flower, ×½; **i,** cross section of ovary, ×3; **j,** apex of style showing marginal stigma, ×1¼; **k,** petaloid stamen with dehiscing half-anther (from bud), ×1¼; **l,** two views of anther with subtending style showing relative positions in bud, ×2; **m,** petaloid stamen (with half-anther), ×½; **n,** petaloid stamen with adnate style, ×½; **o,** lip (inner staminode), ×½; **p,** "lateral" staminodes ("wings"; one outer and one inner staminode), ×½; **q,** large staminode (outer staminode behind stamen), ×½; **r,** immature capsule, ×½; **s,** immature seed embedded in aril-like hairs, ×2; **t,** dehiscing capsule with persistent calyx, ×½; **u,** dehisced capsule (seeds exposed), ×½; **v,** seed, ×3; **w,** longitudinal section of seed, ×3. For flower with labeled staminodes, corresponding floral diagram, and floral formula, see Fig. 4: 5a–c.

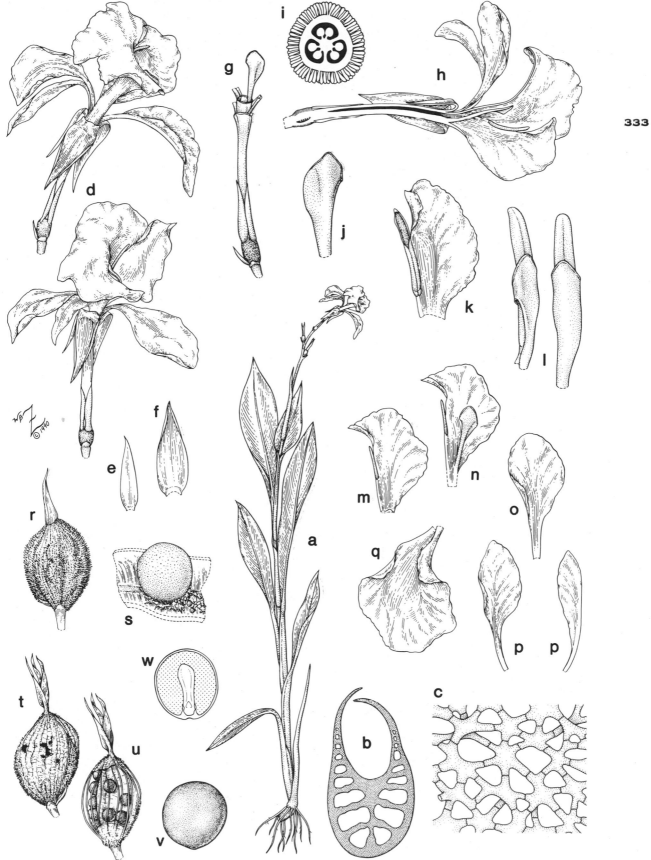

nately veined leaves with open basal sheaths; showy, red to yellow, asymmetric, epigynous flowers; petaloid, conspicuous androecium of 1 partially fertile stamen (with 1-locular anther) and several staminodes (1 enlarged into a lip); petaloid style with decurrent stigma; an irregularly dehiscent, warty capsule as the fruit type; and perispermous seeds embedded in aril-like hairs. Pollen grains with very poorly developed exine. Tissues with tannins and calcium oxalate crystals. Anatomical features: secretory canals (containing mucilage) in the rhizomes and stems; silica bodies adjacent to vascular bundles; and a single row of air canals (abaxial to main vascular bundles; Fig. 156: b) with transverse diaphragms of stellate cells (Fig. 156: c) in the leaf axis ("petiole" and midvein; Tomlinson 1961, 1962, 1969).

Genus/species: *Canna* (25–55 spp.); a monotypic family

Distribution: Primarily tropical and subtropical America; also widely dispersed (probably introduced) in the Old World tropics.

U.S./Canadian representatives: *Canna* (8 spp.)

Economic plants and products: Edible starch (purple-arrowroot, Queensland-arrowroot) from rhizomes of *Canna edulis*. Hundreds of *Canna* hybrids (canna-lilies), such as the *Canna × generalis* group, cultivated as ornamentals.

Commentary: The Cannaceae are a monotypic family, best known by the familiar garden flower "*Canna × generalis*" (a large group of complex hybrid lineages derived from several parent species, such as *C. indica*). The circumscription of *Canna* is variable, with different authorities citing as few as 7 to as many as 60 species in the genus. Numerous nomenclatural problems are associated with the hybrids and overdescribed species (Rogers 1984). The genus *Canna* has been divided into a varying number of sections and subgenera primarily based on the morphology of the staminodes. Several authors have suggested that the genus is exclusively neotropical in origin and has been introduced to other tropical areas, but other botanists claim that certain species actually are native to the tropics of Africa and Asia. Due to the similar floral morphology, the family has been considered close to the Marantaceae, and earlier authors often included both taxa as tribes within one family.

The general vegetative and floral characters that the Cannaceae share with other members of the Zingiberales are summarized in the discussion under the Zingiberaceae and in Table 20; the commentary under the Marantaceae summarizes the similarities between these two sister groups. The plants of *Canna* prefer moist, humus-rich substrates in forests or along riversides, as well as swampy places. The leaves taper to a sheathing base (Fig. 156: a), but the petiole is not well differentiated. Superficially, the vegetative portion of the plant resembles that of the Zingiberaceae or Marantaceae, but it lacks ligules and pulvini (Tomlinson 1961).

As in the other families of the Zingiberales, the most conspicuous part of the flower of *Canna* is the modified, petaloid androecium, which in this case comprises one stamen plus one to four (often three or four) staminode(s). Figure 4: 5a,b graphically depicts the relative positions of the androecial elements of *Canna flaccida*. The single functional stamen (Figs. 4: 5a,b—label no. 5; 156: k,m) represents the uppermost (median) element of the original inner whorl of the biseriate androecium and is only partially fertile, bearing a unilocular, bisporangiate anther (Fig. 156: k). This long and linear "half-anther" is lateral along one edge of the fleshy, petaloid stamen, often inserted far below the apex. One staminode (the lip or labellum; Figs. 4: 5b—label no. 4; 156: o), often wider than the others and recurved, also is generally interpreted as a member of the original inner whorl. In some species the lip is the only staminode present, but most flowers have at least two additional staminodes (sometimes referred to as "wings"; Figs. 4: 5b—label nos. 1,3; 156: p) that are erect but vary in morphology. A fourth staminode, when present (as in *C. flaccida*; Figs. 4: 5b—label no. 2; 156: q), is inserted behind the stamen.

The conspicuous, red to yellow flowers attract various insect pollinators (probably bees) with the nectar secreted by septal nectaries and accumulated around the base of the floral tube. In *Canna flaccida*, the species native to the United States, the flowers are fragrant and open at dusk. The fleshy and petaloid style is flat and encircles the anther in bud (Fig. 156: l). The stigma is decurrent along the apical margin of the style (Fig. 156: j). As in the Marantaceae, pollen is shed and adheres to one side of the style before the flower opens, either directly on the stigma (presumably resulting in self-pollination) or proximal to the stigma (a pollen presentation mechanism requiring a pollinator; Knuth 1909; Dahlgren et al. 1985). When the flower opens, the staminodial lip functions as a landing platform for the pollinator, which first encounters the protruding stigma and dusts it with pollen from another flower. The insect may then contact new pollen already on the style.

The tuberculate ovary develops into a bristly capsule that releases the round, dark seeds when the pericarp disintegrates (Fig. 156: r,t,u). The seeds are embedded in hairs (derived from the funicle) that often are not regarded as true arils (Fig. 156: s); these tufts of hairs remain in the capsule after the seeds fall out.

REFERENCES CITED

Dahlgren, R. M. T., H. T. Clifford, and P. F. Yeo. 1985. *The families of the monocotyledons*, p. 369. Springer-Verlag, Berlin.

Knuth, P. 1909. *Handbook of flower pollination*. Trans. J. R. A. Davis. Vol. 3, *Goodenovieae to Cyadeae*, p. 423. Oxford University Press, Oxford.

Rogers, G. K. 1984. The Zingiberales (Cannaceae, Marantaceae, and Zingiberaceae) in the southeastern United States. *J. Arnold Arbor.* 65:5–55.

Tomlinson, P. B. 1961. The anatomy of *Canna*. *J. Linn. Soc., Bot.* 56:467–473.

——. 1962. Phylogeny of the Scitamineae—Morphological and anatomical considerations. *Evolution* 16:192–213.

——. 1969. *Commelinales—Zingiberales*, vol. 3 of *Anatomy of the monocotyledons*, ed. C. R. Metcalfe, pp. 300, 365–373. Oxford University Press, Oxford.

MARANTACEAE
Arrowroot Family

Perennial herbs, with tuberous rhizome; stems unbranched. **Leaves** simple, entire, alternate or mostly basal, distichous, large, oblong to broadly elliptic, pinnately veined (prominent midrib plus arching-convergent secondary veins), often variegated, rolled in bud; petiole long, with pulvinus at petiole-blade junction, with open basal sheath. **Inflorescence** determinate, cymose and often appearing spicate, racemose, paniculate, or capitate, usually subtended by spathe-like bracts, terminal (sometimes on a scape). **Flowers** irregular, usually arranged in mirror-image pairs on the inflorescence, perfect, epigynous, showy, each subtended by 1 or 2 bract(s). **Calyx** of 3 sepals, distinct, subequal, imbricate. **Corolla** of 3 petals, basally connate and adnate to the androecium and style (forming a tube), unequal with 1 often enlarged and hood-like, variously colored, ephemeral, imbricate. **Androecium** of 1 stamen (partially fertile and partially staminodial) and usually 3 or 4 variable staminodes with 1 staminode forming a hood over the style, basically biseriate, basally connate and adnate to corolla, petaloid, conspicuous; anther lateral on margin of petaloid stamen, 1-locular, dehiscing longitudinally, introrse. **Gynoecium** of 1 pistil, basically 3-carpellate with 2 carpels often reduced to absent; ovary inferior, basically 3-locular with often only 1 locule developed, with septal nectaries; ovules 1 in each fertile locule, erect, anatropous to campylotropous, placentation axile and often appearing basal; style 1, stout, apically expanded, lobed, and involute, with a specialized area for pollen deposition, curved and enclosed by hooded staminode in bud (twisted and recurved when released); stigma terminal, in depression between flaps at style apex, wet. **Fruit** a loculicidal capsule, a berry, or dry and indehiscent; seeds subglobose to ellipsoid, arillate, operculate; endosperm very reduced to absent; perisperm copious, starchy, mealy, penetrated by 1 or 2 canals; embryo linear, strongly curved to plicate.

Family characterization: Rhizomatous, herbaceous plants; distichous, pinnately veined leaves with a pulvinus at the petiole-blade junction and open basal sheath; epigynous, asymmetric flowers arranged in mirror-image pairs enclosed by conspicuous bracts; petaloid, conspicuous androecium of 1 partially fertile stamen (with 1-locular anther) and several staminodes (2 modified for the explosive release of pollen); 3-carpellate ovary with usually only 1 fertile locule containing a solitary, basal ovule; peculiar, specialized style and stigma; and arillate seeds with perisperm penetrated by canals. Pollen grains with very poorly developed exine. Flavonoids and tannins present. Tissues with calcium oxalate crystals. Anatomical features: silica bodies (in all green parts); and a single row of air canals (abaxial to main vascular bundles) with transverse diaphragms of at least some stellate cells in the leaf axis (petiole and midvein, but not well developed in the pulvinus; Tomlinson 1961, 1969).

Genera/species: 30/450

Distribution: Pantropical, and most diverse in the American tropics; sometimes extending into subtropical (or infrequently warm-temperate) regions. Typically in moist forest habitats or swamps.

Major genera: *Calathea* (150–300 spp.) and *Ischnosiphon* (35 spp.)

U.S./Canadian representatives: *Calathea* (2 spp.), *Thalia* (2 spp.), and *Maranta* (1 sp.)

Economic plants and products: West Indian arrowroot (maranta starch) from rhizomes of *Maranta arundinacea*. Ornamental plants (species of about 16 genera), including *Calathea* (maranta), *Ctenanthe*, *Maranta* (arrowroot, prayer plant), and *Thalia*.

Commentary: The Marantaceae have been divided into two tribes based on the number of fertile locules in the ovary (Rogers 1984), but generally the subfamilial divisions are unresolved (Andersson 1981). The family has been considered closely allied with the Cannaceae, and cladistic analyses (Kirchoff 1988; Kress 1990) support their status as sister taxa (but see also the conflicting results of Smith et al. 1993). The two families share the character states of the asym-

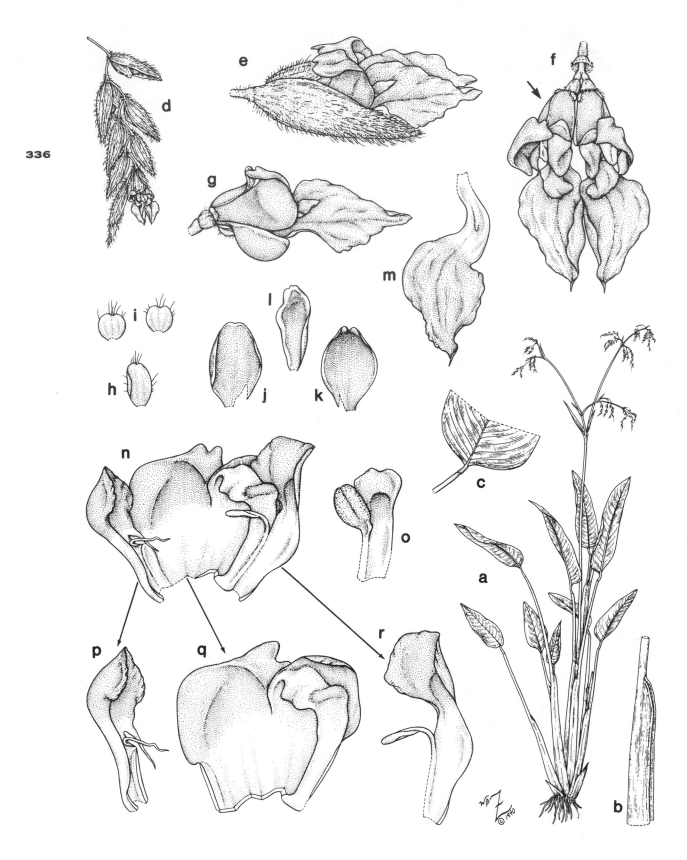

336

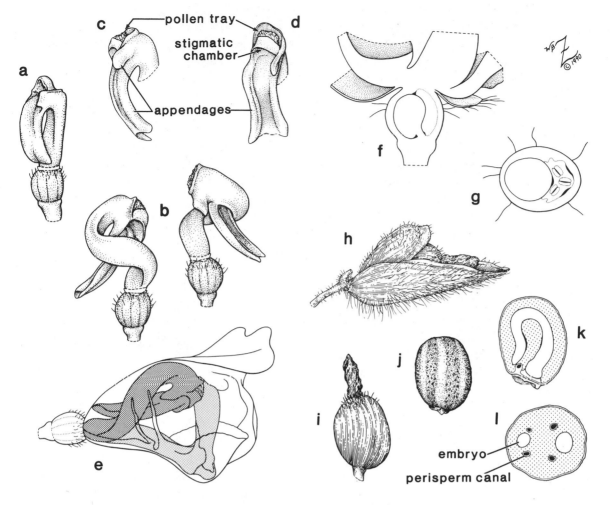

Figure 158 [above]. Marantaceae (continued). *Thalia geniculata*: **a,** pistil from bud (style curved), ×5; **b,** two views of mature pistil from "triggered" flower (style tightly coiled), ×5; **c,** lateral view of stylar apex, ×5; **d,** front view of stylar apex, ×5; **e,** diagrammatic representation of the style-stigma structure within the inner androecium (see Fig. 157: n, p–r) showing the untriggered (light screen) vs. triggered (dark screen) positions, ×5; **f,** longitudinal section of ovary (and perianth base), ×10; **g,** cross section of ovary (note three septal nectaries and two empty [aborted] locules), ×12; **h,** fruit (enclosed by bracts and persistent floral parts), ×2½; **i,** utricle-like fruit (subtending bracts removed), ×2½; **j,** seed, ×2½; **k,** longitudinal section of seed, ×3½; **l,** cross section of seed (note perisperm canals), ×4. All flower parts (**a–g**) from flower on right side of cymule (indicated by an arrow in Fig. 157: f). For photographs and detailed diagrams of the pollination mechanism of *Thalia geniculata*, see Classen-Bockhoff (1991), an elaborate study of the entire sequence.

Figure 157 [facing page]. Marantaceae. *Thalia geniculata*: **a,** habit, ×1/20; **b,** sheathing leaf base, ×¼; **c,** petiole-leaf blade junction showing pulvinus, ×¼; **d,** portion of inflorescence, ×⅝; **e,** two-flowered cymule and subtending bracts (lateral view), ×2½; **f,** two-flowered cymule (top view, subtending bracts removed; right-sided flower indicated by arrow), ×2½; **g,** flower from right side of cymule, ×2¾; **h,** abaxial sepal, ×5; **i,** lateral sepals, ×5; **j,** lateral petal (outer), ×2¾; **k,** lateral petal (inner), ×2¾; **l,** adaxial petal, ×2 ¾; **m,** outer staminode, ×2¾; **n,** inner whorl of androecium (two staminodes and one petaloid stamen), ×4; **o,** petaloid stamen with half-anther (from bud), ×5; **p,** cucullate staminode with "trigger" appendages, ×5; **q,** callose staminode, ×5; **r,** petaloid stamen (half-anther dehisced), ×5. All flower parts (**h–r**) from flower on right side of cymule (indicated by an arrow in f).

metric flowers, half-anther, and secondary pollen presentation mechanism (discussed below).

The general vegetation and floral features that the Marantaceae have in common with the other families of the Zingiberales are summarized in the commentary on the Zingiberaceae and in Table 20. The plants of the Marantaceae are a typical component of understory vegetation in wet, tropical regions, especially the margins of rain forests and clearings. Some species also occur in other moist habitats, such as swamps. Stems, leaves, and bracts typically are pubescent, and the leaves are well differentiated into a basal sheath (Fig. 157: b), petiole, and blade. The plants are easily recognizable due to the characteristic pulvinus, a terete swollen portion of the leaf axis between the blade and petiole (Fig. 157: c). This structure functions in movements of the leaves (see the commentary on the Fabaceae).

As in other members of the Zingiberales, the terminal inflorescence is subtended by conspicuous bracts (Fig. 157: d,e). The ultimate unit of the variable inflorescence is a two-flowered cymule enclosed by bracts (Fig. 157: e,f; Andersson 1976). These two irregular flowers are mirror images of each other (Fig. 157: f).

As is typical of the Zingiberales, the modified and petaloid androecium of the Marantaceae (one stamen plus usually three or four staminodes) is the most conspicuous part of the flower (see Andersson 1981). As in the Cannaceae, only one element of the original inner whorl is functional—a petaloid, partially fertile stamen bearing an elongate, unilocular, bisporangiate anther ("half-anther") along one margin (Fig. 157: o). The other two petaloid elements of the inner whorl are hood-like staminodes functioning in the complex pollination mechanism of the flower (see the discussion below). The "cucullate staminode," which forms a hood over the style and stigma before anthesis, is appendaged with one (e.g., *Maranta*) or two (*Thalia*; Fig. 157: p) marginal process(es). The fleshy and firm "callose staminode" is concave and lobed and bears a conspicuous thickening (callus) on the inner surface (Fig. 157: q). The outer whorl of the androecium is represented by one (*Thalia*; Fig. 157: m) or two (*Maranta*) large and variously shaped staminode(s). In *Thalia* this conspicuous staminode is basally constricted and protrudes from the subtending cymule bract (Fig 157: e,f).

The androecium and style form a complex "trigger mechanism" for insect pollination (Andersson 1981; Classen-Bockhoff 1991). Insects (generally bees) seek the nectar (from septal nectaries; Fig. 158: f,g) accumulated at the base of the tube formed by the fusion of the corolla, androecium, and style. The stout and somewhat hooked style is expanded and lobed at the oblique apex, with the receptive stigma situated in a depression between the lobes (Fig. 158: c,d). In the bud, the half-anther deposits pollen onto a specialized (nonreceptive) subapical cavity on the outer curved part of the style (Fig. 158: a,c,d). The spring-like style is confined under tension within the hood of the cucullate, appendaged staminode on the lower side of the horizontal flower (Fig. 158: e), and the firm callose staminode helps to brace this sensitive mechanism, especially during discharge. As the bee reaches for the nectar between these two staminodes, it pushes the appendage(s) (the "trigger") on the margin of the cucullate staminode: the released style springs from the hood and elastically coils downward (Fig. 158: b,e), with an audible "snap." During this abrupt motion, the stigmatic part of the style scrapes pollen (from another flower) from the bee's body (effecting cross-pollination), and the insect is then dusted with new pollen (held in the special region at the initial apex of the style). Each flower evidently is limited to one such insect visit, since the triggered, coiled style blocks the entrance to the flower.

As is characteristic of the other families in the Zingiberales, mature marantaceous seeds have perisperm, a nutritive tissue that develops from the nucellus. A unique feature of perisperm in this family, however, is the occurrence of hollow channels ("perisperm canals" or "perisperm channels"; Fig. 158: l) that develop within the nucellus from intrusions of apparent conductive tissue from the chalaza, the region of an ovule or seed where the nucellus joins the integuments (i.e., where the integuments originate); the channels result from the subsequent degeneration of these chalazal cells as the seed ripens (Grootjen 1983).

REFERENCES CITED

Andersson, L. 1976. The synflorescence of Marantaceae. Organization and descriptive terminology. *Bot. Not.* 129: 39–48.

———. 1981. The neotropical genera of Marantaceae. Circumscription and relationships. *Nordic J. Bot.* 1:218–245.

Classen-Bockhoff, R. 1991. Untersuchungen zur Konstruktion des Bestäubungsapparates von *Thalia geniculata* (Marantaceen). *Bot. Acta* 104:183–193.

Grootjen, C. J. 1983. Development of ovule and seed in Marantaceae. *Acta Bot. Neerl.* 32:69–86.

Kirchoff, B. K. 1988. Floral ontogeny and evolution in the ginger group of the Zingiberales. In *Aspects of floral development*, ed. P. Leins, S. C. Tucker, and P. K. Endress, pp. 45–56. J. Cramer, Berlin.

Kress, W. J. 1990. The phylogeny and classification of the Zingiberales. *Ann. Missouri Bot. Gard.* 77:698–721.

Rogers, G. K. 1984. The Zingiberales (Cannaceae, Marantaceae, and Zingiberaceae) in the southeastern United States. *J. Arnold Arbor.* 65:5–55.

Smith, J. F., W. J. Kress, and E. A. Zimmer. 1993. Phylogenetic analysis of the Zingiberales based on *rbc*L sequences. *Ann. Missouri Bot. Gard.* 80:620–630.

Tomlinson, P. B. 1961. Morphological and anatomical characteristics of the Marantaceae. *J. Linn. Soc., Bot.* 58:55–78.

——. 1969. *Commelinales—Zingiberales*, vol. 3 of *Anatomy of the monocotyledons*, ed. C. R. Metcalfe, pp. 300, 374–389. Oxford University Press, Oxford.

XYRIDACEAE
Yellow-eyed-grass Family

Perennial or sometimes annual herbs growing in damp habitats, scapose, with short and sometimes bulbous rhizomes; roots fibrous. **Leaves** simple, entire, alternate, distichous, often equitant, basally tufted, narrow and grass-like, flat, terete, or filiform, parallel-veined, with an open basal sheath. **Inflorescence** indeterminate, capitate or spicate, cone-like with spirally arranged and closely imbricated bracts, terminal on a long scape. **Flowers** slightly zygomorphic, perfect, hypogynous, small, sessile, each in the axil of a stiff or coriaceous bract. **Calyx** of 3 sepals, distinct, unequal with the 2 lower (lateral) sepals chaffy, keeled, and persistent, and the inner (interior) sepal larger, membranous, enveloping the corolla (as a hood), and fugacious. **Corolla** of 3 petals, distinct or basally connate, clawed, ephemeral, usually yellow, marcescent. **Androecium** of 3 stamens (opposite the corolla lobes) and usually 3 staminodes (bifid, plumose, or bearded with moniliform hairs), epipetalous; filaments distinct, usually short and flattened; anthers dehiscing longitudinally, extrorse or latrorse. **Gynoecium** of 1 pistil, 3-carpellate; ovary superior, 1-locular or basally 3-locular; ovules few to numerous, orthotropous to anatropous, placentation parietal (1-locular ovary) or free central/basal (in ovary partitioned at base); style 1, trifid; stigmas 3, truncated. **Fruit** a 3-valved loculicidal capsule, enveloped by persistent corolla tube; seeds numerous, minute, usually apiculate and longitudinally striate; endosperm copious, mealy, starchy, proteinaceous; embryo small, lenticular or shield-shaped, apical.

Family characterization: Scapose herbs often growing in damp habitats; basally tufted, narrow leaves; flowers in the axils of stiff bracts congested into cone-like heads; calyx of 2 chaffy, keeled sepals and 1 membranous, hood-like sepal; ephemeral yellow corolla with spreading lobes; androecium of 3 stamens and often 3 plumose staminodes; and 3-valved capsule enclosed by persistent corolla.

Genera/species: *Xyris* (240–250 spp.), *Abolboda* (21 spp.), *Orectanthe* (2 spp.), *Achlyphila* (1 sp.), and *Aratitiyopea* (1 sp.) (Kral 1992)

Distribution: Widespread in tropical and subtropical areas, with relatively few representatives in temperate regions; especially diverse in the southeastern United States, tropical America, and southern Africa.

U.S./Canadian representatives: *Xyris* (23 spp.) (see Kral 1966, 1983)

Economic plants and products: A few *Xyris* species (yellow-eyed-grass) cultivated as ornamental plants for pools and aquaria.

Commentary: Most of the species in the family belong to the genus *Xyris*. In North America, the 23 species are confined to the eastern United States and Canada, particularly the southeastern United States (Kral 1983). The Xyridaceae generally are considered closely allied with the Eriocaulaceae, which they resemble in habit and some floral and embryological features (Dahlgren et al. 1985).

In the field, *Xyris* plants (Fig. 159: a) are unmistakable with their scapose habit, distichous grass-like leaves, cone-like inflorescences, and usually yellow flowers. The spicate to capitate inflorescence (Fig. 159: b) probably represents a condensed, complex panicle (Dahlgren et al. 1985). Generally, the flowers are borne singly in the upper bracts of the inflorescence, and the lower bracts are sterile. The tough bracts often are shiny and darkly colored (brownish) with a scarious margin. The unusual sepals are markedly dimorphic. The enlarged hood-like median sepal (Fig. 159: e) is membranous and caducous; the two boat-like lateral sepals (Fig. 159: d) are scarious (margins often ciliate) and persist in the infructescence (Fig. 159: o). The morphology of the lateral sepals provides important diagnostic features for determining species.

The corolla is ephemeral, with the flower usually open for only a few hours; typically, only one expanded flower is seen on each inflorescence (Fig. 159: b). No nectar is produced. Wind-pollination appears to be prevalent, although pollen-collecting bees occasionally visit the flowers (Kral 1983), evidently gathering the pollen that accumulates on the plumose staminodes (Fig. 159: i; Dahlgren et al. 1985).

REFERENCES CITED

Dahlgren, R. M. T., H. T. Clifford, and P. F. Yeo. 1985. *The families of monocotyledons*, pp. 388–391. Springer-Verlag, Berlin.

Kral, R. 1966. *Xyris* (Xyridaceae) of the continental United States and Canada. *Sida* 2:177–260.

Figure 159. Xyridaceae. *Xyris platylepis*: **a,** habit, ×⅓; **b,** inflorescence, ×3; **c,** two views of bud (flower subtended by two lateral sepals and one interior median sepal), ×6; **d,** one lateral sepal, ×6; **e,** median (interior) sepal before (top) and after (bottom) anthesis, ×6; **f,** flower, ×3; **g,** expanded corolla and androecium, ×3; **h,** stamen, ×6; **i,** staminode, ×6; **j,** pistil, ×5; **k,** longitudinal section of flower, ×5; **l,** cross section of ovary near apex, ×12; **m,** cross section of ovary at base, ×12; **n,** longitudinal section of ovary, ×12; **o,** capsule with persistent lateral sepals (persistent corolla removed), ×6; **p,** two views of capsule (persistent perianth removed), ×6; **q,** seed, ×30.

———. 1983. The Xyridaceae in the southeastern United States. *J. Arnold Arbor.* 64:421–429.

———. 1992. A treatment of American Xyridaceae exclusive of *Xyris. Ann. Missouri Bot. Gard.* 79:819–885.

COMMELINACEAE
Spiderwort Family

Annual or perennial herbs, sometimes climbing, succulent; roots fibrous or sometimes tuberous; stems nodose, soft and mucilaginous, or plant acaulescent. **Leaves** simple, entire, alternate, parallel-veined, flat or sharply folded (V-shaped in cross section), soft, with membranous, closed, basal sheath. **Inflorescence** determinate, cymose, or flower solitary, terminal or axillary, subtended by boat-shaped spathe or foliaceous bracts. **Flowers** actinomorphic or zygomorphic, perfect, hypogynous, showy. **Calyx** of 3 sepals, distinct, imbricate. **Corolla** of 3 petals, distinct, often clawed, equal or unequal (then one petal much reduced), ephemeral, deliquescent, commonly pink, blue, or purple, imbricate and crumpled in bud. **Androecium** of 6 stamens or of 3 stamens and 3 staminodes; filaments distinct, often fringed with long, brightly colored moniliform hairs; anthers basifixed, dehiscing by longitudinal slits, introrse. **Gynoecium** of 1 pistil, 3-carpellate; ovary superior, 3-locular; ovules 1 to few in each locule, orthotropous, placentation axile; style 1; stigma 1, capitate or trifid. **Fruit** a 3-valved loculicidal capsule, thin-walled; seeds crowded, large in relation to capsule, with conspicuous disc-like cap (embryotega) on the seed coat; endosperm copious, mealy; embryo small, beneath embryotega.

Family characterization: Succulent herbs with nodose, mucilaginous stems; leaves with tubular, closed, basal sheaths; 3 delicate, colorful petals; often fringed filaments; a loculicidal capsule as the fruit type; and seed coat with embryotega. Anatomical feature: mucilage cells/canals with raphides (needle-like bundles of calcium oxalate crystals; Tomlinson 1966).

Genera/species: 40/605

Distribution: Widespread in tropical and subtropical regions; also some representatives in warm temperate areas.

Major genera: *Commelina* (150–230 spp.) and *Aneilema* (60–100 spp.)

U.S./Canadian representatives: 13 genera/49 spp.; largest genera: *Tradescantia* and *Commelina*

Economic plants and products: Some weedy plants, such as *Commelina* and *Murdannia* spp. Ground cover plants grown in frost-free areas, and houseplants (species of 20 genera), including *Commelina* (dayflower), *Rhoeo* (oyster plant), *Setcreasea* (purple-queen), *Tradescantia* (spiderwort), and *Zebrina* (wandering-Jew).

Commentary: The Commelinaceae are divided into various subfamilies or tribes primarily based on floral symmetry and androecial characters, discussed briefly below (see also Tucker 1989, and Faden and Hunt 1991, for summaries). Variations of the cymose inflorescence also have been stressed in the literature as reliable subfamilial and generic characters (Woodson 1942). The circumscription of many genera (e.g., *Tradescantia* complex), however, varies considerably according to the authority (Brenan 1966).

The plants usually are recognizable in the field by the more or less succulent habit with jointed, mucilaginous stems. The soft, somewhat fleshy leaves are dilated at the base into a tubular sheath (Fig. 160: 1a; 2a,b). The inflorescence may be subtended by conspicuous bracts, as in *Commelina* (Fig. 160: 1b), which has a relatively large, folded spathe. As noted above, certain floral features also vary in the family. The zygomorphic flowers of *Commelina*, for example, comprise two large petals plus one very reduced petal (Fig. 160: 1c,d) and an androecium of three dimorphic stamens (Fig. 160: 1e,f) plus three staminodes (Fig. 160: 1g). In *Callisia* (Fig. 160: 2c) and *Tradescantia*, the flowers are actinomorphic, with three equal petals and six fertile stamens that may have moniliform hairs on the filaments (Fig. 160: 2d,e).

Although easily recognizable in the field, the deliquescent commelinaceous flowers are difficult to study from herbarium specimens. The entomophilous flowers with colorful petals and often fringed filaments are attractive to insects (especially bees), which visit to collect pollen (Vogel 1978). Marked heteromorphism of the androecium, as in *Commelina*, represents a visible division between food (staminodes) and fertile (stamens) elements. The staminodes, which produce little or no pollen, often have conspicuous, enlarged, yellow connectives (Fig. 160: 1g). Self-pollination may occur in certain genera when the stigma and stamens come into contact with each other as the flower withers (Owens 1981).

REFERENCES CITED

Brenan, J. P. M. 1966. The classification of the Commelinaceae. *J. Linn. Soc., Bot.* 59:349–395.

Faden, R. B., and D. R. Hunt. 1991. The classification of the Commelinaceae. *Taxon* 40:19–31.

Owens, S. J. 1981. Self-incompatibility in the Commelinaceae. *Ann. Bot.* (Oxford) 47:567–581.

Tomlinson, P. B. 1966. Anatomical data in the classification of Commelinaceae. *J. Linn. Soc., Bot.* 59:371–395.

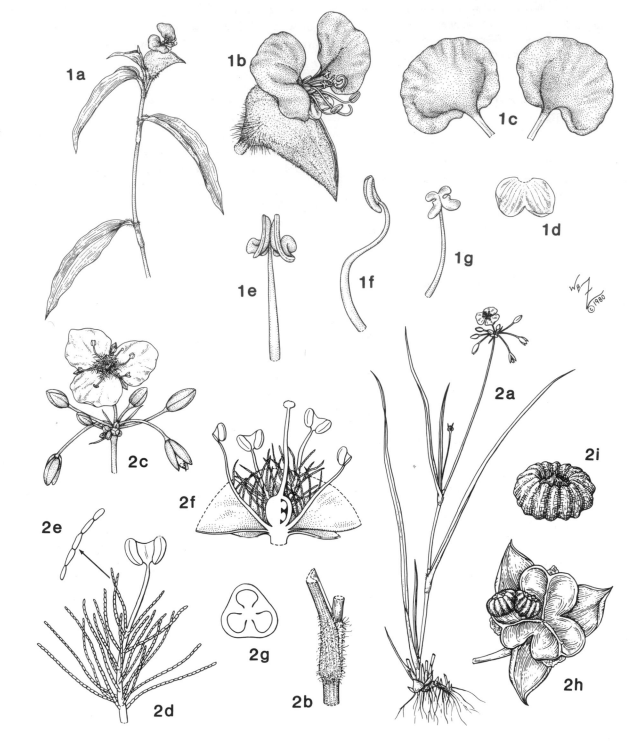

Figure 160. Commelinaceae. 1, *Commelina erecta*: **a,** habit, ×½; **b,** inflorescence (cyme subtended by folded spathe), ×2; **c,** lateral petals, ×2½; **d,** median petal, ×2½; **e,** median stamen, ×6; **f,** lateral stamen, ×6; **g,** staminode, ×6. **2,** *Callisia graminea* (*Cuthbertia graminea*): **a,** habit, ×⅖; **b,** node with sheathing leaf base, ×2¼; **c,** inflorescence, ×1½; **d,** stamen, ×9; **e,** detail of moniliform staminal hair, ×25; **f,** longitudinal section of flower, ×6; **g,** cross section of ovary, ×12; **h,** capsule (with persistent calyx), ×6; **i,** seed, ×12.

Tucker, G. C. 1989. The genera of Commelinaceae in the southeastern United States. *J. Arnold Arbor.* 70:97–130.

Vogel, S. 1978. Evolutionary shifts from reward to deception in pollen flowers. In *The pollination of flowers by insects*, ed. A. J. Richards, pp. 89–96. Linn. Soc. Symp. Ser. No. 6. Academic Press, London.

Woodson, R. E. 1942. Commentary on the North American genera of Commelinaceae. *Ann. Missouri Bot. Gard.* 29: 141–154.

ERIOCAULACEAE
Pipewort Family

Annual or usually perennial herbs growing in damp habitats, scapose; roots fibrous or spongy. **Leaves** simple, entire, alternate, basally tufted, narrow and grass-like, sometimes sheathing at base. **Inflorescence** indeterminate, capitate, subtended by involucre of chaffy bracts, terminal on a scape enclosed basally by spathe-like sheath. **Flowers** actinomorphic to slightly zygomorphic, imperfect (plants usually monoecious), hypogynous, minute, sessile or shortly stalked, each in the axil of a chaffy bract. **Perianth** usually biseriate, somewhat differentiated into calyx and corolla, chaffy, often fringed with hairs. **Calyx** of 2 or (more often) 3 sepals, distinct or connate, usually valvate. **Corolla** of 2 or (more often) 3 petals, distinct or variously connate, sometimes unequal, cupular, frequently with nectariferous glands (near apex on adaxial side), sometimes reduced to hairs in carpellate flowers, frequently whitish, usually valvate. **Androecium** of 4 or (more often) 6 (biseriate) stamens or occasionally 2 or 3 (uniseriate) stamens (then opposite the petals), epipetalous and sometimes arising from a raised stalk (androphore formed by basally fused petals); filaments distinct, often unequal; anthers basifixed or dorsifixed (then often versatile), 1- or 2-locular, dehiscing longitudinally, introrse; staminodes sometimes present in carpellate flowers. **Gynoecium** of 1 pistil, 3- to (less often) 2-carpellate, stipitate (raised on a gynophore); ovary superior, 2- or 3-locular; ovules 1 in each locule, orthotropous, pendulous, placentation axile-apical; style 1, bifid or trifid, with each branch simple or forked, persistent; stigmas slender, papillose; rudimentary pistil (gland-like) usually present in staminate flowers. **Fruit** a loculicidal capsule, membranous; seeds ellipsoidal; endosperm copious, mealy, starchy; embryo minute, undifferentiated, lenticular, apical.

Family characterization: Scapose herbs typically growing in damp habitats; basally tufted, narrow leaves; minute, 2- or 3-merous, imperfect flowers congested into whitish to greyish heads subtended by chaffy involucre of bracts; fringed, chaffy perianth;

stipitate ovary with a solitary apical ovule in each locule; branched style; membranous, loculicidal capsule; and seeds with tiny, undifferentiated embryos. Pollen grains with spiral apertures ("spiraperturate"; Fig. 161: p). Tissues with calcium oxalate crystals. Anatomical features: vascular bundles of stems amphivasal (xylem surrounding the phloem) and arranged in one ring (reduced stem) or several rings (well-developed aerial stem; Tomlinson 1969); and tenuinucellate ovules.

Genera/species: 9/1,175

Distribution: Pantropical, with relatively few representatives in temperate regions; especially diverse in South America.

Major genera: *Paepalanthus* (485 spp.), *Eriocaulon* (400 spp.), *Syngonanthus* (195–200 spp.), and *Leiothrix* (65 spp.)

U.S./Canadian representatives: *Eriocaulon* (10 spp.), *Lachnocaulon* (5 spp.), and *Syngonanthus* (1 sp.)

Economic plants and products: Dried inflorescences of *Syngonanthus* species used as "everlasting flowers" in floral arrangements. Species of *Eriocaulon* (pipewort) occasionally cultivated as ornamentals.

Commentary: The Eriocaulaceae are divided into two or three tribes primarily based on anther and tepal morphology (Kral 1989), and genera generally are distinguished by microscopic characters of the minute flowers (Kral 1966). Most of the species in the family belong to three neotropical genera: *Paepalanthus*, *Eriocaulon*, and *Syngonanthus*.

The family generally is considered to be closely allied with the Xyridaceae, which they resemble in habit: clustered, grass-like, basal leaves and a terminal head on a long scape that much exceeds the leaves (Fig. 161: a). The plants typically inhabit wet soils or are emergent from shallow water. Of our species, those of *Eriocaulon* tend to occur in much wetter areas than those of *Lachnocaulon* and *Syngonanthus*, which prefer drier, acidic and sandy substrates.

The Eriocaulaceae often have been referred to as the "Compositae of the monocots" due to the minute flowers (florets) crowded in tight spirals on involucrate, button-like heads (Fig. 161: b). The heads in this case, however, are whitish or greyish, never colorful. The involucre consists of numerous, chaffy, imbricate bracts ("phyllaries") that vary considerably in coloration and pubescence. As in the Asteraceae, the flat to conical inflorescence axis ("receptacle"; Fig. 161: c) may be glabrous or variously hairy or chaffy. The imperfect, more or less sessile flowers are minute, and

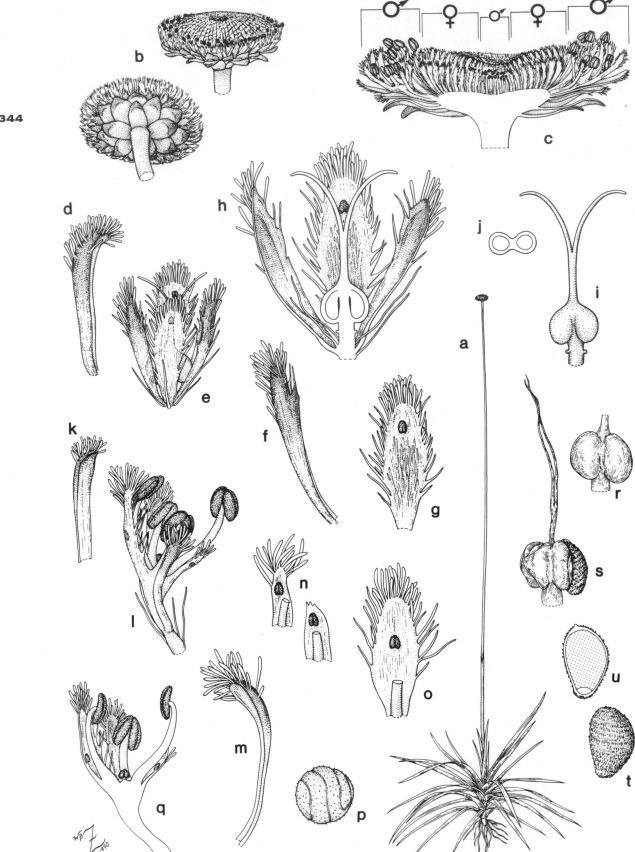

individual flowers are difficult to discern on the densely packed heads. As in many of the Asteraceae, the flowers are each subtended by a scarious and scale-like bract (often with apical hairs; Fig. 161: d,k). The chaffy, biseriate perianth, typically whitish and fringed with hairs, is only somewhat differentiated into calyx and corolla (Fig. 161: f,g,m–o). In the staminate flowers of some species (e.g., *Syngonanthus*), the petals may be fused basally, forming an androphore; in the carpellate flowers of *Lachnocaulon*, the petals are modified into tufts of hairs. Sometimes marginal flowers in a head have an "outer" enlarged petal—appearing somewhat like the ray florets in a heterogamous head of the Asteraceae. Staminate and carpellate flowers usually occur in the same inflorescence, with one sex typically at the periphery of the head and the other filling the center (see Fig. 161: c). As in the Asteraceae, the flowers mature from the outside toward the center (i.e., peripheral flowers open first).

Pollination in the family has been reported as probably anemophilous and/or entomophilous, and self-pollination within the heads presumably is common. In most species, the mature anthers and styles are well exserted. Conspicuous nectar glands occur on the inner, subapical surface of *Eriocaulon* petals (Fig. 161: g,n,o). Although Robertson (1927) reports visits of various bees, flies, and butterflies, Kral (1989) suggests that most of our species of *Eriocaulon* usually are wind pollinated due to the rarity of these insect visits.

REFERENCES CITED

Kral, R. 1966. Eriocaulaceae of continental North America north of Mexico. *Sida* 2:285–332.

———. 1989. The genera of Eriocaulaceae in the southeastern United States. *J. Arnold Arbor.* 70:131–142.

Robertson, C. 1927. Florida flowers and insects. *Trans. Acad. Sci. St. Louis* 25:277–324.

Tomlinson, P. B. 1969. *Commelinales—Zingiberales*, vol. 3 of *Anatomy of the monocotyledons*, ed. C. R. Metcalfe, pp. 146–192. Oxford University Press, Oxford.

JUNCACEAE
Rush Family

Annual or perennial grass-like herbs growing in damp habitats, with erect or creeping sympodial rhizomes; roots fibrous; stems scapose, terete, solid. **Leaves** simple, entire, alternate, 3-ranked, basally tufted, flat to cylindrical, often narrow and grass-like, parallel-veined, with open basal sheath. **Inflorescence** determinate, cymose and often appearing paniculate, corymbose, or capitate, terminal. **Flowers** actinomorphic, perfect or sometimes imperfect (then plants dioecious or monoecious), hypogynous, small and inconspicuous. **Perianth** of usually 6 tepals, distinct, often membranous and chaffy, greenish, reddish-brown, to black, imbricate. **Androecium** of 3 to 6 stamens (opposite the tepals); filaments distinct; anthers basifixed, dehiscing by longitudinal slits, introrse; pollen grains in tetrads. **Gynoecium** of 1 pistil, 3-carpellate; ovary superior, 1- or 3-locular; ovules 1 to numerous in each locule, anatropous, placentation axile or parietal; style(s) 1 or 3, sometimes very short; stigmas 3, often brush-like. **Fruit** a 3-valved loculicidal capsule; seeds small, often appendaged; endosperm copious, fleshy; embryo minute, straight, surrounded by endosperm.

Family characterization: Grass-like herbs of damp habitats; inconspicuous anemophilous flowers in congested cymose inflorescences; 6 chaffy tepals; tetradenous pollen; 3-valved capsule; and appendaged seeds. Anatomical features: aerenchyma tissue in the stems and leaves; and a large pith composed of stellate cells (Fig. 162: 2b; Cutler 1969).

Genera/species: 8/300

Distribution: Generally widespread in damp habitats; particularly frequent in temperate and montane regions.

Major genus: *Juncus* (225–300 spp.)

U.S./Canadian representatives: *Juncus* (107 spp.) and *Luzula* (23 spp.)

Figure 161. Eriocaulaceae. *Eriocaulon compressum*: **a,** habit, ×¼; **b,** two views of inflorescence (head subtended by bracts), ×2¼; **c,** longitudinal section of head showing relative positions of staminate and carpellate flowers and order of development, ×5; **d,** bract subtending carpellate flower, ×18; **e,** carpellate flower, ×18; **f,** sepal from carpellate flower, ×20; **g,** petal from carpellate flower (petals subequal), ×20; **h,** longitudinal section of carpellate flower, ×25; **i,** pistil, ×25; **j,** cross section of ovary, ×25; **k,** bract subtending staminate flower, ×12; **l,** staminate flower, ×12; **m,** sepal from staminate flower, ×18; **n,** range in morphology of smaller petal of staminate flowers (from different inflorescences), ×18; **o,** larger petal from staminate flower, ×18; **p,** pollen grain (spiraperturate), ×630; **q,** longitudinal section of staminate flower, ×12; **r,** capsule before dehiscence, ×22; **s,** dehiscing capsule, ×22; **t,** seed, ×26; **u,** longitudinal section of seed, ×26.

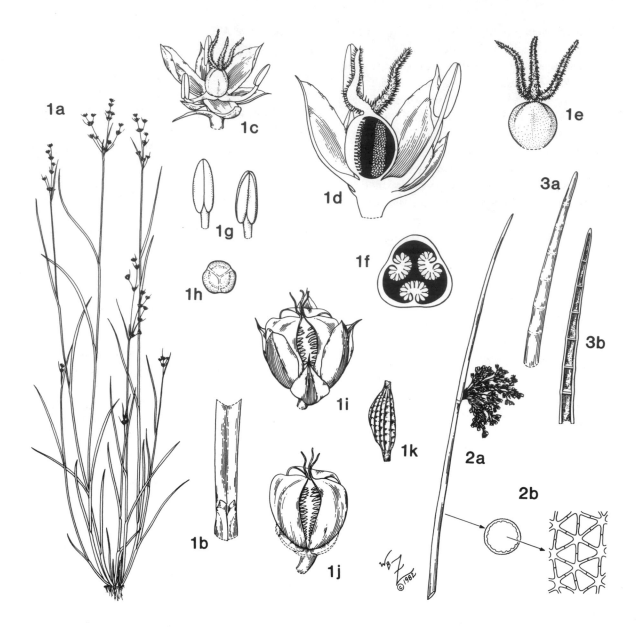

Figure 162. Juncaceae. 1, *Juncus marginatus*: **a,** habit, ×¼; **b,** sheathing leaf base, ×3; **c,** flower, ×8; **d,** longitudinal section of flower, ×15; **e,** pistil, ×12; **f,** cross section of ovary, ×22; **g,** two views of anther, ×15; **h,** pollen tetrad, ×270; **i,** loculicidal capsule with persistent perianth, ×10; **j,** loculicidal capsule (perianth removed), ×10; **k,** seed, ×45. **2,** *Juncus effusus*: **a,** inflorescence, ×½; **b,** cross section of stem, ×3, with detail of stellate cells of pith, ×100. **3,** *Juncus scirpoides*: **a,** leaf, ×3; **b,** longitudinal section of leaf, ×3.

Economic plants and products: Weaving materials from *Juncus* spp. (rush). Ornamental plants: species of *Juncus* and *Luzula*.

Commentary: The 225 to 300 species of *Juncus* constitute most of the family. The rushes often are confused with the sedges (Cyperaceae) and the grasses (Poaceae), which they resemble superficially in habit (Fig. 162: 1a). Upon closer examination, the three groups actually are strikingly different in vegetative, floral, and fruit characters; the stem and leaves, in particular, provide obvious distinguishing field characters (see Table 21).

As in the Poaceae and Cyperaceae, the anatomical features of rush leaves often have taxonomic significance (see Stebbins and Khush 1961). In particular, rush leaves show a wide range of form in transverse section (Adamson 1925): the blade may be flat and grass-like (*Juncus marginatus*), V-shaped or arc-shaped, circular and solid with a pith like the stem, to circular and hollow with numerous transverse septa (*J. scirpoides*; Fig. 162: 3a,b). In many species, such as *J. effusus*, the terminal inflorescence is subtended by a cylindrical leaf, which forms an apparent continuation of the stem (Fig. 162: 2a). Sometimes the leaves are very reduced (e.g., *J. effusus*).

The inconspicuous star-like flowers (Fig. 162: 1c), with reduced perianths and feathery stigmas, are adapted for anemophily. However, certain species of *Juncus* and *Luzula* with white or reddish perianths may attract insect pollinators. Self-pollination usually is prevented by protogyny (Ertter 1986).

Like the sedges and grasses, the rushes should be collected with infructescences for identification to species: the shape and size of the mature capsule and ripe seeds often are diagnostic. The minute seeds of most *Juncus* species probably are wind dispersed. However, some species of *Juncus* and *Luzula* have seeds that become attached to animals with a sticky material, and the seeds of certain *Luzula* species have caruncles, which attract ants (Dahlgren et al. 1985).

REFERENCES CITED

Adamson, R. S. 1925. On the leaf structure of *Juncus*. *Ann. Bot.* (Oxford) 39:599–612.

Cutler, D. F. 1969. *Juncales*, vol.4 of *Anatomy of the monocotyledons*, ed. C. R. Metcalfe, pp. 17–77. Oxford University Press, Oxford.

Dahlgren, R. M. T., H. T. Clifford, and P. F. Yeo. 1985. *The families of the monocotyledons*, p. 406. Springer-Verlag, Berlin.

Ertter, B. 1986. The *Juncus triformis* complex. *Mem. New York Bot. Gard.* 39:1–90.

Stebbins, G. L., and G. S. Khush. 1961. Variation in the organization of the stomatal complex in the leaf epidermis of monocotyledons and its bearing on their phylogeny. *Amer. J. Bot.* 48:51–59.

CYPERACEAE
Sedge Family

Perennial or sometimes annual grass-like herbs growing in damp to marshy habitats, with creeping sympodial rhizomes; roots fibrous; stems often clustered and unbranched (except for the inflorescences), scapose, triangular in cross section, with solid internodes. **Leaves** simple, entire, alternate, basally tufted or cauline or 3-ranked, narrow and grass-like, parallel-veined, with closed basal sheath. **Inflorescence** indeterminate, of small spikes (spikelets) in spicate, racemose, paniculate, or umbellate arrangements, subtended by bracts, terminal. **Flowers** actinomorphic, perfect or imperfect (then plants usually monoecious), hypogynous, minute and inconspicuous, each subtended by a chaffy bract. **Perianth** absent, or sometimes of varying numbers of bristles, scales, or hairs, persistent. **Androecium** of usually 3 stamens; filaments distinct; anthers basifixed, dehiscing longitudinally, introrse. **Gynoecium** of 1 pistil, 2- or 3-carpellate; ovary superior, 1-locular, sometimes enveloped by a persistent sac-like bract (perigynium); ovule solitary, anatropous, erect, placentation basal; style single, bifid, or trifid, sometimes with persistent thickened base (tubercle); stigmas often feathery. **Fruit** an achene, lenticular (2 carpels), trigonous (3 carpels), to sometimes terete; endosperm mealy or fleshy; embryo small, surrounded by endosperm.

Family characterization: Perennial rhizomatous herbs of wet habitats; caespitose, solid, three-sided stems; 3-ranked linear leaves with closed basal sheath; spikelets arranged in various inflorescences; reduced, anemophilous flowers each subtended by a chaffy bract; reduced or absent perianth; basifixed anthers; and an achene as the fruit type. Single pollen grains ("pseudomonads") represent a degraded tetrad (three of the four nuclei resulting from the tetrad division degenerate as the wall of the pollen mother cell develops into the exine; Kern 1974). Anatomical features: aerenchyma tissue in the stems and leaves; and silica bodies in the epidermal cells of the leaves (Metcalfe 1971).

Genera/species: 146/5,315

Distribution: Generally widespread on barren soil or in damp habitats; most abundant in subarctic to temperate regions.

Major genera: *Carex* (1,000–2,000 spp.), *Cyperus* (550–600 spp.), *Fimbristylis* (250–300 spp.), *Scirpus* (200–300 spp.), *Rhynchospora* (200 spp.), *Scleria* (200 spp.), and *Eleocharis* (150–200 spp.)

U.S./Canadian representatives: 26 genera/959 spp.; largest genera: *Carex*, *Cyperus*, and *Rhynchospora*

Economic plants and products: Edible tubers from species of *Cyperus* (chufa) and *Eleocharis* (Chinese water-chestnut). Weaving materials from stems and leaves of *Cyperus* and *Scirpus*. Several weedy species, such as *Cyperus* spp. Ornamental plants for pools: species of *Carex*, *Cyperus*, and *Eleocharis*.

Commentary: The Cyperaceae generally are divided into subfamilies, genera, and even species primarily based on features of the inflorescence. Fruit characters also are particularly important for generic and specific identification (Tucker 1987). The sedges are distinguished from the rushes (Juncaceae) and grasses (Poaceae), which they resemble superficially, in Table 21.

The reduced flowers may be perfect or imperfect. Usually the perianth is absent, but some genera, such as *Rhynchospora* (Fig. 163: 3), are characterized by a

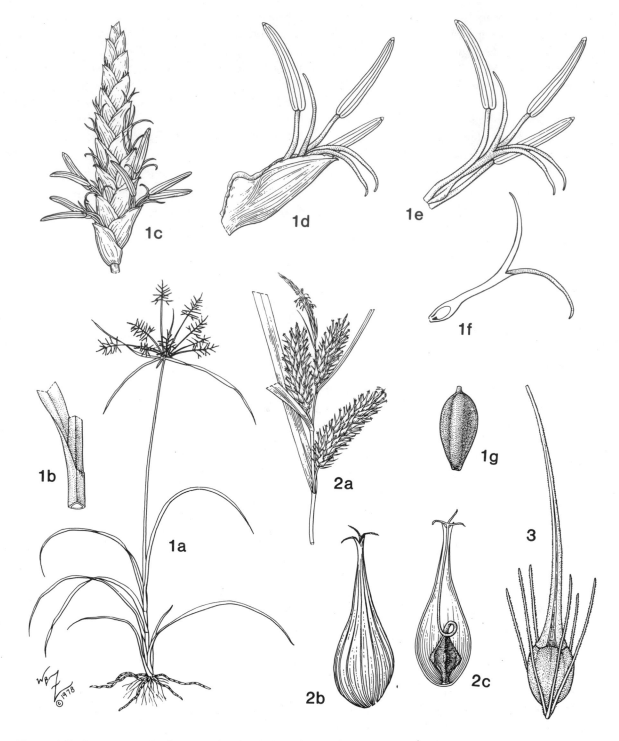

Figure 163. Cyperaceae. 1, *Cyperus esculentus*: **a,** habit, ×⅕; **b,** node showing closed leaf sheath, ×1¼; **c,** spikelet, ×6; **d,** flower and subtending bract, ×12; **e,** flower, ×12; **f,** longitudinal section of pistil, ×12; **g,** achene, ×12. **2,** *Carex lupulina*: **a,** inflorescence, ×⅖; **b,** fruiting perigynium, ×3; **c,** achene with half of subtending perigynium removed, ×3. **3,** *Rhynchospora careyana*: achene with persistent perianth (bristles), ×4.

reduced perianth of bristles or scales (Blaser 1941). The inconspicuous flowers (Fig. 163: 1e), each subtended by a chaffy bract (glume, scale; Fig. 163: 1d), are spirally or distichously arranged into small spikes (spikelets; Fig. 163: 1c). The complexity and diversity

of the sedge inflorescence morphology have long been analyzed in relation to the taxonomy and phylogeny of the family (see Holttum 1948; Koyama 1961; and Dahlgren et al. 1985). Sometimes a single terminal spikelet comprises the entire inflorescence (*Eleocharis*),

TABLE 21. Traditional chart of characters used to separate the Juncaceae, Cyperaceae, and Poaceae—all plants with a "grass-like" habit. Only the most common character states for each family are listed; exceptions are not included.

CHARACTER	JUNCACEAE (RUSHES)	CYPERACEAE (SEDGES)	POACEAE (GRASSES)
GENERA/SPECIES	8/300	146/5,315	650–785/10,000
HABITAT	wet areas	wet areas or sterile soils	dry to moist areas
STEM CROSS SECTION	terete	triangular	terete or ellipsoid
INTERNODES	solid, with large pith	usually solid	usually hollow, or less commonly solid
NODES	not jointed	not jointed	jointed
LEAF RANKS	3	3	2
LEAF BLADE	flat to terete	flat	flat
LEAF SHEATH	open	closed	open and with ligule
INFLORESCENCE	basically cymose, and often congested	arranged in spikelets	arranged in spikelets
NUMBER OF BRACTS SUBTENDING EACH FLOWER	2 or more	1 (glume, scale)	usually 2 (palea and lemma)
PERIANTH	usually 6 chaffy tepals	absent, or reduced to a varying number of bristles or scales	reduced to 2 (or sometimes 3) lodicules
ANTHER ATTACHMENT	basifixed	basifixed	basifixed, but deeply sagittate and appearing versatile
POLLEN	in tetrads	single, but each grain ("pseudomonad") representing a degraded tetrad	single
FRUIT TYPE	loculicidal capsule	achene	caryopsis (grain)
EMBRYO	surrounded by endosperm	embedded in base of endosperm	outside of endosperm

but usually numerous spikelets are arranged in larger spicate, paniculate, or umbellate inflorescences, as in *Cyperus* (Fig. 163: 1a). Usually an involucre of keeled bracts (prophylls) occurs at the base of the inflorescence branches.

One of the most specialized inflorescence types occurs in *Carex*, a genus that includes one-third to one-half of the species of the family. The staminate and carpellate flowers often are arranged in separate inflorescences on the same plant (Fig. 163: 2a). Superficially, the carpellate inflorescence appears to be a simple spikelet; actually, the complex inflorescence is composed of one-flowered spikelets arranged in a compound spike. Each carpellate spikelet (a single carpellate flower and its reduced secondary axis) is enclosed by a modified, sac-like prophyll (perigynium), which persists in fruit (Fig. 163: 2b,c).

As in the grasses, the flowers of the Cyperaceae characteristically are adapted for anemophily, with large anthers, long filaments, and prominent stigmas. A few sedges with showy bracts, such as some warm-temperate to tropical *Rhynchospora* (formerly segregated as *Dichromena*), attract insect pollinators (Leppik 1955).

REFERENCES CITED

Blaser, H. W. 1941. Studies in the morphology of the Cyperaceae. I. Morphology of the flowers. B. Rhynchosporoid genera. *Amer. J. Bot.* 28:832–838.

Dahlgren, R. M. T., H. T. Clifford, and P. F. Yeo. 1985. *The families of the monocotyledons*, pp. 407–418. Springer-Verlag, Berlin.

Holttum, R. E. 1948. The spikelet in Cyperaceae. *Bot. Rev.* (Lancaster) 14:525–541.

Kern, J. H. 1974. Cyperaceae. *Fl. Males. Bull.* 7:435–753.

Koyama, T. 1961. Classification of the family Cyperaceae (1). *J. Fac. Sci. Univ. Tokyo, Sect. 3, Bot.* 8:37–81.

Leppik, E. E. 1955. *Dichromena ciliata*, a noteworthy entomophilous plant among Cyperaceae. *Amer. J. Bot.* 42:455–458.

Metcalfe, C. R. 1971. *Cyperaceae*, vol. 5 of *Anatomy of the monocotyledons*, ed. C. R. Metcalfe, pp. 13–28. Oxford University Press, Oxford.

Tucker, G. C. 1987. The genera of Cyperaceae in the southeastern United States. *J. Arnold Arbor.* 68:361–445.

POACEAE OR GRAMINEAE
Grass Family

Annual or usually perennial herbs, occasionally woody, with rhizomes or stolons, often forming dense tufts or mats; roots fibrous; stems erect to creeping, solitary or caespitose, sometimes branched, usually terete, often with hollow internodes and jointed (swollen) nodes. **Leaves** simple, entire, alternate, 2-ranked, cauline and also crowded at plant base, parallel-veined, with open basal sheath and appendage (ligule) present at sheath-blade junction. **Inflorescence** indeterminate, of small spikes (spikelets) in spicate, racemose, or paniculate arrangements, usually terminal. **Flowers** obscurely zygomorphic, perfect or occasionally imperfect (then plants usually monoecious), hypogynous, small and inconspicuous, typically subtended by 2 bracts (lemma and palea). **Perianth** represented by 1 to 3 minute scales (lodicules), hyaline or fleshy. **Androecium** of usually 1 to 3 stamens; filaments distinct, long and flexuous; anthers basifixed but deeply sagittate and appearing versatile, dehiscing by longitudinal slits, latrorse. **Gynoecium** of 1 pistil, 2- or 3-carpellate; ovary superior, 1-locular; ovule solitary, anatropous, adnate to one side of carpel; styles 2 or sometimes 3, terminal or lateral; stigma plumose or sometimes papillose. **Fruit** a caryopsis (grain); seed adnate to pericarp; endosperm copious, usually hard; embryo outside of endosperm.

Family characterization: Tufted, herbaceous plants with rhizomes or stolons; terete stems with hollow internodes and jointed nodes; 2-ranked linear leaves with a ligule and an open basal sheath; spikelets arranged in various inflorescences; anemophilous, inconspicuous flowers subtended by 2 bracts (palea and lemma); reduced perianth of scales (lodicules); deeply sagittate, functionally versatile anthers; 2 feathery stigmas; and a caryopsis as the fruit type. Silica present (deposited throughout the plant). Anatomical features: several distinctive modifications of the leaf (discussed below).

Genera/species: 650–785/±10,000 (see Watson and Dallwitz 1992)

Distribution: Cosmopolitan; comprising the climax vegetation in many areas with low annual rainfall (e.g., plains, steppes).

Major genera: *Poa* (500 spp.), *Panicum* (370 spp.), *Festuca* (360 spp.), *Eragrostis* (350 spp.), *Paspalum* (320 spp.), *Stipa* (300 spp.), *Aristida* (290 spp.), *Calamagrostis* (230 spp.), *Agrostis* (220 spp.), *Digitaria* (220 spp.), *Muhlenbergia* (160 spp.), *Sporobolus* (160 spp.), and *Elymus* (150 spp.) (Watson and Dallwitz 1992); the various species estimates cited in the literature are due to extensive hybridization, polyploidy, and apomixis in the family.

U.S./Canadian representatives: 231 genera/1,490 spp.; largest genera: *Poa, Panicum, Muhlenbergia, Eragrostis, Paspalum,* and *Bromus*

Economic plants and products: Food crops (staple cereals): *Avena* (oats), *Hordeum* (barley), *Oryza* (rice), *Secale* (rye), *Triticum* (wheat), and *Zea* (corn). Sugar and molasses from *Saccharum* (sugarcane) and *Sorghum*. Forage and fodder grasses: *Agropyron, Andropogon, Brachiaria, Digitaria, Festuca, Panicum, Pennisetum,* and *Setaria*. Alcoholic beverages from *Oryza* (saké); *Saccharum* (rum); and *Hordeum, Secale,* and *Zea* (whiskeys). Paper and building materials from several, such as *Bambusa* (bamboo). Essential oils from *Cymbopogon* (lemongrass) and *Vetiveria* (vetiver oil). Ornamental plants grown mainly as turf grasses (species of 100+ genera), including *Bambusa* (bamboo), *Cortaderia* (pampas grass), *Cynodon* (Bermuda grass), *Festuca* (fescue), *Paspalum* (Bahia grass), *Poa* (bluegrass), *Stenotaphrum* (St. Augustine grass), and *Zoysia*.

Commentary: The Poaceae probably comprise the most economically important group of angiosperms (see the preceding section). This large and complex family has been divided into two to twelve subfamilies and as many as fifty to sixty tribes. Recent treatments (e.g., Gould and Shaw 1983; Clayton and Renvoize 1986; Ellis 1986; Watson and Dallwitz 1992) usually accept five or six subfamilies; Table 22 summarizes some major distinctions between these taxa. In the

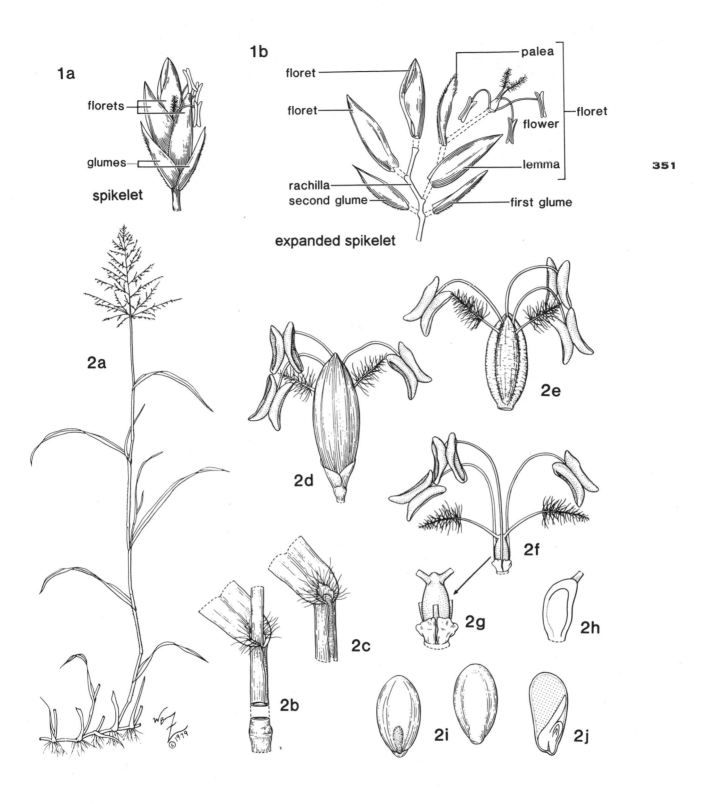

Figure 164. Poaceae. 1, *Eragrostis spectabilis*: **a,** spikelet, ×12; **b,** expanded spikelet with one expanded floret, ×12. **2,** *Panicum maximum*: **a,** habit, ×1/16; **b,** node showing sheathing leaf base, ×1¼; **c,** leaf base with ligule, ×1¼; **d,** spikelet, ×12; **e,** floret (flower plus palea and lemma), ×12; **f,** flower, ×12; **g,** ovary and lodicules, ×25; **h,** longitudinal section of ovary, ×25; **i,** two views of caryopsis, ×12; **j,** longitudinal section of caryopsis, ×12.

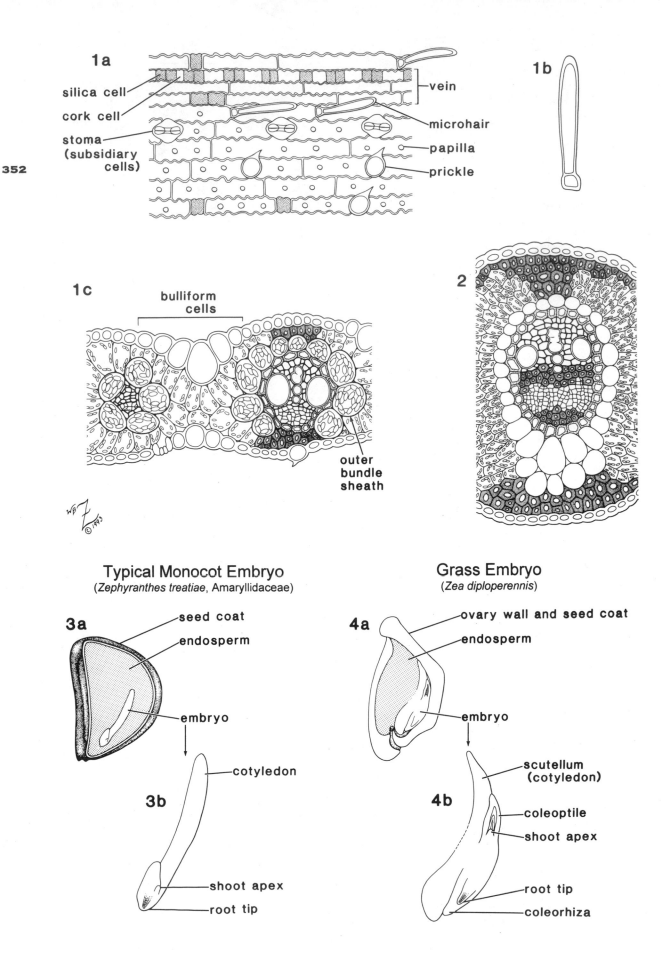

1a

silica cell
cork cell
stoma (subsidiary cells)

vein
microhair
papilla
prickle

352

1b

1c

bulliform cells

outer bundle sheath

2

Typical Monocot Embryo
(*Zephyranthes treatiae*, Amaryllidaceae)

3a

seed coat
endosperm

embryo

3b

cotyledon

shoot apex
root tip

Grass Embryo
(*Zea diploperennis*)

4a

ovary wall and seed coat
endosperm

embryo

4b

scutellum (cotyledon)

coleoptile
shoot apex

root tip
coleorhiza

past, genera, tribes, and subfamilies were classified primarily on the basis of spikelet structure and arrangement. More recent intrafamilial concepts are based on additional cytological, anatomical, and physiological data, such as chromosome number, leaf anatomy, and embryo morphology. In particular, numerous studies have concentrated on the specific type of leaf anatomy (structure of the mesophyll and bundle sheath) associated with each of the two photosynthetic pathways (C_3 or C_4) occurring within the family (Fig. 165: 1c, 2; see Brown 1977, Hattersley 1986, and the discussion of Kranz anatomy in the commentary on the Portulacaceae). The rushes (Juncaceae) and the sedges (Cyperaceae), characterized by grass-like habits and small flowers, are compared to the Poaceae in Table 21.

The grasses, superficially similar in habit and morphology, actually have diverse floral and vegetative structure. Like the orchids, the grasses have a complete set of terminology for vegetative and floral parts. For example, the leaves exhibit many interesting morphological and anatomical features (detailed in Metcalfe 1960 and Ellis 1976, 1979, 1986). A characteristic grass leaf has a flattened blade with an open basal sheath, which closely envelops the stem (Fig. 164: 2b); a ligule, represented by a thin membrane or a fringe of hairs, occurs along the inner surface of the blade-sheath junction (Fig. 164: 2c). Grass blades have well-developed sclerenchyma (seen in cross section), which usually occurs between the larger vascular bundles and the epidermis (Fig. 165: 1c, 2). Bands of enlarged epidermal cells with thin anticlinal walls (bulliform cells) participate in involution and folding movements of the leaves (Fig. 165: 1c). The epidermis, as seen in surface view, consists of long and short cells arranged in longitudinal rows or zones (Fig. 165: 1a); some of these cells may have suberized walls (cork cells) or may be filled with silica bodies of various shapes (silica cells). The morphology of the silica bodies provides subfamilial, tribal, and generic characters. The morphology of minute epidermal hairs, called microhairs, also has diagnostic value; these hairs generally are two-celled, with a relatively thick-walled basal cell and a thin-walled, delicate terminal cell (Fig. 165: 1b).

Grass flowers and inflorescences also are variable and complex (Clifford 1986). The inconspicuous flowers (Fig. 164: 1b, 2f) basically consist of a reduced perianth of one to three, membranous or fleshy scales (lodicules; Fig. 164: 2g), one to three stamens, and a three-carpellate pistil with two stigmas. This entire flower usually is subtended by two protective bracts, the palea (uppermost) and the lemma (Fig. 164: 1b, 2e). The palea, which arises from the base of the flower, usually is membranous, two-nerved, and two-keeled. The lemma encloses the flower and the palea until anthesis and varies in morphology. A flower with subtending palea and lemma composes a unit termed a "floret" (Fig. 164: 1b, 2e). One to many sessile florets are arranged distichously on an axis (rachilla), forming a small spike (spikelet). Each spikelet characteristically bears two sterile modified floral bracts (glumes) at the base (Fig. 164: 1b, 2d). The "first glume" refers to the lower bract; the "second glume," to the upper. The spikelets, the basic inflorescence units, themselves are arranged into large spikes, racemes, or panicles.

Recognizing various modifications of spikelet structure is important for grass identification. For example, the flower, either of the glumes, and/or the palea may be reduced or absent, resulting in difficult spikelet identification. In approximately one-half of the grasses, the lowest spikelet in an inflorescence is represented by a sterile lemma that resembles the second glume. Texture, shape, and nerve number of the lemma and glumes also are important taxonomically. In addition, the glumes, lemma, and palea are sometimes awned.

The reduced flowers are anemophilous, although pollen-gathering insects have been reported for some grass species (Soderstrom and Calderón 1971; Terrell and Batra 1984). At anthesis, the increased turgidity of the lodicules exerts pressure upon the bases of the lemma and palea and forces them apart, followed by the exsertion of the stigmas and anthers (Conner 1986). The large anthers on long, slender filaments are functionally versatile and move freely in the wind (Fig. 164: 2d–f). The feathery stigmas (Fig. 164: 2d–f) are well adapted for catching the wind-borne pollen.

Figure 165. Poaceae (continued). 1, *Panicum maximum*: **a,** abaxial surface of leaf, ×370; **b,** microhair from leaf, ×750; **c,** cross section of leaf showing bulliform cells and vascular bundles with Kranz anatomy (enlarged outer bundle sheath cells with chloroplasts; C_4 photosynthetic pathway), ×330. **2,** *Cortaderia selloana*: cross section of leaf (C_3 plant: note outer bundle leaf cells lacking chloroplasts), ×225. **3,** *Zephyranthes treatiae* (Amaryllidaceae), as an example of a typical monocot embryo morphology: **a,** longitudinal section of seed, ×7½; **b,** longitudinal section of embryo, ×18. **4,** *Zea diploperennis*: **a,** longitudinal section of caryopsis, ×5; **b,** longitudinal section of embryo, ×10.

TABLE 22. Some major morphological differences between the subfamilies of the Poaceae. Characters are from Gould and Shaw (1983), who emphasize features of North American taxa; Campbell (1985); Dahlgren et al. (1985); Ellis (1986); and Watson and Dallwitz (1992). Pooideae, Panicoideae, and Chloridoideae are the major subfamilies in the United States. "Disarticulation" refers to the breaking apart of the rachilla at maturity; "reduced floret" refers to those that are sterile or staminate. The C_4 photosynthetic pathway is associated with Kranz anatomy (see the discussion under Portulacaceae).

CHARACTER	POOIDEAE	PANICOIDEAE	CHLORIDOIDEAE	BAMBUSOIDEAE	ARUNDINOIDEAE
GENERA/SPECIES	153/3,275	207/3,290	145/1,360	91/1,110	55/645
DISTRIBUTION	mainly temperate, with a large proportion adapted to cool/cold climates	comprise the majority of grasses of subtropical-tropical regions	most diverse in warm and semiarid regions (e.g., southwestern United States)	mainly tropical to subtropical areas	widespread; mainly in southern hemisphere
REPRESENTATIVE GENERA	Avena, Bromus, Festuca, Hordeum, Poa, Secale, Triticum	Andropogon, Digitaria, Eriochloa, Pennisetum, Panicum, Paspalum, Saccharum, Setaria, Sorghastrum, Sorghum, Stenotaphrum, Zea	Aristida, Chloris, Eragrostis, Muhlenbergia, Sporobolus, Zoysia	Arundinaria, Bambusa, Oryza, Pharus, Phyllostachys	Arundo, Chasmanthium, Cortaderia, Danthonia, Phragmites
HABIT	annual or perennial, herbaceous	annual or perennial, herbaceous	annual or perennial, mostly herbaceous	usually perennial, woody or herbaceous	usually perennial, herbaceous and mostly large and suffrutescent
LIGULE	usually unfringed membrane	fringe of hairs or fringed membrane	fringed membrane or fringe of hairs	fringed or unfringed membrane; often abaxial + adaxial ligules	fringed membrane, cartilaginous, or fringe of hairs
BICELLULAR MICROHAIRS	absent	elongated, finger-like; cells usually subequal	inflated; spherical or clavate; cells equal to subequal; hairs sometimes also single-celled	elongated, finger-like; cells subequal	elongated, finger-like, or sometimes absent; cells variable

Character					
SILICA CELL SHAPE	oblong or nodular	cross-, saddle-, or dumbbell-shaped	usually saddle- or cross-shaped	variable; often saddle-shaped	variable; often dumbbell-shaped
BUNDLE SHEATH	double	usually single	usually double	double	double
PHOTOSYNTHETIC PATHWAY	C_3	C_3 or C_4	C_4	C_3	usually C_3
SPIKELET COMPOSITION	1 to several florets (reduced florets, when present, are above fertile florets)	1 fertile floret + 1 reduced floret below (perfect or imperfect)	1 to several florets (reduced florets, when present, are usually above fertile florets)	1 to many florets (compound spikelets common)	1 to many florets (perfect or imperfect)
SPIKELET DISARTICULATION RELATIVE TO GLUMES	usually above	below	above	usually above (also between florets)	usually above
LODICULES	2 usually elongate, pointed; fleshy with membranous tips	2 short, truncate; fleshy	2 cuneate; fleshy	usually 3 rounded, ciliate; membranous	2 shape variable; fleshy or membranous, ± ciliate
STAMEN NUMBER	1 to 3	usually 1 to 3	1 to 3	3, or sometimes 6 or more	1 to 3
STYLE NUMBER	2	usually 2	2	1 to 3	2
EMBRYO SIZE IN RELATION TO ENDOSPERM	small	large	large	small	small to moderately large
BASE CHROMOSOME NUMBER	usually 7	usually 9 or 10	usually 9 or 10	10, 11, or 12	variable

The grass fruit (Figs. 164: 2i,j; 165: 4a,b) is unusual because some of the pericarp tissue (ovary wall) becomes united with the seed coat during development. The resulting dry, indehiscent, one-seeded fruit is called a caryopsis or, in common terminology, a grain. The small seed-like fruits often are shed enclosed by the palea and lemma, and sometimes by the glumes and/or various other parts of the infructescence. Teeth, bristles, and awns on the bracts aid in dispersal by wind or animals; wind vibration of the awn(s) may cause a burrowing action that embeds the fruit in the soil (Peart 1979).

The structural details and morphological variation of the complex grass embryo have been important in grass systematics (Reeder 1957). The embryo of other monocotyledons (e.g., Fig. 165: 3a,b) is composed of a single axis, terminated by the cotyledon at one end, and by the radicle (embryonic root) at the other. The unique grass embryo (Fig. 165: 4a,b), however, consists of a relatively large, well-developed cotyledon (scutellum) appressed to the endosperm, which bears the embryonic plant laterally. The plumule (shoot) and radicle, each subtended by a protective sheath, terminate the axis at opposing ends. A modified leaf, the coleoptile, sheaths the embryonic shoot, and a structure called the coleorhiza shields the included root. Some grass embryos are characterized by an additional scale-like outgrowth, the epiblast, situated on the outside of the embryo, opposite the scutellar node.

REFERENCES CITED

Brown, W. V. 1977. The Kranz syndrome and its subtypes in grass systematics. *Mem. Torrey Bot. Club* 23:1–97.

Campbell, C. S. 1985. The subfamilies and tribes of Gramineae (Poaceae) in the southeastern United States. *J. Arnold Arbor.* 66:123–199.

Clayton, W. D., and S. A. Renvoize. 1986. Genera graminum. *Kew Bull.*, Addit. Ser. 13:1–389.

Clifford, H. T. 1986. Spikelet and floral morphology. In *Grass systematics and evolution*, ed. T. R. Soderstrom, K. W. Hilu, C. S. Campbell, and M. E. Barkworth, pp. 21–30. Smithsonian Institution Press, Washington, D.C.

Conner, H. E. 1986. Reproductive biology in the grasses. In *Grass systematics and evolution*, ed. T. R. Soderstrom, K. W. Hilu, C. S. Campbell, and M. E. Barkworth, pp. 117–132. Smithsonian Institution Press, Washington, D.C.

Dahlgren, R. M. T., H. T. Clifford, and P. F. Yeo. 1985. *The families of the monocotyledons*, pp. 425–453. Springer-Verlag, Berlin.

Ellis, R. P. 1976. A procedure for standardizing comparative leaf anatomy in the Poaceae. I. The leaf-blade as viewed in transverse section. *Bothalia* 12:65–109.

——. 1979. Ibid., II. The epidermis as seen in surface view. *Bothalia* 12:641–671.

——. 1986. A review of comparative leaf blade anatomy in the systematics of the Poaceae: The past twenty-five years. In *Grass systematics and evolution*, ed. T. R. Soderstrom, K. W. Hilu, C. S. Campbell, and M. E. Barkworth, pp. 3–10. Smithsonian Institution Press, Washington, D.C.

Gould, F. W., and R. B. Shaw. 1983. *Grass systematics*, pp. 93–134. Texas A & M University Press, College Station.

Hattersley, P. W. 1986. Variations in photosynthetic pathway. In *Grass systematics and evolution*, ed. T. R. Soderstrom, K. W. Hilu, C. S. Campbell, and M. E. Barkworth, pp. 49–64. Smithsonian Institution Press, Washington, D.C.

Metcalfe, C. R. 1960. *Gramineae*, vol. 1 of *Anatomy of the monocotyledons*, ed. C. R. Metcalfe, pp. xvii–xxxi, xli–l. Oxford University Press, Oxford.

Peart, M. H. 1979. Experiments on the biological significance of the morphology of seed-dispersal units in grasses. *J. Ecol.* 67:843–863.

Reeder, J. R. 1957. The embryo in grass systematics. *Amer. J. Bot.* 44:756–768.

Soderstrom, T. R., and C. E. Calderón. 1971. Insect pollination in tropical rain forest grasses. *Biotropica* 3:1–16.

Terrell, E. E., and S. W. T. Batra. 1984. Insects collect pollen of eastern wildrice, *Zizania aquatica* (Poaceae). *Castanea* 49:31–34.

Watson, L., and M. J. Dallwitz. 1992. *The grass genera of the world*, pp. 1–3, 45–54. C.A.B. International, Wallingford, England.

GLOSSARY

The following definitions apply to the angiosperms in this work; these same terms may have different, more general, or more specific interpretations in other texts. Many terms relating either to special features defined in this text or to basic botany (or morphology) are not repeated here. For more thorough treatments, refer to the glossaries in taxonomy texts such as Lawrence (1951), Cronquist (1968, 1988), Radford et al. (1974), and Benson (1979), as well as to more inclusive botanical dictionaries such as Stearn (1966), Cook (1968), Little and Jones (1980), Allaby (1992), and Huxley (1992). My use of inflorescence terms generally follows Rickett (1944, 1955), and the fruit definitions are from Judd (1985). The definitions of terms concerning the principles of plant systematics are based on Simpson (1961), Wiley (1981), and Mayr and Ashlock (1991). Synonyms and commonly used alternative spellings (from other texts) are listed parenthetically. The drawings accompanying these definitions are to the same scale as in the family plates from which they are taken.

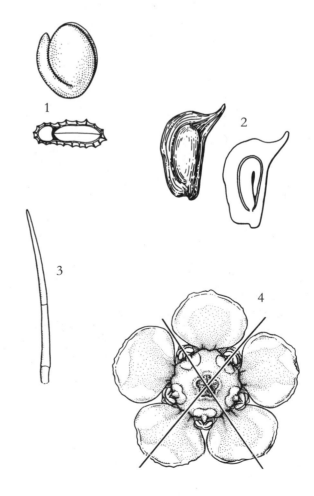

abaxial (= dorsal): Facing away from the axis (e.g., the underside of a leaf).

acaulescent: Stemless or apparently stemless (main stem underground).

accrescent: Increasing in size with age, after flowering.

accumbent: (Of cotyledons), having the edges adjacent to the radicle. (**1**. *Rorippa nasturtium-aquaticum* [Brassicaceae]: embryo [top] and cross section of seed showing accumbent cotyledons [bottom])

achene (akene): A more or less small, indehiscent, dry fruit with a thin and close-fitting wall surrounding the single seed. (**2**. *Sagittaria lancifolia* [Alismataceae]: achene [left] and longitudinal section [right])

achlorophyllous: Lacking chlorophyll.

acicular: Needle-shaped or pointed. (**3**. *Utricularia foliosa* [Lentibulariaceae]: "trigger bristle" from insectivorous bladder)

actinomorphic (= radially symmetrical, regular): Divisible into equal halves by two or more planes. (**4**. *Euonymus americanus* [Celastraceae]: top view of flower showing two [of several] planes dividing the flower into equal halves)

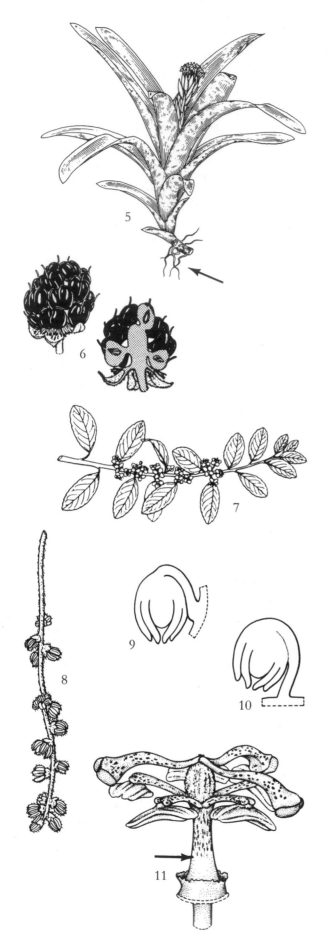

adaxial (= ventral): Facing toward the axis (e.g., the upper side of a leaf).

adherent: Touching closely (more or less adhesively), but not fused to, a dissimilar organ (see also **connivent**).

adnate: United to an unlike organ or part.

adventitious roots: Secondary roots appearing from any part of the plant (e.g., stem or leaf) other than the root system. (**5.** *Billbergia pyramidalis* [Bromeliaceae]: plant with adventitious roots)

aerenchyma: Parenchymatous tissue with particularly large intercellular spaces (air spaces); commonly found in tissues of wetland plants.

aestivation: The arrangement of perianth parts in the bud.

aggregate fruit: The product of several separate carpels of a single gynoecium that stick together or are connate (see also **syncarp**). (**6.** *Rubus trivialis* [Rosaceae]: aggregate fruit of drupelets [left] and longitudinal section [right])

alkaloid: A general term for any of a group of cyclic, nitrogenous bases occurring in a limited number of plants (and animals); many are powerful drugs (e.g., caffeine, cocaine, nicotine).

alternate: (Of structures, such as leaves), arranged singly at different heights along the stem (i.e., one leaf per node). (**7.** *Ilex vomitoria* [Aquifoliaceae]: carpellate flowering branch with alternate leaves)

ament (= catkin): Any pendent, flexuous, and spikelike inflorescence of reduced, often anemophilous, flowers. (**8.** *Quercus geminata* [Fagaceae]: staminate ament)

amphitropous ovule: An ovule whose funicle (stalk) is curved around it so that the ovule tip and funicle base are near each other. (**9.** amphitropous ovule [diagrammatic])

anatropous ovule: An ovule whose body is fully inverted and adnate to the funicle, so that the micropyle is basal, adjoining the funicle. (**10.** anatropous ovule [diagrammatic])

androecium: A collective term for the "male" reproductive units (stamens) of the flower.

androgynophore: A special stalk supporting both stamens and pistil above the point of attachment of the perianth. (**11.** *Passiflora incarnata* [Passifloraceae]: gynoecium and androecium on androgynophore)

anemophilous: Wind-pollinated.

annual: A plant with a life cycle completed within one year.

anther: The pollen-producing apical portion of the stamen. (**12.** *Poncirus trifoliata* [Rutaceae]: longitudinal section of flower [pistil intact])

anthesis: Generally, the time when the flower expands or opens.

anthocarp: The accessory perianth tube plus enclosed fruit (achene or utricle) of the Nyctaginaceae. (**13.** *Boerhavia diffusa* [Nyctaginaceae]: fruit [left] and longitudinal section [right])

anthocyanins: A group of water-soluble flavonoid pigments, ranging in color from blue to violet to red; widespread among angiosperms.

apical placentation: Attachment of ovules at the distal (apical) end of the ovary. (**14.** *Morus rubra* [Urticaceae]: longitudinal section of carpellate flower showing apical placentation of ovule)

apiculate: Terminated by a short, sharp, flexible point.

apocarpous: Having the carpels separate. (**15.** *Cocculus carolinus* [Menispermaceae]: apocarpous gynoecium [left] and longitudinal section of carpellate flower [right])

apomixis: Generally, asexual reproduction (i.e., without meiosis or syngamy); more specifically, also used to denote vegetative reproduction in which part of a plant becomes detached and develops into a separate individual.

apomorphy: A derived character state.

apopetalous: Having the petals separate.

appressed (adpressed): Closely and flatly pressed against another structure.

arcuate: Arching or curved.

areole: Generally, a small, clearly marked space; in the Cactaceae, a specialized cushion-like area on the stem that bears a tuft of hairs, spines, and/or glochids. (**16.** *Opuntia humifusa* [Cactaceae]: portion of plant [left] with detail of stem showing areole [right])

aril: A large and specialized outgrowth of the seed, growing out of the funicle at the hilum (attachment area of seed to funicle). (**17.** *Guaiacum sanctum* [Zygophyllaceae]: longitudinal section of seed enclosed by aril)

artificial classification: A classification based on a few convenient and conspicuous characters without attention to those that indicate relationship.

auricle: An ear-shaped projection. (**18.** *Cabomba caroliniana* [Cabombaceae]: inner tepal with two basal auricles)

awn: A long bristle. (**19.** *Bidens alba* [Asteraceae]: achene with two apical awns)

axil of leaf: The upper angle between a leaf and the stem. (**20.** *Conocarpus erectus* [Combretaceae]: node with leaf)

axile placentation: Attachment of ovules at or near the center of a compound ovary, on the inner angle

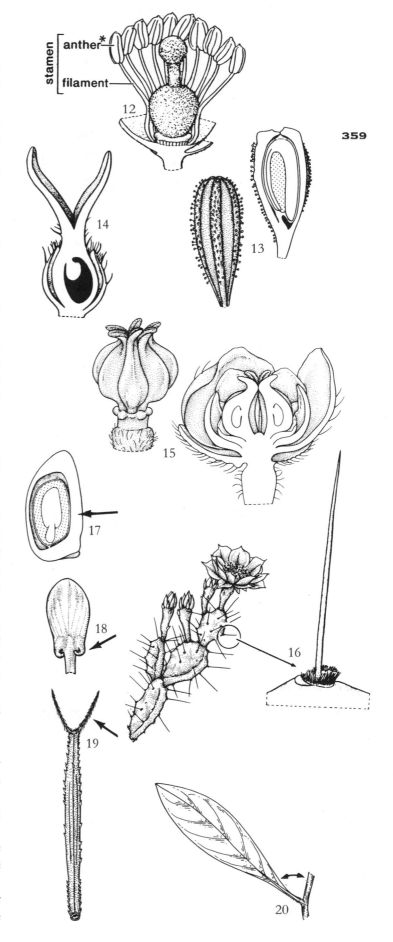

359

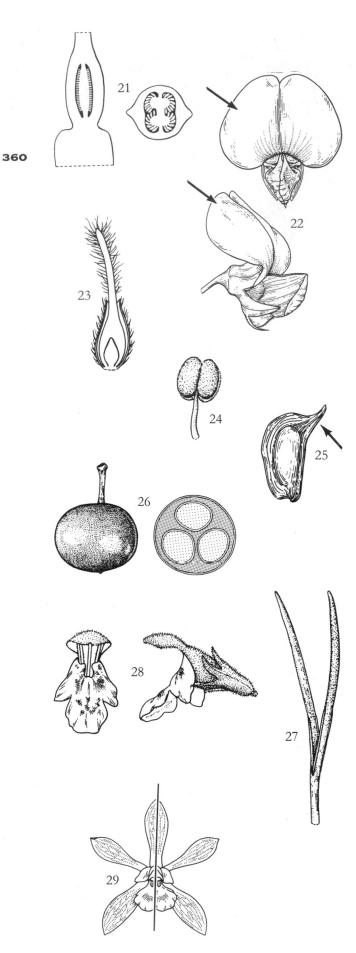

360

formed by the septa (partitions). (**21**. *Campsis radicans* [Bignoniaceae]: longitudinal [left] and cross [right] sections of ovary)

axillary: In the axil.

baccate: Fleshy or pulpy.

banner (= standard): Usually the uppermost and largest petal of the papilionaceous flowers present in many of the Fabaceae (Faboideae). (**22**. *Crotalaria spectabilis* [Fabaceae]: two views of flower)

basal placentation: Attachment of ovules at the proximal (basal) end of the ovary. (**23**. *Boehmeria cylindrica* [Urticaceae]: longitudinal section of carpellate flower showing basal placentation of ovule)

basifixed: Attached at the base. (**24**. *Nyssa ogeche* [Cornaceae]: stamen with filament attached at anther base)

basipetal: Developing in a longitudinal plane from an apical or distal point toward the base.

beak: A long, prominent, and substantial point; in particular, applied to prolongations of pistils/fruits. (**25**. *Sagittaria lancifolia* [Alismataceae]: achene with pronounced beak)

berry: An indehiscent, fleshy fruit with few to many seeds (rarely, a single seed); the flesh may be more or less homogeneous, or heterogeneous with the outer portion more firm or leathery; septa are present in some. (**26**. *Smilax auriculata* [Smilacaceae]: berry [left] and cross section of berry with three seeds [right])

betacyanins: A group of nitrogenous, water-soluble pigments, ranging in color from blue to violet to red; found in particular plants, such as Amaranthaceae.

betalains: A general term for nitrogenous, water-soluble pigments (including betacyanins), ranging in color from red, blue, and violet to yellow and orange.

bicollateral bundle: A vascular bundle having phloem on two sides of the xylem.

biennial: A plant with a life cycle completed in two years.

bifid: Two-cleft. (**27**. *Carphephorus corymbosus* [Asteraceae]: apex of bifid style)

bilabiate: Two-lipped; often applied to the corolla or calyx. (**28**. *Stachys floridana* [Lamiaceae]: two views of flower with bilabiate corolla)

bilaterally symmetrical (= zygomorphic): Divisible into equal halves in one plane only, usually along an anterior-posterior line. (**29**. *Encyclia tampensis* [Orchidaceae]: front view of flower with line dividing the flower into equal halves)

biseriate: In two whorls.

bract: A much-reduced leaf often associated with flowers. (**30.** *Costus pulverulentus* [Costaceae]: cone-like inflorescence with conspicuous, imbricate bracts)

bracteole (= bractlet): A secondary bract.

bractlet: See **bracteole.**

bristle: A stiff hair.

budding: A form of vegetative reproduction (apomixis) in which a new individual develops as an outgrowth of a mature plant. (**31.** *Wolffia columbiana* [Lemnaceae]: dorsal view of thallus with vegetative bud)

bulb: A thickened underground bud (in a resting state) composed of a reduced stem and scales (leaf bases). (**32.** *Allium canadense* [Alliaceae]: plant [left] and detail of base showing bulb [right])

bulbil: A small bulb or bulb-like structure, often formed in a leaf axil, which separates from the parent plant and functions as a vegetative propagule. (**33.** *Dioscorea bulbifera* [Dioscoreaceae]: node with aerial bulbil)

bulliform cell: An enlarged epidermal cell participating in the folding movements of the leaves of the Poaceae. (**34.** *Panicum maximum* [Poaceae]: cross section of leaf through an area with bulliform cells)

C$_3$ photosynthesis: A pathway of photosynthesis in which compounds with three carbon atoms are the immediate products of CO_2 fixation: the Calvin-Benson cycle, the part of photosynthesis in which sugar is made from CO_2; the process occurs in all plants.

C$_4$ photosynthesis: A pathway of photosynthesis in which compounds with four carbon atoms are the immediate products of CO_2 fixation; characteristic of tropical plants (such as some Poaceae) growing in high light intensities, high temperatures, and aridity.

caducous: Falling off very early.

caespitose (cespitose): Growing in tufts or mats.

calyptra (= operculum): A cap, cover, or lid.

calyx: The outermost whorl of the typical perianth, composed of the sepals.

campanulate: Bell-shaped. (**35.** *Boerhavia diffusa* [Nyctaginaceae]: flower [upper portion of perianth campanulate])

campylotropous ovule: An ovule curved by uneven growth so that the micropyle is near the funicle, but the ovule is not adnate to the funicle. (**36.** campylotropous ovule [diagrammatic])

capitate: Generally, of a structure, enlarged and swollen at the tip; also, of an inflorescence, with flowers in a dense cluster (head).

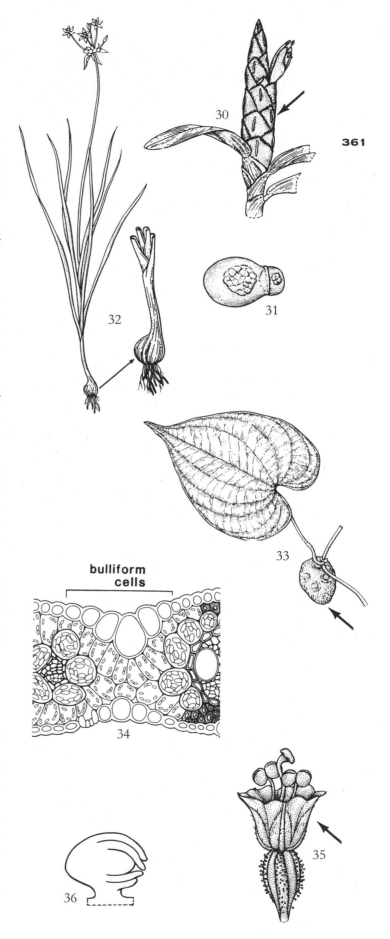

361

bulliform cells

34

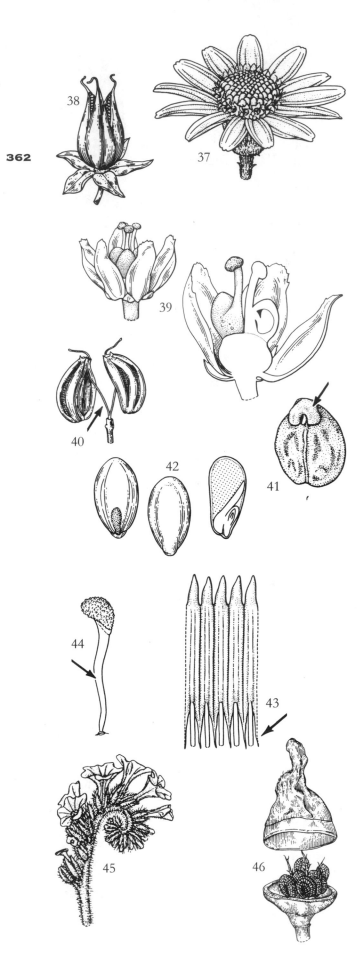

capitulum (= head): A compact inflorescence composed of a very short (often discoid) axis and usually sessile flowers. (**37.** *Senecio glabellus* [Asteraceae]: inflorescence)

capsule: A dry fruit derived from a two- to many-carpellate gynoecium that opens (in various ways) to release few to many seeds. (**38.** *Hypericum myrtifolium* [Clusiaceae]: capsule [with persistent calyx])

carpel: The foliar, ovule-bearing unit of a flower that forms either all (simple pistil) or part of a pistil (compound pistil).

carpellate flower (= pistillate flower): A "female flower," one that has a functional gynoecium and lacks a functional androecium. (**39.** *Zanthoxylum clava-herculis* [Rutaceae]: carpellate flower [left] and longitudinal section [right])

carpophore: The wiry stalk (primarily of carpellary origin) that supports each part of the dehiscing schizocarp; applied to fruits of some of the Apiaceae, but similar structures occur in fruits of other families, such as the Sapindaceae. (**40.** *Cicuta mexicana* [Apiaceae]: dehiscing schizocarp)

caruncle (= oil body): A fleshy outgrowth from the integuments situated at the attachment point (hilum) of a seed. (**41.** *Croton argyranthemus* [Euphorbiaceae]: seed)

caryopsis (= grain): A more or less small, indehiscent, dry fruit with a thin wall surrounding and fused to the single seed. (**42.** *Panicum maximum* [Poaceae]: two views of caryopsis [left and center] and longitudinal section [right])

catkin: See **ament**.

caudate: Bearing a tail-like appendage. (**43.** *Gnaphalium obtusifolium* [Asteraceae]: expanded androecium [abaxial surface] showing caudate anther bases)

caudicle: A portion of the pollinium of some orchids that usually is slender and composed of viscin with some pollen grains. (**44.** *Platanthera ciliaris* [Orchidaceae]: pollinium with well-developed caudicle)

cauliflorous: Producing flowers that appear to grow directly upon woody branches or trunks.

cauline: Pertaining or belonging to the stem.

chaffy: Thin, dry, and membranous.

chasmogamous flowers: Flowers that open normally.

circinate: Coiled at the tip, with the apex nearest the center of the coil. (**45.** *Heliotropium amplexicaule* [Boraginaceae]: circinate inflorescence)

circumscissile capsule (= pyxis): A capsule that dehisces by a horizontal line around the fruit, the top coming off as a lid. (**46.** *Portulaca pilosa* [Portulacaceae]: dehiscing circumscissile capsule)

cladistics: A taxonomic theory in which the relationships of taxa are determined entirely on the basis of shared derived character states (i.e., recency of common ancestry).

cladode: See **phylloclade**.

cladogram: A graphic representation (phylogenetic tree) based on the principles of cladism; i.e., a branching diagram showing sister group relationships.

classification: The delimitation, ordering, and ranking of taxa.

clavate: Club-shaped. (**47.** *Yucca flaccida* [Agavaceae]: stamen with clavate filament)

clawed: (Of a petal, sepal, or tepal), narrowed abruptly basally into a long, slender stalk. (**48.** *Raphanus raphanistrum* [Brassicaceae]: clawed petal)

cleistogamous flowers: Flowers that are small, closed, self-fertilized. (**49.** *Helianthemum corymbosum* [Cistaceae]: cleistogamous flower)

coherent: Touching closely (more or less adhesively), but not fused to, a similar organ (see also **connivent**).

coleoptile: The sheath enclosing the shoot of the embryo in the Poaceae. (**50.** See **coleorhiza**.)

coleorhiza: The sheath enclosing the radicle (embryonic root) of the embryo in the Poaceae. (**50.** *Zea diploperennis* [Poaceae]: longitudinal section of embryo)

colleter: A multicellular hair producing a sticky secretion, often occurring on the adaxial surface of the sepals/calyx lobes, stipules, and petiole bases (and adjacent stem surfaces) of certain plants, such as those of the Loganiaceae. (**51.** *Cephalanthus occidentalis* [Rubiaceae]: adaxial side of stipule [left] with detail showing colleters [right])

columella: The elongated floral axis that supports the carpels in certain flowers/fruits (e.g., Geraniaceae). (**52.** *Geranium carolinianum* [Geraniaceae]: dehisced schizocarp)

column: The structure formed by the union of stamens, style, and stigma present in many of the Orchidaceae. (**53.** *Encyclia tampensis* [Orchidaceae]: lateral view of column and ovary [top] and underside of column [bottom])

coma: A tuft of hairs. (**54.** *Salix caroliniana* [Salicaceae]: dehiscing capsule releasing comose seeds)

combretaceous hair: A thick-walled, unicellular hair with sharply pointed apex and thickened bulbous base that appears bicellular due to an optical illusion caused by a conical or concave cellulose membrane in the cell wall; found in Combretaceae and Cistaceae. (**55.** *Conocarpus erectus* [Combretaceae]: combretaceous hair from leaf)

commissure: The face along which adjoining car-

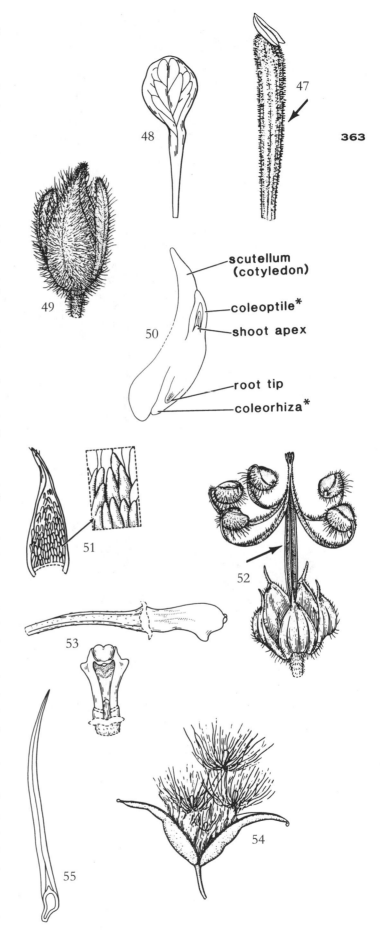

363

scutellum (cotyledon)

coleoptile*

shoot apex

root tip

coleorhiza*

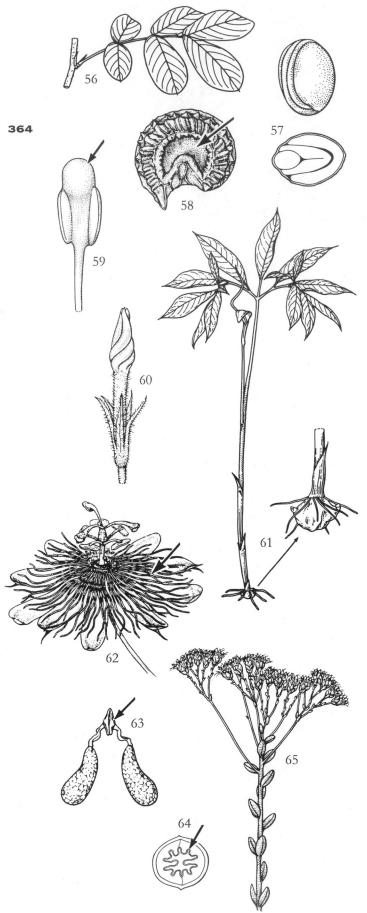

pels (some Apiaceae) or stigmas or style branches (Brassicaceae) are appressed.

comose: Having a tuft of hairs.

compound leaf: A leaf composed of two or more leaflets. (**56**. *Senna obtusifolia* [Fabaceae]: leaf)

conduplicate: Folded together lengthwise; in particular, of cotyledons, having dorsal sides folded lengthwise over the radicle and more or less enclosing it. (**57**. *Raphanus raphanistrum* [Brassicaceae]: embryo [top] and cross section of seed showing conduplicate cotyledons [bottom])

condyle: The adaxial ingrowth of the endocarp that appears as a conspicuous depression on the seeds of the Menispermaceae. (**58**. *Cocculus carolinus* [Menispermaceae]: pyrene)

connate: United to a similar organ or part.

connective: The tissue connecting the two locules of an anther. (**59**. *Begonia cucullata* [Begoniaceae]: stamen with pronounced connective)

connivent: Coming together or converging with, but not fused to, a similar or dissimilar organ, the parts often arching; generally, used synonymously for both **adherent** and **coherent**.

contorted: Twisted.

contractile root: A specialized type of root, often found in bulbous plants, that contracts in length and pulls the bulb (or shoot) deeper into the soil.

convolute aestivation: An arrangement of perianth parts (sepal, petal, tepal, or lobe) in bud in which one edge of a part overlaps the next part while its other edge or margin is overlapped by the preceding part. (**60**. *Phlox drummondii* [Polemoniaceae]: bud showing corolla with convolute aestivation)

coriaceous: Leathery.

corm: A solid bulb-like part of the stem, usually subterranean, sometimes bearing small scale leaves. (**61**. *Arisaema dracontium* [Araceae]: plant [left] and detail of base showing corm [right])

corolla: The inner whorl (or several whorls) of the typical perianth, composed of the petals.

corona: Generally, any appendage between (or from) the corolla and/or the androecium. (**62**. *Passiflora incarnata* [Passifloraceae]: flower [corona filiform])

corpusculum: A connection ("gland") between the arms of pollinia present in many of the Apocynaceae. (**63**. *Asclepias humistrata* [Apocynaceae]: pollinia)

corrugated: Irregularly folded or wrinkled. (**64**. *Carya glabra* [Juglandaceae]: cross section of nut showing corrugated cotyledons)

corymbose: (Of an inflorescence), short, broad, and flat-topped. (**65**. *Carphephorus corymbosus* [Asteraceae]: corymbose inflorescence)

cosmopolitan: Distributed more or less worldwide.

costapalmate: (Of a palmate leaf), having a petiole that continues through the blade as a distinct midrib. (**66.** *Sabal palmetto* [Arecaceae]: leaf)

cotyledon: The primary leaf, or one of two primary leaves, in the embryo. (**67.** *Myrcianthes fragrans* [Myrtaceae]: longitudinal section of seed)

crassulacean acid metabolism (CAM): A variant of the C_4 photosynthetic pathway; characteristic of most succulent plants, such as the Cactaceae.

cruciform: Arranged diagonally (in a cross). (**68.** *Raphanus raphanistrum* [Brassicaceae]: top view of flower showing cruciform arrangement of petals)

cucullate: Hooded or hood-shaped. (**69.** *Xyris platylepis* [Xyridaceae]: cucullate median sepal)

cupule: A cup-like structure at the base of some fruits. (**70.** *Quercus geminata* [Fagaceae]: nut [acorn] subtended by scaly cupule)

cyanogenic glycosides: Certain types of toxic glycosides that liberate hydrogen cyanide when acted upon by appropriate enzymes; widely distributed among angiosperms, often occurring in seeds.

cyathium: A type of specialized inflorescence of some of the Euphorbiaceae in which the reduced unisexual flowers are congested within a bracteate envelope. (**71.** *Euphorbia heterophylla* [Euphorbiaceae]: cyathium with [top] and without [bottom] involucre)

cyme: Generally, a determinate, compound, and frequently more or less flat-topped inflorescence; the basic cymose unit is a three-flowered cluster composed of a peduncle bearing a terminal flower and, below it, two bracts with each bract subtending a lateral flower. (**72.** cyme [diagrammatic])

cystolith: A crystalline concretion (usually calcium carbonate) contained within a specialized cell (lithocyst). (**73.** *Boehmeria cylindrica* [Urticaceae]: surface of dried leaf showing cystoliths [circular raised areas; left] and cross section of fresh leaf showing cystolith in epidermal cell [right])

deciduous: Falling off at the end of each growing season.

decompound: More-than-once compound (or divided).

decurrent: Extending downward—as when the leaf bases extend down and are adnate to the stem, or when stigmas extend down the style. (**74.** *Platanus occidentalis* [Platanaceae]: carpel with decurrent stigma)

decussate: (Of opposite leaves), alternating in pairs at right angles. (**75.** *Stachys floridana* [Lamiaceae]: upper portion of plant showing decussate leaves)

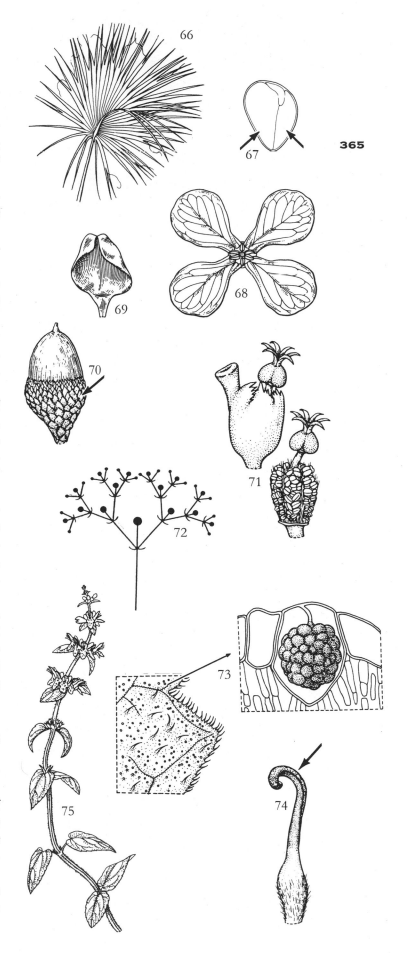

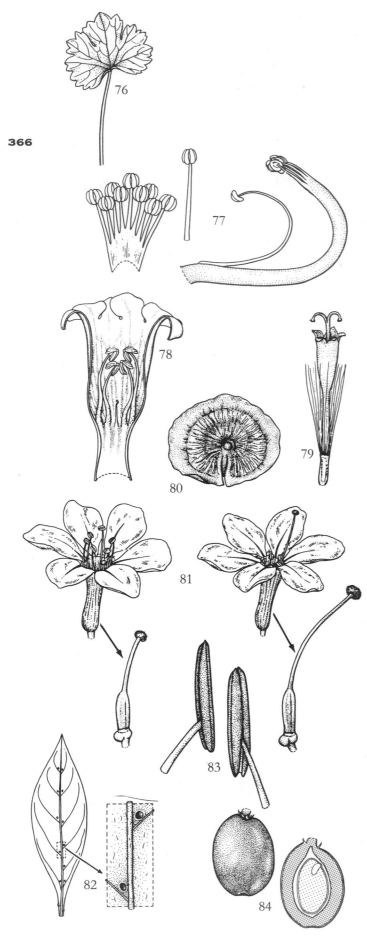

dehiscence: The process of splitting open at maturity.

deliquescent: (Of perianth parts), quickly becoming semiliquid.

dentate: Having a toothed margin with the teeth projecting at right angles. (**76.** *Ranunculus muricatus* [Ranunculaceae]: leaf)

determinate inflorescence: An inflorescence in which the terminal or central flower opens first, resulting in the cessation of primary axis elongation.

diadelphous: Having stamens arranged in two bundles (e.g., many legumes with "9 + 1" stamens). (**77.** *Centrosema virginianum* [Fabaceae]: expanded androecium [left] and configuration as in flower [right])

didynamous: Having stamens in two pairs, the pairs being of unequal lengths. (**78.** *Campsis radicans* [Bignoniaceae]: expanded corolla and androecium)

dimorphic heterostyly: See **distyly**.

dioecious: Bearing staminate and carpellate flowers on different plants.

disc floret (disk floret): A flower with a tubular corolla present in many of the Asteraceae. (**79.** *Senecio glabellus* [Asteraceae]: disc floret)

discoid: Disc-shaped; thin, flat, and circular. (**80.** *Lachnanthes caroliniana* [Haemodoraceae]: seed)

distichous: Two-ranked (e.g., leaves).

distinct: Separate from parts in the same series.

distyly (= dimorphic heterostyly): The occurrence of two different mature style lengths, relative to other parts of the flower, in various plants of the same species; an outcrossing mechanism found in several angiosperm families, such as the Rubiaceae. (**81.** *Lythrum alatum* [Lythraceae]: short-styled flower + pistil [left]; long-styled flower + pistil [right])

domatium (*pl.* domatia): A small cavity, pouch, or merely a tuft of hairs found on the vegetative organs (such as leaves) of certain woody dicotyledons and frequently inhabited by mites, ants, or other minute arthropods. (**82.** *Conocarpus erectus* [Combretaceae]: abaxial surface of leaf [left] and detail showing domatia [right])

dorsal: See **abaxial**; also, in thallose plants (e.g., *Lemna*) the dorsal surface refers to the upper surface, away from the substrate.

dorsifixed: Attached at the back. (**83.** *Hemerocallis fulva* [Hemerocallidaceae]: two views of stamen with filament attached at anther back)

drupe: An indehiscent, fleshy fruit in which the outer part is more or less soft and the center has one or more hard stones (pyrenes) enclosing the seeds. (**84.** *Serenoa repens* [Arecaceae]: drupe [left] and longitudinal section [right])

echinulate: Covered with small bristles.

elastic: Readily dehiscing with a spring.

elliptical: Oval in outline, with the widest point at or about the middle. (**85.** elliptical leaf [diagrammatic])

emarginate: Having a shallow notch at the apex. (**86.** *Silene antirrhina* [Caryophyllaceae]: flower showing emarginate petals)

embryotega: An outgrowth on the seed coat (present in many of the Commelinaceae). (**87.** *Callisia graminea* [Commelinaceae]: seed with embryotega)

enation: An epidermal outgrowth.

endosperm: The starch- and oil-containing tissue of many seeds, derived from the triple fusion nucleus of the embryo sac.

ensiform: Sword-shaped.

entire: Having a margin that lacks any toothing or division. (**88.** *Clusia rosea* [Clusiaceae]: leaf)

entomophilous: Insect-pollinated.

ephemeral: Lasting for a day or less.

epiblast: A scale-like projection on the outside of the embryo in some Poaceae.

epicalyx: An involucre (whorl of bracts) outside the true calyx. (**89.** *Hibiscus incanus* [Malvaceae]: flower bud)

epigynous flower: A flower with perianth and stamens apparently arising upon the ovary (actually growing from the edge of the floral cup, which is adnate to the ovary; ovary inferior). (**90.** *Cornus florida* [Cornaceae]: longitudinal section of flower)

epipetalous: Arising from the corolla. (**91.** *Campsis radicans* [Bignoniaceae]: expanded corolla and epipetalous androecium)

epitepalous: Arising from the tepals. (**92.** *Pontederia cordata* [Pontederiaceae]: expanded perianth [in two sections] and epitepalous stamens)

equitant: (Of leaves), flattened in two overlapping ranks (or rows), as in *Iris*; such leaves often are also sharply folded along their midribs. (**93.** *Sisyrinchium atlanticum* [Iridaceae]: young plant)

ericoid leaf: A small, thick, needle-like leaf with margins that are inrolled abaxially.

erose: Having a margin that is eroded or jagged. (**94.** *Costus speciosus* [Costaceae]: front view of flower showing lip with erose margins)

essential oils: See **ethereal oils**.

ethereal oils (= essential oils): A group of aromatic, oily plant products containing terpenes and often also other compounds (e.g., benzenoid compounds).

exserted: Projecting beyond another structure (as, for example, stamens projecting from a perianth tube). (**95.** *Plantago lanceolata* [Plantaginaceae]: flower with exserted stamens ["male stage"])

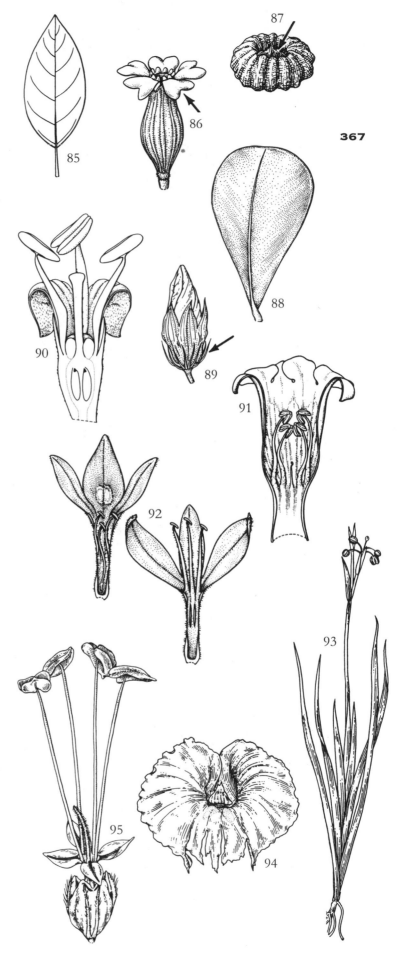

367

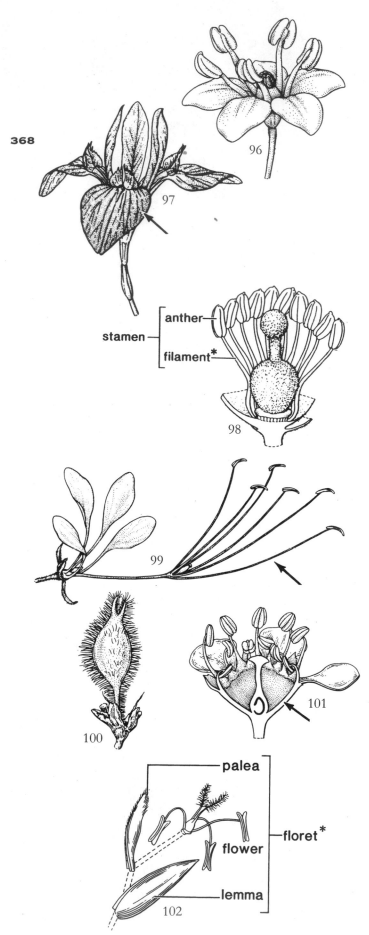

368

stamen { **anther**
filament*

palea

floret*

flower

lemma

exstipulate: Lacking stipules.

extrorse: (Of anther dehiscence), with the locule openings facing outward from the center of the flower. (**96.** *Sambucus canadensis* [Adoxaceae]: flower with extrorse anthers)

falls: In an *Iris* flower, the outer whorl of tepals, which are spreading or deflexed and often broader than the inner tepals. (**97.** *Iris hexagona* [Iridaceae]: flower)

fasciculate: Congested into bundles or clusters.

fenestration: A window-like opening or perforation.

fetid (foetid): Having a very strong, offensive, and rancid/putrid odor.

filament: The stalk of a stamen. (**98.** *Poncirus trifoliata* [Rutaceae]: longitudinal section of flower [pistil intact])

filiform: Thread-like; long and slender. (**99.** *Cleome gynandra* [Capparaceae]: flower showing filiform filaments)

fimbriate: Fringed. (**100.** *Ulmus alata* [Ulmaceae]: samara with fimbriate margin)

flabellate: Fan-like.

floral cup (= hypanthium): The cup-like structure, usually derived from the fusion of the perianth bases and androecium, on which the perianth and stamens are seemingly borne. (**101.** *Prunus serotina* [Rosaceae]: longitudinal section of flower)

florets: Small individual flowers, especially those of the Poaceae (flower plus palea and lemma) and Asteraceae. (**102.** *Eragrostis spectabilis* [Poaceae]: expanded floret)

flower: The reproductive structure of the angiosperms (and gnetopsids), consisting of a determinate shoot bearing sporophylls (modified leaves bearing sporangia); a complete flower includes the perianth, androecium, and gynoecium.

foliaceous: Leaf-like.

follicle: A dry to (less commonly) fleshy fruit derived from a single carpel that opens along a single suture. (**103.** *Aquilegia canadensis* [Ranunculaceae]: dehiscing aggregate of follicles)

free: Separate from unlike organs or parts.

free-central placentation: Attachment of ovules to a free-standing central column in a compound unilocular ovary (i.e., no septa are present). (**104.** *Stellaria media* [Caryophyllaceae]: longitudinal section of flower [bottom] and cross section of ovary [top])

fruit: A matured ovary with associated accessory parts.

fugacious (fugaceous): Falling or withering away very early.

funicle (= funiculus, *pl.* **funiculi):** The stalk of an ovule. (**105.** ovule [diagrammatic])

funnelform: Funnel-shaped. (**106.** *Datura stramonium* [Solanaceae]: flower with funnelform corolla)

fusiform: Spindle-shaped; tapering at both ends from a swollen middle.

geniculate: Bent like a knee. (**107.** *Rhexia mariana* [Melastomataceae]: geniculate stamen)

gibbose: Swollen on one side. (**108.** *Corydalis micrantha* [Papaveraceae]: flower with gibbose upper petal)

glabrous: Lacking hair.

glaucous: Covered with a whitish substance that rubs off (waxy bloom).

globose: Spheroid.

glochid: A minute barbed bristle (of many cacti), often occurring in tufts. (**109.** *Opuntia humifusa* [Cactaceae]: detail of areole [left] and one glochid [right])

glucosides: A group of chemical compounds containing the cyclic form of glucose.

glucosinolates (= mustard oils): Certain sulfurous plant glycosides with a sharp and irritating odor that occur in particular plants, such as the Brassicaceae.

glume: A small, chaffy bract; in particular, a sterile bract at the base of a grass spikelet. (**110.** *Eragrostis spectabilis* [Poaceae]: spikelet)

glycosides: A group of chemical compounds that yield a sugar when hydrolyzed; many are powerful drugs (e.g., digitalis).

grain: See **caryopsis**.

gynobasic style: A style that arises directly from the receptacle and appears to be inserted at the base of the ovary. (**111.** *Stachys floridana* [Lamiaceae]: ovary [with disc; left] and longitudinal section showing gynobasic style [right])

gynoecium: A collective term for the "female" reproductive units (pistils) of the flower.

gynophore: A special stalk supporting a pistil above the point of attachment of the androecium and perianth. (**112.** *Capparis cynophallophora* [Capparaceae]: pistil on gynophore)

gynostegium: The organ formed by the adnation of stamens and stigma present in many of the Apocy-

369

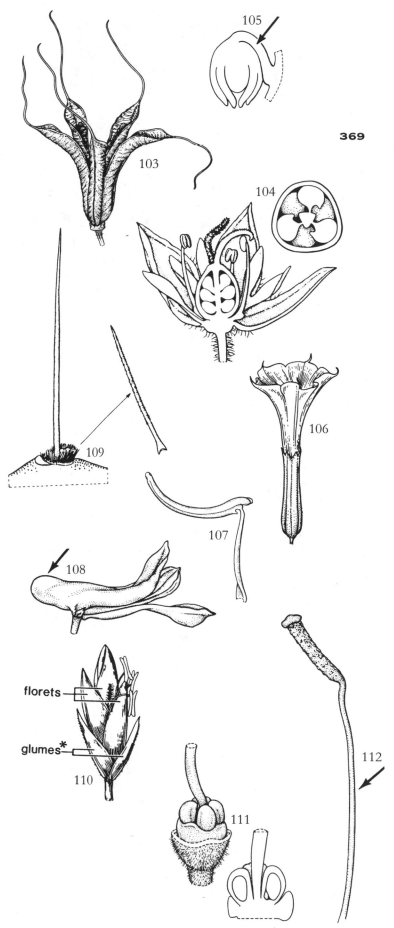

florets

glumes*

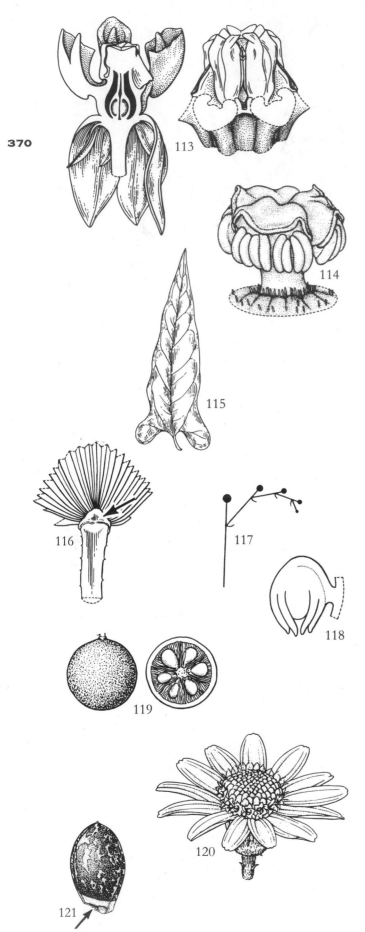

naceae. (**113**. *Asclepias humistrata* [Apocynaceae]: longitudinal section of flower [left; note position of gynostegium] and gynostegium [right; "hoods" and "horns" removed])

gynostemium: The organ formed by the adnation of stamens and style in the Aristolochiaceae. (**114**. *Aristolochia serpentaria* [Aristolochiaceae]: gynostemium from flower at anthesis)

habit: The general appearance of a plant.

hastate: Arrowhead-shaped, with widely divergent basal lobes. (**115**. *Aristolochia serpentaria* [Aristolochiaceae]: leaf)

hastula: The woody ligule of many palms. (**116**. *Serenoa repens* [Arecaceae]: base of leaf showing hastula)

haustorium (*pl.* haustoria): The absorbing organ (often root-like) of a parasitic plant, which connects it to the host.

head: See **capitulum**.

helicoid cyme: A coiled cyme in which the lateral branches develop from the same side of the main axis. (**117**. helicoid cyme [diagrammatic])

hemitropous (hemianatropous) ovule: An ovule in which the funicle is more curved than in an anatropous ovule. (**118**. hemitropous ovule [diagrammatic])

herb: A plant that does not develop persistent woody tissue above the ground and either dies at the end of the growing season or overwinters by means of underground organs (e.g., bulbs, rhizomes, corms).

hesperidium: A leathery-skinned berry with several to many partitions. (**119**. *Poncirus trifoliata* [Rutaceae]: hesperidium [left] and cross section [right])

heterogamous head: A head with more than one type of floret present in many of the Asteraceae. (**120**. *Senecio glabellus* [Asteraceae]: inflorescence)

heteropolar pollen grain: A pollen grain with an aperture on only one of its two faces (distal or proximal).

heterostyly (*adj.* heterostylous): The occurrence of two (distyly) or three (tristyly) different mature style lengths, relative to other parts of the flower, in various plants of the same species; an outcrossing mechanism found in several angiosperm families (see **distyly** and **tristyly**).

hilum: The scar on a seed indicating the point of attachment of the funicle (stalk). (**121**. *Sideroxylon lanuginosum* [Sapotaceae]: seed with well-developed hilum)

hirsute: Having coarse hairs rough to the touch.

hispid: Having rigid or stiff hairs.

homogamous flower: A flower whose stigma is receptive to pollen at the same time that pollen is shed from the anthers of the same flower.

homogamous head: A head with only one kind of floret, present in some of the Asteraceae. (**122.** *Vernonia gigantea* [Asteraceae]: inflorescence)

hood: A hood-like outgrowth of the filaments present in many of the Apocynaceae. (**123.** See **horn.**)

horn: A projecting outgrowth of the filaments present in many of the Apocynaceae. (**123.** *Asclepias humistrata* [Apocynaceae]: two views of flower)

husk: An outer covering of some fruits, usually derived from the perianth and/or involucre. (**124.** *Carya glabra* [Juglandaceae]: nut subtended by dehiscing husk [left] and nut with half of husk removed [right])

hyaline: Transparent, thin, and membranous.

hypanthium: See **floral cup.**

hypogynous flower: A flower with perianth and stamens arising from below the ovary (ovary superior). (**125.** *Guaiacum sanctum* [Zygophyllaceae]: longitudinal section of flower)

imbricate: Overlapping, like the shingles on a roof. (**126.** *Costus pulverulentus* [Costaceae]: inflorescence with imbricate bracts)

imperfect flower: A flower lacking either a functional androecium or a functional gynoecium. (**127.** *Zanthoxylum clava-herculis* [Rutaceae]: staminate [left] and carpellate [right] flowers)

incumbent: (Of cotyledons), having dorsal sides parallel to the radicle. (**128.** *Capsella bursa-pastoris* [Brassicaceae]: embryo [top] and cross section of seed showing incumbent cotyledons [bottom])

indeterminate inflorescence: An inflorescence in which the lowermost or outermost flower opens first, with the primary axis often elongating as the flowers develop; usually no terminal flower is produced.

induplicate: Folded inwards (V-shaped in cross section).

inferior: (Of an ovary), adnate to the floral cup and thus appearing to be below the perianth. (**129.** *Cornus florida* [Cornaceae]: longitudinal section of flower)

inflexed: Curved or bent inward, toward the axis.

inflorescence: The arrangement of flowers on the floral axis, or a flower cluster.

infructescence: The arrangement of fruits on the floral axis, or a fruiting cluster.

integument: The outer coating (or coatings) of an ovule that becomes the seed coat. (**130.** ovule [diagrammatic])

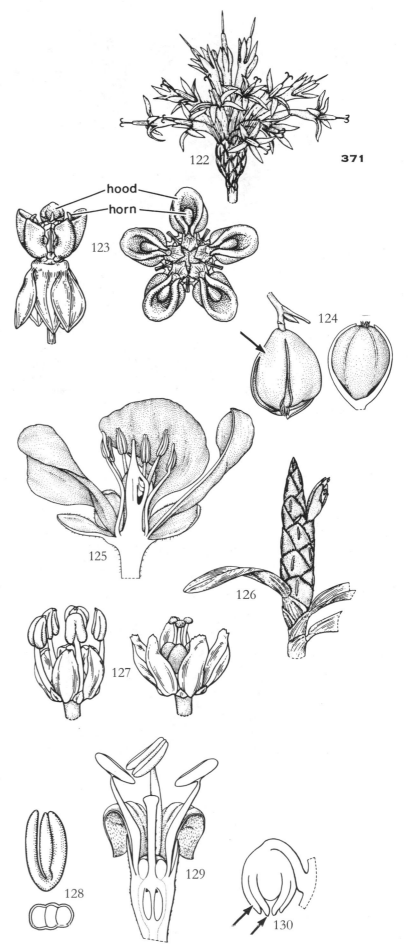

371

hood

horn

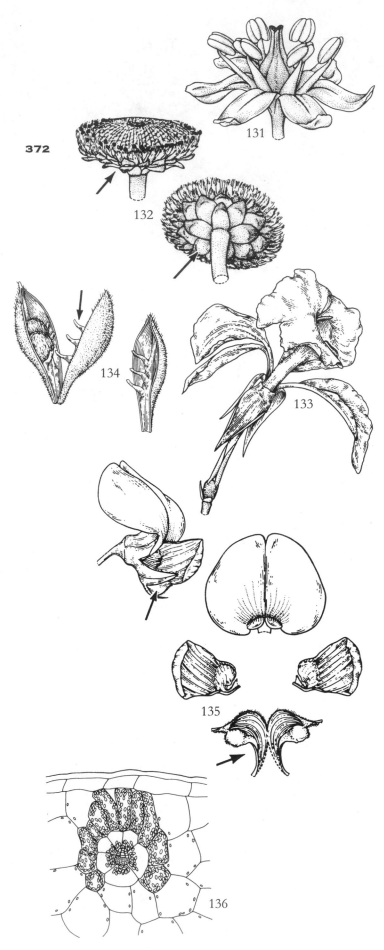

internode: The portion of the stem between two nodes.

intraxylary phloem (= internal phloem): Primary phloem that is located internally from the primary xylem (as compared to its typical position, external to the primary xylem).

introrse: (Of anther dehiscence), with the locule openings facing inward toward the center of the flower. (**131**. *Asparagus densiflorus* [Asparagaceae]: flower with introrse anthers)

inulin: A food reserve polysaccharide (of fructans) that takes the place of starch in certain plants, such as the Asteraceae.

involucre: A series of bracts surrounding a flower or inflorescence. (**132**. *Eriocaulon compressum* [Eriocaulaceae]: two views of inflorescence)

involute: Rolled inward or toward the upper side.

iridoid compounds: A group of bitter-tasting monoterpenoid glucosides.

irregular: Not divisible into equal halves. (**133**. *Canna flaccida* [Cannaceae]: flower)

jaculator (= retinaculum): A hook-like appendage on the funicle of certain ovules (in the Acanthaceae), which aids in the expulsion of seeds from the fruit. (**134**. *Ruellia caroliniensis* [Acanthaceae]: dehiscing capsule [left] and one valve of capsule showing jaculators [seeds ejected; right])

keel: The two front (often lower) petals of the papilionaceous flower present in many of the Fabaceae (Faboideae). (**135**. *Crotalaria spectabilis* [Fabaceae]: lateral view of flower [top] and expanded corolla [bottom])

Kranz anatomy: The wreath-like arrangement of mesophyll cells around a ring of large bundle sheath cells, forming two concentric layers surrounding the vascular bundle of leaves with the C_4 photosynthetic pathway. (**136**. *Portulaca pilosa* [Portulacaceae]: cross section of leaf showing vascular bundle and concentric mesophyll cells)

labellum: The lip of an orchid corolla. (**137**. *Encyclia tampensis* [Orchidaceae]: front view of flower [top] and lip [bottom])

lacerate: Torn; cut irregularly.

laciniate: Irregularly slashed into narrow ribbon-like segments or lobes. (**138**. *Platanthera ciliaris* [Orchidaceae]: portion of lip [spur removed] showing laciniate margin)

lamellate placentation: Attachment of ovules to plate-like lamellae within the ovary; a modification of parietal placentation.

laminar: Blade-like; flat and expanded.

lanceolate: Lance-shaped; much longer than broad, widening above the base and then tapering to the apex. (**139.** *Alstroemeria pulchella* [Alstroemeriaceae]: leaf)

latex: A colorless to (more often) white, yellow, or reddish sap of some plants.

laticifer (*adj.* laticiferous): A cell or cell series producing and containing the fluid latex.

latrorse: (Of anther dehiscence), with the locule openings located on the sides of the anther. (**140.** *Begonia cucullata* [Begoniaceae]: stamen)

legume: A usually dry fruit derived from a single carpel that opens along two longitudinal sutures. (**141.** *Chamaecrista fasciculata* [Fabaceae]: dehiscing legume)

lemma: In many of the Poaceae, the lower of the two bracts immediately subtending the flower. (**142.** *Eragrostis spectabilis* [Poaceae]: expanded floret)

lenticels: Spongy areas in the cork of stems (and other plant parts) that allow an interchange of gases between the interior and exterior of the stem.

lenticular: Lens-shaped; biconvex. (**143.** *Amaranthus hybridus* [Amaranthaceae]: seed)

lepidote: Covered with small scales. (**144.** *Croton argyranthemus* [Euphorbiaceae]: carpellate flower showing lepidote pistil and perianth)

level of universality: The level at which a character evolves from an ancestral to a derived state, defining a monophyletic group inclusive only at that point.

liana: A climbing woody vine.

ligulate floret: Generally, a flower with a strap-like corolla present in many of the Asteraceae (then = **ray floret**); more specifically, it is sometimes restricted to a perfect, fertile floret (then the strap-like corolla limb is typically five-toothed). (**145.** *Pyrrhopappus carolinianus* [Asteraceae]: ligulate floret)

ligule: A projection from the top of the leaf sheath present in the Poaceae, Arecaceae, and a few other families. (**146.** *Panicum maximum* [Poaceae]: leaf base with ligule on adaxial surface)

limb: The expanded flat part of a petal or a sympetalous corolla. (**147.** *Catharanthus roseus* [Apocynaceae]: flower)

limen: The lobed rim or cup at or above the base of the androgynophore in *Passiflora*. (**148.** *Passiflora incarnata* [Passifloraceae]: flower with perianth [and corona] removed)

linear: Long and narrow, with the sides nearly parallel.

lip: One of the (usually two) parts of an unequally di-

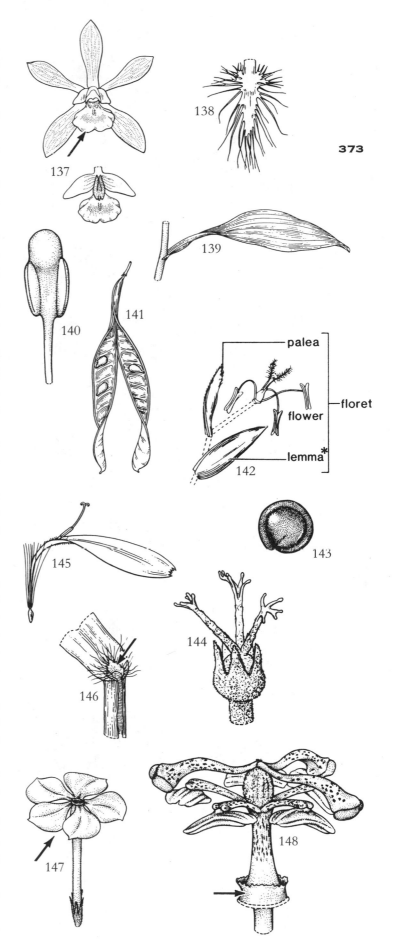

373

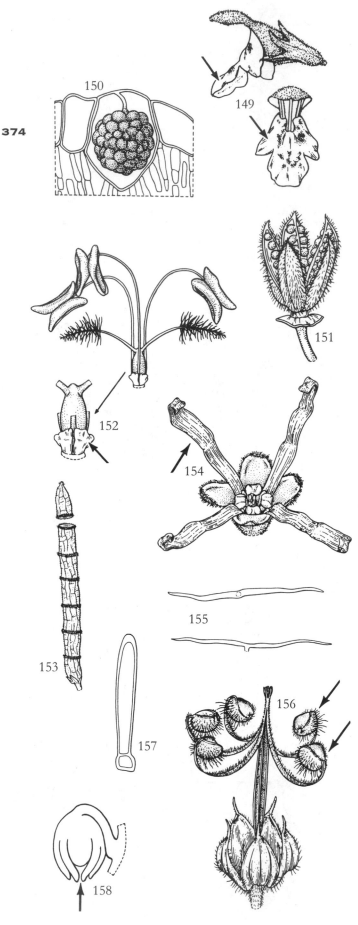

vided calyx or corolla. (**149**. *Stachys floridana* [Lamiaceae]: two views of flower with large lower lip)

lithocyst: A cell containing a cystolith. (**150**. *Boehmeria cylindrica* [Urticaceae]: cross section of leaf showing lithocyst with cystolith)

locule: Generally, a compartment or cavity of an organ such as an ovary or anther.

loculicidal capsule: A capsule that splits between the septa and into the locules (chambers) of the ovary. (**151**. *Hibiscus incanus* [Malvaceae]: capsule)

lodicules: One to three small scale-like structures at the base of a grass flower that represent a modified perianth. (**152**. *Panicum maximum* [Poaceae]: flower [top] and detail of flower base showing ovary and lodicules [bottom])

loment: A usually dry fruit derived from a single carpel that breaks transversely into one-seeded segments. (**153**. *Alysicarpus vaginalis* [Fabaceae]: dehiscing loment)

lorate: Strap-shaped. (**154**. *Hamamelis virginiana* [Hamamelidaceae]: flower showing lorate petals)

malpighian hair: A unicellular, two-armed hair composed of two horizontal branches and a short central stalk (Y- or T-shaped) but sometimes appearing simple due to reduction of one arm; found in the Malpighiaceae and other families, such as the Sapotaceae. (**155**. *Malpighia glabra* [Malpighiaceae]: two views of malpighian hair)

mamelon: The large placenta-like structure filling the ovary locule and containing two embryo sacs (female gametophytes) in the Viscaceae.

mangrove: Generally applied to several groups of tropical trees that grow in tidally flooded ground along coastal banks; more specifically, applied to those characterized by branches that spread and send down roots, thus forming multiple trunks and causing a thick growth.

marcescent: Withering, with the remains persisting.

mericarp: A segment (carpel) of a schizocarp. (**156**. *Geranium carolinianum* [Geraniaceae]: dehisced schizocarp [seeds expelled from mericarps])

-merous: Suffix denoting parts or numbers of each kind or series in a flower.

mesocarp: The middle layer of the pericarp (ovary wall).

microhairs: Minute, thin-walled epidermal hairs, usually two-celled, found in many of the Poaceae. (**157**. *Panicum maximum* [Poaceae]: microhair from leaf)

micropyle: An opening between the integuments of an ovule through which the pollen tube usually enters the nucellus. (**158**. ovule [diagrammatic])

monad: Generally, one of several individuals that are free from each other rather than attached in groups; specifically, single pollen grains.

monadelphous: Having stamens united into one group by the connation of their filaments. (**159**. *Hibiscus incanus* [Malvaceae]: androecium [filaments of stamens united into a tube] and gynoecium)

moniliform: Appearing like a string of beads. (**160**. *Callisia graminea* [Commelinaceae]: stamen [right] and detail of moniliform hair [left])

monocarpic: Describes plants that die after flowering and fruiting once (e.g., *Agave*, many Bromeliaceae, many bamboos, some Arecaceae); usually applied to perennials, but technically also applicable to annuals and biennials.

monocolpate: (Of a pollen grain), having a single elongate germinal furrow. (**161**. *Magnolia virginiana* [Magnoliaceae]: pollen grain)

monoecious: Bearing staminate and carpellate flowers on the same plant.

monograph: A specialized treatise on a single taxon.

monophyletic taxon: A taxon that contains all descendants of the most recent common ancestor of that group.

monopodial stem: A stem in which the height is increased due to the action of a single apical meristem.

monotypic taxon: A taxon that contains only one immediate subordinate taxon (e.g., a genus with only one species).

mucilage: A slimy and/or sticky substance.

multilacunar node: A node in a stem that has numerous gaps and numerous leaf traces related to one leaf. (**162**. *Magnolia virginiana* [Magnoliaceae]: cross section of node showing vascular tissue)

multiple fruit: A fruit formed from several flowers (and associated parts) more or less coalesced into a single structure with a common axis (see also **syncarp**).

multiseriate: Many-layered.

muricate: Rough due to small hard protuberances. (**163**. *Aristolochia serpentaria* [Aristolochiaceae]: lateral view of seed with muricate surface)

mustard oils: See **glucosinolates**.

mycorrhiza: The symbiotic association of fungi and the roots of higher plants.

naked (= nude): (Of a flower), lacking a perianth. (**164**. *Myrica cerifera* [Myricaceae]: carpellate flower)

native species: Species that occur naturally in an area.

natural taxon: A taxon whose members are phylogenetically related (i.e., share a unique common ancestor).

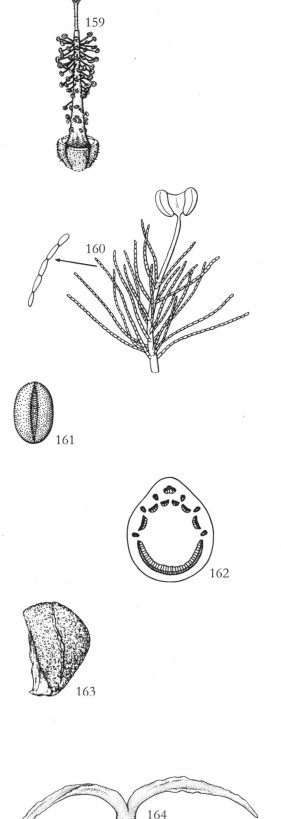

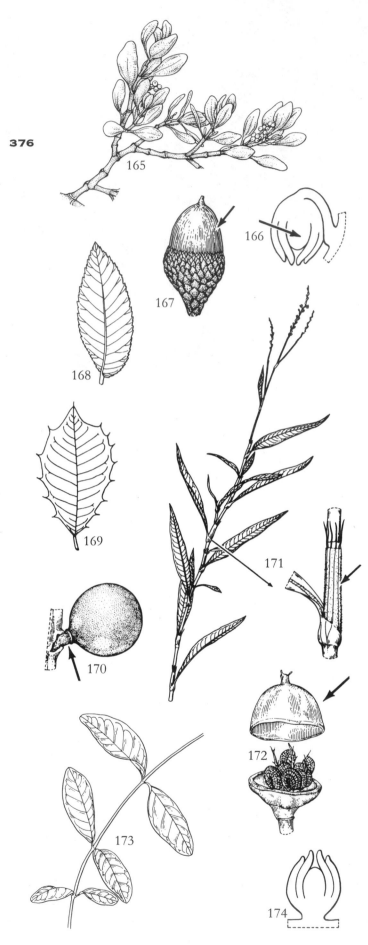

376

naturalized species: Species originally imported from another region that now maintain themselves, reproducing and spreading, without further human intervention.

nectar: The sweet, sugary liquid secreted by certain flowers, which is attractive to insects and other pollinators.

nectariferous: Producing nectar.

nectary: A nectar-secreting gland.

node: The portion of the stem where a leaf arises.

nodose: Knobby. (**165.** *Phoradendron leucarpum* [Viscaceae]: portion of carpellate flowering plant)

nucellus: The inner part (megasporangium) of an ovule in which the embryo develops. (**166.** ovule [diagrammatic])

nude: See **naked.**

nut: A more or less large, indehiscent, dry fruit with a thick and bony wall surrounding the single seed. (**167.** *Quercus geminata* [Fagaceae]: nut [acorn] with subtending cupule)

oblique: Having the sides unequal or asymmetrical at the base. (**168.** *Ulmus alata* [Ulmaceae]: leaf)

obovate: Ovate, but with the narrower end toward the attachment. (**169.** *Ilex opaca* [Aquifoliaceae]: leaf)

obovoid: A three-dimensional object that is egg-shaped.

obturator: A small protuberance on an ovule/seed that is an outgrowth of the placenta. (**170.** *Koelreuteria elegans* [Sapindaceae]: seed with well-developed obturator)

ocrea (ochrea): A nodal sheath or tube formed by the fusion of two stipules present in members of the Polygonaceae. (**171.** *Polygonum punctata* [Polygonaceae]: upper portion of plant [left] and detail of node with ocrea [right])

oil body: See **caruncle.**

operculate: Having a lid or cap.

operculum: A lid or cap (= calyptra). (**172.** *Portulaca pilosa* [Portulacaceae]: dehiscing pyxis). Also, the innermost whorl of the corona of *Passiflora,* which is membranous and functions as a cover over the nectar-secreting chamber.

opposite: (Of structures, such as leaves), occurring in pairs at the same level and on opposite sides of the axis (i.e., two leaves per node). (**173.** *Lonicera japonica* [Caprifoliaceae]: vegetative portion of plant showing opposite leaves)

orthotropous ovule: An erect ovule with the micropyle at the apex and the hilum at the base. (**174.** orthotropous ovule [diagrammatic])

ovary: The basal portion of the pistil that bears the

ovules. (**175**. *Poncirus trifoliata* [Rutaceae]: longitudinal section of flower [pistil intact])

ovate: Egg-shaped in outline, with the broader end below the middle and nearer the point of attachment. (**176**. *Hypericum tetrapetalum* [Clusiaceae]: leaf)

ovule: A structure in seed plants enclosing the female gametophyte and composed of the nucellus, one or two integuments, and the funicle; it differentiates into the seed after fertilization. (**177**. *Helianthemum corymbosum* [Cistaceae]: longitudinal section of pistil from cleistogamous flower)

P-type plastid: A type of sieve element plastid, designated as the "protein type," containing protein (deposited as crystalline bodies) and often also starch; it occurs in certain families.

pale: A receptacular chaffy bract subtending a floret in some of the Asteraceae. (**178**. *Carphephorus corymbosus* [Asteraceae]: floret [right] and subtending pale [left])

palea: In many of the Poaceae, the upper of the two bracts immediately subtending the flower. (**179**. *Eragrostis spectabilis* [Poaceae]: expanded floret)

palmate: (Of leaves and/or venation), having lobes, divisions, or veins that arise at the same point (digitate). (**180**. *Acer rubrum* [Sapindaceae]: leaf showing palmate lobing and venation)

paniculate: (Of an inflorescence), in general, loosely and greatly branched (ultimate units may be of various types). (**181**. *Yucca flaccida* [Agavaceae]: inflorescence)

papilionaceous: Having the "butterfly-like" corolla type (with standard, wings, and keel) of the subfamily Faboideae (Papilionoideae) of the Fabaceae. (**182**. *Crotalaria spectabilis* [Fabaceae]: two views of flower)

papilla (*pl.* papillae): A minute, rounded projection. (**183**. *Canna flaccida* [Cannaceae]: cross section of ovary showing papillose surface)

papillose: Covered with papillae (minute, rounded projections).

pappus: The specialized and reduced calyx of the Asteraceae composed of hairs, bristles, awns, or scales. (**184**. *Gnaphalium obtusifolium* [Asteraceae]: disc floret)

parallel venation: Venation in which veins lie more or less parallel to the leaf margins. (**185**. *Alstroemeria pulchella* [Alstroemeriaceae]: leaf)

paraphyletic taxon: A taxon that contains some, but not all, descendants of the most recent common ancestor of that group.

parietal placentation: Attachment of ovules on the walls (or extrusions of the wall) within a simple or

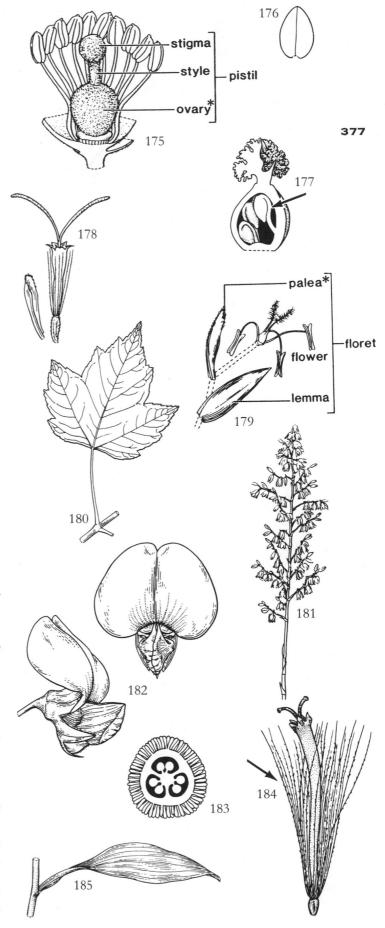

377

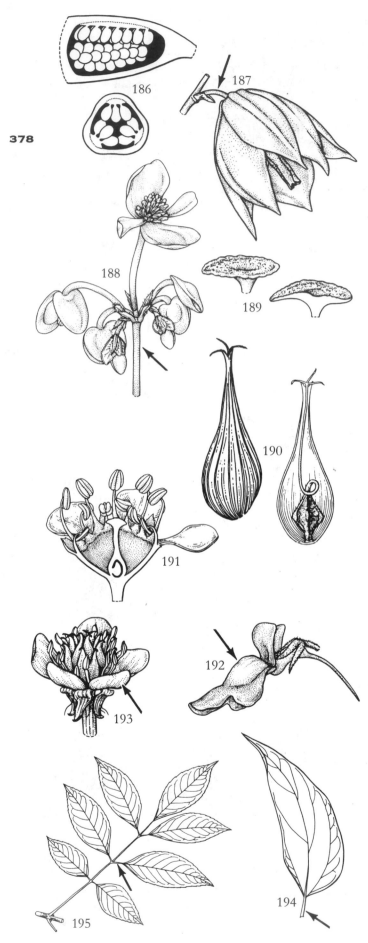

378

compound ovary. (**186.** *Viola septemloba* [Violaceae]: longitudinal [top] and cross [bottom] sections of ovary)

pectinate: Comb-like, as in leaves with very close and narrow divisions.

pedicel: The stalk of a flower. (**187.** *Yucca flaccida* [Agavaceae]: flower)

peduncle: The stalk of a flower cluster. (**188.** *Begonia cucullata* [Begoniaceae]: inflorescence)

pellucid: Clear and almost transparent in transmitted light.

peltate: Umbrella-like; attached to the stalk near the center of the lower surface. (**189.** *Peperomia magnoliifolia* [Piperaceae]: bract [left] and longitudinal section of bract [right])

pepo: The specialized berry of many of the Cucurbitaceae, characterized by a hard or leathery rind (epicarp) and fleshy inner tissue (with no septa).

perennial: A plant with a life cycle of more than two years.

perfect flower: A flower with both functional androecium and functional gynoecium.

perianth: A collective term for the calyx and corolla (or tepals).

pericarp: The wall of a fruit.

perigynium: The modified sac-like bract surrounding the achene of *Carex*. (**190.** *Carex lupulina* [Cyperaceae]: fruiting perigynium [left] and achene with half of subtending perigynium removed [right])

perigynous flower: A flower with perianth and stamens arising from a floral cup that is not adnate to the ovary. (**191.** *Prunus serotina* [Rosaceae]: longitudinal section of flower)

perisperm: Seed storage tissue similar to endosperm but derived from the nucellus.

persistent: Remaining attached.

personate corolla: A bilabiate (two-lipped) corolla whose lower lip has a conspicuous rounded projection that closes the throat. (**192.** *Linaria canadensis* [Scrophulariaceae]: lateral view of flower)

petal: One member of the inner floral envelope(s) (corolla) of a typical flower; it is usually colored and more or less showy. (**193.** *Ranunculus muricatus* [Ranunculaceae]: flower)

petiolate: Having a petiole.

petiole: The stalk of a leaf. (**194.** *Celtis laevigata* [Ulmaceae]: leaf)

petiolule: The stalk of a leaflet. (**195.** *Fraxinus caroliniana* [Oleaceae]: compound leaf)

phenetic gap: The gap of dissimilarity separating taxa, according to phenetics.

phenetics: A taxonomic theory in which the relationships of taxa are determined by the calculation of overall similarity.

phyllary: An involucral bract present in many of the Asteraceae. (**196.** *Senecio glabellus* [Asteraceae]: underside of inflorescence)

phylloclade (= cladode): A flattened, leaf-like stem that functions as a leaf. (**197.** *Asparagus densiflorus* [Asparagaceae]: portion of plant [left] and phylloclade [right])

phyllode: A flattened, leaf-like petiole (with no blade) that functions as a leaf.

phylogeny: The study of the evolutionary history of organisms.

phytomelan: An opaque and brittle, charcoal-like substance that forms a shiny black crust on certain seeds (e.g., those of the Agavaceae and many of the Amaryllidaceae and Alliaceae).

pinna (*pl.* pinnae): A primary division or leaflet of a pinnately compound leaf. (**198.** *Fraxinus caroliniana* [Oleaceae]: compound leaf)

pinnate leaf: A leaf in which there are more than three leaflets arranged in two rows along a common axis. (**199.** *Tribulus cistoides* [Zygophyllaceae]: leaf)

pinnate venation: Venation consisting of a central midvein with many secondary veins emerging on both sides to form a feather-like pattern. (**200.** *Ulmus alata* [Ulmaceae]: leaf)

pistil: The ovule-bearing organ of the flower, typically composed of at least an ovary, a style, and a stigma (thus formed from one or more carpels). (**201.** *Poncirus trifoliata* [Rutaceae]: longitudinal section of flower [pistil intact])

pistillate flower: See **carpellate flower.**

placenta (*pl.* placentae): The place or part in the ovary where the ovules are attached.

placentation: The arrangement of ovules within the ovary.

plesiomorphy: An ancestral character state.

pleurogram: A fine, U-shaped (Mimosoideae) or oval/circular (Caesalpinioideae) line or groove in the seed coat on both sides of the seed. (**202.** *Mimosa quadrivalis* [Fabaceae]: seed with pleurogram)

plicate: Folded into pleats, as in a fan. (**203.** *Ipomoea pandurata* [Convolvulaceae]: expanded plicate corolla with epipetalous androecium)

plumose: (Of a hair), feather-like and compound. (**204.** *Panicum maximum* [Poaceae]: flower showing plumose styles)

pollen grains: The male gametophytes (microgametophytes, which contain the male generative cells) developed within the microspore wall; borne in the anthers of the flower.

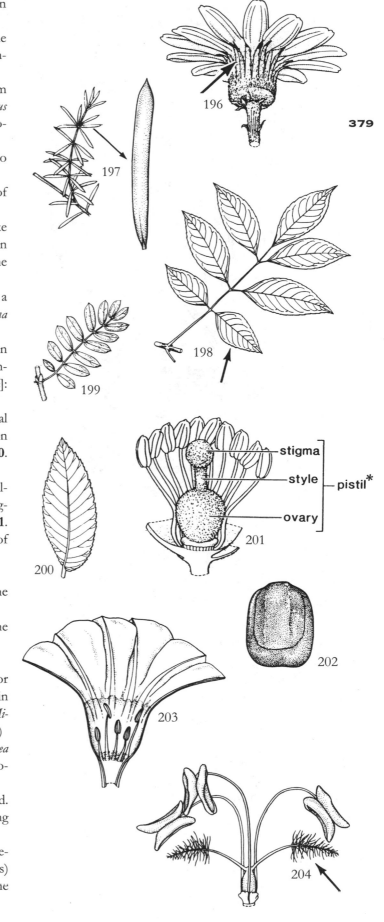

379

stigma

style — pistil*

ovary

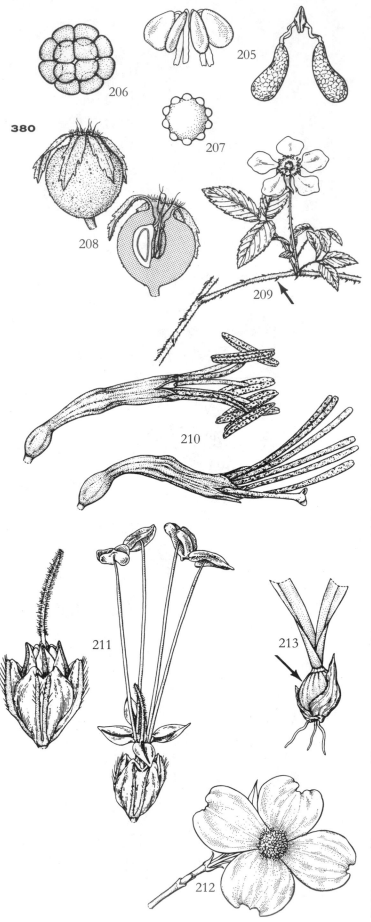

380

pollinium: The coherent, waxy pollen mass produced by many members of the Orchidaceae and Apocynaceae; it is transported as a unit during pollination. (**205**. *Encyclia tampensis* [Orchidaceae; left] and *Asclepias humistrata* [Apocynaceae; right]: pollinia)

polyad: Pollen grains united in a group of more than four (tetrad). (**206**. *Albizia julibrissin* [Fabaceae]: pollen polyad)

polycolporate: (Of a pollen grain), having many (compound) apertures. (**207**. *Polygala lutea* [Polygalaceae]: pollen grain)

polygamodioecious: Functionally dioecious, but bearing a few flowers of the opposite sex or a few perfect flowers on the same plant.

polygamous: Bearing perfect and imperfect flowers on the same plant.

polyphyletic taxon: A taxon that contains descendants of two or more ancestral sources.

pome: An indehiscent fleshy fruit whose outer part is more or less soft and whose center has papery or cartilaginous structures enclosing the seeds. (**208**. *Crataegus uniflora* [Rosaceae]: pome [top] and longitudinal section [bottom])

precocious: (Of flowers), appearing early in the season (before the leaves).

prickle: A sharp, pointed outgrowth from the epidermis or cortex. (**209**. *Rubus trivialis* [Rosaceae]: flowering branch)

prophyll: A small bract.

protandrous (proterandrous) flower: A flower whose anthers shed their pollen before the stigma of the same flower is receptive. (**210**. *Manfreda virginica* [Agavaceae]: flower at anthesis [top; anthers mature, stigma unexpanded]; older flower [bottom; anthers fallen off, stigma mature])

protogynous (proterogynous) flower: A flower whose stigma is receptive to pollen before pollen is shed from the anthers of the same flower. (**211**. *Plantago lanceolata* [Plantaginaceae]: young flower [left; "female stage," with corolla lobes and stamens unexpanded]; older flower [right; "male stage," with stamens and corolla fully expanded])

pseudanthium: A cluster of small or reduced flowers that collectively simulate a single flower. (**212**. *Cornus florida* [Cornaceae]: inflorescence [clustered individual flowers surrounded by large petaloid bracts])

pseudobulb: The thickened, bulb-like stem of certain orchids. (**213**. *Encyclia tampensis* [Orchidaceae]: base of plant)

pseudomonomerous: Appearing to be one-parted.

pseudostem: An apparently thick and strong "stem"

formed by the tightly overlapping leaf bases enclosing the actual weak stem in several of the Zingiberales (Heliconiaceae, Musaceae, Zingiberaceae). (**214**. *Alpinia speciosa* [Zingiberaceae]: cross section of stem plus overlapping leaf sheaths ["pseudostem"])

pubescent: Covered with hairs.

pulvinus: An enlargement of the petiole (or petiolule) base at its point of attachment to the stem, or of the petiole (or petiolule) apex at its point of attachment to the blade. (**215**. *Senna obtusifolia* [Fabaceae]: leaf [left] with detail showing pulvinus at petiole base [right])

punctate: Dotted with translucent or colored glands, dots, or depressions; also used to describe a structure that is small and roughly circular. (**216**. *Hypericum tetrapetalum* [Clusiaceae]: leaf [left] with detail showing punctate surface [right])

pyrene: A stone ("pit") of a drupe composed of a seed and a bony endocarp (inner wall of the fruit). (**217**. *Cocculus carolinus* [Menispermaceae]: longitudinal section of drupelet [top] and whole pyrene [bottom])

pyxis: See **circumscissile capsule**.

racemose: (Of an inflorescence), having stalked flowers (or small flower clusters) arranged along an elongated central axis. (**218**. *Encyclia tampensis* [Orchidaceae]: inflorescence)

rachilla: In many of the Poaceae and Cyperaceae, the axis that bears the florets. (**219**. *Eragrostis spectabilis* [Poaceae]: expanded spikelet)

radially symmetrical: See **actinomorphic**.

radicle: The embryonic root of an embryo or a germinating seed.

raphe: The portion of the funicle adnate to the integument (often appearing as a seam on the seed coat), typically present in anatropous and hemitropous ovules. (**220**. anatropous ovule [diagrammatic])

raphide: A needle-like crystal of calcium oxalate.

ray floret: Generally, a flower with a strap-like corolla present in many of the Asteraceae (then = **ligulate floret**); more specifically, it is sometimes restricted to sterile or carpellate florets (then the strap-like corolla limb is two- or three-toothed). (**221**. *Senecio glabellus* [Asteraceae]: ray floret)

receptacle: The more or less enlarged or elongated apex of the flower axis, which bears some or all of the flower parts. (**222**. *Magnolia virginiana* [Magnoliaceae]: longitudinal section of flower showing elongated receptacle)

recurved: Bent or curved downward or backward.

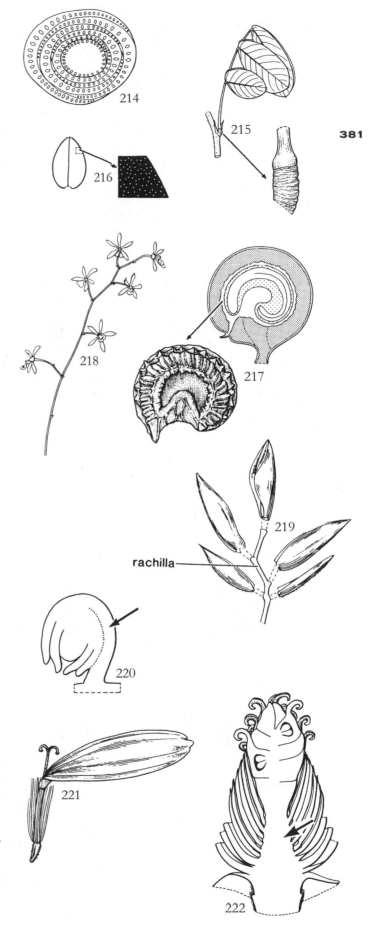

rachilla

381

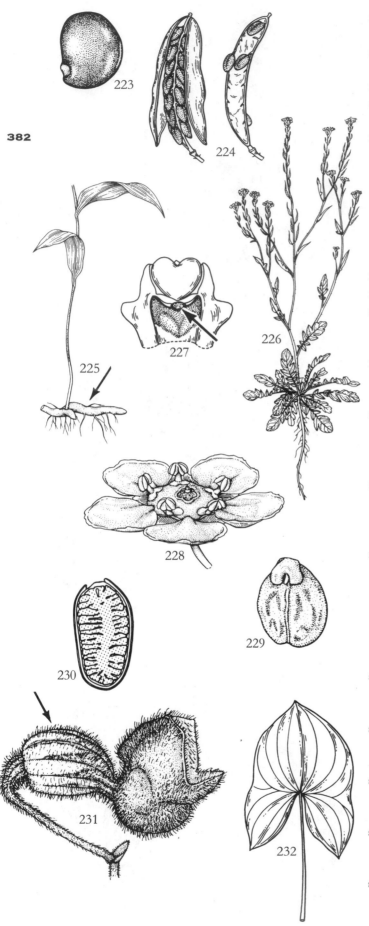

382

reduplicate: Folded downward (Λ-shaped in cross section).

reflexed: Recurved or bent downward, away from the axis.

regular: See **actinomorphic**.

reniform: Kidney-shaped. (**223**. *Phytolacca americana* [Phytolaccaceae]: seed)

replum: A persistent, thickened rim (representing the placentae) of the ovary/fruit bearing the ovules/seeds present in many of the Brassicaceae and some of the Capparaceae; traditionally, it is often restricted to the membranous septum (plus rim) of the brassicaceous ovary/fruit. (**224**. *Descurainia pinnata* [Brassicaceae]: dehiscing silique [left] and replum with a few attached seeds [right])

resupinate: Twisted 180°; turned upside down.

retinaculum (*pl.* retinacula): See **jaculator**.

rhizome: An underground stem that usually persists from season to season. (**225**. *Polygonatum biflorum* [Asparagaceae]: lower portion of plant)

rosette: A circular cluster of leaves, usually close to the ground. (**226**. *Raphanus raphanistrum* [Brassicaceae]: habit with basal rosette of leaves)

rostellum: The beak- or strap-like portion of the stigma of an orchid, specialized to take part in the transfer of pollen. (**227**. *Encyclia tampensis* [Orchidaceae]: detail of column underside)

rotate corolla: A sympetalous corolla with a flat, circular, and spreading limb at right angles to a very short tube. (**228**. *Euonymus americanus* [Celastraceae]: lateral view of flower)

ruminate: Mottled in appearance, with dark and light zones. (**229**. *Croton argyranthemus* [Euphorbiaceae]: seed with mottled seed coat.) Also, used to describe endosperm that is irregularly grooved and ridged. (**230**. *Asimina incana* [Annonaceae]: longitudinal section of seed with ruminate endosperm)

saccate: Bag-shaped; pouched. (**231**. *Aristolochia serpentaria* [Aristolochiaceae]: flower with saccate perianth base)

sagittate: Arrowhead-shaped; triangular, with the basal lobes pointing downward. (**232**. *Sagittaria latifolia* [Alismataceae]: leaf)

salverform corolla: A sympetalous corolla with a slender tube and an abruptly expanded, flat limb. (**233**. *Catharanthus roseus* [Apocynaceae]: flower)

samara: A winged, indehiscent, more or less dry fruit containing a single seed. (**234**. *Betula nigra* [Betulaceae]: samara)

saponins: A group of toxic plant glycosides characterized by foaming when they are shaken with water; they are widely distributed among angiosperms.

sarcotesta: A fleshy seed coat. (**235**. *Magnolia virginiana* [Magnoliaceae]: seed [with sarcotesta; left] and longitudinal section of seed [right])

scabrous: Rough to the touch.

scalariform perforation plate: In vessel members of the xylem, a type of multiperforate plate in which elongated perforations are arranged parallel to one another so that the cell wall bars between them form a ladder-like pattern.

scales: A general term for small, often dry, appressed leaves, bracts, or hairs.

scapose: (Of a plant), bearing flowers or inflorescences on a leafless peduncle that arises from the ground. (**236**. *Pinguicula lutea* [Lentibulariaceae]: habit)

scarious: Dry, thin, membranous, non-green, and translucent.

schizocarp: A fruit derived from a two- to many-carpellate gynoecium that splits into two or more one-seeded segments. (**237**. *Croton argyranthemus* [Euphorbiaceae]: three-lobed schizocarp before dehiscence [bottom] and one dehiscing segment of schizocarp with released seed [top])

scorpioid cyme: A coiled cyme in which the apparently lateral flowers (or branches) develop alternately on opposite sides of the main axis. (**238**. scorpioid cyme [diagrammatic])

scurfy: Having flake-like particles (small scales) on the surface.

scutellum: The relatively large, specialized cotyledon of the Poaceae embryo, which bears the embryonic plant laterally. (**239**. *Zea diploperennis* [Poaceae]: longitudinal sections of caryopsis [left] and embryo [right])

seed: The product of the ovule after fertilization, comprising the embryo with its surrounding food reserves and protective coverings.

sensu lato: "In the broad sense": with a wide or general interpretation (abbreviated *s.l.*).

sensu stricto: "In the narrow sense": with a restricted interpretation (abbreviated *s.s.*).

sepal: One member of the outer floral envelope(s) (calyx) of a typical flower; it is usually green and foliaceous. (**240**. *Ranunculus muricatus* [Ranunculaceae]: flower)

septicidal capsule: A capsule that splits along the septa (partitions between the locules). (**241**. *Hypericum myrtifolium* [Clusiaceae]: capsule [with persistent calyx])

septifragal capsule: A capsule that splits longitudi-

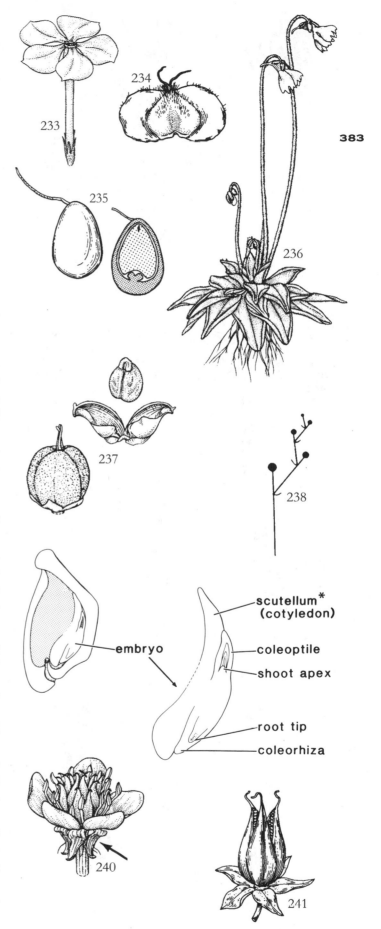

383

scutellum* (cotyledon)

embryo

coleoptile

shoot apex

root tip

coleorhiza

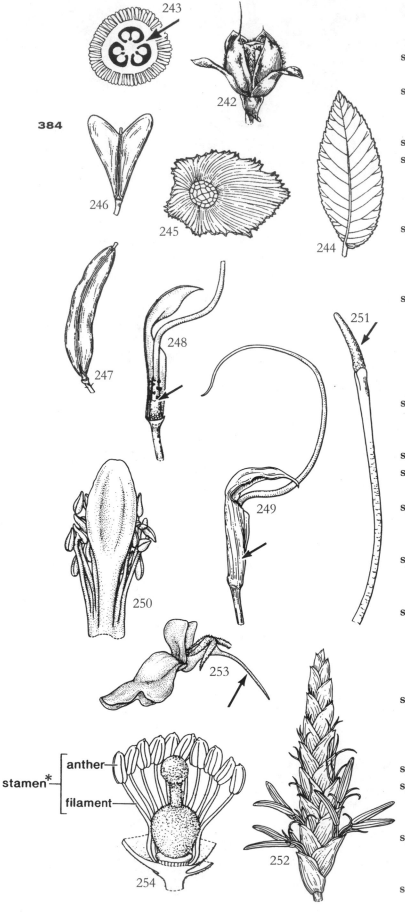

384

stamen*
anther
filament

nally so that the valves break away from the septa (partition walls). (**242**. *Ipomoea pandurata* [Convolvulaceae]: capsule)

septum (*pl.* septa): A partition or cross wall. (**243**. *Canna flaccida* [Cannaceae]: cross section of ovary)

serrate: Having a saw-toothed margin with the teeth pointing forward. (**244**. *Ulmus alata* [Ulmaceae]: leaf)

sessile: Lacking a stalk.

shield cells: The cells of the expanded apical portion of the peltate hairs present on the leaves of many of the Bromeliaceae. (**245**. *Tillandsia recurvata* [Bromeliaceae]: shield cell from leaf)

shrub: A medium-sized (less than 10 m) perennial plant lacking a well-developed main trunk due to the well-developed side branches that arise from near the base.

silicle: A dry fruit of many of the Brassicaceae, derived from a two-carpellate gynoecium in which the two valves split away from a persistent partition (around the rim of which the seeds are attached); it is usually not more than twice as long as wide. (**246**. *Capsella bursa-pastoris* [Brassicaceae]: silicle)

silique: Same as **silicle**, but usually at least twice as long as wide. (**247**. *Descurainia pinnata* [Brassicaceae]: silique)

simple leaf: A leaf not divided into leaflets.

sister group: One of two related groups produced by the dichotomous branching of a cladogram.

spadix: A spike with a thickened fleshy axis. (**248**. *Arisaema dracontium* [Araceae]: inflorescence with half of subtending spathe removed)

spathe: The large bract surrounding or subtending an inflorescence. (**249**. *Arisaema dracontium* [Araceae]: inflorescence with subtending spathe)

spatulate (spathulate): Spoon-shaped; oblong or somewhat rounded, with the basal end long and tapered. (**250**. *Tilia americana* [Tiliaceae]: adaxial view of petaloid spatulate staminode and attached stamens)

spicate: (Of an inflorescence), having sessile flowers arranged singly or in contracted clusters along a central axis. (**251**. *Orontium aquaticum* [Araceae]: spicate inflorescence)

spike: See **spicate**.

spikelet: The small bracteate spike of many of the Poaceae and Cyperaceae. (**252**. *Cyperus esculentus* [Cyperaceae]: spikelet)

spur: A sac-like projection from a flower part (usually a petal or sepal). (**253**. *Linaria canadensis* [Scrophulariaceae]: lateral view of flower)

stamen: The pollen-bearing organ of the flower, composed of a filament (stalk) and anther (pollen sacs). (**254**. *Poncirus trifoliata* [Rutaceae]: longitudinal section of flower [pistil intact])

staminate flower: A "male flower," one that has a functional androecium and lacks a functional gynoecium. (**255**. *Zanthoxylum clava-herculis* [Rutaceae]: staminate flower [left] and longitudinal section [right])

staminode: A sterile stamen. (**256**. *Canna flaccida* [Cannaceae]: petaloid staminode)

standard: See **banner**; also, in an *Iris* flower, one of the inner whorl of tepals that are erect or ascending and often are narrower than the outer tepals. (**257**. *Iris hexagona* [Iridaceae]: flower)

stellate: Star-shaped; with radiating branches. (**258**. *Tilia americana* [Tiliaceae]: stellate hair from leaf)

steroids: A group of compounds including sterols (solid cyclic alcohol compounds), found in plant and animal tissues.

stigma: The apical part of the pistil that is receptive to pollen. (**259**. *Poncirus trifoliata* [Rutaceae]: longitudinal section of flower [pistil intact])

stipe: A strap of tissue derived from the column that connects the orchid pollinia to the viscidium.

stipule: One of a pair of appendages at the base of the petiole at the point of attachment to the stem. (**260**. *Begonia cucullata* [Begoniaceae]: node with leaf and stipules)

stolon: A runner; a horizontal stem at or below the ground surface that produces a new plant at its tip.

striate: Having fine longitudinal lines, channels, or ridges. (**261**. *Nuphar luteum* [Nymphaeaceae]: striate pistil)

stylar canal: In a few angiosperms, a canal (or one of several canals) lined with specialized tissue, passing through the center of the style. (**262**. *Lyonia lucida* [Ericaceae]: cross section of style showing stylar canal)

style: The more or less elongated portion of the pistil between the ovary and the stigma. (**263**. *Poncirus trifoliata* [Rutaceae]: longitudinal section of flower [pistil intact])

stylopodium: A nectariferous enlargement at the style bases present in many of the Apiaceae. (**264**. *Oxypolis filiformis* [Apiaceae]: flower at anthesis [left] and mature gynoecium with petals removed [right])

subulate: Awl-shaped; tapering from base to apex. (**265**. *Carex lupulina* [Cyperaceae]: subulate fruiting perigynium)

succulent: Fleshy, soft, and thick.

suffrutescent: Somewhat shrubby: woody at the base, with herbaceous shoots produced perennially.

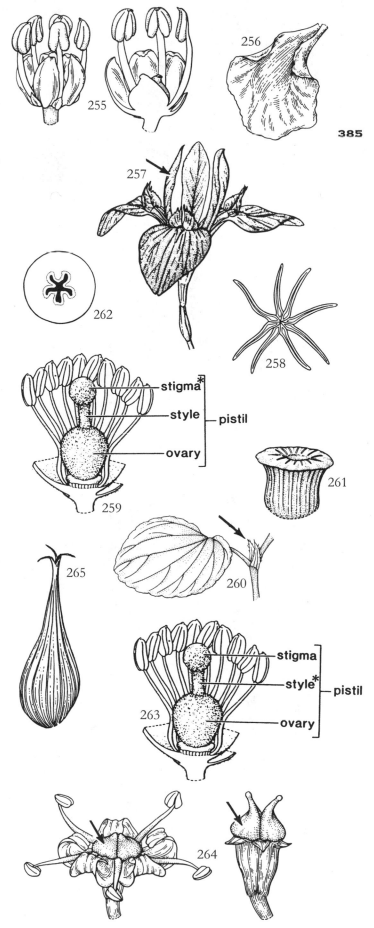

385

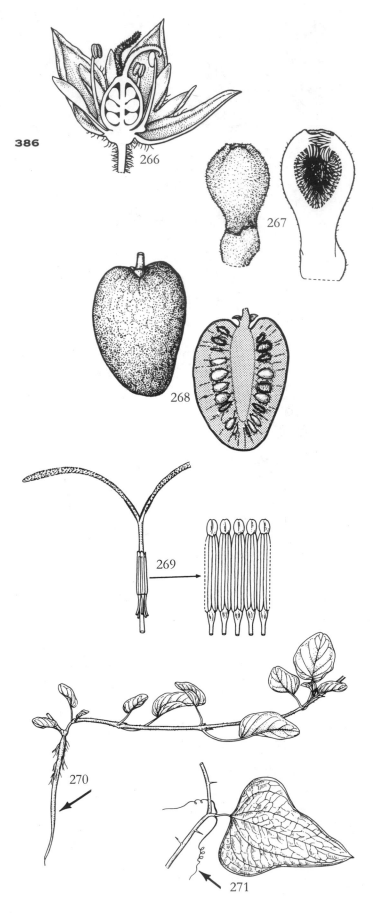

386

superior: (Of an ovary), situated above the point of attachment of the perianth and androecium and free from them. (**266**. *Stellaria media* [Caryophyllaceae]: longitudinal section of flower)

suture: A line of joining; a seam.

syconium: In *Ficus*, the enlarged, hollow, and flasklike structure that bears the flowers and fruits along the inner surface. (**267**. *Ficus aurea* [Urticaceae]: syconium [left] and longitudinal section [right])

sympetalous: Having the petals united, at least at the base.

symplesiomorphy: A shared ancestral character state.

sympodial stem: A stem in which the apex aborts or becomes reproductive, with the vegetative growth being continued in each instance by a lateral branch.

synapomorphy: A shared derived character state.

syncarp: A fruit formed from one (= aggregate fruit) or several (= multiple fruit) flower(s) and associated parts, more or less coalesced into a single structure. (**268**. *Annona glabra* [Annonaceae]: syncarp [formed by fusion of carpels and receptacle; left] and longitudinal section [right])

syncarpous: Having the carpels united.

syngenesious: Having stamens connate by their anthers to form a cylinder. (**269**. *Carphephorus corymbosus* [Asteraceae]: androecium surrounding expanded style branches [left] and expanded androecium [right])

synsepalous: Having the sepals united, at least at the base.

systematics: The scientific study of the kinds and diversity of organisms and of any and all relationships among them.

tannic acids: See **tannins**.

tannins (= tannic acids): A group of yellowish, astringent, bitter-tasting phenolic compounds; they are widespread among angiosperms.

taproot: The primary root, when markedly larger than the others. (**270**. *Boerhavia diffusa* [Nyctaginaceae]: basal portion of plant with taproot)

taxon: A taxonomic group that is distinguished by name and ranked in a definite category.

taxonomy: The theory and practice of classifying organisms.

tendril: An elongated, twining segment of a leaf, stem, or inflorescence by which a plant clings to its support. (**271**. *Smilax auriculata* [Smilacaceae]: leaf with tendrils)

tenuinucellate: Having a nucellus that consists of a single layer of cells.

tepal: One member of those perianths that are not clearly differentiated into a typical calyx and corolla. (**272.** *Sisyrinchium atlanticum* [Iridaceae]: flower)

terete: Approximately circular in cross section.

terminal: Occurring at the tip or apical end.

testa: The seed coat (hardened mature integuments). (**273.** *Hypericum myrtifolium* [Clusiaceae]: seed [left] and longitudinal section [right])

tetrad (*adj.* tetradenous): A group of four; specifically, pollen grains united in a group of four. (**274.** *Mimosa strigillosa* [Fabaceae]: pollen tetrad)

tetradynamous: Having an androecium of six stamens, with the four inner ones longer than the outer two. (**275.** *Raphanus raphanistrum* [Brassicaceae]: androecium and gynoecium)

thallus (*pl.* thalli): A plant body, often flat and leaf-like, that is not clearly differentiated into roots, stems, and leaves. (**276.** *Lemna obscura* [Lemnaceae]: lateral [left] and top [right] views of a clone of three thalli)

thorn: A pointed, reduced branch. (**277.** *Crataegus uniflora* [Rosaceae]: flowering branch)

toothed: Having a margin with alternating small projections and indentations. (**278.** *Ulmus alata* [Ulmaceae]: leaf)

translator arm: One of the elongated structures connecting the pollinia of adjacent anthers present in many of the Apocynaceae. (**279.** *Asclepias humistrata* [Apocynaceae]: pollinia)

tree: A large, perennial, woody plant, usually having a single branched and woody main trunk with few or no branches arising from the base.

trichome: A hair or bristle.

trifid: Three-cleft. (**280.** *Xyris platylepis* [Xyridaceae]: pistil showing trifid style)

trifoliolate: With three leaflets. (**281.** *Toxicodendron radicans* [Anacardiaceae]: leaf)

trigonous: Three-angled. (**282.** *Cyperus esculentus* [Cyperaceae]: node showing trigonous stem)

trilacunar node: A node in a stem that has three leaf gaps related to one leaf. (**283.** *Ostrya virginiana* [Betulaceae]: cross section of node showing vascular tissue)

trimorphic heterostyly: See **tristyly**.

tristyly (= trimorphic heterostyly): The occurrence of three different mature style lengths, relative to other parts of the flower, in various plants of the same species; an outcrossing mechanism found only in the Lythraceae, Oxalidaceae, and Ponte-

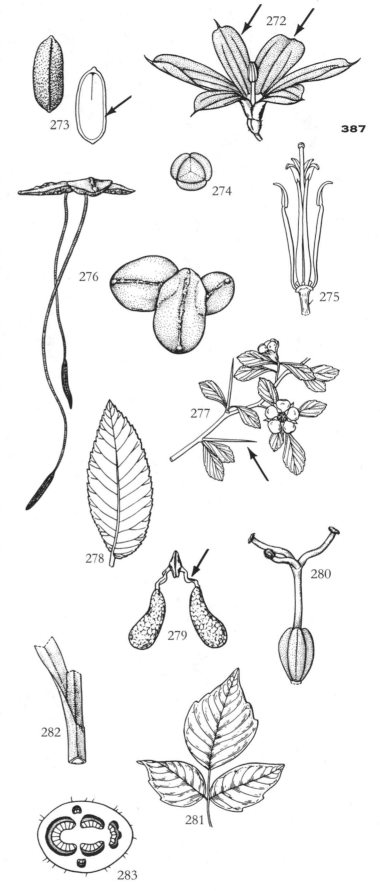

387

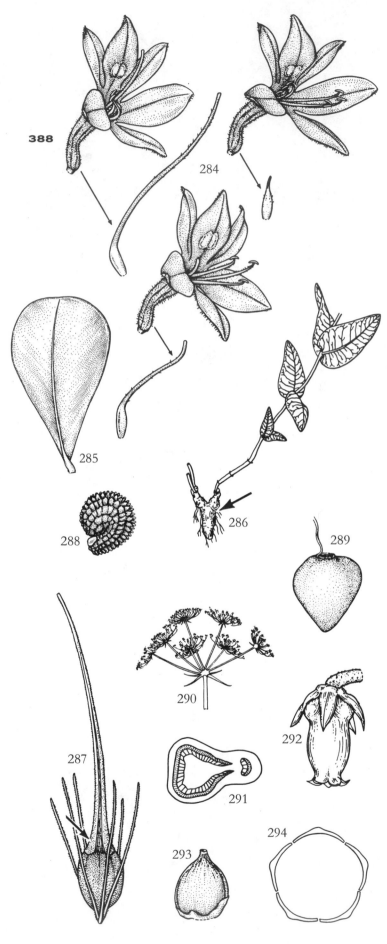

deriaceae. (**284**. *Pontederia cordata* [Pontederiaceae]: long-styled flower + pistil [left], medium-styled flower + pistil [bottom], short-styled flower + pistil [right])

truncate: Having a base or apex that ends abruptly, as though squarely cut off. (**285**. *Clusia rosea* [Clusiaceae]: leaf)

tuber: A thickened, short, underground part of the stem that functions as a storage area for reserve food. (**286**. *Asclepias humistrata* [Apocynaceae]: lower portion of plant with tuber)

tubercle: The thickened, persistent style base on the achenes of some of the Cyperaceae; also, a rounded protrusion (bump) on a surface. (**287**. *Rhynchospora careyana* [Cyperaceae]: achene with well-developed tubercle)

tuberculate: Covered with tubercles (rounded protrusions). (**288**. *Portulaca pilosa* [Portulacaceae]: seed)

turbinate: Top-shaped; more or less resembling an inverted cone. (**289**. *Helianthemum corymbosum* [Cistaceae]: seed)

umbel: An inflorescence composed of several branches that radiate from almost the same point and are terminated by single flowers or secondary umbels. (**290**. *Cicuta mexicana* [Apiaceae]: longitudinal section of umbel with secondary umbels)

unilacunar node: A node in a stem that has one leaf gap related to one leaf. (**291**. *Persea borbonia* [Lauraceae]: cross section of node showing vascular tissue)

uniseriate: In one series or whorl.

unitegmic ovule: An ovule with a single integument.

urceolate: Urn-shaped; more or less globular, with a contracted mouth and small, flaring lobes. (**292**. *Lyonia lucida* [Ericaceae]: flower with urceolate corolla)

utricle: A more or less small, indehiscent, dry fruit with a thin wall (bladder-like) that is loose and free from the single seed. (**293**. *Lemna minor* [Lemnaceae]: utricle)

valvate aestivation: An arrangement of perianth parts (sepal, petal, tepal, or lobe) in bud in which the parts meet by the edges without overlapping. (**294**. *Mimosa quadrivalvis* [Fabaceae]: cross section of corolla in bud)

velamen: The absorbent and protective root epidermis of epiphytic orchids. (**295**. *Encyclia tampensis* [Orchidaceae]: cross section of aerial root with detail showing velamen and portion of cortex)

ventral: See **adaxial**; also, in thallose plants (e.g., *Lemna*) the ventral surface refers to the lower surface, next to the substrate.

versatile: (Of an anther), attached to its filament at the middle and usually moving freely. (**296**. *Oenothera laciniata* [Onagraceae]: lateral view of stamen)

verticil: A whorl.

vestiture: The covering on a surface; e.g., hairs or scales.

villose (villous): Shaggy; densely covered with long and soft hairs.

vine: A woody or herbaceous plant with a long, slender, more or less flexible stem that either trails on the ground or is supported by twining and climbing.

viscidium: A sticky part of the rostellum of some orchids that serves to attach the pollinia to the pollinator. (**297**. *Platanthera ciliaris* [Orchidaceae]: pollinium [left] and detail of caudicle base showing viscidium [right])

viscin: An elastic and somewhat viscid material.

vitta: An aromatic oil or resin canal in the pericarp of many of the Apiaceae. (**298**. *Cicuta mexicana* [Apiaceae]: schizocarp [left] and cross section of schizocarp [right])

viviparous: (Of seeds), germinating while still attached to the parent plant; also used to describe a plant that produces such seeds. (**299**. *Rhizophora mangle* [Rhizophoraceae]: portion of branch with one viviparous fruit)

whorl: A group of leaves or other structures arranged in a circle at a single node (e.g., three or more leaves per node). (**300**. *Medeola virginiana* [Liliaceae]: middle portion of plant showing whorled leaves)

wings: Thin, often membranous extensions of a structure. (**301**. *Acer rubrum* [Sapindaceae]: winged schizocarp.) Also, the lateral pair of petals of a papilionaceous corolla present in many of the Fabaceae (Faboideae). (**302**. *Crotalaria spectabilis* [Fabaceae]: front view of flower [left] and expanded corolla [right])

xerophyte (*adj.* xerophytic, xerophilous): A plant adapted to dry and arid conditions.

zygomorphic: See **bilaterally symmetrical**.

REFERENCES CITED

Allaby, M. (ed.). 1992. *The concise Oxford dictionary of botany.* Oxford University Press, Oxford.

Benson, L. 1979. *Plant classification*, pp. 857–877. D.C. Heath, Lexington, Mass.

Cook, J. G. 1968. *ABC of plant terms.* Merrow, Watford, England.

Cronquist, A. 1968. *The evolution and classification of flowering plants*, pp. 375–384. Houghton Mifflin, Boston.

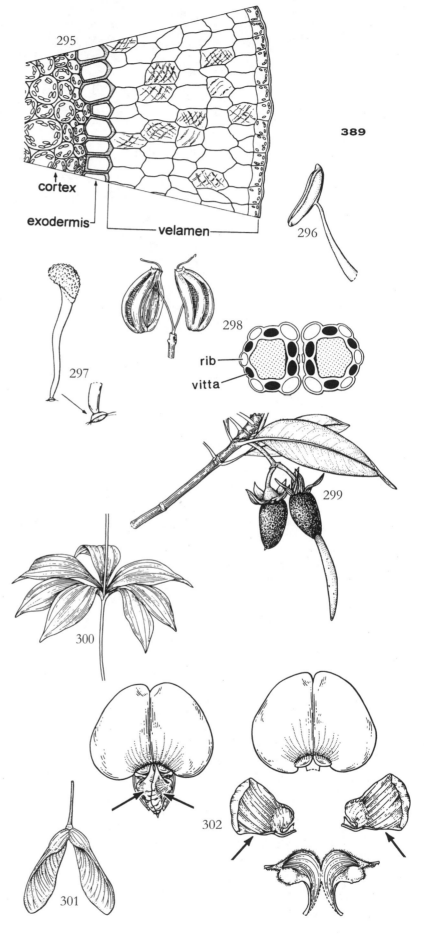

389

————. 1988. *The evolution and classification of flowering plants*, 2d ed., pp. 519–531. New York Botanical Garden, Bronx.

Huxley, A., ed. 1992. *The new Royal Horticultural Society dictionary of gardening*. Vol. 1, *A–C*, pp. xxvii–xlviii. Macmillan, London.

Judd, W. S. 1985. A revised traditional/descriptive classification of fruits for use in floristics and teaching. *Phytologia* 58:233–242.

Lawrence, G. H. M. 1951. *Taxonomy of vascular plants*, pp. 737–775. Macmillan, New York.

Little, R. J., and C. E. Jones. 1980. *A dictionary of botany*. Van Nostrand Reinhold, New York.

Mayr, E., and P. D. Ashlock. 1991. *Principles of systematic zoology*, 2d ed., pp. 407–433. McGraw-Hill, New York.

Radford, A. E., W. C. Dickison, J. R. Massey, and C. R. Bell. 1974. *Vascular plant systematics*, pp. 79–284. Harper and Row, New York.

Rickett, H .W. 1944. The classification of inflorescences. *Bot. Rev.* (Lancaster) 10:187–231.

————. 1955. Materials for a dictionary of botanical terms. III. Inflorescences. *Bull. Torrey Bot. Club* 82:419–445.

Simpson, G. G. 1961. *Principles of animal taxonomy*, pp. 1–28, 107–146. Columbia University Press, New York.

Stearn, W. T. 1966. *Botanical Latin*. Hafner, New York.

Wiley, E. O. 1981. *Phylogenetics*, pp. 1–20, 70–92, 126–130, 240–276. John Wiley and Sons, New York.

390

APPENDIX A

FAMILIES IN THIS BOOK ACCORDING TO CRONQUIST'S CLASSIFICATION SYSTEM

The classification systems used in many textbooks and floras conform closely to that of Cronquist (1968, 1981, 1988). Listed below are the families included in this book arranged according to Cronquist (1988), whose circumscriptions of several taxa differ considerably from those of Thorne (1992), the system followed in this book. Some indication of these differences, but not every variance, is given here. Generally, Cronquist's family circumscriptions are more restrictive (for example, the Ericaceae *s.s.*, with the Pyrolaceae and Monotropaceae maintained as segregates); in such cases, the major segregate families also are listed here. However, a few families, such as the Liliaceae, are interpreted much more broadly by Cronquist; Thorne's most important segregate taxa (mentioned in the text) are then included parenthetically.

Division Magnoliophyta

Class Magnoliopsida

  Subclass Magnoliidae

    Order Magnoliales
    (1) Magnoliaceae
    (2) Annonaceae

    Order Laurales
    (3) Lauraceae

    Order Piperales
    (4) Piperaceae

    Order Aristolochiales
    (5) Aristolochiaceae

    Order Illiciales
    (6) Illiciaceae

    Order Nymphaeales
    (7) Nelumbonaceae
    (8) Nymphaeaceae
    (9) Cabombaceae

    Order Ranunculales
    (10) Ranunculaceae
    (11) Menispermaceae

Order Papaverales
(12) Papaveraceae
(13) Fumariaceae

Subclass Hamamelidae

Order Hamamelidales
(14) Platanaceae
(15) Hamamelidaceae

Order Urticales
(16) Ulmaceae
(17) Cannabaceae
(18) Moraceae
(19) Cecropiaceae
(20) Urticaceae

Order Juglandales
(21) Juglandaceae

Order Myricales
(22) Myricaceae

Order Fagales
(23) Fagaceae
(24) Betulaceae

Subclass Caryophyllidae

Order Caryophyllales (Centrospermae)
(25) Phytolaccaceae
(26) Nyctaginaceae
(27) Cactaceae
(28) Chenopodiaceae
(29) Amaranthaceae
(30) Portulacaceae
(31) Caryophyllaceae

Order Polygonales
(32) Polygonaceae

Subclass Dilleniidae

Order Theales
(33) Theaceae
(34) Clusiaceae

Order Malvales
(35) Tiliaceae
(36) Sterculiaceae
(37) Bombacaceae
(38) Malvaceae

Order Nepenthales
(39) Sarraceniaceae

Order Violales
(40) Cistaceae
(41) Violaceae
(42) Turneraceae
(43) Passifloraceae

(44) Cucurbitaceae
(45) Begoniaceae

Order Salicales
(46) Salicaceae

Order Capparales
(47) Capparaceae
(48) Brassicaceae

Order Ericales
(49) Ericaceae
(50) Pyrolaceae
(51) Monotropaceae

Order Ebenales
(52) Sapotaceae

Order Primulales
(53) Myrsinaceae

Subclass Rosidae

Order Rosales
(54) Hydrangeaceae
(55) Grossulariaceae (incl. Escalloniaceae, Iteaceae)
(56) Rosaceae

Order Fabales
(57) Mimosaceae
(58) Caesalpiniaceae
(59) Fabaceae

Order Myrtales
(60) Sonneratiaceae
(61) Lythraceae
(62) Myrtaceae
(63) Punicaceae
(64) Onagraceae
(65) Melastomataceae
(66) Combretaceae

Order Rhizophorales
(67) Rhizophoraceae

Order Cornales
(68) Cornaceae

Order Santalales
(69) Viscaceae

Order Celastrales
(70) Celastraceae
(71) Hippocrateaceae
(72) Aquifoliaceae (incl. *Phelline, Sphenostemon*)

Order Euphorbiales
(73) Euphorbiaceae

Order Rhamnales
(74) Rhamnaceae

392

(75) Leeaceae
(76) Vitaceae

Order Polygalales
(77) Malpighiaceae
(78) Polygalaceae

Order Sapindales
(79) Sapindaceae
(80) Hippocastanaceae
(81) Aceraceae
(82) Anacardiaceae
(83) Meliaceae
(84) Rutaceae
(85) Zygophyllaceae

Order Geraniales
(86) Oxalidaceae
(87) Geraniaceae

Order Apiales
(88) Araliaceae
(89) Apiaceae

Subclass Asteridae

Order Gentianales
(90) Loganiaceae (*sensu lato*)
(91) Gentianaceae (*sensu stricto*)
(92) Apocynaceae
(93) Asclepiadaceae

Order Solanales
(94) Solanaceae
(95) Convolvulaceae
(96) Cuscutaceae
(97) Polemoniaceae

Order Lamiales
(98) Boraginaceae
(99) Verbenaceae (*sensu lato* and incl. Avicennia-
    ceae, Phrymaceae)
(100) Lamiaceae (*sensu stricto*)

Order Plantaginales
(101) Plantaginaceae

Order Scrophulariales
(102) Oleaceae
(103) Scrophulariaceae
(104) Orobanchaceae
(105) Acanthaceae
(106) Bignoniaceae
(107) Lentibulariaceae

Order Campanulales
(108) Campanulaceae

Order Rubiales
(109) Rubiaceae

Order Dipsacales
(110) Caprifoliaceae (incl. *Sambucus, Viburnum*)
(111) Adoxaceae

Order Asterales
(112) Asteraceae

Class Liliopsida

Subclass Alismatidae

Order Alismatales
(113) Limnocharitaceae
(114) Alismataceae

Order Hydrocharitales
(115) Hydrocharitaceae

Order Najadales
(116) Najadaceae

Subclass Arecidae

Order Arecales
(117) Arecaceae

Order Arales
(118) Araceae
(119) Lemnaceae

Subclass Commelinidae

Order Commelinales
(120) Xyridaceae
(121) Commelinaceae

Order Eriocaulales
(122) Eriocaulaceae

Order Juncales
(123) Juncaceae

Order Cyperales
(124) Cyperaceae
(125) Poaceae

Order Typhales
(126) Sparganiaceae
(127) Typhaceae

Subclass Zingiberidae

Order Bromeliales
(128) Bromeliaceae

Order Zingiberales
(129) Strelitziaceae
(130) Heliconiaceae
(131) Musaceae
(132) Zingiberaceae
(133) Costaceae

(134) Cannaceae

(135) Marantaceae

Subclass Liliidae

Order Liliales

(136) Pontederiaceae

(137) Haemodoraceae

(138) Liliaceae (incl. Alliaceae, Alstroemeriaceae, Amaryllidaceae, Asparagaceae *s.l.*, Hemerocallidaceae, Melanthiaceae)

(139) Iridaceae

(140) Agavaceae (incl. Dracaenaceae *s.l.*)

(141) Smilacaceae (incl. Philesiaceae, Ripogonaceae)

(142) Dioscoreaceae (incl. Trochopodaceae)

Order Orchidales

(143) Orchidaceae

REFERENCES CITED

Cronquist, A. 1968. *The evolution and classification of flowering plants*. Houghton Mifflin, Boston.

———. 1981. *An integrated system of classification of flowering plants*. Columbia University Press, New York.

———. 1988. *The evolution and classification of flowering plants*, 2d ed., pp. 503–517. New York Botanical Garden, Bronx.

Thorne, R. F. 1992. Classification and geography of the flowering plants. *Bot. Rev.* (Lancaster) 58:225–348.

Appendix B

Family Summary Chart

This chart summarizes morphological features of the 130 families included in this book. For easy reference, the families are listed in alphabetical order. Distributions and character states are here generalized from the most common situations cited in the more detailed family diagnoses and family characterization sections. For families with imperfect flowers, the floral summaries for carpellate vs. staminate flowers are not listed; refer to the family descriptions for the condition (sterile/rudimentary or absent) of the androecium and gynoecium of these flowers. Generally, carpellate flowers frequently have staminodes, while the gynoecium of staminate flowers is rudimentary to often absent. The abbreviations and symbols used to represent commonly used terms are listed at the end of the chart; most, however, are self-explanatory in the context of the particular structures described.

Family	**Genera/ spp.**	**Distribution**	**Vegetative:** habit; leaves	**Floral summary:** symmetry; gender; calyx, corolla/-tepals-; androecium; gynoecium	**Fruit type**	**Other important morphological features**
Acanthaceae	256/2,770	pantrop; few temp	h, vi, shr; simp, ent, opp	z; ⚥; (4 or 5), (5); (2) + (2) or 2 ± 2*; (2)	caps	cysto, bilab cor; epipet stam; unus anth; jaculators; muc sds
Adoxaceae	5/243	N Hem; few Austr, mont SA	h, shr, sm tr; simp/comp, ent/ser, opp, ± stip	a; ⚥; (4 or 5), (4 or 5); 4 or 5; (2–5)	dru	umb; redu cal lob; rotate cor; epipet stam; redu sty
Agavaceae	8/300	subtrop-trop; few temp	h, shr-tr; simp, alt/bas, par-v	a to ± z; ⚥; -(6)-; 6; (3)	caps, ber	lrg pl; sec gr; rhiz; succ fibr lvs; lrg term inflor; petald tep; blk sds
Alismataceae	16/100	cosmo; esp temp-trop N Hem	h; simp, ent, bas, par-v, op sh	a; ⚥ / ♂, ♀; 3, 3; 6–∞; 6–∞	aggr ach	aqu; milky sap; rhiz/tub; ephem pet; spir carp; crv embr; no endo
Alliaceae	30/720	cosmo	h; simp, ent, alt/bas, sh	a or z; ⚥; -(6)-; (6); (3)	caps	bulb; scap; umb + brc; usu blk sds
Alstroemeriaceae	4/160	Cen + SA	h; simp, ent, alt, par-v	a or ± z; ⚥; -6-; 6; (3)	caps	rhiz; twis lvs; umb + brc; oft spot tep
Amaranthaceae	65/850	trop-temp	h, shr; simp, ent, alt/opp	a; ⚥ / ♂, ♀; -3–5-; (5); (2 or 3)	utr, ach, pyx	anom sec gr; scari tep; enations; 1-loc ovy; 1 bas ovu; peris; crv embr
Amaryllidaceae	50/860	warm temp-trop	h; simp, ent, alt/bas, par-v, sh	a to ± z; ⚥; -(6)-; (6); (3)	caps, ber	scap; bulb; umb + brc; petald tep; ± corona; blk sds
Anacardiaceae	70/600	pantrop; few subtrop-temp Eurasia + NA	tr, shr, wo vi; comp/simp, ent/ser, alt	a; ♂, ♀ / ⚥; (5), (5); 5 or 10; (3)	dru	resin; nect disc; 1-loc ovy + 1 ovu

FAMILY	GENERA/ SPP.	DISTRI- BUTION	VEGETATIVE: habit; leaves	FLORAL SUMMARY: symmetry; gender; calyx, corolla/-tepals-; androecium; gynoecium	FRUIT TYPE	OTHER IMPORTANT MORPHOLOGICAL FEATURES
Annona- ceae	132/2,300	trop; few temp	tr, shr; simp, ent, alt	a; ⚥; 3, 6; ∞; ∞	aggr- syn ber	arom; biser pet; lam stam; monocol poll; rum endo
Apiaceae (Umbel- liferae)	460/4,250	cosmo; esp n temp	h, shr, tr; pin/palm comp, alt/bas, sh, ± stip	a; ⚥/ ♂, ♀; 0 or 5, 5; 5; (2 or 5)	sch, dru	oft arom; grvd stem; umb; redu cal; nect disc on ovy; 1 pend ovu/loc
Apocyna- ceae	355/3,700	pantrop; few temp	h, wo vi, shr, tr; simp, ent, opp/who, ± stip	a; ⚥; 5, (5); (5); (2)	aggr fol	interx phl; milky sap; tub; appen cor + andr; poll tetr; ± polli; ± gynostegium; dis ovy + conn sty; lrg stig; comose sds
Aquifolia- ceae	1/400	trop-temp	tr, shr; simp, ent/ser, alt, stip	a; ♂, ♀ / ⚥; 4 or 5, 4 or 5; 4 or 5; (4 – 7)	dru	acid soils; 1 pend ovu/loc; 4 pyr
Araceae	104/2,500	pantrop	h, vi; simp/pin-palm comp, ent, alt/bas, par/ pin/palm-v, sh	a; ⚥ / ♂, ♀; -0 or (4 – 6)-; (1 – 6); 1–(3)	ber	usu wet areas; rhiz/tub; pungent sap; spadix + spathe; anth pores
Arecaceae (Palmae)	200/2,780	trop- subtrop; also warm temp	tr, shr, wo vi; simp/pin-palm comp, ent, alt, palm/pin-v, sh	a; ♂, ♀ / ⚥; (3), (3); (6); 1–3 or (3)	dru, nutl	sec gr; term lrg plic lvs; lrg inflor; prophyll; monocol poll
Aristolo- chia- ceae	8/400	pantrop (excl Austr); few temp	wo vi, shr, h; simp, ent/lob, alt, palm-v	z or a; ⚥; (1 – 3), 0; 6–∞; (4 – 6)	caps	arom; bitter/peppery yel sap; fetid flr; campan- trumpet cal; ± gynostemium
Aspara- gaceae	26/240	N Hem + OW trop; few SA	h, vi, shr; simp, ent, alt/ opp/ who/ bas, par-v	a; ⚥ / ♂, ♀; -(6)-; 6; (3)	ber	rhiz; usu phylloclades; oft redu lvs; petald tep; epitep stam; ± blk sds
Astera- ceae (Compos- itae)	1,160/ 19,085	cosmo	h, shr, tr; simp, pin/palm lob/dis, alt/opp/bas	a or z; ⚥ / ♂, ♀; 0 or pappus, (5); (5); (2)	ach	taprt/tub; head + invol; flo- rets; epipet stam; appen anth; 1 bas ovu; no endo; poll pres mech
Begonia- ceae	2/920	trop- subtrop (excl Austr)	h, shr; simp, ent/ser, palm lob, alt, palm-v, stip	± z; ♂, ♀; -2-5-; ∞; (3)	caps	succ; jtd stems; rhiz/tub; wgd ovy; lob plac; twis stig; operc sds; no endo
Betula- ceae	6/157	n + s temp, mont trop	tr, shr; simp, ser, alt, stip	a; ♂, ♀; -0 or 1-6-; 1 – 4; (2)	nut, nutl, sam	monoe; redu flr; catk; frt + brc

Family	Genera/ spp.	Distri-bution	Vegetative: habit; leaves	Floral summary: symmetry; gender; calyx, corolla/-tepals-; androecium; gynoecium	Fruit type	Other important morphological features
Bignonia-ceae	113/800	trop-subtrop; few temp	tr, shr, wo vi; simp/palm-pin comp, ent/ser, opp/who	z; ⚥; ⑤, ⑤; 2 + 2 + 1*; ②	caps, her, pod	anom sec gr; campan/bilab cor; epipet stam; wgd sds; no endo
Bombaca-ceae	20/180	trop	lrg tr; simp/palm comp, ent/lob, alt, palm-v, stip	a; ⚥; ˌ5ˌ, 5; ⑸ ‒∞⹂; ⟨2 ‒ 5⟩	caps, pod	± epical; ± stamin; 1-loc anth; aril
Boragina-ceae	117/2,400	cosmo; esp temp-subtrop	h, shr, tr; simp, ent, alt	a to ± z; ⚥; ˌ5ˌ, ⑤; 5; ②	sch, dru	scabr/hisp; rhiz/taprt; 1-sided crv cyme; infolded cor; epipet stam; ± no endo
Brassica-ceae (Cru-ciferae)	376/3,200	temp-cool N Hem	h; simp, pin-lob, alt/bas	a; ⚥; 4, 4; 2 + 4; ②	silq, silc	acrid sap; clw pet in cross; parie plac; false septa + replum; crv embr
Bromelia-ceae	51/1,520	trop-warm temp Amer	h; simp, ent/ser, alt/spir/bas, par-v, sh	a to ± z; ⚥; ˌ3ˌ, ˌ3ˌ, ˌ6ˌ; ③	ber, caps	epiph; scap; pelt scales; epipet stam; spir twis stig; plum sds
Cabom-baceae	2/8	trop-warm temp	h; simp, ent/div, alt/opp/who	a; ⚥; 3, 3; 3‒∞; 2‒∞	aggr nutl	rhiz; usu monocol poll; operc sds; peris; ± no endo
Cactaceae	93/1,488	NA, Cent Amer, SA	h, vi, shr, tr; simp, alt, redu	a to ± z; ⚥; ‒∞‒; ∞; ⟨3‒∞⟩	ber	succ; spines; spir tep + andr; hypan; 1-loc ovy; parie plac
Campan-ulaceae	65/2,000	cosmo; esp temp-subtrop + mont trop	h, shr; simp, ent/pin-div, alt	a or z; ⚥; ⑤, ⑤; ⑸; ⟨2‒5⟩	caps, ber	oft milky sap; poll pres mech
Canna-ceae	1/25‒55	trop-subtrop Amer; also OW trop	h; simp, ent, spir, pin-v, op sh	i; ⚥; 3, ③; ½+½*+3* or ½*+4*; ⟨3⟩	caps	rhiz; petald andr + sty; stamin lip; decur stig; "aril" hrs; peris; poll pres mech
Cappara-ceae	46/930	trop-subtrop; few temp	shr, h, tr; simp/palm comp, alt, ± stip	a or z; ⚥; 2‒6, 2‒6; 6‒∞; ②	ber, silq	gynop; 1-loc ovy; frame-like replum
Caprifolia-ceae	12/450	temp-subtrop N Hem	shr, wo vi, h; simp, ent/ser, opp	z; ⚥; ⟨4 or 5⟩, ⟨4 or 5⟩; 4 or 5; ⟨2‒5⟩	ber, dru, caps, ach	constr tubu cor; epipet stam; vers anth; aborting carp

FAMILY	**GENERA/ SPP.**	**DISTRI- BUTION**	**VEGETATIVE:** habit; leaves	**FLORAL SUMMARY:** symmetry; gender; calyx, corolla/-tepals-; androecium; gynoecium	**FRUIT TYPE**	**OTHER IMPORTANT MORPHOLOGICAL FEATURES**
Caryo- phylla- ceae	70/1,750	n temp; few s temp, mont trop, Arctic	h; simp, ent, opp	a; ⚥ / ♂, ♀; (4 or 5), 4 or 5; 5–10; (2 – 5)	caps, utr	jtd stems; conn lvs; 1-loc ovy; fr-cen plac; peris; crv embr
Celas- traceae	55/855	trop- subtrop; few temp	tr, shr, wo vi; simp, ent/ser, opp/alt, ± stip	a; ⚥ / ♂, ♀; (4 or 5), 4 or 5; (3 – 5); (2 – 5)	caps, dru	conspic nect disc; aril
Cheno- podia- ceae	104/1,510	cosmo	h, shr; simp, ent/lob/ ser, alt	a; ⚥ / ♂, ♀; -(5)-; 5; (2 or 3)	ach, utr, pyx	anom sec gr; ± succ; 1-loc ovy; 1 bas ovu; peris; crv embr
Cistaceae	8/200	n temp	h, shr; simp, ent, opp, ± stip	a; ⚥; 3 ± 2, 3 or 5; ∞; (3 or 5)	caps	tufted hrs; redu lvs; ephem pet; 1-loc gyn; parie plac; crv embr
Clusia- ceae (Guttif- erae)	45/1,010	trop; many temp	tr, shr, h; simp, ent, opp/who	a; ⚥ / ♂, ♀; 2–10, 4–12; (4 –∞); (3 – 5)	caps, ber, dru	resin sap; punc lvs; fasc stam; no endo
Combre- taceae	20/600	trop; some subtrop	tr, shr, wo vi; simp, ent, alt/ opp/who	a to ± z; ⚥ / ♂, ♀; 4 or 5, 0 or 4 or 5; 4–10; (2 – 5)	dru	interx phl; paired gld on lvs; domatia; unus hrs; hypan; 1-loc ovy; long funic
Comme- lina- ceae	40/605	trop- subtrop; few warm-temp	h; simp, ent, alt, par-v, clo sh	a or z; ⚥; 3, 3; 6 or 3 + 3*; ③	caps	succ; muc; jtd stems; ephem pet; ± fringed stam; embryotega
Convol- vula- ceae	59/1,830	trop- subtrop; some temp	vi, shr, tr; simp, ent/lob/ div, alt	a; ⚥; (5), ⑤; 5; ②	caps	milky sap; interx phl; rhiz/ tub; salv/fun plic cor; epipet stam; axile plac; lrg embr; condup cot
Corna- ceae	6/78	cosmo; esp n temp	tr, shr; simp, ent/ser, opp/alt	a; ⚥ / ♂, ♀; 4 or 5, 4 or 5; 4–10; ②	dru	redu sep; nect disc on ovy; 1 pend ovu/loc; grvd pyr
Costaceae	4/150	pantrop; esp trop Amer	h; simp, ent, spir, pin-v, clo sh	z; ⚥; ③, ③; 1 + ⑤*; ③	caps	rhiz; spir stem; lrg lvs; lig; nect brc; cone inflor; petald andr; stamin lip; anth encl sty; aril; peris
Cucurbi- taceae	118/825	pantrop- subtrop; few temp-cool	vi (h/wo); simp/comp, palm-lob, alt, palm-v	a; ♂, ♀; ⑤, ⑤; (1 – 5); ③	ber, caps	scabr; 5-ang stems; interx phl; vb in 2 rings; tendr; unus stam; lrg parie plac
Cyper- aceae	146/5,315	cosmo; esp subarctic- temp	h; simp, ent, alt/bas, par-v, clo sh	a; ⚥ / ♂, ♀; -0 or bristles/scales-; 3; (2 or 3)	ach	wet areas; rhiz; 3-ang stems; spikel; redu flr; 1 brc/flr

FAMILY	GENERA/ SPP.	DISTRI- BUTION	VEGETATIVE: habit; leaves	FLORAL SUMMARY: symmetry; gender; calyx, corolla/-tepals-; androecium; gynoecium	FRUIT TYPE	OTHER IMPORTANT MORPHOLOGICAL FEATURES
Diosco- reaceae	5/625	trop- subtrop; few n temp	vi; simp, ent, alt, palm + net-v	a; ♂, ♀; -⑥-; 6; ③	caps	vb in 2 rings; rhiz/tub; twis petio; pulv; 3-wgd frt; wgd sds
Dracaena- ceae	9/230	trop- subtrop	h, shr; simp, alt/bas/ api, par-v, ± sh	a; ⚥/♂, ♀; -⑥-; 6; ③	ber, nut	lrg pl; sec gr; rhiz; epitep stam
Ericaceae	99/2,245	temp; also Arctic, mont trop	shr, h; simp, ent/ser, alt	a to z; ⚥; ④ or ⑤, ④ or ⑤; 8 or 10; ④ or ⑤	caps, ber, dru	acid soils; campan-urc cor; appen stam; anth pores; tetr poll; depr api ovy
Eriocaula- ceae	9/1,175	pantrop; few temp	h; simp, ent, alt/bas, ± sh	a to ± z; ♂, ♀; ② or ③, ② or ③; 2–6; ② or ③	caps	wet areas; scap; heads + scari invol; chaffy peria; epipet stam; unus poll; gynop; 1 ovu/loc; undif embr
Escal- lonia- ceae	15/200	mont + temp S Hem; few NA	shr, tr; simp, ser, alt	a; ⚥; ④ or ⑤, 4 or 5; 5; ①–⑥	caps, ber	gldr teeth on lvs
Euphor- biaceae	307/7,030	trop; many temp	tr, shr, h, vi; simp/comp, ent/lob/ser, alt/opp/who, stip	a; ♀, ♂; 0 or 5, 0 or ⑤; ①–∞; ③	sch	milky/resin sap; redu flr; oft cyathia; extra-stam gld; carun
Fabaceae (Legumi- nosae)	630/18,000	cosmo	h, shr, tr; comp, ent, alt, stip	z or a; ⚥; ⑤, ⑤; ⑩–∞; 1	legume, loment	bact nod; pulv; perig flr; hypan; parie plac; no endo
Fagaceae	9/700–800	n temp + mont trop (excl trop + s Afr)	tr, shr; simp, ent/ser/ lob, alt, stip	a; ♂, ♀; -3-8-; 4–∞; ③	nut	monoe; redu flr; catk; frt + invol
Gentiana- ceae	83/965	cosmo; esp temp- subtrop + mont trop	h, shr, sm tr; simp, ent, opp	a; ⚥; ④ or ⑤, ④ or ⑤; 4 or 5; ②	caps	wgd stems; interx phl; sess conn lvs; epipet stam; 1-loc ovy; usu parie plac
Gerania- ceae	14/775	temp- subtrop; few trop	h, shr; palm-pin comp/ simp, ent/ser, opp/alt, palm-v, stip	a to ± z; ⚥; 5, 5; ⑩; ⑤	sch	arom; gldr hrs; rhiz; biser stam; lob frt; no endo
Haemo- dora- ceae	14/100	S Hem, esp Austr; 1 sp in e NA	h; simp, alt/bas, par-v, sh	a to z; ⚥; -⑥-; 3 or 6; ③	caps	wet areas; rhiz/stol; red sap; condup conn lvs; woolly inflor; tep encl epitep stam; lrg pelt plac

FAMILY	GENERA/ SPP.	DISTRI- BUTION	VEGETATIVE: habit; leaves	FLORAL SUMMARY: symmetry; gender; calyx, corolla/-tepals-; androecium; gynoecium	FRUIT TYPE	OTHER IMPORTANT MORPHOLOGICAL FEATURES
Hamame- lida- ceae	30/120	temp- subtrop; discon	tr, shr; simp, ser/palm- lob, alt, stip	a to z; ♀/♂, ♀; 0 or (4 or 5), 0 or 4 or 5; 4 + 4* or ∞; ②̲	caps	stel hrs; redu flr; ½-inf ovy; divergent hard sty; wo frt
Heli- conia- ceae	1/250	Amer trop	h; simp, ent/torn, dist, pin-v, op sh	z; ♂; -⑤ + 1-; 5 + 1*; ③̄	dru-sch	rhiz; pseudostem; boat- shaped brc; resup flr; 1 ov/loc; peris
Hemero- callida- ceae	1/16	mainly temp Asia + s Eur	h; simp, ent, bas, par-v, sh	z; ♂; -⑥-; 6; ③̲	caps	rhiz; ± strk tep; epitep stam; blk sds
Hydran- geaceae	17/250	temp- subtrop N Hem; few w SA	shr, sm tr, wo vi; simp, ser/lob, opp	a; ♂; (4 or 5), 4 or 5; 8 -∞; (2-5)̄	caps	hypan; nect disc on ovy; dis sty
Hydro- chari- taceae	17/130	trop; few temp	h; simp, ent/ser, alt/opp/ who/bas, par-v, sh	a; ♂, ♀; 3, 0 or 3; (2 -∞); (3-6)̄	ber- caps	aqu; rhiz; spathe; 1-loc ovy; parie plac
Illiciaceae	1/37	trop mont, temp NA + se Asia	shr, sm tr; simp, ent, alt	a; ♂; -∞-; ∞; 5-∞	aggr fol	arom; spir flr parts; ± lam stam; condup sty; star frt
Iridaceae	77/1,655	cosmo	h; simp, ent, alt/bas, par-v, op sh	a or z; ♂; -⑥-; ③; ③̄	caps	rhiz/bulb/corm; petald tep
Juglanda- ceae	8/59	temp- subtrop N Hem, mont SA + Asia	tr; pin comp, ser, alt	a; ♂, ♀; -0 or 4-; 3-∞; ②̄	dru-nut	arom; resin; ± catk; redu flr; 1 bas ovu; frt + husk; no endo; corrugated cot
Juncaceae	8/300	cosmo; esp temp and mont	h; simp, ent, alt/bas, par-v, op sh	a; ♀/♂, ♀; -6-; 3 or 6; ③̲	caps	wet areas; rhiz; sm flr; scari tep; tetr poll; appen sds
Lamia- ceae (Labiatae)	258/6,970	cosmo; esp temp	h, shr, tr; simp/pin-palm comp, ser, opp/who	z or ± a; ♂; ⑤, ⑤; 2 + 2 or 2; ②̲	sch, dru	arom; squ stems; cymes; ± verticils; ± bilab cor; epipet stam; false septa + 4-loc ovy; lateral ovu; term/gynob sty; no endo
Lauraceae	31/2,490	trop- subtrop	tr, shr; simp, ent/lob, alt/opp	a; ♀/♂, ♀; -⑥-; 12; 1̲	dru, ber	arom; who stam; anth valves; 1 ovu; cupule; no endo
Lemna- ceae	4/28	cosmo	h (thallus); none	± z; ♂, ♀; -0-; 1; 1̲	utr	aqu; redu pl; ± flr; asex repro

FAMILY	GENERA/SPP.	DISTRIBUTION	VEGETATIVE: habit; leaves	FLORAL SUMMARY: symmetry; gender; calyx, corolla/-tepals-; androecium; gynoecium	FRUIT TYPE	OTHER IMPORTANT MORPHOLOGICAL FEATURES
Lentibulariaceae	4/170	cosmo	h; simp, alt/bas, ent/div	z; ♀⚥; ⸜4 or 5⸝, ⑤; 2 ± 2*; ②	caps	wet/aqu; insectiv; scap; gldr lvs; bilab cor; epipet stam; 1-loc anth; fr-cen plac; unequ stig; no endo; undif embr
Liliaceae	22/485	temp N Hem	h; simp, ent, alt/bas, par-v, sh	a to ± z; ⚥; -3 + 3-; 6; ③	caps, ber	bulb/rhiz; petald spot tep; undif embr
Loganiaceae	17/480	trop-subtrop; few temp	wo vi, shr, tr, h; simp, ent, opp, interpet stip	a to ± z; ⚥; (4 or 5), (4 or 5); 4 or 5; ②	caps, ber	interx phl; colleters; epipet stam
Lythraceae	25/460	subtrop-trop; few temp	h, shr; simp, ent, opp/who, ± stip	a or ± z; ⚥; 4–8, 4–8; 4–∞; (2 – 6)	caps, ber	interx phl; perig flr; persistent hypan; clw pet; unequ stam
Magnoliaceae	7/220	temp-trop se Asia + e NA; few SA	tr, shr; simp, ent, alt, stip	a; ⚥; -6–∞-; ∞; ⊙⊙	aggr: fol, sam	arom; spir flr parts; long recep; lam stam; mono-col poll; condup carp
Malpighiaceae	66/1,200	pantrop; few subtrop	wo vi, shr, sm tr; simp, ent, opp, stip	z; ⚥; ⸜5⸝, 5; ⑩; ③	sam, sch, dru, caps	anom sec gr; malp hrs; abax oil gld on sep; unequ clw pet; unequ sty
Malvaceae	75/1,000+	cosmo; esp trop Amer	h, shr, sm tr; simp, ent/ser/lob, alt, palm-v, stip	a; ⚥; ⸜5⸝, 5; ∞; (2 –∞)	caps, sch	muc sap; stel hrs; epical; 1-loc anth; lrg poll; carp in ring
Marantaceae	30/450	pantrop; few subtrop-warm temp	h; simp, ent, alt/bas/dist, pin-v, op sh	i; ⚥; 3, ③; ½+½*+3* or ½*+4*; ③̄	caps, ± nut, ber	rhiz; pulv; mirror-image flrs; conspic brc; petald andr; 1 bas ovu; unus stig-sty; aril; peris + canals; poll pres mech
Melanthiaceae	25/155	temp-boreal N Hem; few SA	h; simp, ent, spir/bas, par-v, op sh	a; ⚥; -6-; 6; ③̲	caps, aggr fol	rhiz; oft wgd sds
Melastomataceae	244/3,360	pantrop	h, shr, tr; simp, ent/ser, opp, ± par-v	± z; ⚥; 4 or 5, 4 or 5; 8 or 10; (3 – 5)	caps, ber	interx phl; urc hypan; bent stam; appen anth + pores
Meliaceae	51/550	trop-subtrop; few temp	shr, tr; pin comp, ent/ser, alt	a; ⚥ ± ♂, ♀; ⸜4 or 5⸝, ⸜4 or 5⸝; (8 –10); (2 – 5)	caps, dru, ber	arom; api appen andr
Menispermaceae	65/350	trop rainf; few subtrop-temp	wo vi; simp, ent/palm lob, alt, palm-v	a; ♂, ♀; 6, 6; 6; 3 or 6	aggr: dru, nutl	anom sec gr; dioe; biser cal, cor, andr; crv embr

Family	Genera/spp.	Distribution	Vegetative: habit; leaves	Floral summary: symmetry; gender; calyx, corolla/-tepals-; androecium; gynoecium	Fruit type	Other important morphological features
Musaceae	2/42	OW trop	h-tr; simp, ent/torn, spir, pin-v, op sh	z; ♂, ♀; -(5)+1-; 5 + 1*; (3)	caps, ber	muc; pseudostem; conspic brc; peris
Myrica-ceae	3/45	temp-subtrop (excl Austr)	shr, sm tr; simp, ent/ser, alt	a; ♂, ♀; -0-; 2–8; (2)	dru, nutl	xeric/acid soils; arom; bact nod; yel punc lvs; redu flr; 1-loc ovy; 1 bas ovu; waxy frt
Myrsina-ceae	33/1,000	trop-warm temp	tr, shr, wo vi; simp, ent/ser, alt	a; ☿/♂, ♀; 5, (5); 5; (4 or 5)	dru	resin dotted/strk; punc lvs; epipet stam; 1-loc ovy; lrg plac
Myrtaceae	144/3,000	trop-subtrop	tr, shr; simp, ent, opp/alt	a; ☿; 4 or 5, 0 or 4 or 5; ∞; (2–5)	ber, caps	arom; interx phl; punc lvs; hypan
Nelum-bona-ceae	1/2	e Asia + e NA	h; simp, ent, alt/bas	a; ☿; 2, ∞; ∞; ∞	aggr nut	aqu; scap; rhiz/tub; pelt lvs; carp embed in recep; 1 ovu/carp; no endo
Nyctagi-naceae	30/290	trop-subtrop; few temp	h, shr, tr; simp, ent, opp	a; ☿/♂, ♀; -(5)-; 5; 1	ach	anom sec gr; conspic brc; 1 bas ovu; accr peria encl frt; peris
Nym-phaea-ceae	6/62	trop–n cold temp	h; simp, ent, alt/bas, palm/pin-v	a; ☿; 4 or 6, 8–∞; ∞; (5–∞)	± ber	aqu; scatt vb; milky sap; scap; rhiz; lrg lvs; lam stam; monocol poll; lrg stig; operc sds; peris
Oleaceae	29/600	± cosmo	tr, shr, wo vi; simp/pin comp, ent, opp	a; ☿/♂, ♀; (4), 0 or (4); 2; (2)	dru, caps, sam	punc lvs; epipet stam
Onagra-ceae	18/650	temp-subtrop	h, shr, tr; simp, ent/ser/lob, alt/opp/who, ± stip	a or z; ☿; 4, 4; 4 or 8; (4)	caps, ber, nutl	interx phl; hypan; unus poll + viscin; no endo
Orchida-ceae	775/19,500	cosmo; esp trop	h; simp, ent, alt, par-v, sh	z; ☿; 3, 3/-6-; 1 or 2; (3)	caps	resup flr; lip; column; tetr poll in polli; parie plac; num ovu; min sds; undif embr; no endo
Oxalida-ceae	6/890	pantrop; also temp	h, shr; pin/palm comp, ent, alt/bas/api	a; ☿; 5, 5; (10); (5)	caps	acrid sap; rhiz/tub; pulv; di/tristyly; elastic aril
Papavera-ceae	42/660	n temp–subtrop; few s Afr + e Austr	h, vi, shr, tr; simp/comp, pin/palm lob/dis, alt/bas	a or z; ☿; 2 or 3, 4 or 6; 3 + 3 or ∞; (2–∞)	caps, nut	acrid sap; cadu sep; biser cor; 1-loc ovy; parie plac

FAMILY	GENERA/ SPP.	DISTRI- BUTION	VEGETATIVE: habit; leaves	FLORAL SUMMARY: symmetry; gender; calyx, corolla/-tepals-; androecium; gynoecium	FRUIT TYPE	OTHER IMPORTANT MORPHOLOGICAL FEATURES
Passiflora-ceae	18/630	trop-warm temp	vi (h/wo); simp/comp, ent/ser/palm lob, alt, palm-v, stip	a; ⚥; 5, 5, 5; ③	ber, caps	tendr; gldr petio; perig flr; ± complex corona; androgyn; parie plac; aril
Phytolac-caceae	14/97	trop-subtrop	h, shr, tr; simp, ent, alt	a; ⚥; -4 or 5-; 4–10; 1–∞	ber, dru, sam, ach	± succ; anom sec gr; nect disc + andr; 1 ovu/loc; colored sap in frt; aril; peris; crv embr
Pipera-ceae	8/2,000	pantrop	h, shr, tr; simp, ent, alt, palm/pin-v, ± stip	a or z; ⚥/ ♂, ♀; -0-; 2–6; 1, 3, or 4	dru	arom; jtd stems; scatt/ringed vb; redu flr; pelt brc; spadix; 1 ovu; peris; undif embr
Plantagi-naceae	3/220	temp; also mont trop	h; simp, ent/ser, alt/bas, par-v, sh	a to ± z; ⚥; 4, ④; 4; ②	caps	scap; phyllodes; spikes/heads; scari flr; salv cor; epipet stam; vers anth; muc sds
Platana-ceae	1/6–9	temp-subtrop N Hem	tr; simp, ser, palm-lob, alt, palm-v, stip	a; ♂, ♀; 3–7, 0 or 3–7; 3–7; 5–9	aggr ach	monoe; candelabra hrs; peeling bark; petio cap over axil bud; redu flr; flr + frt ball; ach with bristles
Poaceae (Gramin-eae)	650–785/ 10,000	cosmo	h; simp, ent, alt/bas, par-v, op sh	± z; ⚥/ ♂, ♀; -1–3-; 1–3; 2 or 3	grain	jtd stems; rhiz/stol; lig; spikel; redu flr; palea + lemma; lodicules; plum stig
Pole-monia-ceae	16/320	temp NA + Eur, mon-tane SA	h, shr, sm tr, vi; simp/comp, ent/div, alt/opp	a; ⚥; ⑤, ⑤; 5; ③	caps	noxious odor; salv/fun plic cor; epipet stam; muc sds
Polygala-ceae	15/800	cosmo (excl NZ)	h, vi, shr; simp, ent, alt, ± stip	z; ⚥; 2 + 3, 2+1 or ⑤; ⑧; ②	caps	taprt; butterfly-like flr; 2 petald sep; redu pet; 1-loc anth + pores; di-morphic lob sty
Polygona-ceae	49/1,100	n temp; few trop-subtrop	h, shr, vi; simp, ent, alt, stip	a; ⚥; -5 or 6-; 6–9; 2 or 3	ach	jtd stems; ocrea; 1-loc ovy; 1 bas ovu
Ponte-deria-ceae	8/30	subtrop-trop; few n temp	h; simp, ent, opp/ who/bas, par-v, sh	a to z; ⚥; -⑥-; 6; ③	caps, nutl	aqu; rhiz/stol; term inflor + spathe; petald tep; unequ epitep stam
Portulaca-ceae	19/500	warm temp	h, shr; simp, ent, opp/ alt/bas, stip	a; ⚥; -4 – 6-; 4–∞; 2 – 9	caps	succ; taprt; 2 brc; 1-loc ovy; bas fr-cen plac; brn sty; peris; crv embr

APPENDIX B CONTINUED

FAMILY	GENERA/ SPP.	DISTRIBUTION	VEGETATIVE: habit; leaves	FLORAL SUMMARY: symmetry; gender; calyx, corolla/-tepals-; androecium; gynoecium	FRUIT TYPE	OTHER IMPORTANT MORPHOLOGICAL FEATURES
Ranunculaceae	46/1,900	temp-boreal N Hem	h, shr, vi; simp/comp-dis, ser/lob, alt, sh	a or z; ⚥ / ♂, ♀; 5, 0 or 5; ∞; ∞	aggr: fol, ber, ach	scatt/ringed vb; rhiz/tub; spir stam + carp
Rhamnaceae	45/850	temp-trop	tr, shr, wo vi; simp, ent/ser, alt, stip	a; ⚥ / ♂, ♀; 4 or 5, 4 or 5; 4 or 5; (2 or 3)	dru	perig flr; hypan; conspic nect disc; concave pet encl epipet stam; 1 ovu/loc
Rhizophoraceae	12/84	trop-subtrop rainf, shorelines	shr, tr; simp, ent, opp, interpet stip	a; ⚥; 4 or 5, 4 or 5; 8 or 10; (2 – 4)	ber, dru	mangrove; punc lvs; hrs on pet; parti anth + flap; viviparous + grn embr
Rosaceae	100/2,000	cosmo; esp temp	tr, shr, h; simp/palm-pin comp, ser, alt/bas, stip	a; ⚥; 5, 5; ∞; 1-(2 – 5)–∞	pome, dru; aggr: ach, fol, dru	thorns/prickles; perig-epig flr; hypan; clw pet; ± no endo
Rubiaceae	500–600/ 9,000	trop-subtrop; some temp	tr, shr, wo vi, h; simp, ent, opp/who, interpet stip	a; ⚥; (4 or 5), (4 or 5); 4 or 5; (2)	caps, sch, ber, dru	colleters; oft distyly; epipet stam
Rutaceae	154/925	trop-subtrop; many warm temp	shr, tr; simp/palm-pin comp, ent, alt/opp	a; ⚥ / ♂, ♀; 4 or 5, 4 or 5; 4 – 10; (2 – 5)	ber, dru, caps, sam, sch, aggr: fol, dru	arom; punc lvs; nect disc; lob gyn
Salicaceae	3/530	± cosmo (excl Austr); esp n temp + subarctic	tr, shr; simp, ser, alt, stip	± a; ♂, ♀; -0-; 2 – ∞; (2 – 4)	caps	dioe; redu flr + brc; catk; redu peria = disc/gld; comose sds
Sapindaceae	147/2,215	trop-subtrop; some n temp	tr, shr, wo vi; pinn-palm comp/simp, ser, alt/opp, ± stip	a to z; ♀, ♂ / ⚥; 4 or 5, 4 or 5 or 0; 4–10; (2 or 3)	caps, nut, ± ber, sch	conspic nect disc; 1 or 2 ovu/loc; no funicle; no endo; crv embr; radicular pocket
Sapotaceae	53/1,100	pantrop	tr, shr; simp, ent, alt	a; ⚥; 4 – 8, (4 – 8); 4–10 + 4* or 5*; (4 – 5)	ber	milky sap; malp hrs; appen cor; epipet stam; thick testa; lrg hilum
Sarraceniaceae	3/14	e NA, OR, n CA, n SA	h; simp, alt/bas	a; ⚥; 5, 5; ∞; (5)	caps	wet areas; insectiv; rhiz; tubu lvs; pelt sty
Scrophulariaceae	220/3,000	cosmo; esp n temp	h, shr; simp, ent/lob, opp/alt	z; ⚥; 4 or 5, (4 or 5); 2 + 2 ± 1*; (2)	caps	epipet stam; num sm sds

FAMILY	GENERA/ SPP.	DISTRI-BUTION	VEGETATIVE: habit; leaves	FLORAL SUMMARY: symmetry; gender; calyx, corolla/-tepals-; androecium; gynoecium	FRUIT TYPE	OTHER IMPORTANT MORPHOLOGICAL FEATURES
Smilaca-ceae	3/310	trop-subtrop; some n temp	vi (h/± wo); simp, ent/ser/ lob, alt, palm + net-v	a; ♂, ♀; -̣6̣-; ̣6̣; ③	ber	dioe; prickles; rhiz/tub; tendr; umb; ± 1-loc anth; hard endo
Solana-ceae	76/2,900	cosmo	h, shr, tr; simp, ± pin div, alt	a; ♀; ⑤, ⑤; 5; ②	ber, caps	interx phl; plic salv/fun cor; epipet stam; axile plac
Sterculia-ceae	60/700	pantrop; few warm temp	tr, shr, h; simp/palm comp, ent/ ser/palm lob, alt, palm-v, stip	a to z; ♀/ ♂, ♀; ⑤, 0 or 5; 5 + 5*; ⑤ or ⑤	caps, sch, ber, aggr fol	muc sap; stel hrs; gldr hrs on adax sep; androgyn
Strelitzia-ceae	3/7	trop SA + s Afr	h-tr; simp, ent/torn, dist, pin-v, op sh	z; ♀; 3, ②+1; 5±1*; ③̄	caps	± wo stem; rhiz; boat-like brc; aril; peris
Theaceae	28/500	trop-subtrop; few warm temp	tr, shr; simp, ser/ent, alt	a; ♀; ̣5̣, ̣5̣; ∞; ③–5̲	caps	conspic brc; ± fasc stam; wo frt
Tiliaceae	49/450	trop-subtrop; few temp	tr, shr, h; simp, palm lob/ ser, alt, palm-v, stip	a; ♀; ̣5̣, 5 or 0; ∞+5+ *; ②–5̲	nut, caps, sch, dru	stel hrs; ± fasc stam; androgyn
Trilliaceae	2/50	temp N Hem	h; simp, ent, who, net-v	a; ♀; 3–8, 3–8; 6; ③	ber, caps	rhiz
Turnera-ceae	8/120	trop-subtrop Amer + Afr; few warm temp	h, shr; simp, ent/ser/ lob, alt, ± stip	a; ♀; 5, 5; 5; ③	caps	varied hrs; extrafloral nect; distyly; hypan; 1-loc ovy; parie plac; fimbriate stig; aril
Typha-ceae	2/30	± cosmo (excl some trop)	h; simp, ent, alt/bas, par-v, sh	a; ♂, ♀; -3 or 4 or bristles/scales-; ①–6̲; ②	fol, dru	aqu; rhiz; unisexual spike/head; 1-loc ovy; 1 ovu; decur stig
Ulmaceae	15/200	temp-trop N Hem	tr, shr; simp, ent/ser, alt, stip	a to ± z; ♀/ ♂, ♀; -④–8̲-; 4–8; ②	nutl, sam, dru	± muc sap; redu flr; ± endo
Urtica-ceae	98/2,475	trop, subtrop, + temp	tr, shr, h; simp, ent/ser/ lob, alt/opp, pin/palm-v, stip	a; ♂, ♀; -0 or ③–5̲-; 3–5; ②̄	ach, dru; oft syn	milky/watery sap; cysto; redu flr; 1-loc ovy; 1 ovu

FAMILY	GENERA/ SPP.	DISTRI- BUTION	VEGETATIVE: habit; leaves	FLORAL SUMMARY: symmetry; gender; calyx, corolla/-tepals-; androecium; gynoecium	FRUIT TYPE	OTHER IMPORTANT MORPHOLOGICAL FEATURES
Verbena- ceae	36/1,035	trop- subtrop; few temp	h, shr, tr, vi; simp, ent/lob/ ser, opp/who	z; ⚥; ⑤, ⑤; 2 + 2; ②	dru, sch	arom; squ stems; indet inflor; salv ± bilab cor; epipet stam; false septa + 4-loc ovy; marginal ovu; no endo
Violaceae	22/900	cosmo	h, shr, sm tr; simp, ± lob, alt/bas, stip	z to a; ⚥; 5, 5; ⑤; ③	caps	spurred pet + anth; parie plac; unus sty/stig; carun
Viscaceae	8/440	cosmo; esp trop- subtrop	h, shr; simp, ent, opp, par-v	a; ♂, ♀; -3 or 4-; 3 or 4; 3 or 4	ber	hemiparasitic; brittle; haustoria; jtd stems; monoe; redu flr; undif ovu; viscid frt; no testa; grn embr
Vitaceae	13/735	pantrop; few temp	wo vi; simp/pin-palm comp, ± palm lob, alt, palm-v, stip	a; ⚥/♂, ♀; 4 or 5, 4 or 5; 4 or 5; ②	ber	tendr; jtd stems; punc lvs; inflor opp lvs; stam opp pet; grvd sds
Xyrida- ceae	5/270	trop- subtrop; few temp	h; simp, ent, alt/bas, par-v, op sh	± z; ⚥; 3, 3; 3 + 3*; ③	caps	wet areas; scap; rhiz; cone- like heads; scari cal; ephem cor; plum epipet stam
Zingi- bera- ceae	50/1,000	pantrop; esp se Asia	h; simp, ent, alt/bas, par-v, op sh	z; ⚥; ③, ③; 1 + ②* + 2*; ③	caps, ber	arom; rhiz; lrg lvs; lig; pseudostem; conspic brc; stamin lip; 2 gld on ovy; anth encl sty; aril; peris
Zygophyl- laceae	28/255	trop- subtrop; some temp	shr, h, tr; even-pin comp, ent, opp, stip	a to ± z; ⚥; 5, 5; 10; ⑤	caps, sch	xeric; jtd stems; conspic nect disc; clw pet; biser stam + basal gld; wgd ovy

| FAMILY | GENERA/ SPP. | DISTRI- BUTION | VEGETATIVE: habit; leaves | FLORAL SUMMARY: symmetry; gender; calyx, corolla/-tepals-; androecium; gynoecium | FRUIT TYPE | OTHER IMPORTANT MORPHOLOGICAL FEATURES |

Symbols (Floral summary)

☿—perfect flowers

♂, ♀—staminate and carpellate flowers (imperfect flowers)

⊕—floral parts fused

⊍—floral parts basally fused

⊓—floral parts apically fused

⊕—floral parts fused or distinct

⊍—floral parts basally fused or distinct

⊓—floral parts apically fused or distinct

*—staminode(s)

#̲—inferior ovary

#̅—superior ovary

∞—numerous

Abbreviations

a—actinomorphic

abax—abaxial

accr—accrescent

ach—achene

adax—adaxial

Afr—Africa

aggr—aggregate fruit

alt—alternate

Amer—American

andr—androecium

androgyn—androgynophore

ang—angled

anom sec gr—anomalous secondary growth

anth—anther(s)

appen—appendaged, appendages

api—apical

aqu—aquatic

arom—aromatic

asex—asexual

Austr—Australia

axil—axillary

bact nod—bacterial nodules

bas—basal

ber—berry

bilab—bilabiate

biser—biseriate

blk—black

brc—bract(s)

brn—branched, branches

cadu—caducous

cal—calyx

campan—campanulate

caps—capsule

carp—carpel(s)

carun—caruncle(s)

catk—catkins

Cen (Amer)—Central (America)

clo—closed

clw—clawed

comp—compound

condup—conduplicate

conn—connate

conspic—conspicuous

constr—constricted

cor—corolla

cosmo—cosmopolitan

cot—cotyledons

crv—curved

cysto—cystoliths

decur—decurrent

depr—depressed

dioe—dioecious

dis—distinct

discon—discontinuous

dist—distichous

div—divided

dru—drupe

e—east(ern)

embed—embedded

embr—embryo

encl—enclosing

endo—endosperm

ent—entire

ephem—ephemeral

epical—epicalyx

epig—epigynous

epipet—epipetalous

epiph—epiphytic

epitep—epitepalous

esp—especially

Eur—Europe

excl—excluding

fasc—fascicles, fasciculate

fibr—fibrous

fir—flower(s)

fol—follicle(s)

fr-cen—free-central

frt—fruit

fun—funnelform

funic—funicles

furr—furrowed

gld—gland(s)

gldr—glandular

grn—green

grvd—grooved

gyn—gynoecium

gynob—gynobasic

gynop—gynophore

h—herbs, herbaceous

hisp—hispid

hrs—hairs

hypan—hypanthium

i—irregular

inconsp—inconspicuous

indet—indeterminate

inf—inferior

inflor—inflorescence

insectiv—insectivorous

interpet—interpetiolar

interx phl—interxylary phloem

invol—involucre

jtd—jointed

lam—laminar

lig—ligule

lob—lobed, lobes

loc—locule(s), locular

lrg—large

lvs—leaves

malp hrs—Malpighian hairs

min—minute

monocol—monocolpate

monoe—monoecious

mont—montane

muc—mucilaginous

n—north(ern)

N Hem—Northern Hemisphere

NA—North America(n)

nect—nectary(ies), nectariferous

num—numerous

nutl—nutlet

NZ—New Zealand

oft—often

op—open

operc—operculate, operculum

opp—opposite

ovu—ovule(s)

ovy—ovary

OW—Old World

palm—palmate(ly)

pantrop—pantropical

par—parallel

parie—parietal

parti—partitions, partitioned

pelt—peltate

pend—pendulous

peria—perianth

perig—perigynous

peris—perisperm

pet—petal(s)

petald—petaloid

petio—petiole(s)

pin—pinnate(ly)

pl—plant(s)

plac—placentae, placentation

plic—plicate

plum—plumose

pod—indehiscent pod

poll—pollen

poll pres mech—pollen presentation mechanism

polli—pollinia

pulv—pulvini

punc—punctate

pyr—pyrene(s)

pyx—pyxis
rainf—rain forest(s)
recep—receptacle
redu—reduced
repro—reproduction
resin—resinous
resup—resupinate
rhiz—rhizome(s)
rum—ruminate
s—south(ern)
S Hem—Southern Hemisphere
SA—South America(n)
salv—salverform
sam—samara(s)
scabr—scabrous
scap—scapose
scari—scarious
scatt—scattered
sch—schizocarp
sds—seeds
se—southeast(ern)
sec gr—secondary growth
sep—sepal(s)
ser—serrate
sess—sessile

sh—sheath(ing)
shr—shrubs
silc—silicle
silq—silique
simp—simple
sm—small
spikel—spikelets
spir—spiral(ly)
spot - spotted
squ—square
stam—stamen(s)
stamin—staminode(s), staminodial
stel—stellate
stig—stigma(s)
stip—stipulate, stipules
stol—stolon(s)
strk—streaked
sty—style(s)
subtrop—subtropics, subtropical
succ—succulent
syn—syncarp
taprt—taproot(s)
temp—temperate
tendr—tendrils
tep—tepal(s)

term—terminal
tetr—tetrads
tr—trees
trop—tropics, tropical
tub—tuber(s)
tubu—tubular
twis—twisted
umb—umbel(late)
undif—undifferentiated
unequ—unequal
unus—unusual
urc—urceolate
usu—usually
utr—utricle
v—venation, veined
vb—vascular bundle(s)
vers—versatile
vi—vines
w—west(ern)
wgd—winged
who—whorled
wo—woody
yel—yellow
z—zygomorphic

INDEX

This index includes the scientific (Latin) and vernacular names ("common names") of all taxa mentioned in the text, figure legends, charts, and appendices, as well as plant products and main topics emphasized in the discussions. The Latin name and the primary page number(s) for each of the 130 families covered in the book appear in **boldface**; pages with illustrations are listed in *italic* under the appropriate species name.

416

418

423

426